Human Pathogenic Fungi

Molecular Biology and Pathogenic Mechanisms

Edited by

Derek J. Sullivan and Gary P. Moran

Division of Oral Biosciences
School of Dental Science and Dublin Dental University Hospital
Trinity College Dublin
Dublin
Ireland

Caister Academic Press

Caister Academic Press
Norfolk, UK

www.caister.com

British Library Cataloguing-in-Publication Data
A catalogue record for this book is available from the British Library

ISBN: 978-1-908230-44-7 (hardback)
ISBN 978-1-908230-66-9 (ebook)

Cover design adapted from Figures 7.1 and 9.2

Contents

Contributors

Célia Maria de Almeida Soares
Laboratório de Biologia Molecular
Instituto de Ciências Biológicas
Universidade Federal de Goiás
Goiânia, Goiás
Brazil

cmasoares@gmail.com

Alexandre M. Bailão
Laboratório de Biologia Molecular
Instituto de Ciências Biológicas
Universidade Federal de Goiás
Goiânia, Goiás
Brazil

alexandre.bailao@gmail.com

Elaine Bignell
Institute for Inflammation and Repair
Manchester Fungal Infection Group
University of Manchester
Manchester
UK

elaine.bignell@manchester.ac.uk

Clayton L. Borges
Laboratório de Biologia Molecular
Instituto de Ciências Biológicas
Universidade Federal de Goiás
Goiânia, Goiás
Brazil

clbluiz2@gmail.com

Axel A. Brakhage
Molecular and Applied Microbiology
Leibniz Institute for Natural Product Research and Infection Biology
Hans Knöll Institute
Jena
Germany; and
Institute of Microbiology
Friedrich Schiller University
Jena
Germany

axel.brakhage@hki-jena.de

Vincent Bruno
Department of Microbiology and Immunology
Institute for Genome Sciences
University of Maryland School of Medicine
Baltimore, MD
USA

vbruno@som.umaryland.edu

Radames J.B. Cordero
Department of Microbiology and Immunology
Albert Einstein College of Medicine
Bronx, NY
USA

rada.cordero@gmail.com

Sean Doyle
Department of Biology
National University of Ireland Maynooth
Maynooth
County Kildare
Ireland

sean.doyle@nuim.ie

David A. Fitzpatrick
Genome Evolution Laboratory
Department of Biology
The National University of Ireland Maynooth
Maynooth
County Kildare
Ireland

david.fitzpatrick@nuim.ie

Lorna Gallagher
Department of Biology
National University of Ireland Maynooth
Maynooth
County Kildare
Ireland

lorna.gallagher@nuim.ie

Rocío García-Rodas
National Centre for Microbiology
Instituto de Salud Carlos III
Madrid
Spain

rocio.rodas@gmail.com

Gregory M. Gauthier
Department of Medicine
Division of Infectious Diseases
University of Wisconsin-Madison
Madison, WI
USA

gmg@medicine.wisc.edu

Reinhard Guthke
Systems Biology/Bioinformatics
Leibniz Institute for Natural Product Research and Infection Biology
Hans Knöll Institute
Jena
Germany

reinhard.guthke@hki-jena.de

Fabian Horn
Systems Biology/Bioinformatics
Leibniz Institute for Natural Product Research and Infection Biology
Hans Knöll Institute
Jena
Germany

fabian.horn@hki-jena.de

Bernhard Hube
Department of Microbial Pathogenicity Mechanisms
Leibniz Institute for Natural Product Research and Infection Biology
Hans Knöll Institute
Jena
Germany

bernhard.hube@hki-jena.de

Andrew H. Limper
Division of Pulmonary and Critical Care
Thoracic Diseases Research Unit
Department of Internal Medicine
Mayo Clinic College of Medicine
Mayo Clinic
Rochester, MN
USA

limper.andrew@mayo.edu

Donna M. MacCallum
Aberdeen Fungal Group
School of Medical Sciences
Institute of Medical Sciences
University of Aberdeen
Aberdeen
UK

d.m.maccallum@abdn.ac.uk

Bernard Mignon
Veterinary Mycology
Department of Infectious and Parasitic Diseases
Faculty of Veterinary Medicine
University of Liège
Liège
Belgium

bmignon@ulg.ac.be

Michel Monod
Department of Dermatology
Centre Hospitalier Universitaire Vaudois
Lausanne
Switzerland

Michel.Monod@chuv.ch

David L. Moyes
Department of Oral Immunology
Clinical and Diagnostic Science Group
King's College London Dental Institute
King's College
London
UK

david.moyes@kcl.ac.uk

Julian R. Naglik
Department of Oral Immunology
Clinical and Diagnostic Science Group
King's College London Dental Institute
King's College
London
UK

Julian.Naglik@kcl.ac.uk

Karen A. O'Hanlon
Department of Biology
National University of Ireland Maynooth
Maynooth
County Kildare
Ireland

karen.a.ohanlon@nuim.ie

Grainne O'Keeffe
Department of Biology
National University of Ireland Maynooth
Maynooth
County Kildare
Ireland

grainne.okeeffe@nuim.ie

Rosely Maria Zancopé Oliveira
Laboratório de Micologia
Instituto de Pesquisa Clínica Evandro Chagas
Fundação Oswaldo Cruz
Manguinhos, Rio de Janeiro
Brazil

rosely.zancope@ipec.fiocruz.br

Rebecca A. Owens
Department of Biology
National University of Ireland Maynooth
Maynooth
County Kildare
Ireland

rebecca.owens@nuim.ie

Juliana A. Parente
Laboratório de Biologia Molecular
Instituto de Ciências Biológicas
Universidade Federal de Goiás
Goiânia, Goiás
Brazil

juparente@gmail.com

Maristela Pereira
Laboratório de Biologia Molecular
Instituto de Ciências Biológicas
Universidade Federal de Goiás
Goiânia, Goiás
Brazil

maristelaufg@gmail.com

Jakrapun Pupaibool
Division of Infectious Diseases
Mayo Clinic College of Medicine
Rochester, MN
USA

a-jkp@hotmail.com

Chad A. Rappleye
Department of Microbiology
Department of Microbial Infection and Immunity
Ohio State University
Columbus, OH
USA

rappleye.1@osu.edu

Jonathan P. Richardson
Department of Oral Immunology
Clinical and Diagnostic Science Group
King's College London Dental Institute
King's College
London
UK

jonathan.richardson@kcl.ac.uk

Peter Staib
Department of Research and Development
Kneipp-Werke
Würzburg
Germany

peter.staib@kneippe.de

Vito Valiante
Molecular and Applied Microbiology
Leibniz Institute for Natural Product Research and Infection Biology
Hans Knöll Institute
Jena
Germany

vito.valiante@hki-jena.de

Duncan Wilson
Department of Microbial Pathogenicity Mechanisms
Leibniz Institute for Natural Product Research and Infection Biology
Hans Knöll Institute
Jena
Germany

duncan.wilson@abdn.ac.uk

Oscar Zaragoza
National Centre for Microbiology
Instituto de Salud Carlos III
Madrid
Spain

ozaragoza@isciii.es

Preface

Human Pathogenic Fungi: New Technologies and New Insights into Pathogenesis

Infections caused by fungi are a significant contributor to morbidity and mortality in humans. These infections range from comparatively innocuous superficial skin diseases caused by dermatophytes to invasive life-threatening infections caused by species such as *Candida albicans, Cryptococcus neoformans* and *Aspergillus fumigatus,* amongst others. Since many of the diseases caused by fungi are opportunistic in nature, the pathogenesis of fungal infections is complex and our understanding of how these microorganisms cause disease has lagged behind bacterial and viral pathogens. However, the application of new genomic and other molecular techniques in recent years has revolutionized the field of fungal pathogenesis. This has resulted in a dramatic improvement in our understanding of how fungi cause disease in humans and the aim of this book is to summarize recent developments in this rapidly evolving area.

The book is divided into two sections; the first includes chapters detailing the state of the art tools and approaches currently in use at the forefront of comparative biology of pathogenic fungi; and the second an in-depth analysis of the current understanding of the pathogenic mechanisms of specific fungal pathogens. Each chapter has been written by a high profile author(s) who is a recognized leader in their field. In the first section, Chapter 1 summarizes how next generation sequencing techniques are revolutionizing our understanding of how fungi cause disease. Meanwhile, Chapter 2 describes how these same techniques have led to the sequencing of over 100 fungal genomes, including those of the most important human pathogens, and how these data are being used to elucidate the evolution of fungal virulence. How the application of high throughput data-driven systems biology can impact on our understanding of fungal–host interactions is discussed in Chapter 3, while Chapter 4 examines specifically what comparative transcriptomics can tell us about how the fungus and host cells interact and respond during infection. Chapter 5 describes the fungal infection models currently in use and how these have helped enhance our understanding of the processes involved in human disease while Chapter 6 provides an overview of our current understanding of how the host recognizes colonizing and infecting fungi and how host immune defences prevent and resolve infections. The second section includes chapters (Chapters 7–14) that specifically provide a self-contained and detailed analysis of recent advances in the study of the pathogenesis of the major fungal pathogens of humans, including *Candida, Aspergillus, Cryptococcus, Dermatophytes, Histoplasma, Blastomyces, Pneumocystis* and *Paracoccidioides.*

Current Books of Interest

Acanthamoeba: Biology and Pathogenesis (2nd edition)	2015
Microarrays: Current Technology, Innovations and Applications	2014
Metagenomics of the Microbial Nitrogen Cycle: Theory, Methods and Applications	2014
Pathogenic *Neisseria*: Genomics, Molecular Biology and Disease Intervention	2014
Proteomics: Targeted Technology, Innovations and Applications	2014
Biofuels: From Microbes to Molecules	2014
Applied RNAi: From Fundamental Research to Therapeutic Applications	2014
Halophiles: Genetics and Genomes	2014
Molecular Diagnostics: Current Research and Applications	2014
Phage Therapy: Current Research and Applications	2014
Bioinformatics and Data Analysis in Microbiology	2014
The Cell Biology of Cyanobacteria	2014
Pathogenic *Escherichia coli*: Molecular and Cellular Microbiology	2014
Campylobacter Ecology and Evolution	2014
Burkholderia: From Genomes to Function	2014
Myxobacteria: Genomics, Cellular and Molecular Biology	2014
Next-generation Sequencing: Current Technologies and Applications	2014
Omics in Soil Science	2014
Applications of Molecular Microbiological Methods	2014
Mollicutes: Molecular Biology and Pathogenesis	2014
Genome Analysis: Current Procedures and Applications	2014
Bacterial Toxins: Genetics, Cellular Biology and Practical Applications	2013
Bacterial Membranes: Structural and Molecular Biology	2014
Cold-Adapted Microorganisms	2013
Fusarium: Genomics, Molecular and Cellular Biology	2013
Prions: Current Progress in Advanced Research	2013
RNA Editing: Current Research and Future Trends	2013
Real-Time PCR: Advanced Technologies and Applications	2013
Microbial Efflux Pumps: Current Research	2013
Cytomegaloviruses: From Molecular Pathogenesis to Intervention	2013
Oral Microbial Ecology: Current Research and New Perspectives	2013
Bionanotechnology: Biological Self-assembly and its Applications	2013
Real-Time PCR in Food Science: Current Technology and Applications	2013
Bacterial Gene Regulation and Transcriptional Networks	2013

Full details at www.caister.com

Part I

Recent Advances in the Pathogenesis of Human Fungal Infections

Understanding Fungal Pathogenesis with High-throughput Sequencing

1

Vincent Bruno

Abstract

The invention of massively parallel DNA sequencing and the subsequent development of RNA-seq, ChIP-seq and microbiome sequencing have radically changed the landscape of molecular and systems biology by allowing researchers to address biological problems with a sensitivity and scope that was not previously possible. This chapter discusses how each of these high-throughput sequencing (HTS)-based approaches is being used to address biological questions within the field of human fungal pathogenesis. As the cost of sequencing becomes more affordable, the amount of sequence depth coverage that will be attainable will increase significantly allowing researchers to apply these genomic techniques to more *in vivo* models and clinical samples for which a great deal of coverage is required. Integrated approaches, which combine multiple HTS-based experiments, will provide necessary insight into the complex interactions between the host, microbiota and fungal pathogens.

Introduction

High-throughput sequencing (HTS) refers to the recent development of DNA sequencing methods that came into use about three decades after the capillary sequencing method was first developed by Sanger (Bentley *et al.*, 2008; Margulies *et al.*, 2005; Sanger *et al.*, 1977). As a result, DNA sequencing has advanced well beyond the simple determination of genomic DNA sequences to identify open reading frames (ORFs) and polymorphisms, and now allows basic researchers to study other aspects of the central dogma, such as transcription and translation. In contrast to the Sanger method that uses chain-termination inhibitors to sequence one DNA molecule at a time, HTS platforms use highly parallelized reactions that enable the simultaneous sequencing of millions of DNA molecules. While these advances have made the sequencing of nucleic acids more efficient and cost-effective, they have also paved the way for new experimental approaches that allow systematic and comprehensive exploration of molecular mechanisms involved in every sub-field of biology.

This chapter will provide an overview of the various aspects of fungal pathogenesis that have benefited, or have the potential to benefit, from the advancements in DNA sequencing technologies by focusing the discussion on the 3 most widely used HTS-based applications that have been developed: RNA-seq, ChIP-seq, and sequencing that targets the ribosomal DNA (rDNA) locus. For a detailed description and comparison of the technical aspects of each of the different HTS platforms, the reader is referred to two extensive review articles (Liu *et al.*, 2012; Pareek *et al.*, 2011). While HTS has also led to a rapid increase in the number of genome sequences of fungal pathogens, the impact that this has had on the sub-field will be discussed in Chapters 2 and 3.

RNA-seq

The ability to cause disease largely depends on a pathogen's ability to alter its transcriptome in response to different environmental stimuli to ensure survival in different host niches. For example, in *Candida albicans*, a complex transcriptional circuitry ensures that morphogenesis between ovoid yeast cells and elongated filamentous cells, a process tightly linked to virulence, takes place in response to specific stimuli (Biswas *et al.*, 2007). Changes in the transcriptional network also ensure that *C. albicans* can grow in tissues with vastly different pH values and survive stresses believed to be inflicted upon them by cells of the innate immune system (Bensen *et al.*, 2004; Chiranand *et al.*, 2008; Enjalbert *et al.*, 2006; Hromatka *et al.*, 2005; Lorenz and Fink, 2001; Rubin-Bejerano *et al.*, 2003). In *Cryptococcus neoformans*, multiple signal transduction pathways ensure the precise transcriptional regulation of the genes required for biosynthesis of its polysaccharide capsule, which serves several functions in promoting virulence (O'Meara and Alspaugh, 2012). Consequently, there is great interest in the fungal pathogenesis community to obtain a detailed understanding of the transcriptomes of this class of pathogens.

The transcriptome can be defined as the complete set of RNA molecules (products of transcription) that are generated by an organism. A complete knowledge of the transcriptome would contain a catalogue of (i) all the different species of RNAs (e.g. mRNAs, non-coding RNAs, small RNAs), (ii) the structure of each RNA molecule (e.g. 5′ boundaries, 3′ boundaries, splicing patterns), and (iii) information regarding the abundance of each of the species. The task of completely defining the transcriptome for an organism is complicated by the fact that transcriptomes are, as mentioned above, dynamic entities that are constantly changing in response to the extracellular environment.

RNA-seq (deep-sequencing of cDNA) provides a largely unbiased method to comprehensively and systematically define the transcriptome of an organism in a manner that is significantly more sensitive than microarray hybridization approaches (Wang *et al.*, 2009). This approach was first used to identify novel transcribed regions in *Saccharomyces cerevisiae*, *Schizosaccharomyces pombe*, bacteria, humans, and plants (Biswas *et al.*, 2007; Emrich *et al.*, 2007; Mi *et al.*, 2008; Nagalakshmi *et al.*, 2008; Weber *et al.*, 2007; Wilhelm *et al.*, 2008). The data obtained using RNA-seq is also routinely used to identify new splicing events and to quantify gene expression from cells grown under different experimental conditions or cell types (Marioni *et al.*, 2008; Mortazavi *et al.*, 2008; Sultan *et al.*, 2008; Trapnell *et al.*, 2009; Wang *et al.*, 2010b). I describe here several studies that have utilized RNA-seq to increase our understanding of fungal pathogenesis.

Notably, there is no 'standard' method of analysis for RNA-seq data, but rather several different bioinformatics tools that have been developed. For a detailed description and comparison of the technical aspects of each of the different RNA-seq analysis tools, the reader is referred to (Rapaport *et al.*, 2013).

Candida

RNA-seq has significantly improved genome annotation by allowing investigators to identify several features that were overlooked by initial annotation efforts. Early versions of the *C. albicans* genome annotation were almost exclusively based on gene prediction algorithms that detect protein-coding regions. Two initial studies that performed RNA-seq on *C. albicans* grown under an extensive variety of *in vitro* conditions led to the identification of more than 1000 novel transcriptionally active regions (TARs), the identification of several novel splice junctions and the determination of 5′ and 3′ untranslated regions (UTRs) for the majority of protein-coding transcripts (Bruno *et al.*, 2010; Tuch *et al.*, 2010). Many of the novel TARs are potential non-coding RNAs based on the absence of ORFs predicted to encode proteins with significant similarity to known proteins. These two studies analysed *C. albicans* undergoing the white-opaque transition, the yeast-hyphal transition, as well as exposure to cell wall-damaging agents, oxidative stress, nitrosative stress and growth at different pHs (Bruno *et al.*, 2010; Tuch *et al.*, 2010). Two subsequent studies performed RNA-seq on cells grown in biofilms and across the yeast-hyphal transition further increased the number of novel TARs (Grumaz *et al.*, 2013; Nobile *et al.*, 2012). These four studies combined provide a comprehensive view of the *C. albicans* transcriptome by allowing detection of genes that might only be expressed under one condition as well as providing experimental validation for novel transcripts that are expressed in more than one growth condition. The annotation of novel transcripts from each study are coordinated and catalogued in the *Candida* Genome Database (Arnaud *et al.*, 2005) and can be visualized in the GBrowse tool at (http://www.candidagenome.org/).

RNA-seq is increasingly being applied to study straightforward differential gene expression. For *C. albicans* it has been used to identify changes in gene expression in response to a variety of *in vitro* growth conditions including the yeast-hyphal transition (Bruno *et al.*, 2010; Grumaz *et al.*, 2013; Hnisz *et al.*, 2012), white-opaque switching (Tuch *et al.*, 2010), biofilm formation (Desai *et al.*, 2013; Nobile *et al.*, 2012), and treatment with agents that induce cell wall damage, oxidative stress, nitrosative stress (Bruno *et al.*, 2010), and varying pHs (Bruno *et al.*, 2010). It has also proven useful to compare the transcriptomes of different strains grown under the same condition (e.g. wild-type vs. mutant or matched clinical isolates) (Desai *et al.*, 2013; Dhamgaye *et al.*, 2012; Guida *et al.*, 2011; Sun *et al.*, 2013). For all of the comparisons that were previously performed using microarray experiments, the deep-sequencing approach allowed the identification of many gene expression changes that were previously overlooked. Furthermore, the sequencing approach also allowed the investigators to survey the expression of the TARs that were not represented in most *C. albicans* microarrays.

Only 2 studies to date have employed RNA-seq to address biological questions for non-albicans *Candida* species. Guida et al. applied the method to cells grown in media with varying temperatures, nutrients, and oxygen concentrations (Guida *et al.*, 2011). The data served two purposes. First, it was used to improve the genome annotation of *Candida parapsilosis* and in doing so the investigators identified the first examples of 3′ UTR introns in an organism belonging to the Saccharomycotina subphylum (Guida *et al.*, 2011). Second, it was used to characterize the transcriptional profile of *C. parapsilosis* grown in normoxic and hypoxic conditions and to establish the role of a transcription factor, Upc2, in the response (Guida *et al.*, 2011).

Grumaz *et al.* (2013) performed a comparative transcriptomic study by examining the transcriptomes of both *C. albicans* and *C. dubliniensis* across the yeast to hyphal transition. This innovative study design sought to address the apparent differences in morphogenesis and virulence between the two organisms despite their close evolutionary relationship (Butler *et al.*, 2009). The investigators defined a core subset of 84 homologues that are hyphae specific in both species as well as a set of 42 that are specific to *C. albicans*. The authors concluded that the hyphal transition in each organism is governed by a species-specific genetic repertoire.

Interestingly, the only *C. albicans* RNA-seq study to examine small RNAs happened to be the first one and was a large, multiple-species endeavour that provided the first strong evidence that RNA silencing occurs in budding yeast, including *C. albicans* (Drinnenberg *et al.*, 2009). This study firmly established the existence of small RNAs in the *C. albicans* transcriptome by performing RNA-seq on the small RNA population that results from biochemically selecting for the hallmarks of Dicer products, specifically 5′ monophosphates and 3′ hydroxyl groups (Drinnenberg *et al.*, 2009).

Aspergillus

Of the over 250 species of filamentous fungi (Geiser *et al.*, 2007) that make up the genus *Aspergillus*, the opportunistic human pathogens *A. fumigatus* and *A. flavus* are the most highly associated with fungal infection of humans (Marr *et al.*, 2002; Morgan *et al.*, 2005) and, to date, are the only two human pathogenic species from the *Aspergillus* genus to have been the subject of RNA-seq studies. The first such study to examine *A. fumigatus* compared the transcriptomes of cells grown planktonically or in biofilms (Gibbons *et al.*, 2012). The resulting analysis indicated that 3728 genes (representing 38% of the known transcripts) were differentially expressed between the two growth conditions. Interestingly, they observed that the biofilm up-regulated genes were more likely to be physically located in the subtelomeric regions of the chromosomes while no such spatial correlation existed for the biofilm down-regulated genes (Gibbons *et al.*, 2012). In the same organism, Muller *et al.* (2012) focused on defining the set of genes whose expression is controlled by the central mitogen activated protein kinase (MpkA). In doing so they identified about 3-fold more MpkA-dependent genes than were identified using microarray technology on the same experimental design. In addition, the data were also used to identify 185 novel transcripts and to define 5′ and 3′ untranslated regions (UTRs) thereby updating the genome annotation (Muller *et al.*, 2012).

The first two RNA-seq studies to feature *A. flavus* focused on the biology of aflatoxin production (Lin *et al.*, 2013; Yu *et al.*, 2011). The first addressed the observation that aflatoxin production occurs optimally at 30°C while the optimal temperature for fungal growth is 37°C. To this end, RNA-seq was performed on cells grown at each of these temperatures. The analysis revealed 1153 genes whose expression was differentially expressed between the two conditions. Among these genes preferentially expressed at 30°C were 11 of the 55 known secondary metabolite clusters, including aflatoxin biosynthetic genes (Yu *et al.*, 2011). The second study sought to identify genes whose expression change in response to treatment with 5-azacytidine (5-AC), a compound that is known to inhibit aflatoxin production (Lin *et al.*, 2013). RNA-seq analysis on cells treated with 5-AC or left untreated identified 240 differentially expressed genes. The data generated in this second study were also used for annotation purposes and allowed the identification of 1025 novel putative protein-coding genes (> 50 codons) and 107 potential non-coding RNA transcripts (Lin *et al.*, 2013).

In a departure from the majority of the fungal RNA-seq studies that have measured gene expression of one or two strains across one or two conditions, Elmore *et al.* (2012) used RNA-seq data to test the effect of polymorphisms in tandem repeat elements on phenotypic variation. In this case, RNA-seq was used as a sensitive transcriptional output to search for a correlation between copy number polymorphisms in promoters and transcriptional output in eight different isolates each of *A. flavus* and *A. oryzae* (Elmore *et al.*, 2012). The examination of 143 tandem repeats located in promoters revealed that while in some cases the polymorphisms might contribute to variation in transcript abundance, that tandem repeats are not likely to be major regulations of variation in gene expression (Elmore *et al.*, 2012).

Cryptococcus neoformans

To date, there have been only two reports that have addressed the biology of *Cryptococcus neoformans* using RNA-seq (Haynes *et al.*, 2011; Wang *et al.*, 2010a). Wang *et al.* (2010a) deep-sequenced the small RNA population and identified many siRNAs that mapped to transposable elements, providing evidence that RNAi might repress transposable elements in *C. neoformans*, a phenomenon that was known to occur in other organisms (Slotkin and Martienssen, 2007). Haynes *et al.* (2011) used RNA-seq to identify transcriptional targets for three different genes that are required for capsule biosynthesis. This study is discussed in more detail below.

Multiple species RNA-seq

All of the studies mentioned above were performed on pure cultures containing one species. The highly sensitive and probe-independent nature of RNA-seq, combined with depth of sequencing coverage that is now available, allows the potential for the transcriptomes of multiple species to be assayed simultaneously (Wang *et al.*, 2009; Westermann *et al.*, 2012), specifically in the context of a microbial infection or a polymicrobial biofilm. To this end, Tierney *et al.* used RNA-seq to simultaneously define the response of both *C. albicans* and murine dendritic cells over the course of an *in vitro* infection (Tierney *et al.*, 2012). While the overall depth of sequence coverage was low, the strength of this study lies in the combination of deep-sequencing and functional validation of the novel interspecies interactions. Although this work was the first successful application of RNA-seq to the simultaneous analysis fungal and host transcriptomes, significantly greater sequencing depth is required to secure a more complete view of host–pathogen interactions.

ChIP-seq

The major paradigm regarding the mechanism of transcriptional regulation is the binding of protein factors to the promoter regions upstream of genes to activate or repress transcription. Adding to the complexity of this transcriptional regulation is the function of chromatin modification enzymes that allow or block access of proteins to DNA by virtue of their ability to control how tightly the chromatin is compacted. A detailed cataloguing of where and when DNA-binding proteins bind, as well as the modification state of the chromatin across the genome at different times, is essential to understanding the transcriptional circuitry which governs so many aspects of cellular life. Unfortunately, DNA sequence features alone cannot predict the precise genomic locations where these proteins bind. This information can only be obtained by more functional assays. Chromatin immunoprecipitation (ChIP) coupled with high throughput sequencing (ChIP-seq) has become the standard experimental approach for not only identifying the binding sites of specific proteins, but also for determining the states of chromatin modification across entire genomes (Barski *et al.*, 2007; Robertson *et al.*, 2007; Wang *et al.*, 2008). ChIP allows an investigator to enrich for DNA sequences that are associated with specific DNA-binding proteins or chromatin modification states of cells grown under any condition (Solomon *et al.*, 1988). The identities of the bound DNA sequences are then determined by either microarray hybridization (referred to as ChIP-chip) or HTS (Horak *et al.*, 2002; Robertson *et al.*, 2007). ChIP-seq has several distinct advantages over the microarray-based predecessor, ChIP-chip, in that it offers higher resolution, lower background noise, a higher dynamic range, greater genomic coverage that is not limited by the probe set and a much lower requirement for input ChIP DNA (Park, 2009). These significant benefits over array based assays have made ChIP-seq the standard assay for identifying DNA-binding sites on a global scale.

Despite its earlier development compared to RNA-seq, there have been far fewer studies that have used ChIP-seq analysis to address biological questions of human fungal pathogens. To date, there have only been two such reports (Haynes *et al.*, 2011; Hnisz *et al.*, 2012). In each of the studies, the investigators set out to elucidate the molecular mechanism by which a transcription factor governed virulence by identifying the genomic regions that each transcription factor directly occupied.

The first study focused on how the *ADA2* gene of *Cryptococcus neoformans* controls polysaccharide capsule formation (Haynes *et al.*, 2011). During experiments to characterize the transcriptome associated with capsule formation, Haynes *et al.* identified a previously uncharacterized gene whose expression correlated positively with capsule size. This particular gene had strong homology to the *ADA2* gene of *S. cerevisiae*, a member of the SAGA complex that mediates the acetylation of lysine 9 on histone H3 (H3K9) (Grant *et al.*, 1997). After a series of molecular genetic experiments to establish the role of Cn*ADA2* in capsule formation and that it was required for the majority of the H3K9 acetylation in *C. neoformans*, the authors set out to identify Cn*ADA2* targets by performing ChIP-seq. Since they had established that Cn*ADA2* functions to acetylate H3K9, the authors performed the chromatin immunoprecipitation with an antibody against acetylated H3K9 in both wild-type and *ada2Δ* cells grown under capsule-inducing conditions. Using this approach, the genomic regions that displayed H3K9 acetylation in the wild-type cells but not in the ada2Δ cells were considered to be direct Ada2p targets. The authors also performed RNA-seq on wild-type and ada2Δ cells to identify genes whose expression required a functional *ADA2* gene. Comparing these two lists of targets revealed a significant overlap, a finding that is consistent with the function of the SAGA complex in activation of gene expression. This study was an elegant demonstration of the use of ChIP-seq to differentiate direct and indirect transcriptional regulation.

In a subsequent study, Hnisz et al. (2012) set out to identify the molecular mechanism by which SET3C, the defining member of the SET3 histone deacetylase complex, governs hyphal morphogenesis in *C. albicans*, a connection that the group established in a previous study (Hnisz *et al.*, 2010). To accomplish this, the authors first generated a strain that produced a functional epitope-tagged version of Set3C and then identified the binding sites using ChIP-seq targeted against the myc epitope. By performing the experiments on cells growing as yeast or under hyphae-inducing conditions, they were able to identify three different classes of binding sites: (i) those that are occupied specifically during yeast phase growth, (ii) those that are occupied specifically during hyphal growth and (iii) those that are occupied constitutively. RNA-seq analysis of cells growing exponentially in rich media (YPD) established that SET3C bound to the promoters of highly transcribed genes supporting the idea that SET3C functions to activate gene expression, a finding that is consistent with what has been observed for the *S. cerevisiae* homologue (Wang *et al.*, 2002). A subsequent systemic analysis, comparing RNA-seq and ChIP-seq data across the yeast-to-hypha transition, revealed that SETC3 binding correlates highly with transcriptional activation and that absence of binding correlated highly with transcriptional repression. Unexpectedly, RNA-seq analysis of a strain harbouring a homozygous deletion in SET3 was still able to efficiently induce a proper transcriptional response to hyphal induction leaving the investigators without a reasonable explanation for why set3Δ/Δ cells are hyperfilamentous. A more focused analysis of the data indicated that four known transcriptional regulators of morphogenesis (*NRG1*, *EFG1*, *TEC1* and *BRG1*) had altered expression in a set3Δ/Δ mutant compared to wild-type and three are directly bound by SET3. Taken together, these genomic analyses combined with more focused follow-up experiments demonstrate that the SET3 complex governs hyphal morphogenesis by adjusting the transient expression of specific transcription factors following hyphal induction. This study is significant because it is the first to dissect how a chromatin-modifying enzyme controls fungal morphogenesis and demonstrates that this is accomplished by controlling transcriptional kinetics via a key morphogenesis circuit.

Mycobiome sequencing

The human microbiota is the complex collection of microorganisms that inhabit the many niches of the human body and outnumber the host cells by approximately one order of magnitude (Turnbaugh *et al.*, 2007). The influence of the microbiota on human health and disease has been well established (Dethlefsen *et al.*, 2007; Ley *et al.*, 2006; Mazmanian *et al.*, 2005). However, these studies have focused solely on the bacterial component of the microbiota and hence the term 'microbiota' is often used to refer to the composition of the bacterial community in a given environment. Overall, much less is known about the fungal component of the microbiota, or mycobiome, since there have been far fewer studies that have focused on it. Unlike its bacterial counterpart, the role of the human mycobiome in human health and disease has not been firmly established, but it is sure to be relevant as a reservoir for opportunistic fungi that become pathogens when the host becomes immunocompromised or the bacterial microbiota is altered by broad-spectrum antibiotic use.

Historically, the vast majority of the mycobiome surveys have used either growth on selective media, microscopy, biochemical detection of fungal molecules, or Sanger sequencing of PCR products to detect and quantify the fungal species in a clinical sample. However, these techniques have some serious limitations due to their inability to detect non-culturable species, relative lack of sensitivity or reliance on genome sequence data that, until recently, has been lacking.

HTS sequencing has become the standard approach to comprehensively characterize the mycobiome of clinical samples because of its ability to reveal the breadth, depth, and diversity of the fungal species that inhabit the human host. The preferred targets of the HTS-based approaches are the 28S, 18S or 5.8S rRNA genes as well as the internal transcribed spacer regions (ITS1 and ITS2). The species-specific regions of these loci are amplified with primers that bind to the highly conserved flanking regions and deep-sequencing is used to identify each specific sequence present in the heterogeneous amplification mixture. Only a handful of studies have applied the HTS to the characterization of mycobiomes, but the combined efforts have sampled from several distinct niches within the host. Described below are several studies that aimed to characterize the mycobiome in different anatomical contexts.

Oral cavity

Ghannoum *et al.* (2010) were the first to apply high-throughput sequencing to characterize the mycobiome at any human body site. They established the 'basal' oral mycobiome by pyrosequencing the ITS regions of samples isolated from the oral rinses of 20 healthy individuals. Their analysis revealed that the distribution of fungal species varied significantly between individuals. Interestingly, at least 20% of the participants contained at least one member from each of the four most common pathogenic fungi: *Candida, Aspergillus, Fusarium* and *Cryptococcus.* The incidence of *Candida* carriage in this study was much higher than previously thought (75% vs. 40%), a discrepancy that is most likely attributed to the superior sensitivity of the HTS-based approach. Across all of the samples, they identified a total of 101 fungal species representing 74 different culturable and 11 different non-culturable genera, with each individual containing between 9 and 23 different species. This study has set the stage for future detailed characterizations of different mycobiomes in the context of both health and disease.

Lungs

The lungs, which were previously thought to be free of microbial life in healthy individuals, are now thought to contain a bacterial microbiota that is altered during disease (Dickson *et al.*, 2013). Apart from the biology of *Pneumocystis* spp., almost nothing is known regarding the lung mycobiome (Chabe, 2011). Charlson and colleagues surveyed the mycobiomes of both bronchoalveolar lavage fluid and oropharyngeal wash fluid by pyrosequencing the ITS-1 region of samples collected from recipients of lung transplants as well as healthy controls (Charlson *et al.*, 2012). Consistent with the study by Ghannoum et al. (2010), the fungal distribution of the oral washes of the healthy controls was similar to the oral rinses described above. The authors found very little ITS1 amplification in the lavage fluid of healthy individuals. Pyrosequencing of lavage fluid and oral rinses from the lung transplant recipients revealed an increased abundance of *Candida* spp. and *Aspergillus* spp. These results are not surprising given that the lung transplant recipients were taking both antibiotics and immunosuppressants, both of which predispose individuals to fungal disease.

Skin

The human skin microbiota is of great medical interest since changes in the community composition of the skin microbiota may be connected to the onset and persistence of several dermatological diseases (Kong and Segre, 2012). Findley et al. (2013) performed the most extensive HTS-based mycobiome study to date during their analysis of the human skin mycobiome. To explore the composition of the fungal community on the skin, they performed pyrosequencing of the ITS1 regions from 10 healthy individuals at 14 different skin sites. Relatively little diversity was observed at 11 of the 14 body sites and these fungal populations were dominated by organisms of the *Malassezia* genus. In contrast, significant fungal diversity was observed in the samples taken from three foot sites – plantar heel, toenail and toe web. This study was especially comprehensive in its efforts to address the potential relationship between bacteria and fungi by including a parallel analysis of the bacterial microbiota through pyrosequencing of the 16S regions. Measurements of fungal and bacterial diversity at each site revealed a lack of correlation between the two communities indicating that the forces that shape each community are more complex then just depending on other microbes. The merit of this study is that it provides a strong framework for future studies addressing the interaction between fungal and bacterial organisms in the context of any disease.

Gut

To date, there are no published studies that have characterized the human gut mycobiome with high-throughput sequencing, but one particular study of the murine gut has increased our understanding of the relationship between the mycobiome of this host niche and disease (Iliev *et al.*, 2012). As part of a large multilayered study that established the relationship between inflammatory bowel disease (IBD) and Dectin-1, the key host receptor for the recognition of fungi, Iliev *et al.* (2012) used deep-sequencing of the ITS1 and ITS2 loci to define the mouse intestinal mycobiome. The analysis identified over 100 different known fungal species and over 100 unannotated fungal sequences. Interestingly, the overwhelming majority of the sequences

(97.3%) belonged to just 10 fungal species with 65% belonging to *Candida tropicalis* (Iliev *et al.*, 2012). The authors also examined how the fungal community structure differed between health and inflammation in the context of a murine model of induced colitis. Indeed, the intestinal mycobiome did vary between the two states, but only in the mice that lacked the fungal receptor, Dectin-1. The inflamed state had a higher proportion of opportunistic pathogens, specifically *Candida* and *Trichosporon*, and a lower proportion of non-pathogenic *Saccharomyces* species. Taken together, these results suggest that the host's ability to sense and respond to the presence of fungi is important for maintenance of intestinal health.

A comprehensive genomic approach to studying fungal disease: vaginal microbiota

Vulvovaginal candidiasis (VVC) is an example of how high-throughput sequencing has significant potential to increase our understanding of disease by leading to a more detailed understanding of fungal interactions with the bacterial microbiota and the host.

The key criteria for the diagnosis of VVC is the demonstrable presence of *Candida* and inflammation in the vaginal mucosa (Sobel *et al.*, 1998). While many different species of *Candida* have been implicated in the diagnosis of VVC, nearly all events are attributed to isolates of *Candida albicans*, leaving only a small portion of cases that are caused by non-*albicans* species (Hettiarachchi *et al.*, 2010; Sobel, 2007). The clinical manifestations of VVC resulting from *C. albicans* are indistinguishable from those caused by non-*albicans* species (Sobel, 2007).

A comprehensive body of work overwhelmingly supports the notion that neither cell-mediated immunity nor *Candida*-specific antibodies are responsible for host defence against VVC (Fidel, 2007). Consistent with these findings, there is no greater frequency of VVC in HIV-positive women compared to HIV-negative women despite the incidence of oral candidiasis being higher (Leigh *et al.*, 2001; Schuman *et al.*, 1998). There is, however, one additional key player that should not be neglected in conversations about VVC: the vaginal microbiota.

The vaginal microbiota is a community of diverse organisms that exists in a mutualistic relationship with the host (Ma *et al.*, 2012). For many years scientists have sought to understand the functions of the vaginal microbial community and determine whether these microorganisms provide a barrier of protection from invading pathogens (Sobel, 1999). Efforts to explore this scientific question have yielded many interesting findings. In fact, certain types of vaginal microbiota have shown a potential role in preventing bacterial vaginosis, HIV infection, and urinary tract infections (Donders *et al.*, 2000; Gupta *et al.*, 1998; Lai *et al.*, 2009; Pybus and Onderdonk, 1999). Not surprisingly then, disturbances or perturbations in the vaginal ecosystem, characterized by changes in microbial composition and/or increased in vaginal pH, have been associated with greater susceptibility to vaginal infections and other obstetric outcomes (Gray *et al.*, 1997; Taha *et al.*, 1998; van De Wijgert *et al.*, 2000; Wiesenfeld *et al.*, 2003).

The Human Microbiome Project aims to characterize and understand the functions of microbial communities of microorganisms that colonize the human body, focusing primarily on bacteria (Peterson *et al.*, 2009; Turnbaugh *et al.*, 2007). To date, several studies have provided evidence that the microbiota plays a role in influencing host processes, such as immunity, obesity and development (Dethlefsen *et al.*, 2007; Ley *et al.*, 2006; Mazmanian *et al.*, 2005). A community wide effort to study the importance of the microbiota across many body sites has fuelled the investigation of vaginal communities and their impact on vaginal health and disease. For example, the activity of lactic acid-producing bacteria in the vagina appears to be a fundamental element of defence against pathogens (Boskey *et al.*, 2001). This function is mainly performed by *Lactobacillus* spp., mainly *Lactobacillus crispatus*, *L. gasseri*, *L. jensenii* and *L. iners*. These species dominate in healthy vaginal bacterial communities (Antonio *et al.*, 1999; Boskey *et al.*, 2001; Juarez Tomas *et al.*, 2003). Additionally, certain strains of these *Lactobacillus* species produce bactericidal molecules and serve as a physical and chemical barrier to infection by competitively outgrowing pathogens within the vagina (Kaewsrichan *et al.*, 2006; Klebanoff *et al.*, 1991; Voravuthikunchai *et al.*, 2006).

To date, all of the studies that have examined the composition of the vaginal microbiota of VVC patients have relied on culture-based methods or molecular fingerprinting methods. These methods are severely limited in their ability to accurately determine community structure as many species remain unculturable leaving the precise nature of the interactions between fungal species and the microbial community unclear. However, following advances in high-throughput sequencing technologies, cultivation-independent approaches have become the new standard for accurately cataloguing the members of microbial communities (Zhou *et al.*, 2004). It is now possible to describe the bacterial composition of the community in fine detail by selective sequencing of the 16S rRNA gene, an important taxonomic marker universally present in all bacteria and archaea (Fredricks *et al.*, 2005; Linhares *et al.*, 2010; Verstraelen *et al.*, 2004).

A large cross-sectional study of nearly 400 asymptomatic women has provided the most comprehensive characterization of the vaginal microbiota (Ravel *et al.*, 2011) and concluded that vaginal bacterial communities can be categorized into five distinct community state types, which differ in species composition, diversity and relative abundance (Ravel *et al.*, 2011). A subsequent study by the same group followed 32 healthy women longitudinally and demonstrated that the bacterial communities of some women varied greatly over short periods of time, while the community structure of others remained relatively unchanged, suggesting that variations in community composition or diversity could in some cases equate with health (Gajer *et al.*, 2012) (Fig. 1.1).

A detailed analysis of bacterial community composition using 16S rRNA deep sequencing has not been performed in the context of VVC, but promises to be more sensitive and accurate than the methods used previously. In light of the dynamic nature of the microbial communities in healthy women, our understanding of the relationship between VVC and bacterial community structure would benefit greatly from longitudinal studies that survey the same individual at several time points prior to, during and following treatment for a VVC episode. It should be noted that genetic diversity in a bacterial population does not necessarily indicate functional diversity in the microbial community as evidenced by in-depth analysis of several human associated microbiota (Human Microbiome Project Consortium, 2012). In other words, very similar biochemical and metabolic landscapes of the vaginal mucosa can be achieved by many different variations of bacterial community composition. The functional diversity in vaginal swabs collected over the time course of infection can be addressed by RNA-seq.

As discussed above, RNA-seq provides the ability to analyse gene expression in a manner that requires no *a priori* knowledge of the species composition in a sample (Wang *et al.*, 2009). Such an analysis performed on longitudinal samples promises to be invaluable to our understanding of VVC because it allows quantification of RNA transcripts derived from the fungal pathogen, the bacteria that make up the vaginal microbiome, and the plethora of host cells that are present in the vaginal mucosa. This integrative understanding of pathogenesis, host response and microbial influence are critical to our ability to improve routine gynaecological care and treatment of vaginal infections. As VVC remains one of the most common fungal infections, it is essential that we revise our scientific approach to understand this important and costly health problem.

It should be noted that the above scenario is meant to serve as one detailed example of how HTS can be used to obtain a comprehensive understanding of fungal disease. This general approach can, and should, be applied to any fungal disease for which the appropriate longitudinal samples can be obtained.

Future directions

While the works described in this chapter were all innovative in their application of HTS to address questions within the field of fungal pathogenesis, collectively, these studies have only scratched the surface in realizing the full potential of what the approach has to offer the field of fungal pathogenesis. As the cost of sequencing becomes

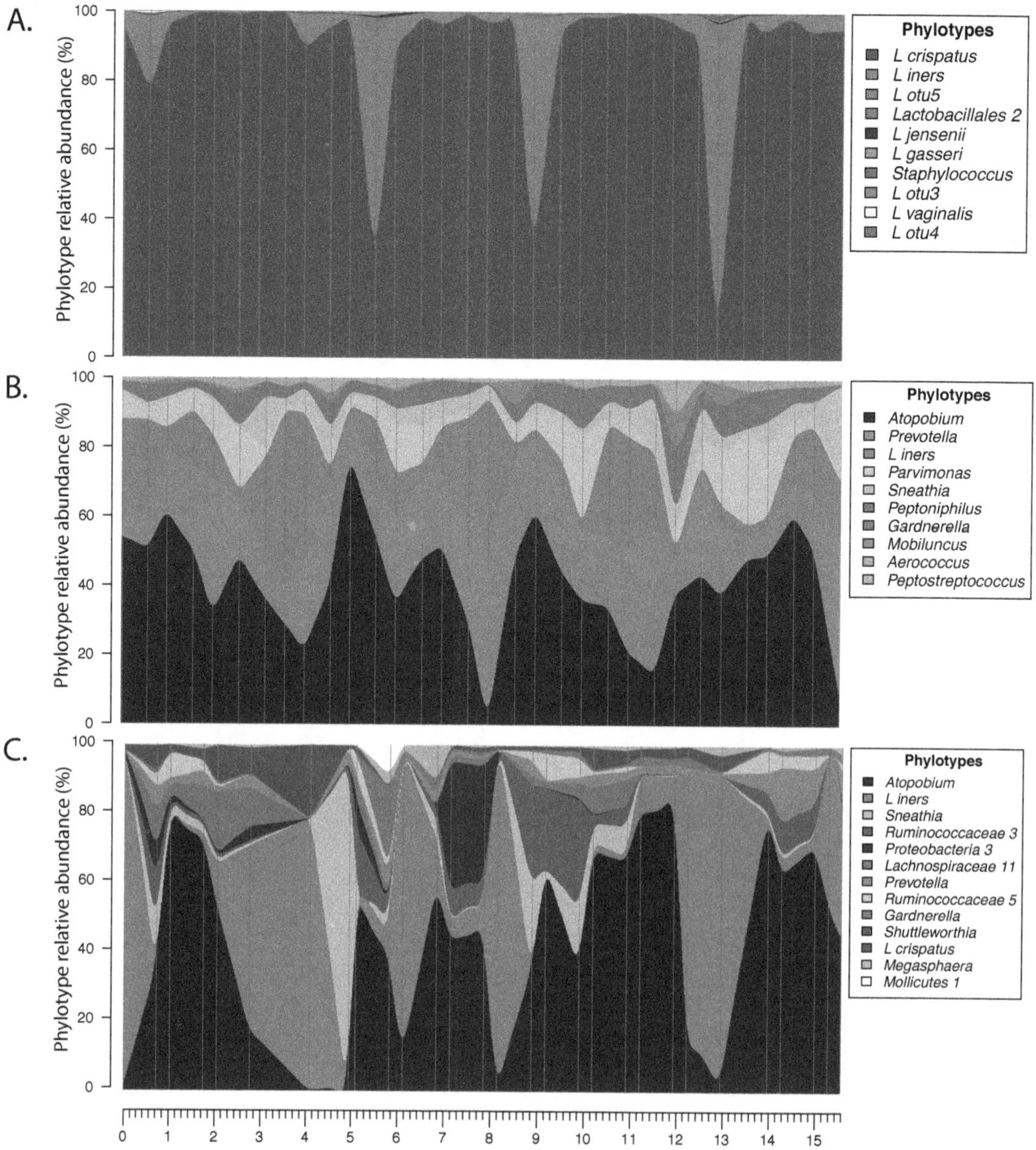

Figure 1.1 Temporal dynamics of vaginal bacterial communities from three different women over 16 weeks. Samples were collected twice-weekly and the community composition was determined by HTS of 16S rRNA gene sequences. The abundances of particular bacteria are depicted as interpolated bar graphs. All phylotypes are coloured as indicated on the right of each bar graph. (A) Example of a highly stable community dominated by *L. crispatus*. (B) Example of a moderately stable community with no or very little *Lactobacillus* sp. (C) Example of an unstable community. Graphs represent sequencing data generated by Gajer *et al.* (2012). Reproduced from Bradford *et al.* (2013) with kind permission from Springer Science Business Media B.V. A colour version of this figure is available in the plate section at the back of the book.

more affordable, the amount of sequence depth coverage that will be attainable will increase significantly allowing researchers to apply RNA-seq to problems that require significant sensitivity.

For example, RNA-seq will increasingly be used to analyse the transcriptional response of multiple interacting species such as in the context of *in vivo* infection models or clinical samples where the amount of RNA derived from the host or associated microbiota far exceeds the amount of RNA produced by the pathogen of interest. Such an approach will also increase our understanding of polymicrobial biofilms. These types of studies will not only allow a comprehensive gene expression profile of multiple species at once but will also allow the identification of transcripts that might have been overlooked by *in vitro* studies because they are only be expressed in a specific biological context. RNA-seq will also be useful to comprehensively explore allele-specific gene expression. This will be particularly useful to study the diploid fungal pathogens (e.g. several *Candida* species) that display significant heterozygosity throughout the genome.

The scope of mycobiome sequencing can also be expanded to include samples collected longitudinally over time from the same individual to learn about how the fungal community structure varies over time in both healthy and diseased individuals. In closing, a comprehensive understanding of fungal infection will require a multifaceted approach which includes all of the HTS-based approaches described in this chapter as well as genetics, molecular techniques and pathogenesis models to provide the necessary functional validation of what the genomic approaches uncover.

References

Antonio, M.A., Hawes, S.E., and Hillier, S.L. (1999). The identification of vaginal *Lactobacillus* species and the demographic and microbiologic characteristics of women colonized by these species. J. Infect. Dis. *180*, 1950–1956.

Arnaud, M.B., Costanzo, M.C., Skrzypek, M.S., Binkley, G., Lane, C., Miyasato, S.R., and Sherlock, G. (2005). The Candida Genome Database (CGD), a community resource for *Candida albicans* gene and protein information. Nucleic Acids Res. 33, D358–363.

Barski, A., Cuddapah, S., Cui, K., Roh, T.Y., Schones, D.E., Wang, Z., Wei, G., Chepelev, I., and Zhao, K. (2007). High-resolution profiling of histone methylations in the human genome. Cell *129*, 823–837.

Bensen, E.S., Martin, S.J., Li, M., Berman, J., and Davis, D.A. (2004). Transcriptional profiling in *Candida albicans* reveals new adaptive responses to extracellular pH and functions for Rim101p. Mol. Microbiol. *54*, 1335–1351.

Bentley, D.R., Balasubramanian, S., Swerdlow, H.P., Smith, G.P., Milton, J., Brown, C.G., Hall, K.P., Evers, D.J., Barnes, C.L., Bignell, H.R., *et al.* (2008). Accurate whole human genome sequencing using reversible terminator chemistry. Nature *456*, 53–59.

Biswas, S., Van Dijck, P., and Datta, A. (2007). Environmental sensing and signal transduction pathways regulating morphopathogenic determinants of *Candida albicans*. Microbiol. Mol. Biol. Rev. *71*, 348–376.

Boskey, E.R., Cone, R.A., Whaley, K.J., and Moench, T.R. (2001). Origins of vaginal acidity: high D/L lactate ratio is consistent with bacteria being the primary source. Hum. Reprod. *16*, 1809–1813.

Bradford, L.L., Ravel, J., and Bruno, V. (2013). Understanding vulvovaginal candidiasis through a community genomics approach. Curr. Fungal Infect. Rep. 7, 126–131.

Bruno, V.M., Wang, Z., Marjani, S.L., Euskirchen, G.M., Martin, J., Sherlock, G., and Snyder, M. (2010). Comprehensive annotation of the transcriptome of the human fungal pathogen *Candida albicans* using RNA-seq. Genome Res. *20*, 1451–1458.

Butler, G., Rasmussen, M.D., Lin, M.F., Santos, M.A., Sakthikumar, S., Munro, C.A., Rheinbay, E., Grabherr, M., Forche, A., Reedy, J.L., *et al.* (2009). Evolution of pathogenicity and sexual reproduction in eight *Candida* genomes. Nature *459*, 657–662.

Chabe, M., Aliouat-Denis, C.M., Delhaes, L., Aliouat el, M., Viscogliosi, E., and Dei-Cas, E. (2011). Pneumocystis: from a doubtful unique entity to a group of highly diversified fungal species. FEMS Yeast Res. *11*, 2–17.

Charlson, E.S., Diamond, J.M., Bittinger, K., Fitzgerald, A.S., Yadav, A., Haas, A.R., Bushman, F.D., and Collman, R.G. (2012). Lung-enriched organisms and aberrant bacterial and fungal respiratory microbiota after lung transplant. Am. J. Respir. Crit. Care Med. *186*, 536–545.

Chiranand, W., McLeod, I., Zhou, H., Lynn, J.J., Vega, L.A., Myers, H., Yates, J.R., 3rd, Lorenz, M.C., and Gustin, M.C. (2008). *CTA4* transcription factor mediates induction of nitrosative stress response in *Candida albicans*. Eukaryot. Cell 7, 268–278.

Desai, J.V., Bruno, V.M., Ganguly, S., Stamper, R.J., Mitchell, K.F., Solis, N., Hill, E.M., Xu, W., Filler, S.G., Andes, D.R., *et al.* (2013). Regulatory role of glycerol in *Candida albicans* biofilm formation. Mbio *4*, e00637–00612.

Dethlefsen, L., McFall-Ngai, M., and Relman, D.A. (2007). An ecological and evolutionary perspective on human-microbe mutualism and disease. Nature *449*, 811–818.

Dhamgaye, S., Bernard, M., Lelandais, G., Sismeiro, O., Lemoine, S., Coppee, J.Y., Le Crom, S., Prasad, R., and Devaux, F. (2012). RNA sequencing revealed novel actors of the acquisition of drug resistance in *Candida albicans*. BMC Genomics *13*, 396.

Dickson, R.P., Erb-Downward, J.R., and Huffnagle, G.B. (2013). The role of the bacterial microbiome in lung disease. Expert Rev. Respir. Med. 7, 245–257.

Donders, G.G., Bosmans, E., Dekeersmaecker, A., Vereecken, A., Van Bulck, B., and Spitz, B. (2000). Pathogenesis of abnormal vaginal bacterial flora. Am. J. Obstet. Gynecol. *182*, 872–878.

Drinnenberg, I.A., Weinberg, D.E., Xie, K.T., Mower, J.P., Wolfe, K.H., Fink, G.R., and Bartel, D.P. (2009). RNAi in budding yeast. Science *326*, 544–550.

Elmore, M.H., Gibbons, J.G., and Rokas, A. (2012). Assessing the genome-wide effect of promoter region tandem repeat natural variation on gene expression. G3 (Bethesda) *2*, 1643–1649.

Emrich, S.J., Barbazuk, W.B., Li, L., and Schnable, P.S. (2007). Gene discovery and annotation using LCM-454 transcriptome sequencing. Genome Res. *17*, 69–73.

Enjalbert, B., Smith, D.A., Cornell, M.J., Alam, I., Nicholls, S., Brown, A.J., and Quinn, J. (2006). Role of the Hog1 stress-activated protein kinase in the global transcriptional response to stress in the fungal pathogen *Candida albicans*. Mol. Biol. Cell *17*, 1018–1032.

Fidel, P.L., Jr. (2007). History and update on host defense against vaginal candidiasis. Am. J. Reprod. Immunol. *57*, 2–12.

Findley, K., Oh, J., Yang, J., Conlan, S., Deming, C., Meyer, J.A., Schoenfeld, D., Nomicos, E., Park, M., Kong, H.H., *et al.* (2013). Topographic diversity of fungal and bacterial communities in human skin. Nature *498*, 367–370.

Fredricks, D.N., Fiedler, T.L., and Marrazzo, J.M. (2005). Molecular identification of bacteria associated with bacterial vaginosis. N. Engl. J. Med. *353*, 1899–1911.

Gajer, P., Brotman, R.M., Bai, G., Sakamoto, J., Schutte, U.M., Zhong, X., Koenig, S.S., Fu, L., Ma, Z.S., Zhou, X., *et al.* (2012). Temporal dynamics of the human vaginal microbiota. Sci. Transl. Med. *4*, 132ra152.

Geiser, D.M., Klich, M.A., Frisvad, J.C., Peterson, S.W., Varga, J., and Samson, R.A. (2007). The current status of species recognition and identification in *Aspergillus*. Stud. Mycol. *59*, 1–10.

Ghannoum, M.A., Jurevic, R.J., Mukherjee, P.K., Cui, F., Sikaroodi, M., Naqvi, A., and Gillevet, P.M. (2010). Characterization of the oral fungal microbiome (mycobiome) in healthy individuals. PLoS Pathog. *6*, e1000713.

Gibbons, J.G., Beauvais, A., Beau, R., McGary, K.L., Latge, J.P., and Rokas, A. (2012). Global transcriptome changes underlying colony growth in the opportunistic human pathogen *Aspergillus fumigatus*. Eukaryot. Cell *11*, 68–78.

Grant, P.A., Duggan, L., Cote, J., Roberts, S.M., Brownell, J.E., Candau, R., Ohba, R., Owen-Hughes, T., Allis, C.D., Winston, F., *et al.* (1997). Yeast Gcn5 functions in two multisubunit complexes to acetylate nucleosomal histones: characterization of an Ada complex and the SAGA (Spt/Ada) complex. Genes Dev. *11*, 1640–1650.

Gray, R.H., Wawer, M.J., Sewankambo, N., and Serwadda, D. (1997). HIV-1 infection associated with abnormal vaginal flora morphology and bacterial vaginosis. Lancet *350*, 1780.

Grumaz, C., Lorenz, S., Stevens, P., Lindemann, E., Schock, U., Retey, J., Rupp, S., and Sohn, K. (2013). Species and condition specific adaptation of the transcriptional landscapes in *Candida albicans* and *Candida dubliniensis*. BMC Genomics *14*, 212.

Guida, A., Lindstadt, C., Maguire, S.L., Ding, C., Higgins, D.G., Corton, N.J., Berriman, M., and Butler, G. (2011). Using RNA-seq to determine the transcriptional landscape and the hypoxic response of the pathogenic yeast *Candida parapsilosis*. BMC Genomics *12*, 628.

Gupta, K., Stapleton, A.E., Hooton, T.M., Roberts, P.L., Fennell, C.L., and Stamm, W.E. (1998). Inverse association of H_2O2-producing lactobacilli and vaginal *Escherichia coli* colonization in women with recurrent urinary tract infections. J. Infect. Dis. *178*, 446–450.

Haynes, B.C., Skowyra, M.L., Spencer, S.J., Gish, S.R., Williams, M., Held, E.P., Brent, M.R., and Doering, T.L. (2011). Toward an integrated model of capsule regulation in *Cryptococcus neoformans*. PLoS Pathog. *7*, e1002411.

Hettiarachchi, N., Ashbee, H.R., and Wilson, J.D. (2010). Prevalence and management of non-*albicans* vaginal candidiasis. Sex. Transm. Infect. *86*, 99–100.

Hnisz, D., Majer, O., Frohner, I.E., Komnenovic, V., and Kuchler, K. (2010). The Set3/Hos2 histone deacetylase complex attenuates cAMP/PKA signaling to regulate morphogenesis and virulence of *Candida albicans*. PLoS Pathog. *6*, e1000889.

Hnisz, D., Bardet, A.F., Nobile, C.J., Petryshyn, A., Glaser, W., Schock, U., Stark, A., and Kuchler, K. (2012). A histone deacetylase adjusts transcription kinetics at coding sequences during *Candida albicans* morphogenesis. PLoS Genet. *8*, e1003118.

Horak, C.E., Mahajan, M.C., Luscombe, N.M., Gerstein, M., Weissman, S.M., and Snyder, M. (2002). GATA-1 binding sites mapped in the β-globin locus by using mammalian chIp-chip analysis. Proc. Natl. Acad. Sci. U.S.A. *99*, 2924–2929.

Hromatka, B.S., Noble, S.M., and Johnson, A.D. (2005). Transcriptional response of *Candida albicans* to nitric oxide and the role of the *YHB1* gene in nitrosative stress and virulence. Mol. Biol. Cell *16*, 4814–4826.

Human Microbiome Project Consortium (2012). Structure, function and diversity of the healthy human microbiome. Nature *486*, 207–214.

Iliev, I.D., Funari, V.A., Taylor, K.D., Nguyen, Q., Reyes, C.N., Strom, S.P., Brown, J., Becker, C.A., Fleshner, P.R., Dubinsky, M., *et al.* (2012). Interactions between commensal fungi and the C-type lectin receptor Dectin-1 influence colitis. Science *336*, 1314–1317.

Juarez Tomas, M.S., Ocana, V.S., Wiese, B., and Nader-Macias, M.E. (2003). Growth and lactic acid production by vaginal *Lactobacillus acidophilus* CRL 1259, and inhibition of uropathogenic *Escherichia coli*. J. Med. Microbiol. *52*, 1117–1124.

Kaewsrichan, J., Peeyananjarassri, K., and Kongprasertkit, J. (2006). Selection and identification of anaerobic lactobacilli producing inhibitory compounds against vaginal pathogens. FEMS Immunol. Med. Microbiol. *48*, 75–83.

Klebanoff, S.J., Hillier, S.L., Eschenbach, D.A., and Waltersdorph, A.M. (1991). Control of the microbial flora of the vagina by H_2O_2-generating lactobacilli. J. Infect. Dis. *164*, 94–100.

Kong, H.H., and Segre, J.A. (2012). Skin microbiome: looking back to move forward. J. Invest. Dermatol. *132*, 933–939.

Lai, S.K., Hida, K., Shukair, S., Wang, Y.Y., Figueiredo, A., Cone, R., Hope, T.J., and Hanes, J. (2009). Human immunodeficiency virus type 1 is trapped by acidic but not by neutralized human cervicovaginal mucus. J. Virol. *83*, 11196–11200.

Leigh, J.E., Barousse, M., Swoboda, R.K., Myers, T., Hager, S., Wolf, N.A., Cutright, J.L., Thompson, J., Sobel, J.D., and Fidel, P.L., Jr. (2001). *Candida*-specific systemic cell-mediated immune reactivities in human immunodeficiency virus-positive persons with mucosal candidiasis. J. Infect. Dis. *183*, 277–285.

Ley, R.E., Turnbaugh, P.J., Klein, S., and Gordon, J.I. (2006). Microbial ecology: human gut microbes associated with obesity. Nature *444*, 1022–1023.

Lin, J.Q., Zhao, X.X., Zhi, Q.Q., Zhao, M., and He, Z.M. (2013). Transcriptomic profiling of *Aspergillus flavus* in response to 5-azacytidine. Fungal Genet. Biol. *56*, 78–86.

Linhares, I.M., Giraldo, P.C., and Baracat, E.C. (2010). [New findings about vaginal bacterial flora]. Rev. Assoc. Med. Bras. *56*, 370–374.

Liu, L., Li, Y., Li, S., Hu, N., He, Y., Pong, R., Lin, D., Lu, L., and Law, M. (2012). Comparison of next-generation sequencing systems. J. Biomed. Biotechnol. *2012*, 251364.

Lorenz, M.C., and Fink, G.R. (2001). The glyoxylate cycle is required for fungal virulence. Nature *412*, 83–86.

Ma, B., Forney, L.J., and Ravel, J. (2012). Vaginal microbiome: rethinking health and disease. Annu. Rev. Microbiol. *66*, 371–389.

Margulies, M., Egholm, M., Altman, W.E., Attiya, S., Bader, J.S., Bemben, L.A., Berka, J., Braverman, M.S., Chen, Y.J., Chen, Z., *et al.* (2005). Genome sequencing in microfabricated high-density picolitre reactors. Nature *437*, 376–380.

Marioni, J.C., Mason, C.E., Mane, S.M., Stephens, M., and Gilad, Y. (2008). RNA-seq: an assessment of technical reproducibility and comparison with gene expression arrays. Genome Res. *18*, 1509–1517.

Marr, K.A., Carter, R.A., Crippa, F., Wald, A., and Corey, L. (2002). Epidemiology and outcome of mould infections in hematopoietic stem cell transplant recipients. Clin. Infect. Dis. *34*, 909–917.

Mazmanian, S.K., Liu, C.H., Tzianabos, A.O., and Kasper, D.L. (2005). An immunomodulatory molecule of symbiotic bacteria directs maturation of the host immune system. Cell *122*, 107–118.

Mi, S., Cai, T., Hu, Y., Chen, Y., Hodges, E., Ni, F., Wu, L., Li, S., Zhou, H., Long, C., *et al.* (2008). Sorting of small RNAs into *Arabidopsis* argonaute complexes is directed by the 5′ terminal nucleotide. Cell *133*, 116–127.

Morgan, J., Wannemuehler, K.A., Marr, K.A., Hadley, S., Kontoyiannis, D.P., Walsh, T.J., Fridkin, S.K., Pappas, P.G., and Warnock, D.W. (2005). Incidence of invasive aspergillosis following hematopoietic stem cell and solid organ transplantation: interim results of a prospective multicenter surveillance program. Med. Mycol. *43* (Suppl 1), S49–58.

Mortazavi, A., Williams, B.A., McCue, K., Schaeffer, L., and Wold, B. (2008). Mapping and quantifying mammalian transcriptomes by RNA-Seq. Nat. Methods *5*, 621–628.

Muller, S., Baldin, C., Groth, M., Guthke, R., Kniemeyer, O., Brakhage, A.A., and Valiante, V. (2012). Comparison of transcriptome technologies in the pathogenic fungus *Aspergillus fumigatus* reveals novel insights into the genome and MpkA dependent gene expression. BMC Genomics *13*, 519.

Nagalakshmi, U., Wang, Z., Waern, K., Shou, C., Raha, D., Gerstein, M., and Snyder, M. (2008). The transcriptional landscape of the yeast genome defined by RNA sequencing. Science *320*, 1344–1349.

Nobile, C.J., Fox, E.P., Nett, J.E., Sorrells, T.R., Mitrovich, Q.M., Hernday, A.D., Tuch, B.B., Andes, D.R., and Johnson, A.D. (2012). A recently evolved transcriptional network controls biofilm development in *Candida albicans*. Cell *148*, 126–138.

O'Meara, T.R., and Alspaugh, J.A. (2012). The *Cryptococcus neoformans* capsule: a sword and a shield. Clin. Microbiol. Rev. *25*, 387–408.

Pareek, C.S., Smoczynski, R., and Tretyn, A. (2011). Sequencing technologies and genome sequencing. J. Appl. Genet. *52*, 413–435.

Park, P.J. (2009). ChIP-seq: advantages and challenges of a maturing technology. Nat. Rev. Genet. *10*, 669–680.

Peterson, J., Garges, S., Giovanni, M., McInnes, P., Wang, L., Schloss, J.A., Bonazzi, V., McEwen, J.E., Wetterstrand, K.A., Deal, C., *et al.* (2009). The NIH human microbiome project. Genome Res. *19*, 2317–2323.

Pybus, V., and Onderdonk, A.B. (1999). Microbial interactions in the vaginal ecosystem, with emphasis on the pathogenesis of bacterial vaginosis. Microbes Infect. *1*, 285–292.

Rapaport, F., Khanin, R., Liang, Y., Pirun, M., Krek, A., Zumbo, P., Mason, C.E., Socci, N.D., and Betel, D. (2013). Comprehensive evaluation of differential gene expression analysis methods for RNA-seq data. Genome Biol. *14*, R95.

Ravel, J., Gajer, P., Abdo, Z., Schneider, G.M., Koenig, S.S., McCulle, S.L., Karlebach, S., Gorle, R., Russell, J., Tacket, C.O., *et al.* (2011). Vaginal microbiome of reproductive-age women. Proc. Natl. Acad. Sci. U.S.A. *108* (Suppl. 1), 4680–4687.

Robertson, G., Hirst, M., Bainbridge, M., Bilenky, M., Zhao, Y., Zeng, T., Euskirchen, G., Bernier, B., Varhol, R., Delaney, A., *et al.* (2007). Genome-wide profiles of STAT1 DNA association using chromatin immunoprecipitation and massively parallel sequencing. Nat. Methods *4*, 651–657.

Rubin-Bejerano, I., Fraser, I., Grisafi, P., and Fink, G.R. (2003). Phagocytosis by neutrophils induces an amino acid deprivation response in *Saccharomyces cerevisiae* and *Candida albicans*. Proc. Natl. Acad. Sci. U.S.A. *100*, 11007–11012.

Sanger, F., Nicklen, S., and Coulson, A.R. (1977). DNA sequencing with chain-terminating inhibitors. Proc. Natl. Acad. Sci. U.S.A. *74*, 5463–5467.

Schuman, P., Sobel, J.D., Ohmit, S.E., Mayer, K.H., Carpenter, C.C., Rompalo, A., Duerr, A., Smith, D.K., Warren, D., and Klein, R.S. (1998). Mucosal Candidal colonization and candidiasis in women with or at risk for human immunodeficiency virus infection. HIV Epidemiology Research Study (HERS) Group. Clin. Infect. Dis. 27, 1161–1167.

Slotkin, R.K., and Martienssen, R. (2007). Transposable elements and the epigenetic regulation of the genome. Nat. Rev. Genet. *8*, 272–285.

Sobel, J.D. (1999). Is there a protective role for vaginal flora? Curr. Infect. Dis. Rep. *1*, 379–383.

Sobel, J.D. (2007). Vulvovaginal candidosis. Lancet *369*, 1961–1971.

Sobel, J.D., Faro, S., Force, R.W., Foxman, B., Ledger, W.J., Nyirjesy, P.R., Reed, B.D., and Summers, P.R. (1998). Vulvovaginal candidiasis: epidemiologic, diagnostic, and therapeutic considerations. Am. J. Obstet. Gynecol. *178*, 203–211.

Solomon, M.J., Larsen, P.L., and Varshavsky, A. (1988). Mapping protein–DNA interactions *in vivo* with formaldehyde: evidence that histone H4 is retained on a highly transcribed gene. Cell *53*, 937–947.

Sun, N., Fonzi, W., Chen, H., She, X., Zhang, L., and Calderone, R. (2013). Azole susceptibility and transcriptome profiling in *Candida albicans* mitochondrial electron transport chain complex I mutants. Antimicrob. Agents Chemother. *57*, 532–542.

Sultan, M., Schulz, M.H., Richard, H., Magen, A., Klingenhoff, A., Scherf, M., Seifert, M., Borodina, T., Soldatov, A., Parkhomchuk, D., *et al.* (2008). A global view of gene activity and alternative splicing by deep sequencing of the human transcriptome. Science *321*, 956–960.

Taha, T.E., Hoover, D.R., Dallabetta, G.A., Kumwenda, N.I., Mtimavalye, L.A., Yang, L.P., Liomba, G.N., Broadhead, R.L., Chiphangwi, J.D., and Miotti, P.G. (1998). Bacterial vaginosis and disturbances of vaginal flora: association with increased acquisition of HIV. AIDS *12*, 1699–1706.

Tierney, L., Linde, J., Muller, S., Brunke, S., Molina, J.C., Hube, B., Schock, U., Guthke, R., and Kuchler, K. (2012). An interspecies regulatory network inferred from simultaneous RNA-seq of *Candida albicans* invading innate immune cells. Front. Microbiol. *3*, 85.

Trapnell, C., Pachter, L., and Salzberg, S.L. (2009). TopHat: discovering splice junctions with RNA-Seq. Bioinformatics *25*, 1105–1111.

Tuch, B.B., Mitrovich, Q.M., Homann, O.R., Hernday, A.D., Monighetti, C.K., De La Vega, F.M., and Johnson, A.D. (2010). The transcriptomes of two heritable cell types illuminate the circuit governing their differentiation. PLoS Genet. *6*, e1001070.

Turnbaugh, P.J., Ley, R.E., Hamady, M., Fraser-Liggett, C.M., Knight, R., and Gordon, J.I. (2007). The human microbiome project. Nature *449*, 804–810.

van De Wijgert, J.H., Mason, P.R., Gwanzura, L., Mbizvo, M.T., Chirenje, Z.M., Iliff, V., Shiboski, S., and Padian, N.S. (2000). Intravaginal practices, vaginal flora disturbances, and acquisition of sexually transmitted diseases in Zimbabwean women. J. Infect. Dis. *181*, 587–594.

Verstraelen, H., Verhelst, R., Claeys, G., Temmerman, M., and Vaneechoutte, M. (2004). Culture-independent analysis of vaginal microflora: the unrecognized association of Atopobium vaginae with bacterial vaginosis. Am. J. Obstet. Gynecol. *191*, 1130–1132.

Voravuthikunchai, S.P., Bilasoi, S., and Supamala, O. (2006). Antagonistic activity against pathogenic bacteria by human vaginal lactobacilli. Anaerobe *12*, 221–226.

Wang, A., Kurdistani, S.K., and Grunstein, M. (2002). Requirement of Hos2 histone deacetylase for gene activity in yeast. Science *298*, 1412–1414.

Wang, X., Hsueh, Y.P., Li, W., Floyd, A., Skalsky, R., and Heitman, J. (2010a). Sex-induced silencing defends the genome of *Cryptococcus neoformans* via RNAi. Genes Dev. *24*, 2566–2582.

Wang, X., Wu, Z., and Zhang, X. (2010b) Isoform abundance inference provides a more accurate estimation of gene expression levels in RNA-seq. J. Bioinform. Comput. Biol. *8* (Suppl. 1), 177–192.

Wang, Z., Zang, C., Rosenfeld, J.A., Schones, D.E., Barski, A., Cuddapah, S., Cui, K., Roh, T.Y., Peng, W., Zhang, M.Q., *et al.* (2008). Combinatorial patterns of histone acetylations and methylations in the human genome. Nat. Genet. *40*, 897–903.

Wang, Z., Gerstein, M., and Snyder, M. (2009). RNA-Seq: a revolutionary tool for transcriptomics. Nat. Rev. Genet. *10*, 57–63.

Weber, A.P., Weber, K.L., Carr, K., Wilkerson, C., and Ohlrogge, J.B. (2007). Sampling the Arabidopsis transcriptome with massively parallel pyrosequencing. Plant Physiol. *144*, 32–42.

Westermann, A.J., Gorski, S.A., and Vogel, J. (2012). Dual RNA-seq of pathogen and host. Nat. Rev. Microbiol. *10*, 618–630.

Wiesenfeld, H.C., Hillier, S.L., Krohn, M.A., Landers, D.V., and Sweet, R.L. (2003). Bacterial vaginosis is a strong predictor of *Neisseria gonorrhoeae* and *Chlamydia trachomatis* infection. Clin Infect Dis *36*, 663–668.

Wilhelm, B.T., Marguerat, S., Watt, S., Schubert, F., Wood, V., Goodhead, I., Penkett, C.J., Rogers, J., and Bahler, J. (2008). Dynamic repertoire of a eukaryotic transcriptome surveyed at single-nucleotide resolution. Nature *453*, 1239–1243.

Yu, J., Fedorova, N.D., Montalbano, B.G., Bhatnagar, D., Cleveland, T.E., Bennett, J.W., and Nierman, W.C. (2011). Tight control of mycotoxin biosynthesis gene expression in *Aspergillus flavus* by temperature as revealed by RNA-Seq. FEMS Microbiol. Lett. *322*, 145–149.

Zhou, X., Bent, S.J., Schneider, M.G., Davis, C.C., Islam, M.R., and Forney, L.J. (2004). Characterization of vaginal microbial communities in adult healthy women using cultivation-independent methods. Microbiology *150*, 2565–2573.

Comparative Genomics and Evolutionary Analyses of Human Fungal Pathogens

2

David A. Fitzpatrick

Abstract

Recent advances in next generation sequencing techniques have dramatically increased the availability of genomic data. Due to their relatively small genome size and importance as human and crop pathogens, over one hundred fungal genomes have been completely sequenced and published to date. This number is expected to increase dramatically over the coming years with individual institutions sequencing their own fungi of interest and the initiation of an ambitious project to sequence 1000 neglected fungal genomes (http://1000.fungalgenomes.org/). Currently a significant proportion of the available genomes are human pathogens as well as closely related non-pathogenic species. This type of data allows us to perform comparative genomic analyses between pathogenic and non-pathogenic species in an effort to uncover molecular mechanisms related to disease. The availability of full genome sequences also allows us to investigate many evolutionary processes in fungi such as horizontal gene transfer, gene fusions/fissions and intron gains and losses. The following chapter will discuss a selection of key comparative genomic analyses that have been performed on a number of the most common human pathogens and also illustrate how complete fungal genomes have helped us understand some of the mechanisms that have shaped fungal evolution.

Introduction

A recent report highlighted the massive effects plant pathogens can have on ecosystems and food sources and in some cases even leading to species extinctions (Fisher *et al.*, 2012). It is also widely accepted that human fungal pathogens can have a serious impact on the health of human populations. Superficial fungal infections affect approximately 25% of the worldwide human population (Havlickova *et al.*, 2008). While the majority of superficial infections are not lethal they can cause serious discomfort such as athlete's foot, ringworm of the scalp, nail infections and thrush to name but a few. The frequency of invasive fungal infections is much lower but a more serious concern as they have very high associated mortality rates and are estimated to be responsible for approximately 1.5 million deaths worldwide annually (Brown *et al.*, 2012). It has also been estimated that 90% of all reported fungal related deaths are caused by species from one of four genera (*Candida, Aspergillus, Pneumocystis* and *Cryptococcus*; Table 2.1) (Brown *et al.*, 2012). Owing to their impact on human health the last decade has seen the publication of many fungal pathogens. The availability of complete genomes has helped us uncover some of the molecular mechanisms that make these fungal species pathogenic. Furthermore, the sequencing of closely related yet non-pathogenic species enables us to perform comparative genomic approaches to uncover the genetic and metabolic mechanisms that may make one species a pathogen while other closely related species relatively harmless.

As well as helping uncover important mechanisms associated with disease and niche specification, the availability of a wide variety of fungal genomes has pushed this kingdom to the forefront of eukaryotic evolutionary studies. In this chapter I will document the comparative

Table 2.1 Estimated statistics of the 10 most significant invasive fungal pathogens. Estimates taken from Brown *et al.* (2012)

Disease	Most common species	Estimated life-threatening infections
Aspergillosis*	*Aspergillus fumigatus*	>200,000
Candidiasis*	*Candida albicans*	>400,000
Cryptococcosis*	*Cryptococcus neoformans*	>1,000,000
Mucormycosis*	*Rhizopus oryzae*	>10,000
*Pneumocystis**	*Pneumocystis jirovecii*	>400,000
Blastomycosis	*Blastomyces dermatitidis*	~3000
Coccidioidomycosis*	*Coccidioides immitis*	~25,000
Histoplasmosis	*Histoplasma capsulatum*	~25,000
Paracoccidioidomycosis*	*Paracoccidioides brasiliensis*	~4000
Penicilliosis	*Penicillium marneffei*	>8000

*Denotes comparative genomic analysis has been performed and is discussed in main body of text.

analyses that have been performed in a number of common human pathogens. Furthermore I will highlight a number of studies that have set out to illustrate how human pathogenic fungal genomes have helped aid a number of evolutionary studies such as the mechanisms of gene fission and fusion, intron evolution, rates of horizontal gene transfer into fungi and fungal phylogenomic studies.

Comparative genomics of *Cryptococcus* species

Cryptococcus species belong to the phylum basidiomycota (Fig. 2.1) and until the late twentieth century would have been considered a minor human pathogen causing unusual mycoses. However an increase in the incidence of immunocompromised patients as a result of HIV and aggressive cancer therapies has resulted in a dramatic increase in the incidence of cryptococcosis.

Initially three varieties of *Cryptococcus neoformans* were recognized (var. *grubii*, var. *neoformans* and var. *gattii*). These varieties are thought to have diverged from one another over 40 million years ago (Xu *et al.*, 2000). Recent analysis has shown that *Cryptococcus neoformans* var. *gattii* is actually an independent species and is now designated *Cryptococcus gattii*. While *C. neoformans* generally causes disease in immunocompromised individuals, *C. gattii* is capable of causing disease in immunocompetent individuals exemplified by *C. gattii* outbreaks on Vancouver island, British Columbia over the last 10 years (Hoang *et al.*, 2004). *C. gattii* and *C. neoformans* share a number of strongly recognized virulence factors such as growth at 37°C (normal human body temperature), a polysaccharide capsule and melanin. The polysaccharide capsule surrounds the *Cryptococcus* cell wall and is known to deplete competent components and interfere with macrophage phagocytosis thereby helping evasion from the host immune response (Bose *et al.*, 2003). Unlike *C. neoformans*, *C. gattii* can also assimilate D-tryptophan and D-proline, use glycine as a sole source of carbon and nitrogen and is resistant to the growth inhibitors canavanine and cycloheximide (Sorrell, 2001). These additional virulence factors may be responsible for the prolonged antifungal treatment required for *C. gattii* strains compared to *C. neoformans* isolates.

In 2005 the genomes of *C. neoformans* JEC21 and B-3501A (serotype D strains) were fully sequenced (Loftus *et al.*, 2005). The size of each genome was 19 Mb and 18.5 Mb respectively spanning 14 chromosomes (Table 2.2). Both genomes were found to be very intron rich containing on average 5.3 and 6.3 introns per gene and cDNA evidence pointed towards alternative transcripts for 4.2% of all genes (Loftus *et al.*, 2005). The vast majority of the ~6500 genes shared by both strains are highly similar (>98% at the nucleotide level). However the phenotypes of both strains differ significantly with B-3501A being more themostable and virulent in animal models than JEC21.

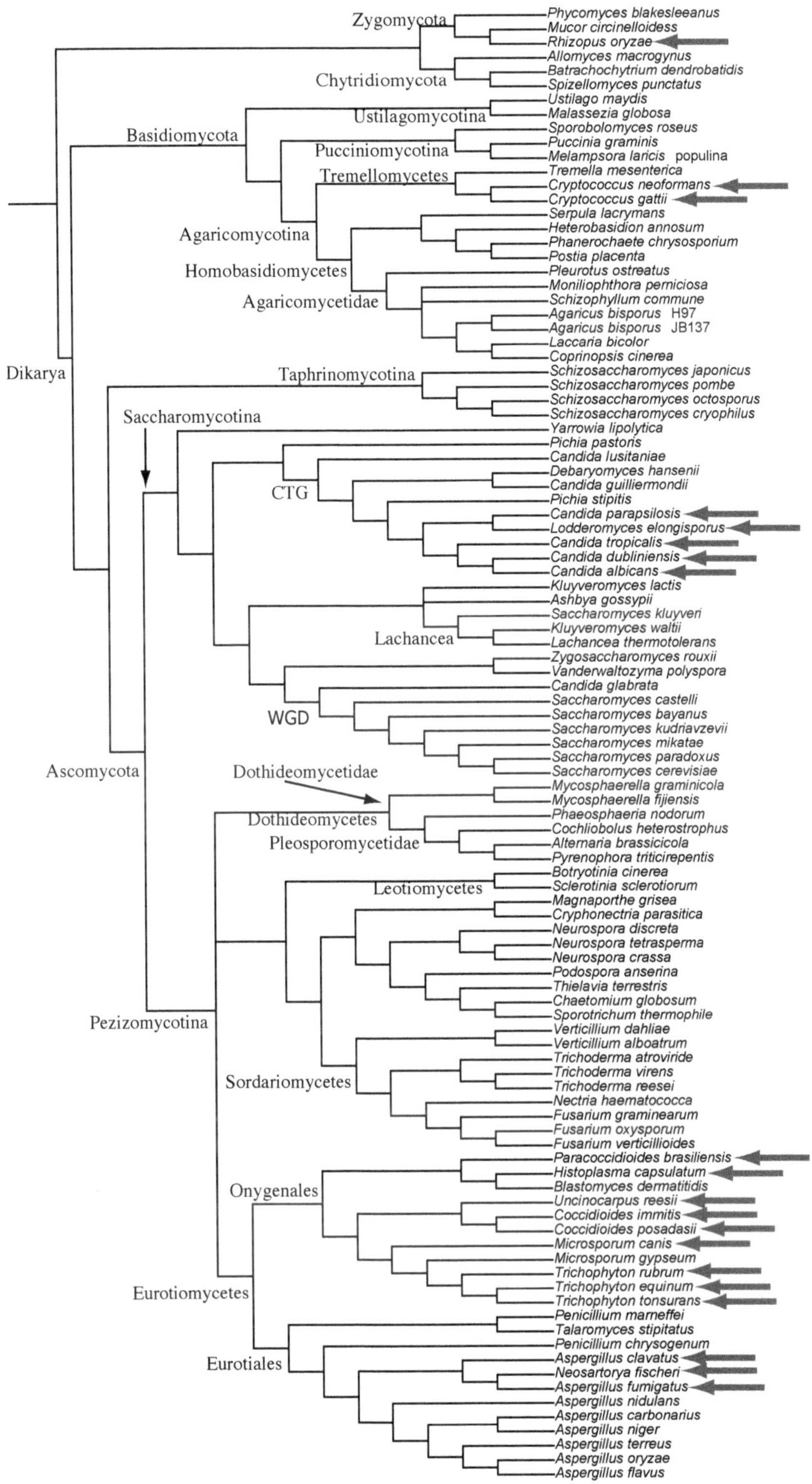

Figure 2.1 Fungal tree of life modified from Medina *et al*. (2011). A number of species discussed in the main text are highlighted with red arrows. A colour version of this figure is available in the plate section at the back of the book.

Table 2.2 Genome statistics for a number of species discussed in this chapter

Organism	Length (Mb)	GC content (%)	No. of genes	Reference
A. fumigatus Af293	28.8	50	9631	Nierman *et al.* (2005)
A. fumigatus A1163	29.2	49	9906	Fedorova *et al.* (2008)
N. fischeri	32.5	49	10,407	Fedorova *et al.* (2008)
A. clavatus	27.8	49	9125	Fedorova *et al.* (2008)
P. lutzii	32.9	42.8	9132	Desjardins *et al.* (2011)
P. brasiliensis Pb03	29.1	44.5	7875	Desjardins *et al.* (2011)
P. brasiliensis Pb18	30.0	44.4	8741	Desjardins *et al.* (2011)
C. gattii WM276	18.4	47.88	6565	D'Souza *et al.* (2011)
C. neoformans JEC21	19.0	48.0	6572	Loftus *et al.* (2005)
C. albicans SC5314	14.3	33.5	6107	Jones *et al.* (2004)
C. dubliniensis	14.6	33.2	5758	Jackson *et al.* (2009)
C. tropicalis	14.5	33.1	6258	Butler *et al.* (2009)
C. parapsilosis	13.1	38.7	5733	Butler *et al.* (2009)
L. elongisporus	15.4	37.0	5802	Butler *et al.* (2009)
C. orthopsilosis	12.6	37.6	5700	Riccombeni *et al.* (2012)
D. hansenii	12.2	36.3	6318	Sherman *et al.* (2004)
C. immitis	28.9	46	10,355	Sharpton *et al.* (2009)
C. posadassi	27.0	46.6	7229	Sharpton *et al.* (2009)
U. reesii	22.3	48.2	7798	Sharpton *et al.* (2009)
H. capsulatum	33.0	42.8	9390	Sharpton *et al.* (2009)
P. jirovecii	8.1	28.4	3898	Cisse *et al.* (2013)
P. jirovecii	6.3	32.5	4591	Hauser *et al.* (2010)
T. rubrum	22.5	48.3	8707	Martinez *et al.* (2012)
T. tonsurans	23.0	48.1	8523	Martinez *et al.* (2012)
T. equinum	24.1	47.3	8679	Martinez *et al.* (2012)
M. canis	23.1	47.5	8915	Martinez *et al.* (2012)
M. gypseum	23.2	48.5	8907	Martinez *et al.* (2012)
T. verrucosum	22.5	48.2	8024	Burmester *et al.* (2011)
A. benhamiae	22.2	48.7	7980	Burmester *et al.* (2011)
R. oryzae	45.3	35.6	17,467	Ma *et al.* (2009)

In an attempt to uncover the genetic mechanisms that may have led to these alternative phenotypes strain specific genes were investigated. Overall small differences were observed with B-3501A having three strain-specific genes, two of which encoded proteins of unknown function and the third a Ras guanosine triphosphatase-activating protein (Loftus *et al.*, 2005). JEC21 contains four additional proteins of unknown function. Genome analysis of both strains identified a range of 30 novel genes involved in capsule biosynthesis and also 50 novel extracellular mannoproteins that are associated with the cell wall (Loftus *et al.*, 2005).

Two *C. gattii* genomes were sequenced in 2011, these included R265, a clinical strain and WM276 the predominant global molecular type (D'Souza *et al.*, 2011). The *C. gattii* WM276 genome is similar in size to *C. neoformans* encompassing 18.4 Mb in 14 chromosomes and containing ~6500 genes. Chromosomal alignments showed that overall nucleotide sequence divergence was higher than expected (~7.6%), similar to what one would observe for closely related but independent species. Overall WM276 was found to contain 154 strain specific genes while R265 contains 108 strain specific genes. The majority of these genes encode hypothetical genes but some notable exceptions included the RNA-induced silencing complex Argonaute genes (Ago1 and Ago2) which are present in WM276 but not in R265 (D'Souza *et al.*, 2011). The authors also examined possible mechanisms of fluconazole resistance in some *C. gattii* strains (i.e. R1413F and R1412F). Both strains showed elevated copy numbers for a number of chromosomes (i.e. chromosomes 9 and 11 and 2,9 and 10 respectively), furthermore chromosome 10 of R1413F appears to be disomic. Disomic chromosomes have been observed in clinically important fluconazole resistance *C. neoformans* isolates previously (Hu *et al.*, 2008; Sionov *et al.*, 2010).

At the whole-genome level *C. gattii* and *C. neoformans* share approximately 87% identity. A number of obvious chromosomal rearrangements are evident between *C. gattii* and *C. neoformans* species particularly between chromosomes 4, 9 and 10. These chromosomal rearrangements may be linked to speciation, as rearrangements can lead to inhibition of chromosomal recombination and pairing (Rieseberg, 2001). Comparing protein-coding genes *C. gattii* (WM276) has 565 genes not found in *C. neoformans* (JEC21), while there are 254 genes exclusive to *C. neoformans* relative to *C. gattii* (D'Souza *et al.*, 2011). An interesting difference included the absence of the isochorismatase gene from *C. neoformans*, this enzyme has a predicted role in siderophore and secondary metabolite biosynthesis (Litwin *et al.*, 1996; May *et al.*, 2001) and deletion in a WM276 mutant led to loss of virulence (D'Souza *et al.*, 2011).

Comparative genomics of *Aspergillus* species

Aspergillus species are Ascomycetes that belong to the Eurotiales order (Fig. 2.1). The *Aspergillus* genus is the most densely sampled in terms of available genome sequences. Currently there are 14 genomes available including pathogens of animals, plants and corals such as *Aspergillus fumigatus*, *Aspergillus terreus*, *Aspergillus flavus* and *Aspergillus sydowii* and species that are important for biotechnological and food production such as *Aspergillus niger*, *Aspergillus oryzae*, *Aspergillus sojae* and *Aspergillus kawachii*. The *Aspergillus* genus is as diverse as the Vertebrate phylum (Rokas *et al.*, 2007). For example, apparently closely related species such as *A. fumigatus* and *A. fischerianus* are as divergent as humans and mice (Fedorova *et al.*, 2008). To date all *Aspergillus* genome sequencing projects have shown that all species contain eight chromosomes with genomes ranging in size from ~27.9 Mb for *A. clavatus* to ~40 Mb for *A. sojae*. Although there appears to be karyotype stability amongst *Aspergillus* species, studies have shown varying chromosomal numbers amongst natural species (Geiser *et al.*, 1996).

The following section will concentrate on *Aspergillus fumigatus*, an organism that is responsible for the highest number of human deaths due to fungi and second only to *Candida albicans* in terms of the number of infections it causes. *A. fumigatus* is both a primary and opportunist pathogen and is a major allergen associated with sinusitis and asthma (Zureik *et al.*, 2002). In immunocompromised patients opportunistic mycelial growth can be observed in pulmonary tissue causing invasive aspergillosis with a mortality rate of ~50%. Postoperative primary pathogenesis of human organs has also been observed (Pasqualotto and Denning, 2006). The medically important genome sequence of *A. fumigatus* Af293 was completed in 2005 (Nierman *et al.*, 2005). Initial genomic analyses showed that this strain is enriched with heat shock responsive genes that may have a role in thermotolerance and also many genes associated with mating and sexual development indicating the potential for this organism to undergo sexual reproduction, a mechanism that was subsequently verified (O'Gorman *et al.*, 2009).

A subsequent study sequenced a second clinically important *A. fumigatus* strain (A1163) as well as two closely related but poorly pathogenic species (i.e. *Neosartorya fischeri* (NRRL181) and *Aspergillus clavatus* (NRRL1)) (Fedorova *et al.*, 2008). A comparative analysis of both *A. fumigatus* strains showed that the A1163 strain had a slightly larger genome than the Af293 strain (29.2 Mb vs. 28.8 Mb) (Fedorova *et al.*, 2008). Unsurprisingly, the two genomes are highly syntenic with one another although a number of translocation events are observed mainly within subtelomeric regions. Overall sequence identity of shared regions is very high (99.8% at the nucleotide level), however, strain-specific regions are evident and they contain 143 and 218 strain-specific genes for Af293 and A1163 respectively. Inspection of these species-specific genes shows a large proportion of them are pseudogenes. Approximately 20% of the Af293 specific genes are the result of two segmental duplication events. One of the strain-specific regions contains an arsenic detoxification cluster while the other contains genes that produce betaine, a product often synthesized under heavy metal stress conditions (Fedorova *et al.*, 2008). Segmental duplication events are thought to contribute to the rapid adaptation of a species and may have been driven by selective pressures from the human host. The vast majority of Af293 and A1163 orthologues were found to be 100% identical at the sequence level; however, 41 orthologous pairs share only between 37% and 95% identity and at least 27 of these are polymorphic in other *A. fumigatus* isolates based on a comparative genomic hybridization study (Nierman *et al.*, 2005). These polymorphic genes may function in heterokaryon incompatibility in *A. fumigatus* as previous studies in filamentous fungi have shown that high levels of variability are associated with heterokaryon (*het*) genes. *het* genes have been shown to trigger cell death when hyphal fusion between two genetically incompatible individuals occurs (Glass and Dementhon, 2006).

Comparing the *N. fischeri* genome to *A. clavatus* and *A. fumigatus* we see that it is ~10% to 15% bigger than the genome of both species (Table 2.2). It also contains a larger number of protein coding genes and transposable genes which are partly responsible for the larger genome size (Fedorova *et al.*, 2008). The *A. clavatus* genome is the smallest *Aspergillus* genome sequenced to date and contains 9125 protein-coding genes (Table 2.2). Despite large differences in genome size the three species show high levels of synteny. Overall the three species share 7514 orthologues (core genes) and each contain 818 (*A. fumigatus*), 1402 (*N. fischeri*) and 1151 (*A. clavatus*) species-specific genes. Interestingly, species-specific genes appear to be approximately 50% shorter than *Aspergillus* core genes, contain fewer introns and display elevated evolutionary rates when compared with orthologues from distantly related fungi (Fedorova *et al.*, 2008). Many of the species-specific genes may be non-functional however, for example only ~32% were found to be expressed in an *A. fumigatus* microarray (Fedorova *et al.*, 2008). The *A. fumigatus*-specific genes were found to be enriched in secondary metabolite biosynthesis and detoxification, carbohydrate transport and catabolism. Interestingly a significant proportion of the species-specific genes show a subtelomeric bias. For example, 50% of the *A. fumigatus*-specific genes are clustered into 13 blocks containing more than 10 *A. fumigatus*-specific genes termed 'species-specific islands' and 68% of the clustered genes are found in subtelomeric regions (300 Kb from telomere ends). The majority of species-specific genes have no orthologues in other fungal species and most likely arose following duplication and divergence of pre-existing genes (Fedorova *et al.*, 2008).

Comparative genomics of *Paracoccidioides* species

Paracoccidioides is a dimorphic fungal pathogen that can lead to systemic mycosis termed paracoccidioidomycosis. Paracoccidioidomycosis requires prolonged antifungal treatment and high rates of complications and relapse are associated with this disease (Martinez, 2010). *Paracoccidioides* is a genus of the Onygenales order (phylum Ascomycota), other dimorphic fungi such as *Coccidioides*, *Blastomyces* and *Histoplasma* also belong to this order (Fig. 2.1) and in total they account for more than one million infections in the United States annually (Saubolle *et al.*, 2007). The Onygenales order also contains non-dimorphic fungi including pathogens and non-pathogens (e.g. *Microsporum* species and *Uncinocarpus* respectively). *Paracoccidioides* infection occurs in mammals after disruption of fungal cells in soil and release and inhalation of microconidia. During infection *Paracoccidioides* species undergo a morphological shift from a non-virulent mycelial form to pathogenic yeast growth forms, this shift is triggered by a change in soil temperature (23°C) to host temperature

(37°C). To date a number of genes associated with pathogenicity in dimorphic fungi have been identified, these include *DRK1* which controls the temperature-dependent dimorphic switch from mycelial to yeast forms and the adhesins *BAD1* and *SOWgp* (Klein and Tebbets, 2007; Nemecek *et al.*, 2006).

A recent sequencing effort successfully sequenced three of the four (S1, PS2, PS3 and Pb01) distinct *Paracoccidioides* lineages (Desjardins *et al.*, 2011). The strains selected were *Paracoccidioides brasiliensis* Pb18 a member of S1, *Paracoccidioides brasiliensis* Pb03 a member of PS2 and Pb01 a clinical strain that has recently been redesignated as a new species (*P. lutzii*; Teixeira *et al.*, 2009). Using a comparative genomics approach the authors set out to identify common features between these three *Paracoccidioides* lineages and also uncover specific gene family expansions and metabolic similarities between dimorphic fungi. Unsurprisingly the *Paracoccidioides brasiliensis* (Pb18 and Pb03) genomes are similar in size (30 and 29.1 Mb respectively) while the *P. lutzii* genome is ~10% larger at 32.9 Mb (Table 2.2). This difference in genome size is the result of expansions of low complexity repeats and transposable elements in the *P. lutzii* genome (Desjardins *et al.*, 2011). All three strains are highly syntenic with one another, furthermore, the *P. brasiliensis* strains share ~96% sequence similarity with one another while *P. lutzii* displays ~90% similarity. The total number of genes found in each species varies from 7875 in *P. brasiliensis* Pb03 to 9132 in *P. lutzii* (Table 2.2).

The evolution of dimorphic gene families was investigated by a comparative analysis of 15 Pezizomycotina species (8 of these belong to the Onygenales order). Within the Onygenales order two distinct phylogenetic lineages are evident (not shown), the first contains the dimorphic pathogens *Paracoccidioides* and *Histoplasma* while the second contains the dimorphic pathogen *Coccidioides* along with the non-dimorphic *Microsporum* and *Uncinocarpus* species (Fig. 2.1). Overall 20,713 orthologous gene clusters were located amongst these species, 995 of these were specific to *Paracoccidioides* species but only ~8% of these had predicted functions associated with them. No cluster was found to be specific to all animal pathogens in this comparative analysis although 33 unique dimorphic clusters were located and included the fungal specific kinase family FunK1 (Desjardins *et al.*, 2011). No functional categories were enriched in *Paracoccidioides* species with respect to their close relative *Histoplasma capsulatum* suggesting that differences in these organisms may be the result of species-specific genes with unknown functions. Overall dimorphic Onygenales species display a significantly reduced repertoire of carbohydrate metabolism genes including glycoside hydrolases, glycoside transferases, carbohydrate esterases and polysaccharide lysases. Onygenales species were also found to have reduced numbers of proteases, specifically cysteine, glutamic, matallo, serine and aspartic proteases. However even though the Onygenales have a reduced number of proteases, their ratio relative to carbohydrate enzymes is much higher indicating that proteins are an important nutrient source to these species (Desjardins *et al.*, 2011).

Initially *Paracoccidioides* was considered a clonal asexual microorganism although a recent study has provided evidence for recombination and in turn a sexual stage in its life cycle (Matute *et al.*, 2006). Genome analysis of the *Paracoccidioides* species provided further evidence for a sexual stage, as two mating idiomorphs and other mating and meiosis specific genes were located. Furthermore, EST sequences proved that genes within the mating loci are actively expressed (Desjardins *et al.*, 2011). The authors also located rapidly evolving genes, as these may be important for virulence. The majority of genes identified encode transcription factors, some of these, for example *RAP1* (Biswas *et al.*, 2003), are known to regulate morphogenesis and pathogenicity in other fungal species and the rapid evolution of transcription factors in *Paracoccidioides* may be linked to dimorphism as a response to host signals or defence mechanisms (Desjardins *et al.*, 2011).

Comparative genomics of *Candida* species

The term *Candida* was originally assigned to yeast species with no known sexual cycle. Today the term covers an array of both sexual and asexual species but provides little information regards their evolutionary relationships. For example, *Candida glabrata* is more closely related to *Saccharomyces cerevisiae* than it is to *Candida albicans* (Fig. 2.1). Alternatively, *Lodderomyces elongisporus* and *Debaryomyces hansenii* are close relatives of *Candida* species (Fig. 2.1). It is now accepted that when discussing *Candida* species we are referring to the monophyletic clade containing *C. albicans* and close relatives, all of these species have the common characteristic in that they translate the CUG codon as serine rather than leucine and are also sometimes termed CTG species (Butler *et al.*, 2009; Fitzpatrick *et al.*, 2006).

A number of *Candida* species are pathogenic yeasts and are the most prevalent cause of opportunistic fungal infections in humans. Currently, *Candida albicans* is the most common causative agent of candidiasis and is responsible for 50–60% of systemic cases. Because of its impact on human health *C. albicans* was the first *Candida* species to have its genome completely sequenced (Jones *et al.*, 2004). Comparative genomic comparisons with the non-pathogenic *Saccharomyces cerevisiae* genome uncovered expansions in genes that encode iron transporters, aspartyl proteinases and telomere associated transcription factors (TLO genes) that may all play an important role in virulence (Jones *et al.*, 2004).

Although *C. albicans* is still the main causative agent of candidiasis a number of other *Candida* species are becoming more prevalent. For example, *C. parapsilosis* has been associated with major outbreaks of infection in neonates (Lupetti *et al.*, 2002b), and individuals receiving central venous catheters (Levin *et al.*, 1998) or parenteral nutrition (Weems *et al.*, 1987). Similarly, *C. tropicalis* is commonly associated with neutropenia and malignancy (Wingard *et al.*, 1979). Recent comparative genomic analyses of *Candida* species have greatly increased our understanding of the genetic mechanisms underlying *Candida* pathogenesis. For example, Butler *et al.*(2009) sequenced six new *Candida* genomes and performed a comparative analysis of eight *Candida* genomes in total (Table 2.2). This analysis included highly pathogenic species (such as *C. albicans, C. tropicalis* and *C. parapsilosis*) to poor pathogens (such as *C. lusitaniae* and *C. guilliermondii*) and rarely pathogenic species such as *Lodderomyces elongisporus* and *Debaryomyces hansenii*. Overall the authors observed large deviations in genome size ranging from 10.6 to 15.5 Mb and GC content ranging from 33% to 45% (Table 2.2). However, despite variations in genome size and indeed phenotype, the predicted number of protein coding genes are very similar ranging from 5733 to 6318 genes, a proteome similar in size to another Saccharomycotina yeast *S. cerevisiae* (Butler *et al.*, 2009). In order to identify genes associated with virulence and pathogenicity, Butler *et al.*(2009) implemented a phylogenomic approach to uncover 21 gene families that are significantly enriched in pathogenic species. These gene families include secreted lipases, ferric reductases, adhesins and oligopeptide transporters that have all been shown to be associated with pathogenicity. Unsurprisingly, a number of cell wall families were found to be enriched within pathogenic *Candida* species, for example the Als-like adhesins (associated with biofilm development (Hoyer *et al.*, 2008)), the Pga30-like family which may be important for *de novo* construction of the cell wall (Castillo *et al.*, 2008) and the Hyr/Iff family, which were found to be evolving rapidly and most likely have a role in evasion of the host immune response (Butler *et al.*, 2009).

Another important comparative *Candida* genome analysis coincided with the publication of the *C. dubliniensis* genome (Jackson *et al.*, 2009). *C. dubliniensis* and *C. albicans* share many phenotypic characteristics, furthermore *C. dubliniensis* is closely related to *C. albicans* and in the past was routinely misidentified as *C. albicans* (Sullivan *et al.*, 1995). However, while *C. albicans* is the most common cause of systemic candidosis, *C. dubliniensis* is only rarely associated with this infection, making it an ideal candidate for a comparative genomic survey to locate virulence specific genes. The main differences observed between *C. albicans* and *C. dubliniensis* genomes are related to specific virulence associated gene families in *C. albicans*. For example, *C. dubliniensis* is missing three hyphal specific genes (*SAP4*, *SAP5* and *ALS3*) (Sullivan *et al.*, 1995). Furthermore, *C. albicans* has noticeable expansions in *IFA* genes (which encode putative transmembrane proteins) and the *TLO* gene family (*C. albicans* has 14 TLO genes while *C. dubliniensis* only contains 2) (Jackson *et al.*, 2009). This comparative analysis also showed that since *C. albicans* and *C. dubliniensis* last shared a common ancestor (approximately 20 million years ago), *C. albicans* has expanded its repertoire of virulence-associated gene families while *C. dubliniensis* has experienced reductive evolution leading to a diminished genetic repertoire. For example, it has lost *HYR*1 a gene associated with hyphal development, similarly, there is a dramatic case of pseudogenization of the *IFA* gene family where 14 of the 21 loci are corrupted by internal frame shifts. This leads to the conclusion that *C. dubliniensis* is a less effective pathogen that has evolved from a pathogenic ancestor.

The genome of *C. orthopsilosis* has also been completely sequenced recently (Riccombeni *et al.*, 2012). *C. orthopsilosis* is closely related to *C. parapsilosis* but is rarely associated with disease whereas *C. parapsilosis* is a serious threat to the immunocompromised and premature neonates. Comparative analysis of these two genomes showed that there is a marked difference in the composition of certain multigene families between both species. For example, *C. parapsilosis* contains 17 Hyr/Iff genes while *C. orthopsilosis* only contains two. Alternatively, relative to *C. albicans*, duplication of FLU1 (confers resistance to the antifungal fluconazole) and MDR1 (expressed in azole resistance isolates) in *C. parapsilosis* and *C. orthopsilosis* may be responsible for increased azole resistance observed in *C. parapsilosis* species (Riccombeni *et al.*, 2012).

Comparative genomics of *Coccidioides* species

Coccidioides is responsible for coccidioidomycosis (valley fever), a condition that can cause fatal disease in immunocompetent mammals. *Coccidioides* infects ~150,000 humans annually, and 5% of these develop a chronic form of coccidioidomycosis (Hector and Laniado-Laborin, 2005). *Coccidioides* is a dimorphic pathogen that is environmentally acquired in North and South American deserts, where it grows as a filamentous soil saprophyte (Fisher *et al.*, 2007). The genus is composed of two closely related species (*C. immitis* and *C. posadasii*) and belongs to the Onygenales order (Fig. 2.1). Members of the Onygenales order are known to commonly associate with animals (Untereiner *et al.*, 2004). In order to elucidate the underlying pathogenic genotypic features of *Coccidioides*, the genome sequences of *C. immitis* and *C. posadasii* have been completed, a comparative genomic analysis against close Onygenales relatives including *Histoplasma capsulatum* (mammalian pathogen) and *Uncinocarpus reesii* (non-pathogenic) has also been performed (Sharpton *et al.*, 2009). These newly sequenced genomes were also compared to 13 distantly related Ascomycota genomes, 12 of which associate with plants. The strength of this type of hierarchical comparative genomic approach is that it has the power to uncover individual genomic changes that may be linked to *Coccidioides'* ability to associate and cause disease in animals (Sharpton *et al.*, 2009).

Overall the genomes of *C. immitis* and *C. posadasii* were found to be of a comparable size (28.9 Mb and 27 Mb respectively (Table 2.2)), the difference in genome size is due to additional repetitive sequences in *C. immitis* (Sharpton *et al.*, 2009). The *Coccidioides* species are unusual in their chromosomal repertoire in that they only have five chromosomes where as most filamentous Ascomycetes have seven or eight chromosomes. There is a high degree of sequence similarity (~93.5%) between *C. immitis* and *C. posadasii* non-repetitive sequences and the mean sequence identity between homologous regions is 98.3%. Non-homologous regions account for 282 and 66 species-specific gene in *C. immitis* and *C. posadasii* respectively. Interestingly these species-specific genes occur in 'genomic islands', while these regions are non-homologous, eight of them share synteny, suggesting that they originate through non-random processes (Sharpton *et al.*, 2009). Molecular clock analysis suggests that *C. immitis* and *C. posadasii* diverged approximately 5.1 million years ago (Sharpton *et al.*, 2009).

Sharpton *et al.* (2009) undertook a phylogenomic approach to locate expansions and contractions in gene families. Their analysis uncovered 13 families that have contracted in the animal associated Onygenales species relative to the Eurotiales plant associated genomes used in their analysis. Interestingly the Onygenales species lacked genes that contain a cellulose binding domain and could account for difference in niches for Eurotiales and Onygenales species (plant vs. animal). Furthermore, Onygenales species are lacking genes associated with the decay of plant material including tannase, cellulase, cutinase, melibiase, pectate lyase and pectinesterase. Only two gene families (i.e. the APH phosphotransferase and subtilisin N domain containing family) were found to be expanded in the Onygenales species. The subtilisin N domain containing family was only found to be expanded in the pathogenic Onygenales species and their extracellular serine protease activity has been implicated in the pathogenicity of several fungal and bacterial species (da Silva *et al.*, 2006).

Overall the vast majority of *Coccidioides* genes were found to be evolving under purifying selection (median dN/dS of 0.25); however, 57 genes do exhibit signatures of positive selection. Five of these genes encode secreted proteins of unknown function and were suggested as putative vaccine targets. Previous functional work has shown that a recombinant divalent vaccine containing *Coccidioides* proline-rich paralogues Prp1 and Prp2 provides better protection than either protein in isolation. Sharpton *et al.* (2009) searched the *C. posadasii* genome and located six additional *Prp* homologues (*Prp*2–8), of these Prp5 has the highest ratio of MHC class II epitopes and shows the greatest activity in IFN gamma ELISPOT assays, indicating they would be ideal candidates for the inclusion in a trivalent vaccine along with Prp1 and Prp2.

Comparative genomics of *Pneumocystis* species

Pneumocystis species belong to the Ascomycota subphylum Taphrinomycotina and are closely related to the fission yeast *Schizosaccharomyces pombe*. *Pneumocystis jirovecii* pneumonia is an AIDS-defining infection of humans and is responsible for mortality in immunocompromised patients (Davis *et al.*, 2008). The lack of a long-term *in vitro* culture method for *P. jirovecii* has seriously hampered our understanding of this organism's biological processes (Hauser *et al.*, 2010). To help uncover some of the possible molecular mechanisms utilized by *P. jirovecii* a recent comparative genomic analysis of *Pneumocystis carinii* was undertaken (Hauser *et al.*, 2010). *P. carinii* infects rats and is closely related to *P. jirovecii*. Using *S. pombe* as a control, ~4000 putative *P. carinii* protein coding genes were predicted in the genome which is approximately 8 Mb in size (Table 2.2). By mapping the proteome of *P. carinii* and *S. pombe* into the KEGG atlas of biochemical pathways (Ogata *et al.*, 1999), the authors observed that genes involved in amino acid metabolism were significantly underrepresented in *P. carinii* (Hauser *et al.*, 2010). Furthermore only 2 of the 54 genes specifically dedicated to the metabolism of the 20 standard amino acids are present in *P. carinii*. This led the authors to conclude that *P. carinii* scavenges amino acids from the host and that it is an obligate parasite. Obligate parasitism of *P. carinii* is a parsimonious conclusion to reach as it has a reduced genome size and gene content relative to other free living Ascomycetes such as *S. cerevisiae* and *S. pombe*. Furthermore, loss of key metabolic pathways is a feature common to other obligate eukaryotic and prokaryotic parasites (Corradi *et al.*, 2010; Messick *et al.*, 2013).

Research into human related *Pneumocystis* disease was greatly advanced in 2013 when the *P. jirovecii* genome was successfully sequenced and assembled from a clinical specimen of a single patient (Cisse *et al.*, 2013). This was a unique accomplishment as it yielded the first *de novo* eukaryotic genome from a metagenome sample. The resultant genome was found to be 8.1 Mb and was predicted to contain 3878 protein-coding genes. As with *P. carinii*, genome mining illustrated that genes specifically dedicated to the synthesis of amino acids are absent and suggests that *P. jirovecii* scavenges amino acids from its host (Cisse *et al.*, 2013). Unsurprisingly 22% of the *P. jirovecii* gene set corresponds to transporters which are necessary for a scavenging lifestyle (Cushion and Stringer, 2010). As well as scavenging amino acids these transporters are also responsible for the uptake of host cholesterol, which is used to build their own membranes and S-adenosylmethionine a methyl donor in many of biological processes and a precursor of polyamines (Cushion and Stringer, 2010). Significantly,

P. jirovecii lacks specific fungal virulence factors such as polyketide synthase clusters and the glyoxylate cycle. Obvious genome reduction coupled with a lack of known virulence factors leads to the conclusion that *P. jirovecii* is normally a non-pathogenic obligate parasite of humans except in immunocompromised individuals (Hauser *et al.*, 2010). It is hoped that the future genome mining of *P. jirovecii* will advance the search for an *in vitro* culture media and lead to the identification of novel vaccine targets.

Comparative genomics of *Rhizopus oryzae*

Rhizopus oryzae belongs to a group of basal fungi known as the Zygomycota (Fig. 2.1). It is a fast growing filamentous fungus and is commonly associated with immunocompromised patients suffering from mucormycosis. *R. oryzae* has an overall mortality rate of more than 50%. While there is a wealth of genomic data available for pathogenic Dikarya species (Ascomycota and Basidiomycota species) there is a noticeable lack of data for basal fungal lineages even though they contain important human pathogens. The *R. oryzae* genome sequence (Ma *et al.*, 2009), was actually the first fungus sequenced from the Zygomycota. As well as shedding light on important pathogenic characteristics of this organism the *R. oryzae* genome also serves as a tool for investigating early fungal evolution. Furthermore, the fact that *R. oryzae* hyphae remain multinucleated and septa free after mitosis means it has the potential to shed light on the evolutionary mechanisms associated with multicellularity.

The *R. oryzae* genome is ~45.3 Mb in length (Table 2.2). It is highly repetitive with approximately 20% of the assembly consisting of transposable elements (TE) with the Ty3/gypsy-like long terminal repeat retrotransposon being the most abundant type of TE (Ma *et al.*, 2009). The genome contains 17,467 protein-coding genes and 79.55% of these are not overlapped by TEs. Interestingly, 45% of the genes that are not overlapped by TEs were found to have paralogues and 17% of these are grouped into two-member gene families which is more than two times greater that what is observed in other fungi such as *S. cerevisiae, Neurospora crassa, Magnaporthe grisea and Ustilago maydis* and led the researchers to investigate whether *R. oryzae* has undergone a whole genome duplication (WGD) similar to what has been observed in *S. cerevisiae* (Ma *et al.*, 2009; Wolfe and Shields, 1997). A number of lines of evidence where located for WGD. Firstly, there are 256 duplicated regions (containing 3–9 duplicated genes) that span ~12% of the genome and have conserved gene order and orientation. Secondly a comparative analysis against the genome of a second Zygomycete (*Phycomyces blakesleeanus*) showed that 84% of the *R. oryzae* paralogous gene pairs had a single gene match in *P. blakesleeanus*. Finally the authors were able to statistically rule out the possibility of independent duplications of the paralogous regions. These findings point to a WGD in an ancestor of *R. oryzae* followed by massive gene loss so that the post WGD genome now contains an extra 648 genes relative to the pre WGD ancestor (Ma *et al.*, 2009). Interestingly the categories of genes retained after the *R. oryzae* WGD differ from those that were retained in *S. cerevisiae*. For example, *S. cerevisiae* has retained glycolytic genes that may permit it to grow in glucose rich niches (Conant and Wolfe, 2007), where as *R. oryzae* may be able to grow rapidly due to the retention of duplicated genes involved in energy generation (Ma *et al.*, 2009).

A number of *R. oryzae* gene families are significantly expanded and some of these may be important in the pathogenic lifestyle of this fungus but also potentially may provide us with reliable diagnostic targets. For example, *R. oryzae* has large numbers of chitin synthase and chitin deacetylase genes that are important enzymes in the synthesis of the *R. oryzae* cell wall. Of the 34 predicted chitin deacetylase genes, 14 are thought to be localized to the cell wall meaning they could be further investigated as potential diagnostic/therapeutic targets. Overall, an expansion of the protease gene families is observed in *R. oryzae* which may be linked to its ability to readily degrade organic material. One expanded protease gene family of note includes the secreted aspartic proteases (*SAP*) which are linked to virulence in a number of pathogenic fungi such as *C. albicans* (Schaller *et al.*, 2005). *SAPs* have also been linked to virulence in pathogenic *Rhizopus* species and have been identified as potential novel vaccine/drug targets (Schoen *et al.*, 2002). The genome analysis of *R. oryzae* showed that components of the ergosterol biosynthetic pathway have been duplicated. This may explain why *R. oryzae* has varying responses to a number of commonly used azoles (antifungals that target this pathway) such as voriconazole and posaconazole similar to what has been observed in a pathogenic strain of *C. albicans* with duplication of the *ERG11* gene (Selmecki *et al.*, 2008).

Comparative genomics of dermatophyte species

The dermatophytes are a group of Ascomycota filamentous fungi. They are found within three genera (*Trichophyton, Epidermophyton* and *Microsporum*) and all belong to the Arthrodermataceae family, which is closely related to the Onygenales family (Fig. 2.1). Initial mitochondrial analyses suggested that due to high levels of gene similarity and order, dermatophytes are a highly conserved closely related lineage of filamentous fungi (Wu *et al.*, 2009). They are highly specialized pathogenic fungi and are the most common cause of superficial mycoses in animals and humans. Dermatophyte infections are rarely fatal but are responsible for irritating skin conditions such as tinea pedis (athlete's foot), tinea capitis (ringworm) and tinea cruris (jock itch).

Two recent sequencing efforts now mean that seven whole-genome dermatophyte sequences are now available (Burmester *et al.*, 2011; Martinez *et al.*, 2012). These include the anthropophiles *Trichophyton rubrum* (causative agent of tinea pedis) and *Trichophyton tonsurans* (causes tinea capitis). The zoophiles *Trichophyton equinum, Trichophyton verrucosum, Arthroderma benhamiae, Microsporum canis* and the geophile *Microsporum gypseum*. The phylogenetic distance and differences in host range make these species ideal candidates for uncovering the genomic basis of pathogenicity within the dermatophytes.

The genomes of all seven dermatophytes are similar in size ranging in size from 22.5 Mb to 24.1 Mb (Table 2.2) and the absence of obvious polymorphism within individual species reinforces the predicted haploid lifestyle of the species sequenced. Overall the dermatophytes display a high level of gene synteny and orthologous genes display high levels of amino acid similarity (Burmester *et al.*, 2011; Martinez *et al.*, 2012), an observation consistent with earlier comparative mitochondrial analyses (Wu *et al.*, 2009). In terms of gene content the number of genes varies from a low of 7980 for *A. benhamiae* to a high of 8915 for *M. canis*. However, a core set of more than 6100 orthologous gene groups between all seven species underlines the close phylogenetic relationships and short divergence time amongst these species. However small differences in gene content can be correlated with changes in niche and host range (Martinez *et al.*, 2012). Approximately 300 of the core orthologous groups are dermatophyte specific and of these 60% have no known functional InterPro (Hunter *et al.*, 2011) domain. Comparative analysis with eleven human associated fungi including representatives from the Onygenales, two *Aspergillus* species and *Coccidioides immitis* uncovered expansions in four dermatophyte gene families that contribute to virulence. These include proteases, kinases, secondary metabolites and LysM proteins (Martinez *et al.*, 2012).

The dermatophyte genomes contain a large number of novel eukaryotic protein kinases, depending on the species they have between 147 and 194 kinases, this compares with only 118 in *Saccharomyces cerevisiae* (Manning *et al.*, 2002). The increase in kinase number is due to orthologue retention and increased duplication to generate paralogues in specific kinase classes (Martinez *et al.*, 2012). Based on elevated rates of Ka/Ks (non-synonymous substitutions per non-synonymous site vs. synonymous substitutions per synonymous site) the dermatophyte novel kinases appear to be evolving quicker than other components of their genomes most probably in response to changes in external factors. Furthermore the increased numbers of novel kinases may provide a mechanism for different dermatophyte species to occupy different ecological niches.

The dermatophytes are also enriched for genes involved in the production of secondary metabolites. Secondary metabolites are compounds that can act as immune response modulators, toxins or signal compounds between fungus and host. Twenty eight of the secondary metabolism genes are found in three or less dermatophyte species and lack orthologues in other species and have been proposed as excellent virulence factors contributing to host range and pathogenesis (Martinez *et al.*, 2012).

In fungi the LysM domain is known to bind to chitin and is thought to aid evasion of the host innate immune response by masking fungal cell wall components and carbohydrates (de Jonge and Thomma, 2009). The number of genes containing LysM domains ranging from nine in *T. verrucosum* to 31 in *M. canis*. It was observed that proteins containing the LysM domains fall into two categories, those that contain LysM domains and the chitin binding domains GH18 and Chitin binding 1 and those that contain LysM domains only. Phylogenetic analysis infers that genes that are missing the GH18 domain are deep branching and are most probably the ancestral state to the LysM-GH18 Chitin binding 1 form (Martinez *et al.*, 2012).

The dermatophytes are known to have expansions in subtilisin (Burmester *et al.*, 2011); these are important proteases for keratin degradation. Dermatophyte genome analysis uncovered other keratin degrading protease expansions across all species examined; in particular the M35 (deuterolysins) and M36 (fungalysins) metalloendopeptidases are present in at least five copies in each species (Martinez *et al.*, 2012).

Using whole-genome sequences to trace the evolution of gene fusion and fission events in the fungal kingdom

Gene fusions are the result of two or more independent open reading frames (ORFs) being joined together. Gene fission is the opposite process where an ORF is split into two or more novel ORFs. Both processes have the potential to increase genetic and enzymatic variation. Initial studies suggested fusions are four times more common than fission events (Kummerfeld and Teichmann, 2005). Authors have argued that fissions are less frequent as they require a number of evolutionary events to occur simultaneously including gain of a new promoter, start codon and stop codon (Stechmann and Cavalier-Smith, 2002).

Due to the availability of many fungal genomes it is possible to investigate the relative rates of fusion and fission in these species. A recent analysis of gene fusions/fissions in 115 fungal genomes identified 63 gene fusions (Leonard and Richards, 2012). A gene ontology (GO) search did not identify a bias towards any GO category in these fused genes. An in-depth functional analysis of the genes that have been fused is yet to be performed. However, looking at a number of common human pathogens we can see that there are no pathogen specific fusions in the CTG clade. There is evidence to show that the last common ancestor of the *Aspergillus* species underwent a gene fusion event giving rise to dihydrolipoamide acetyltransferase although there has been a subsequent reversion (fission) in *Aspergillus fumigatus* (Leonard and Richards, 2012). Similarly, there is also evidence for two independent gene fusions in *C. neoformans* resulting in a 3-methyl-2-oxobutanoate hydroxymethyltransferase and a phosphoribosylaminoimidazole-succinocarboxamide synthase enzyme (Leonard and Richards, 2012).

The authors also uncovered 110 gene fission events, demonstrating that gene fissions occur at a relatively high rate in the fungal kingdom and represent an important source of gene variation. Furthermore the authors also noticed gene fusion counts are affected by homoplasy in the form of reversions (fusion gene subsequently undergoing a fission). This leads to the conclusion that gene fusions are not as stable as previously predicted, this has implications for deep level phylogenetic studies as fusion genes have been suggested as prime phylogenetic markers (Stechmann and Cavalier-Smith, 2002). From an evolutionary standpoint the relative rates of fusion and fission are interesting and give an indication of how organisms generate novel enzymatic function via neofunctionalization. However, at present there is no evidence linking any of the fusion/fission genes to fungal virulence in humans but it is a topic that warrants further study.

Using whole-genome sequences to determine the rates of horizontal gene transfer into pathogenic fungal species

Horizontal gene transfer (HGT) is defined as the exchange and stable integration of genetic material between different strains or species (Doolittle, 1999), it differs from vertical gene transfer, which is the normal transmission of genetic material from parent to offspring. Genome sequencing has shown that HGT is a major evolutionary force in prokaryotic evolution (Eisen, 2000).

The increase in fungal genomic data (including many human pathogens) has permitted investigations into the frequency and putative roles HGT may play in fungal evolution and virulence. For example, it has been shown that *S. cerevisiae* S288C has acquired 13 bacterial genes (less than 1% of its genome) via HGT since it diverged from its close relative *Ashbya gossypii* (Hall *et al.*, 2005). These HGT events have contributed to *S. cerevisiae* ability to synthesize biotin, grow under anaerobic conditions, and to utilize sulfate from several organic sources (Hall *et al.*, 2005). Incidences of HGT have also been linked to virulence in plant associated fungi, the recent acquisition of a toxin gene (ToxA) by *Pyrenophora tritici-repentis* from *Stagonospora nodorum* has resulted in serious *Pyrenophora* infestations of wheat (Friesen *et al.*, 2006).

Initial investigations into human fungal pathogens suggested that while HGT does occur, it is not a frequent event. An analysis of *Candida albicans* and five of its close relatives detected only two incidences of interkingdom HGT from bacteria into the CTG clade, both transfers (a proline racemase and gene involved in phenazine production) were into the emerging pathogen *Candida parapsilosis* (Fitzpatrick *et al.*, 2008). A recent large-scale analysis has uncovered additional interkingdom gene transfers into fungal species (Marcet-Houben and Gabaldon, 2010). Overall this study uncovered 713 bacterial genes in 60 fungal genomes spanning the fungal tree of life and including a number of human pathogens. For example, this analysis located a bacterial arsenate reductase in *Rhizopus oryzae*. The reduction of arsenate to arsenite is a key step in arsenic detoxification thereby potentially enabling *R. oryzae* to detoxify arsenics. This large-scale study also detected aspartate, hydantoin and mandelate bacterial racemases in the human pathogen *Candida glabrata* and a number of *Aspergillus* species, possible roles for these racemases include the assimilation and detoxification of D-isomers of specific amino acids (Marcet-Houben and Gabaldon, 2010). There is also evidence for the transfer of the bacterial peptidoglycan biosynthesis *Mur* operon into a number of *Aspergillus* species including *A. fumigatus*. The function of these genes in fungi is unknown although it has been speculated that they may be involved in degradation of the bacterial cell wall (Marcet-Houben and Gabaldon, 2010).

To date most large-scale analyses have looked for bacterial gene transfers into human fungal pathogens, the reason for this may be due to the fact that bacterial HGT events are easier to detect than eukaryotic transfers. Future large-scale HGT analyses that set out to detect interfungal and intereukaryotic HGT events will further increase our knowledge of the frequency of HGT that occurs in human pathogens.

Evolution of antifungal resistance in human pathogens

The ability of human pathogens to evolve antimicrobial resistance is a well-documented problem. Prokaryotes such as methicillin-resistant *Staphyloccus aureus* (MRSA) are frequently in the news and a serious threat to patient health. Rates of mutation and horizontal gene transfer in bacteria are relatively high and may lead to the emergence of novel bacterial strains of clinical importance. Horizontal gene transfer rates are significantly lower in fungi therefore the genetic mechanisms of antifungal resistance in human fungal pathogens is of great interest.

Antifungal resistance is dependent on the ability of a fungus to generate variable phenotypes. The major source of fungal phenotype variation was thought to be closely linked to individual mutations that effect either protein activity or gene expression (Cowen *et al.*, 2000). For example, mutations in the transcriptional regulator *PDR1* of *S. cerevisiae* lead to the over expression of ATP-binding cassette (ABC) transporters and leads to increased efflux and subsequent resistance to fluconazole (Lupetti *et al.*, 2002a; Sanglard, 2002; Sanglard and Odds, 2002). Furthermore mutations in genes involved in metabolic pathways that may be the targets of antifungals can also have important implications, for example fluconazole resistance is linked to polymorphisms in *ERG11* which codes for lanosterol 14-alpha-demethylase a critical enzyme in the biosynthesis of ergosterol (White *et al.*, 1998). Going forward, the availability of reference genomes for a number of human fungal pathogens and the advent of inexpensive next generation sequencing techniques should permit population genetic studies of antifungal resistance in clinically important pathogens.

Recent analyses have suggested that copy number variation including chromosomal aneuploidy and short segmental variation are important contributors to antifungal resistance (Perepnikhatka *et al.*, 1999; Selmecki *et al.*, 2006). For example, an experimental evolutionary study of *S. cerevisiae* showed that there is an increased frequency of chromosomal (part and whole chromosomes) variation over point mutations indicating the potential fitness benefits associated with these mutations (Zeyl, 2004).

Research in *C. albicans* has shown that a specific aneuploidy, isochromosome 5L, which comprises two copies of the left arm of chromosome 5, occurs at a high frequency soon after fluconazole exposure (Selmecki *et al.*, 2009). The presence of isochromosome 5L is significant as it contains two genes that contribute to fluconazole resistance. These are *ERG11*, and *TAC1* which encodes a transcription factor that activates the multidrug transporters *CDR1* and *CDR2* (Coste *et al.*, 2004).

Using whole-genome sequences of human fungal pathogens to investigate the evolution of fungal introns

Unlike prokaryotes, eukaryotes contain spliceosomal introns that must be removed before translation. To date no universal advantage of introns has been uncovered although some introns are known to contain important regulatory sequences (Duret and Bucher, 1997; Sorek and Ast, 2003), furthermore the presence of multiple introns in a gene may generate diverse proteins *via* alternative splicing. Looking at different lineages of the eukaryotic tree of life we see that the number of introns varies depending on the branch examined. For example, animals such as humans, nematode and fruitfly have 8.1, 4.7 and 3.4 introns per gene (Duret and Bucher, 1997; Sorek and Ast, 2003); in contrast, members of the hemiascomycetes (includes *S. cerevisiae* and *C. albicans*) have very few introns. Owing to the number of fungal genomes available this kingdom has been at the forefront of studies into eukaryotic intron evolution. The diversity in intron density in fungal species adds to their usefulness. For example, *C. albicans* contains 0.07 introns per gene while *C. neoformans* contains 5.5 (Jeffares *et al.*, 2006; Mitrovich *et al.*, 2007).

An analysis of *Magnaporthe grisea, Neurospora crassa, Fusarium graminearum* and an outgroup in *Aspergillus nidulans* identified 3450 strongly supported intron positions amongst these four Pezizomycotina species (Nielsen *et al.*, 2004). Overall the authors observed significant intron gains and losses in *M. grisea, N. crassa* and *F. graminearum* genomes varying from 150 to 250 gains and 150 to 350 losses (Nielsen *et al.*, 2004). *N. crassa* and *F. graminearum* display more than 100 intron lineage specific gains while *M. grisea* has more than 200 intron gains indicating that intron gain is an important evolutionary force in these fungal species, this contrasts with mammalian genes where intron gain is a rare occurrence (Roy *et al.*, 2003).

A subsequent study undertook a multi-kingdom comparative analysis of intron gain and loss in an effort to accurately reconstruct intron evolution throughout the fungal kingdom (Stajich *et al.*, 2007). This analysis included 21 fungal genomes as well as three animal and one plant outgroups. Comparing 1161 sets of orthologous genes the authors inferred that the last common fungal ancestor and fungal–animal ancestor (Opisthokont) was very intron rich and matches or exceeds the highest densities observed in other eukaryotes. Overall a trend of intron loss is observed in all fungal lineages although some intron rich species such as *C. neoformans* have apparently gained many new introns. The authors also observed massive intron loss and very little gain in the hemiascomycetes (Stajich *et al.*, 2007).

Fungal phylogenomics

Owing to their simple morphology, poor fossil record and high degree of biological and physiological diversity evolutionary relationships amongst fungi were poorly understood until recent years (Guarro *et al.*, 1999). Traditional fungal evolutionary studies relied on morphology, sexual states, cell wall composition, cytological testing, ultrastructure and metabolism to infer taxonomic relationships (Guarro *et al.*, 1999). With the advent of DNA sequencing, molecular phylogenetic analyses have revealed that there are at least seven distinct phyla within the fungal kingdom (Hibbett *et al.*, 2007; James *et al.*, 2006a,b) these include the Chytridiomycota, Zygomycota (Glomeromycota), Microsporidia, Blastocladiomycota, Neocallimastigomycota, Ascomycota and Basidiomycota. Initially the majority of fungal molecular phylogenies were derived from the analysis of individual ribosomal genes (Lutzoni *et al.*, 2004). Due to some of the pitfalls associated with single gene phylogenies such as too few sites or compositional biases, phylogenomic approaches that set out to maximize the amount of molecular data utilized such as multigene concatenation or supertree methods have been successful in addressing relationships amongst diverse fungal species and have resulted in a well supported fungal tree of life (FTOL) (Fitzpatrick *et al.*, 2006; James *et al.*, 2006a; Kuramae *et al.*, 2006; Liu *et al.*, 2009; Marcet-Houben and Gabaldon, 2009; Robbertse *et al.*, 2006).

Having a coherent FTOL has many advantages; not only does it infer evolutionary relationships amongst diverse species it also allows us to systematically map metabolic pathways, virulence genes and sex related genes for example onto an evolutionary framework and permits us to visualize the gain or loss of key components that may be linked to disease or niche specification. For example, a recent phylogenomic study examined the phyletic distribution of yeast prion-like proteins in the fungal kingdom and mapped their presence/absence onto the FTOL (Medina *et al.*, 2011). Prions are infectious proteins that have the capability of converting native molecules of the same type into the infectious prion form. This study demonstrated that there is a wide-ranging distribution of potential yeast-like prion orthologues across the FTOL. The most dramatic and restricted orthologue distribution is for Rnq1, where orthologues are only found in 13 species and restricted to a monophyletic clade that contains close relatives of *S. cerevisiae* (Medina *et al.*, 2011). The two most extensively studied yeast prion-forming proteins, Sup35 and Ure2, show a very different distribution in conservation of their prion forming domains. The prion domain in Sup35 is much more widely conserved throughout the FTOL compared to the Ure2 prion domain (Medina *et al.*, 2011). Another recent fungal phylogenomic analysis uncovered the phyletic distribution of two DNA repair proteins, methylphosphotriester (MPT) and O^6-alkylguanine methylphosphotriester (AGT). This analysis found that AGT is ubiquitous across the FTOL but only a select group of fungal species, some of which are major human pathogens contain MPT (O'Hanlon *et al.*, 2012). *In vitro* DNA repair assays in *Aspergillus fumigatus* confirmed the ability of MPT and AGT to repair methylphosphotriester and O^6-methylguanine lesions, furthermore phylogenetic analysis showed that inter-kingdom horizontal gene transfer between a bacterial phylum (Firmicutes) and Ascomycete ancestor introduced the adaptive response into the Fungal kingdom (O'Hanlon *et al.*, 2012).

Future perspectives

Genome sequences for many human fungal pathogens are now available. The availability of these data has helped us understand some of the molecular mechanisms associated with pathogenicity and in some cases have also uncovered a number of novel antifungal targets. To date most comparative genomic studies have compared pathogenic species to closely related but non-pathogenic species to uncover these mechanisms. Going forward, the continued decrease in sequencing costs will make it possible to re-sequence (relative to a reference genome) large numbers of pathogenic and non-pathogenic strains to uncover single nucleotide polymorphisms that may be linked to virulence. A re-sequencing strategy also has the potential to locate duplications, deletions or *de novo* synthesis of virulence associated genes. The availability of numerous human fungal pathogens has also allowed us to address many eukaryotic evolutionary problems such as intron evolution and the frequency of horizontal gene transfer and gene fission and fusion events. Currently there is a sustained effort to sequence neglected fungal lineages by the 1000 fungal genome project. This data promises to shine a light on many neglected fungal lineages and will further aid our understanding of many eukaryotic evolutionary innovations such as multicellularity and sexual reproduction.

References

Biswas, K., Rieger, K.J., and Morschhauser, J. (2003). Functional analysis of *CaRAP1*, encoding the Repressor/activator protein 1 of *Candida albicans*. Gene *307*, 151–158.

Bose, I., Reese, A.J., Ory, J.J., Janbon, G., and Doering, T.L. (2003). A yeast under cover: the capsule of *Cryptococcus neoformans*. Eukaryot. Cell *2*, 655–663.

Brown, G.D., Denning, D.W., Gow, N.A., Levitz, S.M., Netea, M.G., and White, T.C. (2012). Hidden killers: human fungal infections. Sci. Transl. Med. *4*, 165rv113.

Burmester, A., Shelest, E., Glockner, G., Heddergott, C., Schindler, S., Staib, P., Heidel, A., Felder, M., Petzold, A., Szafranski, K., *et al.* (2011). Comparative and functional genomics provide insights into the pathogenicity of dermatophytic fungi. Genome Biol. *12*, R7.

Butler, G., Rasmussen, M.D., Lin, M.F., Santos, M.A.S., Sakthikumar, S., Munro, C.A., Rheinbay, E., Grabherr, M., Forche, A., and Reedy, J.L. (2009). Evolution of pathogenicity and sexual reproduction in eight *Candida* genomes. Nature 459, 657–662.

Castillo, L., Calvo, E., Martinez, A., Ruiz-Herrera, J., Valentin, E., Lopez, J.A., and Sentandreu, R. (2008). A study of the *Candida albicans* cell wall proteome. Proteomics *8*, 3871–3881.

Cisse, O.H., Pagni, M., and Hauser, P.M. (2013). De novo assembly of the *Pneumocystis jirovecii* genome from a single bronchoalveolar lavage fluid specimen from a patient. Mbio *4*, e00428–00412.

Conant, G.C., and Wolfe, K.H. (2007). Increased glycolytic flux as an outcome of whole-genome duplication in yeast. Mol. Syst. Biol. *3*, 129.

Corradi, N., Pombert, J.F., Farinelli, L., Didier, E.S., and Keeling, P.J. (2010). The complete sequence of the smallest known nuclear genome from the microsporidian *Encephalitozoon intestinalis*. Nat. Commun. *1*, 77.

Coste, A.T., Karababa, M., Ischer, F., Bille, J., and Sanglard, D. (2004). *TAC1*, transcriptional activator of *CDR* genes, is a new transcription factor involved in the regulation of *Candida albicans* ABC transporters *CDR1* and *CDR2*. Eukaryot. Cell 3, 1639–1652.

Cowen, L.E., Sanglard, D., Calabrese, D., Sirjusingh, C., Anderson, J.B., and Kohn, L.M. (2000). Evolution of drug resistance in experimental populations of *Candida albicans*. J. Bacteriol. *182*, 1515–1522.

Cushion, M.T., and Stringer, J.R. (2010). Stealth and opportunism: alternative lifestyles of species in the fungal genus *Pneumocystis*. Ann. Rev. Microbiol. *64*, 431–452.

D'Souza, C.A., Kronstad, J.W., Taylor, G., Warren, R., Yuen, M., Hu, G., Jung, W.H., Sham, A., Kidd, S.E., Tangen, K., *et al.* (2011). Genome variation in *Cryptococcus gattii*, an emerging pathogen of immunocompetent hosts. Mbio 2, e00342–00310.

Davis, J.L., Fei, M., and Huang, L. (2008). Respiratory infection complicating HIV infection. Curr. Opin. Infect. Dis. *21*, 184–190.

Desjardins, C.A., Champion, M.D., Holder, J.W., Muszewska, A., Goldberg, J., Bailao, A.M., Brigido, M.M., Ferreira, M.E., Garcia, A.M., Grynberg, M., *et al.* (2011). Comparative genomic analysis of human fungal pathogens causing paracoccidioidomycosis. PLoS Genet. 7, e1002345.

Doolittle, W.F. (1999). Lateral genomics. Trends Cell Biol. *9*, M5–8.

Duret, L., and Bucher, P. (1997). Searching for regulatory elements in human noncoding sequences. Curr. Opin. Struct. Biol. 7, 399–406.

Eisen, J.A. (2000). Assessing evolutionary relationships among microbes from whole-genome analysis. Curr. Opin. Microbiol. 3, 475–480.

Fedorova, N.D., Khaldi, N., Joardar, V.S., Maiti, R., Amedeo, P., Anderson, M.J., Crabtree, J., Silva, J.C., Badger, J.H., and Albarraq, A. (2008). Genomic islands in the pathogenic filamentous fungus *Aspergillus fumigatus*. PLoS Genet. 4, e1000046.

Fisher, F.S., Bultman, M.W., Johnson, S.M., Pappagianis, D., and Zaborsky, E. (2007). *Coccidioides* niches and habitat parameters in the southwestern United States: a matter of scale. Ann. N. Y. Acad. Sci. *1111*, 47–72.

Fisher, M.C., Henk, D.A., Briggs, C.J., Brownstein, J.S., Madoff, L.C., McCraw, S.L., and Gurr, S.J. (2012). Emerging fungal threats to animal, plant and ecosystem health. Nature *484*, 186–194.

Fitzpatrick, D.A., Logue, M.E., Stajich, J.E., and Butler, G. (2006). A fungal phylogeny based on 42 complete genomes derived from supertree and combined gene analysis. BMC Evol. Biol. 6, 99.

Fitzpatrick, D.A., Logue, M.E., and Butler, G. (2008). Evidence of recent interkingdom horizontal gene transfer between bacteria and *Candida parapsilosis*. BMC Evol. Biol. *8*, 181.

Friesen, T.L., Stukenbrock, E.H., Liu, Z., Meinhardt, S., Ling, H., Faris, J.D., Rasmussen, J.B., Solomon, P.S., McDonald, B.A., and Oliver, R.P. (2006). Emergence of a new disease as a result of interspecific virulence gene transfer. Nat. Genet. *38*, 953–956.

Geiser, D.M., Arnold, M.L., and Timberlake, W.E. (1996). Wild chromosomal variants in *Aspergillus nidulans*. Curr. Genet. 29, 293–300.

Glass, N.L., and Dementhon, K. (2006). Non-self recognition and programmed cell death in filamentous fungi. Curr. Opin. Microbiol. *9*, 553–558.

Guarro, J., Gené, J., and Stchigel, A.M. (1999). Developments in fungal taxonomy. Clin. Microbiol. Rev. *12*, 454–500.

Hall, C., Brachat, S., and Dietrich, F.S. (2005). Contribution of horizontal gene transfer to the evolution of *Saccharomyces cerevisiae*. Eukaryot. Cell *4*, 1102–1115.

Hauser, P.M., Burdet, F.X., Cisse, O.H., Keller, L., Taffe, P., Sanglard, D., and Pagni, M. (2010). Comparative genomics suggests that the fungal pathogen *Pneumocystis* is an obligate parasite scavenging amino acids from its host's lungs. PLoS One *5*, e15152.

Havlickova, B., Czaika, V.A., and Friedrich, M. (2008). Epidemiological trends in skin mycoses worldwide. Mycoses *51 (Suppl 4)*, 2–15.

Hector, R.F., and Laniado-Laborin, R. (2005). Coccidioidomycosis--a fungal disease of the Americas. PLoS Med. *2*, e2.

Hibbett, D.S., Binder, M., Bischoff, J.F., Blackwell, M., Cannon, P.F., Eriksson, O.E., Huhndorf, S., James, T., Kirk, P.M., Lücking, R., *et al.* (2007). A higher-level phylogenetic classification of the Fungi. Mycol. Res. *111*, 509–547.

Hoang, L.M., Maguire, J.A., Doyle, P., Fyfe, M., and Roscoe, D.L. (2004). *Cryptococcus neoformans* infections at Vancouver Hospital and Health Sciences Centre (1997–2002): epidemiology, microbiology and histopathology. J. Med. Microbiol. *53*, 935–940.

Hoyer, L.L., Green, C.B., Oh, S.H., and Zhao, X. (2008). Discovering the secrets of the *Candida albicans* agglutinin-like sequence (*ALS*) gene family--a sticky pursuit. Med. Mycol. *46*, 1–15.

Hu, G., Liu, I., Sham, A., Stajich, J.E., Dietrich, F.S., and Kronstad, J.W. (2008). Comparative hybridization reveals extensive genome variation in the AIDS-associated pathogen *Cryptococcus neoformans*. Genome Biol. *9*, R41.

Hunter, S., Jones, P., Mitchell, A., Apweiler, R., Attwood, T.K., Bateman, A., Bernard, T., Binns, D., Bork, P., Burge, S., *et al.* (2011). InterPro in 2011: new developments in the family and domain prediction database. Nucleic Acids Res. *40*, D306–312.

Jackson, A.P., Gamble, J.A., Yeomans, T., Moran, G.P., Saunders, D., Harris, D., Aslett, M., Barrell, J.F., Butler, G., and Citiulo, F. *et al.* (2009). Comparative genomics of the fungal pathogens *Candida dubliniensis* and *Candida albicans*. Genome Res. *19*, 2231–2244.

James, T.Y., Kauff, F., Schoch, C.L., Matheny, P.B., Hofstetter, V., Cox, C.J., Celio, G., Gueidan, C., Fraker, E., and Miadlikowska, J. (2006a). Reconstructing the early evolution of Fungi using a six-gene phylogeny. Nature *443*, 818–822.

James, T.Y., Letcher, P.M., Longcore, J.E., Mozley-Standridge, S.E., Porter, D., Powell, M.J., Griffith, G.W., and Vilgalys, R. (2006b). A molecular phylogeny of the flagellated fungi (Chytridiomycota) and description of a new phylum (Blastocladiomycota). Mycologia *98*, 860–871.

Jeffares, D.C., Mourier, T., and Penny, D. (2006). The biology of intron gain and loss. Trends Genet. *22*, 16–22.

Jones, T., Federspiel, N.A., Chibana, H., Dungan, J., Kalman, S., Magee, B.B., Newport, G., Thorstenson, Y.R., Agabian, N., and Magee, P.T. (2004). The diploid genome sequence of *Candida albicans*. Proc. Natl. Acad. Sci. U.S.A. *101*, 7329–7334.

de Jonge, R., and Thomma, B.P. (2009). Fungal LysM effectors: extinguishers of host immunity? Trends Microbiol. *17*, 151–157.

Klein, B.S., and Tebbets, B. (2007). Dimorphism and virulence in fungi. Curr. Opin. Microbiol. *10*, 314–319.

Kummerfeld, S.K., and Teichmann, S.A. (2005). Relative rates of gene fusion and fission in multi-domain proteins. Trends Genet. *21*, 25–30.

Kuramae, E.E., Robert, V., Snel, B., Weifl, M., and Boekhout, T. (2006). Phylogenomics reveal a robust fungal tree of life. FEMS Yeast Res. *6*, 1213–1220.

Leonard, G., and Richards, T.A. (2012). Genome-scale comparative analysis of gene fusions, gene fissions, and the fungal tree of life. Proc. Natl. Acad. Sci. U.S.A. *109*, 21402–21407.

Levin, A.S., Costa, S.F., Mussi, N.S., Basso, M., Sinto, S.I., Machado, C., Geiger, D.C., Villares, M.C., Schreiber, A.Z., Barone, A.A., and Branchini, M.L. (1998). *Candida parapsilosis* fungemia associated with implantable and semi-implantable central venous catheters and the hands of healthcare workers. Diagn. Microbiol. Infect. Dis. *30*, 243–249.

Litwin, C.M., Rayback, T.W., and Skinner, J. (1996). Role of catechol siderophore synthesis in *Vibrio vulnificus* virulence. Infect. Immun. *64*, 2834–2838.

Liu, Y., Leigh, J.W., Brinkmann, H., Cushion, M.T., Rodriguez-Ezpeleta, N., Philippe, H., and Lang, B.F. (2009). Phylogenomic analyses support the monophyly of Taphrinomycotina, including *Schizosaccharomyces* fission yeasts. Mol. Biol. Evol. *26*, 27–34.

Loftus, B.J., Fung, E., Roncaglia, P., Rowley, D., Amedeo, P., Bruno, D., Vamathevan, J., Miranda, M., Anderson, I.J., and Fraser, J.A. (2005). The genome of the basidiomycetous yeast and human pathogen *Cryptococcus neoformans*. Science *307*, 1321–1324.

Lupetti, A., Danesi, R., Campa, M., Del Tacca, M., and Kelly, S. (2002a). Molecular basis of resistance to azole antifungals. Trends Mol. Med. *8*, 76–81.

Lupetti, A., Tavanti, A., Davini, P., Ghelardi, E., Corsini, V., Merusi, I., Boldrini, A., Campa, M., and Senesi, S. (2002b). Horizontal transmission of *Candida parapsilosis* candidemia in a neonatal intensive care unit. J. Clin. Microbiol. *40*, 2363–2369.

Lutzoni, F., Kauff, F., Cox, C.J., McLaughlin, D., Celio, G., Dentinger, B., Padamsee, M., Hibbett, D., James, T.Y., Baloch, E., *et al.* (2004). Assembling the fungal tree of life: progress, classification, and evolution of subcellular traits. Am. J. Bot. *91*, 1446–1480.

Ma, L.J., Ibrahim, A.S., Skory, C., Grabherr, M.G., Burger, G., Butler, M., Elias, M., Idnurm, A., Lang, B.F., and Sone, T. (2009). Genomic analysis of the basal lineage fungus Rhizopus oryzae reveals a whole-genome duplication. PLoS Genet. *5*, e1000549.

Manning, G., Plowman, G.D., Hunter, T., and Sudarsanam, S. (2002). Evolution of protein kinase signaling from yeast to man. Trends Biochem. Sci. *27*, 514–520.

Marcet-Houben, M., and Gabaldon, T. (2009). The tree versus the forest: the fungal tree of life and the topological diversity within the yeast phylome. PLoS One *4*, e4357.

Marcet-Houben, M., and Gabaldon, T. (2010). Acquisition of prokaryotic genes by fungal genomes. Trends Genet. *26*, 5–8.

Martinez, D.A., Oliver, B.G., Graser, Y., Goldberg, J.M., Li, W., Martinez-Rossi, N.M., Monod, M., Shelest, E., Barton, R.C., Birch, E., *et al.* (2012). Comparative genome analysis of *Trichophyton rubrum* and related dermatophytes reveals candidate genes involved in infection. Mbio 3, e00259–00212.

Martinez, R. (2010). Paracoccidioidomycosis: the dimension of the problem of a neglected disease. Rev. Soc. Bras. Med. Trop. *43*, 480.

Matute, D.R., McEwen, J.G., Puccia, R., Montes, B.A., San-Blas, G., Bagagli, E., Rauscher, J.T., Restrepo, A., Morais, F., Nino-Vega, G., and Taylor, J.W. (2006). Cryptic speciation and recombination in the fungus *Paracoccidioides brasiliensis* as revealed by gene genealogies. Mol. Biol. Evol. *23*, 65–73.

May, J.J., Wendrich, T.M., and Marahiel, M.A. (2001). The dhb operon of Bacillus subtilis encodes the biosynthetic template for the catecholic siderophore 2,3-dihydroxybenzoate-glycine-threonine trimeric ester bacillibactin. J. Biol. Chem. *276*, 7209–7217.

Medina, E.M., Jones, G.W., and Fitzpatrick, D.A. (2011). Reconstructing the fungal tree of life using phylogenomics and a preliminary investigation of the distribution of yeast prion-like proteins in the fungal kingdom. J. Mol. Evol. *73*, 116–133.

Messick, J.B., Santos, A.P., and Guimaraes, A.M. (2013). Complete genome sequences of two hemotropic mycoplasmas, *Mycoplasma haemofelis* strain Ohio2 and *Mycoplasma suis* Strain Illinois. J. Bacteriol. *193*, 2068–2069.

Mitrovich, Q.M., Tuch, B.B., Guthrie, C., and Johnson, A.D. (2007). Computational and experimental approaches double the number of known introns in the pathogenic yeast *Candida albicans*. Genome Res. *17*, 492–502.

Nemecek, J.C., Wuthrich, M., and Klein, B.S. (2006). Global control of dimorphism and virulence in fungi. Science *312*, 583–588.

Nielsen, C.B., Friedman, B., Birren, B., Burge, C.B., and Galagan, J.E. (2004). Patterns of intron gain and loss in fungi. PLoS Biol. *2*, e422.

Nierman, W.C., Pain, A., Anderson, M.J., Wortman, J.R., Kim, H.S., Arroyo, J., Berriman, M., Abe, K., Archer, D.B., and Bermejo, C. (2005). Genomic sequence of the pathogenic and allergenic filamentous fungus *Aspergillus fumigatus*. Nature *438*, 1151–1156.

O'Gorman, C.M., Fuller, H., and Dyer, P.S. (2009). Discovery of a sexual cycle in the opportunistic fungal pathogen *Aspergillus fumigatus*. Nature *457*, 471–474.

O'Hanlon, K.A., Margison, G.P., Hatch, A., Fitzpatrick, D.A., Owens, R.A., Doyle, S., and Jones, G.W. (2012). Molecular characterization of an adaptive response to alkylating agents in the opportunistic pathogen *Aspergillus fumigatus*. Nucleic Acids Res. *40*, 7806–7820.

Ogata, H., Goto, S., Sato, K., Fujibuchi, W., Bono, H., and Kanehisa, M. (1999). KEGG: Kyoto Encyclopedia of genes and genomes. Nucleic Acids Res. *27*, 29–34.

Pasqualotto, A.C., and Denning, D.W. (2006). Postoperative aspergillosis. Clin. Microbiol. Infect. *12*, 1060–1076.

Perepnikhatka, V., Fischer, F.J., Niimi, M., Baker, R.A., Cannon, R.D., Wang, Y.K., Sherman, F., and Rustchenko, E. (1999). Specific chromosome alterations in fluconazole-resistant mutants of *Candida albicans*. J. Bacteriol. *181*, 4041–4049.

Riccombeni, A., Vidanes, G., Proux-Wera, E., Wolfe, K.H., and Butler, G. (2012). Sequence and analysis of the genome of the pathogenic yeast *Candida orthopsilosis*. PLoS One 7, e35750.

Rieseberg, L.H. (2001). Chromosomal rearrangements and speciation. Trends Ecol. Evol. *16*, 351–358.

Robbertse, B., Reeves, J.B., Schoch, C.L., and Spatafora, J.W. (2006). A phylogenomic analysis of the *Ascomycota*. Fungal Genet. Biol. *43*, 715–725.

Rokas, A., Payne, G., Fedorova, N.D., Baker, S.E., Machida, M., Yu, J., Georgianna, D.R., Dean, R.A., Bhatnagar, D., Cleveland, T.E., *et al.* (2007). What can comparative genomics tell us about species concepts in the genus *Aspergillus*? Stud. Mycol. *59*, 11–17.

Roy, S.W., Fedorov, A., and Gilbert, W. (2003). Large-scale comparison of intron positions in mammalian genes shows intron loss but no gain. Proc. Natl. Acad. Sci. U.S.A. *100*, 7158–7162.

Sanglard, D. (2002). Resistance of human fungal pathogens to antifungal drugs. Curr. Opin. Microbiol. *5*, 379–385.

Sanglard, D., and Odds, F.C. (2002). Resistance of *Candida* species to antifungal agents: molecular mechanisms and clinical consequences. Lancet Infect. Dis. *2*, 73–85.

Saubolle, M.A., McKellar, P.P., and Sussland, D. (2007). Epidemiologic, clinical, and diagnostic aspects of coccidioidomycosis. J. Clin. Microbiol. *45*, 26–30.

Schaller, M., Borelli, C., Korting, H.C., and Hube, B. (2005). Hydrolytic enzymes as virulence factors of *Candida albicans*. Mycoses *48*, 365–377.

Schoen, C., Reichard, U., Monod, M., Kratzin, H.D., and Ruchel, R. (2002). Molecular cloning of an extracellular aspartic proteinase from Rhizopus microsporus and evidence for its expression during infection. Med. Mycol. *40*, 61–71.

Selmecki, A., Forche, A., and Berman, J. (2006). Aneuploidy and isochromosome formation in drug-resistant *Candida albicans*. Science *313*, 367–370.

Selmecki, A., Gerami-Nejad, M., Paulson, C., Forche, A., and Berman, J. (2008). An isochromosome confers drug resistance *in vivo* by amplification of two genes, *ERG11* and *TAC1*. Mol. Microbiol. *68*, 624–641.

Selmecki, A.M., Dulmage, K., Cowen, L.E., Anderson, J.B., and Berman, J. (2009). Acquisition of aneuploidy provides increased fitness during the evolution of antifungal drug resistance. PLoS Genet. *5*, e1000705.

Sharpton, T.J., Stajich, J.E., Rounsley, S.D., Gardner, M.J., Wortman, J.R., Jordar, V.S., Maiti, R., Kodira, C.D., Neafsey, D.E., and Zeng, Q. (2009). Comparative genomic analyses of the human fungal pathogens *Coccidioides* and their relatives. Genome Res. *19*, 1722.

Sherman, D., Durrens, P., Beyne, E., Nikolski, M., and Souciet, J.L. (2004). Genolevures: comparative genomics and molecular evolution of hemiascomycetous yeasts. Nucleic Acids Res. *32*, D315–318.

da Silva, B.A., dos Santos, A.L., Barreto-Bergter, E., and Pinto, M.R. (2006). Extracellular peptidase in the fungal pathogen *Pseudallescheria boydii*. Curr. Microbiol. *53*, 18–22.

Sionov, E., Lee, H., Chang, Y.C., and Kwon-Chung, K.J. (2010). *Cryptococcus neoformans* overcomes stress of azole drugs by formation of disomy in specific multiple chromosomes. PLoS Pathog. *6*, e1000848.

Sorek, R., and Ast, G. (2003). Intronic sequences flanking alternatively spliced exons are conserved between human and mouse. Genome Res. *13*, 1631–1637.

Sorrell, T.C. (2001). *Cryptococcus neoformans* variety *gattii*. Med. Mycol. *39*, 155–168.

Stajich, J.E., Dietrich, F.S., and Roy, S.W. (2007). Comparative genomic analysis of fungal genomes reveals intron-rich ancestors. Genome Biol. *8*, R223.

Stechmann, A., and Cavalier-Smith, T. (2002). Rooting the eukaryote tree by using a derived gene fusion. Science *297*, 89–91.

Sullivan, D.J., Westerneng, T.J., Haynes, K.A., Bennett, D.E., and Coleman, D.C. (1995). *Candida dubliniensis* sp. nov.: phenotypic and molecular characterization of a novel species associated with oral candidosis in HIV-infected individuals. Microbiology *141*, 1507–1521.

Teixeira, M.M., Theodoro, R.C., de Carvalho, M.J., Fernandes, L., Paes, H.C., Hahn, R.C., Mendoza, L., Bagagli, E., San-Blas, G., and Felipe, M.S. (2009). Phylogenetic analysis reveals a high level of speciation in the *Paracoccidioides* genus. Mol. Phylogenet. Evol. *52*, 273–283.

Untereiner, W.A., Scott, J.A., Naveau, F.A., Sigler, L., Bachewich, J., and Angus, A. (2004). The Ajellomycetaceae, a new family of vertebrate-associated Onygenales. Mycologia *96*, 812–821.

Weems, J.J., Jr., Chamberland, M.E., Ward, J., Willy, M., Padhye, A.A., and Solomon, S.L. (1987). *Candida parapsilosis* fungemia associated with parenteral nutrition and contaminated blood pressure transducers. J. Clin. Microbiol. *25*, 1029–1032.

White, T.C., Marr, K.A., and Bowden, R.A. (1998). Clinical, cellular, and molecular factors that contribute to antifungal drug resistance. Clin. Microbiol. Rev. *11*, 382–402.

Wingard, J.R., Merz, W.G., and Saral, R. (1979). *Candida tropicalis*: a major pathogen in immunocompromised patients. Ann. Intern. Med. *91*, 539–543.

Wolfe, K.H., and Shields, D.C. (1997). Molecular evidence for an ancient duplication of the entire yeast genome. Nature *387*, 708–713.

Wu, Y., Yang, J., Yang, F., Liu, T., Leng, W., Chu, Y., and Jin, Q. (2009). Recent dermatophyte divergence revealed by comparative and phylogenetic analysis of mitochondrial genomes. BMC Genomics *10*, 238.

Xu, J., Vilgalys, R., and Mitchell, T.G. (2000). Multiple gene genealogies reveal recent dispersion and hybridization in the human pathogenic fungus *Cryptococcus neoformans*. Mol. Ecol. *9*, 1471–1481.

Zeyl, C. (2004). Capturing the adaptive mutation in yeast. Res. Microbiol. *155*, 217–223.

Zureik, M., Neukirch, C., Leynaert, B., Liard, R., Bousquet, J., and Neukirch, F. (2002). Sensitisation to airborne moulds and severity of asthma: cross sectional study from European Community respiratory health survey. Brit. Med. J. *325*, 411–414.

Data-driven Systems Biology of Fungal Infections

3

Fabian Horn, Vito Valiante, Reinhard Guthke and Axel A. Brakhage

Abstract

Recent technical advancements in biotechnology unlocked a new fascinating period of research. High-throughput data can be obtained within biomolecular experiments with reasonable efforts and costs. This enables the investigation of the host–fungal interactions at a system-wide level. The bottleneck in the experiment workflow has shifted from obtaining the data to its proper analysis. The analysis of high-throughput data can be described by the operational protocol of data-driven systems biology. The protocol includes experimental design, data pre-processing, feature selection, computational modelling, and biological evaluation. Throughout this procedure, data need to be analysed within the corresponding biological context. The goal of the analysis pipeline is to obtain model variables and parameters in order to make new hypotheses for further experiments. The intertwined nature of the host–pathogen interaction is also reflected in the complex data analysis where first experiences are currently gathered and new methods developed. Pressing challenges are the integration of different levels of omics data, i.e. modelling should consider multiple molecular levels and the integration of different technologies at each level. Advances in systems biology are expected to help translate *in vivo* knowledge about fungal infections into clinical applications.

Background

Patients suffering from a weakened immune system are at high risk for developing fungal infections. Fungal pathogens can cause a large variety of diseases including allergies, skin and superficial infections, and invasive mycoses (Brakhage, 2005; Brakhage *et al.*, 2010). Moreover, the number of patients with immune compromised systems is rising in the last decades. This is basically due to the increase of medical practices such as chemotherapy and organ or bone marrow transplantation leading to an increased number of patients with impaired immune status. Mortality due to mycoses increased by 400% between the 1980s and the year 2000 in the USA. This rise is associated with the increase of sepsis (incremented by 207%) caused by fungal organisms whereas the number of cases of bacterial sepsis have only increased moderately (Martin *et al.*, 2003). In Europe, there is no similar wide-range study but national clinical reports confirmed this trend.

The total number of estimated fungal species is around 10^6–10^7 (Hawksworth, 2001). Among them, only approximately 100 species are associated with human diseases. The yeast *Candida albicans* and the filamentous fungus *Aspergillus fumigatus* are by far the most important species causing invasive mycoses. Apart from *A. fumigatus*, other species of the genus *Aspergillus* are considered as human pathogens causing adverse effects, e.g. *A. terreus, A. flavus,* and *A. niger* (Brakhage, 2005). Concerning the *Candida* genus, *C. albicans* is most prevalent in clinical samples (50–70%), followed by *C. glabrata* (20–25%). Other *Candida* species, such as *C. tropicalis, C. dubliniensis, C. krusei* and *C. parapsilosis,* were found in clinical samples as well. However, their distribution strongly depends on temporal and geographic influences (overview in Pfaller and Diekema, 2007). Another clinically relevant human-pathogenic fungus is *Cryptococcus neoformans,* which is the most common fungal species associated with AIDS patients. Other fungal species causing invasive mycoses (e.g. *Pneumocystis jirovecii, Zygomycetes* and *Fusarium* species) are recently emerging.

Fungi causing invasive mycoses have common traits during infection. After reaching the host by landing on epidermal layers or via oral cavities, the first crucial step during the infection is to overcome the epithelial barriers and the subsequent colonization of the host. At this stage, innate immunity represents the major defence system. The ability to overcome the host immune system depends on the physiological characteristic of the fungus and on its virulence determinants.

Studies on fungal infections mainly focus on three levels: (i) virulence determinants and physiological characteristics of the fungus; (ii) mechanism of innate immunity and immune effector cells of the human host; (iii) the interaction between the fungus and the host during infection. All these approaches have advantages and disadvantages. One basic issue is the simplification of the system studied during the experiments. This can cause artefacts or even hide important traits. The second problem is how to integrate diverse data types in order to get a complete and more realistic picture of infection. Because of the high level of complexity, it is hard to cover all the relevant aspects concerning the pathobiology of pathogenic organisms with traditional approaches (Fig. 3.1). Thus, it is necessary to describe and understand invasive fungal infection by taking a systems biological approach.

Systems biology

The research field of systems biology promises new ways of investigating the intertwined nature of host–pathogen interactions (Horn *et al.*, 2012). In literature, the term is widely used for emphasizing different aspects.

The first implied aspect is a general naming of an academic field of study. This research field combines life sciences (e.g. biology, medicine, biochemistry), mathematics, information sciences and engineering sciences, thus, bridging the gap

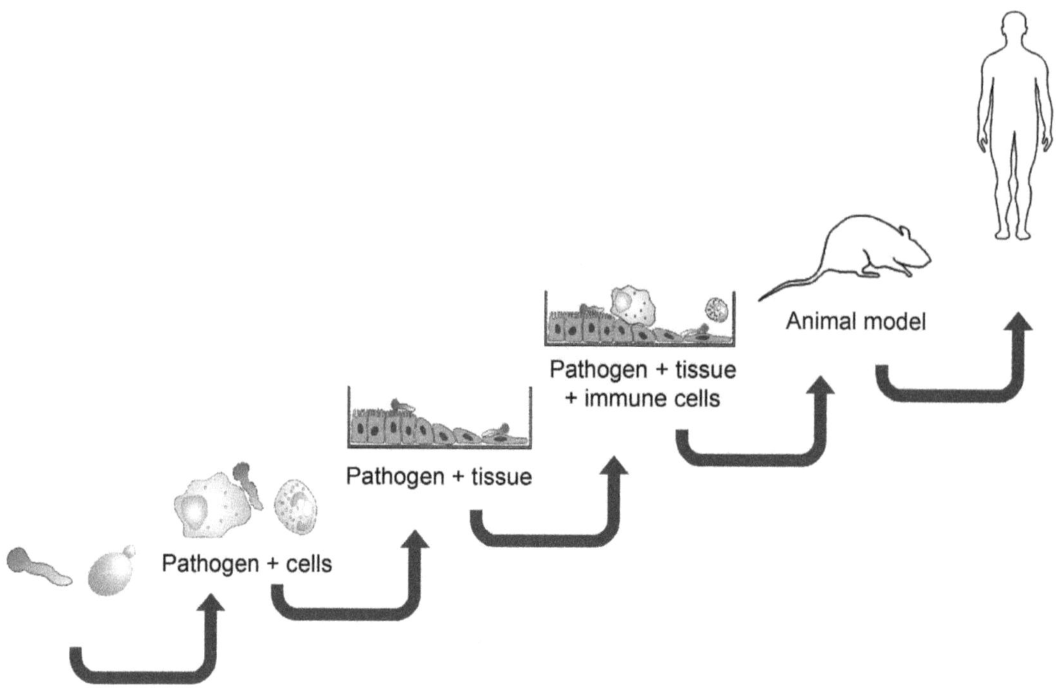

Figure 3.1 Increasing levels of complexity in infection biology studies. A colour version of this figure is available in the plate section at the back of the book.

between computational modelling techniques and biomolecular experimental techniques. Biological systems should be analysed and described by mathematical and/or computational modelling in order to understand complex interrelationships. The description of biological phenomena using mathematics can already be retraced to an area without computer support. The first famous application was the model by Bernoulli (1760). He described the spreading of diseases within a population under the influence of cowpox vaccination. Today, this kind of modelling refers to a distinct field of research called population biology.

Generally, the main work of molecular biological research focuses on time-invariant mechanisms, e.g. the phenotype of gene knockouts or the discovery of new protein interactions. Most processes, particularly the response of species to changing environmental conditions, are not time-invariant but time-dependent. The analysis of time-dependent, dynamic processes is very advanced in the fields of physics and mathematics and the experience gained in these fields should be applied to biological systems. Therefore, the term systems biology can, as a second aspect, emphasize the application of well-known dynamic modelling techniques to biological research.

The basis for these methods is formed by high-throughput experimental data, which become increasingly available due to the rapid advances in biotechnology. The investigation of these genome-wide datasets using a more holistic approach by integration rather than reduction is the third aspect described by systems biology. This paradigm emphasizes that the behaviour of the whole systems and the discovery of its emergent properties is more crucial for understanding of the cells and organisms than the isolated molecular mechanisms and pathways. In fact, systems biology tries to decipher the global systems behaviour while basing its observations on molecular mechanisms, which can be validated by experiments (Gatherer, 2010).

The last aspect is that *systems biology* describes an operational protocol, which describes the interdisciplinary collaboration of 'wet-lab' and 'dry-lab' research, groups, e.g. experimentally working biologists and bioinformaticians (Kitano, 2002). Basically, the workflow consists of data acquisition, data pre-processing, feature selection, and modelling (Fig. 3.2). As a last step, new biological hypotheses are formed which may be tested experimentally in order to gain new biological insights. At this step, new data is gathered,

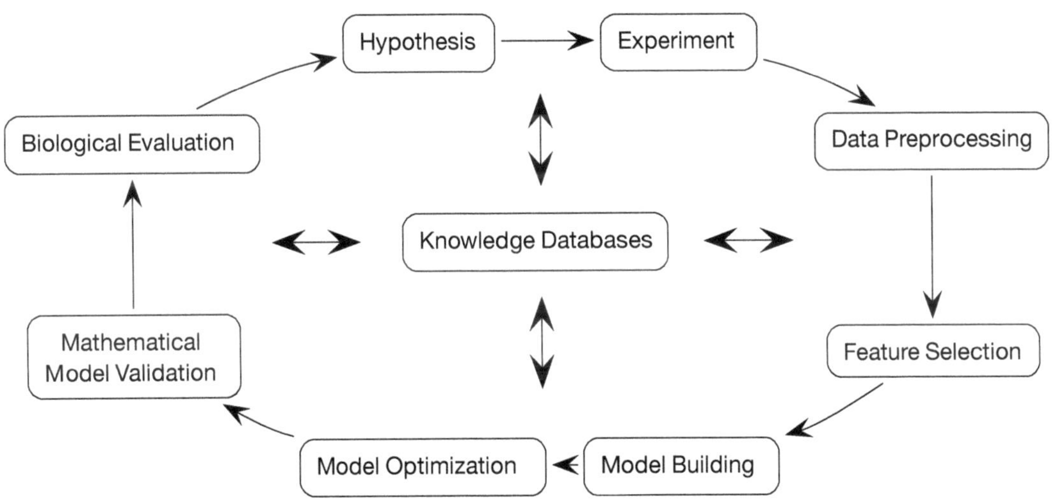

Figure 3.2 Systems biology circle. The experiment is based on working hypotheses and the resulting high-throughput data is pre-processed. Interesting features of the data are selected and can be used for further modelling. The model should be used to make further predictions, which represent new working hypotheses. At each stage during this procedure, prior knowledge and information from the widespread of existing databases is integrated.

thus, the circle is closed and a new iteration of this procedure can be realized.

The application of systems biology to host–pathogen interactions can be classified in respect to the biological perspective investigated. Several studies focus on the modelling of the human immune response mechanisms that represent the host-side (reviewed by Vodovotz *et al.*, 2010). This perspective is of special importance for fungal infections since most affected patients suffer from a compromised immune system or have a genetic disposition for immune system-related diseases. Nevertheless, fungal infections cannot be explained by the host-side alone. Single fungal species have a higher virulence, e.g. *A. fumigatus*, than their close relatives and some fungi can switch from a commensal lifestyle to a pathogenic lifestyle such as *C. albicans*. This shows that the biological system of the fungi plays a critical role within fungal interactions. Thus, other studies and models focus on the determination of the fungal pathogenic traits and try to describe the pathogen perspective of the dynamics of a fungal infection (e.g. Albrecht *et al.*, 2011). The independent investigation of both biological systems cannot really reflect the precise nature of host–pathogen interactions. Neither are immunosuppressed hosts susceptible for any fungal infection, nor is a single pathogen capable of invading any host-species. The host–pathogen interaction is very specific and sensitive and it is also assumed that only the co-evolution of hosts and pathogens has led to complex immune system mechanisms and the immune-evading systems of fungi (Casedevall, 2012). As a consequence, the intertwined nature of infections can only be fully understood if both biological systems were considered at the same time.

The different perspectives and new molecular technologies have led to a diversification of research foci. Research groups and research institutes have specialized on different biological models and distinct aspects of the host–pathogen interaction. The results either represent general mechanisms of pathobiology, e.g. formation of neutrophil exracellular traps (NETs) (Bruns *et al.*, 2010a; McCormick *et al.*, 2010), or they may only reflect specific pathogenic mechanisms, e.g. the hydrophobic rodlet layer of *A. fumigatus* on the surface immunologically silences the airborne conidia (Aimanianda *et al.*, 2009). The application of systems biology to the investigation of fungal pathogens is rather diverse and different reviews emphasizing various aspects have been published. Interested readers are referred to Albrecht *et al.* (2011) for the transcriptome and proteome analysis of *A. fumigatus*, Santamaría *et al.* (2011) for a drug design perspective, Rizzetto and Cavalieri (2011) and Tierney *et al.* (2012a) for the complexity of host–pathogen interactions, and Horn *et al.* (2012) for a survey of current -omics studies and their associated modelling methods including the perspective of image-based modelling methods.

In this chapter, we will focus on the general framework for the analysis of high-throughput data of fungal infections. We give a brief overview about the field and present some published applications focusing on the fungal perspective, in particular on *A. fumigatus* and *C. albicans*. This way, the reader may adapt successful concepts to their own research questions.

Data management and bioinformatics resources

Molecular research and computational modelling is based on information, i.e. experimental data, or existing knowledge, i.e. widely accepted facts about biological systems. In the last couple decades, there has been a revolutionary change for data acquisition methods in biology (Lee, 2008). Before the 1990s, the new frontier was genome sequencing. Genomic era had its hallmark with the complete sequence of the human genome (Lander *et al.*, 2001). However already at the end of the 1980s, scientists tried to identify new methodologies that could be used to study the entire set of transcripts, proteins and metabolites expressed by an organism under certain laboratory conditions. The spreading of this technical knowledge practically started the 'post-genomic' era, laying the foundations for what nowadays is known as systems biology. As a consequence, collecting data during the experimental procedure is not the time-limiting step any more today. The availability of extensive data is increasing at the order of different magnitudes due to the dramatically decreasing costs and time for performing high-throughput experiments (Lee, 2008). In fact, not only is the number of data is increasing, but also the number of different data types. As an example, the recent advancements in RNA-seq technology opens up a new field of applications and the generation of modified protocols (Ekblom and Galindo, 2011). New technologies do not necessarily substitute existing approaches, but several technologies will be used complementary. Studies will continue to integrate data obtained with older technologies and successfully established laboratory technology platforms will not be replaced at once.

The different types of high-throughput data are classified according to the molecular level they describe (Fig. 3.3). As an example, *transcriptomics* contains data, which gives a system-wide view of the transcript abundance in a biological sample. The suffixes *-ome* or *-omics* refer to the obtained high-throughput data and to the associated technologies and all analysis methods connected with these data types. For the investigation of fungal infections, several molecular layers are of high interest and have been the focus of recent studies (reviewed in Horn *et al.*, 2012).

Genome

The genomic information from the host and the pathogen builds the basis for all further molecular and computational analyses. At the beginning, genome sequencing has been very

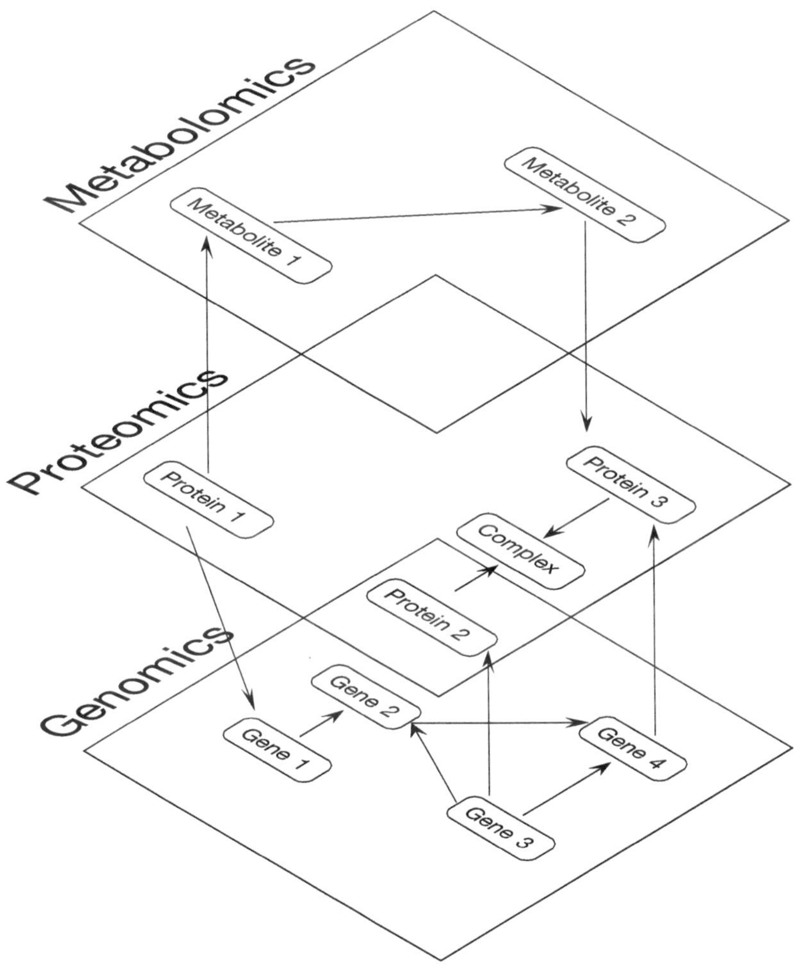

Figure 3.3 Schematic illustration of interactions between different omics-layers.

expensive and only a limited number of species, which are in the focus of a wider research community, have been sequenced. The first one released was the *C. albicans* genome (Jones *et al.*, 2004; Braun *et al.*, 2005), followed by the *C. neoformans* genome (Loftus *et al.*, 2005) and by the *A. fumigatus* genome (Nierman *et al.*, 2005). The first dermatophyte genomes were published in 2011 (Burmester *et al.*, 2011). Currently, we have access to about 45 genomes corresponding to pathogenic fungi (Horn *et al.*, 2012). With the decreasing costs for DNA-sequencing, the number of sequenced fungal species will increase considerably, and, in fact, an ambitious '1000 Fungal Genomes Project' was initiated at the DOE Joint Genome Institute (JGI). This project will provide a broad genome coverage of fungal species while allowing for a dense sampling of selected groups of closely related fungi. Additionally, by checking the NCBI website, it is possible to notice that about 100 genome projects are already in progress, such as the sequencing of several *C. albicans* strains to evaluate *Candida* genome plasticity. As a future perspective, another foundation is also being laid concerning the host-perspective. Genome sequencing of single patients will allow for the discovery of possible genetic dispositions contributing to fungal infections. The genome sequence alone does not allow the understanding of the genetic repertoire of the pathogenic fungus. It rather has to be structurally and functionally annotated in order to find potential regions in the genome, which are important for either infectious traits or the evasion from the immune system. Most gene prediction methods are based on DNA features (e.g. start codon, stop codon, GC content), sequence similarity to closely related species, and specific protein domain patterns (reviewed in Yandell and Ence, 2012). Gene prediction provides us with a variety of information concerning genes, including expression patterns, alternative splicing variants, localization, and putative interaction partners. As an example, commonly applied tools for gene prediction are Augustus (Stanke and Waack, 2003), InterProScan (Zdobnov and Apweiler, 2001), or Blast2GO (Conesa *et al.*, 2005). Predicted functional annotations are usually supplied in standardized data formats where gene ontology (GO) (Ashburner *et al.*, 2000) and functional categories (Ruepp *et al.*, 2004) play an important role for fungal research.

One specific part of the fungal genome annotation is the prediction of secondary metabolite gene clusters – a process referred to as *genome mining* (Scharf and Brakhage, 2013). The gene clusters can be detected with the help of sequence similarity and by identifying specific protein domains necessary for non-ribosomal peptide synthetases (NRPSs) or polyketide synthases (PKSs). These enzymes process a small range of metabolic precursors, e.g. malonyl-CoA and amino acids and derivatives thereof, respectively, into a large range of chemically complex metabolites, e.g. penicillin. Some of the produced secondary metabolite gene clusters may be directly involved in fungal invasion. As an example, gliotoxin is produced by a secondary metabolite gene cluster in *A. fumigatus*, which was identified using bioinformatics methods (Gardiner and Howlett, 2005). Its biosynthesis was recently elucidated and gliotoxin is supposed to suppress the immune system during infection (Scharf *et al.*, 2012). Besides the identification of putative secondary metabolite gene clusters, one main task is to decipher the regulation of the gene clusters, which is eventually responsible for the activation of the respective secondary metabolite pathway (Brakhage, 2013). Thus, the investigation of underlying gene regulatory networks is crucial and there are many bioinformatics tools designed for this purpose. In detail, there are methods for the prediction of transcription factor binding sites. As an example, SiTar is a tool which predicts binding sites only based on the sequence composition without relying on known motifs or sequence alignments of highly conserved promoter sequences (Fazius *et al.*, 2011). Fungal transcription factors are not only newly predicted, but there are also efforts to collect known fungal transcription factors in specialized databases, e.g. the Fungal Transcription Factor Database (Park *et al.*, 2008).

The genomic sequences and their respective annotations are provided through genome portals (e.g. NCBI, Ensembl), by the sequencing institution (e.g. Broad, JGI) or by community data portals (e.g. CADRE, Aspergillus Genome Database, Candida Genome Database, FungiDB).

Table 3.1 Resources of special importance for systems biology of fungal infections. All the resources have been mentioned in the text

Resource	URL
Genome	
1000 Fungal Genome Project	http://1000.fungalgenomes.org
Aspergillus Genome Database	http://www.aspgd.org
Broad Institute	http://www.broadinstitute.org
CADRE	http://www.cadre-genomes.org.uk
Candida Genome Database	http://www.Candidagenome.org
Ensembl Fungi	http://fungi.ensembl.org
Fungal Transcription Factor Database	http://ftfd.snu.ac.kr
FungiDB	http://fungidb.org
FungiFun	http://www.omnifung.hki-jena.de/FungiFun/
JIG	http://www.jig.doe.gov
NCBI	http://www.ncbi.nlm.nih.gov
Transcriptome	
ArrayExpress	http://www.ebi.ac.uk/arrayexpress
Gene Expression Omnibus	http://www.ncbi.nlm.nih.gov/geo/
Interactome	
AsperCyc	http://www.aspercyc.org.uk/
Biomolecular Interaction Network Database	http://bind.ca
Database of Interacting Proteins	http://dip.doe-mbi.ucla.edu
Host-Pathogen Interaction Database	http://agbase.msstate.edu/hpi/main.html
Molecular INTeraction database	http://mint.bio.uniroma2.it/mint/
Pathogen–Host Interactions Database	http://www.phi-base.org
Pathogen–Host Interaction Search Tool	http://www.phisto.org
Search Tool for the Retrieval of Interacting Genes	http://string.embl.de
Collaboration infrastructure	
Omnifung	http://www.omnifung.hki-jena.de
SysMoDB	http://www.sysmo-db.org
Taverna Workflow Management System	http://www.taverna.org.uk
WikiPathways	http://www.wikipathways.org

Besides deepening the taxonomic knowledge, the availability of many released genome sequences allowed the extensive use of comparative genome studies. In the infection biology field, these studies are useful to understand the evolution of virulence determinants among different fungi. One of the best examples is the phenomenon of gene family expansion. Studies of different *Candida* sp. highlighted that cell wall-associated genes, lipases, or GPI anchored proteins are important for host recognition and virulence (Butler *et al.*, 2009; van het Hoog *et al.*, 2007; Dujon, 2010). These genes have been subject to gene duplications and consequent 'family expansion' during evolution. Having more copies of important genes prevents that host genome variations can be eventually used to neutralize virulence determinants of the pathogenic fungi. This mechanism was very well studied in the plant pathogenic fungus *Cladosporium fulvum* (de Wit *et al.*, 2012) but seems to be a wider adaptation mechanism. Another comparative genomic approach led to the identification of different signalling pathways in pathogenic fungi (Rispail *et al.*, 2009). The arrangement of protein kinases and their activation mechanisms seem to be very conserved among fungi. Many of these genes were found to be important for virulence in both plant and human pathogens (Hamel *et al.*, 2012). The aim of these studies is to look for highly conserved core structures involved in signalling, in order to identify suitable intraspecific or interspecific targets for therapeutic intervention.

Transcriptome

The genome is a static feature of the biological system and it cannot be directly used to draw conclusions about the physiological, time-variant state of the organism. Environmental and developmental changes of the fungi can be observed by measuring the transcript abundance of activated genes. Today, several technologies are available for transcriptome studies. Currently, microarrays are still the most common methodology and numerous microarray-based transcriptome studies were performed in the last decade. A great part of these studies focus on transcriptome variations during stress conditions. In *C. albicans*, studies were performed to investigate the mode of action of antifungal agents (e.g. azole derivatives, amphotericin B, echinocandins (Backer *et al.*, 2001; Barker *et al.*, 2004; Liu *et al.*, 2005) but also the regulation of developmental stages (e.g. mating) and physiological stages [e.g. biofilm formation (Doedt *et al.*, 2004; Murillo *et al.*, 2005]. *A. fumigatus* transcriptome history is quite recent. Transcriptomics studied mainly focused on developmental stages, nutrient starvation, biofilm formation, and response to antifungals, e.g. Nierman *et al.* (2005), da Silva Ferreira *et al.* (2006), Schrettl *et al.* (2008), Bruns *et al.* (2010b), Cagas *et al.* (2011) and Jain *et al.* (2011)).

The main remaining challenge in infection biology is to understand transcriptomic changes during infection. Limits for these studies are (i) the development of good protocols in order to increase the quantity and the quality of mRNA prior hybridization, and (ii) the decision on the co-incubation strategy which would ideally be closer to a 'realistic' infection situation, e.g. co-culture of fungi and immune cells or directly infection of tissue. The first problem was partly solved by separating the two different organisms after co-incubation and by enriching the total mRNA with the application of a few reverse-transcription steps using random primers. This approach was successful used to identify differentially regulated genes in *C. albicans* when co-cultured with human effector cells, clarifying the importance of genes involved in nutrient acquisition, oxidative stress response, and metal homeostasis (Lorenz and Fink, 2001; Lorenz *et al.*, 2004; Wilson *et al.*, 2009; Fukuda *et al.*, 2013). A similar approach was also used to study other pathogenic fungi such as *A. fumigatus* (Lessing *et al.*, 2007; Sugui *et al.*, 2008; Morton *et al.*, 2011). Interestingly, transcriptomics approaches highlighted that many genes involved in primary metabolism appeared to be differentially expressed during infection of both human and plant pathogens. This result was obtained by comparing transcriptome data obtained from pathogenic fungi during organ or tissue infections and it suggests that physiological reprogramming remains relatively well conserved among fungal pathogens during infection (Cairns *et al.*, 2010).

All efforts made to develop and improve microarray technology have been already overtaken by the appearance of new techniques based on deep RNA sequencing (RNA-seq). There are several advantages of this new sequencing technology. In principle, it is possible to identify lowly expressed transcripts that could not be detected using microarrays. As RNA-seq does not rely on a hybridization process, it is assumed that there are less false positive results. Host–pathogen interaction studies using microarray analysis were always quite limited. The separation of organism-specific RNA prior the hybridization was hard to achieve. The simultaneous analysis of both interacting partners is becoming more feasible with the recent innovations of RNA-seq analysis and single cell measurements (Westermann *et al.*, 2012). Technically, RNA from different species can be pooled, and then the data obtained can be separated during the analysis by aligning raw sequence data to different genomes. This approach is possible because the programs that process RNA-seq reads automatically discard sequences that do not match with the referring genome. The RNA-seq technique could potentially give us a way of monitoring gene expression profiles from the pathogen and the host simultaneously. However, this promising approach shows also some technical limitations. The coverage is one important parameter to be considered during a RNA-seq experiment. In order to have a good statistical analysis on the differentially regulated genes, it is important to choose the reads length and the reads amount for the analysed genome. Studies of the suitable number of reads required for a specific genome were performed in *Caenorhabditis elegans* (Hillier *et al.*, 2009), but we still lack a suitable formula. Another study reported that 193 million paired end reads, which were uniquely mapped against the *A. fumigatus* genome, were able to identify 80% of the total annotated genes in a wild-type strain, by using a stringent RPKM cut-off of 10 reads per gene (Müller *et al.*, 2012). It is easy to understand that if we pool two organisms with different genome sizes and different numbers of expressed transcripts, the amount of obtained reads will be divided by the two species. One of the limiting factors is the calculation of the fraction of each transcriptome in the total sample. It is necessary to determine the minimal coverage that is still sufficient to provide a good statistical analysis, and actually this can be done only empirically, because we do not know the real number of RNA molecules present in a cell. As an example, let us consider a co-incubation experiment using a pathogenic fungus with around 10^4 transcripts and its human host with around $\sim 2.5 \times 10^4$ transcripts. Under the assumption that the two mRNA pools are equally distributed, and that all the genes are expressed, the coverage for the fungus will be at least twice as large when compared with the coverage obtained for the human host. This situation would be already optimal. Unfortunately, human immune effector cells, which are the most interesting targets for infection studies, appear to have a low transcription capability, turning the experiment totally unbalanced.

Despite considering these technical limitations, deep RNA sequencing is going to be a leading approach in transcriptomics profiling. Recently, first studies investigating fungal infections with the help of RNA-seq data have been published. One example is the study by Müller *et al.* (2012), who compared microarray and RNA-seq data for the investigation of the regulatory role of the mitogen activated protein kinase A encoded by the *mpkA* gene with the help of knock-out mutants. The authors discussed that, in comparison to microarrays, RNA-seq is less error-prone and that it can be used for the detection of novel transcripts and alternative splice variants.

Transcriptome data originating from microarrays is provided via Gene Expression Omnibus (GEO) or ArrayExpress. Currently, RNA-seq data is distributed within the community on the same platform, but the central data storages and distribution does not come without cost on the long run as an enormous numbr of data is to be expected (Sboner *et al.*, 2011). Presumably, a proper data management of next generation sequencing data will become more expensive than its acquisition.

Proteome

The measurement of the abundance of gene transcripts does not allow draw direct conclusions about the protein level, as additional regulatory processes have to be considered. Eventually, the protein abundance of the cells determines the enzymatic activity, the production of cellular components, the signal transduction, and the gene regulation of the cells. Several techniques have been developed in order to identify the available proteins and to measure their quantity. Two-dimensional gel (2D-gel) electrophoresis was invented in the mid-1970s, and was the first approach used to study proteins on a global scale (O'Farrell, 1975; Klose and Kobalz, 1995). Progressively, this technique was coupled with mass spectrometry (MS)-based methods. Spots identified on 2D gel electrophoresis are typically digested and analysed by mass spectrometry (reviewed by Westermeier and Görg, 2011). These methods allow the identification of differences in protein abundance followed by the identification of peptides using mass signatures.

In the last 20 years, proteomics was used to identify differentially regulated proteins in response to various stimuli and conditions. As for transcriptomics, proteomics was mainly used to investigate the physiology of stress responses, the mode of action of antifungals, and regulation of developmental stages in pathogenic fungi. The variation of the different experiments was mainly based on extraction protocols, pH transition, and isolation of specific cellular compartments (e.g. mitochondria and cell wall) (reviewed by Kniemeyer *et al.*, 2011). Of special importance for fungal infections is the investigation of the secretome during infection, i.e. all secreted proteins. There have been studies establishing the composition of the *Aspergillus* core secretome (Vödisch *et al.*, 2009; Wartenberg *et al.*, 2011) and also studies of virulence traits, such as the response to hypoxia (Vödisch *et al.*, 2011). Proteomics studies of the interplay between fungal pathogens and human cells remain a challenging task because of the complexity and the number of methods available for the separation of fungal and immune effector cells. This approach was used to study the interaction between *C. albicans* and murine macrophages that led to the identification of chaperons and stress-related proteins accumulated during pathogenesis (Shin *et al.*, 2005; Martínez-Solano *et al.*, 2006, 2009). The proteomics field is as well subject to continuous technical innovations. Despite its advantages (robustness, separation of intact proteins, high resolution), 2D-gel electrophoresis is biased towards low abundant and hydrophobic proteins, which limits many applications. In the last few years, significant technological advances have been developed in mass spectrometry instruments and LC-MS/MS based techniques. Many of these approaches are so called 'gel free', because they do not apply any pre-separation on 2D-gels. These new approaches were mainly applied to study the *S. cerevisiae* proteome, but they will be sooner or later optimized also for filamentous fungi. The amount of identified proteins may be impressively high and some protocols allow the identification of very low abundant proteins (50 copies/cell) (Picotti *et al.*, 2009). However, the biggest challenge is to perform proteomics approaches that enable to identify post-translational modifications (e.g. phosphorylation, glycosylation, methylation, and ubiquitination), which are supposed to be relevant for the pathobiology of medically relevant fungi (Leach and Brown, 2012). For these approaches, we are still in a premature phase for human pathogenic fungi and many protocols need to be adapted. Of course, like for transcriptomics, the best protocol would be the one that allows the screening of global protein changes in the fungus and in the host at the same time, but there are no established methods so far. With regard to the many technological advantages, which have been emerging, further progress in this direction can be expected soon.

Besides the conclusions directly drawn from the quantity of the protein, the interaction between proteins is of high interest. Such interactome data are stored in data collections without a specific biological focus (e.g. DIP, BIND, MINT, STRING). Moreover, there exist databases specialized on protein interactions which are associated with the infection process (e.g. HPIDB, PHI-base, PHISTO). Such data build the basis for data integration and analysis.

Data integration

Most published studies on fungal infections focus on the analysis of one single high-throughput data type. These datasets represent just a small part of the biological system. An integrative analysis of different -omics data is desirable in order to understand all regulatory interactions, which may occur at several layers, i.e. transcriptome, epigenome and proteome.

The first problem encountered for data integration concerns the differences among the applied methods used. Different groups use different protocols, which naturally contain variations in media composition, used strains, or incubation temperature. All these differences create barriers that make direct comparison complicated. Additionally, there are many manufacture differences in the used technology (e.g. microarray chips, 2D-gel equipment, LC columns) decreasing comparability of experiments. One example is given by the different microarray platforms developed in the last years. Even for the same organism, they use different microarray probe sequences and eventually result in different outputs for the same experimental settings.

Comparisons among different microarray platforms already drove attention to these problems. As different microarray platforms have different advantages and disadvantages, each technology is normally optimized for an application. Therefore, it is not practical to ask the research community to use only one single provider and to adopt the same working procedure. The difference between the transcriptome experiments is also apparent when the microarray technology is compared with the RNA-seq technology. A recent study underlined the technical bias differences among these technologies as well as between different microarray platforms for different samples of *A. fumigatus* wild-type strains (Müller *et al.*, 2012). The comparison of high-throughput data obtained from different -omics technologies is another example demonstrating the challenges of data integration. A study by Albrecht *et al.* (2010a) used transcriptome and proteome data from an experiment investigating the response of *A. fumigatus* to heat shock. Transcripts and proteins were correlated while considering possible time-delays. A median correlation of 0.84 between proteome and transcriptome data was observed. This is noteworthy because many studies only found a low correlation between these two data types although a higher correlation is to be expected (Greenbaum *et al.*, 2003; Gygi *et al.*, 1999). It was hypothesized that proteins without a positive correlation to their respective transcripts are under the regulation of additionally regulatory mechanism. These examples already show that the integration of data from different layers is difficult. Between these levels additional regulations occur which are not only fixed to a single layer but can occur across multiple layers, e.g. the product inhibition of enzymes is a regulatory process shared between the metabolite and protein level (see Fig. 3.3).

In bioinformatics, data integration is a research field itself and, according to Huttenhower and Hofmann (2010), it basically has to fulfil three main tasks: (1) assemble data from repositories, (2) establish an efficient querying methodology, and (3) integrate different experimental data types. These tasks can be best explained by an allegory to well-known internet search engines. In a first step, the content of the webpages is gathered by web-crawlers and it is pre-processed and stored in efficient data structures (search index). In a second step, an efficient and simple querying system is offered. The simple text search box can be used to retrieve a weighted list of results within milliseconds. Lastly, different types of data, i.e. webpages, images or maps, are presented, whereas most current search-engines do not provide an integrated, connected view of all search results but arrange them in different result fields.

The same technology would also be ideal for biological research. This idea is not new, e.g. the Sequence Retrieval System (SRS) was developed shortly after the establishment of sequence databases (Etzold and Argos, 1993) but was not extended to a wider range of data types. The realization of such technical platforms is hampered by the remarkable challenges of biological data. Specific problems for data integration are the amount of experimental data, inconsistent identifiers across different databases, different vocabularies for different research fields, permanently changing programming interfaces, missing standards, and missing enforcement of existing standards (see Goble and Stevens, 2008, for a detailed review).

Different requirements on the infrastructure for data storage have been met with different technical realizations. The goal of data storage systems differs for database which either focus on a biological level or technology (e.g. expression data or protein interaction databases), or which focus on the investigated species (e.g. Aspergillus Genome Database). From a technical point of view, there are differences in the way the underlying data resources are queried and integrated (Köhler, 2004). Technical solutions range from a decentralized federated database systems, i.e. data resources are technically linked with the help integration layers, to centralized data warehouses, i.e. the merging of different databases into a new data resource (Zhang *et al.*, 2009). Frequently, data warehouse systems additionally offer standard data analysis methods and standard visualization methods, e.g. Omnifung – a data warehouse for fungal -omics data – offers the possibility to preprocess and visualize stored raw data (Albrecht *et al.*, 2007). The wide range of different database solutions is completed by the group of community- or project-orientated frameworks. As an example, SysMoDB facilitates a quick communication between collaborators by allowing for a quick exchange of data stored in a broad range of data formats (especially spreadsheets). Other solutions are based on wikipedias (e.g. WikiPathways) or workflow management systems (e.g. Taverna). While the high number of different technologies makes the interplay between them difficult, they ensure the existence of a tailor-made data storage solution for each project.

Data analysis

The generation of experimental data in molecular biology is no longer considered a bottleneck. Nowadays, the number of available data sets is strongly increasing even for smaller research communities. However, the accumulating data does not necessarily mean that new biological insights are gained at the same speed. As an example, in the scope of the ENCODE project it took 35 research groups to make a systematic analysis of 1% of the human genome (ENCODE Project Consortium *et al.*, 2007). Ongoing research tries to identify methods that help analyse genome sequences quicker and to make the predictions more reliable. Thus, the bottleneck during the analysis pipeline shifted from the data acquisition to the integrative data analysis.

Here, we present the operational protocol of data-driven systems biology which focuses on the interpretation of experimental high-throughput data gained from perturbation experiments (Fig. 3.4). The model helps to classify bioinformatics methods and to understand interdependencies between them. The main goal of the analysis is to obtain a manageable amount of biological hypothesis from the high-throughput data, which can be validated experimentally. At each stage, the experimental data is filtered, refined, and it is getting more specific in respect to the research question. One passage in the procedure of data-driven systems biology corresponds to a single run through the classical systems biology cycle (Fig. 3.2). Throughout the procedure, methods are applied which integrate prior knowledge available in public databases. The prior knowledge gets more specific as we proceed towards the precise biological hypotheses. Within this protocol, additional bioinformatics methods, such as sequence analysis and text mining, build the bases for the retrieval of prior knowledge or the validation of new results.

The research question guides every analysis stage but it is mainly incorporated in the experiment design and therefore in the raw data. The obtained data 'drive' the analysis in three directions: first, the data type determines the pre-processing and also limits the possible biological interpretations as only single perspectives of the biological systems are measured; second, the quality and quantity of the data determines

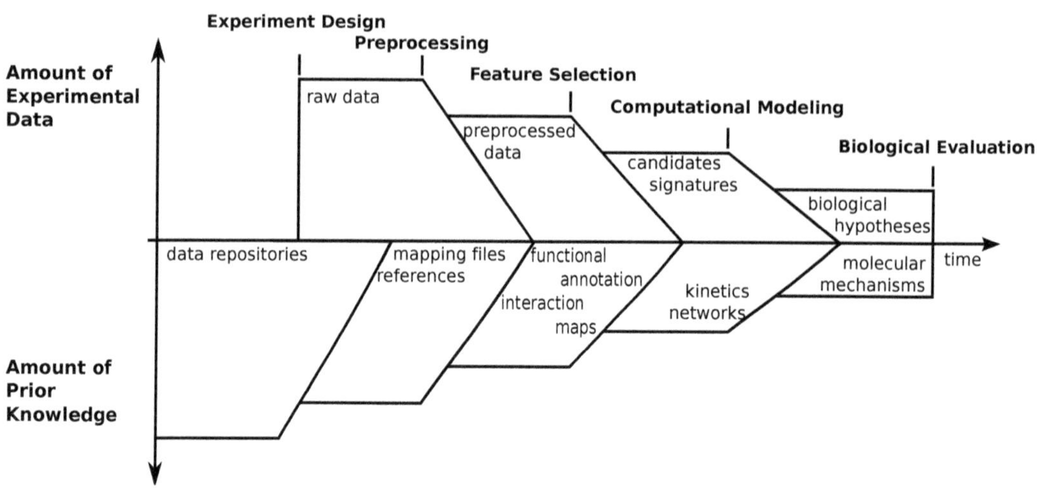

Figure 3.4 Scheme of data- and knowledge-driven analysis consisting of five stages. At each stage, the experiment is refined in order to finally obtain experimentally testable hypotheses. At the same time, the utilized prior knowledge is getting more precise. The experimental validation of new hypothesis leads to new experimental data where the data-driven analysis is applied again.

which mathematical methods are applicable and what statistical power can be reached; third, the biological information contained within the dataset is responsible for the biologically motivated decisions during the analysis. (See Santamaría *et al.* 2011, for an introduction into data-driven systems biology.)

Data-driven systems biology can be divided into five different stages: experiment design and data acquisition, data pre-processing, feature selection, modelling, and biological evaluation. Each single stage needs to meet the required specifications and tailor-made methods should be applied. Thus, existing software needs to be adapted or new solutions need to be found in respect to the research question and the available data. Ideally, the tools should be generically applicable, i.e. the transfer of results to the next stage in the analysis pipeline should be supported. In practice, many tools are inconsistent, use different data file standards, and only a few standard operational protocols exist. A comprehensive comparison of different analysis methods is hard, because in order to estimate the propagation of errors introduced by each method, they should not only be assessed within a single stage but within the framework of the whole workflow. The number of possible combinations of different methods at each stage and the strong influence of the data make this approach not feasible.

Experiment design and data acquisition

The first stage is the experimental design where all the aspects and expertise should be involved. During this fundamental stage, the most appropriate technology is chosen, the number of replicates is determined, and time points and perturbations are defined. The decision depends on the precise research question, the available experience with technologies, and the cost–benefit ratio of possible solutions. Systematic errors concerning the technical and biological variability can significantly vary across different platforms. The choice of subsequent analysis methods and the statistical power of latter approaches are determined by the experimental design. A good review on experimental design was written by Kreutz and Timmer (2009). During this stage, it also needs to be assessed what prior knowledge is available in order to perform more sophisticated analysis methods. If this information is not available, methods based on sequence analysis or text mining need to be considered in order to gain additional prior knowledge.

Data pre-processing

After carrying out the experiments, the data needs to be pre-processed in order to remove systematic biases introduced by the technology or the experiment design. In a first step, a quality control of the obtained data is performed, i.e. the noise level is estimated, technical and biological biases are detected, the reproducibility and the completeness of the data is checked. In the worst case, the obtained quality measures may lead to a rejection of the dataset. Commonly, the observed systematic errors can be minimized with the help of normalization and filtering. As an example, for microarrays it was shown that different normalization and filtering methods have a significant influence on the resulting number of regulated genes (Parrish and Spencer, 2004). A multiple-testing correction is necessary as we are working with high-throughput data and, from a mathematical point of view, thousands of experiments are performed at the same time (Benjamini and Hochberg, 1995). Together with the error correction, the mapping of measured signals to the respective biological objects is performed, e.g. spot intensities from microarrays are mapped to their gene identifiers. This step usually depends on a good genome annotation or an extensive reference database. In more complex experiment designs, missing values need to be imputed to allow a consistent mathematical analysis, e.g. during time-series proteome analysis missing gel spots need to be estimated (Albrecht *et al.*, 2010b).

Data pre-processing strongly depends on the technology platform applied. It may involve extensive manual assistance, especially for technologies where images and videos are processed. Despite the reference to commonly applied pre-processing workflows, there is no general valid standard operation procedure for data pre-processing and the selected methods need to be adapted to the technology platform and the research question.

Feature selection

The starting point for this stage is the pre-processed data, which may consist of many different conditions that investigate thousands of different biological objects, e.g. expression of genes at different time-points. The characteristics of the experimental pre-processed data help to (i) validate the experiment design, (ii) test for inconsistencies and discrepancies in contrast to already published literature, and (iii) prioritize candidates for further analysis. A specific task of feature selection is the determination of data signatures, i.e. characteristic data features capable of discriminating the original perturbation. Those signatures can be used as biomarkers.

From a methodological point of view there exists two complementary approaches for feature selection: confirmatory data analysis and exploratory data analysis (Tukey, 1977). Confirmatory data analysis is used to decide an open, precise working hypothesis. One example is the biomarker detection based on the hypothesis of the existence of data features, which are capable of discriminating different conditions. This kind of analysis is usually performed with the help of statistics, supervised and unsupervised learning methods. Software solutions supporting confirmatory data analysis usually have a defined workflow and, thus, computer assistance is desirable and realizable. The analysis of metabolic pathways also falls in this category. Experimental data is matched to well-known pathways and statements of physiological change can be inferred with the help of a statistical enrichment analysis. For fungi, there exist specialized platforms offering this type of analysis, e.g. FungiFun (Priebe *et al.*, 2011).

A more general data analysis is the exploratory data analysis approach where no clear working hypothesis is given. Thus, the analysis focuses on finding the right question, rather than the right answer (Kelder *et al.*, 2010). The researcher should be capable to follow his intuition and to realize his creativity while exploring the data. Usually, this step is carried out by expert biologists who use external resources for their visual analysis – mainly through web-browser applications. However, this analysis method is restricted by the deficiency of technology frameworks which assist the handling of high-throughput data. A user-friendly data integration framework combining existing knowledge with experimental data is essential for the fast realization of creative exploration strategies. Additionally, the framework should allow work with different molecular levels and different database technologies data integration.

Exploratory data analysis does not only rely on data integration, but also relies on data visualization. Different visualizations of the same dataset are not just a different dump of data but represent different analysis methods. The main task of visualization is to reduce complexity in order to emphasize the underlying patterns As an example, during the pre-processing of microarray data there are different visualization methods especially designed to intuitively detect technical biases, e.g. the separation of signals according to the print-tips (Yang *et al.*, 2002). Currently, biological data is preferably visualized with the help of networks (as reviewed by Pavlopoulos *et al.*, 2011). If data from different data domains, e.g. images and physiological measurements, need to be visualized within an integrated view, more complex approaches need to be applied (Rohn *et al.*, 2011).

Despite the importance of exploratory data analysis, there are only a few tools supporting data integration and visualization without precise and pre-implemented workflows (Kelder *et al.*, 2010). These tools enable the user to explore data structures which integrated existing knowledge and experimental data. Cytoscape (Shannon *et al.*, 2003) and Ondex (Köhler, 2004) are examples of software frameworks that are based on networks offering data integration and visualization methods via an interactive user interface (Bebek *et al.*, 2012). BiologicalNetworks (Kozhenkov *et al.*, 2011) is a similar tool which is specialized for the integration and visualization of studies obtained from host–pathogen studies. In order to achieve a wider usage of these analysis methods in the biological research community, the tools need to improve the integration of diverse data types, the handling of larger networks, the interface usability, the systematic incorporation of user feedback, and the minimization of technical hurdles (Kelder *et al.*, 2010). Within academic projects, a high effort needs to be invested by the developers of such software frameworks in order to insert new methodologies while guaranteeing a high reliability and maintaining the flexibility and usability of these systems.

Modelling

If the quality of the data is high and if the experiment design was arranged accordingly, a mathematical modelling stage is subsequently applied. The main goal of the modelling is to describe the observed experimental data dynamically, to further prioritize candidates, or to predict new, so far unknown interactions within the data.

Different strategies can be employed in order to reach these goals. On the one hand, the top-down modelling strategy decomposes the data into smaller components and tries to infer single knowledge for the data. On the other hand, the bottom-up modelling strategy is based on the knowledge and data for specific model variables of high interest. Hybrid strategies combining both strategies are used to understand the interdependencies within several hierarchical layers (Forst, 2006).

Modelling can occur at different abstraction levels where each level has its own time and spatial scale (Horn *et al.*, 2012). Models are either used to describe the system at the molecular level, the cell level, the tissue level, the organ level, the systemic level, or the population level. These different levels can be combined to multi-scaled models, i.e. the dynamic behaviour of one level can affect a neighbouring abstraction layer (An *et al.*, 2012). Several different methods are available for modelling, e.g. ordinary differential equations, partial differential equations, agent-based models, process algebra, and rule-based models (Ghosh *et al.*, 2011). Each modelling technique includes a parameter set which needs to be estimated with the help of *in vivo* measurements and *in silico* parameter-optimization methods.

There are many studies using modelling techniques on the human host side. They do not only investigate fungal pathogens but also viruses and bacteria (Vodovotz *et al.*, 2010). Modelling techniques should not only be limited to the host perspective of the infection process, but should also include the molecular mechanisms of fungal pathogenicity. There are some studies that have applied modelling techniques to the fungal perspective of an infection (Linde *et al.*, 2012; Altwasser *et al.*, 2012). In the focus of these studies are the elucidation of key virulence traits and their biomolecular characterization. A better annotation of genomes is essential for this purpose and the incorporation of information gained from RNA-seq data into structural gene annotation seems promising (Müller *et al.*, 2012). The application of modelling techniques focusing on the host–pathogen interactions, thus incorporating cross-talk studies between two biological systems, is just at the beginning (Sturdevant *et al.*, 2010; Cottier and Pavelka, 2012; Tierney *et al.*, 2012b). More studies in this direction are to be expected. Additionally, new methods based on spatio-temporal data are developing. The first steps for the application of image-based systems biology to fungal infections by agent-based modelling has been done (An *et al.*, 2009; Tokarski *et al.*, 2012) and have been reviewed by Horn *et al.* (2012). Here, we focus on omics-based systems biology where networks are currently the most common representation.

Network modelling tries to represent biological systems and their interactions with the help of a graph. In this graph, the nodes usually correspond to molecular entities, e.g. genes or proteins, and the edges between them depict a causal relationship, e.g. regulation. These types of networks help to understand the biological system and to put the biological information into a single context. One goal of systems biology is to infer such networks. Thus, causal relationships among molecular entities are reconstructed with respect to the experimental context.

Networks can be constructed solely based on manually curated data as it is usually done for the reconstruction of metabolic networks. As an example, Leach and Brown (2012) developed a model for the heat shock response of *C. albicans* and integrated experimental data into it. The predicted dynamic behaviour could be experimentally validated.

Boolean and Bayesian inference, regression-, information theory- and ordinary differential equation-based network inference are mathematical and computational approaches which utilize experimental high-throughput data for gene regulatory network inference (Ghosh *et al.*, 2011). These methods are capable of predicting new relationships between the molecular entities which have not been described in literature yet. Despite the high amount of experimental data available, the complexity of most methods impose the challenge of dimensionality, i.e. the large number of variables and parameters cannot be optimized on the experimental data alone. Among the approaches to reduce the dimensionality and to guide the network reconstruction procedure is the usage of prior knowledge (Hecker *et al.*, 2009) and to aggregate the variables (e.g. gene expression profiles) by clustering (Guthke *et al.*, 2005; Foteinou *et al.*, 2009).

The incorporation of curated networks into the automatic network reconstruction represents a hybrid approach, which includes the information from experimental data and the knowledge available. In any case, prior knowledge is necessary for the validation of the resulting networks or for the assessment of different reconstruction methods (Prill *et al.*, 2010). For this purpose, it needs to be ensured that the prior knowledge is up-to-date and has a high quality. On this account, expert knowledge is currently preferred over automatically derived knowledge. As an example, text mining is noisy and it is hard to carefully decide the context, i.e. only find disease-specific interactions (Harmston *et al.*, 2010). On the other hand, biological experts may only know a limited part of the research field or miss current parts of the overwhelming amount of literature. One approach to circumvent both drawbacks is the collaborative curation of prior knowledge. Unfortunately, due to missing procedures to acknowledge the researchers' efforts within the community, these approaches are only applied in rare cases (Ghosh *et al.*, 2011).

Gene regulatory networks

In gene regulatory networks, nodes represent genes and edges represent an interaction between them, i.e. repression or activation. The regulation does no represent direct interactions, but symbolizes indirect interaction with the help of, so far unknown, intermediate genes or additional signal transduction. For the reconstruction of gene regulatory networks, the integration of prior knowledge is crucial (Vicente *et al.*, 2012). For this reason, known and predicted transcription factors and their corresponding target genes play an important role. Challenges occurring during gene-regulatory network and available methods have been reviewed by Hecker *et al.* (2009).

For fungal infections, the reconstruction has been applied to *C. albicans* and *A. fumigatus*. As an example, NetGenerator (Guthke *et al.*, 2005; Toepfer *et al.*, 2007; Weber *et al.*, 2013), a tool based on ordinary differentially equations, has been used by Guthke *et al.* (2007) to investigate the heat shock response of *A. fumigatus*. This approach was successfully applied to study the iron acquisition of *A. fumigatus* during infection (Linde *et al.*, 2010, 2012). The authors extended the method by the incorporation of prior knowledge and the utilization of the scale-free network property. NetGenerator was also applied to study RNA-seq data, which have been simultaneously obtained from immune cells and the invading fungus *C. albicans* (Tierney *et al.*, 2012b). This is the first gene regulatory network inference approach to directly study the host–pathogen interaction and which infers an inter-species regulatory network with the help of RNA-seq data simultaneously measured from both host and pathogen cells.

The infection process is usually studied focusing on a single virulent trait. The consideration of multiple traits during modelling is desirable in order to reflect multiple aspects of infection. Consequently, the network inference tools have been extended to allow such multiple experiment stimuli (Weber *et al.*, 2013). It can be assumed that more stimuli also result in a higher number of genes involved in the regulatory response. Owing to a large number of variables and parameters in the model, the inference methods can only infer small-scale networks, which may contain about 30 model variables. It may be assumed that during the infection process a higher number of genes are involved and, therefore, large-scale network inference methods are necessary to get a more complete view. For this purpose, a method based on the least absolute shrinkage and selection operator (LASSO) (van Someren *et al.*, 2002) was adapted to analyse a microarray compendium for *C. albicans* (Altwasser *et al.*, 2012). The resulting network contained more than 6000 genes and the interpretation of the model focused on the hubs, i.e. connected components within the network topology. Nevertheless, the size of the resulting network makes its interpretation challenging. The biological interpretation of regulatory edges in the network is additionally hampered as the predicted interactions only represent indirect interactions. A mechanistic, biomolecular model of the regulation cannot be derived. An approach, called exTILAR allows for a more direct interpretation of the predicted gene regulatory network (Vlaic *et al.*, 2012). The method ensures that all edges within a gene regulatory network correspond to predicted or known direct transcription factor interactions. Thus, the predicted interactions directly represent a new biological hypothesis, which can be experimentally validated.

Protein interaction networks

Nodes in a protein–protein interaction network (PPI) represent proteins whereas the edges between them represent an interaction which may directly imply a possible binding between two proteins. Data for these networks are obtained from existing databases, yeast two-hybrid screens (Xenarios *et al.*, 2000), through text mining of the scientific literature (Kim *et al.*, 2008), or with the help of the bioinformatics prediction of protein binding (Arnold *et al.*, 2012).

There exist different strategies for the investigation of host–pathogen interactions. In a study by Dyer *et al.* (2008), an analysis of a PPI showed that most pathogenic proteins target the same host-protein. This observation can be used in order to predict proteins which are possibly involved in the infection processes (Mukhtar *et al.*, 2011). At the same time, the position of proteins within the PPI was also used for the discovery of potential drug targets (Hase *et al.*, 2009; Zhu *et al.*, 2009) or to predict the protein function based on its interaction partners in the cellular context (Xu and Li, 2006; Sharan *et al.*, 2007). So far, only a few studies exist that have applied PPI modelling methods to the specific context of fungal infections.

Signalling networks

Edges in a signalling network represent the activation of inhibition during a signal transduction, e.g. phosphorylation and ubiquitination. The nodes represent proteins, which may either only have an active or inactive state or, depending on the modelling approach, can have a certain activation level. So far, only a few experimental data for the protein activity are known for fungi although the first steps towards the modelling of genome-scale signalling have been made (Hyduke and Palsson, 2010). Nevertheless, there are some studies where the first attempts have been made to model the signal transduction. Currently, many models try to decipher the signal transduction network of *S. cerevesiae* (Waltermann and Klipp, 2010). A broader range of models is available for the human host where many different signalling pathways of the immune system are known (Vera *et al.*, 2011; Raza *et al.*, 2008). Published studies usually focus on the response of a single signalling pathway to a specific pathogen signal. This response is very sensitive and specific, i.e. it is triggered by signals recognizing pathogens, e.g. pathogen-associated molecular patterns (Hasenberg *et al.*, 2011). There always exists an extensive cross-talk between different signal transduction pathways (Borisov *et al.*, 2009). One example, where multiple signalling pathways are considered during the modelling of the activation of macrophages during infection is a study by Raza *et al.* (2008). A systemic analysis which considers multiple pathogenic signals and their effects on all known signalling pathways of the immune system would help to understand the complexity of the immune response.

Metabolic networks

Metabolic networks contain all possible chemical reactions which occur either by spontaneous reactions or by reactions catalysed with the help of enzymes. Generally, the nodes represent the metabolites and the directed edges show which metabolites are consumed and produced during one reaction. For enzymatic reactions, the responsible enzymes can be mapped to these reactions and also the regulatory genes are included in more sophisticated models, as reviewed in Ruppin *et al.* (2010). Metabolic networks are basically commonly curated with the help of literature. Despite the existence of tools supporting the semi-automatic network reconstruction, e.g. rBioNet (Thorleifsson and Thiele, 2011), this procedure is still labour-intensive and the model reconstruction already helps to identify gaps in our knowledge about the species metabolism (Thiele and Palsson, 2010). Whereas advanced models for humans (Rolfsson *et al.*, 2011) and the bacterium *Escherichia coli* (Feist *et al.*, 2007) exist, the metabolic network reconstruction for fungal pathogens is just at its beginning. First fully automatically produced, uncurated networks are presented with the help of platforms, such as AsperCyc.

There are numerous applications for metabolic networks, which mainly focus on the metabolic engineering of producer strains (reviewed by Ruppin *et al.*, 2010). The metabolic models can be used to integrate experimental data in order to provide a biological context, as was exemplary carried out for *Salmonella* (Bumann, 2009). Metabolic models could also be used in a comparative approach, i.e. models of closely related pathogenic and non-pathogenic fungi should reveal parts of the metabolic network which might be responsible for its pathogenic potential. Such approaches have already been successfully applied to *Pseudonomas* sp. (Oberhardt *et al.*, 2011). In a study by Garcia *et al.* (2011), the authors demonstrated the application of metabolic models to fungal infections with the example of a topology analysis of a sphingolipid biochemical system of *C. neoformans*. Further adaptions of these modelling approaches to fungal infections are expected as soon as further advancements for the automatic metabolic network reconstructions were made.

Biological evaluation

All hypotheses made during the modelling stage of the data-driven systems biology procedure should be experimentally validated in the last stage. This last stage ensures that the findings are not only reproducible at the same molecular level while using different technologies, but also that they affect other molecular levels in order to complete the full picture, as reviewed by Santamaría *et al.* (2011). At this stage, commonly applied experimental technologies are western blotting, ChIP, flow cytometry, ELISA, immunohistochemistry, siRNA, qRT-PCR, northern blotting, and the generation and analysis of gene knock-out mutants. This stage does not only confirm the aforementioned stages in the data-driven systems biology, but it helps to assess the quality of applied computational methods and their according parameter settings. For this purpose, the results need to be put into the research context and biological expertise is inevitable to draw new conclusions. An interdisciplinary cooperation between bioinformaticians and biologists may lead to a re-definition of existing models where the analysis iteratively goes back into the computational analysis with the help of new biological contributions.

Ideally, the obtained results are transferred from the *in vivo* confirmation to the clinical application. With respect to systems biology, the consideration of additional, clinical-relevant model parameters comprises an own research focus where the term *translational systems biology* has been established (Vodovotz *et al.*, 2010). The goal is to connect well-investigated molecular mechanisms with the physiology of the individual patient. For example, it deals with the genetic variability of both the pathogen and the host, drug resistance, the effect of age and gender, diversity of microbial populations, and the inter-dependence of different diseases (Santamaría *et al.*, 2011; Vodovotz *et al.*, 2010).

Acknowledgements

Research in the authors' laboratories is supported by the ERA-NET Programme, by the German Research Foundation (DFG) Excellence Graduate School Jena School for Microbial Communication (JSMC), the International Leibniz Research School for Microbial and Biomolecular Interactions (ILRS) as part of the JSMC, the DFG and the "Pakt für Forschung und Innovation" of the BMBF and the TMBWK.

References

Albrecht, D., Kniemeyer, O., Brakhage, A.A., and Guthke, R. (2007). Integration of transcriptome and proteome data from human-pathogenic fungi by using a data warehouse. J. Integr. Bioinf. *4*, 52.

Albrecht, D., Guthke, R., Brakhage, A.A., and Kniemeyer, O. (2010a). Integrative analysis of the heat shock response in *Aspergillus fumigatus*. BMC Genomics *11*, 32.

Albrecht, D., Kniemeyer, O., Brakhage, A.A., and Guthke, R. (2010b). Missing values in gel-based proteomics. Proteomics *10*, 1202–1211.

Albrecht, D., Kniemeyer, O., Mech, F., Gunzer, M., Brakhage, A.A., and Guthke, R. (2011). On the way toward systems biology of *Aspergillus fumigatus* infection. Int. J. Med. Microbiol. *301*, 453–459.

Altwasser, R., Linde, J., Buyko, E., Hahn, U., and Guthke, R. (2012). Genome-wide scale-free network inference for *Candida albicans*. Front. Microbiol. *3*, 51.

Aimanianda, V., Bayry, J., Bozza, S., Kniemeyer, O., Perruccio, K., Elluru, S.R., Clavaud, C., Paris, S., Brakhage, A.A., Kaveri, S.V., *et al.* (2009). Surface hydrophobin prevents immune recognition of airborne fungal spores. Nature, *460*, 1117–1121.

An, G., Mi, Q., Dutta-Moscato, J., and Vodovotz, Y. (2009). Agent-based models in translational systems biology. Wiley Interdiscip. Rev. Syst. Biol. Med. *1*, 159–171.

An, G., Nieman, G., and Vodovotz, Y. (2012). Computational and systems biology in trauma and sepsis: current state and future perspectives. Int. J. Burns Trauma *2*, 1–10.

Arnold, R., Boonen, K., Sun, M.G.F., and Kim, P.M. (2012). Computational analysis of interactomes: current and future perspectives for bioinformatics approaches to model the host–pathogen interaction space. Methods *57*, 508–518.

Ashburner, M., Ball, C.A., Blake, J.A., Botstein, D., Butler, H., Cherry, J.M., Davis, A.P., Dolinski, K., Dwight, S.S., Eppig, J.T., *et al.* (2000). Gene Ontology: tool for the unification of biology. The Gene Ontology Consortium. Nat. Genet. *25*, 25–29.

Backer, M.D.D., Ilyina, T., Ma, X.J., Vandoninck, S., Luyten, W.H., and Bossche, H.V. (2001). Genomic profiling of the response of *Candida albicans* to itraconazole treatment using a DNA microarray. Antimicrob. Agents Chemother. *45*, 1660–1670.

Barker, K.S., Crisp, S., Wiederhold, N., Lewis, R.E., Bareither, B., Eckstein, J., Barbuch, R., Bard, M., and Rogers, P.D. (2004). Genome-wide expression profiling reveals genes associated with amphotericin B and fluconazole resistance in experimentally induced antifungal resistant isolates of *Candida albicans*. J. Antimicrob. Chemother. *54*, 376–385.

Bebek, G., Koyutürk, M., Price, N.D., and Chance, M.R. (2012). Network biology methods integrating biological data for translational science. Brief. Bioinform. *13*, 446–459.

Benjamini, Y., and Hochberg, Y. (1995). Controlling the false discovery rate: a practical and powerful approach to multiple testing. Journal of the Royal Statistical Society. Series B (Methodological) *57*, 289–300.

Bernoulli, D. (1760). Essai d'une nouvelle analyse de la mortalité causée par la petite vérole et des avantages de l'inoculation pour la prévenir. Mémoires de Mathématique et de Physique. Histoire de l'Académie Royale des Sciences *1*, 1–45.

Borisov, N., Aksamitiene, E., Kiyatkin, A., Legewie, S., Berkhout, J., Maiwald, T., Kaimachnikov, N.P., Timmer, J., Hoek, J.B., and Kholodenko, B.N. (2009). Systems-level interactions between insulin–EGF networks amplify mitogenic signaling. Mol. Syst. Biol. *5*, 256.

Brakhage, A.A. (2005). Systemic fungal infections caused by *Aspergillus* species: epidemiology, infection process and virulence determinants. Curr. Drug. Targets. *6*, 875–886.

Brakhage, A.A. (2013). Regulation of fungal secondary metabolism. Nat. Rev. Microbiol. *11*, 21–32.

Braun, B.R., van Het Hoog, M., d'Enfert, C., Martchenko, M., Dungan, J., Kuo, A., Inglis, D.O., Uhl, M.A., Hogues, H., Berriman, M., *et al.* (2005). A human-curated annotation of the *Candida albicans* genome. PLoS Genet. *1*, 36–57.

Casadevall, A. (2012). Fungi and the rise of mammals. PLoS Pathog. *8*, e1002808.

Bruns, S., Kniemeyer, O., Hasenberg, M., Aimanianda, V., Nietzsche, S., Thywissen, A., Jeron, A., Latgé, J.-P., Brakhage, A.A., and Gunzer, M. (2010a). Production of extracellular traps against *Aspergillus fumigatus in vitro* and in infected lung tissue is dependent on invading neutrophils and influenced by hydrophobin RodA. PLoS Pathog. *6*, e1000873.

Bruns, S., Seidler, M., Albrecht, D., Salvenmoser, S., Remme, N., Hertweck, C., Brakhage, A.A., Kniemeyer, O., and Müller, F.-M.C. (2010b). Functional genomic profiling of *Aspergillus fumigatus* biofilm reveals enhanced production of the mycotoxin gliotoxin. Proteomics *10*, 3097–3107.

Bumann, D. (2009). System-level analysis of *Salmonella* metabolism during infection. Curr. Opin. Microbiol. *12*, 559–567.

Butler, G., Rasmussen, M.D., Lin, M.F., Santos, M.A.S., Sakthikumar, S., Munro, C.A., Rheinbay, E., Grabherr, M., Forche, A., Reedy, J.L., *et al.* (2009). Evolution of pathogenicity and sexual reproduction in eight *Candida* genomes. Nature *459*, 657–662.

Cagas, S.E., Jain, M.R., Li, H., and Perlin, D.S. (2011). Profiling the *Aspergillus fumigatus* proteome in response to caspofungin. Antimicrob. Agents Chemother. *55*, 146–154.

Cairns, T., Minuzzi, F., and Bignell, E. (2010). The host-infecting fungal transcriptome. FEMS Microbiol. Lett. *307*, 1–11.

Conesa, A., Götz, S., García-Gómez, J.M., Terol, J., Talón, M., and Robles, M. (2005). Blast2GO: a universal tool for annotation, visualization and analysis in functional genomics research. Bioinformatics *21*, 3674–3676.

Cottier, F., and Pavelka, N. (2012). Complexity and dynamics of host–fungal interactions. Immunol. Res. *53*, 127–135.

Doedt, T., Krishnamurthy, S., Bockmühl, D.P., Tebarth, B., Stempel, C., Russell, C.L., Brown, A.J.P., and Ernst, J.F. (2004). APSES proteins regulate morphogenesis and metabolism in *Candida albicans*. Mol. Biol. Cell. *15*, 3167–3180.

Dujon, B. (2010). Yeast evolutionary genomics. Nat. Rev. Genet. *11*, 512–524.

Dyer, M.D., Murali, T.M., and Sobral, B.W. (2008). The landscape of human proteins interacting with viruses and other pathogens. PLoS Pathog. *4*, e32.

Ekblom, R., and Galindo, J. (2011). Applications of next generation sequencing in molecular ecology of non-model organisms. Heredity (Edinb) *107*, 1–15.

ENCODE Project Consortium *et al.* (2007). Identification and analysis of functional elements in 1% of the human genome by the ENCODE pilot project. Nature *447*, 799–816.

Etzold, T., and Argos, P. (1993). SRS–an indexing and retrieval tool for flat file data libraries. Comput. Appl. Biosci. *9*, 49–57.

Fazius, E., Shelest, V., and Shelest, E. (2011). SiTaR: a novel tool for transcription factor binding site prediction. Bioinformatics *27*, 2806–2811.

Feist, A.M., Henry, C.S., Reed, J.L., Krummenacker, M., Joyce, A.R., Karp, P.D., Broadbelt, L.J., Hatzimanikatis, V., and Palsson, B.O. (2007). A genome-scale metabolic reconstruction for *Escherichia coli* K-12 MG1655 that accounts for 1260 ORFs and thermodynamic information. Mol. Syst. Biol. *3*, 121.

Forst, C.V. (2006). Host-pathogen systems biology. Drug. Discov. Today *11*, 220–227.

Foteinou, P.T., Calvano, S.E., Lowry, S.F., and Androulakis, I.P. (2009). Modeling endotoxin-induced systemic inflammation using an indirect response approach. Math. Biosci. *217*, 27–42.

Fukuda, Y., Tsai, H.-F., Myers, T.G., and Bennett, J.E. (2013). Transcriptional profiling of *Candida glabrata* during phagocytosis by neutrophils and in the infected mouse spleen. Infect. Immun. *4*, 1325–1333.

Garcia, J., Sims, K.J., Schwacke, J.H., and Poeta, M.D. (2011). Biochemical systems analysis of signaling pathways to understand fungal pathogenicity. Methods. Mol. Biol. *734*, 173–200.

Gardiner, D.M., and Howlett, B.J. (2005). Bioinformatic and expression analysis of the putative gliotoxin biosynthetic gene cluster of *Aspergillus fumigatus*. FEMS Microbiol. Lett. *248*, 241–248.

Gatherer, D. (2010). So what do we really mean when we say that systems biology is holistic? BMC Syst. Biol. *4*, 22.

Ghosh, S., Matsuoka, Y., Asai, Y., Hsin, K.-Y., and Kitano, H. (2011). Software for systems biology: from tools to integrated platforms. Nat. Rev. Genet. *12*, 821–832.

Goble, C., and Stevens, R. (2008). State of the nation in data integration for bioinformatics. J. Biomed. Inform. *41*, 687–693.

Greenbaum, D., Colangelo, C., Williams, K., and Gerstein, M. (2003). Comparing protein abundance and mRNA expression levels on a genomic scale. Genome. Biol. *4*, 117.

Guthke, R., Möller, U., Hoffmann, M., Thies, F., and Töpfer, S. (2005). Dynamic network reconstruction from gene expression data applied to immune response during bacterial infection. Bioinformatics *21*, 1626–1634.

Guthke, R., Kniemeyer, O., Albrecht, D., Brakhage, A., and Moeller, U. (2007). Discovery of gene regulatory networks in *Aspergillus fumigatus*. Lect. Notes Bioinf. *4366*, 22–41.

Gygi, S.P., Rochon, Y., Franza, B.R., and Aebersold, R. (1999). Correlation between protein and mRNA abundance in yeast. Mol. Cell. Biol. *19*, 1720–1730.

Hamel, L.-P., Nicole, M.-C., Duplessis, S., and Ellis, B.E. (2012). Mitogen-activated protein kinase signaling in plant-interacting fungi: distinct messages from conserved messengers. The Plant Cell Online *24*, 1327–1351.

Harmston, N., Filsell, W., and Stumpf, M.P.H. (2010). What the papers say: text mining for genomics and systems biology. Hum. Genomics *5*, 17–29.

Hase, T., Tanaka, H., Suzuki, Y., Nakagawa, S., and Kitano, H. (2009). Structure of protein interaction networks and their implications on drug design. PLoS Comput. Biol. *5*, e1000550.

Hasenberg, M., Behnsen, J., Krappmann, S., Brakhage, A., and Gunzer, M. (2011). Phagocyte responses towards *Aspergillus fumigatus*. Int. J. Med. Microbiol. *301*, 436–444.

Hawksworth, D. (2001). The magnitude of fungal diversity: the 1.5 million species estimate revisited. Mycol. Res. *105*, 1422–1432.

Hecker, M., Lambeck, S., Toepfer, S., van Someren, E., and Guthke, R. (2009). Gene regulatory network inference: data integration in dynamic models-a review. Biosystems *96*, 86–103.

Hillier, L.W., Reinke, V., Green, P., Hirst, M., Marra, M.A., and Waterston, R.H. (2009). Massively parallel sequencing of the polyadenylated transcriptome of *C. elegans*. Genome Res. *19*, 657–666.

Horn, F., Heinekamp, T., Kniemeyer, O., Pollmächer, J., Valiante, V., and Brakhage, A.A. (2012). Systems biology of fungal infection. Front. Microbiol. *3*, 108.

Huttenhower, C., and Hofmann, O. (2010). A quick guide to large-scale genomic data mining. PLoS Comput. Biol. *6*, e1000779.

Hyduke, D.R., and Palsson, B.O. (2010). Towards genome-scale signalling network reconstructions. Nat. Rev. Genet. *11*, 297–307.

Jain, R., Valiante, V., Remme, N., Docimo, T., Heinekamp, T., Hertweck, C., Gershenzon, J., Haas, H., and Brakhage, A.A. (2011). The MAP kinase MpkA controls cell wall integrity, oxidative stress response, gliotoxin production and iron adaptation in *Aspergillus fumigatus*. Mol. Microbiol. *82*, 39–53.

Jones, T., Federspiel, N.A., Chibana, H., Dungan, J., Kalman, S., Magee, B.B., Newport, G., Thorstenson, Y.R., Agabian, N., Magee, P.T., *et al.* (2004). The diploid genome sequence of *Candida albicans*. Proc. Natl. Acad. Sci. U.S.A. *101*, 7329–7334.

Kelder, T., Conklin, B.R., Evelo, C.T., and Pico, A.R. (2010). Finding the right questions: exploratory pathway analysis to enhance biological discovery in large datasets. PLoS Biol. *8*, e1000472.

Kim, S., Shin, S.-Y., Lee, I.-H., Kim, S.-J., Sriram, R., and Zhang, B.-T. (2008). PIE: an online prediction system for protein–protein interactions from text. Nucleic Acids Res. *36*, W411–W415.

Kitano, H. (2002). Computational systems biology. Nature *420*, 206–210.

Klose, J., and Kobalz, U. (1995). Two-dimensional electrophoresis of proteins: an updated protocol and implications for a functional analysis of the genome. Electrophoresis *16*, 1034–1059.

Kniemeyer, O., Schmidt, A.D., Vödisch, M., Wartenberg, D., and Brakhage, A.A. (2011). Identification of virulence determinants of the human pathogenic fungi *Aspergillus fumigatus* and *Candida albicans* by proteomics. Int. J. Med. Microbiol. *301*, 368–377.

Kozhenkov, S., Sedova, M., Dubinina, Y., Gupta, A., Ray, A., Ponomarenko, J., and Baitaluk, M. (2011). Biological Networks–tools enabling the integration of multi-scale data for the host–pathogen studies. BMC Syst. Biol. *5*, 7.

Kreutz, C., and Timmer, J. (2009). Systems biology: experimental design. FEBS J. *276*, 923–942.

Köhler, J. (2004). Integration of life science databases. Drug Discovery Today: BIOSILICO *2*, 61 – 69.

Lander, E.S., Linton, L.M., Birren, B., Nusbaum, C., Zody, M.C., Baldwin, J., Devon, K., Dewar, K., Doyle, M., FitzHugh, W., *et al.* (2001). Initial sequencing and analysis of the human genome. Nature *409*, 860–921.

Leach, M.D., and Brown, A.J.P. (2012). Posttranslational modifications of proteins in the pathobiology of medically relevant fungi. Eukaryot. Cell *11*, 98–108.

Lee, T.-L. (2008). Big data: open-source format needed to aid wiki collaboration. Nature *455*, 461.

Lessing, F., Kniemeyer, O., Wozniok, I., Loeffler, J., Kurzai, O., Haertl, A., and Brakhage, A.A. (2007). The *Aspergillus fumigatus* transcriptional regulator AfYap1 represents the major regulator for defense against reactive oxygen intermediates but is dispensable for pathogenicity in an intranasal mouse infection model. Eukaryot. Cell *6*, 2290–2302.

Linde, J., Wilson, D., Hube, B., and Guthke, R. (2010). Regulatory network modelling of iron acquisition by a fungal pathogen in contact with epithelial cells. BMC Syst. Biol. *4*, 148.

Linde, J., Hortschansky, P., Fazius, E., Brakhage, A.A., Guthke, R., and Haas, H. (2012). Regulatory interactions for iron homoeostatis in *Aspergillus fumigatus* inferred by a Systems Biology approach. BMC Syst. Biol. *6*, 6.

Liu, T.T., Lee, R.E.B., Barker, K.S., Lee, R.E., Wei, L., Homayouni, R., and Rogers, P.D. (2005). Genome-wide expression profiling of the response to azole, polyene, echinocandin, and pyrimidine antifungal agents in *Candida albicans*. Antimicrob. Agents Chemother. *49*, 2226–2236.

Loftus, B.J., Fung, E., Roncaglia, P., Rowley, D., Amedeo, P., Bruno, D., Vamathevan, J., Miranda, M., Anderson, I.J., Fraser, J.A., *et al.* (2005). The genome of the basidiomycetous yeast and human pathogen *Cryptococcus neoformans*. Science *307*, 1321–1324.

Lorenz, M.C., and Fink, G.R. (2001). The glyoxylate cycle is required for fungal virulence. Nature *412*, 83–86.

Lorenz, M.C., Bender, J.A., and Fink, G.R. (2004). Transcriptional response of *Candida albicans* upon internalization by macrophages. Eukaryot. Cell *3*, 1076–1087.

Martin, G.S., Mannino, D.M., Eaton, S., and Moss, M. (2003). The epidemiology of sepsis in the United States from 1979 through 2000. N. Engl. J. Med. *348*, 1546–1554.

Martínez-Solano, L., Nombela, C., Molero, G., and Gil, C. (2006). Differential protein expression of murine macrophages upon interaction with *Candida albicans*. Proteomics *6 (Suppl 1)*, S133–S144.

Martínez-Solano, L., Reales-Calderón, J.A., Nombela, C., Molero, G., and Gil, C. (2009). Proteomics of RAW 264.7 macrophages upon interaction with heat-inactivated *Candida albicans* cells unravel an anti-inflammatory response. Proteomics *9*, 2995–3010.

McCormick, A., Heesemann, L., Wagener, J., Marcos, V., Hartl, D., Loeffler, J., Heesemann, J., and Ebel, F. (2010). NETs formed by human neutrophils inhibit growth of the pathogenic mold *Aspergillus fumigatus*. Microbes Infect. *12*, 928–936.

Morton, C.O., Varga, J.J., Hornbach, A., Mezger, M., Sennefelder, H., Kneitz, S., Kurzai, O., Krappmann, S., Einsele, H., Nierman, W.C., *et al.* (2011). The temporal dynamics of differential gene expression in *Aspergillus fumigatus* interacting with human immature dendritic cells *in vitro*. PLoS One *6*, e16016.

Mukhtar, M.S., Carvunis, A.-R., Dreze, M., Epple, P., Steinbrenner, J., Moore, J., Tasan, M., Galli, M., Hao, T., Nishimura, M.T., *et al.* (2011). Independently evolved virulence effectors converge onto hubs in a plant immune system network. Science *333*, 596–601.

Murillo, L.A., Newport, G., Lan, C.-Y., Habelitz, S., Dungan, J., and Agabian, N.M. (2005). Genome-wide transcription profiling of the early phase of biofilm formation by *Candida albicans*. Eukaryot. Cell *4*, 1562–1573.

Müller, S., Baldin, C., Groth, M., Guthke, R., Kniemeyer, O., Brakhage, A.A., and Valiante, V. (2012). Comparison of transcriptome technologies in the pathogenic fungus *Aspergillus fumigatus* reveals novel insights into the genome and MpkA dependent gene expression. BMC Genomics *13*, 519.

Nierman, W.C., Pain, A., Anderson, M.J., Wortman, J.R., Kim, H.S., Arroyo, J., Berriman, M., Abe, K., Archer, D.B., Bermejo, C., *et al.* (2005). Genomic sequence of the pathogenic and allergenic filamentous fungus *Aspergillus fumigatus*. Nature *438*, 1151–1156.

Oberhardt, M.A., Puchałka, J., Martins dos Santos, V.A.P., and Papin, J.A. (2011). Reconciliation of genome-scale metabolic reconstructions for comparative systems analysis. PLoS Comput. Biol. *7*, e1001116.

O'Farrell, P.H. (1975). High resolution two-dimensional electrophoresis of proteins. J. Biol. Chem. *250*, 4007–4021.

Park, J., Park, J., Jang, S., Kim, S., Kong, S., Choi, J., Ahn, K., Kim, J., Lee, S., Kim, S., *et al.* (2008). FTFD: an informatics pipeline supporting phylogenomic analysis of fungal transcription factors. Bioinformatics *24*, 1024–1025.

Parrish, R.S., and Spencer, H.J. (2004). Effect of normalization on significance testing for oligonucleotide microarrays. J. Biopharm. Stat. *14*, 575–589.

Pavlopoulos, G.A., Secrier, M., Moschopoulos, C.N., Soldatos, T.G., Kossida, S., Aerts, J., Schneider, R., and Bagos, P.G. (2011). Using graph theory to analyze biological networks. BioData Min. *4*, 10.

Pfaller, M.A., and Diekema, D.J. (2007). Epidemiology of invasive candidiasis: a persistent public health problem. Clin. Microbiol. Rev. *20*, 133–163.

Picotti, P., Bodenmiller, B., Mueller, L.N., Domon, B., and Aebersold, R. (2009). Full dynamic range proteome analysis of *S. cerevisiae* by targeted proteomics. Cell *138*, 795–806.

Priebe, S., Linde, J., Albrecht, D., Guthke, R., and Brakhage, A.A. (2011). FungiFun: a web-based application for functional categorization of fungal genes and proteins. Fungal Genet. Biol. *48*, 353–358.

Prill, R.J., Marbach, D., Saez-Rodriguez, J., Sorger, P.K., Alexopoulos, L.G., Xue, X., Clarke, N.D., Altan-Bonnet, G., and Stolovitzky, G. (2010). Towards a rigorous assessment of systems biology models: the DREAM3 challenges. PLoS One 5, e9202.

Raza, S., Robertson, K.A., Lacaze, P.A., Page, D., Enright, A.J., Ghazal, P., and Freeman, T.C. (2008). A logic-based diagram of signalling pathways central to macrophage activation. BMC Syst. Biol. *2*, 36.

Rispail, N., Soanes, D.M., Ant, C., Czajkowski, R., Grünler, A., Huguet, R., Perez-Nadales, E., Poli, A., Sartorel, E., Valiante, V., *et al.* (2009). Comparative genomics of MAP kinase and calcium-calcineurin signalling components in plant and human pathogenic fungi. Fungal Genet. Biol. *46*, 287–298.

Rizzetto, L., and Cavalieri, D. (2011). Friend or foe: using systems biology to elucidate interactions between fungi and their hosts. Trends Microbiol. *19*, 509–515.

Rohn, H., Klukas, C., and Schreiber, F. (2011). Creating views on integrated multidomain data. Bioinformatics *27*, 1839–1845.

Rolfsson, O., Palsson, B.O., and Thiele, I. (2011). The human metabolic reconstruction Recon 1 directs hypotheses of novel human metabolic functions. BMC Syst. Biol. *5*, 155.

Ruepp, A., Zollner, A., Maier, D., Albermann, K., Hani, J., Mokrejs, M., Tetko, I., Güldener, U., Mannhaupt, G., Münsterkötter, M., *et al.* (2004). The FunCat, a functional annotation scheme for systematic classification of proteins from whole genomes. Nucleic Acids Res. *32*, 5539–5545.

Ruppin, E., Papin, J.A., de Figueiredo, L.F., and Schuster, S. (2010). Metabolic reconstruction, constraint-based analysis and game theory to probe genome-scale metabolic networks. Curr. Opin. Biotechnol *21*, 502–510.

Santamaría, R., Rizzetto, L., Bromley, M., Zelante, T., Lee, W., Cavalieri, D., Romani, L., Miller, B., Gut, I., Santos, M., *et al.* (2011). Systems biology of infectious diseases: a focus on fungal infections. Immunobiology *216*, 1212–1227.

Sboner, A., Mu, X.J., Greenbaum, D., Auerbach, R.K., and Gerstein, M.B. (2011). The real cost of sequencing: higher than you think! Genome. Biol. *12*, 125.

Scharf, D.H., and Brakhage, A.A. (2013). Engineering fungal secondary metabolism: a roadmap to novel compounds. J. Biotechnol *163*, 179–183.

Scharf, D.H., Heinekamp, T., Remme, N., Hortschansky, P., Brakhage, A.A., and Hertweck, C. (2012). Biosynthesis and function of gliotoxin in *Aspergillus fumigatus*. Appl. Microbiol. Biotechnol 93, 467–472.

Schrettl, M., Kim, H.S., Eisendle, M., Kragl, C., Nierman, W.C., Heinekamp, T., Werner, E.R., Jacobsen, I., Illmer, P., Yi, H., *et al.* (2008). SreA-mediated iron regulation in *Aspergillus fumigatus*. Mol. Microbiol. *70*, 27–43.

Shannon, P., Markiel, A., Ozier, O., Baliga, N.S., Wang, J.T., Ramage, D., Amin, N., Schwikowski, B., and Ideker, T. (2003). Cytoscape: a software environment for integrated models of biomolecular interaction networks. Genome Res. *13*, 2498–2504.

Sharan, R., Ulitsky, I., and Shamir, R. (2007). Network-based prediction of protein function. Mol. Syst. Biol. *3*, 88.

Shin, Y.-K., Kim, K.-Y., and Paik, Y.-K. (2005). Alterations of protein expression in macrophages in response to *Candida albicans* infection. Mol. Cells *20*, 271–279.

da Silva Ferreira, M.E., Malavazi, I., Savoldi, M., Brakhage, A.A., Goldman, M.H.S., Kim, H.S., Nierman, W.C., and Goldman, G.H. (2006). Transcriptome analysis of *Aspergillus fumigatus* exposed to voriconazole. Curr. Genet. *50*, 32–44.

Stanke, M., and Waack, S. (2003). Gene prediction with a hidden markov model and a new intron submodel. Bioinformatics *19 (Suppl 2)*, ii215–ii225.

Sturdevant, D.E., Virtaneva, K., Martens, C., Bozinov, D., Ogundare, O., Castro, N., Kanakabandi, K., Beare, P.A., Omsland, A., Ohmsland, A., *et al.* (2010). Host–microbe interaction systems biology: lifecycle transcriptomics and comparative genomics. Future Microbiol. *5*, 205–219.

Sugui, J.A., Kim, H.S., Zarember, K.A., Chang, Y.C., Gallin, J.I., Nierman, W.C., and Kwon-Chung, K.J. (2008). Genes differentially expressed in conidia and hyphae of *Aspergillus fumigatus* upon exposure to human neutrophils. PLoS One *3*, e2655.

Thiele, I., and Palsson, B.O. (2010). A protocol for generating a high-quality genome-scale metabolic reconstruction. Nat. Protoc. *5*, 93–121.

Thorleifsson, S.G., and Thiele, I. (2011). rBioNet: A COBRA toolbox extension for reconstructing high-quality biochemical networks. Bioinformatics *27*, 2009–2010.

Tierney, L., Kuchler, K., Rizzetto, L., and Cavalieri, D. (2012a). Systems biology of host–fungus interactions: turning complexity into simplicity. Curr. Opin. Microbiol. *15*, 440–446.

Tierney, L., Linde, J., Müller, S., Brunke, S., Molina, J., Hube, B., Schöck, U., Guthke, R., and Kuchler, K. (2012b). An interspecies regulatory network inferred from simultaneous RNA-seq of *Candida albicans* invading innate immune cells. Front. Microbiol. *3*, 85.

Toepfer, S., Guthke, R., Driesch, D., Woetzel, D., and Pfaff, M. (2007). The NetGenerator algorithm: reconstruction of gene regulatory networks. Lect. Notes Bioinf. *4366*, 119–130.

Tokarski, C., Hummert, S., Mech, F., Figge, M.T., Germerodt, S., Schroeter, A., and Schuster, S. (2012). Agent-based modeling approach of immune defense against spores of opportunistic human pathogenic fungi. Front. Microbiol. *3*, *129*.

Tukey, J.W. (1977). Exploratory data analysis. Series in behavioral science: quantitative methods (Addison-Wesley Publishing Company, Reading, MA).

van het Hoog, M., Rast, T.J., Martchenko, M., Grindle, S., Dignard, D., Hogues, H., Cuomo, C., Berriman, M., Scherer, S., Magee, B.B., *et al.* (2007). Assembly of the *Candida albicans* genome into sixteen supercontigs aligned on the eight chromosomes. Genome. Biol. *8*, R52.

van Someren, E.P., Wessels, L.F.A., Backer, E., and Reinders, M.J.T. (2002). Genetic network modeling. Pharmacogenomics 3, 507–525.

Vera, J., Rateitschak, K., Lange, F., Kossow, C., Wolkenhauer, O., and Jaster, R. (2011). Systems biology of JAK-STAT signalling in human malignancies. Prog. Biophys. Mol. Biol. *106*, 426–434.

Vicente, F.F.R., Lopes, F.M., Hashimoto, R.F., and Cesar, R.M. (2012). Assessing the gain of biological data integration in gene networks inference. BMC Genomics *13 (Suppl 6)*, S7.

Vlaic, S., Schmidt-Heck, W., Matz-Soja, M., Marbach, E., Linde, J., Meyer-Baese, A., Zellmer, S., Guthke, R., and Gebhardt, R. (2012). The extended TILAR approach: a novel tool for dynamic modeling of the transcription factor network regulating the adaption to *in vitro* cultivation of murine hepatocytes. BMC Syst. Biol. *6*, 147.

Vodovotz, Y., Constantine, G., Faeder, J., Mi, Q., Rubin, J., Bartels, J., Sarkar, J., Squires, R.H., Okonkwo, D.O., Gerlach, J., *et al.* (2010). Translational systems approaches to the biology of inflammation and healing. Immunopharmacol. Immunotoxicol. *32*, 181–195.

Vödisch, M., Albrecht, D., Lessing, F., Schmidt, A.D., Winkler, R., Guthke, R., Brakhage, A.A., and Kniemeyer, O. (2009). Two-dimensional proteome reference maps for the human pathogenic filamentous fungus *Aspergillus fumigatus*. Proteomics *9*, 1407–1415.

Vödisch, M., Scherlach, K., Winkler, R., Hertweck, C., Braun, H.-P., Roth, M., Haas, H., Werner, E.R., Brakhage, A.A., and Kniemeyer, O. (2011). Analysis of the *Aspergillus fumigatus* proteome reveals metabolic changes and the activation of the pseurotin A biosynthesis gene cluster in response to hypoxia. J. Proteome. Res. *10*, 2508–2524.

Waltermann, C., and Klipp, E. (2010). Signal integration in budding yeast. Biochem. Soc. Trans. *38*, 1257–1264.

Wartenberg, D., Lapp, K., Jacobsen, I.D., Dahse, H.-M., Kniemeyer, O., Heinekamp, T., and Brakhage, A.A. (2011). Secretome analysis of *Aspergillus fumigatus* reveals Asp-hemolysin as a major secreted protein. Int. J. Med. Microbiol. *301*, 602–611.

Weber, M., Henkel, S.G., Vlaic, S., Guthke, R., van Zoelen, E.J., and Driesch, D. (2013). Inference of dynamical gene-regulatory networks based on time-resolved multi-stimuli multi-experiment data applying NetGenerator V2.0. BMC Syst. Biol. *7*, 1.

Westermann, A.J., Gorski, S.A., and Vogel, J. (2012). Dual RNA-seq of pathogen and host. Nat. Rev. Microbiol. *10*, 618–630.

Westermeier, R., and Görg, A. (2011). Two-dimensional electrophoresis in proteomics. Methods. Biochem. Anal. *54*, 411–439.

Wilson, D., Thewes, S., Zakikhany, K., Fradin, C., Albrecht, A., Almeida, R., Brunke, S., Grosse, K., Martin, R., Mayer, F., *et al.* (2009). Identifying infection-associated genes of *Candida albicans* in the postgenomic era. FEMS Yeast Res. *9*, 688–700.

de Wit, P.J.G.M., van der Burgt, A., Ökmen, B., Stergiopoulos, I., Abd-Elsalam, K.A., Aerts, A.L., Bahkali, A.H., Beenen, H.G., Chettri, P., Cox, M.P., *et al.* (2012). The genomes of the fungal plant pathogens *Cladosporium fulvum* and *Dothistroma septosporum* reveal adaptation to different hosts and lifestyles but also signatures of common ancestry. PLoS Genet. *8*, e1003088.

Xenarios, I., Rice, D.W., Salwinski, L., Baron, M.K., Marcotte, E.M., and Eisenberg, D. (2000). DIP: the database of interacting proteins. Nucleic Acids Res. *28*, 289–291.

Xu, J., and Li, Y. (2006). Discovering disease-genes by topological features in human protein–protein interaction network. Bioinformatics *22*, 2800–2805.

Yandell, M., and Ence, D. (2012). A beginner's guide to eukaryotic genome annotation. Nat. Rev. Genet. *13*, 329–342.

Yang, Y.H., Dudoit, S., Luu, P., Lin, D.M., Peng, V., Ngai, J., and Speed, T.P. (2002). Normalization for cDNA microarray data: a robust composite method addressing single and multiple slide systematic variation. Nucleic Acids Res. *30*, e15.

Zdobnov, E.M., and Apweiler, R. (2001). InterProScan–an integration platform for the signature-recognition methods in InterPro. Bioinformatics *17*, 847–848.

Zhang, Z., Cheung, K.-H., and Townsend, J.P. (2009). Bringing Web 2.0 to bioinformatics. Brief. Bioinform. *10*, 1–10.

Zhu, M., Gao, L., Li, X., Liu, Z., Xu, C., Yan, Y., Walker, E., Jiang, W., Su, B., Chen, X. *et al.* (2009). The analysis of the drug-targets based on the topological properties in the human protein–protein interaction network. J. Drug. Target *17*, 524–532.

Comparative Pathogenesis: Transcriptomic Analyses of Host Cell–Fungus Interactions

4

Elaine Bignell

Abstract

The contribution of host activities to positive, and negative, outcomes of fungal disease is increasingly recognized. Technological advances are permitting scientists to unravel gene-by-gene the complexities of the host–pathogen interaction, an objective which will remain a crucial component of infectious disease research for the foreseeable future, and will likely yield a new generation of immunotherapeutic antifungal entities. The fungal host–pathogen interaction includes a remarkable array of behaviours from pathogen-mediated killing of host cells, to intracellular occupancy and lytic expulsion, behaviours which drive the outcomes of disease by governing pathogen replication, immune evasion, survival, and dissemination. This chapter will explore the fruits of transcriptomic analyses of fungal host–pathogen interactions.

Introduction

The molecular basis of pathogenic strategy presents some of the most compelling research questions posed by infectious disease. Owing to the opportunistic nature of most invasive mycoses, fungal pathogenicity has proven difficult to define in terms of conventional virulence factors. Whilst microbial subversion of normal host-cell processes to create a specialized niche has become common parlance in the field of bacterial pathogenesis, the search for a sensical pathogenic strategy amongst usually soil-dwelling fungi is teleologically flawed. This conundrum is resolvable if one considers host contributions to disease, since host defects and their impact upon immune homeostasis represent a crucial component of most invasive mycoses (Casadevall and Pirofski, 2003, 2009). Indeed the very nature of immune deficits predisposing humans to fungal infection, and widely varying manifestations of fungal disease, have helped medical mycologists to define the cellular and molecular basis of pathogen entry and host defence (Romani, 2011). Given extensive physiological similarities between diverse fungal pathogens it is remarkable that such diversity prevails amongst immune lesions causing the risk of fungal infection. Amongst the major inhaled fungal pathogens several utterly distinct modes of microbial persistence in host cells and tissues can be found. If anything, such observations demonstrate the remarkable stratification of multiple immune defences ensuring that over reliance upon any single defence strategy is abolished. Additionally one can appreciate that the normal functioning of human immunity is manifestly efficient in safe guarding human health against fungal diseases. HIV sufferers are a case in point. In this patient cohort rectification of the immune deficit via antiretroviral therapy, rather than treatment with antifungal drugs, provides the patient with the best chance of surviving fungal infection. Such is the power of an adequate antifungal immune response.

The first encounter between host and pathogen is governed by the route of pathogen entry. Fungi which exist as normally commensal organisms will frequently encounter innate immune cells, and have probably evolved to survive many of these encounters, while fungi which occupy environmental niches are more likely to have developed generic stress tolerance capabilities which become gainfully employed within the mammalian host. Regardless of an origin intrinsic or extrinsic to that of the host niche, infecting fungal pathogens will ultimately be challenged with the full force of the residual innate defences the host can muster. A meta-analytical understanding of what happens next will surely enlighten our understanding of disease and health as herein lies the key to understanding healthy host defences and embellishment of appropriate antifungal defence mechanisms.

Different types of innate immune cells exert different types of antimicrobial functions. Much progress, here discussed, has recently been made in assessing the interaction between cells of the innate immune system and fungal pathogens. For the purposes of this chapter we will consider the three most significant fungal pathogens of man, *Aspergillus fumigatus*, *Candida albicans* and *Cryptococcus neoformans*. As the whole animal in-host fungal transcriptome has previously been reviewed elsewhere (Cairns *et al.*, 2010; Moran *et al.*, 2011) we will omit such studies and focus instead upon fungal responses to specific innate immune cells and the corresponding host responses to fungal challenge (a summary of all studies discussed in this chapter can be found in Table 4.1).

Fungus–epithelial interactions

Several mammalian mucosae provide important portals of fungal entry, the most intensively studied of these being the pulmonary, oral, and intestinal epithelia (Table 4.1). Increasingly, epithelial strata are recognized for their role in orchestrating local immune responses and it is now widely accepted that the latter role is of equal importance to that of the physical prevention of pathogen entry. The pulmonary epithelium provides the first point of human contact for the vast majority of fungal species causing invasive mycoses, as all but *Candida* species are acquired via inhalation of a fungal particle. The oral and gastrointestinal epithelia, both in health and in disease, are those most relevant to the *Candida albicans* host–pathogen interaction and it is interesting to compare the various responses and antifungal activities of these somewhat different host–pathogen interfaces to dissect commonalities and differences in containment, or otherwise, of fungal interlopers.

Candida albicans–epithelial interactions

Significant interactions between *Candida albicans* and epithelial cell types include those occurring at the oral mucosa during oropharyngeal candidiasis, at the gastrointestinal epithelium during commensalization of, or escape from, the digestive tract and at the vaginal epithelia during genitourinary infection and/or carriage. Important landmarks of the interaction of *C. albicans* with host cells include germination of *C. albicans* blastospores to produce hyphae, adherence to and internalization by host cells, and host damage (Naglik *et al.*, 2011).

Adherence of *C. albicans* to epithelial strata prompts internalization of the fungal cell, elicited by binding of the invasin Als3 to N-cadherin and other epithelial surface proteins (Phan *et al.*, 2005, 2007). Park *et al.* (2009) performed a longitudinal time-series analysis of *C. albicans* gene expression during interaction with FaDu oral epithelial cells (Fig. 4.1), surveying gene expression in response to host cell exposure after 45, 90 and 180 minutes of co-culture. Initial experimentation probed the time scale of the physical interaction between blastospores and epithelial cells or polystyrene tissue culture plates, revealing germination of blastospores on both surfaces after 45 minutes of co-incubation and subsequent production of elongated hyphal cells. At each time point of the analysis, and in several distinct *C. albicans* isolates, the hyphae that formed in the presence of host cells were similar in length to those grown on polystyrene. This confirmed that rates of fungal germination and hyphal elongation were unimpeded by the presence of host cells at the time points studied and, moreover, that the observed differential gene expression derived from the interaction with host cells, rather than progressive morphological change. Blastospores adhered to oral epithelial cells within 45 minutes of co-incubation and after 90 minutes 24% of cell-associated organisms had become internalized by host cells. By 3 hours the majority of fungal cells had become internalized (Fig. 4.1).

Relative to blastospores exposed to polystyrene culture vessels, *C. albicans* exposed to oral epithelial cells up-regulated 31, 8 and 12 genes, and down-regulated 211, 54 and 32 genes at 45, 90 and 180 minutes of co-incubation respectively. Initial scrutiny of this dataset posed some inexplicable observations, such as down-regulation, rather than up-regulation of well-characterized adhesion gene functions in response to host cells. Additionally, and unsurprisingly given the small cohort of up-regulated gene functions, gene ontological (GO) analysis was minimally useful in defining functional cohorts impacted by epithelial cells. Amongst the up-regulated gene functions no enrichment of functional classifications could be deciphered whilst down-regulated genes were associated, at all three time points, with adhesion, and at 45 and 90 minutes with ribonucleoprotein complex biogenesis and protein catabolism respectively. Despite these anomalies, a strategy to focus upon uncharacterised up-regulated genes was rewarded with identification of two previously unrecognized mediators of epithelial damage, namely *YCK2*, and *VPS51*, thereby demonstrating that even modest changes in gene expression can have biological significance. A highly interesting finding was that null mutants of both of these genes exhibited increased susceptibility to protamine and human β-defensin, host-derived antimicrobial peptides encountered within the oral niche. This prompted the hypothesis that contact with oral epithelial cells induces a protective reaction that enables *C. albicans* to withstand these peptides. *YCK2* encodes the highly conserved serine/threonine kinase, casein kinase 1 (CK1) and is required for normal hyphal morphology and maintenance of cell polarity, while *VPS51* is predicted to encode a component of the Vps53 complex required for fusion of endosome-derived vesicles with the late Golgi compartment. Null mutants in both genes prompted defects in epithelial damage, as measured by a ^{51}Cr release assay, and in morphogenesis. The *vps51* null mutant produced unbranched hyphae that were shorter than those of the wild-type strain when it was grown on epithelial cells. Furthermore, the mutant also exhibited significantly increased susceptibility to all stressors tested and demonstrated a highly fragmented vacuolar phenotype, as determined by FM4-64-mediated visualization of the vacuolar membrane.

The finding that adhesion-related gene functions were down-regulated was a seemingly counterintuitive one. The authors, raising an important qualification of the dataset, were able to rationalize this finding by hypothesizing that these genes might be more potently stimulated by the reference growth condition, polystyrene, than by exposure to host cells. Clearly the influence of the reference mRNA population upon output of two-colour microarrays is substantial and requires careful consideration during design, and interpretation of, comparative transcriptomic studies. A further pertinent observation along these lines includes that, in the Park *et al.* (2009) experiments, the reference condition (contact with polystyrene) caused the organisms to form hyphae, and thus increased transcript levels that were induced by hyphal formation alone were likewise not detected.

An important aspect of this study is that two distinct fungal isolates, derived from laboratory and clinical sources, were analysed for transcriptomic responses. Responses which are frequently documented are far less likely to arise as artefactual findings and some studies, such as that of Park *et al.* (2009), have been specifically designed to seek core fungal transcriptional responses by repeated experimentation using multiple fungal isolates and host cell types.

Sohn *et al.* (2006) also addressed the differences between host and plastic contact with *C. albicans* cells. Motivated by an interest in adhesion, the authors dissected the time-course and physical appearance of adhering cells, by electron microscopy, as well as implementing an investigation of gene expression. In this study *in vitro* adhesion assays were established using the human colorectal carcinoma cell line Caco-2 and epidermoid vulvo-vaginal A-431 cells. In both instances *C. albicans* was found to adhere very well and to produce hyphae. The transcriptional profiles of *C. albicans* cells adhering to these cell lines were very similar to that of cells adhering to plastic although some genes did differ significantly in response to plastic. The authors demonstrated significant

Table 4.1 Published *in vitro* transcriptomic studies of fungal interactions with mammalian cells. Studies are ordered by pathogen studied, with genotypes as indicated. Where raw data have been deposited in publicly accessible repositories, URLs and/or accession numbers are provided. N/A indicates not applicable

Pathogen/ genotype	Host cell type/ genotype	Test conditions	Biological replicates	Time period	Media	Array/methodology	Data repository/ accession number	Reference
Candida albicans CA14-URA [Sc5314] and 7392	Oral epithelial cells/FaDu	*Candida albicans* plus and minus host cells	3	45, 90 and 180 min	Eagle's minimum essential medium with Earle's balanced salt solution (Irvine Scientific) containing 10% fetal bovine serum, 1 mM pyruvic acid, 2 mM l-glutamine, 0.1 mM nonessential amino acids, 100 IU/ml penicillin, and 100 IU/ml streptomycin	LA Biomed_candida albicans_6266_1.0	GEO: GSE5340	Park *et al.* (2009)
Candida albicans ATCC 90029	Primary human gingival epithelial cells (HGECs)	*Candida albicans* plus and minus gingival epithelial cells		8 h	EpiLife medium with supplement S7 and antibiotics (Penicillin, Streptomycin and Amphotericin B) (Cascade Biologics, Portland, OR, USA)	HG-U133A Affymetrix	N/A	Ikuta *et al.* (2012)
Candida albicans CBS-562	HEp2 epithelial cells	*Candida albicans* plus and minus host cells	3	180 min	DMEM, 2 mM glutamine, 200 IU penicillin, 200 μg streptomycin, 10% FCS	Custom *C. albicans* microarrays spotted using Qiagen ~75 mer oligonucleotides	N/A	Sandovski-Losica *et al.* (2006)
Candida albicans SC5314	Reconstituted human epithelium (RHE) composed of TR146 buccal mucosa cells	RHE infection at 1, 3, 6, 12 and 24 h, relative to common reference sample (YPD)	2 to 5	1, 3, 6, 12 and 24 h	Serum-free maintenance medium (SkinEthic Laboratories)	*C. albicans* microarrays (Eurogentec, Seraing, Belgium)	http://www.galarfungail.org/data.htm	Zakikhany *et al.* (2007)

Candida albicans SC5314	Reconstituted human epithelium (RHE) composed of TR146 buccal mucosa cells	*Candida albicans*plus and minus host cells	3	90 min	MCDB 153 maintenance medium (SkinEthic Laboratories)	*C. albicans* microarrays 70 mer oligoarrays, NRC, Canada	GEO Accession number GSE13318 and GSE13345	Spiering *et al*. (2010)
Candida albicans SC5314	Reconstituted human epithelium (RHE) composed of TR146 buccal mucosa cells	*Candida albicans*plus host cells 0 versus 30 min	3	30 min	MCDB 153 maintenance medium (SkinEthic Laboratories)	*C. albicans* microarrays (Eurogentec, Seraing, Belgium)	GEO Accession number GSE13318 and GSE13345	Spiering *et al*. (2010)
Candida albicans SC5314	Caco-2 ATCC HTB-37 Human clorectal carcinoma cell line	*Candida albicans*plus and minus host cells	1	0, 30, 60, 120, 240	DMEM, 10% heat-inactivated fetal calf serum, 1 mM sodium pyruvate, 1% gentamycin	In-house custom microarrays	Not provided	Sohn *et al*. (2006)
Candida albicans SC5314	Whole human blood	*Candida albicans* plus and minus blood	3 technical repetitions	10, 20, 30 and 60 min	N/A	Nylon membrane arrays http://www.pasteur.fr/recherche/unites/RIF/	http://www.galarfungail.org/data.htm	Fradin *et al*. (2003)
Candida albicans SC5314	Primary human monocytes and neutrophils	*Candida albicans* plus and minus human monocytes or neutrophils		10, 20, 40 60 or 80 min	RPMI medium 1640 with 10% human serum	Custom microarray (Bennet *et al*. (2003))	http://jura.wi.mit.edu/fink_public/neutrophils	Rubin-Bejerano *et al*. (2003)
C. albicans SC5314	Murine macrophage cell line J774A (ATCC TIB-67)	*Candida albicans* plus and minus murine macrophages	2 biological replicates and two technical replicates	1–6 h	RPMI plus 10% FBS	Custom microarray (Bennet *et al*. (2003))	http://www.lorenzlab.org	Lorenz *et al*. (2004)
C. albicans SC5314	Primary human neutrophils	Neutrophils plus and minus live yeast, killed yeast or live hyphae	2 technical repetitions	60 min	N/A	Custom human microarrays AMADID 011412 Agilent	http://www.galarfungail.org/data.htm	Fradin *et al*. (2007)
C. albicans SC5314	THP-1 human monocytic cell line (American Type Culture Collection)	Plus and minus *C.albicans*		3 h	RPMI 1640, 10% fetal calf serum (FCS)	HG-U133A Affymetrix		Barker (2005)

Table 4.1 (Continued)

Pathogen/ genotype	Host cell type/ genotype	Test conditions	Biological replicates	Time period	Media	Array/methodology	Data repository/ accession number	Reference
Candida albicans CA14-URA [Sc5314] and 36082	Primary umbilical cord endothelial cells	*Candida albicans* plus and minus host cells	3	45, 90 and 180 min	M-199 medium (Gibco) supplemented with 10% fetal bovine serum, 10% defined bovine calf serum, and 2 mM l-glutamine with penicillin and streptomycin	LA Biomed_candida albicans_6266_1.0	GEO: GSE5344	Park *et al.* (2009)
Candida albicans ATCC 14053	ATCC CRL1730 cell line: Human umbilical vein endothelial cells (HUVECs) 5e6 cells	Plus and minus high density yeast cells 5e7	3	8 h	RPMI 1640, 2mM L-glutamine, 10% FCS	sentrix HumanRef-8_V3_0_R0 Illumina	EMBL-EBI ArrayExpress database E-TABM-941	Lim *et al.* (2011)
Candida albicans ATCC 14053	ATCC CRL1730 cell line: Human umbilical vein endohelial cells (HUVECs) 5e6 cells	Plus and minus low density hyphal cells 1e6	3	8 h	RPMI 1640, 2mM L-glutamine, 10% FCS		EMBL-EBI ArrayExpress database E-TABM-941	Lim *et al.* (2011)
Candida albicans SC5314	Primary umbilical cord endothelial cells	Plus and minus *Candida albicans*	3	3 and 8 h	M-199 medium with 10% pooled human serum	HG-U133A Affymetrix	GEO: GSE8166	Barker *et al.* (2008)
Candida albicans CAN34 *cph1Δ/cph1Δ efg1Δ/efg1Δ*	Primary umbilical cord endothelial cells		3	3 and 8 h	M-199 medium with 10% pooled human serum	HG-U133A Affymetrix	GEO: GSE8166	Barker *et al.* (2008)
Candida albicans CAN35 *cph1Δ/cph1Δ efg1Δ/efg1::EFG1*	Primary umbilical cord endothelial cells		3	3 and 8 h	M-199 medium with 10% pooled human serum	HG-U133A Affymetrix	GEO: GSE8166	Barker *et al.* (2008)

A. fumigatus ATCC 13073 [GFP]	Epithelial cells/16HBE14o-	Plus and minus conidia (entire monolayer)	4	6 h	DMEM [1:1] 10e5 spores	Host array: Agilent Whole Human Genome Oligo Microarrays in 4644 K format (product number G4112F, design ID 014850) Pathogen array: JCVIGR Aspergillus fumigatus Version 3 microarray slides (Pathogen Functional Genomics Resource Centre, Rockville, MD, USA).	Gene Expression Omnibus (GEO) GSE16628, GSE16630 and GSE16637	Oosthuizen *et al*. (2011)
A. fumigatus ATCC 13073 [GFP]	Epithelial cells/16HBE14o-	Plus and minus conidia (host cells sorted by FACS)	4		serum-free minimal essential medium (MEM)		GEO: GSE16627	Gomez *et al*. (2010)
A. fumigatus AF293	Epithelial cells/ A549	Plus and minus conidia	3	8 h		GeneChip Human Genome U133A 2.0 array and later, the Human Gene 1.0 ST	GEO: GSE 24983–5	Sharon *et al*. (2011)
A. fumigatus *ΔprtT*	Epithelial cells/ A549	Plus and minus conidia	3	8 h		GeneChip Human Genome U133A 2.0 array and later, the Human Gene 1.0 ST	GEO: GSE 24983–5).	Sharon *et al*. (2011)
A. fumigatus AF293	Epithelial cells/ A549	Plus and minus fungal culture filtrate	3	8 h		GeneChip Human Genome U133A 2.0 array and later, the Human Gene 1.0 ST	GEO:GSE 24983–5).	Sharon *et al*. (2011)
A. fumigatus *ΔprtT*	Epithelial cells/ A549	Plus and minus fungal culture filtrate	3	8 h		GeneChip Human Genome U133A 2.0 array and later, the Human Gene 1.0 ST	GEO: GSE 24983–5	Sharon *et al*. (2011)
A. fumigatus clinical isolate B-5233	Primary human neutrophils (healthy and CGD)	Plus and minus *A. fumigatus* spores and hyphae	03-Apr	1.5 h (conidia), 75 min (hyphae)	RPMI/HEPES	Af293 DNA amplicon arrays	Pending	Sugui *et al*. (2008)
A. fumigatus clinical isolate	Primary human monocytes	Moncytes plus and minus *A. fumigatus* conidia (resting, swollen and germinating)	3	3, 6 and 9 h	RPMI 1640, 25 mM HEPES, 10% heat inactivated fetal calf serum (FCS)	HG-U133A Affymetrix	N/A	Loeffler *et al*. (2009)

Table 4.1 (Continued)

Pathogen/ genotype	Host cell type/ genotype	Test conditions	Biological replicates	Time period	Media	Array/methodology	Data repository/ accession number	Reference
A. fumigatus clinical isolate	Primary human monocytes	Monocytes plus and minus ethanol-killed *A. fumigatus* hyphae	3	3, 6 and 9 h	RPMI 1640, 25 mM HEPES, 10% heat inactivated fetal calf serum (FCS)	HG-U133A Affymetrix	N/A	Loeffler *et al.* (2009)
A. fumigatus clinical isolate NIH 4215, ATCC MYA-1163	Primary human monocytes	Monocytes plus and minus *A. fumigatus*conidia	2 technical repetitions for each of 5 donors	0, 2, 4, and 6 h	HBSS plus Ca^{2+} and Mg^{2+}	HG-U133A Affymetrix		Cortez (2006)
Cryptococcus neoformans B3501	Human brain microvascular endothelial cells (HBMECs)	Not stated	Not stated	0, 4, 8, 12, 16, 20, 24 h	Not stated	Affymetrix HU95A	Not stated	Jong *et al.* (2008)
C. neoformans H99	J774A.1 (ATCC TIB-67) macrophage cell line (reticulum sarcoma)	*C. neoformans* H99 plus and minus macrophages	3	2 and 24 h	Dulbeccos' modified Eagle's medium, 10% heat-inactivated fetal calf serum, 1% non-essential amino acids, 100 µg/ml monoclonal antibody 18B7, 50 U/ ml recombinant mouse interferon gamma, 0.3 µg/ ml lipopolysaccharide	Custom see Kraus *et al*, Euk. Cell 3: 1249		Fan *et al.* (2005)

C. neoformans H99	J774A.1 (ATCC TIB-67) macrophage cell line (reticulum sarcoma)	*C. neoformans* H99 plus and minus macrophages	6 h	5% CO2 in DMEM, supplemented with 10% heat-inactivated fetal calf serum, 10% NCTC-106 109 medium and 1% nonessential amino acids. mAb 18B7 was added at 10 μg/ml	Custom	Derengowski *et al*. (2013)
C. neoformans H99	J774A.1 (ATCC TIB-67) macrophage cell line (reticulum sarcoma)	*C. neoformans* H99 plus and minus macrophages	6 h	10% CO2 in Dulbecco's modified Eagle's medium (DMEM) supplemented with 10% heat-inactivated fetal calf serum, 1% non-essential amino acids, 100 mg ml-1 penicillinstreptomycin and 4 mM L-glutamine. Cryptococcus cells were opsonized with monoclonal antibody 18B7 against capsule (1 mg ml-1), and macrophages were treated with recombinant mouse gamma interferon (IFN-gamma) (50 U ml-1) and lipopolysaccharide (0.3 mg ml-1) prior to co-incubation at a MOI of 1:1	Serial analysis of gene expression (SAGE)	Griffiths *et al*. (2012)

DMEM, Dulbecco's modified Eagle medium; FACS, fluorescence-activated cell sorting; FBS, fetal bovine serum; FCS, fetal calf serum; GEO, Gene Expression Omnibus; HBMEC, human brain microvascular endothelial cell; HBSS, Hanks' balanced salt solution; HGEC, human gingival epithelial cell; HUVEC, human umbilical vein endothelial cell; MEM, minimum essential medium; N/A, not applicable; NRC, National Research Council; RPMI, Roswell Park Memorial Institute Medium; RHE, reconstituted human epithelium; SAGE, serial analysis of gene expression; YPD, Yeast Protein Database.

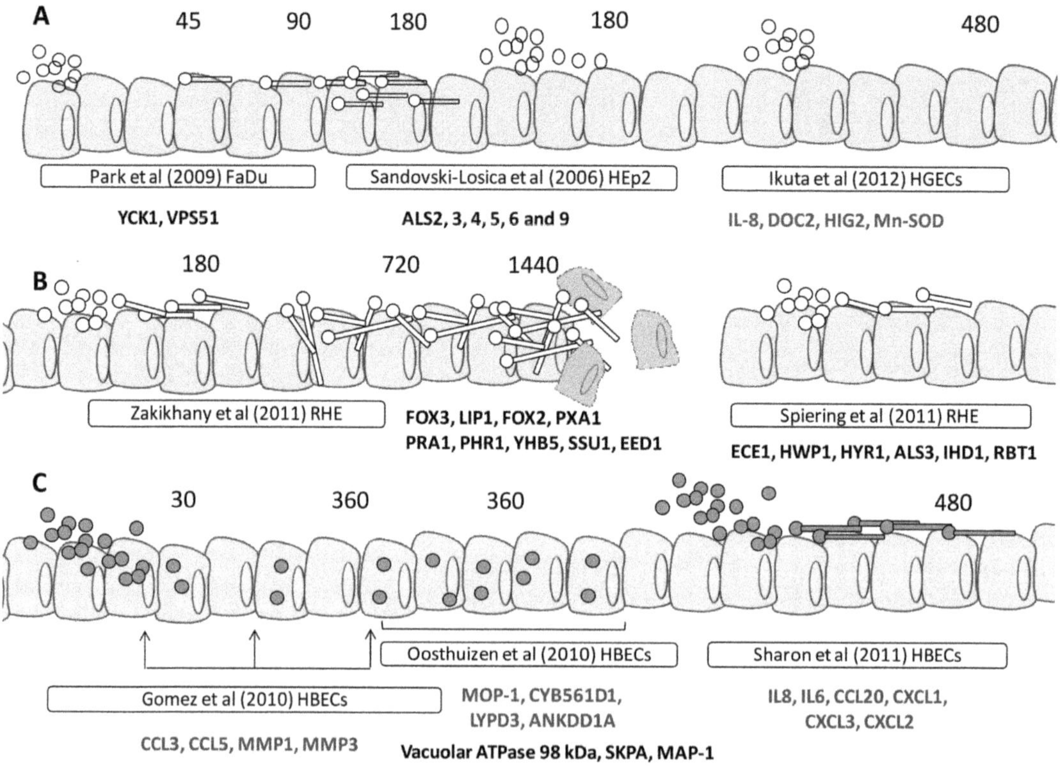

Figure 4.1 Fungus–epithelial interactions subjected to transcriptomic analyses. Overview of studies discussed in this Chapter, including key transcriptomic responses of host (red) and pathogen (black) and morphogenetic responses of fungal cells to host contact. Time-points of analyses (minutes) are indicated above each scheme. Citations and cell types are indicated below each scheme. (A) *Candida*–epithelial interactions (B) *Candida*-reconstituted human epithelium interactions (C) *Aspergillus*–epithelial interactions. Gomez *et al*. (2010) and Oosthuizen *et al*. (2010) studies differ by nature of the host cells studied, respectively targeting only those having internalized *A. fumigatus* conidia or assessing transcriptional responses at the level of the entire monolayer. A colour version of this figure is available in the plate section at the back of the book.

up-regulation of *PGA23*, *PGA7*, *HWP1* and *PRA1* during adhesive growth, as compared to that in suspension culture. With respect to differential behaviour of cells exposed to epithelia or plastic, Sohn *et al.* (2006) discovered that adhesion of fungal cells to plastic surfaces was very rapid and occurred with a higher efficiency. Maximum adhesion, via quantitative analysis, had already occurred by 30 minutes of co-culture. Anchoring to epithelial monolayers of either cell line occurred much more slowly.

Sandovski-Losica *et al.* (2006) compared *C. albicans* gene expression in the presence of Hep2 epithelial cells to gene expression in the absence of host cells (Fig. 4.1) selectively harvesting adherent cells for their analysis. Unfortunately the absence of hypha-specific gene induction observed by Park *et al.* (2009) cannot be illuminated by the findings of this particular study since, in this host–pathogen interaction, *C. albicans* remained in the yeast morphology (Fig. 4.1). This stands in stark contrast to the findings of Park *et al.* (2009), who observed germination just 45 minutes after the initial host–pathogen encounter (Fig. 4.1). It is worth noting the effect that the different host cell lines and fungal isolates used by the different experimenters might contribute to such distinct morphological behaviours, nonetheless observations alluded to by Zakikhany *et al.* (2007) include normal filamentation of *C. albicans* in the presence of Hep2 cells. It is also plausible that the interaction between fungal and host cells studied by Sandovski-Losica was occurring over a much longer timescale and this is substantiated by the finding that several adhesion genes, predominantly *ALS2* and *ALS5*, were up-regulated in the presence of host cells (Fig. 4.1). In order to substantiate the results of microarray analysis a targeted RT-PCR analysis of all *C. albicans* adhesins was performed on the same mRNA samples. This revealed that *ALS2, 3, 4, 5, 6,* and 9 were significantly up-regulated in response to host cells from as early as one hour post infection. A further, possibly more likely explanation for the lack of hyphal cells in the Sandovski-Losica study is that the high multiplicity of infection (effector–target ratio of 10:1) impeded filamentation. In fact other researchers (Lim *et al.*, 2011) have intentionally employed such high MOIs as a means, in the absence of genetic mutation, to lock *C. albicans* cells into the yeast morphology during interaction with vascular endothelial cells.

Ikuta *et al.* (2012) examined primary human gingival epithelial cells (HGECs) after 8 hours of co-incubation with *C. albicans* cells (Fig. 4.1). A large number of host genes were found to have been differentially expressed, and the authors used Ingenuity pathway analysis to identify highly interconnected nodes. This ear-marked IL-8 as a key 'rheostat' in immune responses to *C. albicans*, a finding substantiated by RT-PCR analyses of IL-8, COX2 and ICAM-1 gene expression and immunohistochemical staining of infected tissues from a rat experimental model of infection. Interestingly the transcripts demonstrating the highest levels of differential expression following *C. albicans* challenge were *DOC2* (cell death and apoptosis) and *HIG2* and Mn-*SOD* (response to stress) indicating trauma imposed upon host cells exposed to the pathogen. Compared to other available studies the time-course adopted for this experiment was considerably longer (8 h) and the authors did not report any visual inspection of monolayer integrity or fungal morphogenesis so the extent to which valid comparative conclusions can be reached on the status of the fungal cells or viability of infected host cells is somewhat limited. Overall, the magnitude of up- and down-regulation of genes seems to have been somewhat muted in these experiments.

Reconstituted human epithelium (RHE) has provided the research community with an invaluable tool for probing the host–microbe interface. The RHE infection model (Schaller *et al.*, 2008) permits a quantitative assessment of *C. albicans* virulence, thereby negating the use of whole animals, and reproducibly yields sufficient material for longitudinal transcriptome analysis of host and fungal activities during infection.

(Zakikhany *et al.*, 2007) studied the interaction between RHE and *C. albicans*, describing the series of events occurring over a 24-hour time course of the host–pathogen interaction (Fig. 4.1). Two modes of epithelial penetration were identified, the first occurring via active physical breach of the epithelial barrier and the second via a passive, induced endocytic process. In this infection model, distinct events are observable during the host–pathogen interaction: fungal attachment at 1–3 hours, invasion at 3–12 hours, and tissue destruction at 12–24 hours. 98% of fungal cells had undergone a switch to hyphal morphology within the first hour of host contact. Electron microscopy revealed ruffling of the host cell membrane and fungal penetration of host cells at early (1 h) time points of infection, behaviours which correlated temporally with up-regulation of 164 fungal genes. Multiple hypha-specific genes were in evidence amongst the 38 genes up-regulated across the entire time series of the study. Gene functions having heightened transcript abundance exclusively at the later stages of infection included those involved in utilization of two-carbon energy sources of possibly lipid origin as evidenced by *FOX3*, *PXA1* and *FOX2* and *LIP1* gene up-regulation. A clear signature of response to nitrosative stress was also observable as *YHB5* and *SSU1*, involved in nitric oxide detoxification were found to be up-regulated during the late phase of interaction with epithelial cells. The pH-responsive *PRA1* and *PHR1* genes were also up-regulated during late-phase infection indicating gradual alkalinization of the infectious arena, plausibly linked to lysis of host cells.

Zakikhany *et al.* (2007) made important strides towards demonstrating physiological relevance of their findings by extrapolating the analysis to smear samples isolated from HIV positive sufferers of pseudomembranous candidiasis. Amongst genes found to be up-regulated in both experimental and human infection of oral epithelia were multiple hypha-associated, non-glucose carbon-utilizing, nitrosative stress and alkaline-induced genes, thereby cementing the physiological relevance of the hyphal stage of growth.

Focusing upon genes of previously unidentified function a single ORF, orf19.7561, deletion of which maximally impacted damage elicited by C. *albicans* in RHE, was followed up in a functional analysis. Assigned the name EED1 (epithelial escape and dissemination) the null mutant of this gene grew equivalently to the wild type *in vitro* but would not form hyphal filaments in response to standard hypha-inducing cues. Minimal elongation was achieved in response to 10% serum, a phenotype which might contribute to defective invasion of RHE. After 24 hours Eed1 null mutants had achieved endocytically mediated host cell invasion but had not undergone hyphal elongation, being restricted to the upper portion of the RHE and remaining exclusively in yeast or pseudohyphal morphologies. The authors concluded that once inside human epithelial cells the Eed1 gene product is required to promote epithelial escape and inter-epithelial dissemination. Importantly the major C. *albicans* invasin Als3 remains up-regulated during the early phases of the Eed1 null–host interaction. This might also contribute to the ability of very primitive germlings to undergo induced endocytosis.

Spiering *et al.* (2010) used an RHE infection model to investigate the differences in gene expression of C. *albicans* and C. *dubliniensis*, two closely related species that differ widely in their capacity to cause disease. The time course of C. *albicans* infection was found to be highly similar to that reported by Zakikhany *et al.* (2007) whereby most fungal cells were forming germ tubes within one hour of host contact (Fig. 4.1). Protein synthesis, transport, amino acid metabolism and mitochondrial activities were noted as being enriched amongst up-regulated genes. In common with the findings of Zakikhany *et al.* (2007) multiple hypha-associated genes were up-regulated including *ECE1*, *HWP1*, *HYR1*, *ALS3*, *IHD1*, and *RBT1*. The protein mannosyl transferases *PMT1*, *2*, *4*, and *6* were also up-regulated. Ribsosomal protein genes were significantly up-regulated in both C. *albicans* and C. *dubliniensis*, a common regulatory motif which cannot be attributed to activities required for hyphal growth since C. *dubliniensis* fails to undergo morphogenesis under the conditions tested. One hundred and seventy-six genes were down-regulated by C. *albicans*, the majority of which had unknown functions. In keeping with the findings of Park *et al.* (2009), the Spiering study also found that contact with polycarbonate filters prompted formation of hyphae. Two hundred and twenty-two C. *albicans* genes were up-regulated in response to RHE relative to exposure to plastic. Among these, genes involved in two-carbon metabolism were up-regulated as well as several heat shock genes. Importantly, these latter data relate to gene products specifically up-regulated by both fungal species and representing a core response.

Aspergillus fumigatus–epithelial interactions

Diseases caused by the major mould pathogen of humans, *A. fumigatus*, initiate with inhalation of a fungal spore, which eventually settles in contact with the pulmonary epithelium. The nature, in man, of this initial host–pathogen interaction remains speculative since it occurs long before positive diagnosis of infection becomes possible and likely results from a low infectious inoculum. The prospect of fungal latency has been widely suggested (for an excellent recent review see Osherov, 2012) and garners firm support from multiple observations on internalization of *A. fumigatus* spores. Studies have found that 20–50% of *A. fumigatus* conidia interacting with bronchial or nasal epithelia are internalized into late phagosomes where germination is contained. It is worth noting that these conidia remain viable for up to 20 hours. Externally adherent conidia germinate and ultimately cause damage to cells. The mere presence of ungerminated conidia is sufficient to prompt retraction and loosening of the host epithelium and cytoskeletal depolymerization. This might promote invasive growth.

Two related studies assessed the response of airway epithelial cells to *A. fumigatus* challenge. In the first of these studies Gomez *et al.* (2010) used fluorescence-activated cell sorting to select human bronchial epithelial cells (HBECs) having bound or internalized conidia. Thus the cell population under study had directly interacted with the spores of *A. fumigatus* (Fig. 4.1). The use of a GFP-expressing *A. fumigatus* isolate, and nystatin protection assays, revealed that HBECs internalized up to 50% of spores within 6 hours of co-incubation. This process begins almost immediately following spore contact, whereby 38% of bound conidia become internalized by 30 minutes of co-incubation. In response to spore contact 889 host genes were differentially expressed. Functional analysis of the predicted gene products revealed that HBECs respond to spore challenge by altering expression of repair and inflammatory responses. Pilot analyses revealed that the internalization process did not alter surface expression of E-cadherin – which the authors interpret as meaning that the epithelium was not 'stimulated' by the internalization process. The authors note a relatively modest 'magnitude' of differential gene expression which may, in this instance relate to flattening of the output due to incomplete cell sorting. An alternative hypothesis relates to a non-reactionary interaction with the host cell providing ease of immune escape for the pathogen.

In a subsequent, related, study utilizing the same HBEC cell line Oosthuizen *et al.* (2011) implemented a dual organism transcriptomic analysis of each infected monolayer sample, this time avoiding the enrichment for host cells having contacted the pathogen, and again examining a 6 hour time-point to permit direct comparisons with the prior study (Fig. 4.1). Monolayers were incubated with 10^5 *A. fumigatus* conidia, for an average final multiplicity of infection of one conidium per human cell. Two hundred and fifty-five human genes were found to be differentially expressed. In order to broaden the cohort of potentially relevant human genes the authors used Ingenuity network pathway analysis to capture clustering of differentially expressed genes and to triangulate nodes of interest in the regulatory network. The idea was to use down-stream or co-regulated genes to implicate hubs in the network. Based upon this analysis a number of candidates were selected for follow up by qPCR. Transcript abundance was then examined in two cell populations including the initial cell line of interest and primary airway epithelial cells. Among those tested, IL6 (not up-regulated in the microarray data) was the only transcript significantly up-regulated in both cell populations. Clearly, then, even in the hands of the same researchers, heterogeneity of response is to be expected, and this might be potentiated by the mild transcriptional responses observed in response to fungal challenge. Between these two studies only 17 genes were commonly regulated, none of which could demonstrably be proven as significantly up-regulated in both cultured and primary HBECs.

One hundred and fifty fungal genes were up-regulated in conidia exposed to HBECs and 33 down-regulated genes were identified. Among the most prominent annotated functions were those involved in vacuolar acidification and metallopeptidase activity. To validate the expression levels generated by the microarray, eight genes were chosen for further RT-qPCR analysis. Interestingly, seven out of eight genes tested achieved significance following incubation with at least one host cell type. Genes encoding the vacuolar ATPase 98 kDa subunit, SkpA, and MAP-1 were significantly differentially expressed following incubations with both HBECs and primary airway cells. V-ATPases are structurally conserved proton pumps found in all eukaryotes that function to acidify the lumen of vacuoles. Although the authors note that the utility of up-regulating vacuolar acidification mechanisms in the presence of human airway epithelial cells is unknown, a reasonable hypothesis was postulated which states that that V-ATPase up-regulation would promote cation uptake, modulate fungal pH homeostasis and/or assist in nutrient degradation. A putative sulfur metabolism regulator gene, *skpA*, was also up-regulated in conidia following exposure to both HBECs and primary airway cells, plausibly indicating the requirement for methionine and cysteine-rich proteins during exposure to epithelial cells.

A further study of airway epithelial responses to *A. fumigatus* challenge was implemented by Sharon *et al.* (2011), who sought to dissect host responses to *A. fumigatus* isolates possessing, or lacking, a number of protease gene functions previously found to be necessary for *A. fumigatus*-mediated epithelial damage. The mutant under study, lacking a transcriptional regulator, of protease gene expression, PrtT, is deficient in the production of fungal proteases required for causing host damage. Despite lacking several proteases and therefore being deficient in causing late-stage damage to monolayer cells, the PrtT mutant retains virulence in whole animal models of invasive disease. The PrtT gene product is therefore dispensable for colonization of immunosuppressed mammalian hosts but might lack activities which contribute to the process of disease pathogenesis in immunocompetent hosts, such as exacerbation of allergic-type responses to inhaled fungal spores. A549 cells were incubated in the presence of WT or *ΔprtT* conidia, or culture filtrate derived from WT or *ΔprtT* mycelia (Fig. 4.1). The incubation period for all investigations was 8 hours, a time point at which wild type germinating conidia or fungal culture filtrates had previously been found to evoke actin-cytoskeleton depolymerization and cell rounding, without killing the cells. The response of A549 cells to conidial challenge with wild type or *ΔprtT* conidia was found to be largely concordant. This was noted as being an expected finding since the relevance of fungal proteases to the host–pathogen interaction is most significant at later stages of infection. Wild type germinating conidia (as surveyed at 8 hours of co-incubation) significantly increased the mRNA levels of 94 genes relative to untreated cells. Functional categorization was performed using 'Expander' identifying cytokines, signal transduction (particularly MAPK signalling) and transcription factors as significantly over-represented. Together, the results of this assay suggested that conidial infection induces a strong protective response in infected A549 cells, characterized by the activation of genes participating in intracellular signalling pathways and the secretion of inflammatory cytokines. This would agree with the overall conclusions reached by the Gomez and Oosthuizen studies, albeit that direct comparisons of relevant gene functions is prevented by the heterogeneity of outputs reported in the two studies. Ninety-four genes were flagged as a 'relatively modest' number of genes, compared to the 291 genes up-regulated in response to culture filtrate. This would agree with the proposal that epithelial cells maintain a somewhat inert stance to conidial challenge. The response to culture filtrate was more exuberant in terms of the numbers of genes activated where an altered functional relevance was observed amongst up-regulated gene functions. These included genes encoding proteins involved in signal transduction and MAPK signalling. Stress was identified as a possible response to protease exposure as a large number of gene functions were down-regulated including those involved in amino acid, nitrogen and lipid metabolism suggesting a general shutdown of energy-intensive metabolic processes. A comparison of the effects of wild-type and *ΔprtT* culture filtrates showed that the former significantly increased the mRNA levels of 226 genes relative to the latter. These included genes encoding proteins involved in cell cycle control, microtubules, and the unfolded protein response which suggests that the wild-type culture filtrate-treated cells could be modulating cell cycling and protein synthesis and folding in response to protease activity or other PrtT-dependent secreted proteins in the wild-type culture filtrate. Interestingly, conidial infection activated more cytokine-encoding genes than exposure to culture filtrates.

Fungus–endothelial interactions

The interaction with endothelial strata is relevant to both *C. albicans* and *C. neoformans* infections. During the initiation of haematogenously disseminated candidiasis, *C. albicans* adheres to and invades the endothelial cell lining of the blood vessels. In infections caused by *C. neoformans* the propensity of the pathogen to target the central nervous system requires traversal of the vascular endothelium at the blood–brain barrier.

Candida albicans–endothelial interactions

Barker *et al.* (2008) assessed the interaction of *C. albicans* with vascular endothelial cells to examine events occurring at the commencement of disseminated infection (Fig. 4.2). This study employed two fungal isolates of isogenic origin, one of which (CAN34) lacked the Cph1 and Efg1 transcriptional regulators, thereby being unable to

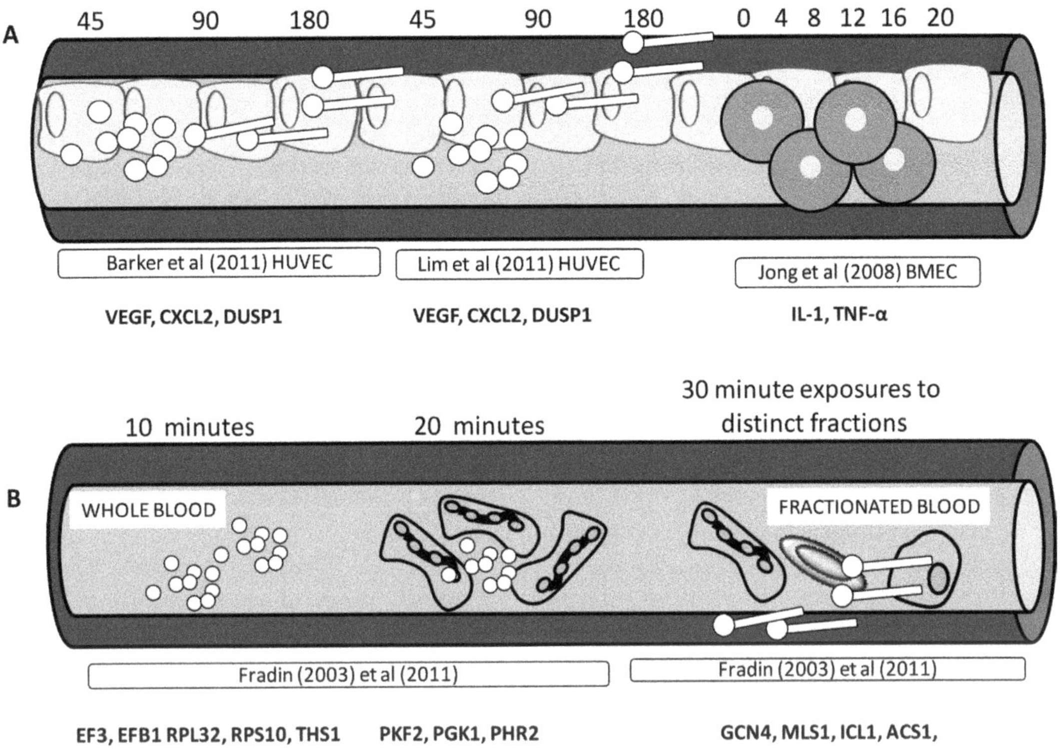

Figure 4.2 Fungus–endothelial and fungus–blood interactions. Overview of studies discussed in this chapter, including key transcriptomic responses of and pathogen (black) and morphogenetic responses of fungal cells to host contact. Time-points of analyses (minutes) are indicated above each scheme. Citations and cell types are indicated below each scheme. (A) *Candida*– and *Cryptococcus*–endothelial interactions (B) *Candida*–blood interactions. A colour version of this figure is available in the plate section at the back of the book.

form hyphae. This was an interesting set of comparisons. Not only did the authors interrogate the response to wild type exposure, in three biological replicates, they also interrogated responses to non-filamenting and reconstituted isolates embedding several degrees of robust control to the datasets. Additionally an investigation of a prototypical response (that to TNF-α exposure) was analysed. Hierarchical clustering of the datasets revealed that, at 3 hours of co-incubation, endothelial cell responses to both wild type (SC5314) and mutant (CAN34) clustered together, indicating similarity of response to these distinct fungal challenges. The same scenario was observable for the 8-hour exposure. Endothelial responses to TNF were similar to each other at 3 and 8 hours, and were most similar to the 8 hour challenge with fungal cells. Comparison to a previous study conducted by the same researchers, and employing THP-1 monocytes, revealed that the responses of these two cell types differ very markedly. Remarkably the number of genes, which were differentially regulated, was very low. The wild-type isolate, by 3 hours post exposure, had up-regulated six genes and down-regulated six genes. This increased to 72 up-regulated and 40 down-regulated at 8 hours post-exposure. Statistical evaluations identified few genes transcending significance thresholds at plus 3 hours (6 and 6 respectively). At 8 hours gene ontological analyses identified inflammation, cell proliferation, and responses to stress or wounding as significant functional trends. CCL3 and CCL4 were noted as being up-regulated in both endothelial cells and THP-1 cells, Barker *et al.* proposed that these two gene functions might play a very important and key role in the host defence. Having confirmed that *C. albicans* elicits secretion of the two proteins, the study was extended to investigation of fungal infection in mice lacking the receptor for both of these proteins, CCR5. Survival and fungal burden following infection of these animals was comparable to that of control mice. The authors noted that CCR5$^{-/-}$ mice are more susceptible to cryptococcosis (Huffnagle *et al.*, 1999) where CCR5 is required for leucocyte recruitment into the brain thereby suggesting an organ-specific role for this molecule. Also CCR5 modulates recruitment of monocytes, not neutrophils, and so neutrophils might still have been normally functioning and masked any effect.

In order to decipher the different responses of *Candida* to epithelial and endothelial cells, Park *et al.* (2009) implemented a parallel analysis of the host–pathogen interaction in response to these two types of host cells. Initial observations identified a conserved morphological response to epithelial and endothelial cells, which differed little from that of the response to polystyrene surfaces. On all three media blastospores germinated within 45 minutes of initial contact, and hyphae extended at similar rates. *Candida* cells were more prone to adhering to epithelial cells, no detectable damage was observed to either type of host cell at 45 minutes, and very few organisms were endocytosed. After 90 minutes 50% of the endothelially incubated cells versus 24% of the epithelially incubated cells had been endocytosed. Epithelial cells suffered significantly less damage. By 180 minutes internalization of fungal cells was largely complete for both cell types. Endothelial cells internalized more quickly but overall took up fewer fungal cells than epithelial cells. Nonetheless epithelial cells suffered 50% less damage. With organisms grown on polystyrene as the reference, the transcript levels of only 29 genes were commonly regulated in response to the two types of host cells and all of these were down-regulated. Using the two cell lines and (ultimately) three fungal isolates, the authors were unable to identify a core response to host exposure. Reasoning that this could be due to the use of laboratory and clinical *C. albicans* isolates the authors resorted to solely documenting the CAI4-URA response to host exposure. This affirmed that strain-specific differences were confounding the data, and revealed over a time-course of three hours that 160 genes exhibited similar behaviours in response to both cell types. In response to endothelial cells the authors found, by microarray analysis, up-regulation of 9 gene functions. The majority of up-regulated gene expression was confirmed by qPCR analysis. So the two types of host cell respond differently – epithelial cells take up *Candida* at a slower rate incurring less damage and the transcriptional response to host contact is markedly different between the two cell types. This resistance to pathogen-mediated damage was proposed to originate from continual exposure of the oral epithelium to *Candida* cells. The authors comment that responses to *Candida* cells, mounted by epithelia – are not great in magnitude – as found for *A. fumigatus*.

Lim *et al.* (2011) took a novel approach to defining genes up-regulated by *Candida* in response to endothelial cells of umbilical vein origin. In order to limit morphogenesis-associated gene expression, variation of population density was utilized to maintain either a yeast (effector to target ratio of 10) or hyphal (effector to target ratio of 0.2) population of host-interacting fungal cells (Fig. 4.2). The working hypothesis was that high density yeast populations in circulating blood might influence the initiation of invasive infection. The study revealed that high density yeast cell populations lead to stronger, broader transcriptional responses than low density hyphal infections.

Concordant with the findings of Park *et al.* (2009), low density infections had yielded cell-penetrating hyphae by one hour post-infection. In the high-density infections the blastospore morphology was largely maintained, with less than 10% of fungal cells converting to pseudohyphae within the 8-hour observation period. Adhesion remained a feature of the low dose infection. Overall functional themes included induction of more genes with higher density infections, predominance of apoptotic pathway genes, immune responses, cell–cell signalling and stress. Two hundred and fifteen genes were differentially regulated in high-density infection, 127 up and 88 down. Of these, 182 were exclusive to the high-density infection. Only 41 genes were induced in the low density infection: 38 up and three down. The results of this study (low dose infection) bear resemblances to those of Barker *et al.* (2008) who found induction of VEGF, CXCL2 and DUSp1. Conditioned medium harvested from high-density fungal cultures was capable, in isolation, of reducing HUVEC viability. This effect, not seen with medium alone, or conditioned medium from low-density cultures suggests that the interaction of host cells with high numbers of blastospore morphotypes evokes release of extracellular molecules, of plausibly either host or pathogen origin. A new hypothesis is formulated on the basis of these data which proposes that high density yeast cell infections, as found perhaps in microabscesses, might cause endothelial breach via apoptosis of endothelial cells.

Cryptococcus–endothelial interactions

Haematogenous spread of cryptococcal cells often culminates with initiation of meningitis. In order to achieve this, the invading pathogen must traverse the blood–brain barrier consisting mainly of brain microvascular endothelial cells (BMECs). Similar to the interaction between epithelial cells and *A. fumigatus* this host–pathogen interaction results in cytoskeletal alterations. Jong *et al.* (2008) documented gene expression at 0, 4, 8, 12, 16 and 20 hours of infection with *C. neoformans* (Fig. 4.2). This identified two major clusters of pathogen-responsive genes differentially regulated early (0–12 h) and late (16–24 hours) in infection. The overall gene expression profile was more profoundly influenced during the early rather than late interaction.

Principal component analysis was implemented to interrogate expression of interferon genes and MHC genes. Expression of IL-1 was observed to have increased slightly, while MCP-1 expression did not change at all. Neither IP-10 nor ICAM-1 signals were identified while TNF-α expression increased at a relatively late stage of the infection.

Fungus–whole blood interactions

The dissemination of fungal elements, from initial sites of infection to remote host niches, likely takes multiple forms. Obvious examples include dendritic cell-mediated passage of *A. fumigatus* spores and hyphae to the mediastinal lymph nodes, cryptococcal invasion of the central nervous system and blood-borne candidaemia. Whole blood, sterile in healthy human beings, represents a complex and threatening milieu to all classes of pathogens being composed of multiple innate immune effectors. Often, murine models of fungal infection are initiated with direct inoculation of the fungal pathogen into the blood stream to mimic disseminated disease. The success of this approach in establishing systemic candidiasis, aspergillosis and cryptococcosis in immunocompetent hosts is testament to the tenacity of these organisms to withstand host immune defences; not only must fungal cells evade killing during passage via peripheral blood but ultimately escape the blood vessel to colonize target organs. How does the pathogen adapt to the biotic and abiotic stresses imposed by such a harsh environment?

Candida–whole blood interactions

In a series of linked studies (Fig. 4.2) Fradin *et al.* (2003) addressed *C. albicans* gene expression likely to be relevant to blood-borne *Candida* infections, initially in response to challenge with whole blood and subsequently with components thereof (Fradin *et al.*, 2005).

An initial approach, combining microarray and cDNA subtraction methodologies (Fradin *et al.*, 2003), identified unique sets of genes expressed at different stages of exposure to human blood. Then, by removing host cells, genes specifically influenced by the various cellular components of blood were identified. The data derived during this study revealed how the fungus adapts to the blood micro-environment providing important insight into dissemination which is an essential step in invasive infections. During the experiments the conversion of blastospores to hyphae was observable, with > 55% of cells producing germ tubes in plasma and blood after 60 minutes. Hypha-specific genes were in evidence amongst those differentially expressed including *HWP1*, *ECE1*, *HYR1*, *SAPs 4, 5* and *6* and *ALS3* and *ALS8*. Fungal cells were found to be associated with leucocytes soon after inoculation into human blood. After 10 minutes 92% of cells were either bound or internalized by leucocytes. After 60 minutes 46% of all cells had been ingested.

Within 60 minutes of incubation with whole blood around 280 genes were found to be up-regulated and more than 240 genes down-regulated. Supplementation with cDNA extraction served to capture transcripts for whom probes were absent from microarrays, and as a means to verify significant results. The gene cohorts were examined by cluster analysis, accounting for those which were differentially expressed at various stages of the infection process. In plasma, gene expression profiles changed slowly from 10 to 30 minutes post exposure, but after 20 minutes clear differences were discernible. In contrast, by 20 minutes exposure to whole blood there was a significant fungal transcriptional response, indicating that the cellular fraction of whole blood exerts the most significant impact upon fungal transcription. When fungal cells were transferred into blood up-regulation of protein biosynthetic gene functions was observed within 10 minutes. This response included elongation factors *EF3* and *EFB1* and ribosomal proteins such as *RPL32* and *RPS10* and tRNA synthetases such as *THS1*. These transcripts subsequently decreased in abundance, possibly reflecting a move from nutrient poor to nutrient-rich conditions. In the early stages of blood exposure glycolysis and the citric acid cycle gene functions were down-regulated, possibly signifying a carbon- and energy-rich phase for the cells. This situation changed with time such that, at 20 minutes post exposure *PFK2* and *PGK1* were > 16- and 41-fold up-regulated, respectively. Despite elevated glycolytic functions there was also evidence of simultaneous glyoxylate gene functions being expressed, so at least some cells were experiencing starvation. Similar to glycolysis these gene functions were first down and then subsequently up-regulated, and again this correlated with presence of host cells, however, the observed up-regulation of both glycolytic and glyoxylate cycle genes is conflicting, the most likely explanation being that multiple populations of metabolically distinct cells contributed to the overall profile.

In this study stress responses were anticipated by the experimenter and quelled by pre-incubating fungal cells with obvious prompts such as alkaline and thermal challenge. Thus, prototypical responses to alkaline stress and temperature flux were not observed upon transfer to blood. In contrast *PRA1* (also alkaline-regulated) was highly down-regulated and *PHR2*, an acid-regulated gene was 20-fold up-regulated. Responses to oxidative stress were evident with heightened abundance of detoxification-associated transcripts mainly detected in blood, not in plasma.

A subsequent analysis of fractionated blood addressed the impact of erythrocyte, polymorphonuclear, and mononuclear cells upon *C. albicans* growth and gene expression (Fradin *et al.*, 2005). Using fresh blood from healthy human donors, analysed in parallel with fractionated components thereof, and infected at a 1:1 ratio of host to pathogen cells, *C. albicans* gene expression was analysed after 30 minutes of co-incubation. Interaction with leucocytes was observed well within this timescale with more than 90% of fungal cells bound or ingested. In whole blood the majority of fungal cells remained in the yeast form, however exposure to plasma, erythrocyte or monocyte fractions failed to restrict the morphology of *C. albicans* to the yeast form resulting in all instances in > 79% germ tube formation. In direct contrast, *C. albicans* cells incubated with PMN fractions remained almost entirely (96%) in the yeast form. Fungal cells survived exposure to all fractions and whole blood.

Adopting a common reference experimental design, with erythrocyte-exposed *C. albicans* mRNA as reference, 1518 genes were identified by transcriptome analysis to become significantly differentially expressed in response to one or more of the test conditions. Comparative analysis of the datasets confirmed vast differences between plasma- and whole blood-responsive transcriptomes and extensive similarities between *C. albicans* whole blood- and PMN-responsive transcriptomes. Exposure to neutrophils prompted reduced expression of genes involved in protein synthesis, ribosomal proteins, and elongation or translation initiation factors, a signature usually indicative of nutrient starvation. Accordingly, genes required for amino acid biosynthesis, including the cross pathway regulator of amino acid biosynthetic genes *GCN4*, were all up-regulated by *C. albicans* soon after exposure to whole blood or PMNs. A switch to two-carbon metabolism was also evident with key enzymes of the glyoxylate pathway (*MLS1, ICL1, ACS1*) being strongly up-regulated. Suppression of filamentation was reflected by down-regulation of multiple hypha-specific genes *including SAP4–6, HYR1, ECE1* and *ALS3*.

The authors surmised that neutrophils significantly impact growth, morphogenesis, oxidative and nutrient stresses and conclude that these observations might derive from internalized fungal cells. Although only 38% of fungal cells had become internalized at the point of analysis this conclusion is further buttressed by the findings of Rubin-Bejerano *et al.* (Rubin-Bejerano *et al.*, 2003), who discovered that *C. albicans* exposed to purified neutrophils and monocytes displayed amino acid-deprived transcriptome signatures only in response to neutrophils.

Fungi, monocytes and neutrophils interactions

Healthy defences against fungal infection include phagocytosis, by macrophages and neutrophils, of fungal cells. Patients suffering qualitative or quantitative deficiencies in these activities are particularly susceptible to systemic disease.

Candida, monocytes and neutrophils

Rubin-Bejerano *et al.* (2003) performed the directly comparative analysis of *S. cerevisiae* and *C. albicans* transcriptomes following phagocytosis by primary human macrophages or neutrophils. This identified a transcriptional signature common to both organisms and involving up-regulation of gene functions required for methionine and arginine biosyntheses. In both organisms, a similar response was evoked upon transfer from amino acid-rich medium to amino acid-deficient medium, thereby suggesting that the internal phagosome of the neutrophil is an amino acid-deficient environment. The response of *S. cerevisiae* to neutrophils was found to be distinct from that to monocytes, where amino acid biosynthetic pathways are not induced. Together with the absence of amino acid biosynthesis amongst the *C. albicans* gene functions documented by Lorenz and Fink (2001) these data demonstrate that a conserved fungal response to the neutrophil phagosome requires amino acid biosynthesis, and that the monocyte and neutrophil phagosome environments differ significantly with respect to available nutrients.

Lorenz *et al.* (2004) assessed the interaction between cells of the murine macrophage cell line, J774A and *C. albicans*, revealing a dramatic reprogramming of transcription that occurs in two successive phases. The initial response to encounter with macrophages is starvation, demonstrated by a shift in metabolism to gluconeogenic growth. Gluconeogenesis is serviced by the production of acetyl coA via fatty acid degradation and translation is emphatically down-regulated. Subsequent to the starvation response, and likely corresponding to phagosomal escape, a resumption of glycolytic growth was observed. As documented by Rubin-Bejerano *et al.* (2003) an up-regulation of arginine biosynthesis was observed following phagocytosis.

The leucocyte component of the innate immune response includes neutrophils and monocytes, the latter of which secrete a wide array of cytokines in response to pathogen encounter, thereby initiating and amplifying the host immune response to infection. In studies addressing bioactive components of whole blood (Fradin *et al.*, 2005) granulocytes had emerged as those cells exerting the most potent effects upon fungal gene expression. In a subsequent study Fradin and co-workers (Fradin *et al.*, 2007) implemented the targeted analysis of neutrophil responses to *C. albicans* contact in order to address whether these host cells mounted a directly microbicidal, rather than immunomodulatory, response to fungal challenge. Using microarrays representing 8500 genes of immune-biased functionality, the response to UV-killed yeasts, live yeasts and live hyphae was assessed. Cluster analysis revealed that the responses to live cells, regardless of morphotype, were most similar, thus viability rather than morphotype is the dominant protagonist of host response. The up-regulated gene cohort was largely devoid of gene products represented amongst granule proteins and a core response of 42 genes, commonly up-regulated regardless of morphotype, was identified. Amongst these an immunomodulatory programme of gene

Table 4.2 Commonalities of host responses to fungal challenge

Pathogen and morphotype	Study	Host	MCP-1	IL-8	IL-6	IL-1a	IL-10	IL-1b	CCL3	CCL4	CXCL2	CXCL3	CCL2	CXCL6	CCL5	ICAM-1	PTX3	UPA	UPAR	Kyn	BAX	Casp1	Casp4	Bcl2
A. fumigatus resting spores	Loeffler *et al*. (2009)	MC															↑							
A. fumigatus swollen spores	Loeffler *et al*. (2009)	MC											↑											
A. fumigatus germinating spores	Loeffler *et al*. (2009)	MC		↑									↑			↑		↑						
A. fumigatus hyphae	Loeffler *et al*. (2009)	MC		↑	↑		↑	↑					↑	↑			↑		↑					
A. fumigatus conidia	Gomez *et al*. (2010)	MC							↑		↑	↑			↑					↑	↑	↑	↑	↑
A. fumigatus conidia	Oosthuizen *et al*. 2011)	EC			↑							↑												
A. fumigatus conidia	Sharon *et al*. (2011)	EC	↑	↑	↑						↑	↑	↑	↑										
A. fumigatus culture filtrate	Sharon *et al*. (2011)	EC		↑									↑											
A. fumigatus conidia	Cortez *et al*. (2006)	MC		↑			↑	↑		↑	↑					↑								
C. albicans	Ikuta *et al*. (2012)	EC		↑	↑																			
C. albicans	Barker *et al*. (2008)	MC	↑	↑			↓		↑															
C. albicans	Barker *et al*. (2008)	VEC		↑		↑			↑	↑	↑	↑	↑											
C. albicans live yeast or hyphae	Fradin *et al*. (2007)	PMN									↑		↑											
C. albicans yeasts	Kim *et al*. (2005)	MC	↑	↑	↑																			

EC, epithelial cell; MC, monocytes; PMN, polymorphonuclear cell; VEC, vascular endothelial cell.

expression was observed, including up-regulation of CCL3, CXCL2, MIP1A and MIP2A genes (Table 4.2).

Kim *et al.* (2005) examined the response of primary human monocytes to *C. albicans* infection, assessing gene expression for 0 to 18 hours in a parallel analysis of infected and uninfected cells. Amongst 42,142 genes represented on the microarray the up-regulation of genes encoding pro-inflammatory cytokines TNFα and IL6 was observed within the first six hours of pathogen encounter. Further pro-inflammatory gene functions were up-regulated with a slight delay, peaking at 4 to 6 hours, including IL8, MIP1–3 and the chemokine receptors CCR1, 5 and 7 and CXCR5 (Table 4.2).

Barker *et al.* (2005) used the THP-1 human monocytic cell line to probe the host response to *C. albicans* exposure. Noting that the yeast to hyphal transition had been completed within 1 hour of host and pathogen co-incubation, monocyte gene expression was interrogated at 3 hours post infection, revealing 131 genes to be differentially regulated, 47 of which were induced and 84 down regulated. Amongst the anti-pathogenic functions favoured by the challenged host cells MIP1A and MIP1B, IL8 and TNF-alpha were up-regulated. The microarray approach failed to detect significance for IL1B induction however this could be demonstrated at all time-points measured by qPCR. The authors proposed that late induction of IL8 might be indicative of its reliance upon earlier regulatory components. In favour of this hypothesis Loeffler *et al.* (2009) found a similar phenomenon following *A. fumigatus* challenge of human monocytes.

Aspergillus fumigatus–monocyte interactions

The interaction between *A. fumigatus* and primary human monocytes has been explored in detail and gene expression of the host response to fungal challenge carefully documented. Based on these studies monocytes have been found to mount differential transcriptional responses to varying growth forms of the fungus. Loeffler *et al.* (2009) studied the interaction between viable conidia and monocytes, profiling the temporal basis of host gene expression at 3 hours (resting conidia), 6 hours (swollen conidia) and 9 hours (germinating conidia). Additionally the response of monocytes to ethanol-killed *A. fumigatus* hyphae was studied at 3, 6 and 9 hours of co-incubation.

In order to minimize noise caused by physical disturbance of the monocytes in human peripheral blood the authors implemented an indirect selection method which labels non-monocytic cells thus stimulation by magnetic beads was avoided. The interaction between host and pathogen was also profiled to correlate gene expression with phagocytosis of conidia in a time-dependent manner. The authors observed that the majority of monocytes phagocytosed conidia within the first 3 hours of co-incubation. Conidia were observed to swell, partly escape phagocytosis and those which escaped phagocytosis germinated and outgrew the monocytes after 9 hours. Interestingly, the uptake of conidia did not induce the expression of cytokines and chemokines, thus supporting immunological inertness of resting spores. In contrast, hyphae induced the expression of multiple genes encoding cytokines and chemokines (Table 4.2). TNF-α, IL-8 and IL-6-encoding genes were up-regulated in a hypha-specific manner. No differential regulation of TLR2/4 was observed after co-incubation with either hyphae or spores. An early phase induction of PTX3 expression was observed commencing at 1 hour and peaking at 3 hours' co-incubation in response to conidia, but not hyphae. The authors found that CCR2 expression was down-regulated by germinating conidia. This chemokine is required for neutrophil recruitment and has been reported as down-regulated in bronchial epithelial cells in response to LPS challenge.

Cortez *et al.* (2006) studied the early response of human monocytes to *A. fumigatus*, assessing host gene expression at 0, 2, 4 and 6 hours of co-incubation. In general terms the functions identified as crucial during this interaction included phagocytosis, endocytosis, cytokine and chemokine signalling (and relevant receptors), cell adhesion and cell–cell communication. Having established a minimum of 80% viability of donor monocytes for all of the five independent donor samples, analysis of phagocytosis rates indicated progressive escalation of phagocytosis rates from 0 to 80% within the studied 6 hours time-frame. During the analysis 1827 genes were differentially expressed, these could be assigned to seven clusters according to their temporal regulation and indicated that stage-specific gene expression occurs in response to conidial exposure. Two hundred and thirty-nine genes could be functionally assigned as having immunological involvement and of these, 125 were up-regulated and 114 were down-regulated. An interesting pattern of chemokine and chemokine receptor gene expression was observed whereby CCL3 and CCL20 were early responders, CXCL2 and CCL4 climbed steadily and CCR3 was a late responder. This differential temporal regulation likely indicates differential responses to swelling and germination of spores within the 6 hours timeframe, although morphogenesis of the fungus was not monitored.

Both of the Loeffler and Cortez studies found that cytokine expression was not evoked immediately upon conidial exposure and phagocytosis. In both studies up-regulation of IL1B was reported, but this did not approach mathematical significance until 3 to 4 hours post exposure. This might be due to the immunological silencing of conidiospores provided by the RodA hydrophobin, which is progressively shed during germination to reveal pathogen associated molecular patterns (Aimanianda *et al.*, 2009). Presumably, given the timescale of phagocytosis measured in both studies (which was remarkably similar), a proportion of the host-mediated recognition occurs inside the phagolysosome.

Aspergillus fumigatus–neutrophil interactions

Sugui *et al.* (2008) analysed fungal transcriptional profiles of *A. fumigatus* conidia and hyphae exposed to neutrophils, deriving host cells either from normal donors or from chronic granulomatous disease (CGD) patients. Upon exposure to either normal or CGD neutrophils, 244 genes were up-regulated in conidia. Several of the gene products are involved in the degradation of fatty acids, peroxisome function and the glyoxylate cycle which suggests that, like *C. albicans* yeast cells, conidia exposed to neutrophils reprogram their metabolism to adjust to the host environment. In addition, the mRNA levels of four genes encoding proteins putatively involved in iron/copper assimilation were found to be higher in conidia and hyphae exposed to normal neutrophils compared to those exposed to CGD neutrophils, indicative of alternative exposures to oxidative stress in healthy or diseased cells. Null mutants constructed of up-regulated genes showed relevant phenotypes *in vitro*; however, none of the mutants showed reduced resistance to neutrophil attack.

Cryptococcus–macrophage interactions

Like *C. albicans* and *A. fumigatus*, *C. neoformans* can be efficiently phagocytosed and killed by macrophages. However, by virtue of its unique polysaccharide capsule this pathogen can also inhibit and evade phagocytosis, and survive and/or escape the phagosome of activated macrophages. The intracellular environment of phagocytic cells, while hostile, is also postulated as a safe-haven from exogenous antifungal effectors such as complement, antibodies, serum, and alveolar immune factors, or even antifungal drugs. It has also been proposed that fungi are transported as in-dwelling passengers thereby accessing a route for *C. neoformans* to enter the bloodstream and lymphatic system, enabling dissemination from the lung to distant organs, including its propensity to invade the central nervous system.

C. neoformans infections initiate via spore inhalation which is later followed by disseminated infection, the interaction with alveolar macrophages and with monocytes are therefore the most relevant. Fan *et al.* (2005) examined the transcriptional response of opsonized *C. neoformans* exposed to LPS and IFNƔ-activated macrophages. Selectively studying the internalized fungal population Fan *et al.* (2005) identified 123 up-regulated pathogen genes, among which 15 were restricted to early phase up-regulation, 35 demonstrated a consistently up-regulated profile and 73 were induced at later time points. Amongst 157 down-regulated gene functions, 17 were relevant to early-phase exposure, 23 consistently down-regulated and 117 genes down-regulated at the later time point. Thus, the majority of pathogen gene regulation was evident at the later time point.

Nineteen genes modulating carbohydrate, amino acid, iron, ammonium, nucleotide, phosphate and fatty acid transport were up-regulated, eight of which were sustained for the duration of macrophage exposure. This signature might indicate nutrient limitation in the phagosome. Also evident were responses to nitrosative and oxidative stress and lipid catabolism. Similar to findings in *C. albicans* an isocitrate lyase-encoding gene function was up-regulated at the 2-hour time point. Both of the *C. neoformans* laccase encoding genes were also up-regulated, indicating a renewed biosynthesis of melanin by internalized cryptococcal cells. Although the capsule biosynthesising *CAP10* was induced, other capsule biogenesis genes were not, an observation which might be explained by their up-regulation in the basal medium used for culture of the reference RNA sample, or indeed by an alternative (e.g. post-translational) mode of regulation. Genes involved in modulation of capsule structure were, however, found to be differentially regulated. Wholesale repression of the translation machinery, including ribosomal proteins, translation initiation factors, elongation factors and tRNA synthetases was observed.

In an almost technically identical infection scenario Griffiths *et al.* (2012) studied gene expression of internalized fungi at a 6 hour timepoint of incubation with macrophages. These authors noted the similarity of their transcriptome data to those described by Fan *et al.* (2005) for example, genes encoding functions for amino acid and phosphate transport, fatty acid and lipid metabolism, translational machinery, and histone and chromatin remodelling were differentially expressed in both studies.

Addressing the evolutionary origins of virulence factors in *C. neoformans*, Derengowski *et al.* (2013) performed a parallel analysis of fungal responses to macrophage exposure and ameobae. Following a 6-hour exposure period, 293 genes were differentially expressed in response to macrophage exposure. Amongst important regulatory signatures the switch from glycolysis to gluconeogenesis and fatty acid catabolism via the glyoxylate cycle were identified. This study found that induction of fructose 1,6-bisphosphate and phoshoenolpyruvate carboxykinase gene functions occurred, being therefore indicative of gluconeogenesis as the favoured route of energy production. This study also found up-regulation of isocitrate lyase and malate synthase suggesting that derivation of acetyl coA from fatty acid beta oxidation drives the glyoxylate cycle. All steps of the fatty acid beta oxidation pathway were found to be up-regulated at the level of transcript in this study.

Perspectives

Interactions between pathogens and cells of the innate immune system are commonplace during infection and commensalism. The growing repertoire of accessible information reveals many similarities between the responses of both host and pathogen during such interactions. Amongst fungal pathogens the induction of two-carbon utilizing metabolism is a common theme occurring in *Candida* cells interacting with both epithelial cells and macrophages and in *A. fumigatus* and *C. neoformans* interactions with neutrophils and monocytes respectively. Where undertaken, directly comparative analyses (such as that conducted by Park *et al.* (2009) who compared epithelial and endothelial cell responses to *C. albicans* challenge), can be highly informative. Park *et al.* (2009) found that the transcript levels of only 29 *C. albicans* genes changed significantly in response to both epithelial and endothelial cells at 45 minutes, and the transcript levels of all of these genes were decreased. Closer analysis revealed that the response of *C. albicans* to oral epithelial cells is substantially different from its response to endothelial cells. *C. albicans* hyphae were endocytosed more slowly by epithelial cells and caused less damage to these cells than endothelial cells. Also, the transcriptional response of *C. albicans* to epithelial cells was different from its response to endothelial cells. Interestingly, the *C. albicans* genes whose transcripts were increased in response to either of these cell types were also different from those that have been reported to be induced by exposure to monocytes or neutrophils. These divergent responses to different types of host cells are likely to occur more widely amongst fungal pathogens and act to promote survival and proliferation in diverse anatomic sites within the host.

References

Aimanianda, V., Bayry, J., Bozza, S., Kniemeyer, O., Perruccio, K., Elluru, S.R., Clavaud, C., Paris, S., Brakhage, A., Kaveri, S.V., Romani, L., and Latge, J.-P. (2009). Surface hydrophobin prevents immune recognition of airborne fungal spores. Nature *46*, 1117–1123.

Barker, K.S., Liu, T., and Rogers, D. (2005). Co-culture of THP-1 human Mononuclear cells with *Candida albicans* results in pronounced changes in host gene expression J. Infect. Dis. *192*, 901–912.

Barker, K.S., Park, H., Phan, Q.T., Xu, L., Homayouni, R., Rogers, P.D., and Filler, S.G. (2008). Transcriptome profile of the vascular endothelial cell response to *Candida albicans*. J. Infect. Dis. *198*, 193–202.

Cairns, T., Minuzzi, F., and Bignell, E. (2010). The host-infecting fungal transcriptome. FEMS Microbiol. Lett. *307*, 1–11.

Casadevall, A., and Pirofski, L.A. (2003). The damage-response framework of microbial pathogenesis. Nat. Rev. Microbiol. *1*, 17–24.

Casadevall, A., and Pirofski, L.A. (2009). Virulence factors and their mechanisms of action: the view from a damage-response framework. J. Water Health 7 (Suppl. 1), S2-S18.

Cortez, K.J., Lyman, C.A., Kottilil, S., Kim, H.S., Roilides, E., Yang, J., Fullmer, B., Lempicki, R., and Walsh, T.J. (2006). Functional genomics of innate host defense molecules in normal human monocytes in response to *Aspergillus fumigatus*. Infect. Immun. *74*, 2353–2365.

Derengowski, L.S., Paes, H.C., Albuquerque, P., Tavares, A.H., Fernandes, L., Silva-Pereira, I., and Casadevall, A. (2013). The transcriptional response of *Cryptococcus neoformans* to ingestion by *Acanthamoeba castellanii* and macrophages provides insights into the evolutionary adaption to the mammalian host. Eukaryot. Cell *12*, 761–774.

Fradin, C., Kretschmar, M., Nichterlein, T., Gaillardin, C., d'Enfert, C., and Hube, B. (2003). Stage-specific gene expression of *Candida albicans* in human blood. Mol. Microbiol. *47*, 1523–1543.

Fradin, C., De Groot, P., MacCallum, D., Schaller, M., Klis, F., Odds, F.C., and Hube, B. (2005). Granulocytes govern the transcriptional response, morphology and proliferation of *Candida albicans* in human blood. Mol. Microbiol. *56*, 397–415.

Fradin, C., Mavor, A.L., Weindl, G., Schaller, M., Hanke, K., Kaufmann, S.H., Mollenkopf, H., and Hube, B. (2007). The early transcriptional response of human granulocytes to infection with *Candida albicans* is not essential for killing but reflects cellular communications. Infect. Immun. *75*, 1493–1501.

Griffiths, E.J., Hu, G., Fries, B., Caza, M., Wang, J., Gsponer, J., Gates-Hollingsworth, M.A., Kozel, T.R., De Repentigny, L., and Kronstad, J.W. (2012). A defect in ATP-citrate lyase links acetyl-CoA production, virulence factor elaboration and virulence in *Cryptococcus neoformans*. Mol. Microbiol. *86*, 1404–1423.

Gomez, P., Hackett, T.L., Moore, M.M., Knight, D.A., and Tebbutt, S.J. (2010). Functional genomics of human bronchial epithelial cells directly interacting with conidia of *Aspergillus* fumigatus. BMC Genomics *11*, 358.

Huffnagle, G.B., McNeil, L.K., McDonald, R.A., Murphy, J.W., Toews, G.B., Maeda, N., and Kuziel, W.A. (1999). Cutting edge: Role of C-C chemokine receptor 5 in organ-specific and innate immunity to *Cryptococcus neoformans*. J. Immunol. *163*, 4642–4646.

Ikuta, T., Bhawal, U.K., Tsushima, K., Aoki, A., Kuboyama, N., and Abiko, Y. (2012). Identification by DNA microarray of genes involved in *Candida albicans*-treated gingival epithelial cells. J. Oral Pathol. Med. *41*, 769–778.

Jong, A., Wu, C., Zhou, W., Chen, H., and Huang, S. (2008). Infectomic analysis of gene expression profiles of human brain microvascular endothelial cells infected with *Cryptococcus neoformans*. J. Biomed. Biotechnol. 2008, Article ID 375620.

Kim, H.S., Choi, E.H., Khan, J., Roilides, E., Francesconi, A., Kasai, M., Sein, T., Schaufele, R.L., Sakurai, K., Son, C.G., Greer, B.T., Chanock, S., Lyman, C.A., and Walsh, T.J. (2005). Expression of genes encoding innate host defense molecules in normal human monocytes in response to *Candida albicans*. Infect. Immun. *73*, 3714–3724.

Kraus, P.R., Boily, M., Giles, S.S., Stajich, J.I., Allen, A., Cox, G.M., Dietrich, F.S., Perfect, J.R., and Heitman, J. (2004). Identification of *Cryptococcus neoformans* temperature-regulated genes with a genomic-DNA microarray. Eukaryot. *Cell* 3, 1249–1260.

Lim, C.S., Rosli, R., Seow, H.F., and Chong, P.P. (2011). Transcriptome profiling of endothelial cells during infections with high and low densities of *C. albicans* cells. Int. J. Med. Microbiol. *301*, 536–546.

Loeffler, J., Haddad, Z., Bonin, M., Romeike, N., Mezger, M., Schumacher, U., Kapp, M., Gebhardt, F., Grigoleit, G.U., Stevanovic, S., Einsele, H., and Hebart, H. (2009). Interaction analyses of human monocytes co-cultured with different forms of *Aspergillus fumigatus*. J Med. Microbiol. *58*, 49–58.

Lorenz, M.C., and Fink, G.R. (2001). The glyoxylate cycle is required for fungal virulence. Nature *412*, 83–86.

Lorenz, M.C., Bender, J.A., and Fink, G.R. (2004). Transcriptional response of *Candida albicans* upon internalization by macrophages. Eukaryot. Cell 3, 1076–1087.

Moran, G.P., Coleman, D.C., and Sullivan, D.J. (2011). Comparative genomics and the evolution of pathogenicity in human pathogenic fungi. Eukaryot. Cell *10*, 34–42.

Naglik, J., Moyes, D.L., Wächtler, B., and Hube, B. (2011). *Candida albicans* interactions with epithelial cells and mucosal immunity. Microbes Infect. *13*, 963–976.

Oosthuizen, J.L., Gomez, P., Ruan, J., Hackett, T.L., Moore, M.M., Knight, D.A., and Tebbutt, S.J. (2011). Dual organism transcriptomics of airway epithelial cells interacting with conidia of *Aspergillus fumigatus*. PLoS. One. *6*, e20527.

Osherov, N. (2012). Interaction of the pathogenic mold *Aspergillus* fumigatus with lung epithelial cells. Front. Microbiol. 3, 346.

Park, H., Liu, Y., Solis, N., Spotkov, J., Hamaker, J., Blankenship, J.R., Yeaman, M.R., Mitchell, A.P., Liu, H., and Filler, S.G. (2009). Transcriptional responses of *Candida albicans* to epithelial and endothelial cells. Eukaryot. Cell *8*, 1498–1510.

Phan, Q.T., Fratti, R.A., Prasadarao, N.V., Edwards, J.E., Jr., and Filler, S.G. (2005). N-cadherin mediates endocytosis of *Candida albicans* by endothelial cells. J. Biol. Chem. *280*, 10455–10461.

Phan, Q.T., Myers, C.L., Fu, Y., Sheppard, D.C., Yeaman, M.R., Welch, W.H., Ibrahim, A.S., Edwards, J.E., Jr., and Filler, S.G. (2007). Als3 is a *Candida albicans* invasin that binds to cadherins and induces endocytosis by host cells. PLoS. Biol. *5*, e64.

Romani, L. (2011). Immunity to fungal infections. Nat. Rev. Immunol. *11*, 275–288.

Rubin-Bejerano, I., Fraser, I., Grisafi, P., and Fink, G.R. (2003). Phagocytosis by neutrophils induces an amino acid deprivation response in *Saccharomyces cerevisiae* and *Candida albicans*. Proc. Natl. Acad. Sci. U.S.A. *100*, 11007–11012.

Sandovsky-Losica, H., Chauhan, N., Calderone, R., and Segal, E. (2006). Gene transcription studies of *Candida albicans* following infection of HEp2 epithelial cells. Med. Mycol. *44*, 329–334.

Schaller, M., Zakikhany, K., Weindl, G., Borelli, C., Korting, H.C., and Hube, B. (2008). Characterization of the interaction between *Candida albicans* and host cells: *In vitro* model using reconstituted human skin and mucosa. Hautarzt *59*, 964–970.

Sharon, H., Amar, D., Levdansky, E., Mircus, G., Shadkchan, Y., Shamir, R., and Osherov, N. (2011). PrtT-regulated proteins secreted by *Aspergillus* fumigatus activate MAPK signaling in exposed A549 lung cells leading to necrotic cell death. PLoS. One. *6*, e17509.

Sohn, K., Senyurek, I., Fertey, J., Konigsdorfer, A., Joffroy, C., Hauser, N., Zelt, G., Brunner, H., and Rupp, S. (2006). An *in vitro* assay to study the transcriptional response during adherence of *Candida albicans* to different human epithelia. FEMS Yeast Res. *6*, 1085–1093.

Zakikhany, K., Naglik, J.R., Schmidt-Westhausen, A., Holland, G., Schaller, M., and Hube, B. (2007). *In vivo* transcript profiling of *Candida albicans* identifies a gene essential for interepithelial dissemination. Cell Microbiol. *9*, 2938–2954.

Animal Models of Human Fungal Infection

5

Donna M. MacCallum

Abstract

Fungal infection is a major cause of morbidity and mortality, with most infections caused by dermatophytes, *Candida* species, *Aspergillus* species, *Cryptococcus neoformans* and *Pneumocystis jirovecii*. Whilst dermatophytes can be considered primary pathogens, the majority of fungal infections are caused by opportunistic fungal species. *Candida* infections tend to originate from commensal sources, but *Aspergillus*, *Cryptococcus* and *Pneumocystis* infections are acquired from the environment. These infections are found in severely ill patients or in immunosuppressed individuals, particularly those with HIV/AIDS. Mortality associated with these infections remains high, partially due to difficulties in diagnosis, but also due to limited therapeutic choices. In order to facilitate development of these badly needed diagnostics and antifungal agents, a greater understanding of fungal pathogenesis is required. Animal models provide opportunities to investigate fungal disease initiation and progression and to evaluate novel antifungal agents. Animal models of dermatophyte, *Candida*, *Aspergillus*, *Cryptococcus* and *Pneumocystis* infections are described and their contribution to our current understanding of these important fungal infections discussed.

Why model fungal infections?

Fungi cause a spectrum of disease including skin and nail infections, superficial mucosal lesions (oral and vaginal), lung infections, central nervous system infections, through to deep-seated and invasive infections. Whilst some fungi, including dermatophytes, are primary pathogens, the majority are opportunistic, relying upon susceptible hosts to allow initiation of infection and disease. This latter group includes the major fungal pathogens worldwide: *Candida* species, *Aspergillus* species, *Cryptococcus neoformans* and *Pneumocystis jirovecii* (Brown et al., 2012).

The most common human fungal infections are superficial skin and nail infections, affecting approximately one quarter of the world's population (Havlickova *et al.*, 2008), oral thrush, affecting approximately 50% of HIV positive individuals (Bendick *et al.*, 2002; Felix and Wray, 1993; Gabler *et al.*, 2008; Schulten *et al.*, 1989), and vaginal thrush, affecting up to three quarters of women (Sobel *et al.*, 1998; Sobel, 2007). However, fungi also have the ability to cause invasive infection, and are responsible for approximately 10% of all bloodstream infections (Wisplinghoff *et al.*, 2004). The major causes of invasive fungal infection worldwide are *Aspergillus* spp., *Candida* spp., *Cryptococcus neoformans* and *Pneumocystis jirovecii* (Brown *et al.*, 2012). In North America, between 2004 and 2008, *Candida* species were responsible for 73% of invasive fungal infection (Azie *et al.*, 2012). *Candida albicans* was the most commonly isolated *Candida* species, accounting for 48% of infections, followed by *Candida glabrata* (25%), *Candida parapsilosis* (13%), *Candida tropicalis* (8%) and *Candida krusei* (3%). *Aspergillus* species accounted for 13% of all invasive infections, with *Aspergillus fumigatus* (54%), *Aspergillus flavus* (7%), *Aspergillus niger* (7%) and *Aspergillus terreus* (3%) the major causes of infections (Azie *et al.*, 2012). Mortality rates for invasive fungal infections range from 20% to 95% (Brown *et al.*, 2012).

With the majority of invasive fungal infections caused by opportunistic fungal species, increasing numbers of patients suffering immunosuppression and undergoing invasive treatments, e.g. for cancers and organ transplants, mean that there is an ever increasing population at risk of invasive fungal infection. However, these fungal infections have non-specific symptoms, leading to difficulties in diagnosis. Coupled with a limited repertoire of antifungal agents, delays in initiation of antifungal therapy and emergence of antifungal drug resistance have resulted in high mortality rates for invasive fungal infection. With an urgent medical need for the development of new and more efficient diagnostics and therapies for fungal infection a greater understanding of fungal pathogenesis is required.

Although a great deal of preliminary research on fungal pathogenesis can be carried out by laboratory experiment, infection modelling requires the involvement of a host organism. It is only in a whole organism that the complex host–fungus interactions that determine whether or not disease will occur can be investigated.

Experimental infection models allow disease development to be followed from the moment that fungal cells are introduced into the host until the host becomes severely ill. A good infection model should be reproducible, relatively easy to set up and should reproduce the major clinical symptoms seen in the human disease. It is also an added advantage if the model is cost-effective. Models which satisfy these conditions allow further in-depth investigation of fungal pathogenesis to be carried out and, subsequently, allow inferences about human disease to be made. This chapter describes and discusses the various infection models used to model major fungal pathogen infections, specifically dermatophyte, *Candida*, *Aspergillus*, *Cryptococcus* and *Pneumocystis* infections.

Dermatophyte infections

Superficial skin and nail infections are the most common human fungal infections, affecting approximately one-quarter of the world's population (Havlickova *et al.*, 2008). These common infections include athletes' foot (tinea pedis), scalp ringworm (tinea tonsurans), other dermatophytosis (ringworm) and *Malassezia*-associated infections. The majority of skin infections are caused by dermatophytes (Foster *et al.*, 2004; Havlickova *et al.*, 2008), although *Candida* species can cause skin infections in certain individuals, including nappy-associated infections in babies and chronic mucocutaneous infections in genetically susceptible individuals. In humans, the fungal species most commonly associated with dermatophytosis are *Trichophyton* spp., particularly *Trichophyton rubrum* and *Trichophyton mentagrophytes*, *Microsporum* spp. and *Epidermophyton floccosum* (Foster *et al.*, 2004; Havlickova *et al.*, 2008).

Skin infection models

The guinea pig (*Cavia porcellus*) has been the most popular model of dermatophytosis as these animals demonstrate similar disease progression and pathology to that seen in humans (Achterman and White, 2012b; Greenberg *et al.*, 1976; Treiber *et al.*, 2001). Most studies have focused on *Trichophyton mentagrophytes* (Chittasobhon and Smith, 1979; Gaughran *et al.*, 1960; Green and Balish, 1979, 1980; Greenberg and Kerbs, 1980; Hernandez *et al.*, 1986; Itoyama *et al.*, 1997; Knight, 1972, 1973; Nagino *et al.*, 2000; Odds *et al.*, 2004; Saunte *et al.*, 2008), and *Microsporum canis* (Cavalcanti *et al.*, 2002; Chittasobhon and Smith, 1979; Mignon *et al.*, 1999; Odds *et al.*, 2004; Saunte *et al.*, 2008), with many fewer studying *Epidermophyton* infection (Chittasobhon and Smith, 1979). However, while this model works well with zoophilic (animal-related) species, it is less effective in modelling infections caused by anthropophilic (restricted to humans) dermatophytes (Achterman and White, 2012a). The guinea pig has also been used to model *Malassezia* infection (Van Cutsem *et al.*, 1990).

In the majority of dermatophyte studies, infection is initiated by direct application of fungal cells or spores to the skin of the guinea pig (Chittasobhon and Smith, 1979; Hunjan and Cronholm, 1979; Mignon *et al.*, 1999; Odds *et al.*, 2004; Saunte *et al.*, 2008). Prior to infection, areas on the back or flanks of the guinea pig are shaved and scarified (gently scraped or abraded). The fungal inoculum is either applied as a saline or water suspension and gently rubbed into the area (Chittasobhon and Smith, 1979; Hunjan and Cronholm, 1979; Saunte *et al.*, 2008) or can be mixed with honey prior to applying to the scarified area (Mignon *et al.*, 1999; Odds *et al.*, 2004). The numbers of spores administered to initiate infection has varied massively, ranging from 1×10^2 to 1×10^8 (Foster *et al.*, 2004; Ghannoum *et al.*, 2004; Greenberg *et al.*, 1976; Grumbt *et al.*, 2011; Itoyama *et al.*, 1997; Staib *et al.*, 2010), possibly reflecting the variation seen between different fungal strains in their ability to cause disease (Hay *et al.*, 1983). The majority of strains use an inoculum towards the higher end of the range mentioned. The infected area is then either left uncovered (Chittasobhon and Smith, 1979; Kerbs and Allen, 1978; Saunte *et al.*, 2008) or covered and sealed (to model occlusion) (Cavalcanti *et al.*, 2002; Saunte *et al.*, 2008); however, studies have found that occlusion makes little difference in establishment of infection (Kerbs and Allen, 1978; Saunte *et al.*, 2008).

The guinea pig infection model has been used to evaluate novel antifungal therapies for dermatophytosis (Gaughran *et al.*, 1960; Ghannoum *et al.*, 2004; Itoyama *et al.*, 1997; Nagino *et al.*, 2000; Niwano *et al.*, 1995; Saunte *et al.*, 2007; Saunte *et al.*, 2008; Uchida and Yamaguchi, 1984, 1994, 1999), prophylaxis (Treiber *et al.*, 2001), vaccine possibilities (Descamps *et al.*, 2003) and to model re-infection by dermatophytes (Chittasobhon and Smith, 1979; Hernandez *et al.*, 1986). By investigating dermatophyte skin and systemic infections van Cutsem and Janssen (1984) proposed that recurrent dermatophyte infections may actually originate from residual deep-seated infection sites within the host (Van Cutsem and Janssen, 1984). The guinea pig has also been used to model athlete's foot (tinea pedis) (Fujita and Matsuyama, 1987; Niwano *et al.*, 1995). In this model the guinea pig foot pad is infected with *Trichophyton* and then covered with a dressing to allow infection to progress (Fujita and Matsuyama, 1987; Niwano *et al.*, 1995).

In comparison to other fungal infections (see below) the mouse has been a much less popular host for fungal skin infection modelling. This is mostly due to the inherent resistance of some mouse strains (C57BL/6, DBA/2, CBA) to dermatophyte infection, although BALB/c and Swiss Webster (CFW) mice are susceptible (Calderon and Hay, 1984a, 1984b; Hay *et al.*, 1983; Sharma *et al.*, 2011). Similar to the guinea pig model, mice are shaved and the skin abraded prior to application of fungal cells or spores to the area (Hay *et al.*, 1983), with oestradiol pretreatment used in some studies to render the mice slightly immunodeficient prior to infection (Odds *et al.*, 2004; Sharma *et al.*, 2011). *Trichophyton mentagrophytes* infection is, again, the most commonly modelled infection in this host (Calderon and Hay, 1984a, 1984b; Hay *et al.*, 1983; Venturini *et al.*, 2012).

Using dermatophyte models, investigators have been able to demonstrate that different strains of the same dermatophyte species vary in their ability to cause infection (Hay *et al.*, 1983). This model has also been used to investigate dermatophyte reinfection following a primary infection. Mice have been used in attempts to model athlete's foot, with fungi administered subcutaneously into the foot pad (Venturini *et al.*, 2012). Swiss Webster mice are susceptible to this infection, but athymic nude BALB/c mice are resistant.

Perhaps the best model of human dermatophytosis is the human infection model. The first report of a human model of dermatophyte infection was published in 1909 (Bloch and Massini, 1909), where a human host was infected with *Microsporum canis*. Since this time, a number of other studies have been carried out using human volunteers (reviewed in Shimamura *et al.*, 2012). The most common genera used in these infections is *Trichophyton* (Aly *et al.*, 1992; Knight, 1972, 1973; Reinhardt *et al.*, 1974), although *Arthroderma* (Staib *et al.*, 2010) and *Malassezia* (Tebo *et al.*, 2000) studies have also been carried out. In these studies the fungus was applied to the ankles, thighs or forearms (Aly *et al.*, 1992; Knight, 1972, 1973; Reinhardt *et al.*, 1974), with infection resolving, without therapy, within 90 days.

A few studies have used less common hosts to model dermatophyte infections, including rats (Green *et al.*, 1983), rabbits (Faergemann, 1979) and cats (DeBoer and Moriello, 1994; Tabart *et al.*, 2007). The mini-host *Galleria mellonella* (wax moth) larvae, which have been used successfully to model other fungal infections (see below), have been evaluated as an alternative host for dermatophyte infection, but were found to be inadequate (Achterman *et al.*, 2011).

Candida infections

Candida species are normally harmless members of the normal human gastrointestinal (GI), oral and vaginal microbial flora (reviewed by MacCallum, 2010). The most common species identified in healthy, human carriage in the GI tract and oral cavity are *Candida albicans*, *Candida parapsilosis*, *Candida glabrata*, *Candida tropicalis* and *Candida krusei* (Belazi *et al.*, 2005; Ben-Aryeh *et al.*, 1995; Bougnoux *et al.*, 2006; Campisi *et al.*, 2002; Kusne *et al.*, 1994; Scanlan and Marchesi, 2008; Thaweboon *et al.*, 2008; Wang *et al.*, 2006) and in the vagina are *C. albicans* and *C. glabrata* (Beltrame *et al.*, 2006; Dan *et al.*, 2006; de Leon *et al.*, 2002; Grigoriou *et al.*, 2006; Paulitsch *et al.*, 2006; Pirotta and Garland, 2006). However, *Candida* species are also a very common cause of human fungal infection, capable of adopting an alternative lifestyle to cause opportunistic infection in hosts with altered physiological or immune responses. Infections caused by *Candida* species range from self-limiting, superficial mucosal lesions (commonly referred to as thrush), chronic and/or recurrent mucosal, skin and nail infections, through to life-threatening invasive or disseminated infection (Havlickova *et al.*, 2008; Kirkpatrick, 1989, 2001; MacCallum, 2010; Odds, 1988).

In humans, the commonest *Candida* infections are superficial infections of the mucosa, skin and nails (Kirkpatrick, 1989, 2001; Kisand *et al.*, 2011; Puel *et al.*, 2010, 2011). Oral thrush is common in both babies and in the elderly, but is also found in HIV-positive individuals and cancer patients (reviewed in MacCallum, 2010; Richardson and Warnock, 1997). Denture stomatitis is also a significant infection, occurring in approximately 60% of denture wearers (Daniluk *et al.*, 2006; Figueiral *et al.*, 2007). The majority of these infections are caused by *C. albicans* (58%), with the remainder caused by *C. parapsilosis*, *C. tropicalis*, *C. glabrata* and *C. krusei* (Lyon *et al.*, 2006; Pires-Goncalves *et al.*, 2007). Vaginal thrush affects approximately 75% of women of child-bearing age (Sobel *et al.*, 1998, 2007), with *C. albicans* the most commonly cause (Asticcioli *et al.*, 2009; Corsello *et al.*, 2003; Guzel *et al.*, 2011; Kennedy and Sobel, 2010; Paulitsch *et al.*, 2006; Sobel, 2007). A further form of candidiasis which affects the mucous membranes, skin and nails is chronic mucocutaneous candidiasis; however, unlike other forms of candidiasis, this infection is due to a genetic defect and almost every infection is caused by *C. albicans* (Kirkpatrick, 1989, 2001; Kisand *et al.*, 2011; Puel *et al.*, 2010, 2011).

The most serious infections caused by *Candida* species are invasive or disseminated infections. *Candida* species cause ~10% of all bloodstream infections and one-fifth of all intensive care unit (ICU) bloodstream infections (Markogiannakis *et al.*, 2008; Orsi *et al.*, 2006; Sarvikivi *et al.*, 2008). These infections have high mortality rates (45–75%) (Brown *et al.*, 2012) due, in part, to difficulties in diagnosing the infection and limited antifungal therapy choices. These infections occur in patients who are already seriously ill, with major risk factors including long term stays in ICU, surgery (especially abdominal surgery) and neutropenia (reviewed by MacCallum, 2010). The majority (90%) of invasive fungal infections are caused by *C. albicans*, *C. glabrata*, *C. tropicalis*, *C. parapsilosis* and *C. krusei* (MacCallum, 2010; Wisplinghoff *et al.*, 2006).

Modelling *Candida* infections

Although larger animals have been used to study *Candida* infections, e.g. macaques (Budtz-Jorgensen, 1971; Steele *et al.*, 1999), piglets (Andrutis *et al.*, 2000), rabbits (Filler *et al.*, 1991; Lyman *et al.*, 1999; Polanco *et al.*, 1999) and guinea pigs (Fransen *et al.*, 1984; Van Cutsem and Thienpont, 1971), the majority of studies have used rodent infection models. Mini-hosts are growing in popularity as *Candida* infection models. *Galleria mellonella* (wax moth) larvae (Brennan *et al.*, 2002; Cotter *et al.*, 2000; Dunphy *et al.*, 2003; Fuchs *et al.*, 2010; Kavanagh and Reeves, 2004) have been used to assay virulence of *C. albicans* strains and have been shown to produce similar results to those found in mouse models (Brennan *et al.*, 2002; Cotter *et al.*, 2000; Dunphy *et al.*, 2003; Fuchs *et al.*, 2010). The fruit fly, *Drosophila melanogaster* (Alarco *et al.*, 2004;

Chamilos *et al.*, 2006, 2009; Glittenberg *et al.*, 2011a, 2011b), and embryonated eggs have also be used as hosts for virulence testing (Gow *et al.*, 2003; Jacobsen *et al.*, 2011), and the zebra fish (*Danio rerio*) is gaining in popularity as a host to investigate and visualize virulence differences and host immune responses in invasive *Candida* infections (Brothers *et al.*, 2011; Chao *et al.*, 2010; Chen *et al.*, 2006; Levraud *et al.*, 2008; Lo *et al.*, 1997; Meeker and Trede, 2008). However, a major disadvantage of mini-host models, with the exception of *G. mellonella*, is the requirement for lower incubation temperatures, which prevents accurate mimicking of human infection.

Modelling *Candida* mucosal infections

The main models of *Candida* oral and vaginal infections have been developed in rats and mice (De Bernardis *et al.*, 1999; Farah *et al.*, 2002; Fidel and Sobel, 1999; Kamai *et al.*, 2001; Naglik *et al.*, 2008; Nett *et al.*, 2010; Rahman *et al.*, 2007; Samaranayake and Samaranayake, 2001). The procedures used in rats and mice are generally similar. However, the rat has the added advantage that denture-associated fungal biofilm formation can also be studied in this host (Nett *et al.*, 2010). Establishment of infection at mucosal sites generally requires treatment with either immunosuppressive agents, oestrogen or antibiotics prior to infection, or the use of germ-free animals (Fidel and Sobel, 1999; Kamai *et al.*, 2001; Samaranayake and Samaranayake, 2001). However, use of the nude (*Foxn1*^{nu}) mouse allows oral infection to be established without any immunosuppression or other pre-treatment (Farah *et al.*, 2002). Greater detail can be found in more extensive reviews of these infection models (De Bernardis *et al.*, 1999; Fidel and Sobel, 1999; Naglik *et al.*, 2008; Samaranayake and Samaranayake, 2001).

In order to assess *Candida* virulence in the murine oral infection model, mice are routinely pre-treated with corticosteroids and *Candida* cells administered into the oral cavity of anaesthetized animals either by applying a *Candida*-soaked cotton bud under the tongue or by applying the inoculum directly onto the teeth, gums and oral cavity (Naglik *et al.*, 2008; Rahman *et al.*, 2007; Samaranayake and Samaranayake, 2001). Virulence in this model is usually determined by fungal organ burden and histopathology.

Both rat and mouse models have been used to compare the virulence of *C. albicans* mutant strains and clinical isolates (Badrane *et al.*, 2008; Chiang *et al.*, 2007; Holbrook *et al.*, 1983; Nobile *et al.*, 2008; Park *et al.*, 2005a). Using these models, *C. albicans* mutant strains which are unable to switch between the yeast and hyphal growth forms were found to be unable to cause oral infection, demonstrating a requirement for yeast-hypha switching in oral infection (Nobile *et al.*, 2008). In addition, protein kinase Ck2 was also shown to be required for oropharyngeal *C. albicans* infections (Chiang *et al.*, 2007).

Mouse and rat models have also been developed to assay *Candida* virulence in vaginal infection. In these models rodents are maintained in pseudo-oestrus to maintain colonization and infection, which mimics pregnancy associated candidiasis (Chen and Kong, 2007; Clemons *et al.*, 2004; De Bernardis *et al.*, 1989; Sobel *et al.*, 1985). In rats, this generally involves surgery to remove the ovaries, with subsequent administration of oestrogen (Sobel *et al.*, 1985). Recently, however, a new rat model has been developed, similar to the mouse model, where oestrus is maintained through administration of oestrogen alone (Carrara *et al.*, 2010), increasing the ease of setting up the infection model. These models allow us to examine single vaginitis episodes; unfortunately a satisfactory model of recurrent, chronic vaginitis is not yet available. The virulence of both genetically modified *C. albicans* strains and clinical isolates have been compared in this model (Bader *et al.*, 2006; Fu *et al.*, 2008; Tavanti *et al.*, 2010; Taylor *et al.*, 2000).

In addition to assessing *C. albicans* virulence, this model has been used to examine virulence of other *Candida* species. As *C. glabrata* has also been associated with human vaginal infection, researchers have used the rat vaginitis model to evaluate the virulence of a *C. glabrata* petite mutant, discovering that the mutant appears more virulent that the parental strain (Ferrari *et al.*, 2011a). In addition, *C. parapsilosis* isolates were also assessed for their ability to cause vaginal infection in the rat model (De Bernardis *et al.*, 1989). In this study only a single isolate, recently obtained from a woman with active vaginal infection, was capable of initiating infection.

A major development in *Candida* virulence testing at mucosal surfaces occurred with the development of a concurrent murine oral and vaginal infection model (Rahman *et al.*, 2007). This model allows both oral and vaginal infections to be initiated in the same host at the same time, greatly reducing the numbers of animals required for virulence assays. A comparison of the virulence of three different *C. albicans* isolates in this model clearly demonstrated that *C. albicans* isolates were not equally virulent, with obvious differences in their ability to initiate mucosal infections (Rahman *et al.*, 2007).

Modelling *Candida* invasive infections

Mouse models of invasive fungal infection have been the most popular method for assessment of *Candida* virulence, although assays have also been carried out in rabbits, guinea pigs and rats. There are two major models of *Candida* invasive infection, the intravenous (i.v.) challenge model and the gastrointestinal (GI) colonization with subsequent dissemination model.

Intravenous challenge model

The mouse i.v. challenge model has been used to study *Candida* virulence since the 1960s and is both well characterized and reproducible (Louria *et al.*, 1963; MacCallum and Odds, 2005; Papadimitriou and Ashman, 1986). *Candida* cells are injected directly into the lateral tail vein, bypassing any requirement of the fungus to cross epithelial and endothelial barriers to gain entry into the bloodstream. In this model, which is similar to human invasive infection occurring with catheter involvement, fungal cells are found in all organs, but disease progresses only in the kidneys and brain, which depends upon inoculum level and mouse strain (Ashman *et al.*, 1996; MacCallum and Odds, 2005; Papadimitriou and Ashman, 1986). As disease progresses sepsis develops, eventually leading to the death of the mouse (MacCallum *et al.* 2005, 2009a; Spellberg *et al.*, 2005). Virulence in this model is determined by monitoring survival of infected mice and/or by quantifying fungal organ burdens at predetermined times post infection. Drug treatments can also be administered to the host to mimic underlying host conditions, e.g. immunosuppression (Anaissie *et al.*, 1993; Bistoni *et al.*, 1984; Calcagno *et al.*, 2003, 2004, 2005; de Bernardis *et al.*, 1990; de Repentigny *et al.*, 1992; Ferrari *et al.*, 2011a, 2011b; Girmenia *et al.*, 1996; Graybill *et al.*, 1995; Jacobsen *et al.*, 2010a; Kamran *et al.*, 2004; Louria *et al.*, 1967; Mellado *et al.*, 2000; Nakayama *et al.*, 2011) or diabetes (Louria *et al.*, 1967), with increased disease seen in response to both treatments.

The i.v. challenge model has allowed the virulence of different *Candida* species to be compared (Arendrup *et al.*, 2002; Bistoni *et al.*, 1984; Brieland *et al.*, 2001; de Repentigny *et al.*, 1992; Hasenclever and Mitchell, 1961; Koga-Ito *et al.*, 2010; Louria *et al.*, 1967; Mellado *et al.*, 2000). *C. albicans* is clearly the most virulent species (Arendrup *et al.*, 2002; Bistoni *et al.*, 1984; de Repentigny *et al.*, 1992; Hasenclever and Mitchell, 1961; Koga-Ito *et al.*, 2010), followed closely by *C. tropicalis* (Arendrup *et al.*, 2002; Bistoni *et al.*, 1984; de Repentigny *et al.*, 1992; Hasenclever and Mitchell, 1961; Koga-Ito *et al.*, 2010). In contrast, *C. krusei* and *C. parapsilosis* were unable to kill the infected animals, even at high inoculum levels, and fungi were eventually cleared from the host (Arendrup *et al.*, 2002; Bistoni *et al.*, 1984; Koga-Ito *et al.*, 2010). In immunosuppressed mice, *C. tropicalis* was able to cause disease, with infection progressing in the kidneys (Bistoni *et al.*, 1984; Fromtling *et al.*, 1987; Graybill *et al.*, 1995; Louria *et al.*, 1967; Mellado *et al.*, 2000). *C. parapsilosis* and *C. krusei* remained unable to initiate progressive infections, even with addition of immunosuppressive treatments (Bistoni *et al.*, 1984; Mellado *et al.*, 2000), although administration of a very high inoculum potentially allows some *C. parapsilosis* isolates to cause disease (Anaissie *et al.*, 1993; de Bernardis *et al.*, 1990).

Within each *Candida* species, clinical isolates showed considerable differences in virulence in the i.v. challenge model. This was true for *C. albicans* (de Repentigny *et al.*, 1992; MacCallum *et al.*, 2009b; Mellado *et al.*, 2000; Sampaio *et al.*, 2010), *C. tropicalis* (de Repentigny *et al.*, 1992; Fromtling *et al.*, 1987; Hasenclever and Mitchell, 1961; Louria *et al.*, 1967; Okawa *et al.*, 2008) and

C. parapsilosis (Cassone *et al.*, 1995; de Bernardis *et al.*, 1990). *C. glabrata* behaves very differently to the other *Candida* species in the mouse model of invasive infection. Although *C. glabrata* is maintained, or tolerated, at high levels in the kidneys of immunocompetent mice, mice did not die and there was little inflammation associated with the fungal cells in the organs (Arendrup *et al.*, 2002; Brieland *et al.*, 2001). Immunosuppression appears to increase virulence of *C. glabrata* in terms of higher fungal organ burdens, but mouse survival is only decreased with some *C. glabrata* isolates (Calcagno *et al.*, 2003, 2004, 2005; Jacobsen *et al.*, 2010a; Kamran *et al.*, 2004). However, because immunosuppression may allow development of invasive disease in *C. glabrata*-infected mice, these treatments have been added to the infection model to compare the virulence of genetically modified *C. glabrata*, with fungal burdens used as the virulence estimate (Ferrari *et al.*, 2011a,b; Kamran *et al.*, 2004; Nakayama *et al.*, 2011). Using immunosuppressed mice this model has demonstrated the importance of hypertonic stress responses, the cell wall integrity pathway and nitrogen starvation responses in *C. glabrata* virulence (Calcagno *et al.*, 2003, 2004, 2005). In addition, a petite mutant, strains expressing hyperactive alleles of the transcription factor gene *PDR1* and the *ace2* null mutant were shown to be more virulent than their parent strains (Ferrari *et al.*, 2009, 2011a; Kamran *et al.*, 2004). However, it should be noted that the hyper-virulent phenotype of the *C. glabrata ace2* null was completely lost in immunocompetent mice (MacCallum *et al.*, 2006).

Gastrointestinal colonisation and dissemination model.

The murine gastrointestinal colonization and dissemination model is believed to be a more accurate reflection of the events occurring in the human patient, with broad spectrum antibiotics allowing fungal overgrowth and later invasive therapies causing mucosal damage. Gut damage allows *Candida* to enter the bloodstream and to disseminate to the internal organs. Gastrointestinal models can either be set up in neonatal or adult mice. Intragastric infection of neonatal mice leads to persistent colonization, without any requirement for pre-treatment of the mice. However, to obtain colonization of adult mice the natural mouse gastrointestinal flora must first be removed by treatment with broad spectrum antibiotics. Adult mice can either be infected by gavage (intragastrically) or orally via their chow or drinking water. Subsequent treatment of *Candida* colonized mice with immunosuppressants and/or drugs which damage the gut wall allow fungal dissemination to occur (reviewed by Cole *et al.*, 1996; Naglik *et al.*, 2008). In the model, fungal colonization is highest in the stomach, caecum and small intestine (Clemons *et al.*, 2006; Mellado *et al.*, 2000; Sandovsky-Losica *et al.*, 1992; Wiesner *et al.*, 2001), reflecting clinical findings seen in human invasive infection. Colonization of the GI tract is routinely monitored by non-invasive faecal fungal counts, and following dissemination *Candida* cells can be cultured from the liver, kidneys and spleen (Clemons *et al.*, 2006; Koh *et al.*, 2008; Sandovsky-Losica *et al.*, 1992). Again, virulence differences can be seen between mouse strains (Cantorna and Balish, 1990). In the mouse model, similar to human patients, there is increased animal-to-animal variation, relative to the intravenous challenge model, requiring higher numbers of animals per group to obtain statistically significant results (Clemons *et al.*, 2006; Koh *et al.*, 2008; Sandovsky-Losica *et al.*, 1992).

Comparison of *Candida* species virulence in this model demonstrated that, although capable of establishing persistent colonization (Mellado *et al.*, 2000), *C. parapsilosis* had lower virulence than *C. albicans* and *C. tropicalis*, and there was little evidence of dissemination from the gut (Kennedy and Volz, 1983; Mellado *et al.*, 2000). In separate studies, *C. tropicalis* appeared more virulent than *C. albicans* in this model, with greater dissemination to the internal organs (Wingard *et al.*, 1980, 1982) and higher mortality rates (de Repentigny *et al.*, 1992; Wingard *et al.*, 1982). However, given the levels of variation observed in other models for the virulence of different strains of different *Candida* species, further isolates would need to be assayed before a definitive conclusion on the relative virulence of the two species could be made.

To date, only a limited number of *C. albicans* mutant strains have been tested in the gastrointestinal colonization and dissemination infection model, with only six mutants identified so far as contributing to virulence in this model (Skrzypek *et al.*, 2010; Szabo and MacCallum, 2011). However, this model was instrumental in demonstrating that the ability to switch between morphological forms may be more important for dissemination from the GI tract (Bendel *et al.*, 2003).

Mouse systemic models have allowed the emergence of drug resistance in an antifungal-treated host to be studied (Andes *et al.*, 2006), with ineffective antifungal dosing regimens allowing the emergence of resistant isolates, which was prevented with effective dosing. In addition, *C. albicans* strains with specific drug resistance mutations have been shown to be more resistant to antifungal therapy in the mouse, with the greatest resistance seen in strains with multiple mutations (MacCallum *et al.*, 2010; Park *et al.*, 2005b).

Systemic infection of knockout mouse strains has also contributed to our knowledge of host susceptibility to *Candida* infection. Pattern recognition receptor knockout mice demonstrated the importance of dectin-1, TLR2 and TLR4 in the recognition and control of systemic fungal infection (reviewed in Netea and Marodi, 2010). In another example, both TNF and IL6 were shown to be critical for normal host responses during disseminated infection, using both the intravenous and gastrointestinal infection models of *Candida* systemic infection (Mencacci *et al.*, 1998). In contrast, some host genes are only required for normal host responses in one model, or the other, e.g. IL-12 is important for the gastrointestinal model but is dispensable for virulence in the intravenous model (Ashman *et al.*, 2011), and the opposite is true for B cell knockout mice (Wagner *et al.*, 1996). Mouse strain background can be important when working with knockout mouse strains as different mouse strains vary in their susceptibility to systemic *Candida* infection (Ashman *et al.*, 1993, 1996; Marquis *et al.*, 1988).

Both mouse models of systemic *Candida* infection have been used to evaluate novel diagnostics prior to clinical trial (Nichterlein *et al.*, 2003; Uno *et al.*, 2007). Evaluation of new diagnostics in a host where systemic infection was reliably induced demonstrated that serological tests for *Candida* mannan and beta-glucan were more sensitive than nested PCR and blood culture for prediction of systemic infection in the mouse (Uno *et al.*, 2007). These tests have been further developed for clinical use, e.g. Platelia® *Candida* mannan antigen sandwich ELISA (Bio-rad Laboratories) and Fungitell® assay (Associates of Cape Cod, Inc.).

Mouse models of systemic *Candida* infection have also played a critical role in the early stages of antifungal drug development (Andes, 2005; Herrera and Guentzel, 1982), allowing *in vivo* antifungal efficacy to be determined. It is important, however, to consider that results obtained for antifungal agents may differ in mice and man, e.g. triazoles are metabolized more quickly in mice than in man due to differences in liver cytochrome P450 enzyme activity (Sugar and Liu, 2000). Inhibition of mouse cytochrome P450 activity increased azole levels and improved infection outcome (MacCallum and Odds, 2002b; Sugar and Liu, 2000), although this was mouse strain dependent (MacCallum and Odds, 2002a).

Potential antifungal antibodies and vaccines have also been evaluated in mouse model of systemic *Candida* infection (Cabezas *et al.*, 2010; Matthews *et al.*, 2003; Spellberg *et al.*, 2006). Mycograb, a human recombinant antibody against fungal HSP90, possessed antifungal activity in the mouse model and showed synergy when used in combination with amphotericin B (Matthews *et al.*, 2003). Mycograb has since become the first anti-*Candida* antibody to reach the clinic (Cabezas *et al.*, 2010). The search for vaccines to prevent life-threatening systemic *Candida* infection in at-risk patients also utilized the mouse infection model to evaluate whether vaccines could protect hosts from subsequent infection. A vaccine based upon Als1 and Als3 protected immunocompromised and immunocompetent mice from systemic candidiasis (Spellberg *et al.*, 2006). This vaccine, NDV-3, has shown success in phase 1 clinical trials (Schmidt *et al.*, 2012) and will be taken forward by NovaDigm for phase II clinical trials in 2013.

Aspergillus infections

Aspergillus exposure occurs through the inhalation of fungal spores from the environment, which can lead to invasive disease in the immunocompromised host (Hope *et al.*, 2005). The main patient groups at risk of invasive aspergillosis are severely immunocompromised individuals, such as those undergoing organ transplants, haematopoietic stem cell transplants, treatment for leukaemia, treatment with new immunosuppressive agents, long term therapy with corticosteroids, advanced AIDS and those with chronic granulomatous disease (CGD) (Baddley, 2011; Guinea *et al.*, 2010; Segal, 2009). Symptoms of infection are non-specific, with patients presenting with fever, cough and shortness of breath (Segal, 2009) and the lung is the most common site of invasive infection. Invasive aspergillosis mortality rates remain high, with 30–95% of individuals dying of this infection (Brown *et al.*, 2012). Again, high mortality rates relate to difficulties in diagnosis and delays in initiation of antifungal therapy.

Models of *Aspergillus* infection

The majority of *Aspergillus* infections are modelled in mice (Arendrup *et al.*, 2010; Bok *et al.*, 2006; Bowman *et al.*, 2001; Chiller *et al.*, 2003; da Silva Ferreira *et al.*, 2006; Denning *et al.*, 1997; Dennis *et al.*, 2006; High and Washburn, 1997; Jahn *et al.*, 1997; Johnson *et al.*, 2000; Kothary *et al.*, 1984; Krappmann *et al.*, 2004; Langfelder *et al.*, 1998; Lessing *et al.*, 2007; Liebmann *et al.*, 2004; Mircescu *et al.*, 2009; Mondon *et al.*, 1996; Mosquera *et al.*, 2001; Nawada *et al.*, 1996; Oakley *et al.*, 1997; Odds *et al.*, 1998; Sheppard *et al.*, 2006a,b; Smith *et al.*, 1994; Steinbach *et al.*, 2004; Willger *et al.*, 2008), with rats (Becker *et al.*, 2000; Kurtz *et al.*, 1995; van Vianen *et al.*, 2006), guinea pigs (Martin *et al.*, 1997; O'Connor *et al.*, 2005; Odds *et al.*, 1998), rabbits (Kirkpatrick *et al.*, 2005), embryonated eggs (Jacobsen *et al.*, 2010b), *Galleria mellonella* (Jackson *et al.*, 2009; Slater *et al.*, 2011) and fruit flies (Lionakis and Kontoyiannis, 2005; Lionakis *et al.*, 2005) also used. Animals can be infected in a number of ways. Many studies have used the intravenous infection route (Arendrup *et al.*, 2010; Bowman *et al.*, 2001; Denning *et al.*, 1997; Dennis *et al.*, 2006; High and Washburn, 1997; Jahn *et al.*, 1997; Johnson *et al.*, 2000; Langfelder *et al.*, 1998; Mosquera *et al.*, 2001; Oakley *et al.*, 1997), with inocula ranging from 4×10^2 to 4×10^6 per mouse. Studies have also used intracranial inoculation (Chiller *et al.*, 2003), but this model is rarely used.

As human infection normally occurs through inhalation of fungal spores, the majority of studies use inhalation methods to infect the host (Aufauvre-Brown *et al.*, 1998; Bok *et al.*, 2006; da Silva Ferreira *et al.*, 2006; Dennis *et al.*, 2006; Dixon *et al.*, 1989; Kothary *et al.*, 1984; Krappmann *et al.*, 2004; Kurtz *et al.*, 1995; Lessing *et al.*, 2007; Liebmann *et al.*, 2004; Mircescu *et al.*, 2009; Mondon *et al.*, 1996; Nawada *et al.*, 1996; Piggott and Emmons, 1960; Sheppard *et al.*, 2004, 2006a,b; Sidransky and Friedman, 1959; Steinbach *et al.*, 2004; van Vianen *et al.*, 2006; White, 1977; Willger *et al.*, 2008). Spores can either be administered directly into the trachea (Dennis *et al.*, 2006; Kothary *et al.*, 1984; Kurtz *et al.*, 1995; Mircescu *et al.*, 2009; Nawada *et al.*, 1996; van Vianen *et al.*, 2006; Willger *et al.*, 2008) or are administered into the animal's nostrils and inhaled (Aufauvre-Brown *et al.*, 1998; Bok *et al.*, 2006; da Silva Ferreira *et al.*, 2006; Dixon *et al.*, 1989; Krappmann *et al.*, 2004; Lessing *et al.*, 2007; Liebmann *et al.*, 2004; Mondon *et al.*, 1996; Steinbach *et al.*, 2004). Inoculum levels used for both infection routes are similar and range from $1 \times 10^4 – 1 \times 10^7$ per mouse (Bhagwat *et al.*, 2010; Cushion *et al.*, 2010; Gigliotti *et al.*, 1998; Ito *et al.*, 2000; Linke *et al.*, 2001; McKinley *et al.*, 2006; Rudner *et al.*, 2007; Saijo *et al.*, 2007; Shellito *et al.*, 1996; Vuk-Pavlovic *et al.*, 2006; Zhang *et al.*, 2010). Inhalation chambers can also be used to reproducibly and easily infect large numbers of animals in a single session using an aerosolized inoculum, with the advantage that the animals do not require anaesthesia prior to infection (Kothary *et al.*, 1984; Piggott and Emmons, 1960; Sheppard *et al.*, 2004, 2006a,b; Sidransky and Friedman, 1959; Steinbach *et al.*, 2004; White, 1977). Infection initiated in this manner has been suggested to be more homogeneous compared to that achieved by intranasal infection (Steinbach *et al.*, 2004). However, there is a potential for contamination of inocula between uses of the apparatus due to the hydrophobic and sticky nature of *Aspergillus* spores, but the model does show good inter-lab reproducibility (Sheppard *et al.*, 2006a).

A range of different mouse strains have been

used to model invasive *Aspergillus* infection, with no evidence of susceptibility differences. Many studies use outbred mice, such as CD-1 (Bok *et al.*, 2006; Chiller *et al.*, 2003; Denning *et al.*, 1997; High and Washburn, 1997; Johnson *et al.*, 2000; Kothary *et al.*, 1984; Krappmann *et al.*, 2004; Mosquera *et al.*, 2001; Oakley *et al.*, 1997; Odds *et al.*, 1998; Smith *et al.*, 1994; Steinbach *et al.*, 2004; Willger *et al.*, 2008), CFW/Swiss Webster (Jahn *et al.*, 1997; Langfelder *et al.*, 1998), OF-1 (Mondon *et al.*, 1996) or NMRI (Arendrup *et al.*, 2010) mice. These have the advantage of being cheaper and more representative of a heterogeneous population (such as human beings), but they generally show more variability and require larger group sizes for experimental comparison. The remaining studies use inbred mouse strains, such as BALB/c (da Silva Ferreira *et al.*, 2006; Lessing *et al.*, 2007; Liebmann *et al.*, 2004; Nawada *et al.*, 1996; Sheppard *et al.*, 2006a,b), C57BL/6 (Mircescu *et al.*, 2009) and DBA/2 (Bowman *et al.*, 2001).

Immunocompetent mice are resistant to inhaled *Aspergillus*, even at high inoculum levels (Dixon *et al.*, 1989; Piggott and Emmons, 1960), making it necessary to immunosuppress animals through administration of cyclophosphamide (Arendrup *et al.*, 2010; Becker *et al.*, 2000; MacCallum *et al.*, 2005; Mosquera *et al.*, 2001; Odds *et al.*, 1998), cortisone acetate (Kurtz *et al.*, 1995; Mondon *et al.*, 1996; Nawada *et al.*, 1996; Sidransky and Friedman, 1959; Smith *et al.*, 1994) or a combination of both (Bok *et al.*, 2006; da Silva Ferreira *et al.*, 2006; Dixon *et al.*, 1989; Kothary *et al.*, 1984; Lessing *et al.*, 2007; Liebmann *et al.*, 2004; Sheppard *et al.*, 2006a; Steinbach *et al.*, 2004; Willger *et al.*, 2008). Cyclophosphamide treatment renders hosts neutropenic and cortisone acetate impairs phagocyte function.

Animal models have been critical in confirming that neutropenia is a major predisposing factor for invasive aspergillosis (Mircescu *et al.*, 2009; Smith *et al.*, 1994; Stephens-Romero *et al.*, 2005), with neutrophil activity critical within the first 6 hours of infection (Mircescu *et al.*, 2009). Surprisingly, alveolar macrophage depletion had little effect on susceptibility to infection (Mircescu *et al.*, 2009). Animal models have also demonstrated that immune infiltrates containing neutrophils, lymphocytes and macrophages are required to prevent fungal spores germinating and invading the lung tissue (Nawada *et al.*, 1996). The requirement for functional phagocytes correlates with findings that mouse strains with NADPH oxidase disorders, which mimic chronic granulomatous disease (CGD) in humans, are more susceptible to infection (Dennis *et al.*, 2006; Willger *et al.*, 2008).

Animal models of invasive aspergillosis have allowed demonstration of variable virulence of *A. fumigatus* strains (Arendrup *et al.*, 2010; Aufauvre-Brown *et al.*, 1998; Denning *et al.*, 1997; Mondon *et al.*, 1996) and have also demonstrated that drug resistant isolates are less virulent than their drug susceptible parental strains (Denning *et al.*, 1997). Models have also been invaluable for assaying the virulence of *A. fumigatus* gene knockout strains (Bok *et al.*, 2006; da Silva Ferreira *et al.*, 2006; Jackson *et al.*, 2009; Jacobsen *et al.*, 2010b; Krappmann *et al.*, 2004; Liebmann *et al.*, 2004; Lionakis and Kontoyiannis, 2005; Lionakis *et al.*, 2005; Slater *et al.*, 2011; Willger *et al.*, 2008). Interestingly, whilst some studies of *Aspergillus fumigatus* virulence in mini-host *Galleria mellonella* have shown good agreement with results in rodent models (Slater *et al.*, 2011) this is not always the case. For conidial colour mutants, virulence is reduced in rodent models (Jahn *et al.*, 1997; Langfelder *et al.*, 1998), yet is increased in *G. mellonella* (Jackson *et al.*, 2009). The reasons for this discrepancy are not yet clear.

A major role for aspergillosis infection models has been in the evaluation of potential new antifungal therapies (Hughes *et al.*, 1994; Kirkpatrick *et al.*, 2005; Kurtz *et al.*, 1995; Lionakis and Kontoyiannis, 2005; Lionakis *et al.*, 2005; MacCallum *et al.*, 2005; Mosquera *et al.*, 2001; Oakley *et al.*, 1997; Sheppard *et al.*, 2006a; van Vianen *et al.*, 2006). Animal models also clearly demonstrated that *in vitro* antifungal susceptibility results do not always correlate with antifungal efficacy *in vivo* (Johnson *et al.*, 2000; Mosquera *et al.*, 2001; Odds *et al.*, 1998). These models have also been vital in evaluating diagnostic methods for monitoring *Aspergillus* infection progression and response to antifungal therapy (Becker *et al.*, 2000; Bowman *et al.*, 2001; Sheppard *et al.*, 2006b), with PCR determination of organ burdens found to be superior to viable count methods (Bowman *et al.*, 2001; Sheppard *et al.*, 2006b), and galactomannan ELISA superior to both (Becker *et al.*, 2000).

Cryptococcus infections

Cryptococcus species cause invasive disease in individuals with AIDS or undergoing prolonged corticosteroid treatment or immunosuppressive therapy (Bicanic and Harrison, 2010; Singh *et al.*, 2007, 2008), with over one million cases reported worldwide every year (Park *et al.*, 2009). In sub-Saharan Africa, *Cryptococcus* causes meningitis in around 45% of AIDS sufferers (Hakim *et al.*, 2000), and over 600,000 deaths every year (Park *et al.*, 2009). Mortality rates are much greater in the developing world, 55–70% in Latin America and sub-Saharan Africa, compared to 15–20% in the USA (Brown *et al.*, 2012). *Cryptococcus* infection is generally acquired through inhalation of air-borne organisms from the environment (Brown *et al.*, 2012), and individuals usually present with fever and headache, progressing to confusion and/or drowsiness over the next 2–4 weeks.

Modelling *Cryptococcus* infection

The mouse has been the most popular host for modelling *Cryptococcus* infection (Barchiesi *et al.*, 2005; Blasi *et al.*, 1992; Capilla *et al.*, 2006; Cauley and Murphy, 1979; Charlier *et al.*, 2005, 2009; Chaturvedi *et al.*, 2005; Cheng *et al.*, 2009; Chretien *et al.*, 2002; Clemons and Stevens, 1998; Cox *et al.*, 2001; Dan *et al.*, 2008; Decken *et al.*, 1998; Feldmesser *et al.*, 2000; Fraser *et al.*, 2005; Fries and Casadevall, 1998; Fries *et al.*, 2001; Gigliotti *et al.*, 1998; Giles *et al.*, 2005; Griffith *et al.*, 2004; Gyetko *et al.*, 1996; Hardison *et al.*, 2010; Hata *et al.*, 1996a,b; Huffnagle *et al.*, 1991b, 1997, 1998; Kawakami *et al.*, 1996, 1997, 2000; Kleinschek *et al.*, 2006, 2010; Lessing *et al.*, 2007; Lortholary *et al.*, 1999; Lovchik and Lipscomb, 1993; Mazzolla *et al.*, 1997; Mody *et al.*, 1988; Mukherjee *et al.*, 1992, 1993, 1994a,b, 1995; Muller *et al.*, 2007; Nielsen *et al.*, 2005; Olszewski *et al.*, 2000; Salkowski and Balish, 1990; Santangelo *et al.*, 2004; Schwarz *et al.*, 2006; Serena *et al.*, 2007; Velagapudi *et al.*, 2009; Wozniak *et al.*, 2006, 2011; Yauch *et al.*, 2004; Zaragoza *et al.*, 2007; Zhou *et al.*, 2007), although the rat (Fries *et al.*, 2005; Goldman *et al.*, 1994, 2000; Hata *et al.*, 1996a,b; Krockenberger *et al.*, 2010; Najvar *et al.*, 1999), rabbit (Alspaugh *et al.*, 1997; Cox *et al.*, 2000, 2001; Perfect *et al.*, 1980) and guinea pig (Diamond *et al.*, 1974; Odds *et al.*, 2000) have also been used. Rabbits are inherently resistant to *Cryptococcus* spp. infection, but can be rendered susceptible through hydrocortisone acetate pre-treatment (Cox *et al.*, 2000, 2001). In recent years, wax moth (*Galleria mellonella*) larvae have also grown in popularity as an alternative model for cryptococcosis (Mukherjee *et al.*, 1992; Velagapudi *et al.*, 2009).

Although various mouse strains have been used to model *Cryptococcus* infections, considerable variability in susceptibility to infection has been found (Decken *et al.*, 1998; Hoag *et al.*, 1995; Huffnagle *et al.*, 1991b, 1998; Mukherjee *et al.*, 1992, 1993; Nielsen *et al.*, 2005; Zaragoza *et al.*, 2007). In models where infection is initiated by direct administration into the lungs, BALB/c mice have generally been found to be resistant to infection whilst C57BL/6 mice are susceptible (Hoag *et al.*, 1995; Huffnagle *et al.*, 1991b; Nielsen *et al.*, 2005; Zaragoza *et al.*, 2007). Other strains, e.g. A/JcR and CBA/J, are also susceptible to infection (Nielsen *et al.*, 2005; Zaragoza *et al.*, 2007). Similar susceptibilities to infection are also found for intraperitoneal (Mukherjee *et al.*, 1992) and intracranial (Huffnagle *et al.*, 1998; Mukherjee *et al.*, 1993) models. However, for the intravenous model there is less agreement about the relative susceptibility of different mouse strains to *Cryptococcus* infection, with two studies finding no difference between various mouse strains (Nielsen *et al.*, 2005; Zaragoza *et al.*, 2007), and another finding that C57BL/6 and 129Sve mice are more resistant than BALB/c mice (Decken *et al.*, 1998). These differences may be due to differing inoculum levels and different *Cryptococcus* strains used for the infection models. The use of different strains for infection models is an important factor as different strains have been found to vary considerably in virulence, both for *C. neoformans* (Barchiesi *et al.*, 2005; Fries and Casadevall, 1998; Huffnagle *et al.*, 1991b; Nielsen *et al.*, 2005) and for *C. gattii* (Chaturvedi *et al.*, 2005; Cheng *et al.*, 2009; Fraser *et al.*, 2005). The availability of mouse knockout strains has also been a major factor in the preference of this host, allowing the contribution of various host genes in virulence and pathogenesis to be evaluated (Dan *et al.*, 2008; Decken *et al.*, 1998; Hardison *et al.*, 2010; Huffnagle *et al.*, 1991a; Huffnagle *et al.*, 1991b; Kawakami *et al.*, 2000; Kleinschek *et al.*, 2006, 2010; Lovchik and Lipscomb, 1993; Muller *et al.*, 2007; Olszewski *et al.*, 2000; Salkowski and Balish, 1990; Wozniak *et al.*, 2011; Yauch *et al.*, 2004; Zhou *et al.*, 2007).

To model inhalation of *Cryptococcus* from the environment, lung infection models have been developed. The inoculum can be administered into the airways in three different ways; intratracheal administration into the top of the trachea (Goldman *et al.*, 1994; Krockenberger *et al.*, 2010; Santangelo *et al.*, 2004), direct administration into the surgically exposed trachea (Fries *et al.*, 2001; Gyetko *et al.*, 1996; Huffnagle *et al.*, 1991a,b, 1997, 1998; Kawakami *et al.*, 1996, 1997, 2000; Mody *et al.*, 1988; Olszewski *et al.*, 2000) or inhalation of fungal cells administered into the nostrils (Barchiesi *et al.*, 2005; Chaturvedi *et al.*, 2005; Cheng *et al.*, 2009; Cox *et al.*, 2000, 2001; Dan *et al.*, 2008; Feldmesser *et al.*, 2000; Fraser *et al.*, 2005; Giles *et al.*, 2005; Griffith *et al.*, 2004; Hardison *et al.*, 2010; Lessing *et al.*, 2007; Muller *et al.*, 2007; Nielsen *et al.*, 2005; Velagapudi *et al.*, 2009; Wozniak *et al.*, 2006, 2011; Yauch *et al.*, 2004; Zaragoza *et al.*, 2007). All methods require anaesthesia of the animals. Development of disease in the lungs of animals infected via the airway differs depending upon the inoculum used and the host chosen. In rats, lungs counts initially remain constant, and then drop as granulomas form and disease resolves (Goldman *et al.*, 1994). In the mouse model, fungi initially remain localized to the lungs, with burdens initially remaining relatively constant (Kawakami *et al.*, 1996, 2010). Lung burdens increase from day 14 post infection (Kawakami *et al.*, 1996; Krockenberger *et al.*, 2010), with some dissemination to the brain seen after 30 days (Chaturvedi *et al.*, 2005). However, administration of a higher inoculum (3×10^7 fungal cells per mouse) leads to 100% mortality between day 10 and day 14 post infection (Santangelo *et al.*, 2004). The importance of both $CD4^+$ and $CD8^+$ T cells in clearance of *Cryptococcus* has been demonstrated in this model (Huffnagle *et al.*, 1991a). In addition, this model has shown that the mouse strains susceptible to *Cryptococcus* have higher levels of *Cryptococcus* replication within their macrophages (Zaragoza *et al.*, 2007) and have also demonstrated that the fungus produces polysaccharide within the macrophage (Feldmesser *et al.*, 2000).

Mouse models have been critical in demonstrating the role of allergic lung inflammation in cryptococcosis (Huffnagle *et al.*, 1997, 1998; Muller *et al.*, 2007; Olszewski *et al.*, 2000). Prevention of an allergic response through neutralization of MIP-1α (Huffnagle *et al.*, 1997) or IL-5 (Huffnagle *et al.*, 1998), or through deletion of IL-13 (Muller *et al.*, 2007), reduced the proinflammatory response and associated pathology. IL-12 was shown to reduce fungal counts and prolong host survival, but only if administered very early in infection (Kawakami *et al.*, 1996). Dexamethasone administration also abrogates the granulomatous response (Goldman *et al.*, 2000). The role of IL-17 has also been investigated, with IL-17A shown to be protective (Hardison *et al.*, 2010; Wozniak *et al.*, 2011). Depletion of this cytokine led to high lung burdens at day 7 (Hardison *et al.*, 2010); however, it has only an early role as there were no differences in lung fungal burdens later in disease (14 days post infection) (Hardison *et al.*, 2010). Surprisingly, the early production of IL-17 appears to be synthesized by neutrophils (Wozniak *et al.*, 2011). IL-23p19$^{-/-}$ knockout mice showed reduced survival to intraperitoneally administered *Cryptococcus* due to reduced IL-17 expression (Kleinschek *et al.*, 2006). However, administration of IL-23 to infected mice enhanced the pro-inflammatory response. The importance of IL-12 was also demonstrated in the intravenous model, where IL-12$^{-/-}$ mice were found to be more susceptible to infection, but administration of IL-12 to infected mice reduced lung burdens and granuloma formation (Decken *et al.*, 1998). The importance of the proinflammatory response in infection also affects the ability of *C. gattii* to cause infection, and strains of lower virulence are less able to induce a proinflammatory response (Cheng *et al.*, 2009).

As *Cryptococcus* causes meningitis, models have also been developed where the fungus is administered directly into the cranium (Alspaugh *et al.*, 1997; Blasi *et al.*, 1992, 1993; Cox *et al.*, 2000, 2001; Fries *et al.*, 2005; Goldman *et al.*, 2000; Hata *et al.*, 1996a,b; Mazzolla *et al.*, 1997; Mukherjee *et al.*, 1993; Najvar *et al.*, 1999; Perfect *et al.*, 1980; Serena *et al.*, 2007; Zhou *et al.*, 2007). The intracranial infection model shows inoculum-dependent disease progression (Blasi *et al.*, 1992), with high inocula leading to high fungal brain burdens, massive damage and death (Blasi *et al.*, 1993; Mukherjee *et al.*, 1993). Dissemination to the spleen, kidneys and lungs can also be observed (Blasi *et al.*, 1993; Perfect *et al.*, 1980). In the rabbit model, where animals are rendered susceptible to infection with hydrocortisone acetate pre-treatment, similar brain burdens are achieved at day 14 post infection using a range of inocula (10^2–10^7 per animal) (Perfect *et al.*, 1980).

A simple model of the haematogenous spread of the fungus is achieved via intravenous administration of the fungus (Barchiesi *et al.*, 2005; Capilla *et al.*, 2006; Charlier *et al.*, 2005, 2009; Chaturvedi *et al.*, 2005; Chretien *et al.*, 2002; Clemons and Stevens, 1998; Cox *et al.*, 2000; Decken *et al.*, 1998; Diamond *et al.*, 1974; Kleinschek *et al.*, 2006; Lortholary *et al.*, 1999; Lovchik and Lipscomb, 1993; Mody *et al.*, 1988; Mukherjee *et al.*, 1994a,b; Odds *et al.*, 2000; Salkowski and Balish, 1990; Schwarz *et al.*, 2006; Yauch *et al.*, 2004; Zaragoza *et al.*, 2007) or intraperitoneal administration (Cauley and Murphy, 1979; Fries and Casadevall, 1998; Goldman *et al.*, 1994; Kleinschek *et al.*, 2006, 2010; Mukherjee *et al.*, 1992, 1995; Yauch *et al.*, 2004). These infection models are technically easier than the lung administration models as they do not require animals to be anaesthetized prior to infection. In both intravenous and intraperitoneal infections a chronic infection develops (Decken *et al.*, 1998; Kleinschek *et al.*, 2006). Five minutes after intravenous administration yeasts are found in the brain, spleen and lungs, with rapid increases in fungal burden seen after this time (Charlier *et al.*, 2005). In the brains fungal counts increase for the first 2 weeks, then decrease as infection resolves (depending upon the inoculum level used) (Chretien *et al.*, 2002). In T-cell deficient hosts, skin lesions can also been seen (Salkowski and Balish, 1990).

The intravenous and intraperitoneal models have been particularly important in demonstrating the importance of the macrophage in *Cryptococcus* infections (Charlier *et al.*, 2009; Chretien *et al.*, 2002; Lortholary *et al.*, 1999; Mukherjee *et al.*, 1995). *Cryptococcus* cells are phagocytosed by macrophages, but surprisingly depletion of macrophages reduces disease (Charlier *et al.*, 2009). This can be partially explained by the fact that fungal cells are seen circulating in the bloodstream within macrophages (Lortholary *et al.*, 1999), which may represent a way of evading host defences and of dissemination. Granulomas in the brain were also found to mostly contain macrophages (Chretien *et al.*, 2002). An antibody against capsular glucuronoxylomannan (GXM) was found to enhance macrophage anti-*Cryptococcus* activity (Mukherjee *et al.*, 1995). The importance of complement was also demonstrated in the intravenous model, where it was shown to be important in clearance of this fungus from the lung (Lovchik and Lipscomb, 1993) and in survival in a guinea pig model (Diamond *et al.*, 1974).

Modelling infection has allowed the evaluation of various mutant *Cryptococcus* strains, and has demonstrated the importance of urokinase (Gyetko *et al.*, 1996), urease (Olszewski *et al.*, 2004), phospholipase B (Cox *et al.*, 2001; Santangelo *et al.*, 2004), UDP-glucose dehydrogenase (Griffith *et al.*, 2004), a G-protein subunit homologue (Alspaugh *et al.*, 1997), superoxide dismutase (Giles *et al.*, 2005) and the capsule (Fromtling *et al.*, 1982) in infection. Animal models have been important in demonstrating that phenotypic switching may play a role in virulence, with mucoid cells more virulent that smooth cells (Fries *et al.*, 2001, 2005) and that spores are fully virulent (Velagapudi *et al.*, 2009). Mucoid cells have been shown to resist phagocytosis and to induce a more destructive inflammatory response (Fries *et al.*, 2001).

Pneumocystis jirovecii infections

Pneumocystis jirovecii (previously known as *Pneumocystis carinii*) causes pneumonia in humans. This was the most common AIDS-defining disease in North America until the advent of highly active antiretroviral therapy (HAART)

(Stringer, 1996). *P. jirovecii* is an unusual fungus, an obligate intracellular pathogen which is difficult to culture *in vitro* (Stringer, 1996), although there has been some success in discovering media conditions allowing limited *in vitro* culture (Cushion and Ebbets, 1990). In addition, *P. jirovecii* is not susceptible to standard antifungals (Cushion and Stringer, 2010; Stringer, 1996), possibly due to the presence of cholesterol rather than ergosterol in its cell membrane (Stringer, 1996). Infected individuals present with a cough, shortness of breath and fever (McKinnell *et al.*, 2012) and mortality ranges from 30% to 80% (Brown *et al.*, 2012; Carmona and Limper, 2011; Fisk *et al.*, 2003).

Modelling *Pneumocystis* infection

Because of the obligate intracellular nature of *Pneumocystis* spp., and their host specificities (Stringer *et al.*, 2002), the majority of models have used *Pneumocystis murina* to model infection in rodents, with most studies carried out in mice (Bhagwat *et al.*, 2010; Chen *et al.*, 1992; Fuchs *et al.*, 2011; Gigliotti *et al.*, 1998; Ito *et al.*, 2000; Kunz *et al.*, 1995; Linke *et al.*, 2001; McKinley *et al.*, 2006; Powles *et al.*, 1992; Rudner *et al.*, 2007; Saijo *et al.*, 2007; Shellito *et al.*, 1996; Vuk-Pavlovic *et al.*, 2006; Zhang *et al.*, 2010) and rats (Bartlett *et al.*, 1988; Boylan and Current, 1992; Choukri *et al.*, 2011; Cushion *et al.*, 2010; Hughes, 1982; Limper *et al.*, 1997). As an alternative, the ferret has also served as a model of this infection (Stokes *et al.*, 1987), but the mini-host *Galleria mellonella* has proved resistant to *P. murina* infection (Fuchs *et al.*, 2011).

Most laboratory rats and mice have latent *Pneumocystis* infections, which can reactivate (Boylan and Current, 1992; Frenkel *et al.*, 1966) when animals are treated with corticosteroids (Chen *et al.*, 1992; Cushion *et al.*, 2010; Hughes, 1982, Huges *et al.* 1983; Linke *et al.*, 2001; Powles *et al.*, 1992). This is the basis of the 'dirty mouse/rat' model, where animals are immunosuppressed and develop a naturally acquired infection (Chen *et al.*, 1992; Frenkel *et al.*, 1966; Hughes, 1982; Hughes *et al.*, 1983; Kunz *et al.*, 1995; Linke *et al.*, 2001; Powles *et al.*, 1992). The alternative is to use organisms from the lungs of diseased animals as the infectious inoculum. Rodents can be infected intranasally (Bhagwat *et al.*, 2010; Cushion *et al.*, 2010; Ito *et al.*, 2000; Saijo *et al.*, 2007; Vuk-Pavlovic *et al.*, 2006) or intratracheally (Bartlett *et al.*, 1988; Boylan and Current, 1992; Choukri *et al.*, 2011; Gigliotti *et al.*, 1998; Linke *et al.*, 2001; McKinley *et al.*, 2006; Rudner *et al.*, 2007; Shellito *et al.*, 1996; Vuk-Pavlovic *et al.*, 2006; Zhang *et al.*, 2010) and all strains of mice (Ito *et al.*, 2000; Powles *et al.*, 1992; Rudner *et al.*, 2007; Steinbach *et al.*, 2004) and rats (Boylan and Current, 1992; Choukri *et al.*, 2011; Cushion *et al.*, 2010; Hughes, 1982; Limper *et al.*, 1997) appear equally susceptible once immunosuppressed.

Animal models have had a vital role in demonstrating that co-housed infected mice can transmit infection to their cage mates (Dumoulin *et al.*, 2000; Powles *et al.*, 1992; Walzer *et al.*, 1977) and have provided evidence that the cyst form of the fungus is the transmissible form (Cushion *et al.*, 2010) and that animals release *Pneumocystis* into the air as lung counts increase during disease progression (Choukri *et al.*, 2011).

Animal models of *Pneumocystis* infection have been used to evaluate potential drug therapies (Boylan and Current, 1992; Hughes *et al.*, 1994; Ito *et al.*, 2000) and vaccines (Gigliotti *et al.*, 1998), but the main use has been to investigate the contribution of the immune system to infection susceptibility (Chen *et al.*, 1992; Gigliotti *et al.*, 1998; Kelly *et al.*, 2013; Limper *et al.*, 1997; McKinley *et al.*, 2006; Rudner *et al.*, 2007; Shellito *et al.*, 1990, 1996; Vuk-Pavlovic *et al.*, 2006; Zhang *et al.*, 2010). Infected immunocompetent mice clear *Pneumocystis* infection within 3 weeks (Shellito *et al.*, 1996), with early peaks in IL-23 expression (Rudner *et al.*, 2007). Neutralization of IL-23 or use of IL-23$^{-/-}$ mice leads to higher lung burdens and lower IL-17 expression. In addition, administration of anti-IL-17 to infected animals also increases lung burdens (Rudner *et al.*, 2007).

Both macrophages (Limper *et al.*, 1997) and T cells (Gigliotti *et al.*, 1998; McKinley *et al.*, 2006; Shellito *et al.*, 1990; Shellito *et al.*, 1996; Vuk-Pavlovic *et al.*, 2006; Zhang *et al.*, 2010) are important in resistance to *Pneumocystis* infection. Depletion of macrophages impaired clearance of the fungus (Limper *et al.*, 1997) and depletion of T cells (Bhagwat *et al.*, 2010; Gigliotti *et al.*, 1998), or specifically CD4$^+$ T cells (Shellito et al., 1990; Shellito et al., 1996; Vuk-Pavlovic et al., 2006; Zhang et al., 2010), increased susceptibility to infection, with evidence of increased inflammatory infiltrates. Discontinuation of T cell depletion allows mice to recover and clear infection (Gigliotti et al., 1998). Regulatory T cells (CD4$^+$ CD25$^+$ FoxP3$^+$) have also been demonstrated to have a role in *Pneumocystis* infection, with depletion leading to an enhanced proinflammatory reaction and injury in the lungs (McKinley et al., 2006). Therefore, in *Pneumocystis* infection control of excessive lung inflammation appears essential to prevent lung injury.

Outlook

Animal models of fungal infection provide vital opportunities to investigate disease progression, and to evaluate potential antifungal therapies and novel diagnostic tools. The majority of models use mice, which are cheap, easy to handle and can be genetically manipulated to investigate the contribution of host factors in disease susceptibility and progression. However, it should be remembered that although the immune systems of mice and man are similar there are differences which could affect the translation of results from mouse models to human infection (Gibbons and Spencer, 2011; Jiang et al., 2010; Mestas and Hughes, 2004; Rehli, 2002). Of particular importance to fungal infections are differences in proportions of neutrophils and lymphocytes in the blood, complement receptor expression and T cell differentiation, to name but a few (reviewed in Mestas and Hughes, 2004). However, as long as the researcher bears these limitations in mind, fungal infection models will continue to provide insights into fungal pathogenesis and will enable development of badly needed diagnostics and antifungal drugs.

Acknowledgements

DMM would like to apologize to all of the researchers whose studies were not included in this chapter owing to a lack of space. The research of DMM is currently supported by grants from the National Centre for the Replacement, Refinement and Reduction of Animals in Research (NC3Rs), the Wellcome Trust and STRIFE (an EC grant).

References

Achterman, R.R., and White, T.C. (2012a). Dermatophyte virulence factors: identifying and analyzing genes that may contribute to chronic or acute skin infections. Int. J. Microbiol. *2012*, 358305.

Achterman, R.R., and White, T.C. (2012b). A foot in the door for dermatophyte research. PLoS Pathog. *8*, e1002564.

Achterman, R.R., Smith, A.R., Oliver, B.G., and White, T.C. (2011). Sequenced dermatophyte strains: growth rate, conidiation, drug susceptibilities, and virulence in an invertebrate model. Fungal Genet. Biol. *48*, 335–341.

Alarco, A.M., Marcil, A., Chen, J., Suter, B., Thomas, D., and Whiteway, M. (2004). Immune-deficient *Drosophila melanogaster*: a model for the innate immune response to human fungal pathogens. J. Immunol. *172*, 5622–5628.

Alspaugh, J.A., Perfect, J.R., and Heitman, J. (1997). *Cryptococcus neoformans* mating and virulence are regulated by the G-protein alpha subunit GPA1 and cAMP. Genes Dev. *11*, 3206–3217.

Aly, R., Maibach, H.I., Ho, I., and Abrams, B.B. (1992). Simplified experimental human dermatophytosis model. Int. J. Dermatol. *31*, 122–125.

Anaissie, E., Hachem, R., K-Tin-U, C., Stephens, L.C., and Bodey, G.P. (1993). Experimental hematogenous candidiasis caused by *Candida krusei* and *Candida albicans*: species differences in pathogenicity. Infect. Immun. *61*, 1268–1271.

Andes, D. (2005). Use of an animal model of disseminated candidiasis in the evaluation of antifungal therapy. Methods Mol. Med. *118*, 111–128.

Andes, D., Lepak, A., Nett, J., Lincoln, L., and Marchillo, K. (2006). *In vivo* fluconazole pharmacodynamics and resistance development in a previously susceptible *Candida albicans* population examined by microbiologic and transcriptional profiling. Antimicrob. Agents Chemother. *50*, 2384–2394.

Andrutis, K.A., Riggle, P.J., Kumamoto, C.A., and Tzipori, S. (2000). Intestinal lesions associated with disseminated candidiasis in an experimental animal model. J. Clin. Microbiol. *38*, 2317–2323.

Arendrup, M., Horn, T., and Frimodt-Moller, N. (2002). *In vivo* pathogenicity of eight medically relevant *Candida* species in an animal model. Infection *30*, 286–291.

Arendrup, M.C., Mavridou, E., Mortensen, K.L., Snelders, E., Frimodt-Moller, N., Khan, H., Melchers, W.J., and Verweij, P.E. (2010). Development of azole resistance in *Aspergillus fumigatus* during azole therapy associated with change in virulence. PLoS One *5*, e10080.

Ashman, R.B., Bolitho, E.M., and Papadimitriou, J.M. (1993). Patterns of resistance to *Candida albicans* in inbred mouse strains. Immunol. Cell Biol. *71 (Pt 3)*, 221–225.

Ashman, R.B., Fulurija, A., and Papadimitriou, J.M. (1996). Strain-dependent differences in host response to *Candida albicans* infection in mice are related to organ susceptibility and infectious load. Infect. Immun. *64*, 1866–1869.

Ashman, R.B., Vijayan, D., and Wells, C.A. (2011). IL-12 and related cytokines: function and regulatory implications in *Candida albicans* infection. Clin. Dev. Immunol. *2011*, 686597.

Asticcioli, S., Sacco, L., Daturi, R., Matti, C., Nucleo, E., Zara, F., and Pagani, L. (2009). Trends in frequency and *in vitro* antifungal susceptibility patterns of *Candida* isolates from women attending the STD outpatients clinic of a tertiary care hospital in Northern Italy during the years 2002–2007. New Microbiol. *32*, 199–204.

Aufauvre-Brown, A., Brown, J.S., and Holden, D.W. (1998). Comparison of virulence between clinical and environmental isolates of *Aspergillus fumigatus*. Eur. J. Clin. Microbiol. Infect. Dis. *17*, 778–780.

Azie, N., Neofytos, D., Pfaller, M., Meier-Kriesche, H.U., Quan, S.P., and Horn, D. (2012). The PATH (Prospective Antifungal Therapy) Alliance(R) registry and invasive fungal infections: update 2012. Diagn. Microbiol. Infect. Dis. *73*, 293–300.

Baddley, J.W. (2011). Clinical risk factors for invasive aspergillosis. Med. Mycol. *49* (Suppl. 1), S7–S12.

Bader, T., Schroppel, K., Bentink, S., Agabian, N., Kohler, G., and Morschhauser, J. (2006). Role of calcineurin in stress resistance, morphogenesis, and virulence of a *Candida albicans* wild-type strain. Infect. Immun. *74*, 4366–4369.

Badrane, H., Nguyen, M.H., Cheng, S., Kumar, V., Derendorf, H., Iczkowski, K.A., and Clancy, C.J. (2008). The *Candida albicans* phosphatase Inp51p interacts with the EH domain protein Irs4p, regulates phosphatidylinositol-4,5-bisphosphate levels and influences hyphal formation, the cell integrity pathway and virulence. Microbiology *154*, 3296–3308.

Barchiesi, F., Cogliati, M., Esposto, M.C., Spreghini, E., Schimizzi, A.M., Wickes, B.L., Scalise, G., and Viviani, M.A. (2005). Comparative analysis of pathogenicity of *Cryptococcus neoformans* serotypes A, D and AD in murine cryptococcosis. J. Infect. *51*, 10–16.

Bartlett, M.S., Fishman, J.A., Queener, S.F., Durkin, M.M., Jay, M.A., and Smith, J.W. (1988). New rat model of *Pneumocystis carinii* infection. J. Clin. Microbiol. *26*, 1100–1102.

Becker, M.J., de Marie, S., Willemse, D., Verbrugh, H.A., and Bakker-Woudenberg, I.A. (2000). Quantitative galactomannan detection is superior to PCR in diagnosing and monitoring invasive pulmonary aspergillosis in an experimental rat model. J. Clin. Microbiol. *38*, 1434–1438.

Belazi, M., Velegraki, A., Fleva, A., Gidarakou, I., Papanaum, L., Baka, D., Daniilidou, N., and Karamitsos, D. (2005). Candidal overgrowth in diabetic patients: potential predisposing factors. Mycoses *48*, 192–196.

Beltrame, A., Matteelli, A., Carvalho, A.C., Saleri, N., Casalini, C., Capone, S., Patroni, A., Manfrin, M., and Carosi, G. (2006). Vaginal colonization with *Candida* spp. in human immunodeficiency virus-infected women: a cohort study. Int. J. STD AIDS *17*, 260–266.

Ben-Aryeh, H., Blumfield, E., Szargel, R., Laufer, D., and Berdicevsky, I. (1995). Oral *Candida* carriage and blood group antigen secretor status. Mycoses *38*, 355–358.

Bendel, C.M., Hess, D.J., Garni, R.M., Henry-Stanley, M., and Wells, C.L. (2003). Comparative virulence of *Candida albicans* yeast and filamentous forms in orally and intravenously inoculated mice. Crit. Care Med. *31*, 501–507.

Bendick, C., Scheifele, C., and Reichart, P.A. (2002). Oral manifestations in 101 Cambodians with HIV and AIDS. J. Oral Pathol. Med. *31*, 1–4.

de Bernardis, F., Morelli, L., Ceddia, T., Lorenzini, R., and Cassone, A. (1990). Experimental pathogenicity and acid proteinase secretion of vaginal isolates of *Candida parapsilosis*. J. Med. Vet. Mycol. *28*, 125–137.

Bhagwat, S.P., Wright, T.W., and Gigliotti, F. (2010). Anti-CD3 antibody decreases inflammation and improves outcome in a murine model of *Pneumocystis* pneumonia. J. Immunol. *184*, 497–502.

Bicanic, T., and Harrison, T. (2010). *Cryptococcus*: spectrum of disease and treatment. In Pathogenic Yeasts, Ashbee, R., Bignell, EM. eds (Springer, Berlin) pp. 146–165.

Bistoni, F., Vecchiarelli, A., Cenci, E., Sbaraglia, G., Perito, S., and Cassone, A. (1984). A comparison of experimental pathogenicity of *Candida* species in cyclophosphamide-immunodepressed mice. Sabouraudia *22*, 409–418.

Blasi, E., Barluzzi, R., Mazzolla, R., Mosci, P., and Bistoni, F. (1992). Experimental model of intracerebral infection with *Cryptococcus neoformans*: roles of phagocytes and opsonization. Infect. Immun. *60*, 3682–3688.

Blasi, E., Barluzzi, R., Mazzolla, R., and Bistoni, F. (1993). Differential host susceptibility to intracerebral infections with *Candida albicans* and *Cryptococcus neoformans*. Infect. Immun. *61*, 3476–3481.

Bloch, F., and Massini, R. (1909). Studien uber immuntat und ubermpfindlichkeit bei hyphomyzetenerkrankunzen. Z Med Mikrobiol Immunol *13*, 68–85.

Bok, J.W., Chung, D., Balajee, S.A., Marr, K.A., Andes, D., Nielsen, K.F., Frisvad, J.C., Kirby, K.A., and Keller, N.P. (2006). GliZ, a transcriptional regulator of gliotoxin biosynthesis, contributes to *Aspergillus fumigatus* virulence. Infect. Immun. *74*, 6761–6768.

Bougnoux, M.E., Diogo, D., Francois, N., Sendid, B., Veirmeire, S., Colombel, J.F., Bouchier, C., Van Kruiningen, H., d'Enfert, C., and Poulain, D. (2006). Multilocus sequence typing reveals intrafamilial transmission and microevolutions of *Candida albicans* isolates from the human digestive tract. J. Clin. Microbiol. *44*, 1810–1820.

Bowman, J.C., Abruzzo, G.K., Anderson, J.W., Flattery, A.M., Gill, C.J., Pikounis, V.B., Schmatz, D.M., Liberator, P.A., and Douglas, C.M. (2001). Quantitative PCR assay to measure *Aspergillus fumigatus* burden in a murine model of disseminated aspergillosis: demonstration of efficacy of caspofungin acetate. Antimicrob. Agents Chemother. *45*, 3474–3481.

Boylan, C.J., and Current, W.L. (1992). Improved rat model of *Pneumocystis carinii* pneumonia: induced laboratory infections in *Pneumocystis*-free animals. Infect. Immun. *60*, 1589–1597.

Brennan, M., Thomas, D.Y., Whiteway, M., and Kavanagh, K. (2002). Correlation between virulence of *Candida albicans* mutants in mice and *Galleria mellonella* larvae. FEMS Immunol. Med. Microbiol. *34*, 153–157.

Brieland, J., Essig, D., Jackson, C., Frank, D., Loebenberg, D., Menzel, F., Arnold, B., DiDomenico, B., and Hare, R. (2001). Comparison of pathogenesis and host immune responses to *Candida glabrata* and *Candida albicans* in systemically infected immunocompetent mice. Infect. Immun. *69*, 5046–5055.

Brothers, K., Newman, Z., and Wheeler, R. (2011). Live imaging of disseminated candidiasis in zebrafish reveals role of phagocyte oxidase in limiting filamentous growth. Eukaryot. Cell. *10*, 932–944.

Brown, G.D., Denning, D.W., Gow, N.A., Levitz, S.M., Netea, M.G., and White, T.C. (2012). Hidden killers: human fungal infections. Sci. Transl. Med. *4*, 165rv13.

Budtz-Jorgensen, E. (1971). Denture stomatitis. IV. An experimental model in monkeys. Acta Odontol. Scand. *29*, 513–526.

Cabezas, J., Albaina, O., Montanez, D., Sevilla, M.J., Moragues, M.D., and Ponton, J. (2010). Potential of anti-*Candida* antibodies in immunoprophylaxis. Immunotherapy *2*, 171–183.

Calcagno, A.M., Bignell, E., Warn, P., Jones, M.D., Denning, D.W., Muhlschlegel, F.A., Rogers, T.R., and Haynes, K. (2003). *Candida glabrata STE12* is required for wild-type levels of virulence and nitrogen starvation induced filamentation. Mol. Microbiol. *50*, 1309–1318.

Calcagno, A.M., Bignell, E., Rogers, T.R., Canedo, M., Muhlschlegel, F.A., and Haynes, K. (2004). *Candida glabrata* Ste20 is involved in maintaining cell wall integrity and adaptation to hypertonic stress, and is required for wild-type levels of virulence. Yeast *21*, 557–568.

Calcagno, A.M., Bignell, E., Rogers, T.R., Jones, M.D., Muhlschlegel, F.A., and Haynes, K. (2005). *Candida glabrata* Ste11 is involved in adaptation to hypertonic stress, maintenance of wild-type levels of filamentation and plays a role in virulence. Med. Mycol. *43*, 355–364.

Calderon, R.A., and Hay, R.J. (1984a). Cell-mediated immunity in experimental murine dermatophytosis. I. Temporal aspects of T-suppressor activity caused by *Trichophyton quinckeanum*. Immunology *53*, 457–464.

Calderon, R.A., and Hay, R.J. (1984b). Cell-mediated immunity in experimental murine dermatophytosis. II. Adoptive transfer of immunity to dermatophyte infection by lymphoid cells from donors with acute or chronic infections. Immunology *53*, 465–472.

Campisi, G., Pizzo, G., Milici, M.E., Mancuso, S., and Margiotta, V. (2002). Candidal carriage in the oral cavity of human immunodeficiency virus-infected subjects. Oral Surg. Oral Med. Oral Pathol. Oral Radiol. Endod. *93*, 281–286.

Cantorna, M.T., and Balish, E. (1990). Mucosal and systemic candidiasis in congenitally immunodeficient mice. Infect. Immun. *58*, 1093–1100.

Capilla, J., Maffei, C.M., Clemons, K.V., Sobel, R.A., and Stevens, D.A. (2006). Experimental systemic infection with *Cryptococcus neoformans* var. *grubii* and *Cryptococcus gattii* in normal and immunodeficient mice. Med. Mycol. *44*, 601–610.

Carmona, E.M., and Limper, A.H. (2011). Update on the diagnosis and treatment of *Pneumocystis* pneumonia. Ther. Adv. Respir. Dis. *5*, 41–59.

Carrara, M.A., Donatti, L., Damke, E., Svidizinski, T.I., Consolaro, M.E., and Batista, M.R. (2010). A new model of vaginal infection by *Candida albicans* in rats. Mycopathologia *170*, 331–338.

Cassone, A., De Bernardis, F., Pontieri, E., Carruba, G., Girmenia, C., Martino, P., Fernandez-Rodriguez, M., Quindos, G., and Ponton, J. (1995). Biotype diversity of *Candida parapsilosis* and its relationship to the clinical source and experimental pathogenicity. J. Infect. Dis. *171*, 967–975.

Cauley, L.K., and Murphy, J.W. (1979). Response of congenitally athymic (nude) and phenotypically normal mice to *Cryptococcus neoformans* infection. Infect. Immun. *23*, 644–651.

Cavalcanti, J.N., Guerra, J.L., Gambale, W., Corrêa, B., and Paula, C.R. (2002). Histopathologic and mycologic aspects of experimental infection of guinea pigs with *Microsporum canis*. Braz. J. Vet. Res. Anim. Sci. *39*, 238–243.

Chamilos, G., Lionakis, M.S., Lewis, R.E., Lopez-Ribot, J.L., Saville, S.P., Albert, N.D., Halder, G., and Kontoyiannis, D.P. (2006). *Drosophila melanogaster* as a facile model for large-scale studies of virulence mechanisms and antifungal drug efficacy in *Candida* species. J. Infect. Dis. *193*, 1014–1022.

Chamilos, G., Nobile, C.J., Bruno, V.M., Lewis, R.E., Mitchell, A.P., and Kontoyiannis, D.P. (2009). *Candida albicans* Cas5, a regulator of cell wall integrity, is required for virulence in murine and toll mutant fly models. J. Infect. Dis. *200*, 152–157.

Chao, C.C., Hsu, P.C., Jen, C.F., Chen, I.H., Wang, C.H., Chan, H.C., Tsai, P.W., Tung, K.C., Wang, C.H., Lan, C.Y., and Chuang, Y.J. (2010). Zebrafish as a model host for *Candida albicans* infection. Infect. Immun. *78*, 2512–2521.

Charlier, C., Chretien, F., Baudrimont, M., Mordelet, E., Lortholary, O., and Dromer, F. (2005). Capsule structure changes associated with *Cryptococcus neoformans* crossing of the blood–brain barrier. Am. J. Pathol. *166*, 421–432.

Charlier, C., Nielsen, K., Daou, S., Brigitte, M., Chretien, F., and Dromer, F. (2009). Evidence of a role for monocytes in dissemination and brain invasion by *Cryptococcus neoformans*. Infect. Immun. *77*, 120–127.

Chaturvedi, S., Ren, P., Narasipura, S.D., and Chaturvedi, V. (2005). Selection of optimal host strain for molecular pathogenesis studies on *Cryptococcus gattii*. Mycopathologia *160*, 207–215.

Chen, C.G., Yang, Y.L., Cheng, H.H., Su, C.L., Huang, S.F., Chen, C.T., Liu, Y.T., Su, I.J., and Lo, H.J. (2006). Non-lethal *Candida albicans cph1/cph1 efg1/efg1* transcription factor mutant establishing restricted zone of infection in a mouse model of systemic infection. Int. J. Immunopathol. Pharmacol. *19*, 561–565.

Chen, W., Havell, E.A., and Harmsen, A.G. (1992). Importance of endogenous tumor necrosis factor alpha and gamma interferon in host resistance against *Pneumocystis carinii* infection. Infect. Immun. *60*, 1279–1284.

Chen, Z., and Kong, X. (2007). Study of *Candida albicans* vaginitis model in Kunming mice. J. Huazhong Univ. Sci. Technolog Med. Sci. *27*, 307–310.

Cheng, P.Y., Sham, A., and Kronstad, J.W. (2009). *Cryptococcus gattii* isolates from the British Columbia cryptococcosis outbreak induce less protective inflammation in a murine model of infection than *Cryptococcus neoformans*. Infect. Immun. *77*, 4284–4294.

Chiang, L.Y., Sheppard, D.C., Bruno, V.M., Mitchell, A.P., Edwards, J.E.,Jr, and Filler, S.G. (2007). *Candida albicans* protein kinase CK2 governs virulence during oropharyngeal candidiasis. Cell. Microbiol. *9*, 233–245.

Chiller, T.M., Sobel, R.A., Luque, J.C., Clemons, K.V., and Stevens, D.A. (2003). Efficacy of amphotericin B or itraconazole in a murine model of central nervous system *Aspergillus* infection. Antimicrob. Agents Chemother. *47*, 813–815.

Chittasobhon, N., and Smith, J.M. (1979). The production of experimental dermatophyte lesions in guinea pigs. J. Invest. Dermatol. *73*, 198–201.

Choukri, F., Aliouat el, M., Menotti, J., Totet, A., Gantois, N., Garin, Y.J., Bergeron, V., Dei-Cas, E., and Derouin, F. (2011). Dynamics of *Pneumocystis carinii* air shedding during experimental pneumocystosis. J. Infect. Dis. *203*, 1333–1336.

Chretien, F., Lortholary, O., Kansau, I., Neuville, S., Gray, F., and Dromer, F. (2002). Pathogenesis of cerebral *Cryptococcus neoformans* infection after fungemia. J. Infect. Dis. *186*, 522–530.

Clemons, K.V., and Stevens, D.A. (1998). Comparison of fungizone, Amphotec, AmBisome, and Abelcet for treatment of systemic murine cryptococcosis. Antimicrob. Agents Chemother. *42*, 899–902.

Clemons, K.V., Spearow, J.L., Parmar, R., Espiritu, M., and Stevens, D.A. (2004). Genetic susceptibility of mice to *Candida albicans* vaginitis correlates with host estrogen sensitivity. Infect. Immun. *72*, 4878–4880.

Clemons, K.V., Gonzalez, G.M., Singh, G., Imai, J., Espiritu, M., Parmar, R., and Stevens, D.A. (2006). Development of an orogastrointestinal mucosal model of candidiasis with dissemination to visceral organs. Antimicrob. Agents Chemother. *50*, 2650–2657.

Cole, G.T., Halawa, A.A., and Anaissie, E.J. (1996). The role of the gastrointestinal tract in hematogenous candidiasis: from the laboratory to the bedside. Clin. Infect. Dis. *22 (Suppl 2)*, S73–88.

Corsello, S., Spinillo, A., Osnengo, G., Penna, C., Guaschino, S., Beltrame, A., Blasi, N., and Festa, A. (2003). An epidemiological survey of vulvovaginal candidiasis in Italy. Eur. J. Obstet. Gynecol. Reprod. Biol. *110*, 66–72.

Cotter, G., Doyle, S., and Kavanagh, K. (2000). Development of an insect model for the *in vivo* pathogenicity testing of yeasts. FEMS Immunol. Med. Microbiol. *27*, 163–169.

Cox, G.M., Mukherjee, J., Cole, G.T., Casadevall, A., and Perfect, J.R. (2000). Urease as a virulence factor in experimental cryptococcosis. Infect. Immun. *68*, 443–448.

Cox, G.M., McDade, H.C., Chen, S.C., Tucker, S.C., Gottfredsson, M., Wright, L.C., Sorrell, T.C., Leidich, S.D., Casadevall, A., Ghannoum, M.A., and Perfect, J.R. (2001). Extracellular phospholipase activity is a virulence factor for *Cryptococcus neoformans*. Mol. Microbiol. *39*, 166–175.

Cushion, M.T., and Ebbets, D. (1990). Growth and metabolism of *Pneumocystis carinii* in axenic culture. J. Clin. Microbiol. *28*, 1385–1394.

Cushion, M.T., and Stringer, J.R. (2010). Stealth and opportunism: alternative lifestyles of species in the fungal genus *Pneumocystis*. Annu. Rev. Microbiol. *64*, 431–452.

Cushion, M.T., Linke, M.J., Ashbaugh, A., Sesterhenn, T., Collins, M.S., Lynch, K., Brubaker, R., and Walzer, P.D. (2010). Echinocandin treatment of *Pneumocystis* pneumonia in rodent models depletes cysts leaving trophic burdens that cannot transmit the infection. PLoS One *5*, e8524.

Dan, M., Segal, R., Marder, V., and Leibovitz, A. (2006). *Candida* colonization of the vagina in elderly residents of a long-term-care hospital. Eur. J. Clin. Microbiol. Infect. Dis. *25*, 394–396.

Dan, J.M., Kelly, R.M., Lee, C.K., and Levitz, S.M. (2008). Role of the mannose receptor in a murine model of *Cryptococcus neoformans* infection. Infect. Immun. *76*, 2362–2367.

Daniluk, T., Tokajuk, G., Stokowska, W., Fiedoruk, K., Sciepuk, M., Zaremba, M.L., Rozkiewicz, D., Cylwik-Rokicka, D., Kedra, B.A., Anielska, I., Gorska, M., and Kedra, B.R. (2006). Occurrence rate of oral *Candida albicans* in denture wearer patients. Adv. Med. Sci. *51 (Suppl 1)*, 77–80.

De Bernardis, F., Lorenzini, R., Morelli, L., and Cassone, A. (1989). Experimental rat vaginal infection with *Candida parapsilosis*. FEMS Microbiol. Lett. *53*, 137–141.

De Bernardis, F., Lorenzini, R., and Cassone, A. (1999). Rat model of *Candida* vaginal infection. In Handbook of animal models of infection: Experimental models in antimicrobial chemotherapy, O. Zak, and M.A. Sande, eds. (Academic Press, New York) pp. 735–740.

DeBoer, D.J., and Moriello, K.A. (1994). Development of an experimental model of *Microsporum canis* infection in cats. Vet. Microbiol. *42*, 289–295.

Decken, K., Kohler, G., Palmer-Lehmann, K., Wunderlin, A., Mattner, F., Magram, J., Gately, M.K., and Alber, G. (1998). Interleukin-12 is essential for a protective Th1 response in mice infected with *Cryptococcus neoformans*. Infect. Immun. *66*, 4994–5000.

Denning, D.W., Venkateswarlu, K., Oakley, K.L., Anderson, M.J., Manning, N.J., Stevens, D.A., Warnock, D.W., and Kelly, S.L. (1997). Itraconazole resistance in *Aspergillus fumigatus*. Antimicrob. Agents Chemother. *41*, 1364–1368.

Dennis, C.G., Greco, W.R., Brun, Y., Youn, R., Slocum, H.K., Bernacki, R.J., Lewis, R., Wiederhold, N., Holland, S.M., Petraitiene, R., Walsh, T.J., and Segal, B.H. (2006). Effect of amphotericin B and micafungin combination on survival, histopathology, and fungal burden in experimental aspergillosis in the p47phox-/- mouse model of chronic granulomatous disease. Antimicrob. Agents Chemother. *50*, 422–427.

Descamps, F.F., Brouta, F., Vermout, S.M., Willame, C., Losson, B.J., and Mignon, B.R. (2003). A recombinant 31.5 kDa keratinase and a crude exo-antigen from *Microsporum canis* fail to protect against a homologous experimental infection in guinea pigs. Vet. Dermatol. *14*, 305–312.

Diamond, R.D., May, J.E., Kane, M.A., Frank, M.M., and Bennett, J.E. (1974). The role of the classical and alternate complement pathways in host defenses against *Cryptococcus neoformans* infection. J. Immunol. *112*, 2260–2270.

Dixon, D.M., Polak, A., and Walsh, T.J. (1989). Fungus dose-dependent primary pulmonary aspergillosis in immunosuppressed mice. Infect. Immun. *57*, 1452–1456.

Dumoulin, A., Mazars, E., Seguy, N., Gargallo-Viola, D., Vargas, S., Cailliez, J.C., Aliouat, E.M., Wakefield, A.E., and Dei-Cas, E. (2000). Transmission of *Pneumocystis carinii* disease from immunocompetent contacts of infected hosts to susceptible hosts. Eur. J. Clin. Microbiol. Infect. Dis. *19*, 671–678.

Dunphy, G.B., Oberholzer, U., Whiteway, M., Zakarian, R.J., and Boomer, I. (2003). Virulence of *Candida albicans* mutants toward larval *Galleria mellonella* (Insecta, Lepidoptera, Galleridae). Can. J. Microbiol. *49*, 514–524.

Faergemann, J. (1979). Experimental tinea versicolor in rabbits and humans with *Pityrosporum orbiculare*. J. Invest. Dermatol. 72, 326–329.

Farah, C.S., Elahi, S., Drysdale, K., Pang, G., Gotjamanos, T., Seymour, G.J., Clancy, R.L., and Ashman, R.B. (2002). Primary role for CD4(+) T lymphocytes in recovery from oropharyngeal candidiasis. Infect. Immun. *70*, 724–731.

Feldmesser, M., Kress, Y., Novikoff, P., and Casadevall, A. (2000). *Cryptococcus neoformans* is a facultative intracellular pathogen in murine pulmonary infection. Infect. Immun. *68*, 4225–4237.

Felix, D.H., and Wray, D. (1993). The prevalence of oral candidiasis in HIV-infected individuals and dental attenders in Edinburgh. J. Oral Pathol. Med. *22*, 418–420.

Ferrari, S., Ischer, F., Calabrese, D., Posteraro, B., Sanguinetti, M., Fadda, G., Rohde, B., Bauser, C., Bader, O., and Sanglard, D. (2009). Gain of function mutations in CgPDR1 of *Candida glabrata* not only mediate antifungal resistance but also enhance virulence. PLoS Pathog. *5*, e1000268.

Ferrari, S., Sanguinetti, M., De Bernardis, F., Torelli, R., Posteraro, B., Vandeputte, P., and Sanglard, D. (2011a). Loss of mitochondrial functions associated with azole resistance in *Candida glabrata* results in enhanced virulence in mice. Antimicrob. Agents Chemother. *55*, 1852–1860.

Ferrari, S., Sanguinetti, M., Torelli, R., Posteraro, B., and Sanglard, D. (2011b). Contribution of *CgPDR1*-regulated genes in enhanced virulence of azole-resistant *Candida glabrata*. PLoS One *6*, e17589.

Fidel, P.L.,Jr, and Sobel, J.D. (1999). Murine models of *Candida* vaginal infections. In Handbook of animal models of infection, O. Zak, and M.A. Sande, eds. (Academic Press, New York) pp. 741–748.

Figueiral, M.H., Azul, A., Pinto, E., Fonseca, P.A., Branco, F.M., and Scully, C. (2007). Denture-related stomatitis: identification of aetiological and predisposing factors – a large cohort. J. Oral Rehabil. *34*, 448–455.

Filler, S.G., Crislip, M.A., Mayer, C.L., and Edwards, J.E.,Jr. (1991). Comparison of fluconazole and amphotericin B for treatment of disseminated candidiasis and endophthalmitis in rabbits. Antimicrob. Agents Chemother. 35, 288–292.

Fisk, D.T., Meshnick, S., and Kazanjian, P.H. (2003). *Pneumocystis carinii* pneumonia in patients in the developing world who have acquired immunodeficiency syndrome. Clin. Infect. Dis. *36*, 70–78.

Foster, K.W., Ghannoum, M.A., and Elewski, B.E. (2004). Epidemiologic surveillance of cutaneous fungal infection in the United States from 1999 to 2002. J. Am. Acad. Dermatol. *50*, 748–752.

Fransen, J., Van Cutsem, J., Vandesteene, R., and Janssen, P.A. (1984). Histopathology of experimental systemic candidosis in guinea-pigs. Sabouraudia *22*, 455–469.

Fraser, J.A., Giles, S.S., Wenink, E.C., Geunes-Boyer, S.G., Wright, J.R., Diezmann, S., Allen, A., Stajich, J.E., Dietrich, F.S., Perfect, J.R., and Heitman, J. (2005). Same-sex mating and the origin of the Vancouver Island *Cryptococcus gattii* outbreak. Nature *437*, 1360–1364.

Frenkel, J.K., Good, J.T., and Shultz, J.A. (1966). Latent *Pneumocystis* infection of rats, relapse, and chemotherapy. Lab. Invest. *15*, 1559–1577.

Fries, B.C., and Casadevall, A. (1998). Serial isolates of *Cryptococcus neoformans* from patients with AIDS differ in virulence for mice. J. Infect. Dis. *178*, 1761–1766.

Fries, B.C., Taborda, C.P., Serfass, E., and Casadevall, A. (2001). Phenotypic switching of *Cryptococcus neoformans* occurs *in vivo* and influences the outcome of infection. J. Clin. Invest. *108*, 1639–1648.

Fries, B.C., Lee, S.C., Kennan, R., Zhao, W., Casadevall, A., and Goldman, D.L. (2005). Phenotypic switching of *Cryptococcus neoformans* can produce variants that elicit increased intracranial pressure in a rat model of cryptococcal meningoencephalitis. Infect. Immun. 73, 1779–1787.

Fromtling, R.A., Shadomy, H.J., and Jacobson, E.S. (1982). Decreased virulence in stable, acapsular mutants of *Cryptococcus neoformans*. Mycopathologia 79, 23–29.

Fromtling, R.A., Abruzzo, G.K., and Giltinan, D.M. (1987). *Candida tropicalis* infection in normal, diabetic, and neutropenic mice. J. Clin. Microbiol. 25, 1416–1420.

Fu, Y., Luo, G., Spellberg, B.J., Edwards, J.E.,Jr, and Ibrahim, A.S. (2008). Gene overexpression/

suppression analysis of candidate virulence factors of *Candida albicans*. Eukaryot. Cell. 7, 483–492.

Fuchs, B.B., Eby, J., Nobile, C.J., El Khoury, J.B., Mitchell, A.P., and Mylonakis, E. (2010). Role of filamentation in *Galleria mellonella* killing by *Candida albicans*. Microbes Infect. *12*, 488–496.

Fuchs, B.B., Bishop, L.R., Kovacs, J.A., and Mylonakis, E. (2011). *Galleria mellonella* are resistant to *Pneumocystis murina* infection. Mycopathologia *171*, 273–277.

Fujita, S., and Matsuyama, T. (1987). Experimental tinea pedis induced by non-abrasive inoculation of *Trichophyton mentagrophytes* arthrospores on the plantar part of a guinea pig foot. J. Med. Vet. Mycol. *25*, 203–213.

Gabler, I.G., Barbosa, A.C., Velela, R.R., Lyon, S., and Rosa, C.A. (2008). Incidence and anatomic localization of oral candidiasis in patients with AIDS hospitalized in a public hospital in Belo Horizonte, MG, Brazil. J. Appl. Oral Sci. *16*, 247–250.

Gaughran, E.R., Grutter, F.H., and Swartz, H.E. (1960). Parenteral administration of griseofulvin in experimental dermatophyte infections. Arch. Dermatol. *81*, 690–693.

Ghannoum, M.A., Hossain, M.A., Long, L., Mohamed, S., Reyes, G., and Mukherjee, P.K. (2004). Evaluation of antifungal efficacy in an optimized animal model of *Trichophyton mentagrophytes*-dermatophytosis. J. Chemother. *16*, 139–144.

Gibbons, D.L., and Spencer, J. (2011). Mouse and human intestinal immunity: same ballpark, different players; different rules, same score. Mucosal Immunol. *4*, 148–157.

Gigliotti, F., Wiley, J.A., and Harmsen, A.G. (1998). Immunization with *Pneumocystis carinii* gpA is immunogenic but not protective in a mouse model of *P. carinii* pneumonia. Infect. Immun. *66*, 3179–3182.

Giles, S.S., Batinic-Haberle, I., Perfect, J.R., and Cox, G.M. (2005). *Cryptococcus neoformans* mitochondrial superoxide dismutase: an essential link between antioxidant function and high-temperature growth. Eukaryot. Cell. *4*, 46–54.

Girmenia, C., Martino, P., De Bernardis, F., Gentile, G., Boccanera, M., Monaco, M., Antonucci, G., and Cassone, A. (1996). Rising incidence of *Candida parapsilosis* fungemia in patients with hematologic malignancies: clinical aspects, predisposing factors, and differential pathogenicity of the causative strains. Clin. Infect. Dis. *23*, 506–514.

Glittenberg, M., Kounatidis, I., Christensen, D., Kostov, M., Kimber, S., Roberts, I., and Ligoxygakis, P. (2011a). Pathogen and host factors are needed to provoke a systemic host response to gastrointestinal infection of *Drosophila* larvae by *Candida albicans*. Dis. Model. Mech. *4*, 515–525.

Glittenberg, M.T., Silas, S., Maccallum, D.M., Gow, N.A., and Ligoxygakis, P. (2011b). Wild-type *Drosophila melanogaster* as an alternative model system for investigating the pathogenicity of *Candida albicans*. Dis. Model. Mech. *4*, 504–514.

Goldman, D., Lee, S.C., and Casadevall, A. (1994). Pathogenesis of pulmonary *Cryptococcus neoformans* infection in the rat. Infect. Immun. *62*, 4755–4761.

Goldman, D.L., Lee, S.C., Mednick, A.J., Montella, L., and Casadevall, A. (2000). Persistent *Cryptococcus neoformans* pulmonary infection in the rat is associated with intracellular parasitism, decreased inducible nitric oxide synthase expression, and altered antibody responsiveness to cryptococcal polysaccharide. Infect. Immun. *68*, 832–838.

Gow, N.A., Knox, Y., Munro, C.A., and Thompson, W.D. (2003). Infection of chick chorioallantoic membrane (CAM) as a model for invasive hyphal growth and pathogenesis of *Candida albicans*. Med. Mycol. *41*, 331–338.

Graybill, J.R., Najvar, L.K., Holmberg, J.D., and Luther, M.F. (1995). Fluconazole, D0870, and flucytosine treatment of disseminated *Candida tropicalis* infections in mice. Antimicrob. Agents Chemother. *39*, 924–929.

Green, F.,3rd, and Balish, E. (1979). Suppression of *in vitro* lymphocyte transformation during an experimental dermatophyte infection. Infect. Immun. *26*, 554–562.

Green, F.,3rd, and Balish, E. (1980). *Trichophyton mentagrophytes* dermatophytosis in germfree guinea pigs. J. Invest. Dermatol. *75*, 476–480.

Green, F.,3rd, Weber, J.K., and Balish, E. (1983). The thymus dependency of acquired resistance to *Trichophyton mentagrophytes* dermatophytosis in rats. J. Invest. Dermatol. *81*, 31–38.

Greenberg, J.H., and Kerbs, S. (1980). Cutaneous basophilic hypersensitivity response to fungal antigens in guinea pigs. J. Invest. Dermatol. *74*, 26–28.

Greenberg, J.H., King, R.D., Krebs, S., and Field, R. (1976). A quantitative dermatophyte infection model in the guinea pig--a parallel to the quantitated human infection model. J. Invest. Dermatol. *67*, 704–708.

Griffith, C.L., Klutts, J.S., Zhang, L., Levery, S.B., and Doering, T.L. (2004). UDP-glucose dehydrogenase plays multiple roles in the biology of the pathogenic fungus *Cryptococcus neoformans*. J. Biol. Chem. *279*, 51669–51676.

Grigoriou, O., Baka, S., Makrakis, E., Hassiakos, D., Kapparos, G., and Kouskouni, E. (2006). Prevalence of clinical vaginal candidiasis in a university hospital and possible risk factors. Eur. J. Obstet. Gynecol. Reprod. Biol. *126*, 121–125.

Grumbt, M., Defaweux, V., Mignon, B., Monod, M., Burmester, A., Wostemeyer, J., and Staib, P. (2011). Targeted gene deletion and *in vivo* analysis of putative virulence gene function in the pathogenic dermatophyte *Arthroderma benhamiae*. Eukaryot. Cell. *10*, 842–853.

Guinea, J., Torres-Narbona, M., Gijon, P., Munoz, P., Pozo, F., Pelaez, T., de Miguel, J., and Bouza, E. (2010). Pulmonary aspergillosis in patients with chronic obstructive pulmonary disease: incidence, risk factors, and outcome. Clin. Microbiol. Infect. *16*, 870–877.

Guzel, A.B., Ilkit, M., Akar, T., Burgut, R., and Demir, S.C. (2011). Evaluation of risk factors in patients with vulvovaginal candidiasis and the value of chromID *Candida* agar versus CHROMagar *Candida* for recovery and presumptive identification of vaginal yeast species. Med. Mycol. *49*, 16–25.

Gyetko, M.R., Chen, G.H., McDonald, R.A., Goodman, R., Huffnagle, G.B., Wilkinson, C.C., Fuller, J.A., and

Toews, G.B. (1996). Urokinase is required for the pulmonary inflammatory response to *Cryptococcus neoformans*. A murine transgenic model. J. Clin. Invest. *97*, 1818–1826.

Hakim, J.G., Gangaidzo, I.T., Heyderman, R.S., Mielke, J., Mushangi, E., Taziwa, A., Robertson, V.J., Musvaire, P., and Mason, P.R. (2000). Impact of HIV infection on meningitis in Harare, Zimbabwe: a prospective study of 406 predominantly adult patients. Aids *14*, 1401–1407.

Hardison, S.E., Wozniak, K.L., Kolls, J.K., and Wormley, F.L.,Jr. (2010). Interleukin-17 is not required for classical macrophage activation in a pulmonary mouse model of *Cryptococcus neoformans* infection. Infect. Immun. *78*, 5341–5351.

Hasenclever, H.F., and Mitchell, W.O. (1961). Pathogenicity of *C. albicans* and *C. tropicalis*. Sabouraudia *1*, 16–21.

Hata, K., Kimura, J., Miki, H., Toyosawa, T., Moriyama, M., and Katsu, K. (1996a). Efficacy of ER-30346, a novel oral triazole antifungal agent, in experimental models of aspergillosis, candidiasis, and cryptococcosis. Antimicrob. Agents Chemother. *40*, 2243–2247.

Hata, K., Kimura, J., Miki, H., Toyosawa, T., Nakamura, T., and Katsu, K. (1996b). *In vitro* and *in vivo* antifungal activities of ER-30346, a novel oral triazole with a broad antifungal spectrum. Antimicrob. Agents Chemother. *40*, 2237–2242.

Havlickova, B., Czaika, V.A., and Friedrich, M. (2008). Epidemiological trends in skin mycoses worldwide. Mycoses *51 (Suppl 4)*, 2–15.

Hay, R.J., Calderon, R.A., and Collins, M.J. (1983). Experimental dermatophytosis: the clinical and histopathologic features of a mouse model using *Trichophyton quinckeanum* (mouse favus). J. Invest. Dermatol. *81*, 270–274.

Hernandez, A.D., Reece, R.E., and Zucker, A.H. (1986). *Trichophyton mentagrophytes* spores differ from mycelia in their ability to induce pustules and activate complement. J. Invest. Dermatol. *87*, 683–687.

Herrera, C., and Guentzel, M.N. (1982). Mice with persistent gastrointestinal *Candida albicans* as a model for antifungal therapy. Antimicrob. Agents Chemother. *21*, 51–53.

High, K.P., and Washburn, R.G. (1997). Invasive aspergillosis in mice immunosuppressed with cyclosporin A, tacrolimus (FK506), or sirolimus (rapamycin). J. Infect. Dis. *175*, 222–225.

Hoag, K.A., Street, N.E., Huffnagle, G.B., and Lipscomb, M.F. (1995). Early cytokine production in pulmonary *Cryptococcus neoformans* infections distinguishes susceptible and resistant mice. Am. J. Respir. Cell Mol. Biol. *13*, 487–495.

Holbrook, W.P., Sofaer, J.A., and Southam, J.C. (1983). Experimental oral infection of mice with a pathogenic and a non-pathogenic strain of the yeast *Candida albicans*. Arch. Oral Biol. *28*, 1089–1091.

Hope, W.W., Walsh, T.J., and Denning, D.W. (2005). The invasive and saprophytic syndromes due to *Aspergillus* spp. Med. Mycol. *43 (Suppl 1)*, S207–238.

Huffnagle, G.B., Yates, J.L., and Lipscomb, M.F. (1991a). Immunity to a pulmonary *Cryptococcus neoformans* infection requires both CD4+ and CD8+ T cells. J. Exp. Med. *173*, 793–800.

Huffnagle, G.B., Yates, J.L., and Lipscomb, M.F. (1991b). T cell-mediated immunity in the lung: a *Cryptococcus neoformans* pulmonary infection model using SCID and athymic nude mice. Infect. Immun. *59*, 1423–1433.

Huffnagle, G.B., Strieter, R.M., McNeil, L.K., McDonald, R.A., Burdick, M.D., Kunkel, S.L., and Toews, G.B. (1997). Macrophage inflammatory protein-1alpha (MIP-1alpha) is required for the efferent phase of pulmonary cell-mediated immunity to a *Cryptococcus neoformans* infection. J. Immunol. *159*, 318–327.

Huffnagle, G.B., Boyd, M.B., Street, N.E., and Lipscomb, M.F. (1998). IL-5 is required for eosinophil recruitment, crystal deposition, and mononuclear cell recruitment during a pulmonary *Cryptococcus neoformans* infection in genetically susceptible mice (C57BL/6). J. Immunol. *160*, 2393–2400.

Hughes, W.T. (1982). Natural mode of acquisition for *de novo* infection with *Pneumocystis carinii*. J. Infect. Dis. *145*, 842–848.

Hughes, W.T., Bartley, D.L., and Smith, B.M. (1983). A natural source of infection due to *Pneumocystis carinii*. J. Infect. Dis. *147*, 595.

Hughes, W.T., Killmar, J.T., and Oz, H.S. (1994). Relative potency of 10 drugs with anti-*Pneumocystis carinii* activity in an animal model. J. Infect. Dis. *170*, 906–911.

Hunjan, B.S., and Cronholm, L.S. (1979). An animal model for cell-mediated immune responses to dermatophytes. J. Allergy Clin. Immunol. *63*, 361–369.

Ito, M., Nozu, R., Kuramochi, T., Eguchi, N., Suzuki, S., Hioki, K., Itoh, T., and Ikeda, F. (2000). Prophylactic effect of FK463, a novel antifungal lipopeptide, against *Pneumocystis carinii* infection in mice. Antimicrob. Agents Chemother. *44*, 2259–2262.

Itoyama, T., Uchida, K., Yamaguchi, H., and Fujita, S. (1997). Therapeutic effects of omoconazole nitrate on experimental tinea pedis, an intractable dermatophytosis, in guinea-pigs. J. Antimicrob. Chemother. *40*, 441–444.

Jackson, J.C., Higgins, L.A., and Lin, X. (2009). Conidiation color mutants of *Aspergillus fumigatus* are highly pathogenic to the heterologous insect host *Galleria mellonella*. Plos One *4*, e4224.

Jacobsen, I.D., Brunke, S., Seider, K., Schwarzmuller, T., Firon, A., d'Enfert, C., Kuchler, K., and Hube, B. (2010a). *Candida glabrata* persistence in mice does not depend on host immunosuppression and is unaffected by fungal amino acid auxotrophy. Infect. Immun. *78*, 1066–1077.

Jacobsen, I.D., Grosse, K., Slesiona, S., Hube, B., Berndt, A., and Brock, M. (2010b). Embryonated eggs as an alternative infection model to investigate *Aspergillus fumigatus* virulence. Infect. Immun. *78*, 2995–3006.

Jacobsen, I.D., Grosse, K., Berndt, A., and Hube, B. (2011). Pathogenesis of *Candida albicans* Infections in the Alternative Chorio-Allantoic Membrane

Chicken Embryo Model Resembles Systemic Murine Infections. PLoS One *6*, e19741.

Jahn, B., Koch, A., Schmidt, A., Wanner, G., Gehringer, H., Bhakdi, S., and Brakhage, A.A. (1997). Isolation and characterization of a pigmentless-conidium mutant of *Aspergillus fumigatus* with altered conidial surface and reduced virulence. Infect. Immun. *65*, 5110–5117.

Jiang, X., Shen, C., Yu, H., Karunakaran, K.P., and Brunham, R.C. (2010). Differences in innate immune responses correlate with differences in murine susceptibility to *Chlamydia muridarum* pulmonary infection. Immunology *129*, 556–566.

Johnson, E.M., Oakley, K.L., Radford, S.A., Moore, C.B., Warn, P., Warnock, D.W., and Denning, D.W. (2000). Lack of correlation of *in vitro* amphotericin B susceptibility testing with outcome in a murine model of *Aspergillus* infection. J. Antimicrob. Chemother. *45*, 85–93.

Kamai, Y., Kubota, M., Kamai, Y., Hosokawa, T., Fukuoka, T., and Filler, S.G. (2001). New model of oropharyngeal candidiasis in mice. Antimicrob. Agents Chemother. *45*, 3195–3197.

Kamran, M., Calcagno, A.M., Findon, H., Bignell, E., Jones, M.D., Warn, P., Hopkins, P., Denning, D.W., Butler, G., Rogers, T., Muhlschlegel, F.A., and Haynes, K. (2004). Inactivation of transcription factor gene *ACE2* in the fungal pathogen *Candida glabrata* results in hypervirulence. Eukaryot. Cell. *3*, 546–552.

Kavanagh, K., and Reeves, E.P. (2004). Exploiting the potential of insects for *in vivo* pathogenicity testing of microbial pathogens. FEMS Microbiol. Rev. *28*, 101–112.

Kawakami, K., Tohyama, M., Xie, Q., and Saito, A. (1996). IL-12 protects mice against pulmonary and disseminated infection caused by *Cryptococcus neoformans*. Clin. Exp. Immunol. *104*, 208–214.

Kawakami, K., Qureshi, M.H., Zhang, T., Okamura, H., Kurimoto, M., and Saito, A. (1997). IL-18 protects mice against pulmonary and disseminated infection with *Cryptococcus neoformans* by inducing IFN-gamma production. J. Immunol. *159*, 5528–5534.

Kawakami, K., Koguchi, Y., Qureshi, M.H., Miyazato, A., Yara, S., Kinjo, Y., Iwakura, Y., Takeda, K., Akira, S., Kurimoto, M., and Saito, A. (2000). IL-18 contributes to host resistance against infection with *Cryptococcus neoformans* in mice with defective IL-12 synthesis through induction of IFN-gamma production by NK cells. J. Immunol. *165*, 941–947.

Kelly, M.N., Zheng, M., Ruan, S., Kolls, J., D'Souza, A., and Shellito, J.E. (2013). Memory CD4+ T cells are required for optimal NK cell effector functions against the opportunistic fungal pathogen *Pneumocystis murina*. J. Immunol. *190*, 285–295.

Kennedy, M.A., and Sobel, J.D. (2010). Vulvovaginal caused by non-*albicans Candida* species: new insights. Curr. Infect. Dis. Rep. *12*, 465–470.

Kennedy, M.J., and Volz, P.A. (1983). Dissemination of yeasts after gastrointestinal inoculation in antibiotic-treated mice. Sabouraudia *21*, 27–33.

Kerbs, S., and Allen, A.M. (1978). Effect of occlusion on *Trichophyton mentagrophytes* infections in guinea pigs. J. Invest. Dermatol. *71*, 301–304.

Kirkpatrick, C.H. (1989). Chronic mucocutaneous candidiasis. Eur. J. Clin. Microbiol. Infect. Dis. *8*, 448–456.

Kirkpatrick, C.H. (2001). Chronic mucocutaneous candidiasis. Pediatr. Infect. Dis. J. *20*, 197–206.

Kirkpatrick, W.R., Vallor, A.C., McAtee, R.K., Ryder, N.S., Fothergill, A.W., Rinaldi, M.G., and Patterson, T.F. (2005). Combination therapy with terbinafine and amphotericin B in a rabbit model of experimental invasive aspergillosis. Antimicrob. Agents Chemother. *49*, 4751–4753.

Kisand, K., Lilic, D., Casanova, J.L., Peterson, P., Meager, A., and Willcox, N. (2011). Mucocutaneous candidiasis and autoimmunity against cytokines in APECED and thymoma patients: Clinical and pathogenetic implications. Eur. J. Immunol. *41*, 1517–1527.

Kleinschek, M.A., Muller, U., Brodie, S.J., Stenzel, W., Kohler, G., Blumenschein, W.M., Straubinger, R.K., McClanahan, T., Kastelein, R.A., and Alber, G. (2006). IL-23 enhances the inflammatory cell response in *Cryptococcus neoformans* infection and induces a cytokine pattern distinct from IL-12. J. Immunol. *176*, 1098–1106.

Kleinschek, M.A., Muller, U., Schutze, N., Sabat, R., Straubinger, R.K., Blumenschein, W.M., McClanahan, T., Kastelein, R.A., and Alber, G. (2010). Administration of IL-23 engages innate and adaptive immune mechanisms during fungal infection. Int. Immunol. *22*, 81–90.

Knight, A.G. (1972). A review of experimental human fungus infections. J. Invest. Dermatol. *59*, 354–358.

Knight, A.G. (1973). Human models for *in vivo* and *in vitro* assessment of topical antifungal compounds. Br. J. Dermatol. *89*, 509–514.

Koga-Ito, C.Y., Komiyama, E.Y., De Paiva Martins, C.A., Vasconcellos, T.C., Cardoso Jorge, A.O., Carvalho, Y.R., Do Prado, R.F., and Balducci, I. (2010). Experimental systemic virulence of oral *Candida dubliniensis* isolates in comparison with *Candida albicans, Candida tropicalis and Candida krusei*. Mycoses.

Koh, A.Y., Kohler, J.R., Coggshall, K.T., Van Rooijen, N., and Pier, G.B. (2008). Mucosal Damage and neutropenia are required for *Candida albicans* dissemination. PLoS Pathog. *4*, e35.

Kothary, M.H., Chase, T.,Jr, and Macmillan, J.D. (1984). Correlation of elastase production by some strains of *Aspergillus fumigatus* with ability to cause pulmonary invasive aspergillosis in mice. Infect. Immun. *43*, 320–325.

Krappmann, S., Bignell, E.M., Reichard, U., Rogers, T., Haynes, K., and Braus, G.H. (2004). The *Aspergillus fumigatus* transcriptional activator CpcA contributes significantly to the virulence of this fungal pathogen. Mol. Microbiol. *52*, 785–799.

Krockenberger, M.B., Malik, R., Ngamskulrungroj, P., Trilles, L., Escandon, P., Dowd, S., Allen, C., Himmelreich, U., Canfield, P.J., Sorrell, T.C., and Meyer, W. (2010). Pathogenesis of pulmonary *Cryptococcus gattii* infection: a rat model. Mycopathologia *170*, 315–330.

Kunz, S., Junker, U., Blaser, J., Joos, B., Meyer, B., Zak, O., and O'Reilly, T. (1995). The scid mouse as an experimental model for the evaluation of anti-*Pneumocystis carinii* therapy. J. Antimicrob. Chemother. *36*, 137–155.

Kurtz, M.B., Bernard, E.M., Edwards, F.F., Marrinan, J.A., Dropinski, J., Douglas, C.M., and Armstrong, D. (1995). Aerosol and parenteral pneumocandins are effective in a rat model of pulmonary aspergillosis. Antimicrob. Agents Chemother. *39*, 1784–1789.

Kusne, S., Tobin, D., Pasculle, A.W., Van Thiel, D.H., Ho, M., and Starzl, T.E. (1994). *Candida* carriage in the alimentary tract of liver transplant candidates. Transplantation 57, 398–402.

Langfelder, K., Jahn, B., Gehringer, H., Schmidt, A., Wanner, G., and Brakhage, A.A. (1998). Identification of a polyketide synthase gene (*pksP*) of *Aspergillus fumigatus* involved in conidial pigment biosynthesis and virulence. Med. Microbiol. Immunol. *187*, 79–89.

de Leon, E.M., Jacober, S.J., Sobel, J.D., and Foxman, B. (2002). Prevalence and risk factors for vaginal *Candida* colonization in women with type 1 and type 2 diabetes. BMC Infect. Dis. *2*, 1.

Lessing, F., Kniemeyer, O., Wozniok, I., Loeffler, J., Kurzai, O., Haertl, A., and Brakhage, A.A. (2007). The *Aspergillus fumigatus* transcriptional regulator AfYap1 represents the major regulator for defense against reactive oxygen intermediates but is dispensable for pathogenicity in an intranasal mouse infection model. Eukaryot. Cell. *6*, 2290–2302.

Levraud, J.P., Colucci-Guyon, E., Redd, M.J., Lutfalla, G., and Herbomel, P. (2008). *In vivo* analysis of zebrafish innate immunity. Methods Mol. Biol. *415*, 337–363.

Liebmann, B., Muhleisen, T.W., Muller, M., Hecht, M., Weidner, G., Braun, A., Brock, M., and Brakhage, A.A. (2004). Deletion of the *Aspergillus fumigatus* lysine biosynthesis gene *lysF* encoding homoaconitase leads to attenuated virulence in a low-dose mouse infection model of invasive aspergillosis. Arch. Microbiol. *181*, 378–383.

Limper, A.H., Hoyte, J.S., and Standing, J.E. (1997). The role of alveolar macrophages in *Pneumocystis carinii* degradation and clearance from the lung. J. Clin. Invest. *99*, 2110–2117.

Linke, M.J., Harris, C.E., Korfhagen, T.R., McCormack, F.X., Ashbaugh, A.D., Steele, P., Whitsett, J.A., and Walzer, P.D. (2001). Immunosuppressed surfactant protein A-deficient mice have increased susceptibility to *Pneumocystis carinii* infection. J. Infect. Dis. *183*, 943–952.

Lionakis, M.S., and Kontoyiannis, D.P. (2005). Fruit flies as a minihost model for studying drug activity and virulence in *Aspergillus*. Med. Mycol. *43 (Suppl 1)*, S111–114.

Lionakis, M.S., Lewis, R.E., May, G.S., Wiederhold, N.P., Albert, N.D., Halder, G., and Kontoyiannis, D.P. (2005). Toll-deficient *Drosophila* flies as a fast, high-throughput model for the study of antifungal drug efficacy against invasive aspergillosis and *Aspergillus* virulence. J. Infect. Dis. *191*, 1188–1195.

Lo, H.J., Kohler, J.R., DiDomenico, B., Loebenberg, D., Cacciapuoti, A., and Fink, G.R. (1997). Nonfilamentous *C. albicans* mutants are avirulent. Cell *90*, 939–949.

Lortholary, O., Improvisi, L., Nicolas, M., Provost, F., Dupont, B., and Dromer, F. (1999). Fungemia during murine cryptococcosis sheds some light on pathophysiology. Med. Mycol. 37, 169–174.

Louria, D.B., Brayton, R.G., and Finkel, G. (1963). Studies on the pathogenesis of experimental *Candida albicans* infections in mice. Sabouraudia 2, 271–283.

Louria, D.B., Buse, M., Brayton, R.G., and Finkel, G. (1967). The pathogenesis of *Candida tropicalis* infections in mice. Sabouraudia 5, 14–25.

Lovchik, J.A., and Lipscomb, M.F. (1993). Role for C5 and neutrophils in the pulmonary intravascular clearance of circulating *Cryptococcus neoformans*. Am. J. Respir. Cell Mol. Biol. *9*, 617–627.

Lyman, C.A., Gonzalez, C., Schneider, M., Lee, J., and Walsh, T.J. (1999). Effects of the hematoregulatory peptide SKandF 107647 alone and in combination with amphotericin B against disseminated candidiasis in persistently neutropenic rabbits. Antimicrob. Agents Chemother. *43*, 2165–2169.

Lyon, J.P., da Costa, S.C., Totti, V.M., Munhoz, M.F., and de Resende, M.A. (2006). Predisposing conditions for *Candida spp.* carriage in the oral cavity of denture wearers and individuals with natural teeth. Can. J. Microbiol. *52*, 462–467.

MacCallum, D.M. (2010). *Candida* infections and modelling disease. In Pathogenic Yeasts, The Yeast Handbook, H.R. Ashbee, and E. Bignell, eds. (Springer-Verlag), pp. 41–67.

MacCallum, D.M., and Odds, F.C. (2002a). Efficacy of parenteral itraconazole against disseminated *Candida albicans* infection in two mouse strains. J. Antimicrob. Chemother. *50*, 225–229.

MacCallum, D.M., and Odds, F.C. (2002b). Influence of grapefruit juice on itraconazole plasma levels in mice and guinea pigs. J. Antimicrob. Chemother. *50*, 219–224.

MacCallum, D.M., and Odds, F.C. (2005). Temporal events in the intravenous challenge model for experimental *Candida albicans* infections in female mice. Mycoses *48*, 151–161.

MacCallum, D.M., Whyte, J.A., and Odds, F.C. (2005). Efficacy of caspofungin and voriconazole combinations in experimental aspergillosis. Antimicrob. Agents Chemother. *49*, 3697–3701.

MacCallum, D.M., Findon, H., Kenny, C.C., Butler, G., Haynes, K., and Odds, F.C. (2006). Different consequences of *ACE2* and *SWI5* gene disruptions for virulence of pathogenic and nonpathogenic yeasts. Infect. Immun. *74*, 5244–5248.

MacCallum, D.M., Castillo, L., Brown, A.J.P., Gow, N.A.R., and Odds, F.C. (2009a). Early-expressed chemokines predict kidney immunopathology in experimental disseminated *Candida albicans* infections. Plos One *4*, e6420.

MacCallum, D.M., Castillo, L., Nather, K., Munro, C.A., Brown, A.J., Gow, N.A., and Odds, F.C. (2009b).

Property differences among the four major *Candida albicans* strain clades. Eukaryot. Cell. *8*, 373–387.

MacCallum, D.M., Coste, A., Ischer, F., Jacobsen, M.D., Odds, F.C., and Sanglard, D. (2010). Genetic dissection of azole resistance mechanisms in *Candida albicans* and their validation in a mouse model of disseminated infection. Antimicrob. Agents Chemother. *54*, 1476–1483.

Markogiannakis, H., Pachylaki, N., Samara, E., Kalderi, M., Minettou, M., Toutouza, M., Toutouzas, K.G., Theodorou, D., and Katsaragakis, S. (2008). Infections in a surgical intensive care unit of a university hospital in Greece. Int. J. Infect. Dis. *13*, 145–153.

Marquis, G., Montplaisir, S., Pelletier, M., Auger, P., and Lapp, W.S. (1988). Genetics of resistance to infection with *Candida albicans* in mice. Br. J. Exp. Pathol. *69*, 651–660.

Martin, M.V., Yates, J., and Hitchcock, C.A. (1997). Comparison of voriconazole (UK-109,496) and itraconazole in prevention and treatment of *Aspergillus fumigatus* endocarditis in guinea pigs. Antimicrob. Agents Chemother. *41*, 13–16.

Matthews, R.C., Rigg, G., Hodgetts, S., Carter, T., Chapman, C., Gregory, C., Illidge, C., and Burnie, J. (2003). Preclinical assessment of the efficacy of mycograb, a human recombinant antibody against fungal HSP90. Antimicrob. Agents Chemother. *47*, 2208–2216.

Mazzolla, R., Barluzzi, R., Brozzetti, A., Boelaert, J.R., Luna, T., Saleppico, S., Bistoni, F., and Blasi, E. (1997). Enhanced resistance to *Cryptococcus neoformans* infection induced by chloroquine in a murine model of meningoencephalitis. Antimicrob. Agents Chemother. *41*, 802–807.

McKinley, L., Logar, A.J., McAllister, F., Zheng, M., Steele, C., and Kolls, J.K. (2006). Regulatory T cells dampen pulmonary inflammation and lung injury in an animal model of *Pneumocystis* pneumonia. J. Immunol. *177*, 6215–6226.

McKinnell, J.A., Cannella, A.P., Kunz, D.F., Hook 3rd, E.W., Moser, S.A., Miller, L.G., Baddley, J.W., and Pappas, P.G. (2012). *Pneumocystis* pneumonia in hospitalized patients: a detailed examination of symptoms, management, and outcomes in human immunodeficiency virus (HIV)-infected and HIV-uninfected persons. Transplant Infectious Disease *14*, 510–518.

Meeker, N.D., and Trede, N.S. (2008). Immunology and zebrafish: spawning new models of human disease. Dev. Comp. Immunol. *32*, 745–757.

Mellado, E., Cuenca-Estrella, M., Regadera, J., Gonzalez, M., Diaz-Guerra, T.M., and Rodriguez-Tudela, J.L. (2000). Sustained gastrointestinal colonization and systemic dissemination by *Candida albicans*, *Candida tropicalis* and *Candida parapsilosis* in adult mice. Diagn. Microbiol. Infect. Dis. *38*, 21–28.

Mencacci, A., Cenci, E., Bistoni, F., Bacci, A., Del Sero, G., Montagnoli, C., Fe d'Ostiani, C., and Romani, L. (1998). Specific and non-specific immunity to *Candida albicans*: a lesson from genetically modified animals. Res. Immunol. *149*, 352–61; discussion 517–519.

Mestas, J., and Hughes, C.C. (2004). Of mice and not men: differences between mouse and human immunology. J. Immunol. *172*, 2731–2738.

Mignon, B.R., Leclipteux, T., Focant, C., Nikkels, A.J., Pierard, G.E., and Losson, B.J. (1999). Humoral and cellular immune response to a crude exo-antigen and purified keratinase of *Microsporum canis* in experimentally infected guinea pigs. Med. Mycol. *37*, 123–129.

Mircescu, M.M., Lipuma, L., van Rooijen, N., Pamer, E.G., and Hohl, T.M. (2009). Essential role for neutrophils but not alveolar macrophages at early time points following *Aspergillus fumigatus* infection. J. Infect. Dis. *200*, 647–656.

Mody, C.H., Toews, G.B., and Lipscomb, M.F. (1988). Cyclosporin A inhibits the growth of *Cryptococcus neoformans* in a murine model. Infect. Immun. *56*, 7–12.

Mondon, P., De Champs, C., Donadille, A., Ambroise-Thomas, P., and Grillot, R. (1996). Variation in virulence of *Aspergillus fumigatus* strains in a murine model of invasive pulmonary aspergillosis. J. Med. Microbiol. *45*, 186–191.

Mosquera, J., Warn, P.A., Morrissey, J., Moore, C.B., Gil-Lamaignere, C., and Denning, D.W. (2001). Susceptibility testing of *Aspergillus flavus*: inoculum dependence with itraconazole and lack of correlation between susceptibility to amphotericin B *in vitro* and outcome *in vivo*. Antimicrob. Agents Chemother. *45*, 1456–1462.

Mukherjee, J., Scharff, M.D., and Casadevall, A. (1992). Protective murine monoclonal antibodies to *Cryptococcus neoformans*. Infect. Immun. *60*, 4534–4541.

Mukherjee, J., Pirofski, L.A., Scharff, M.D., and Casadevall, A. (1993). Antibody-mediated protection in mice with lethal intracerebral *Cryptococcus neoformans* infection. Proc. Natl. Acad. Sci. U.S.A. *90*, 3636–3640.

Mukherjee, J., Zuckier, L.S., Scharff, M.D., and Casadevall, A. (1994a). Therapeutic efficacy of monoclonal antibodies to *Cryptococcus neoformans* glucuronoxylomannan alone and in combination with amphotericin B. Antimicrob. Agents Chemother. *38*, 580–587.

Mukherjee, S., Lee, S., Mukherjee, J., Scharff, M.D., and Casadevall, A. (1994b). Monoclonal antibodies to *Cryptococcus neoformans* capsular polysaccharide modify the course of intravenous infection in mice. Infect. Immun. *62*, 1079–1088.

Mukherjee, S., Lee, S.C., and Casadevall, A. (1995). Antibodies to *Cryptococcus neoformans* glucuronoxylomannan enhance antifungal activity of murine macrophages. Infect. Immun. *63*, 573–579.

Muller, U., Stenzel, W., Kohler, G., Werner, C., Polte, T., Hansen, G., Schutze, N., Straubinger, R.K., Blessing, M., McKenzie, A.N., Brombacher, F., and Alber, G. (2007). IL-13 induces disease-promoting type 2 cytokines, alternatively activated macrophages and allergic inflammation during pulmonary infection of mice with *Cryptococcus neoformans*. J. Immunol. *179*, 5367–5377.

Nagino, K., Shimohira, H., Ogawa, M., Uchida, K., and Yamaguchi, H. (2000). Comparison of the therapeutic efficacy of oral doses of fluconazole and itraconazole in a guinea pig model of dermatophytosis. J. Infect. Chemother. *6*, 41–44.

Naglik, J.R., Fidel, P.L.,Jr, and Odds, F.C. (2008). Animal models of mucosal *Candida* infection. FEMS Microbiol. Lett. *283*, 129–139.

Najvar, L.K., Bocanegra, R., and Graybill, J.R. (1999). An alternative animal model for comparison of treatments for cryptococcal meningitis. Antimicrob. Agents Chemother. *43*, 413–414.

Nakayama, H., Ueno, K., Uno, J., Nagi, M., Tanabe, K., Aoyama, T., Chibana, H., and Bard, M. (2011). Growth defects resulting from inhibiting *ERG20* and *RAM2* in *Candida glabrata*. FEMS Microbiol. Lett. *317*, 27–33.

Nawada, R., Amitani, R., Tanaka, E., Niimi, A., Suzuki, K., Murayama, T., and Kuze, F. (1996). Murine model of invasive pulmonary aspergillosis following an earlier stage, noninvasive *Aspergillus* infection. J. Clin. Microbiol. *34*, 1433–1439.

Netea, M.G., and Marodi, L. (2010). Innate immune mechanisms for recognition and uptake of *Candida* species. Trends Immunol. *31*, 346–353.

Nett, J.E., Marchillo, K., Spiegel, C.A., and Andes, D.R. (2010). Development and validation of an *in vivo Candida albicans* biofilm denture model. Infect. Immun. *78*, 3650–3659.

Nichterlein, T., Buchheidt, D., Hein, A., Becker, K.P., Mosbach, K., and Kretschmar, M. (2003). Comparison of glucan detection and galactomannan enzyme immunoassay in gastrointestinal and systemic murine candidiasis. Diagn. Microbiol. Infect. Dis. *46*, 103–108.

Nielsen, K., Cox, G.M., Litvintseva, A.P., Mylonakis, E., Malliaris, S.D., Benjamin, D.K.,Jr, Giles, S.S., Mitchell, T.G., Casadevall, A., Perfect, J.R., and Heitman, J. (2005). *Cryptococcus neoformans* alpha strains preferentially disseminate to the central nervous system during coinfection. Infect. Immun. *73*, 4922–4933.

Niwano, Y., Tabuchi, T., Kanai, K., Hamaguchi, H., Uchida, K., and Yamaguchi, H. (1995). Short-term topical therapy of experimental tinea pedis in guinea pigs with lanoconazole, a new imidazole antimycotic agent. Antimicrob. Agents Chemother. *39*, 2353–2355.

Nobile, C.J., Solis, N., Myers, C.L., Fay, A.J., Deneault, J.S., Nantel, A., Mitchell, A.P., and Filler, S.G. (2008). *Candida albicans* transcription factor Rim101 mediates pathogenic interactions through cell wall functions. Cell. Microbiol. *10*, 2180–2196.

Oakley, K.L., Morrissey, G., and Denning, D.W. (1997). Efficacy of SCH-56592 in a temporarily neutropenic murine model of invasive aspergillosis with an itraconazole-susceptible and an itraconazole-resistant isolate of *Aspergillus fumigatus*. Antimicrob. Agents Chemother. *41*, 1504–1507.

O'Connor, L., Lahiff, S., Casey, F., Glennon, M., Cormican, M., and Maher, M. (2005). Quantification of *ALS1* gene expression in *Candida albicans* biofilms by RT-PCR using hybridisation probes on the LightCycler. Mol. Cell. Probes *19*, 153–162.

Odds, F.C. (1988). *Candida* and candidosis (Bailliere Tindall, London).

Odds, F.C., Van Gerven, F., Espinel-Ingroff, A., Bartlett, M.S., Ghannoum, M.A., Lancaster, M.V., Pfaller, M.A., Rex, J.H., Rinaldi, M.G., and Walsh, T.J. (1998). Evaluation of possible correlations between antifungal susceptibilities of filamentous fungi *in vitro* and antifungal treatment outcomes in animal infection models. Antimicrob. Agents Chemother. *42*, 282–288.

Odds, F.C., Oris, M., Van Dorsselaer, P., and Van Gerven, F. (2000). Activities of an intravenous formulation of itraconazole in experimental disseminated *Aspergillus*, *Candida*, and *Cryptococcus* infections. Antimicrob. Agents Chemother. *44*, 3180–3183.

Odds, F., Ausma, J., Van Gerven, F., Woestenborghs, F., Meerpoel, L., Heeres, J., Vanden Bossche, H., and Borgers, M. (2004). *In vitro* and *in vivo* activities of the novel azole antifungal agent r126638. Antimicrob. Agents Chemother. *48*, 388–391.

Okawa, Y., Miyauchi, M., and Kobayashi, H. (2008). Comparison of pathogenicity of various *Candida tropicalis* strains. Biol. Pharm. Bull. *31*, 1507–1510.

Olszewski, M.A., Huffnagle, G.B., McDonald, R.A., Lindell, D.M., Moore, B.B., Cook, D.N., and Toews, G.B. (2000). The role of macrophage inflammatory protein-1 alpha/CCL3 in regulation of T cell-mediated immunity to *Cryptococcus neoformans* infection. J. Immunol. *165*, 6429–6436.

Olszewski, M.A., Noverr, M.C., Chen, G.H., Toews, G.B., Cox, G.M., Perfect, J.R., and Huffnagle, G.B. (2004). Urease expression by *Cryptococcus neoformans* promotes microvascular sequestration, thereby enhancing central nervous system invasion. Am. J. Pathol. *164*, 1761–1771.

Orsi, G.B., Scorzolini, L., Franchi, C., Mondillo, V., Rosa, G., and Venditti, M. (2006). Hospital-acquired infection surveillance in a neurosurgical intensive care unit. J. Hosp. Infect. *64*, 23–29.

Papadimitriou, J.M., and Ashman, R.B. (1986). The pathogenesis of acute systemic candidiasis in a susceptible inbred mouse strain. J. Pathol. *150*, 257–265.

Park, B.J., Wannemuehler, K.A., Marston, B.J., Govender, N., Pappas, P.G., and Chiller, T.M. (2009). Estimation of the current global burden of cryptococcal meningitis among persons living with HIV/AIDS. Aids 23, 525–530.

Park, H., Myers, C.L., Sheppard, D.C., Phan, Q.T., Sanchez, A.A., E Edwards, J., and Filler, S.G. (2005a). Role of the fungal Ras-protein kinase A pathway in governing epithelial cell interactions during oropharyngeal candidiasis. Cell. Microbiol. 7, 499–510.

Park, S., Kelly, R., Kahn, J.N., Robles, J., Hsu, M.J., Register, E., Li, W., Vyas, V., Fan, H., Abruzzo, G., *et al.* (2005b). Specific substitutions in the echinocandin target Fks1p account for reduced susceptibility of rare laboratory and clinical *Candida* sp. isolates. Antimicrob. Agents Chemother. *49*, 3264–3273.

Paulitsch, A., Weger, W., Ginter-Hanselmayer, G., Marth, E., and Buzina, W. (2006). A 5-year (2000–2004) epidemiological survey of *Candida* and non-*Candida* yeast species causing vulvovaginal candidiasis in Graz, Austria. Mycoses *49*, 471–475.

Perfect, J.R., Lang, S.D., and Durack, D.T. (1980). Chronic cryptococcal meningitis: a new experimental model in rabbits. Am. J. Pathol. *101*, 177–194.

Piggott, W.R., and Emmons, C.W. (1960). Device for inhalation exposure of animals to spores. Proc. Soc. Exp. Biol. Med. *103*, 805–806.

Pires-Goncalves, R.H., Miranda, E.T., Baeza, L.C., Matsumoto, M.T., Zaia, J.E., and Mendes-Giannini, M.J. (2007). Genetic relatedness of commensal strains of *Candida albicans* carried in the oral cavity of patients' dental prosthesis users in Brazil. Mycopathologia *164*, 255–263.

Pirotta, M.V., and Garland, S.M. (2006). Genital *Candida* species detected in samples from women in Melbourne, Australia, before and after treatment with antibiotics. J. Clin. Microbiol. *44*, 3213–3217.

Polanco, A., Mellado, E., Castilla, C., and Rodriguez-Tudela, J.L. (1999). Detection of *Candida albicans* in blood by PCR in a rabbit animal model of disseminated candidiasis. Diagn. Microbiol. Infect. Dis. *34*, 177–183.

Powles, M.A., McFadden, D.C., Pittarelli, L.A., and Schmatz, D.M. (1992). Mouse model for *Pneumocystis carinii* pneumonia that uses natural transmission to initiate infection. Infect. Immun. *60*, 1397–1400.

Puel, A., Picard, C., Cypowyj, S., Lilic, D., Abel, L., and Casanova, J.L. (2010). Inborn errors of mucocutaneous immunity to *Candida albicans* in humans: a role for IL-17 cytokines? Curr. Opin. Immunol. *22*, 467–474.

Puel, A., Cypowyj, S., Bustamante, J., Wright, J.F., Liu, L., Lim, H.K., Migaud, M., Israel, L., Chrabieh, M., Audry, M., *et al.* (2011). Chronic mucocutaneous candidiasis in humans with inborn errors of interleukin-17 immunity. Science *332*, 65–68.

Rahman, D., Mistry, M., Thavaraj, S., Challacombe, S.J., and Naglik, J.R. (2007). Murine model of concurrent oral and vaginal *Candida albicans* colonization to study epithelial host–pathogen interactions. Microbes Infect. *9*, 615–622.

Rehli, M. (2002). Of mice and men: species variations of Toll-like receptor expression. Trends Immunol. *23*, 375–378.

Reinhardt, J.H., Allen, A.M., Gunnison, D., and Akers, W.A. (1974). Experimental human *Trichophyton mentagrophytes* infections. J. Invest. Dermatol. *63*, 419–422.

de Repentigny, L., Phaneuf, M., and Mathieu, L.G. (1992). Gastrointestinal colonization and systemic dissemination by *Candida albicans* and *Candida tropicalis* in intact and immunocompromised mice. Infect. Immun. *60*, 4907–4914.

Richardson, M.D., and Warnock, D.W. (1997). Superficial candidosis. In Fungal infection: diagnosis and management (Blackwell Science, London) pp. 78–93.

Rudner, X.L., Happel, K.I., Young, E.A., and Shellito, J.E. (2007). Interleukin-23 (IL-23)-IL-17 cytokine axis in murine *Pneumocystis carinii* infection. Infect. Immun. *75*, 3055–3061.

Saijo, S., Fujikado, N., Furuta, T., Chung, S.H., Kotaki, H., Seki, K., Sudo, K., Akira, S., Adachi, Y., Ohno, N., *et al.* (2007). Dectin-1 is required for host defense against *Pneumocystis carinii* but not against *Candida albicans*. Nat. Immunol. *8*, 39–46.

Salkowski, C.A., and Balish, E. (1990). Pathogenesis of *Cryptococcus neoformans* in congenitally immunodeficient beige athymic mice. Infect. Immun. *58*, 3300–3306.

Samaranayake, Y.H., and Samaranayake, L.P. (2001). Experimental oral candidiasis in animal models. Clin. Microbiol. Rev. *14*, 398–429.

Sampaio, P., Santos, M., Correia, A., Amaral, F.E., Chavez-Galarza, J., Costa-de-Oliveira, S., Castro, A.G., Pedrosa, J., and Pais, C. (2010). Virulence attenuation of *Candida albicans* genetic variants isolated from a patient with a recurrent bloodstream infection. PLoS One *5*, e10155.

Sandovsky-Losica, H., Barr-Nea, L., and Segal, E. (1992). Fatal systemic candidiasis of gastrointestinal origin: an experimental model in mice compromised by anticancer treatment. J. Med. Vet. Mycol. *30*, 219–231.

Santangelo, R., Zoellner, H., Sorrell, T., Wilson, C., Donald, C., Djordjevic, J., Shounan, Y., and Wright, L. (2004). Role of extracellular phospholipases and mononuclear phagocytes in dissemination of cryptococcosis in a murine model. Infect. Immun. *72*, 2229–2239.

Sarvikivi, E., Lyytikainen, O., Vaara, M., and Saxen, H. (2008). Nosocomial bloodstream infections in children: an 8-year experience at a tertiary-care hospital in Finland. Clin. Microbiol. Infect. *14*, 1072–1075.

Saunte, D.M., Simmel, F., Frimodt-Moller, N., Stolle, L.B., Svejgaard, E.L., Haedersdal, M., Kloft, C., and Arendrup, M.C. (2007). *In vivo* efficacy and pharmacokinetics of voriconazole in an animal model of dermatophytosis. Antimicrob. Agents Chemother. *51*, 3317–3321.

Saunte, D.M., Hasselby, J.P., Brillowska-Dabrowska, A., Frimodt-Moller, N., Svejgaard, E.L., Linnemann, D., Nielsen, S.S., Haedersdal, M., and Arendrup, M.C. (2008). Experimental guinea pig model of dermatophytosis: a simple and useful tool for the evaluation of new diagnostics and antifungals. Med. Mycol. *46*, 303–313.

Scanlan, P.D., and Marchesi, J.R. (2008). Micro-eukaryotic diversity of the human distal gut microbiota: qualitative assessment using culture-dependent and -independent analysis of faeces. Isme j. *2*, 1183–1193.

Schmidt, C.S., White, C.J., Ibrahim, A.S., Filler, S.G., Fu, Y., Yeaman, M.R., Edwards, J.E.,Jr, and Hennessey, J.P.,Jr. (2012). NDV-3, a recombinant alum-adjuvanted vaccine for *Candida* and *Staphylococcus aureus*, is safe and immunogenic in healthy adults. Vaccine *30*, 7594–7600.

Schulten, E.A., ten Kate, R.W., and van der Waal, I. (1989). Oral manifestations of HIV infection in 75 Dutch patients. J. Oral Pathol. Med. *18*, 42–46.

Schwarz, P., Dromer, F., Lortholary, O., and Dannaoui, E. (2006). Efficacy of amphotericin B in combination with flucytosine against flucytosine-susceptible or flucytosine-resistant isolates of *Cryptococcus neoformans* during disseminated murine cryptococcosis. Antimicrob. Agents Chemother. *50*, 113–120.

Segal, B.H. (2009). Aspergillosis. N. Engl. J. Med. *360*, 1870–1884.

Serena, C., Pastor, F.J., Marine, M., Rodriguez, M.M., and Guarro, J. (2007). Efficacy of voriconazole in a murine model of cryptococcal central nervous system infection. J. Antimicrob. Chemother. *60*, 162–165.

Sharma, B., Kumar, P., and Joshi, S.C. (2011). Topical treatment of dermatophytic lesion on mice (*Mus musculus*) Model. Indian J. Microbiol. *51*, 217–222.

Shellito, J., Suzara, V.V., Blumenfeld, W., Beck, J.M., Steger, H.J., and Ermak, T.H. (1990). A new model of *Pneumocystis carinii* infection in mice selectively depleted of helper T lymphocytes. J. Clin. Invest. *85*, 1686–1693.

Shellito, J.E., Kolls, J.K., Olariu, R., and Beck, J.M. (1996). Nitric oxide and host defense against *Pneumocystis carinii* infection in a mouse model. J. Infect. Dis. *173*, 432–439.

Sheppard, D.C., Rieg, G., Chiang, L.Y., Filler, S.G., Edwards, J.E.,Jr, and Ibrahim, A.S. (2004). Novel inhalational murine model of invasive pulmonary aspergillosis. Antimicrob. Agents Chemother. *48*, 1908–1911.

Sheppard, D.C., Graybill, J.R., Najvar, L.K., Chiang, L.Y., Doedt, T., Kirkpatrick, W.R., Bocanegra, R., Vallor, A.C., Patterson, T.F., and Filler, S.G. (2006a). Standardization of an experimental murine model of invasive pulmonary aspergillosis. Antimicrob. Agents Chemother. *50*, 3501–3503.

Sheppard, D.C., Marr, K.A., Fredricks, D.N., Chiang, L.Y., Doedt, T., and Filler, S.G. (2006b). Comparison of three methodologies for the determination of pulmonary fungal burden in experimental murine aspergillosis. Clin. Microbiol. Infect. *12*, 376–380.

Shimamura, T., Kubota, N., and Shibuya, K. (2012). Animal model of dermatophytosis. J. Biomed. Biotechnol. *2012*, 125384.

Sidransky, H., and Friedman, L. (1959). The effect of cortisone and antibiotic agents on experimental pulmonary aspergillosis. Am. J. Pathol. *35*, 169–183.

da Silva Ferreira, M.E., Kress, M.R., Savoldi, M., Goldman, M.H., Hartl, A., Heinekamp, T., Brakhage, A.A., and Goldman, G.H. (2006). The akuB(KU80) mutant deficient for nonhomologous end joining is a powerful tool for analyzing pathogenicity in *Aspergillus fumigatus*. Eukaryot. Cell. *5*, 207–211.

Singh, N., Alexander, B.D., Lortholary, O., Dromer, F., Gupta, K.L., John, G.T., del Busto, R., Klintmalm, G.B., Somani, J., Lyon, G.M., *et al.* (2007). *Cryptococcus neoformans* in organ transplant recipients: impact of calcineurin-inhibitor agents on mortality. J. Infect. Dis. *195*, 756–764.

Singh, N., Dromer, F., Perfect, J.R., and Lortholary, O. (2008). Cryptococcosis in solid organ transplant recipients: current state of the science. Clin. Infect. Dis. *47*, 1321–1327.

Skrzypek, M.S., Arnaud, M.B., Costanzo, M.C., Inglis, D.O., Shah, P., Binkley, G., Miyasato, S.R., and Sherlock, G. (2010). New tools at the *Candida* Genome Database: biochemical pathways and full-text literature search. Nucleic Acids Res. *38*, D428–432.

Slater, J.L., Gregson, L., Denning, D.W., and Warn, P.A. (2011). Pathogenicity of *Aspergillus fumigatus* mutants assessed in *Galleria mellonella* matches that in mice. Med. Mycol. 49 (Suppl. 1), S107–113.

Smith, J.M., Tang, C.M., Van Noorden, S., and Holden, D.W. (1994). Virulence of *Aspergillus fumigatus* double mutants lacking restriction and an alkaline protease in a low-dose model of invasive pulmonary aspergillosis. Infect. Immun. *62*, 5247–5254.

Sobel, J.D. (2007). Vulvovaginal candidosis. Lancet *369*, 1961–1971.

Sobel, J.D., Muller, G., and McCormick, J.F. (1985). Experimental chronic vaginal candidosis in rats. Sabouraudia *23*, 199–206.

Sobel, J.D., Faro, S., Force, R.W., Foxman, B., Ledger, W.J., Nyirjesy, P.R., Reed, B.D., and Summers, P.R. (1998). Vulvovaginal candidiasis: epidemiologic, diagnostic, and therapeutic considerations. Am. J. Obstet. Gynecol. *178*, 203–211.

Spellberg, B., Ibrahim, A.S., Edwards, J.E.,Jr, and Filler, S.G. (2005). Mice with disseminated candidiasis die of progressive sepsis. J. Infect. Dis. *192*, 336–343.

Spellberg, B.J., Ibrahim, A.S., Avanesian, V., Fu, Y., Myers, C., Phan, Q.T., Filler, S.G., Yeaman, M.R., and Edwards, J.E.,Jr. (2006). Efficacy of the anti-*Candida* rAls3p-N or rAls1p-N vaccines against disseminated and mucosal candidiasis. J. Infect. Dis. *194*, 256–260.

Staib, P., Zaugg, C., Mignon, B., Weber, J., Grumbt, M., Pradervand, S., Harshman, K., and Monod, M. (2010). Differential gene expression in the pathogenic dermatophyte *Arthroderma benhamiae in vitro* versus during infection. Microbiology *156*, 884–895.

Steele, C., Ratterree, M., and Fidel, P.L.,Jr. (1999). Differential susceptibility of two species of macaques to experimental vaginal candidiasis. J. Infect. Dis. *180*, 802–810.

Steinbach, W.J., Benjamin, D.K.,Jr, Trasi, S.A., Miller, J.L., Schell, W.A., Zaas, A.K., Foster, W.M., and Perfect, J.R. (2004). Value of an inhalational model of invasive aspergillosis. Med. Mycol. *42*, 417–425.

Stephens-Romero, S.D., Mednick, A.J., and Feldmesser, M. (2005). The pathogenesis of fatal outcome in murine pulmonary aspergillosis depends on the neutrophil depletion strategy. Infect. Immun. *73*, 114–125.

Stokes, D.C., Gigliotti, F., Rehg, J.E., Snellgrove, R.L., and Hughes, W.T. (1987). Experimental *Pneumocystis carinii* pneumonia in the ferret. Br. J. Exp. Pathol. *68*, 267–276.

Stringer, J.R. (1996). *Pneumocystis carinii*: what is it, exactly? Clin. Microbiol. Rev. *9*, 489–498.

Stringer, J.R., Beard, C.B., Miller, R.F., and Wakefield, A.E. (2002). A new name (*Pneumocystis jiroveci*) for *Pneumocystis* from humans. Emerg. Infect. Dis. *8*, 891–896.

Sugar, A.M., and Liu, X.P. (2000). Effect of grapefruit juice on serum voriconazole concentrations in the mouse. Med. Mycol. *38*, 209–212.

Szabo, E.K., and MacCallum, D.M. (2011). The contribution of mouse models to our understanding of systemic candidiasis. FEMS Microbiol. Lett. *320*, 1–8.

Tabart, J., Baldo, A., Vermout, S., Nusgens, B., Lapiere, C., Losson, B., and Mignon, B. (2007). Reconstructed interfollicular feline epidermis as a model for *Microsporum canis* dermatophytosis. J. Med. Microbiol. *56*, 971–975.

Tavanti, A., Campa, D., Arancia, S., Hensgens, L.A., de Bernardis, F., and Senesi, S. (2010). Outcome of experimental rat vaginitis by *Candida albicans* isolates with different karyotypes. Microb. Pathog. *49*, 47–50.

Taylor, B.N., Fichtenbaum, C., Saavedra, M., Slavinsky, J., Swoboda, R., Wozniak, K., Arribas, A., Powderly, W., and Fidel, P.L.,Jr. (2000). *In vivo* virulence of *Candida albicans* isolates causing mucosal infections in people infected with the human immunodeficiency virus. J. Infect. Dis. *182*, 955–959.

Tebo, J.M., Datta, S., Kishore, R., Kolosov, M., Major, J.A., Ohmori, Y., and Hamilton, T.A. (2000). Interleukin-1-mediated stabilization of mouse KC mRNA depends on sequences in both 5′- and 3′-untranslated regions. J. Biol. Chem. *275*, 12987–12993.

Thaweboon, S., Thaweboon, B., Srithavaj, T., and Choonharuangdej, S. (2008). Oral colonization of *Candida* species in patients receiving radiotherapy in the head and neck area. Quintessence Int. 39, e52–57.

Treiber, A., Pittermann, W., and Schuppe, H.C. (2001). Efficacy testing of antimycotic prophylactics in an animal model. Int. J. Hyg. Environ. Health *204*, 239–243.

Uchida, K., and Yamaguchi, H. (1984). Assessment of *in vivo* activity of bifonazole against dermatophytic infection in guinea pigs on the basis of the amount of a specific fungal cell wall component chitin in the infected skin. Dermatologica *169 (Suppl 1)*, 47–49.

Uchida, K., and Yamaguchi, H. (1994). Effectiveness of oral treatment with terbinafine in a guinea pig model of tinea pedis. Jpn. J. Antibiot. *47*, 1401–1406.

Uchida, K., and Ymaguchi, H. (1999). Therapeutic efficacy of a topical antifungal solution preparation formulated with pyrrolnitrin and clotrimazole in combination (Pyroace W) in guinea pig model of tinea pedis. Jpn. J. Antibiot. *52*, 68–74.

Uno, K., Sugiura, S., Konishi, M., Yasuda, Y., Mikasa, K., and Kita, E. (2007). Evaluation of diagnostic methods for *Candida albicans* translocation in a mouse model: seminested polymerase chain reaction, blood culture, and serological assays. J. Infect. Chemother. *13*, 196–203.

Van Cutsem, J., and Thienpont, D. (1971). Experimental cutaneous *Candida albicans* infection in guinea-pigs. Sabouraudia *9*, 17–20.

Van Cutsem, J., and Janssen, P.A. (1984). Experimental systemic dermatophytosis. J. Invest. Dermatol. *83*, 26–31.

Van Cutsem, J., Van Gerven, F., Fransen, J., Schrooten, P., and Janssen, P.A. (1990). The *in vitro* antifungal activity of ketoconazole, zinc pyrithione, and selenium sulfide against *Pityrosporum* and their efficacy as a shampoo in the treatment of experimental pityrosporosis in guinea pigs. J. Am. Acad. Dermatol. *22*, 993–998.

Velagapudi, R., Hsueh, Y.P., Geunes-Boyer, S., Wright, J.R., and Heitman, J. (2009). Spores as infectious propagules of *Cryptococcus neoformans*. Infect. Immun. *77*, 4345–4355.

Venturini, J., Alvares, A.M., Camargo, M.R., Marchetti, C.M., Fraga-Silva, T.F., Luchini, A.C., and Arruda, M.S. (2012). Dermatophyte–host relationship of a murine model of experimental invasive dermatophytosis. Microbes Infect. *14*, 1144–1151.

van Vianen, W., de Marie, S., ten Kate, M.T., Mathot, R.A., and Bakker-Woudenberg, I.A. (2006). Caspofungin: antifungal activity *in vitro*, pharmacokinetics, and effects on fungal load and animal survival in neutropenic rats with invasive pulmonary aspergillosis. J. Antimicrob. Chemother. *57*, 732–740.

Vuk-Pavlovic, Z., Mo, E.K., Icenhour, C.R., Standing, J.E., Fisher, J.H., and Limper, A.H. (2006). Surfactant protein D enhances *Pneumocystis infection* in immune-suppressed mice. Am. J. Physiol. Lung Cell. Mol. Physiol. *290*, L442–449.

Wagner, R.D., Vazquez-Torres, A., Jones-Carson, J., Warner, T., and Balish, E. (1996). B cell knockout mice are resistant to mucosal and systemic candidiasis of endogenous origin but susceptible to experimental systemic candidiasis. J. Infect. Dis. *174*, 589–597.

Walzer, P.D., Schnelle, V., Armstrong, D., and Rosen, P.P. (1977). Nude mouse: a new experimental model for *Pneumocystis carinii* infection. Science *197*, 177–179.

Wang, J., Ohshima, T., Yasunari, U., Namikoshi, S., Yoshihara, A., Miyazaki, H., and Maeda, N. (2006). The carriage of *Candida* species on the dorsal surface of the tongue: the correlation with the dental, periodontal and prosthetic status in elderly subjects. Gerodontology 23, 157–163.

White, L.O. (1977). Germination of *Aspergillus fumigatus* conidia in the lungs of normal and cortisone-treated mice. Sabouraudia *15*, 37–41.

Wiesner, S.M., Jechorek, R.P., Garni, R.M., Bendel, C.M., and Wells, C.L. (2001). Gastrointestinal colonization by *Candida albicans* mutant strains in antibiotic-treated mice. Clin. Diagn. Lab. Immunol. *8*, 192–195.

Willger, S.D., Puttikamonkul, S., Kim, K.H., Burritt, J.B., Grahl, N., Metzler, L.J., Barbuch, R., Bard, M., Lawrence, C.B., and Cramer, R.A.,Jr. (2008). A sterol-regulatory element binding protein is required for cell polarity, hypoxia adaptation, azole drug resistance, and virulence in *Aspergillus fumigatus*. PLoS Pathog. *4*, e1000200.

Wingard, J.R., Dick, J.D., Merz, W.G., Sandford, G.R., Saral, R., and Burns, W.H. (1980). Pathogenicity of *Candida tropicalis* and *Candida albicans* after gastrointestinal inoculation in mice. Infect. Immun. *29*, 808–813.

Wingard, J.R., Dick, J.D., Merz, W.G., Sandford, G.R., Saral, R., and Burns, W.H. (1982). Differences in virulence of clinical isolates of *Candida tropicalis* and *Candida albicans* in mice. Infect. Immun. *37*, 833–836.

Wisplinghoff, H., Bischoff, T., Tallent, S.M., Seifert, H., Wenzel, R.P., and Edmond, M.B. (2004). Nosocomial bloodstream infections in US hospitals: analysis of 24,179 cases from a prospective nationwide surveillance study. Clin. Infect. Dis. *39*, 309–317.

Wisplinghoff, H., Seifert, H., Wenzel, R.P., and Edmond, M.B. (2006). Inflammatory response and clinical course of adult patients with nosocomial bloodstream infections caused by *Candida* spp. Clin. Microbiol. Infect. *12*, 170–177.

Wozniak, K.L., Vyas, J.M., and Levitz, S.M. (2006). *In vivo* role of dendritic cells in a murine model of pulmonary cryptococcosis. Infect. Immun. *74*, 3817–3824.

Wozniak, K.L., Hardison, S.E., Kolls, J.K., and Wormley, F.L. (2011). Role of IL-17A on resolution of pulmonary *C. neoformans* infection. PLoS One *6*, e17204.

Yauch, L.E., Mansour, M.K., Shoham, S., Rottman, J.B., and Levitz, S.M. (2004). Involvement of CD14, toll-like receptors 2 and 4, and MyD88 in the host response to the fungal pathogen *Cryptococcus neoformans in vivo*. Infect. Immun. *72*, 5373–5382.

Zaragoza, O., Alvarez, M., Telzak, A., Rivera, J., and Casadevall, A. (2007). The relative susceptibility of mouse strains to pulmonary *Cryptococcus neoformans* infection is associated with pleiotropic differences in the immune response. Infect. Immun. *75*, 2729–2739.

Zhang, C., Wang, S.H., Liao, C.P., Shao, S., Lasbury, M.E., Durant, P.J., and Lee, C.H. (2010). Down-regulation of PU.1 leads to decreased expression of Dectin-1 in alveolar macrophages during *Pneumocystis* pneumonia. Infect. Immun. *78*, 1058–1065.

Zhou, Q., Gault, R.A., Kozel, T.R., and Murphy, W.J. (2007). Protection from direct cerebral *Cryptococcus* infection by interferon-gamma-dependent activation of microglial cells. J. Immunol. *178*, 5753–5761.

Host Responses to Fungal Infection

6

David L. Moyes, Jonathan P. Richardson and Julian R. Naglik

Abstract

The three major opportunistic fungal pathogens of humans are *Candida albicans, Aspergillus fumigatus* and *Cryptococcus neoformans*. These pathogens primarily infect immunocompromised individuals either as a result of immunodeficiency (e.g. neutropenia, HIV infection) or intervention therapy (e.g. leukaemia, organ transplant, cancer). This highlights the importance of host immune defences in controlling or preventing fungal infections. In health, these fungal pathogens initially interact with mucosal surfaces which may lead to colonization and establishment of commensalism, but in the immunocompromised these fungi can cause serious superficial infections and may disseminate to cause life-threatening systemic infections. To combat these fungal infections, the host utilizes vast communication network of cells, proteins and chemical signals distributed in blood and tissues, which constitute innate and adaptive immunity. In this chapter we will review how the host recognizes these fungi, the events induced by fungal cells, and the host immune defences that ultimately resolve the infections during health. The overview will primarily target *C. albicans*, the most common fungal pathogen of humans.

Host recognition of fungi

Recognition of microbes such as fungi by a host organism occurs via several mechanisms. Adaptive immune cells recognize specific antigenic moieties, developing specific responses. In contrast, innate immune recognition of fungi occurs through recognition of Pathogen Associated Molecular Patterns (PAMPs). PAMPs are molecules or motifs that are shared between different groups of microbes. Unlike antigens, individual PAMPs are not specific to a single microbial species, but rather are shared between many different species. These PAMPs are recognized by host pattern recognition receptors (PRRs) providing a pre-programmed method of microbial recognition, allowing instant, automatic recognition of general microbial components. The majority of fungal PAMPs are cell wall associated and include β-glucans, *N*- and *O*-linked mannans and phosphomannans (Netea *et al.*, 2008). They are detected by three main families of PRR: the toll-like receptors (TLRs), the C-type lectin receptors (CLRs) and the nucleotide-binding domain, leucine-rich receptors (NLRs). These receptors are expressed either on the surface, in endosomes or in the cytoplasm of most cells in the host and their activation by PAMPs leads to triggering of intracellular signalling pathways such as the nuclear factor kappa B (NF-κB) and mitogen-activated protein kinase (MAPK) pathways. This in turn leads to the up-regulation of gene expression for a variety of genes involved in combating infection, including cytokines, chemokines, antimicrobial peptides and inflammatory mediators.

Toll-like receptors

The Toll-like receptor (TLR) family is an evolutionarily conserved group of leucine repeat-rich PRRs. These PRRs respond to a variety of bacterial, viral and fungal PAMPs as well as some endogenous factors released when host cells are injured. The human genome encodes for 10 different genes (TLR1–10) whilst the murine genome encodes 12 (TLR1–9, 11–13). These receptors are the most heavily researched PRRs and as a result are currently the best characterized of these receptors. They are type I transmembrane proteins that are present either on the plasma membrane or in the endosomal compartment. They consist of an extracellular domain comprised of leucine-rich repeated domain that recognizes the target PAMP, along with a Toll/interleukin-1 receptor (TIR)-domain containing cytoplasmic domain that transmits the activation signal. The extracellular domain of these proteins recognizes a variety of PAMPs, including lipopolysaccharide (LPS), peptidoglycan, triacylated proteins and modified nucleic acids (unmethylated CpG rich DNA, double and single stranded RNA). Recognition of target PAMP ligands results in dimerization and triggering of a signalling cascade through interaction of the cytoplasmic TIR domains with one or more of four adapter proteins: Myeloid differentiation primary response gene (88) (MyD88), TIR-domain-containing adapter-inducing interferon-β (TRIF), MyD88-adapter-like (MAL) and TRIF–related adaptor molecule (TRAM). This cascade ultimately leads to activation of the main signalling pathways, including NF-kB and MAPK pathways. This in turn leads to activation and nuclear localization of transcription factors including activating protein 1 (AP-1), NF-κB and interferon regulatory factor (IRF)-3 and -7. The net result of this is to induce gene expression and secretion of a range of defensive proteins including cytokines, chemokines, antimicrobial peptides and other inflammatory mediators – all of which function to trigger an innate and adaptive immune response.

Several TLR receptors play a major role in immune cell recognition of fungi, including TLR2, 3, 4 and 9 (Table 6.1), and their role has best been characterized for *C. albicans*. Details for the other medically important fungi are less well characterized, although it seems that TLR4 recognizes *C. neoformans* glucuronoxylomannan, TLR3 recognizes *A. fumigatus* conidia and TLR9 recognizes of all three of the major medically important fungi. Although mice lacking the TLR signalling adapter protein MyD88 are highly susceptible to fungal infection (Bourgeois *et al.*, 2010; Netea *et al.*, 2008), the exact role of the individual TLR receptors in combating fungal disease is less clear. This may be due to difference in study design where different fungal species, morphotypes and routes of infection have been used. As a consequence, studies in various TLR knock-out mice have revealed significant differences in the putative role of the different TLRs in systemic or mucosal immunity against fungal infections. Furthermore, specific TLRs such as TLR2, 4, 6 and 9 show variation in their role depending on which arm of the innate immune response they engage with. However, they can appear to promote adaptive responses by facilitating antigen presentation in DCs (Blander and Medzhitov, 2006). In other cases, such as TLR3 in *A. fumigatus* infection of epithelial cells, these receptors are involved in activating specific cell types (Carvalho *et al.*, 2012). Nevertheless, human studies indicate that a polymorphism in TLR4 (Asp299Gly) is associated with susceptibility to systemic candidiasis (Van der Graaf *et al.*, 2006) and pulmonary aspergillosis (Bochud *et al.*, 2008), and a polymorphism in the promoter of TLR9 (T-1237C) is associated with allergic bronchopulmonary aspergillosis (Carvalho *et al.*, 2008). Despite the ambiguous data, it is clear that TLRs play a key, albeit complex, role in antifungal immunity.

C-type lectin receptors

This family of carbohydrate binding receptors is a superfamily of over 1000 heterogeneous proteins characterized by the presence of an extracellular carbohydrate-recognition domain (CRD) or in a broader sense a C-type lectin-like domain (CTLD) (Zelensky and Gready, 2005). They have become the subject of intense study as their role in anti-fungal immunity has begun to be deciphered and several key CLRs have now been demonstrated to play a role in this process, with critical functions in binding, uptake and destruction of fungi.

Dectin-1

Dendritic cell-associated C-type lectin-1, dectin-1 (also known as CLEC7a) is one of a number of CLRs recently identified as playing a major role in fungal recognition by the host immune system (Brown, 2006). It is a type II transmembrane protein and belongs to a sub-group of the CLRs termed the natural killer (NK) receptor-like

Table 6.1 Different PRRs that recognize fungal PAMPs

Receptor	Location	Ligand	Fungus
TLR			
2	Surface	Phospholipomannan	*C. albicans*, *A. fumigatus*, *C. neoformans*
2/6	Surface	Phospholipomannan	*C. albicans*, *A. fumigatus*, *C. neoformans*
3	Endosomal	dsRNA	*A. fumigatus*
4	Surface/ cytoplasmic	*O*-linked mannan; glucuronoxylomannan	*C. albicans*, *A. fumigatus*, *C. neoformans*
9	Endosomal	Unmethylated CpG DNA	*C. albicans*, *A. fumigatus*, *C. neoformans*
Lectins			
Dectin-1	Surface	β(1,3)-glucan	*C. albicans*, *A. fumigatus*, *C. neoformans*
Dectin-2	Surface	*N*-linked high mannose; α-mannans	*C. albicans*
Mannose Receptor	Surface	*N*-linked mannan	*C. albicans*, *A. fumigatus*, *C. neoformans*
DC-SIGN	Surface	*N*-linked mannan; terminal di-mannose; internal mannose	*C. albicans*, *A. fumigatus*, *C. neoformans*
Mincle	Surface	*N*-linked α-mannosyl structures	*C. albicans*
Galectin-3	Cytoplasmic/ extracellular	β-mannosides and β-galactosides	*C. albicans*
Pentraxin-3	Body fluids	Galactomannan	*C. albicans*, *A. fumigatus*, *C. neoformans*
NLR			
NLRP3	Cytoplasmic	Variable	*C. albicans*, *A. fumigatus*, *C. neoformans*
NLRC4	Cytoplasmic	Variable	*C. albicans*
Other receptors			
CD14	Surface	Glucuronoxylomannan	*A. fumigatus*, *C. neoformans*
CR3	Surface	Mannose, β-glucan, *N*-acetylglucosamine	*C. albicans*, *A. fumigatus*, *C. neoformans*
Lactosylceramide	Surface	β-Glucan	*C. albicans*, *C. neoformans*
SP-A	Extracellular	Monosaccharides including mannose, fucose and glucose	*C. albicans*, *A. fumigatus*, *C. neoformans*
SP-D	Extracellular	Monosaccharides including mannose, fucose and glucose	*C. albicans*, *A. fumigatus*, *C. neoformans*
Mannose-binding lectin	Extracellular	Monosaccharides including mannose, fucose and glucose	*C. albicans*, *A. fumigatus*, *C. neoformans*

CLRs. The target ligands of this receptor are β(1,3)-glucan polymers which constitute a major part of fungal cell walls. The intracellular region of this protein contains a modified immunoreceptor tyrosine-based activation motif (ITAM) containing a single tyrosine residue instead of the usual two (hence the terms hemITAM or hemi-ITAM). Activation of the receptor leads to phosphorylation of this domain and phosphorylation of spleen tyrosine kinase (SYK) leading to the formation and activation of a B-cell lymphoma/leukaemia 10 – caspase-associated recruitment domain 9

– mucosa associated lymphoid tissue lymphoma translocation gene 1 (Bcl10-CARD9-MALT1) complex. This leads to activation of both the canonical and non-canonical NF-κB pathways (Gringhuis *et al.*, 2009) as well as nuclear factor of activated T-cells (NFAT) pathway (Reid *et al.*, 2009). Dectin-1 can also induce signalling via Raf-1 in a SYK-dependent fashion (Gringhuis *et al.*, 2009) and is associated with phospholipase C and A2 activation (Plato *et al.*, 2013).

A key consequence of dectin-1 engagement is the induction of phagocytosis (Herre *et al.*, 2004; Underhill *et al.*, 2005). A unique feature of dectin-1 is its ability to be either activated or suppressed by its target ligand. To fully activate this receptor, cells need to be treated with insoluble particles of β-glucan – treatment of this receptor by soluble β-glucan has been shown to block activation of this receptor. This is due to the need to form a 'phagocytic synapse' whereby suppressive phosphatases are excluded from a central concentration of dectin-1 receptor molecules, allowing phosphorylation of the intracellular hemITAM motif (Goodridge *et al.*, 2011). Dectin-1 can also synergise with both TLR2 and TLR4, inducing tumour necrosis factor (TNF) α, interleukin (IL) 10, transforming growth factor (TGF) β and DC maturation (Brown *et al.*, 2003; Dillon *et al.*, 2006; Gantner *et al.*, 2003).

Dectin-1 plays a key role in anti-fungal activity given that a major constituent of fungal cell walls are β(1,3)-glucan polymers, which is strongly immunoactivatory. This might be why some fungi have developed surface structures to 'mask' themselves from the immune system. For instance, *Histoplasma capsulatum* masks its β(1,3)-glucan with a layer of α(1,3)-glucan (Rappleye *et al.*, 2007) and *C. albicans* hyphae β(1,3)-glucan is masked by layers of *N*- and *O*-linked mannoproteins to prevent detection by dectin-1. However, in the yeast form of *C. albicans*, although *N*- and *O*-linked mannoproteins are present, the underlying β-glucan layer can become exposed at the budding scar, allowing recognition by dectin-1. Thus, dectin-1 could primarily be involved in the recognition of yeast/conidial morphological forms of fungi. Also, the β-glucan that is detectable in hyphal cell walls seems to be significantly different and may not be susceptible to detection by dectin-1, possibly due to differences in its structure, compared with yeast/conidial β-glucan.

Knockout mice deficient in dectin-1 have given mixed results in systemic fungal infection models, showing both no difference and increased mortality (Nakamura *et al.*, 2007; Saijo *et al.*, 2007; Taylor *et al.*, 2007), depending on the study and the fungal organism being used to infect. However, a role for dectin-1 is supported given that CARD9 knock-out mice are susceptible to fungal infection (Gross *et al.*, 2006). Notably, in mice dectin-1 does not appear to play a role in controlling gastrointestinal colonization of *C. albicans* (Vautier *et al.*, 2012). However, in humans a stop codon mutation (Tyr238X) in dectin-1 is known to be associated with an increased risk of developing mucocutaneous fungal infections with increased oral and gastrointestinal colonization (Plantinga *et al.*, 2009). Finally, a recent study has identified a role for dectin-1 in maintenance of health. Dectin-$1^{-/-}$ mice show increased severity during induced colitis, but this severity can be reversed by the application of fluconazole to clear the fungal mycobiota. Histologically, these animals show extensive invasion by fungi of the underlying tissue that is not evident in wild-type mice (Iliev *et al.*, 2012). Early clinical data appears to suggest that a sub-group of ulcerative colitis patients with particularly aggressive disease show a common single nucleotide polymorphism (rs2078178) in dectin-1, potentially demonstrating a need for a functional dectin-1 receptor to maintain the mycobiota in a commensal state.

Dectin-2

Although referred to as dectin-2 (CLEC6a), this protein has significant differences to dectin-1. It is a type II transmembrane protein, but unlike dectin-1 it lacks an intracellular signalling domain (Graham and Brown, 2009). In order to transmit a signal after ligand ligation, dectin-2 dimerizes with the Fc receptor gamma chain (FcRγ) (Sato *et al.*, 2006). Dectin-2 recognizes high mannose structures that are common to many fungi and binds to hyphae with higher affinity than to yeast (McGreal *et al.*, 2006), which may explain why dectin-2 deficient mice are susceptible to *C. albicans* infection (Robinson *et al.*, 2009; Saijo *et al.*, 2010). Dectin-2 can induce several cytokines/

chemokines through multiple signalling pathways including SYK, PKCδ, CARD9-Bcl10-Malt1, NF-κB and MAPK, and can also activate the NLRP3 inflammasome and respiratory burst (Kerscher *et al.*, 2013; Sato *et al.*, 2006).

Mannose receptor

Also known as CD206, mannose receptor (MR) is predominantly a macrophage and DC receptor that binds several carbohydrate residues, including branched *N*-linked mannans, fucose, *N*-acetylglucosamine and glucose (Taylor *et al.*, 2005b). As a result, it recognizes a wide range of different fungal, bacterial and viral pathogens (McKenzie *et al.*, 2007; Taylor *et al.*, 2005a). This type I transmembrane protein lacks conventional signalling domains within its intracellular domain although ligation by a ligand still results in a variety of cellular responses, including phagocytosis, promotion of antigen presentation to T cells and induction of cytokine production (Cambi *et al.*, 2008; Lam *et al.*, 2007). Notably, deficiency of this receptor does not appear to confer susceptibility to *C. albicans* infection like it does to *C. neoformans* (Dan *et al.*, 2008) although minor changes in fungal burdens can be observed (Lee *et al.*, 2003). Interestingly, ligation can also lead to enhanced shedding of this lectin from the cell. These roles have been ascribed to several different fungi although curiously, it is also capable of playing an inhibitory role in the production of cytokines in response to *Pneumocystis carinii* (Zhang *et al.*, 2005).

Mincle

As with Dectin-2, macrophage-inducible CTL (Mincle; CLEC4e) is a type II transmembrane protein that transmits its signal in conjunction with the FcRγ adaptor protein (Graham and Brown, 2009). Mincle is expressed by monocytes, macrophages, neutrophils, myeloid DCs and some B cell subsets but not T cells, plasmacytoid DCs and NK cells (Kerscher *et al.*, 2013). Mincle predominantly binds carbohydrates containing α-mannans (Lee *et al.*, 2011; Yamasaki *et al.*, 2009) and recognizes *C. albicans* (Bugarcic *et al.*, 2008; Vijayan *et al.*, 2012; Wells *et al.*, 2008), *Malessezia* spp. (Yamasaki *et al.*, 2009) and *Fonsecaea pedrosi*, the causative agent of chromoblastomycosis (Sousa Mda *et al.*, 2011). Like dectin-2, Mincle is not thought to be required for phagocytosis (Wells *et al.*, 2008) but does stimulate cytokine/chemokine production via SYK, PKCδ, CARD9-Bcl10-Mat1t, NF-κB and MAPK (Kerscher *et al.*, 2013; Yamasaki *et al.*, 2008). Although Mincle-induced responses are MyD88 independent, it may synergize with TLRs to induce inflammatory cytokine production and the respiratory burst (Lee *et al.*, 2012).

DC-SIGN

Also known as CD209, DC-SIGN is expressed on endothelium and DCs. It is another type II transmembrane receptor with an uncertain role in anti-fungal immunity (Willment and Brown, 2008). It recognizes *N*-linked mannans to induce IL-6 production and although its role in the endocytosis and uptake of pathogens to promote antigen presentation is well documented (Cambi *et al.*, 2008; Lam *et al.*, 2007), its role in phagocytosis is questionable (Cambi *et al.*, 2003; Cambi *et al.*, 2008).

Nod-like receptors

The Nod-Like receptor (NLR) family PRRs are intracellular proteins that detect PAMPs present in the cytoplasm of the cell. As with other PRRs, they recognize microbial products but they also recognize host cellular danger signals or alarmins (Martinon *et al.*, 2009). Currently 23 human and 34 mouse NLRs have been identified (Kumar *et al.*, 2011) and members of this PRR family usually associate with two other proteins: apoptosis-associated speck-like protein containing a CARD (ASC) and pro- cysteine-dependent aspartate-directed protease 1 (pro-caspase-1), to form a multimeric protein complex usually termed the inflammasome. The main function of the inflammasome is to convert pro-caspase-1 into active caspase-1 which processes the immature pro-IL-1β and pro-IL-18 into mature IL-1β and IL-18 (Bryant and Fitzgerald, 2009). Inflammasomes incorporating NACHT, LRR and PYD domains-containing protein 3 (NLRP3) (Joly *et al.*, 2009) and NLR family CARD domain-containing protein 4 (NLRC4) (Tomalka *et al.*, 2011) have recently been identified as being activated by *C. albicans* resulting in the production of IL-1β.

Notably, mice lacking IL-1 receptor type 1 (IL-1RI), IL-18 or caspase-1 have contrasting susceptibility patterns to fungal infections (van de Veerdonk *et al.*, 2008), whereas mice lacking NLRP3 appear susceptible to candidiasis (Gross *et al.*, 2009). Other studies have also identified a crucial role for NLRP3 alongside TLR2 and dectin-1 in preventing dissemination of *C. albicans* in a murine model of oral infection (Hise *et al.*, 2009). Consistent with a role for NLRP3 in epithelial protection (Zaki *et al.*, 2010), defective NLRP3 activation increases *C. albicans* colonization in the gut and exacerbates Crohn's disease (Rehaume *et al.*, 2010). Nevertheless, the full extent of any potential role for NLRs in host antifungal responses is not currently fully understood.

Host responses to fungi

There are several different cell types in the host that recognize fungal cells. Given the different times and conditions under which these cells encounter fungi, it is unsurprising that they mount differing responses with varying degrees of importance. These cells fall broadly into two categories – the haemopoietic and non-haemopoietic cells. Haemopoietic cells, including myeloid and lymphoid cells play a central role in combating microbial infections. As such, they are central players in removing fungal invaders at multiple sites. Foremost among these cells are the neutrophils, which have a dominant role in fungal clearance. However, both macrophages and DCs have a key role to play, not just in directly combating fungi, but also in activating and informing subsequent adaptive immune responses. These responses are largely coordinated by T helper cells and with the discovery of new phenotypes of these cells, our understanding of how fungal infections are combated and how the adaptive and innate immune systems link up and work together has massively expanded. Although long neglected, over the last decade our appreciation of the role of non-haemopoietic cells, and particularly epithelial and endothelial cells has rapidly expanded. Far from simply acting as barriers, we are now beginning to appreciate the key role that these cells play in identifying fungal pathogens and orchestrating protective responses to clear these microbes. Although cell-mediated responses form the bulk of protective responses to a fungal invader, we should not neglect the growing body of information indicating important roles for humoral immune responses such as complement and antibodies. As well as acting directly, these responses also act to boost the efficacy of cell-mediated responses, playing a vital role in clearing the fungal invader. In the following sections, we will look at the various aspects of host immunity by cell type and compartment to identify their role in host responses to the fungal invader.

Epithelial cells

Although immune cells are of great importance in host responses to infections, epithelial cells that comprise mucosal surfaces across which fungi invade play a crucial role in protective responses. Epithelial cells are the first point of contact with the vast majority of fungal cells and provide the surface that fungi will initially adhere to. As such, they need to be able to discriminate between commensal, colonizing fungi and invasive, pathogenic fungi. For many fungal infections, invasion and cell interactions are isolated to mucosal surfaces and it is only in the case of disseminated infections that systemic immunity may come in to play. Until recently, the role epithelial cells played in host–fungal interactions was limited to the opinion that they provided an anchorage point for colonization and a food source for invading fungi. However, recent studies have dramatically changed our view of these epithelial cells and their place in host responses to fungi (Moyes and Naglik, 2011; Naglik and Moyes, 2011; Naglik *et al.*, 2011; Zhu and Filler, 2010).

Adhesion and invasion

Whether a fungal cell is colonizing for commensal growth or invading host surfaces, the first stage in a naturally occurring infection is the interaction with host epithelial cells. Central to these events is the ability of the fungus to adhere to these cells. This process is key to all subsequent events, whilst invasion, damage and escape are all activities associated with infection. The varied nature of the fungal cell wall both between different species and in different growth stages within a species indicate that these events are complex and multifactorial, including non-specific events such as hydrophobic interactions and specific ligand–receptor interactions. In the case of *C. albicans*, this adhesion is thought to involve interactions between host receptors and *Candida* adhesins that vary as the morphological state changes (Chaffin, 2008). Much of our knowledge of the initial contact and adhesion between fungus and host epithelial cells is based on *in vitro* studies utilizing yeast/conidia. Although yeast–epithelial interactions do not represent the full repertoire of interaction events, they are likely to comprise the majority of initial interactions due to the sessile nature of hyphal fungal forms. It is important to recognize, however, that direct host-host contact could conceivably result in 'hyphal transfer' rather than 'yeast transfer', particularly for fungi such as *C. albicans*. These interactions lead to responses from both the host cell and the fungal cell. In dimorphic (*Candida* spp.) or filamentous (*Aspergillus* spp.) fungal cells, contact with epithelial cells stimulates germination and germ-tube/hypha formation (Osherov, 2012; Zakikhany *et al.*, 2007). These changes in fungal gene expression and morphology result in a two-way relationship, both improving fungal adhesion and preparing the epithelial cell and host for the presence and potential invasion by the fungus.

As well as changes and responses in the fungus, the interaction and adhesion of fungal and epithelial cells triggers numerous ligand-receptor mediated responses in the host epithelial cell. In *C. albicans* infections, these responses cover a wide range of events. Initially, these events trigger host cytoskeletal reorganization to form membrane processes that engulf the surface associated hyphae (but not yeast), resulting in their uptake by the cell in a process referred to as induced endocytosis (Moreno-Ruiz *et al.*, 2009; Park *et al.*, 2005; Phan *et al.*, 2005; Wachtler *et al.*, 2011; Zhu *et al.*, 2012). Recently, detection of the fungal adhesin agglutinin-like sequence 3 (Als3) by epithelial cell receptors epidermal growth factor (EGFR) and human epidermal growth factor receptor 2 (Her2) has been shown to be a key event in this process (Zhu *et al.*, 2012) In contrast, *A. fumigatus* conidia can be endocytosed (Botterel *et al.*, 2008; Gomez *et al.*, 2010) and appears to require dectin-1 (Han *et al.*, 2011). Induced endocytosis is far from unique among fungi and has been ascribed to a variety of bacteria as well, including *Salmonella*, *Shigella* and *Yersinia* spp. (Goosney *et al.*, 1999; Isberg, 1996). The key feature of induced endocytosis is that it is host driven, requiring an active response by the epithelial cell in response to interaction with the fungal cell, which is not required to be viable. The ability of fungal/epithelial cell interactions to induce endocytic uptake appears to be a function of its location. For example, uptake of *C. albicans* by oral epithelial cells involves induced endocytosis, whereas uptake by gut epithelial cells does not (Dalle *et al.*, 2010).

Apoptosis

Subsequent to the uptake of hyphae by epithelial cells, the fungi then proceed to an invasion phase, during which there is extensive expansion and epithelial penetration. The net result of this phase is damage to the mucosal tissues and a concomitant release of inflammatory mediators including cytokines and damage- or danger-associated molecular pattern (DAMP) molecules. The damage caused during infection and invasion occurs either through fungal-driven necrosis or by cell-induced apoptosis. This apoptotic mechanism is a naturally occurring cellular process whereby activation of a series of internal biochemical pathways drives the breakdown of a cell into discrete membrane-bound apoptotic bodies that can then be disposed of by scavenging phagocytic cells. Unlike necrosis, which is almost always detrimental, apoptosis can be a beneficial process, aiding in combating invasion and infection. It is a common response to pathogens across both the plant and animal kingdoms (Philpott *et al.*, 2001).

The relative importance of this mechanism in pathogen invasion and host defence can be underlined by the number of pathogens that have evolved ways to interfere with host control of this mechanism. For example, *C. albicans* has been reported to inactivate anti-apoptotic proteins in macrophages and neutrophils as well as forcing the activation of the cellular caspases that are central to apoptosis (Ibata-Ombetta *et al.*, 2003; Rotstein *et al.*, 2000). The exact role of apoptosis in the epithelial cell response to fungal infection is less clear. It has been reported that *C. albicans* induces early apoptotic events such as annexin V and intranucleosomal chromosomal degradation, driven by caspase activation, although this process often stalls before moving to late apoptotic events (Villar *et al.*, 2012; Villar and Zhao, 2010). This activity was limited to the first 12 hours of infection, with no role in the later stages of an infection. The importance of fungal-epithelial cell interactions in this process is underlined by the finding that these apoptotic events can be avoided by preventing the epithelial cell from endocytosing the fungal hypha. However, little is known regarding the mechanism of these events or the precise role they play in *Candida* infections.

Fungal cells are not only involved in inducing apoptosis. They also play a role in suppressing apoptosis. Although *C. albicans* induces apoptosis in oral epithelial cells, other fungi have the opposite effect. It has been reported that infection of airway epithelial cells (type II pneumocytes) by *A. fumigatus* suppresses apoptosis (Berkova *et al.*, 2006). Thus, although apoptosis is a key response of host epithelial cells to fungi, whether the response is to induce or inhibit apoptosis depends largely on the cell location and the fungal species, and potentially on other factors such as fungal burdens. The fungal factors that have been proposed to induce/inhibit epithelial apoptosis are not currently identified, although several candidates have been proposed based on conflicting experimental data. One of these factors is the *C. albicans* quorum sensing molecule, farnesol (Scheper *et al.*, 2008). However, it should be noted that for most of these apoptotic effects, contact and uptake are required, so the likelihood of a secreted molecule like farnesol being a major initiator is small. More likely is the observation that cell wall *O*- and *N*-linked mannans are the main inducing factors (Wagener *et al.*, 2012) with secreted components modulating the strength or quality of response. These components could be fungal, such as farnesol, or host, such as nitric oxide produced by activated epithelial cells (Hosogi *et al.*, 2008).

Epithelial cell involvement in host immunity

A key attribute of many fungal pathogens is that they are not obligate pathogens. Rather they are opportunistic commensal organisms that only infect and invade when there are perturbations in host defences. Although normally non-pathogenic, fungi are a major cause for concern in immunocompromised individuals, indicating the importance of immune responses in preventing fungal infections in a healthy host or restricting them to the commensal form at mucosal surfaces. Given the number of microbes resident on the body's various surfaces, it is essential that the host is in some way able to discriminate between microbes that are commensal or pathogenic. In recent years, it has become increasingly clear that epithelial cells play a critical role in this discrimination process. It has now been well established that epithelial cells produce a variety of cytokines and chemokines in response to fungal infection, with the exact combination depending on the fungus and epithelial cell type concerned. For example, infection of oral epithelial cells with *C. albicans* results in the production of the cytokines G-CSF, GM-CSF, IL-1α, IL-1β, IL-6 and the chemokines RANTES, IL-8 and CCL20 (Moyes *et al.*, 2010, 2011; Schaller *et al.*, 2002). Equally, infection of airway epithelial cells with *A. fumigatus* or culture filtrates of *A. fumigatus* grown cells induces cytokine production including IL-8, IL-6 and monocyte chemotactic protein-1 (MCP-1) (Balloy *et al.*, 2008; Osherov, 2012; Sharon *et al.*, 2011). In addition, epithelial cells are widely known to produce anti-microbial peptides such as β-defensins and cathelicidin (LL-37) in response to fungal infection (Diamond *et al.*, 2008). However, until recently, it has been unclear whether these mediators are produced in response to the fungus or in response to damage caused by the fungus. What is evident given the commensal nature of many fungi is that there is a mechanism that allows the mucosal surfaces to discriminate between fungi that are colonizing and fungi that are invading or causing disease. The cellular mechanisms by which this occurs are only now beginning to be elucidated but it is becoming clear that epithelial cells play a key role in these discrimination events. For dimorphic fungi such as *Candida spp*, pathogenicity is usually associated with hypha formation (Moyes *et al.*, 2010), with those species or strains not producing hyphae in culture conditions being unable to produce strong effector responses in epithelial cells (Moyes *et al.*,

2010, 2012; Schaller *et al.*, 2002). For other fungi, it can be assumed that there will be similar shifts in their external profile allowing for their recognition as pathogens rather than commensals. Thus, there are several questions to be answered relating to immunity to fungi: (i) What is the role of epithelial cells in these discriminatory processes? (ii) What is the nature of these discriminatory mechanisms and how do they lead to protection? (iii) What are the fungal moieties and cellular receptors involved in discrimination?

Epithelial cell discrimination mechanisms

The importance of epithelial cells in initial responses to fungal infection is becoming increasingly apparent, in particular with recent observations about the role of epithelial cells in mucosal candidiasis. It has been shown that epithelial cells can communicate with neutrophils in a cross talk mechanism to induce their own protection (Weindl *et al.*, 2007). Infection of epithelial cells by *C. albicans* leads to epithelial-induced activation of neutrophils which then induce epithelial derived protective mechanisms – all without the neutrophils coming in to contact with the fungus. This implies that epithelial cells directly recognize this fungus and produce specific responses to it. Recent work has now demonstrated that epithelial cells are capable of rapidly detecting *Candida* in any form and of discriminating between the yeast and hyphal forms of this fungus (Moyes *et al.*, 2010, 2012). This discrimination is independent of the viability of the fungus, indicating that it is not a feature of the damage process of epithelial cells, but rather is a result of specific recognition of a fungus. The recognition of the fungus results in activation of intracellular signalling pathways in oral epithelial cells – specifically, the NF-κB, phosphatidylinositide 3-kinase (PI3K) and MAPK pathways (Moyes *et al.*, 2010, 2013). This recognition appears to be tailored to the state of the fungus. The presence of *Candida* yeast results in the activation of NF-κB and PI3K signalling along with weak, transient activation of all three MAPK pathways. This drives the brief activation of the transcription factor c-Jun via the c-Jun N-terminal kinase (JNK) and extracellular signal-regulated kinase (ERK) pathways. However, on the appearance of *Candida* hyphae, a much stronger activation of MAPK signalling is triggered with sustained activation of the transcription factor c-Fos via the p38 pathway. Blocking the activity of c-Fos (but not c-jun) results in the loss of production of most cytokines and inflammatory mediators, demonstrating a key role for c-Fos in epithelial cell responses to *Candida* hyphae. Moreover, this discriminatory response is dependent on the fungal burden encountered by the epithelial cell, with a minimum level required to activate the response. Thus, epithelial cells are able to discriminate between a colonizing and invading form of a fungus. This discriminatory process appears to be common to different epithelial cell types. Despite vaginal epithelial cells showing different initial recognition characteristics and a different cytokine/chemokine secretion profile, the key components of the discriminatory pathway are the same (p38/c-Fos), indicating a commonality of response (Moyes *et al.*, 2011). Thus, this mechanism represents a 'danger response' mechanism allowing epithelial surfaces to remain quiescent in the presence of *Candida* yeast or low burdens of hyphae common in colonizing fungi but allows a swift, specific and strong response to dangerous levels of invasive hyphae common in disease pathologies. The p38 and PI3K pathways may also be important in airway epithelial cell cytokine responses (IL-8 production) to *A. fumigatus* whilst secreted proteases in culture filtrates appear to activate JNK and ERK1/2 pathways, which may be associated with damage induction (Balloy *et al.*, 2008; Fekkar *et al.*, 2012; Osherov, 2012). However, it is currently unclear whether these pathways discriminate between conidia and germinative forms or between commensal and pathogenic *A. fumigatus*. Thus, it is clear that epithelial cells may play a critical role in generating appropriate host responses to fungal presence.

Epithelial cell detection of fungus

The mechanisms by which epithelial cells detect fungi are currently not well elucidated. Although several innate receptors for detecting fungi have been identified (see above), whether any of these receptors are involved in the recognition of fungi by epithelial cells is unclear. A role for TLRs has been much mooted with data both in support and in contradiction of this role. For example, in oral epithelial cells recognition of yeast cells seems to be via conventional fungal PAMPs whereas recognition of hyphae is independent of either these PAMPs or of the main PRRs, TLR2, 4 or dectin-1 (Moyes *et al.*, 2010). There are seemingly conflicting studies indicating a potential role for TLRs in recognition of several different *Candida* species, however, here it appears that the recognition by TLRs is a secondary event which acts to induce a protective or inflammatory response in the local epithelium (Bahri *et al.*, 2010; Weindl *et al.*, 2007). Likewise, airway epithelial cell recognition of *A. fumigatus* conidia contains both TLR-dependent and -independent aspects. Although IL-8 appears to be independent of TLR activation (Balloy *et al.*, 2008), other aspects of the epithelial response are dependent on TLR3/TRIF signalling (de Luca *et al.*, 2010). Recently, non-PRRs have been reported that recognize *C. albicans* – most notably Her2. In this instance, rather than a generic fungal PAMP, the receptor detects specific proteins (Als3 and Ssa1) which then trigger the induced endocytic process (Zhu *et al.*, 2012).

Epithelial induced protective responses

Despite advances in our knowledge of the role of different aspects of the immune response in combating fungal infection, the role of the epithelium and epithelial cells in general is only now beginning to be elucidated. Detection of fungi by epithelial cells leads to a variety of protective responses, including production of cytokines to recruit and modulate both innate and adaptive immune cells and responses. Other key responses include the production of anti-microbial compounds, including the S100 alarmins. These peptides can play a key role in host anti-fungal responses and have been shown to be critical in governing the vaginal responses to *C. albicans* infections (Yano *et al.*, 2010, 2012a,b). Whatever responses are generated by epithelial cells in response to fungal presence and infection, it is clear that there is likely to be a significant amount of crosstalk between epithelial cells and other cells of the immune system – both local and systemic. This crosstalk fulfils two functions, acting to both maintain normal physiological conditions when the fungi are recognized as commensal organisms and to elicit/initiate a clearance response when the fungi are recognized as pathogenic. The production of cytokines and chemokines by epithelial cells in response to invasive fungal infection results in the recruitment, differentiation and activation of various different immune cells, including neutrophils, monocyte/macrophages DCs and T cells. This process leads to the generation of tailored immune responses with the aim of clearing fungal infections via both the innate and adaptive immune pathways (see below).

Endothelial cells

The data for endothelial cell responses to fungi are much sparser than for epithelial cells but some information has been reported. Endothelial cells constituting the lining of the vasculature represent the next barrier for fungal infections – both to move beyond local infections to systemic infections and to move from the blood into local tissue sites post-dissemination. Both of these events lead to a significant increase in the mortality associated with infection. The endothelium is not merely a passive, physical barrier, but functions as an integral component of the innate immune system. Endothelial cells are capable of producing a variety of pro-inflammatory and pro-coagulant responses to fungal and other microbial infections, and are also the target for suppression and evasion by invading microbes. They mount tailored responses to fungal pathogens, depending on the morphology and species encountered.

As with epithelial cells, one of the first interaction events between fungi and endothelial cells is uptake and invasion. Once again, this process can be an actively induced endocytic process whereby the fungal cells induce their own endocytosis by the endothelial cells (Filler *et al.*, 1996; Lopes *et al.*, 2004) and is key for downstream events such as cytokine and chemokine induction as well as damage. As endothelial cells are polarized, encountering the fungus on different surfaces can lead to differences in the response mounted by the endothelial cell. For example, *Aspergillus* induces greater levels of pro-inflammatory cytokines and responses (increased E-selectin, tissue factor etc.) when encountered from the abluminal surface but induces greater damage when encountered from the luminal surface (Kamai *et al.*, 2009). In general, endothelial cell contact with fungi results in the increase in expression of cytokines such as TNFα, IL-1α and IL-1β as well as the chemokine IL-8. They also show increases in integrins and selectins including E-selectin, intercellular adhesion molecule-1 (ICAM-1) and vascular cell adhesion molecule-1 (VCAM-1) (Filler *et al.*, 1996; Orozco *et al.*, 2000), all of which function as leucocyte adhesion molecules. In common with epithelial cells, many of these responses are driven by activation of the NF-κB and p38 MAPK signalling pathways in response to fungal recognition (Müller *et al.*, 2007), with at least part of this recognition occurring via the TLR/MyD88 pathway.

As well as activating host responses via endothelial cells, some fungi can induce non-responsiveness via endothelial cells. For example, endocytosis of *C. neoformans* by endothelial cells does not result in the production of cytokines or chemokines (Lortholary *et al.*, 1999; Mozaffarian *et al.*, 2000). Importantly, this effect is an active process, as subsequent stimulation of these cells with activatory cytokines such as TNFα or interferon gamma (IFNγ) no longer results in the production of chemokines either. This inhibition of chemokine production by endocytosis of *C. neoformans* occurs by rendering endothelial cells quiescent as, unlike other species, no NF-κB signalling is induced, explaining the lack of downstream effector responses and the fact that no cell death or damage is induced (Bernasconi *et al.*, 2010). However, there must be some degree of recognition and activation of endothelial cells to enable the cryptococcal cells to be endocytosed.

Innate immune responses

The first stage of any immune response is the initiation of innate immunity. These non-specific responses activate immediately to recognize and respond to a microbe in a pre-programmed fashion. The importance of innate immune responses in fungal and particularly *Candida* infections can be seen by the fact that depletion of neither CD4 nor CD8 T cells in mice (removing the adaptive immune system) results in an increase in colonization and infection (Ashman *et al.*, 2003). As such, we can see the importance of innate immunity in preventing fungal infections from gaining a foothold in the host. The innate immune response is made up of a series of humoral (complement cascade) and cellular (neutrophil, macrophage) components that act in concert to generate a watchful, protective screen preventing the vast majority of pathogens from establishing an invasive infection. Further, it has become increasingly apparent that these responses, far from operating in isolation, can function to activate the more specific adaptive immunity as well as acting in concert with both adaptive immunity and other homeostatic processes to provide further protection.

Complement

A key component of innate immunity, the complement cascade plays a key role in host responses to fungal cells. Invasion of a host by fungal cells results in a rapid increase in synthesis and release of complement components. Activation of the cascade occurs by all three known pathways (classical, alternative, mannose-binding lectin (MBL)) with no obvious case for a dominant pathway in most cases (Speth *et al.*, 2008). Given the preponderance of mannoproteins on the cell wall surface of fungal species such as *Candida* spp., these fungi are particularly good at activating the mannose-binding lectin pathway and it has been demonstrated that this pathway is important in opsonising fungi for subsequent phagocytosis (Brouwer *et al.*, 2008) and other complement functions (Neth *et al.*, 2000). Owing to the thickness of the fungal cell wall, activation of the membrane attack complex (MAC) is not a likely method of killing fungi. However, the rapid deposition of complement components on the cell wall readily opsonises them for uptake and destruction by neutrophils, macrophages and DCs that may have been recruited by complement anaphylatoxins. For many fungi, interactions between activated C3 and the complement receptor CR3 is key for the adherence and uptake of fungal cells by phagocytes (Forsyth and Mathews, 1996; Sturtevant and Latge, 1992). After phagocytosis of fungal cells, the host cells activate an oxidative burst that leads to fungal killing, a process which can be blocked by blocking the C3–CR3 interaction with monoclonal antibodies. These C3–CR3 interactions are important for other functions, including inhibition of hyphal growth and production of cytokines by lymphocytes (Forsyth and Mathews, 2002). As well as the complement activating activity, these proteins can also exert direct effects. For example, MBL has been shown to directly inhibit growth of *Candida* (Ip and Lau, 2004) as well as enhancing the release of TNFα from *Candida* infected monocytes (Ghezzi *et al.*, 1998). The anaphylatoxin C3a released from C3 during complement activation has also been observed to have direct anti-fungal activity, independently of its chemotactic activity (Sonesson *et al.*, 2007).

Neutrophils

Neutrophils are the key effector component of cellular innate immunity and appear to have a dual role in anti-fungal responses. First they directly kill and clear the infecting fungal cells and, secondly, they indirectly mediate fungal protection via epithelial cell activation. Neutrophils can directly kill both conidial and hyphal fungal cells by one of two mechanisms. The first involves the classical phagocytosis of fungal cells, degranulation and oxidative killing (Martinez *et al.*, 2009). Killing of *Aspergillus* hyphae involves production and secretion of reactive oxidative intermediaries, which go on to rapidly destroy these fungal cells (Latge, 1999). The second killing mechanism is via the recently discovered neutrophil extracellular traps (NETs) (Mosser and Edwards, 2008). NETs are formed in a specialized method of neutrophil cell death termed NETosis. In this process, the neutrophil 'explodes', releasing a web of chromatin fibrils coated with the contents of the neutrophil, such as serine proteases, antimicrobial peptides and other microbicidal compounds. So far, NETs have been shown to be involved in killing *C. albicans* (Mosser and Edwards, 2008) and *A. fumigatus* (Arora *et al.*, 2011), although *Cryptococcus* spp. have been proposed to have developed resistance mechanisms.

Recent work has highlighted the degree to which neutrophils communicate with epithelial cells. As well as receiving recruitment and activation signals from epithelial cells, neutrophils can also communicate back to epithelial cells and induce resistance to fungal killing without coming into direct contact with the fungal cells (Weindl *et al.*, 2007). This finding leads to the hypothesis of a circular loop mechanism for combating fungal infection, whereby epithelial cells sense a shift from commensalism (colonization) to pathogenicity (invasion/damage), thus recruiting in and activating neutrophils which secrete TNFα in response, leading to activation of TLR4 and LL-37-dependent protective mechanisms in epithelial cells. This protective mechanism is independent of direct contact between fungal cells and neutrophils. It is important to note the time frame in which TLR4 is utilized by epithelial cells in protecting against fungal infection. Whilst not used for the original detection of fungal invasion (Moyes *et al.*, 2010), it plays a critical role later in infection in subsequent neutrophil-induced protection (Weindl *et al.*, 2007).

The role of neutrophils in combating fungal infection is location specific. Although they play a key role in protecting oral epithelium, as indicated above, there is evidence to suggest that this role is not ubiquitous. Indeed, several studies have demonstrated that in vaginal epithelium, they play no obvious protective role and can even be detrimental by causing many of the symptoms associated with *Candida* infections (Fidel, 2005). However, interestingly, activation and recruitment of neutrophils in vaginal tissue is still initiated by epithelial cells (Yano *et al.*, 2010), underlining the importance of epithelial cells and the role that cross talk with neutrophils plays in anti-fungal responses.

Macrophages

Macrophages are key cells in any immune response. They are a bridge between innate and adaptive immunity, able to act as innate, phagocytic cells and also as antigen presenting cells capable of activating T cells. Although they are capable of responding in a non-activated state, upon activation, they become highly responsive and capable of a wide range of functions. Upon activation, macrophages differentiate into two phenotypically and functionally diverse subsets, called M1 and M2 (Martinez *et al.*, 2009). Which subset they differentiate into depends on the cytokine milieu in which they are activated. When in the presence of high levels of the cytokine IFNγ, they differentiate into the classical M1 phenotype. However, in the presence of Th2 cytokines, in particular IL-4 and IL-13, they develop into the alternatively activated M2 phenotype. The two different phenotypes have differing functions. The classical M1 macrophages are microbicidal and proinflammatory, whilst the alternatively activated M2 macrophages are more involved in wound healing and extracellular matrix remodelling, producing a cocktail of cytokines which aid in these processes. Fungicidal M1 macrophages synthesize the reactive nitrogen species, NO through the action of inducible NO synthase (iNOS) to directly kill phagocytosed fungal cells, along with TNFα and the chemokines CXCL9 and CXCL10 (Mosser and Edwards, 2008). These chemokines are ligands for the CXCR3 receptor expressed by Th1 T cells and NK cells, attracting these cells to the site of infection. Thus, M1 macrophages are generally fungicidal by two mechanisms – both the direct killing via NO synthesis, but also by recruiting Th1 and NK cells and potentially by activating anti-fungal epithelial cell responses. In contrast, M2 macrophages promote fungal persistence within the macrophage, providing an ideal hiding place. Differentiation to the M2 phenotype results in an up-regulation of the MR (CD206) resulting in increased phagocytosis of fungi (Stein *et al.*, 1992). In conjunction, there is also increased expression in the arginase-1 (Arg1) gene which competes with iNOS for the same substrate, thus reducing the levels of fungicidal NO produced within these cells (Hesse *et al.*, 2001). This is further exacerbated by reduced levels of TNFα production. Thus, macrophages play a critical role in the host resistance to fungi but, again, this largely depends on the fungal strain interacting with the macrophage (Tavanti *et al.*, 2006).

The importance of the cytokine milieu and thus the macrophage polarization that it drives can clearly be seen in the changes in susceptibility seen with cytokine-deficient mice. Mice deficient in the key Th1 and M1-driving cytokine, IFNγ, show increased levels of Th2 cytokines that drive an M2 phenotype. Thus, IFNγ deficient mice show increased levels of M2 polarized macrophages, which results in a lack of clearance and increased lung pathology during *C. neoformans* infections (Arora *et al.*, 2005). Conversely, infection with an IFNγ expressing strain of *C. neoformans* results in preferential polarization to M1 phenotype along with elevated Th1 and Th17 cytokine responses, driving a rapid resolution of the infection (Hardison *et al.*, 2010). However, the host cytokine milieu is not the only factor to affect host macrophage polarization. Several groups have reported the ability of fungi and in particular *Cryptococcus* spp. to influence the polarization of macrophages to a more fungal-friendly M2 phenotype (Wozniak *et al.*, 2012).

Adaptive immune responses

Adaptive immune responses are activated after recruitment of macrophages and particularly DCs during innate immunity. DCs are recruited to the site of infection in response to epithelial-produced CCL20 and β-defensin 2. These chemo-attractants act as ligands for the CCR6 chemokine receptor expressed on mucosal-homing DCs and T cells. Once recruited, DCs will recognize fungal cells and their PAMPs through a variety of PRRs including TLR2/4, dectin-1, dectin-2, DC-SIGN and MR. Recognition is followed by ingestion of the fungal cells, DC activation, and trafficking to the local lymph nodes where the DCs will present processed fungal antigens to naïve and memory T cells, thereby initiating adaptive immunity. However, different DC subsets can be activated via distinct signalling pathways following PRR to shape T cell responses against fungal infections (Bonifazi *et al.*, 2009; Huang *et al.*, 2009). Inflammatory (myeloid) DCs initiate antifungal T helper (Th)17 and Th2 cell responses via TLR-MYD88 pathways, whereas tolerogenic (plasmacytoid) DCs activate Th1 and T regulatory (Treg) cells via TRIF (Romani, 2011). The exact 'flavour' of T cell response is governed by the cytokine milieu the cells are exposed to during activation: IL-12/IFNγ for Th1, IL-4 for Th2, IL-1β/IL-6/IL-23 for Th17, IL-2/TGFβ for Treg. Furthermore, signal transducer and activator of transcription 3 (STAT3) is involved in determining canonical or non-canonical activation of NF-κB and thus the expression of indoleamine 2,3-dioxygenase (IDO), a key enzyme that controls DC plasticity and function. These functionally distinct pathways in DCs ultimately affect the balance between effector Th cells and Treg cells and are probably exploited by fungi to promote commensalism or infection (Romani, 2011).

The role of different T cell phenotypes during fungal infection has been much debated over recent years. Although there is ongoing debate as to the roles of Th1 and Th17 phenotypes, the Th2 phenotype is widely accepted as allowing increased growth and dissemination for most, if not all, fungal species. Evidence from both murine models and human clinical studies, looking at immunocompromised patients (such as AIDS and immunosuppressive therapies) suggests a role for cell-mediated immunity and specifically the Th1 phenotype in combating oral and gastrointestinal *C. albicans* infections (Del Sero *et al.*, 1999; Fidel, 1999; Naglik *et al.*, 2008; Schaller *et al.*, 2004). Indeed, in AIDS patients a high proportion (50%) develop oropharyngeal candidiasis (OPC) (Klein *et al.*, 1984). These patients have low $CD4^+$ T cell levels that track and correlate with the incidence of OPC, indicating the importance of these T cells in host defence against oral infections (Glocker and Grimbacher, 2010). During murine *Candida* gut infections, fungal clearance was found to correlate with the levels of IFNγ and IL-5 producing T cells in Peyer's patches and mesenteric lymph nodes. In these cases, neutralization of the Th2 response by blocking IL-4 activity resulted in better fungal clearance, coinciding with an enhanced Th1 response (Cenci *et al.*, 1995). Thus, Th1 responses were for a long time regarded as protective against fungal infection. This was further supported by the finding that T cell deficient mice are susceptible to OPC but can be protected using adoptive transfer of $CD4^+$ T cells, and further, the Th1 cytokines IFNγ and TNFα were identified in the surrounding tissue (Farah *et al.*, 2002). Despite these findings, it is clear that as with epithelial cell responses, the role of the different phenotypes is location specific. Although Th1 cellular immunity is generated during cases of vulvovaginal candidiasis, this does not appear to be protective. Instead, local mucosal responses governed by γδ T cells and DCs have been implicated as being the predominant mechanism for anti-*Candida* immunity at this site (Fidel, 2007).

Our view of the role and importance of the different T cell phenotypes during fungal infections all changed with the identification of a novel T cell phenotype in addition to the standard Th1/Th2 model. This new phenotype was identified as an IL-17 secreting phenotype and termed Th17 (Harrington *et al.*, 2006). These cells are induced by a combination of IL-6, IL-1β and TGFβ on naïve cells and further matured or re-activated on stimulation with the IL-12-like cytokine IL-23. They produce IL-17A, IL-17F and IL-22 and are now regarded as playing a major role in combating extracellular infections and autoimmunity. Production of IL-17 appears to be a key event in the protection of a host against many mucosal fungal infections, including *C. albicans.* Th17 cytokines have several functions that will theoretically promote fungal immunity. For example, IL-17A and F stimulate a variety of cells to produce chemokines that recruit neutrophils, and activate neutrophils, improving their ability to home to the site of infection and to clear fungal cells upon their arrival. IL-22 also plays a major role in maintaining epithelial cell barrier function.

Data regarding the role of the Th17 cytokines, IL-17 and IL-22, and their role in anti-fungal immunity is conflicting. The role of IL-17 in activating neutrophils would suggest that this cytokine and Th17 responses in general play an important role in clearing fungal infections and this is supported by mouse data demonstrating that neutralization of IL-17 in early responses to *A. fumigatus* results in an increased fungal burden (Werner *et al.*, 2009). However, other studies suggest that IL-17 also decreases killing and clearance of *A. fumigatus* (Zelante *et al.*, 2007). In addition, *A. fumigatus* is a poor inducer of Th17 responses in humans, possibly by inducing the host to down-regulate its own Th17 responses (Chai *et al.*, 2010). Although this may suggest a limited role for IL-17 in anti-*Aspergillus* responses, it is possible that IL-17 produced by γδ T cells and innate lymphoid cells could still be playing a role in these infections.

In contrast, there are more solid data regarding a potential role for Th17 and IL-17 in anti-*Candida* defence. A role for Th17 and IL-17 in host defence against *C. albicans* first came to light in 2004 when IL-17 receptor knockout mice were shown to be more susceptible to systemic *C. albicans* infection than wild-type animals (Huang *et al.*, 2004). Whilst there are reports that suggest that Th17 immunity may exacerbate *C. albicans* infection (De Luca *et al.*, 2007; Zelante *et al.*, 2007), most data are consistent with a generally protective role for IL-17 during systemic infection. There is also evidence to support a positive role for IL-17 during mucosal infections. Mice lacking Th1 cytokines (e.g. IFNγ) that inhibit Th17 immunity have been shown to be resistant to oral infections, whilst those lacking components of the Th17-driving IL-23 cytokine show increased susceptibility to infection (Farah *et al.*, 2006). Furthermore, IL-17AR deficient mice have increased susceptibility to oropharyngeal candidiasis (Conti *et al.*, 2009) and production of IL-17 and the Th17-driving cytokine IL-23 appears essential in preventing skin infections (Kagami *et al.*, 2010). However, a role for Th17 cells in vaginal candidiasis is still unclear given that one study showed that IL-17 and IL-23 were required to reduce fungal burdens in a mouse model (Pietrella *et al.*, 2011), whereas another study showed that the acute neutrophil response mediated by S100 alarmins during vaginal *Candida* infections was independent of the Th17-pathway (Yano *et al.*, 2012a). Irrespective, the role of Th17 cells and their associated cytokines in immunity to mucosal fungal infections is gradually becoming clearer.

Cell-mediated immunity is also important in combating cryptococcal infection, with immunocompromised individuals showing increased prevalence and severity of infection (Levitz, 1991; Shoham and Levitz, 2005). $CD4^+$ Th1 immunity is known to play a role in host clearance of this fungus (Yauch *et al.*, 2006); however, $CD8^+$ T cells are important in clearing cryptococcal infections. Deficiency in $CD8^+$ T cells results in decreased leucocyte and $CD4^+$ T cell infiltration at the site of infection, along with a reduction in the Th1 cytokine IFNγ (Huffnagle *et al.*, 1994). Further, depleting the levels of $CD8^+$ T cells results in both reduced pulmonary clearance of *C. neoformans* and reduced delayed-type hypersensitivity responses associated with infection resolution without affecting the ability of lymphocytes to respond to antigen (Mody *et al.*, 1993). In this protective response, it is clear that IFNγ plays a key role as it can compensate for a lack of either $CD4^+$ or $CD8^+$ cells, allowing survival and a protective recall response on subsequent challenge with the fungus (Wozniak *et al.*, 2011b). A role for the Th17 phenotype in cryptococcal infections is less clear. Although IL-17 levels increase after infection, depletion of this cytokine had minimal effects on clearance or production of other cytokines. Further, the source of IL-17 during these infections appears to be neutrophils rather than Th17 cells (Wozniak *et al.*, 2011a)

Although originally thought to be exclusively produced by Th17 T cells, it is now known that IL-17 is produced by a variety of innate cell types as well, including γδ T cells, NKT cells, lymphoid tissue inducer (LTi) cells and macrophages (Cua and Tato, 2010). There is also some evidence to suggest neutrophils produce this cytokine in response to dectin-1 stimulation by *A fumigatus* during lung infections (Werner *et al.*, 2011). However, the role of IL-17 production from these cells during fungal infections remains to be determined.

The importance of Th17 and IL-17 in antifungal immunity is evident when evaluating individuals with specific autoimmune conditions. These patients have lost their ability to mount a normal Th17 response via a variety of mechanisms, either through failure to activate signalling events in Th17 maturation (STAT3 mutations), inability to signal via IL-17 (IL-17 receptor deficiencies) or production of neutralizing antibodies to IL-17 (Lilic, 2012; Naglik *et al.*, 2011). Patients with these genetic abnormalities have all been described as showing a reduction in Th17 effector responses and a concurrent increase in mucosal fungal infections. Many patients have a condition similar to chronic mucocutaneous candidiasis, with persistent infections of skin and mucosal surfaces, although, interestingly, there is no increase in systemic infections.

Antibodies and fungal defence

The final part of adaptive anti-fungal immunity is the production of anti-fungal antibodies. For many years, antibody-mediated immunity was thought to be of no real consequence in anti-fungal defence. Recent work using new experimental methods, however, has provided us with a better picture in which antibody-mediated immunity has begun to take an important place in our view of host anti-fungal responses. Antibodies are capable of mediating several different effects, these being protective, non-protective or pathogenic (acting to enhance disease). Typically, methods of antibody-mediated immunity are regarded as being neutralization (viral or toxin), opsonisation and antibody-dependent cellular cytotoxicity (ADCC). Although neutralization is not thought to play a significant role in anti-fungal immunity, it is possible that the other two may play some role. With the ability of *C. neoformans*, via its polysaccharide capsule, to avoid phagocytosis and opsonisation through complement (Zaragoza *et al.*, 2003), the ability of antibodies to bind and opsonise this fungus is of critical importance in clearing infections (Schlageter and Kozel, 1990). Likewise, activation of complement via classical pathway activation has also been implicated as a key process in anti-fungal immunity (Han *et al.*, 2001; Shapiro *et al.*, 2002). It is now clear that it is not sufficient for a host to simply phagocytose a fungus as many fungi (e.g. *C. neoformans*) have developed mechanisms to avoid phagosomal killing by preventing maturation of the phagolysosome. However, it is possible that antibodies can be generated to overcome this mechanism. For instance, monoclonal antibodies to *H. capsulatum* have been identified that promote phagolysosomal maturation, neatly circumventing this mechanism (Shi *et al.*, 2008).

As well as these indirectly mediated effects, antibodies also have direct effects on the interaction of fungi with their host. Their ability to bind and 'neutralize' targets enables them inhibit a variety of fungal actions. Notable among these are the potential to inhibit biofilm formation (Martinez and Casadevall, 2005), bind to adhesins (Brena *et al.*, 2007) and to remove polysaccharide and other components released from the cell wall. As well as being neutralizing, antibodies can also affect the fungus directly by interfering with metabolic processes by preventing nutrient acquisition (Brena *et al.*, 2011). Binding of antibodies to fungal cells can also affect other aspects of fungal life cycle. Several reports have implicated antibodies as being able to inhibit replication of fungal cells and even induce cell death (Moragues *et al.*, 2003; Rosas *et al.*, 2001) although the mechanisms by which this occurs are unclear.

Summary

Host responses to fungi are as varied as the fungi themselves. What is clear is that there are a variety of fungal PAMPs and antigens which enable responses that can be generated specifically to fungi. These responses vary depending on the site and cell type that encounter the fungal cells and change depending on the morphological status of the fungus upon contact. There are a variety of detection mechanisms utilized by different host cells, some of which enable the host to discriminate between commensal and pathogenic fungi. Over the last decade, the importance of non-haemopoietic cells in these responses has begun to be elucidated giving us a more complete picture of the complex network of responses between host and fungus as well as a better understanding of how previously established responses work to prevent the invasion and damage associated with pathogenic fungal infections. With this clearer picture, our understanding of the complex interactions between host and microbe and the role of both host and fungal factors in maintaining a balance between them has taken a seismic leap forward.

References

Aksoy, E., Vanden Berghe, W., Detienne, S., Amraoui, Z., Fitzgerald, K.A., Haegeman, G., Goldman, M., and Willems, F. (2005). Inhibition of phosphoinositide 3-kinase enhances TRIF-dependent NF-kappa B activation and IFN-beta synthesis downstream of Toll-like receptor 3 and 4. Eur. J. Immunol. *35*, 2200–2209.

Araki, N., Johnson, M.T., and Swanson, J.A. (1996). A role for phosphoinositide 3-kinase in the completion of macropinocytosis and phagocytosis by macrophages. J. Cell. Biol. *135*, 1249–1260.

Arora, S., Hernandez, Y., Erb-Downward, J.R., McDonald, R.A., Toews, G.B., and Huffnagle, G.B. (2005). Role of IFN-gamma in regulating T2 immunity and the development of alternatively activated macrophages during allergic bronchopulmonary mycosis. J. Immunol. *174*, 6346–6356.

Arora, S., Olszewski, M.A., Tsang, T.M., McDonald, R.A., Toews, G.B., and Huffnagle, G.B. (2011). Effect of cytokine interplay on macrophage polarization during chronic pulmonary infection with *Cryptococcus neoformans*. Infect. Immun. *79*, 1915–1926.

Ashman, R.B., Papadimitriou, J.M., Fulurija, A., Drysdale, K.E., Farah, C.S., Naidoo, O., and Gotjamanos, T. (2003). Role of complement C5 and T lymphocytes in pathogenesis of disseminated and mucosal candidiasis in susceptible DBA/2 mice. Microb. Pathogen. *34*, 103–113.

Ather, J.L., Alcorn, J.F., Brown, A.L., Guala, A.S., Suratt, B.T., Janssen-Heininger, Y.M., and Poynter, M.E. (2010). Distinct functions of airway epithelial nuclear factor-kappaB activity regulate nitrogen dioxide-induced acute lung injury. Am. J. Respir. Cell. Mol. Biol. *43*, 443–451.

Bahri, R., Curt, S., Saidane-Mosbahi, D., and Rouabhia, M. (2010). Normal human gingival epithelial cells sense *C. parapsilosis* by toll-like receptors and module its pathogenesis through antimicrobial peptides and pro-inflammatory cytokines. Med. Inflamm. *2010*, Article ID 940383.

Balloy, V., Sallenave, J.M., Wu, Y., Touqui, L., Latge, J.P., Si-Tahar, M., and Chignard, M. (2008). *Aspergillus fumigatus*-induced interleukin-8 synthesis by respiratory epithelial cells is controlled by the phosphatidylinositol 3-kinase, p38 MAPK, and ERK1/2 pathways and not by the toll-like receptor-MyD88 pathway. J. Biol. Chem. *283*, 30513–30521.

Barrenschee, M., Lex, D., and Uhlig, S. (2010). Effects of the TLR2 agonists MALP-2 and Pam3Cys in isolated mouse lungs. PloS One 5, e13889.

Benson, S.A., and Ernst, J.D. (2009). TLR2-dependent inhibition of macrophage responses to IFN-gamma is mediated by distinct, gene-specific mechanisms. PloS One 4, e6329.

Berkova, N., Lair-Fulleringer, S., Femenia, F., Huet, D., Wagner, M.C., Gorna, K., Tournier, F., Ibrahim-Granet, O., Guillot, J., Chermette, R., *et al.* (2006). *Aspergillus fumigatus* conidia inhibit tumour necrosis factor- or staurosporine-induced apoptosis in epithelial cells. Internat. Immunol. *18*, 139–150.

Bernasconi, E., Favre, L., Maillard, M.H., Bachmann, D., Pythoud, C., Bouzourene, H., Croze, E., Velichko, S., Parkinson, J., Michetti, P., *et al.* (2010). Granulocyte-macrophage colony-stimulating factor elicits bone marrow-derived cells that promote efficient colonic mucosal healing. Inflamm. Bowel. Dis. *16*, 428–441.

Bianchi, M.E. (2007). DAMPs, PAMPs and alarmins: all we need to know about danger. J. Leuk. Biol. *81*, 1–5.

Blander, J.M., and Medzhitov, R. (2006). On regulation of phagosome maturation and antigen presentation. Nat. Immunol. 7, 1029–1035.

Bochud, P.Y., Chien, J.W., Marr, K.A., Leisenring, W.M., Upton, A., Janer, M., Rodrigues, S.D., Li, S., Hansen, J.A., Zhao, L.P., *et al.* (2008). Toll-like receptor 4 polymorphisms and aspergillosis in stem-cell transplantation. N. Engl. J. Med. *359*, 1766–1777.

Bonifazi, P., Zelante, T., D'Angelo, C., De Luca, A., Moretti, S., Bozza, S., Perruccio, K., Iannitti, R.G., Giovannini, G., Volpi, C., *et al.* (2009). Balancing inflammation and tolerance *in vivo* through dendritic cells by the commensal *Candida albicans*. Mucos. Immunol. 2, 362–374.

Botterel, F., Gross, K., Ibrahim-Granet, O., Khoufache, K., Escabasse, V., Coste, A., Cordonnier, C., Escudier, E., and Bretagne, S. (2008). Phagocytosis of *Aspergillus fumigatus* conidia by primary nasal epithelial cells *in vitro*. BMC Microbiol. *8*, 97.

Bourgeois, C., Majer, O., Frohner, I.E., Tierney, L., and Kuchler, K. (2010). Fungal attacks on mammalian hosts: pathogen elimination requires sensing and tasting. Curr. Opin. Microbiol. *13*, 401–408.

Brena, S., Omaetxebarria, M.J., Elguezabal, N., Cabezas, J., Moragues, M.D., and Ponton, J. (2007). Fungicidal monoclonal antibody C7 binds to *Candida albicans* Als3. Infect. Immun. *75*, 3680–3682.

Brena, S., Cabezas-Olcoz, J., Moragues, M.D., Fernandez de Larrinoa, I., Dominguez, A., Quindos, G., and Ponton, J. (2011). Fungicidal monoclonal antibody C7 interferes with iron acquisition in *Candida albicans*. Antimicrob. Agents Chemother. *55*, 3156–3163.

Brouwer, N., Dolman, K.M., van Houdt, M., Sta, M., Roos, D., and Kuijpers, T.W. (2008). Mannose-binding lectin (MBL) facilitates opsonophagocytosis of yeasts but not of bacteria despite MbL binding. J. Immunol. *180*, 4124–4132.

Brown, G.D. (2006). Dectin-1: a signalling non-TLR pattern-recognition receptor. Nat. Rev. Immunol. *6*, 33–43.

Brown, G.D., Herre, J., Williams, D.L., Willment, J.A., Marshall, A.S., and Gordon, S. (2003). Dectin-1 mediates the biological effects of beta-glucans. J. Exp. Med. *197*, 1119–1124.

Bryant, C., and Fitzgerald, K.A. (2009). Molecular mechanisms involved in inflammasome activation. Trends Cell Biol. *19*, 455–464.

Bugarcic, A., Hitchens, K., Beckhouse, A.G., Wells, C.A., Ashman, R.B., and Blanchard, H. (2008). Human and mouse macrophage-inducible C-type lectin (Mincle) bind *Candida albicans*. Glycobiology *18*, 679–685.

Cambi, A., Gijzen, K., de Vries l, J., Torensma, R., Joosten, B., Adema, G.J., Netea, M.G., Kullberg, B.J., Romani, L., and Figdor, C.G. (2003). The C-type lectin DC-SIGN (CD209) is an antigen-uptake receptor for *Candida albicans* on dendritic cells. Eur. J. Immunol. *33*, 532–538.

Cambi, A., Netea, M.G., Mora-Montes, H.M., Gow, N.A., Hato, S.V., Lowman, D.W., Kullberg, B.J., Torensma, R., Williams, D.L., and Figdor, C.G. (2008). Dendritic cell interaction with *Candida albicans* critically depends on N-linked mannan. J. Biol. Chem. *283*, 20590–20599.

Carvalho, A., Pasqualotto, A.C., Pitzurra, L., Romani, L., Denning, D.W., and Rodrigues, F. (2008). Polymorphisms in toll-like receptor genes and susceptibility to pulmonary aspergillosis. J. Infect. Dis. *197*, 618–621.

Carvalho, A., De Luca, A., Bozza, S., Cunha, C., D'Angelo, C., Moretti, S., Perruccio, K., Iannitti, R.G., Fallarino, F., Pierini, A., *et al.* (2012). TLR3 essentially promotes protective class I-restricted memory CD8(+) T-cell responses to *Aspergillus fumigatus* in hematopoietic transplanted patients. Blood *119*, 967–977.

Cenci, E., Mencacci, A., Spaccapelo, R., Tonnetti, L., Mosci, P., Enssle, K.H., Puccetti, P., Romani, L., and Bistoni, F. (1995). T helper cell type 1 (Th1)- and Th2-like responses are present in mice with gastric candidiasis but protective immunity is associated with Th1 development. J. Infect. Dis. *171*, 1279–1288.

Chaffin, W.L. (2008). *Candida albicans* cell wall proteins. Microbiol. Mol. Biol. Rev. 72, 495–544.

Chai, L.Y., van de Veerdonk, F., Marijnissen, R.J., Cheng, S.C., Khoo, A.L., Hectors, M., Lagrou, K., Vonk,

A.G., Maertens, J., Joosten, L.A., *et al.* (2010). Anti-*Aspergillus* human host defence relies on type 1 T helper (Th1), rather than type 17 T helper (Th17), cellular immunity. Immunology *130*, 46–54.

Conti, H.R., Shen, F., Nayyar, N., Stocum, E., Sun, J.N., Lindemann, M.J., Ho, A.W., Hai, J.H., Yu, J.J., Jung, J.W., *et al.* (2009). Th17 cells and IL-17 receptor signaling are essential for mucosal host defense against oral candidiasis. J. Exp. Med. *206*, 299–311.

Cua, D.J., and Tato, C.M. (2010). Innate IL-17-producing cells: the sentinels of the immune system. Nat. Rev. Immunol. *10*, 479–489.

Dalle, F., Wachtler, B., L'Ollivier, C., Holland, G., Bannert, N., Wilson, D., Labruere, C., Bonnin, A., and Hube, B. (2010). Cellular interactions of *Candida albicans* with human oral epithelial cells and enterocytes. Cell Microbiol. *12*, 248–271.

Dan, J.M., Kelly, R.M., Lee, C.K., and Levitz, S.M. (2008). Role of the mannose receptor in a murine model of *Cryptococcus neoformans* infection. Infect. Immun. *76*, 2362–2367.

De Luca, A., Montagnoli, C., Zelante, T., Bonifazi, P., Bozza, S., Moretti, S., D'Angelo, C., Vacca, C., Boon, L., Bistoni, F., *et al.* (2007). Functional yet balanced reactivity to *Candida albicans* requires TRIF, MyD88, and IDO-dependent inhibition of Rorc. J. Immunol. *179*, 5999–6008.

Del Sero, G., Mencacci, A., Cenci, E., d'Ostiani, C.F., Montagnoli, C., Bacci, A., Mosci, P., Kopf, M., and Romani, L. (1999). Antifungal type 1 responses are up-regulated in IL-10-deficient mice. Microb. Infect. *1*, 1169–1180.

Diamond, G., Beckloff, N., and Ryan, L.K. (2008). Host defense peptides in the oral cavity and the lung: similarities and differences. J. Dent. Res. *87*, 915–927.

Dillon, S., Agrawal, S., Banerjee, K., Letterio, J., Denning, T.L., Oswald-Richter, K., Kasprowicz, D.J., Kellar, K., Pare, J., van Dyke, T., *et al.* (2006). Yeast zymosan, a stimulus for TLR2 and dectin-1, induces regulatory antigen-presenting cells and immunological tolerance. J. Clin. Invest. *116*, 916–928.

Farah, C.S., Gotjamanos, T., Seymour, G.J., and Ashman, R.B. (2002). Cytokines in the oral mucosa of mice infected with *Candida albicans*. Oral Microbiol. Immunol. *17*, 375–378.

Farah, C.S., Hu, Y., Riminton, S., and Ashman, R.B. (2006). Distinct roles for interleukin-12p40 and tumour necrosis factor in resistance to oral candidiasis defined by gene-targeting. Oral Microbiol. Immunol. *21*, 252–255.

Fekkar, A., Balloy, V., Pionneau, C., Marinach-Patrice, C., Chignard, M., and Mazier, D. (2012). Secretome of human bronchial epithelial cells in response to the fungal pathogen *Aspergillus fumigatus* analyzed by differential in-gel electrophoresis. J. Infect. Dis. *205*, 1163–1172.

Fidel, P.L., Jr. (1999). Host defense against oropharyngeal and vaginal candidiasis: Site-specific differences. Revist. Iberoamerican. Micolog. *16*, 8–15.

Fidel, P.L., Jr. (2005). Immunity in vaginal candidiasis. Curr. Opin. Infect. Dis. *18*, 107–111.

Fidel, P.L., Jr. (2007). History and update on host defense against vaginal candidiasis. Am. J. Reprod. Immunol. *57*, 2–12.

Filler, S.G., Pfunder, A.S., Spellberg, B.J., Spellberg, J.P., and Edwards, J.E., Jr. (1996). *Candida albicans* stimulates cytokine production and leukocyte adhesion molecule expression by endothelial cells. Infect. Immun. *64*, 2609–2617.

Forsyth, C.B., and Mathews, H.L. (1996). Lymphocytes utilize CD11b/CD18 for adhesion to *Candida albicans*. Cell. Immunol. *170*, 91–100.

Forsyth, C.B., and Mathews, H.L. (2002). Lymphocyte adhesion to *Candida albicans*. Infect. Immun. *70*, 517–527.

Gantner, B.N., Simmons, R.M., Canavera, S.J., Akira, S., and Underhill, D.M. (2003). Collaborative induction of inflammatory responses by dectin-1 and Toll-like receptor 2. J. Exp. Med. *197*, 1107–1117.

Ghezzi, M.C., Raponi, G., Angeletti, S., and Mancini, C. (1998). Serum-mediated enhancement of TNF-alpha release by human monocytes stimulated with the yeast form of *Candida albicans*. J. Infect. Dis. *178*, 1743–1749.

Glocker, E., and Grimbacher, B. (2010). Chronic mucocutaneous candidiasis and congenital susceptibility to *Candida*. Curr. Opin. Allerg. Clin. Immunol. *10*, 542–550.

Gomez, P., Hackett, T.L., Moore, M.M., Knight, D.A., and Tebbutt, S.J. (2010). Functional genomics of human bronchial epithelial cells directly interacting with conidia of *Aspergillus fumigatus*. BMC Genomics *11*, 358.

Goodridge, H.S., Reyes, C.N., Becker, C.A., Katsumoto, T.R., Ma, J., Wolf, A.J., Bose, N., Chan, A.S., Magee, A.S., Danielson, M.E., *et al.* (2011). Activation of the innate immune receptor Dectin-1 upon formation of a 'phagocytic synapse'. Nature *472*, 471–475.

Goosney, D.L., Knoechel, D.G., and Finlay, B.B. (1999). Enteropathogenic *E. coli*, *Salmonella*, and *Shigella*: masters of host cell cytoskeletal exploitation. Emerg. Infect. Dis. *5*, 216–223.

Graham, L.M., and Brown, G.D. (2009). The Dectin-2 family of C-type lectins in immunity and homeostasis. Cytokine *48*, 148–155.

Gringhuis, S.I., den Dunnen, J., Litjens, M., van der Vlist, M., Wevers, B., Bruijns, S.C., and Geijtenbeek, T.B. (2009). Dectin-1 directs T helper cell differentiation by controlling noncanonical NF-kappaB activation through Raf-1 and Syk. Nat. Immunol. *10*, 203–213.

Gross, O., Gewies, A., Finger, K., Schafer, M., Sparwasser, T., Peschel, C., Forster, I., and Ruland, J. (2006). Card9 controls a non-TLR signalling pathway for innate antifungal immunity. Nature *442*, 651–656.

Gross, O., Poeck, H., Bscheider, M., Dostert, C., Hannesschlager, N., Endres, S., Hartmann, G., Tardivel, A., Schweighoffer, E., Tybulewicz, V., *et al.* (2009). Syk kinase signalling couples to the Nlrp3 inflammasome for anti-fungal host defence. Nature *459*, 433–436.

Han, X., Yu, R., Zhen, D., Tao, S., Schmidt, M., and Han, L. (2011). beta-1,3-Glucan-induced host phospholipase D activation is involved in *Aspergillus fumigatus* internalization into type II human pneumocyte A549 cells. PloS One *6*, e21468.

Han, Y., Kozel, T.R., Zhang, M.X., MacGill, R.S., Carroll, M.C., and Cutler, J.E. (2001). Complement is essential for protection by an IgM and an IgG3 monoclonal antibody against experimental, hematogenously disseminated candidiasis. J. Immunol. *167*, 1550–1557.

Hardison, S.E., Ravi, S., Wozniak, K.L., Young, M.L., Olszewski, M.A., and Wormley, F.L., Jr. (2010). Pulmonary infection with an interferon-gamma-producing *Cryptococcus neoformans* strain results in classical macrophage activation and protection. Am. J. Pathol. *176*, 774–785.

Harrington, L.E., Mangan, P.R., and Weaver, C.T. (2006). Expanding the effector CD4 T-cell repertoire: the Th17 lineage. Curr. Opin. Immunol. *18*, 349–356.

Herre, J., Gordon, S., and Brown, G.D. (2004). Dectin-1 and its role in the recognition of beta-glucans by macrophages. Mol. Immunol. *40*, 869–876.

Hesse, M., Modolell, M., La Flamme, A.C., Schito, M., Fuentes, J.M., Cheever, A.W., Pearce, E.J., and Wynn, T.A. (2001). Differential regulation of nitric oxide synthase-2 and arginase-1 by type 1/type 2 cytokines *in vivo:* granulomatous pathology is shaped by the pattern of L-arginine metabolism. J. Immunol. *167*, 6533–6544.

Hise, A.G., Tomalka, J., Ganesan, S., Patel, K., Hall, B.A., Brown, G.D., and Fitzgerald, K.A. (2009). An essential role for the NLRP3 inflammasome in host defense against the human fungal pathogen *Candida albicans*. Cell Host Microbe *5*, 487–497.

Hosogi, S., Iwasaki, Y., Yamada, T., Komatani-Tamiya, N., Hiramatsu, A., Kohno, Y., Ueda, M., Arimoto, T., and Marunaka, Y. (2008). Effect of inducible nitric oxide synthase on apoptosis in *Candida*-induced acute lung injury. Biomed Res *29*, 257–266.

Huang, H., Ostroff, G.R., Lee, C.K., Wang, J.P., Specht, C.A., and Levitz, S.M. (2009). Distinct patterns of dendritic cell cytokine release stimulated by fungal beta-glucans and toll-like receptor agonists. Infect. Immun. *77*, 1774–1781.

Huang, W., Na, L., Fidel, P.L., and Schwarzenberger, P. (2004). Requirement of interleukin-17A for systemic anti-*Candida albicans* host defense in mice. J. Infect. Dis. *190*, 624–631.

Huffnagle, G.B., Lipscomb, M.F., Lovchik, J.A., Hoag, K.A., and Street, N.E. (1994). The role of CD4+ and CD8+ T cells in the protective inflammatory response to a pulmonary cryptococcal infection. J. Leuk. Biol. *55*, 35–42.

Ibata-Ombetta, S., Idziorek, T., Trinel, P.A., Poulain, D., and Jouault, T. (2003). *Candida albicans* phospholipomannan promotes survival of phagocytosed yeasts through modulation of bad phosphorylation and macrophage apoptosis. J. Biol. Chem. *278*, 13086–13093.

Iliev, I.D., Funari, V.A., Taylor, K.D., Nguyen, Q., Reyes, C.N., Strom, S.P., Brown, J., Becker, C.A., Fleshner, P.R., Dubinsky, M., *et al.* (2012). Interactions between commensal fungi and the C-type lectin receptor Dectin-1 influence colitis. Science *336*, 1314–1317.

Ip, W.K., and Lau, Y.L. (2004). Role of mannose-binding lectin in the innate defense against *Candida albicans*: enhancement of complement activation, but lack of opsonic function, in phagocytosis by human dendritic cells. J. Infect. Dis. *190*, 632–640.

Isberg, R.R. (1996). Uptake of enteropathogenic *Yersinia* by mammalian cells. Curr. Top. Microbiol. Immunol. *209*, 1–24.

Joly, S., Ma, N., Sadler, J.J., Soll, D.R., Cassel, S.L., and Sutterwala, F.S. (2009). Cutting edge: *Candida albicans* hyphae formation triggers activation of the Nlrp3 inflammasome. J. Immunol. *183*, 3578–3581.

Kagami, S., Rizzo, H.L., Kurtz, S.E., Miller, L.S., and Blauvelt, A. (2010). IL-23 and IL-17A, but not IL-12 and IL-22, are required for optimal skin host defense against *Candida albicans*. J. Immunol. *185*, 5453–5462.

Kamai. Y., Lossinsky, A.S., Liu, H., Sheppard, D.C., and Filler, S.G. (2009). Polarized response of endothelial cells to invasion by *Aspergillus fumigatus*. Cell. Microbiol. *11*, 170–182.

Kerscher, B., Willment, J.A., and Brown, G.D. (2013). The Dectin-2 family of C-type lectin-like receptors: an update. Internat. Immunol. *25*, 271–277.

Klein, R.S., Harris, C.A., Small, C.B., Moll, B., Lesser, M., and Friedland, G.H. (1984). Oral candidiasis in high-risk patients as the initial manifestation of the acquired immunodeficiency syndrome. N. Engl. J. Med. *311*, 354–358.

Kumar, H., Kawai, T., and Akira, S. (2011). Pathogen recognition by the innate immune system. Int. Rev. Immunol. *30*, 16–34.

Lam, J.S., Huang, H., and Levitz, S.M. (2007). Effect of differential N-linked and O-linked mannosylation on recognition of fungal antigens by dendritic cells. PloS One *2*, e1009.

Latge, J.P. (1999). *Aspergillus fumigatus* and aspergillosis. Clinical Microbiol. Rev. *12*, 310–350.

Lee, R.T., Hsu, T.L., Huang, S.K., Hsieh, S.L., Wong, C.H., and Lee, Y.C. (2011). Survey of immune-related, mannose/fucose-binding C-type lectin receptors reveals widely divergent sugar-binding specificities. Glycobiology *21*, 512–520.

Lee, S.J., Zheng, N.Y., Clavijo, M., and Nussenzweig, M.C. (2003). Normal host defense during systemic candidiasis in mannose receptor-deficient mice. Infect. Immun. *71*, 437–445.

Lee, W.B., Kang, J.S., Yan, J.J., Lee, M.S., Jeon, B.Y., Cho, S.N., and Kim, Y.J. (2012). Neutrophils Promote Mycobacterial Trehalose Dimycolate-Induced Lung Inflammation via the Mincle Pathway. PLoS Pathog. *8*, e1002614.

Levitz, S.M. (1991). The ecology of *Cryptococcus neoformans* and the epidemiology of cryptococcosis. Rev. Infectious Dis. *13*, 1163–1169.

Lilic, D. (2012). Unravelling fungal immunity through primary immune deficiencies. Curr. Opin. Microbiol. *15*, 420–426.

Lopes Bezerra, L.M., and Filler, S.G. (2004). Interactions of *Aspergillus fumigatus* with endothelial cells: internalization, injury, and stimulation of tissue factor activity. Blood *103*, 2143–2149.

Lortholary, O., Improvisi, L., Nicolas, M., Provost, F., Dupont, B., and Dromer, F. (1999). Fungemia during murine cryptococcosis sheds some light on pathophysiology. Med. Mycol. *37*, 169–174.

de Luca, A., Bozza, S., Zelante, T., Zagarella, S., D'Angelo, C., Perruccio, K., Vacca, C., Carvalho, A., Cunha, C., Aversa, F., *et al.* (2010). Non-hematopoietic cells contribute to protective tolerance to *Aspergillus* fumigatus via a TRIF pathway converging on IDO. Cell Mol. Immunol.*7*, 459–470.

Martinez, F.O., Helming, L., and Gordon, S. (2009). Alternative activation of macrophages: an immunologic functional perspective. Ann. Rev Immunol. *27*, 451–483.

Martinez, L.R., and *Casadevall*, A. (2005). Specific antibody can prevent fungal biofilm formation and this effect correlates with protective efficacy. Infect. Immun. *73*, 6350–6362.

Martinon, F., Mayor, A., and Tschopp, J. (2009). The inflammasomes: guardians of the body. Ann. Rev. Immunol. *27*, 229–265.

McGreal, E.P., Rosas, M., Brown, G.D., Zamze, S., Wong, S.Y., Gordon, S., Martinez-Pomares, L., and Taylor, P.R. (2006). The carbohydrate-recognition domain of Dectin-2 is a C-type lectin with specificity for high mannose. Glycobiology *16*, 422–430.

McKenzie, E.J., Taylor, P.R., Stillion, R.J., Lucas, A.D., Harris, J., Gordon, S., and Martinez-Pomares, L. (2007). Mannose receptor expression and function define a new population of murine dendritic cells. J. Immunol. *178*, 4975–4983.

Mody, C.H., Chen, G.H., Jackson, C., Curtis, J.L., and Toews, G.B. (1993). Depletion of murine CD8+ T cells *in vivo* decreases pulmonary clearance of a moderately virulent strain of *Cryptococcus neoformans*. J. Lab. Clin. Med. *121*, 765–773.

Moragues, M.D., Omaetxebarria, M.J., Elguezabal, N., Sevilla, M.J., Conti, S., Polonelli, L., and Ponton, J. (2003). A monoclonal antibody directed against a *Candida albicans* cell wall mannoprotein exerts three anti-*C. albicans* activities. Infect. Immun. *71*, 5273–5279.

Moreno-Ruiz, E., Galan-Diez, M., Zhu, W., Fernandez-Ruiz, E., d'Enfert, C., Filler, S.G., Cossart, P., and Veiga, E. (2009). *Candida albicans* internalization by host cells is mediated by a clathrin-dependent mechanism. Cell. Microbiol. *11*, 1179–1189.

Mosser, D.M., and Edwards, J.P. (2008). Exploring the full spectrum of macrophage activation. Nat. Rev. Immunol. *8*, 958–969.

Moyes, D.L., and Naglik, J.R. (2011). Mucosal immunity and *Candida albicans* infection. Clin. Dev. Immunol. *2011*, 346307.

Moyes, D.L., Runglall, M., Murciano, C., Shen, C., Nayar, D., Thavaraj, S., Kohli, A., Islam, A., Mora-Montes, H., Challacombe, S.J., *et al.* (2010). A biphasic innate immune MAPK response discriminates between the yeast and hyphal forms of *Candida albicans* in epithelial cells. Cell Host Microbe *8*, 225–235.

Moyes, D.L., Murciano, C., Runglall, M., Islam, A., Thavaraj, S., and Naglik, J.R. (2011). *Candida albicans* yeast and hyphae are discriminated by MAPK signaling in vaginal epithelial cells. PloS One *6*, e26580.

Moyes, D.L., Murciano, C., Runglall, M., Kohli, A., Islam, A., and Naglik, J.R. (2012). Activation of MAPK/c-Fos induced responses in oral epithelial cells is specific to *Candida albicans* and *Candida dubliniensis* hyphae. Med. Microbiol. Immunol. *201*, 93–101.

Moyes, D.L., Shen, C., Murciano, C., Runglall, M., Richardson, J.P., Arno, M., Aldecoa-Otalora, E., and Naglik, J.R. (2013). Protection against epithelial damage during *Candida albicans* infection is mediated by PI3K/Akt and mTOR signaling. J. Infect. Dis. [Epub ahead of print].

Mozaffarian, N., Casadevall, A., and Berman, J.W. (2000). Inhibition of human endothelial cell chemokine production by the opportunistic fungal pathogen *Cryptococcus neoformans*. J. Immunol. *165*, 1541–1547.

Müller, V., Viemann, D., Schmidt, M., Endres, N., Ludwig, S., Leverkus, M., Roth, J., and Goebeler, M. (2007). *Candida albicans* triggers activation of distinct signalling pathways to establish a proinflammatory gene expression program in primary human endothelial cells. J. Immunol. *179*, 8435–8445.

Naglik, J.R., and Moyes, D. (2011). Epithelial cell innate response to *Candida albicans*. Adv. Dent. Res. *23*, 50–55.

Naglik, J.R., Fidel, P.L., Jr., and Odds, F.C. (2008). Animal models of mucosal *Candida* infection. FEMS Microbiol. Lett. *283*, 129–139.

Naglik, J.R., Moyes, D.L., Wachtler, B., and Hube, B. (2011). *Candida albicans* interactions with epithelial cells and mucosal immunity. Microb. Infect. *13*, 963–976.

Nakamura, K., Kinjo, T., Saijo, S., Miyazato, A., Adachi, Y., Ohno, N., Fujita, J., Kaku, M., Iwakura, Y., and Kawakami, K. (2007). Dectin-1 is not required for the host defense to *Cryptococcus neoformans*. Microbiol. Immunol. *51*, 1115–1119.

Netea, M.G., Brown, G.D., Kullberg, B.J., and Gow, N.A. (2008). An integrated model of the recognition of *Candida albicans* by the innate immune system. Nat. Rev. Microbiol. *6*, 67–78.

Neth, O., Jack, D.L., Dodds, A.W., Holzel, H., Klein, N.J., and Turner, M.W. (2000). Mannose-binding lectin binds to a range of clinically relevant microorganisms and promotes complement deposition. Infect. Immun. *68*, 688–693.

Orozco, A.S., Zhou, X., and Filler, S.G. (2000). Mechanisms of the proinflammatory response of endothelial cells to *Candida albicans* infection. Infect. Immun. *68*, 1134–1141.

Osherov, N. (2012). Interaction of the pathogenic mold *Aspergillus fumigatus* with lung epithelial cells. Front. Microbiol. 3, 346.

Park, H., Myers, C.L., Sheppard, D.C., Phan, Q.T., Sanchez, A.A., E. Edwards., J., and Filler, S.G. (2005). Role of the fungal Ras-protein kinase A pathway in governing epithelial cell interactions during oropharyngeal candidiasis. Cell Microbiol. *7*, 499–510.

Phan, Q.T., Fratti, R.A., Prasadarao, N.V., Edwards, J.E., Jr., and Filler, S.G. (2005). N-cadherin mediates endocytosis of *Candida albicans* by endothelial cells. J. Biol. Chem. *280*, 10455–10461.

Philpott, D.J., Girardin, S.E., and Sansonetti, P.J. (2001). Innate immune responses of epithelial cells following infection with bacterial pathogens. Curr. Opin. Immunol. *13*, 410–416.

Pietrella, D., Rachini, A., Pines, M., Pandey, N., Mosci, P., Bistoni, F., d'Enfert, C., and Vecchiarelli, A. (2011). Th17 cells and IL-17 in protective immunity to vaginal candidiasis. PloS One *6*, e22770.

Plantinga, T.S., van der Velden, W.J., Ferwerda, B., van Spriel, A.B., Adema, G., Feuth, T., Donnelly, J.P., Brown, G.D., Kullberg, B.J., Blijlevens, N.M., *et al.* (2009). Early stop polymorphism in human DECTIN-1 is associated with increased *Candida* colonization in hematopoietic stem cell transplant recipients. Clin. Infect. Dis. *49*, 724–732.

Plato, A., Willment, J.A., and Brown, G.D. (2013). C-type lectin-like receptors of the dectin-1 cluster: ligands and signaling pathways. Int. Rev. Immunol. *32*, 134–156.

Rappleye, C.A., Eissenberg, L.G., and Goldman, W.E. (2007). *Histoplasma capsulatum* alpha-(1,3)-glucan blocks innate immune recognition by the beta-glucan receptor. Proc. Natl. Acad. Sci. U.S.A. *104*, 1366–1370.

Rehaume, L.M., Jouault, T., and Chamaillard, M. (2010). Lessons from the inflammasome: a molecular sentry linking *Candida* and Crohn's disease. Trends Immunol. *31*, 171–175.

Reid, D.M., Gow, N.A., and Brown, G.D. (2009). Pattern recognition: recent insights from Dectin-1. Curr. Opin. Immunol. *21*, 30–37.

Robinson, M.J., Osorio, F., Rosas, M., Freitas, R.P., Schweighoffer, E., Gross, O., Verbeek, J.S., Ruland, J., Tybulewicz, V., Brown, G.D., *et al.* (2009). Dectin-2 is a Syk-coupled pattern recognition receptor crucial for Th17 responses to fungal infection. J. Exp. Med. *206*, 2037–2051.

Romani, L. (2011). Immunity to fungal infections. Nat. Rev. Immunol. *11*, 275–288.

Rosas, A.L., Nosanchuk, J.D., and Casadevall, A. (2001). Passive immunization with melanin-binding monoclonal antibodies prolongs survival of mice with lethal *Cryptococcus neoformans* infection. Infect. Immun. *69*, 3410–3412.

Rotstein, D., Parodo, J., Taneja, R., and Marshall, J.C. (2000). Phagocytosis of *Candida albicans* induces apoptosis of human neutrophils. Shock *14*, 278–283.

Saijo, S., Fujikado, N., Furuta, T., Chung, S.H., Kotaki, H., Seki, K., Sudo, K., Akira, S., Adachi, Y., Ohno, N., *et al.* (2007). Dectin-1 is required for host defense against *Pneumocystis* carinii but not against *Candida albicans*. Nat. Immunol. *8*, 39–46.

Saijo, S., Ikeda, S., Yamabe, K., Kakuta, S., Ishigame, H., Akitsu, A., Fujikado, N., Kusaka, T., Kubo, S., Chung, S.H., *et al.* (2010). Dectin-2 recognition of alpha-mannans and induction of Th17 cell differentiation is essential for host defense against *Candida albicans*. Immunity 32, 681–691.

Sato, K., Yang, X.L., Yudate, T., Chung, J.S., Wu, J., Luby-Phelps, K., Kimberly, R.P., Underhill, D., Cruz, P.D., Jr., and Ariizumi, K. (2006). Dectin-2 is a pattern recognition receptor for fungi that couples with the Fc receptor gamma chain to induce innate immune responses. J. Biol. Chem. *281*, 38854–38866.

Schaller, M., Mailhammer, R., Grassl, G., Sander, C.A., Hube, B., and Korting, H.C. (2002). Infection of human oral epithelia with *Candida* species induces cytokine expression correlated to the degree of virulence. J. Invest. Derm. *118*, 652–657.

Schaller, M., Boeld, U., Oberbauer, S., Hamm, G., Hube, B., and Korting, H.C. (2004). Polymorphonuclear leukocytes (PMNs) induce protective Th1-type cytokine epithelial responses in an *in vitro* model of oral candidosis. Microbiology *150*, 2807–2813.

Scheper, M.A., Shirtliff, M.E., Meiller, T.F., Peters, B.M., and Jabra-Rizk, M.A. (2008). Farnesol, a fungal quorum-sensing molecule triggers apoptosis in human oral squamous carcinoma cells. Neoplasia *10*, 954–963.

Schlageter, A.M., and Kozel, T.R. (1990). Opsonization of *Cryptococcus neoformans* by a family of isotype-switch variant antibodies specific for the capsular polysaccharide. Infect. Immun. *58*, 1914–1918.

Shapiro, S., Beenhouwer, D.O., Feldmesser, M., Taborda, C., Carroll, M.C., Casadevall, A., and Scharff, M.D. (2002). Immunoglobulin G monoclonal antibodies to *Cryptococcus neoformans* protect mice deficient in complement component C3. Infect. Immun. *70*, 2598–2604.

Sharon, H., Amar, D., Levdansky, E., Mircus, G., Shadkchan, Y., Shamir, R., and Osherov, N. (2011). PrtT-regulated proteins secreted by *Aspergillus fumigatus* activate MAPK signaling in exposed A549 lung cells leading to necrotic cell death. PloS One *6*, e17509.

Shi, L., Albuquerque, P.C., Lazar-Molnar, E., Wang, X., Santambrogio, L., Gacser, A., and Nosanchuk, J.D. (2008). A monoclonal antibody to *Histoplasma capsulatum* alters the intracellular fate of the fungus in murine macrophages. Eukaryot. Cell 7, 1109–1117.

Shoham, S., and Levitz, S.M. (2005). The immune response to fungal infections. B. J. Haematol. *129*, 569–582.

Sonesson, A., Ringstad, L., Nordahl, E.A., Malmsten, M., Morgelin, M., and Schmidtchen, A. (2007). Antifungal activity of C3a and C3a-derived peptides against *Candida*. Biochim. Biophys. ACTA *1768*, 346–353.

Sousa Mda, G., Reid, D.M., Schweighoffer, E., Tybulewicz, V., Ruland, J., Langhorne, J., Yamasaki, S., Taylor, P.R., Almeida, S.R., and Brown, G.D. (2011). Restoration of pattern recognition receptor costimulation to treat chromoblastomycosis, a chronic fungal infection of the skin. Cell Host Microbe 9, 436–443.

Speth, C., Rambach, G., Wurzner, R., and Lass-Florl, C. (2008). Complement and fungal pathogens: an update. Mycoses *51*, 477–496.

Stein, M., Keshav, S., Harris, N., and Gordon, S. (1992). Interleukin 4 potently enhances murine macrophage mannose receptor activity: a marker of alternative immunologic macrophage activation. J. Exp. Med. *176*, 287–292.

Sturtevant, J.E., and Latge, J.P. (1992). Interactions between conidia of *Aspergillus fumigatus* and human complement component C3. Infect. Immun. *60*, 1913–1918.

Tavanti, A., Campa, D., Bertozzi, A., Pardini, G., Naglik, J.R., Barale, R., and Senesi, S. (2006). *Candida albicans* isolates with different genomic backgrounds display a differential response to macrophage infection. Microb. Infect. *8*, 791–800.

Taylor, P.R., Gordon, S., and Martinez-Pomares, L. (2005a). The mannose receptor: linking homeostasis and immunity through sugar recognition. Trends Immunol. *26*, 104–110.

Taylor, P.R., Martinez-Pomares, L., Stacey, M., Lin, H.H., Brown, G.D., and Gordon, S. (2005b). Macrophage receptors and immune recognition. Ann. Rev. Immunol. *23*, 901–944.

Taylor, P.R., Tsoni, S.V., Willment, J.A., Dennehy, K.M., Rosas, M., Findon, H., Haynes, K., Steele, C., Botto, M., Gordon, S., *et al.* (2007). Dectin-1 is required for beta-glucan recognition and control of fungal infection. Nat. Immunol. *8*, 31–38.

Tomalka, J., Ganesan, S., Azodi, E., Patel, K., Majmudar, P., Hall, B.A., Fitzgerald, K.A., and Hise, A.G. (2011). A novel role for the NLRC4 inflammasome in mucosal defenses against the fungal pathogen *Candida albicans*. PLoS Pathog. *7*, e1002379.

Underhill, D.M., Rossnagle, E., Lowell, C.A., and Simmons, R.M. (2005). Dectin-1 activates Syk tyrosine kinase in a dynamic subset of macrophages for reactive oxygen production. Blood *106*, 2543–2550.

Van der Graaf, C.A., Netea, M.G., Morre, S.A., Den Heijer, M., Verweij, P.E., Van der Meer, J.W., and Kullberg, B.J. (2006). Toll-like receptor 4 Asp299Gly/Thr399Ile polymorphisms are a risk factor for *Candida* bloodstream infection. Eur. Cytokine Netw. *17*, 29–34.

Vautier, S., MacCallum, D.M., and Brown, G.D. (2012). C-type lectin receptors and cytokines in fungal immunity. Cytokine *58*, 89–99.

van de Veerdonk, F.L., Kullberg, B.J., van der Meer, J.W., Gow, N.A., and Netea, M.G. (2008). Host–microbe interactions: innate pattern recognition of fungal pathogens. Curr. Opin. Microbiol. *11*, 305–312.

Vijayan, D., Radford, K.J., Beckhouse, A.G., Ashman, R.B., and Wells, C.A. (2012). Mincle polarizes human monocyte and neutrophil responses to *Candida albicans*. Immunol. Cell. Biol. *90*, 889–895.

Villar, C.C., and Zhao, X.R. (2010). *Candida albicans* induces early apoptosis followed by secondary necrosis in oral epithelial cells. Mol. Oral Microbiol. *25*, 215–225.

Villar, C.C., Chukwuedum Aniemeke, J., Zhao, X.R., and Huynh-Ba, G. (2012). Induction of apoptosis in oral epithelial cells by *Candida albicans*. Mol. Oral Microbiol. *27*, 436–448.

Wachtler, B., Wilson, D., Haedicke, K., Dalle, F., and Hube, B. (2011). From attachment to damage: defined genes of *Candida albicans* mediate adhesion, invasion and damage during interaction with oral epithelial cells. PloS One *6*, e17046.

Wagener, J., Weindl, G., de Groot, P.W., de Boer, A.D., Kaesler, S., Thavaraj, S., Bader, O., Mailander-Sanchez, D., Borelli, C., Weig, M., *et al.* (2012). Glycosylation of *Candida albicans* cell wall proteins is critical for induction of innate immune responses and apoptosis of epithelial cells. PloS One *7*, e50518.

Weindl, G., Naglik, J.R., Kaesler, S., Biedermann, T., Hube, B., Korting, H.C., and Schaller, M. (2007). Human epithelial cells establish direct antifungal defense through TLR4-mediated signaling. J. Clin. Invest. *117*, 3664–3672.

Wells, C.A., Salvage-Jones, J.A., Li, X., Hitchens, K., Butcher, S., Murray, R.Z., Beckhouse, A.G., Lo, Y.L., Manzanero, S., Cobbold, C., *et al.* (2008). The macrophage-inducible C-type lectin, mincle, is an essential component of the innate immune response to *Candida albicans*. J. Immunol. *180*, 7404–7413.

Werner, J.L., Metz, A.E., Horn, D., Schoeb, T.R., Hewitt, M.M., Schwiebert, L.M., Faro-Trindade, I., Brown, G.D., and Steele, C. (2009). Requisite role for the dectin-1 beta-glucan receptor in pulmonary defense against *Aspergillus fumigatus*. J. Immunol. *182*, 4938–4946.

Werner, J.L., Gessner, M.A., Lilly, L.M., Nelson, M.P., Metz, A.E., Horn, D., Dunaway, C.W., Deshane, J., Chaplin, D.D., Weaver, C.T., *et al.* (2011). Neutrophils produce interleukin 17A (IL-17A) in a dectin-1- and IL-23-dependent manner during invasive fungal infection. Infect. Immun. *79*, 3966–3977.

Willment, J.A., and Brown, G.D. (2008). C-type lectin receptors in antifungal immunity. Trends Microbiol. *16*, 27–32.

Wozniak, K.L., Hardison, S.E., Kolls, J.K., and Wormley, F.L. (2011a). Role of IL-17A on resolution of pulmonary *C. neoformans* infection. PloS One *6*, e17204.

Wozniak, K.L., Young, M.L., and Wormley, F.L., Jr. (2011b). Protective immunity against experimental pulmonary cryptococcosis in T cell-depleted mice. Clin. Vacc. Immunol. *18*, 717–723.

Wozniak, K.L., Hardison, S., Olszewski, M., and Wormley, F.L., Jr. (2012). Induction of protective immunity against cryptococcosis. Mycopathologia *173*, 387–394.

Yamasaki, S., Ishikawa, E., Sakuma, M., Hara, H., Ogata, K., and Saito, T. (2008). Mincle is an ITAM-coupled activating receptor that senses damaged cells. Nat. Immunol. *9*, 1179–1188.

Yamasaki, S., Matsumoto, M., Takeuchi, O., Matsuzawa, T., Ishikawa, E., Sakuma, M., Tateno, H., Uno, J., Hirabayashi, J., Mikami, Y., *et al.* (2009). C-type lectin Mincle is an activating receptor for pathogenic fungus, *Malassezia*. Proc. Natl. Acad. Sci. U.S.A. *106*, 1897–1902.

Yano, J., Lilly, E., Barousse, M., and Fidel, P.L., Jr. (2010). Epithelial cell-derived S100 calcium-binding proteins as key mediators in the hallmark acute neutrophil response during *Candida* vaginitis. Infect. Immun. *78*, 5126–5137.

Yano, J., Kolls, J.K., Happel, K.I., Wormley, F., Wozniak, K.L., and Fidel, P.L., Jr. (2012a). The acute neutrophil response mediated by S100 alarmins during vaginal *Candida* infections is independent of the Th17-pathway. PloS One *7*, e46311.

Yano, J., Noverr, M.C., and Fidel, P.L., Jr. (2012b). Cytokines in the host response to *Candida* vaginitis: Identifying a role for non-classical immune mediators, S100 alarmins. Cytokine *58*, 118–128.

Yauch, L.E., Lam, J.S., and Levitz, S.M. (2006). Direct inhibition of T-cell responses by the *Cryptococcus* capsular polysaccharide glucuronoxylomannan. PLoS Pathog. *2*, e120.

Zaki, M.H., Boyd, K.L., Vogel, P., Kastan, M.B., Lamkanfi, M., and Kanneganti, T.D. (2010). The NLRP3 inflammasome protects against loss of epithelial integrity and mortality during experimental colitis. Immunity *32*, 379–391.

Zakikhany, K., Naglik, J.R., Schmidt-Westhausen, A., Holland, G., Schaller, M., and Hube, B. (2007). *In vivo* transcript profiling of *Candida albicans* identifies a gene essential for interepithelial dissemination. Cell Microbiol. *9*, 2938–2954.

Zaragoza, O., Taborda, C.P., and Casadevall, A. (2003). The efficacy of complement-mediated phagocytosis of *Cryptococcus neoformans* is dependent on the location of C3 in the polysaccharide capsule and involves both direct and indirect C3-mediated interactions. Eur. J. Immunol. *33*, 1957–1967.

Zelante, T., De Luca, A., Bonifazi, P., Montagnoli, C., Bozza, S., Moretti, S., Belladonna, M.L., Vacca, C., Conte, C., Mosci, P., *et al.* (2007). IL-23 and the Th17 pathway promote inflammation and impair antifungal immune resistance. Eur. J. Immunol. *37*, 2695–2706.

Zelensky, A.N., and Gready, J.E. (2005). The C-type lectin-like domain superfamily. FEBS J. *272*, 6179–6217.

Zhang, J., Tachado, S.D., Patel, N., Zhu, J., Imrich, A., Manfruelli, P., Cushion, M., Kinane, T.B., and Koziel, H. (2005). Negative regulatory role of mannose receptors on human alveolar macrophage proinflammatory cytokine release *in vitro*. J. Leuk. Biol. *78*, 665–674.

Zhu, W., and Filler, S.G. (2010). Interactions of *Candida albicans* with epithelial cells. Cell Microbiol. *12*, 273–282.

Zhu, W., Phan, Q.T., Boontheung, P., Solis, N.V., Loo, J.A., and Filler, S.G. (2012). EGFR and HER2 receptor kinase signaling mediate epithelial cell invasion by *Candida albicans* during oropharyngeal infection. Proc. Natl. Acad. Sci. U.S.A. *109*, 14194–14199.

Part II

Pathogenic Mechanisms of Human Fungal Pathogens

A Molecular Update on the Pathogenesis of Candidiasis

7

Duncan Wilson and Bernhard Hube

Abstract

In contrast to most of the human pathogenic fungi described in this book, pathogenic species of the genus *Candida* do not normally exist in ecological niches such as soil or compost. Instead, these fungi have evolved in close association with warm-blooded animals, such as humans. For most people, at most times, these yeasts exist as harmless members of the microbiome. However, some species are capable of causing devastating infections in their human hosts. In this chapter we summarize and discuss important recent developments that have been made in our understanding of the molecular basis of *Candida* pathogenicity, with an emphasis on the dominant pathogenic species, *Candida albicans*.

Introduction: the clinical relevance of *Candida* species

Although the genus *Candida* encompasses some 150 species, only a fraction of these are commonly associated with human disease. However, the breadth of diseases caused by these few *Candida* species (collectively referred to as candidiasis) is enormous. They include superficial mucosal infections which, although generally not life-threatening, are extremely common: 75% of all women suffer from at least one episode of vulvovaginal candidiasis (VVC) and (without the intervention of highly active anti-retroviral therapy) a staggering 90% of HIV patients developed oral candidiasis (Ruhnke, 2002; Sobel, 2007; Vazquez and Sobel, 2002). More seriously, certain individuals are predisposed to developing invasive candidiasis – where the fungus gains access to the bloodstream, or other normally sterile anatomical sites. *Candida* species are by far the most common cause of invasive mycosis in humans and, in the past decade, have been recognized as the third most common cause of nosocomial bloodstream infection. Compounding this problem is the severity of invasive candidiasis: the mortality rate of disseminated candidiasis can be as high as 40%, even when patients are treated with first-line anti-fungal therapy (Perlroth *et al.*, 2007).

Candida infections are predominantly caused by *C. albicans*. However, non-*albicans Candida* species (especially *C. glabrata*, *C. parapsislosis* and *C. tropicalis*) now account for a significant proportion of infections. A number of other *Candida* species also cause infections in humans, but less frequently (Pfaller *et al.*, 2010).

C. albicans, *C. dubliniensis*, *C. tropicalis* and *C. parapsilosis* are related to each other, and all belong to the CUG clade of fungi. These species uniquely decode the CUG codon as serine instead of leucine (Butler *et al.*, 2009; Santos *et al.*, 2011). Interestingly, *C. glabrata*, the second most common cause of invasive candidiasis, is distantly related to *C. albicans*, belonging to the same (whole genome duplication) clade as the brewer's yeast, *Saccharomyces cerevisiae* (Brunke and Hube, 2012; Dujon *et al.*, 2004).

Because *C. albicans* is the most common cause of fungal infections in humans, the vast majority of studies have focused on this species. In the following sections we will discuss recent insights into the molecular basis of *Candida* pathogenicity mechanisms and how these influence the interactions of these fungi with their hosts.

Morphogenesis

C. albicans is a polymorphic fungus, as it can reversibly switch between several morphological forms, including yeast, chlamydospores, pseudohyphae and hyphae (Sudbery *et al.*, 2004). Of the other pathogenic *Candida* species, only *C. dubliniensis* is known to form true hyphae (Sullivan *et al.*, 1995). *C. glabrata*, *C. parapsilosis* and *C. tropicalis*, on the other hand, grow as yeast or pseudohyphae, but true hyphal growth has never been observed for these species and it is generally accepted that non-*albicans Candida* species (with the exception of *C. dubliniensis*) are unable to undergo the yeast to true hypha transition (Odds, 1985).

Despite the fact that most pathogenic *Candida* species do not form true hyphae, the yeast to hypha transition of *C. albicans* is considered to be a key virulence attribute of this fungus. Indeed, dimorphism is possibly the most intensively investigated aspects of *C. albicans* biology.

The role of hyphae in pathogenicity is intuitive, as the nature of hyphal growth permits mechanical penetration through tissue; moreover, hyphae are often observed in patient biopsies and in the organs of experimentally infected animals (Fig. 7.1) (Szabo and MacCallum, 2011). The concept of hypha formation as a virulence factor gained a molecular foothold in 1997 when two groups demonstrated that that non-filamentous gene deletion mutants were avirulent in mouse models of disseminated candidiasis (Leberer *et al.*, 1997; Lo *et al.*, 1997). Indeed, a plethora of subsequent

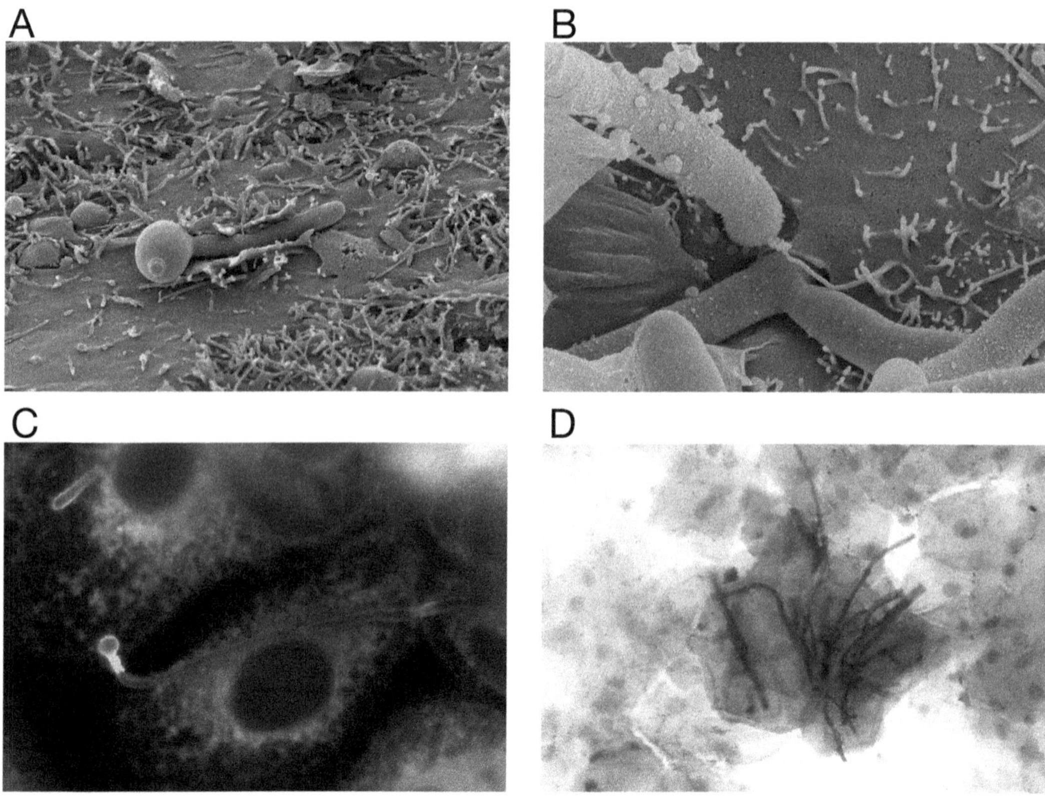

Figure 7.1 *Candida albicans*–epithelial interactions. (A and B) Scanning electron microscopy images of *C. albicans* infecting human oral epithelial cells, 1 hour (A) and 24 hours (B) post infection. Note the presence of epithelial pseudopod-like structures beginning to engulf the *C. albicans* germ tube (A) (Zakikhany K, Holland G, Özel M, with permission). (C) Differential staining reveals inter-epithelial invasion by *C. albicans* hyphae. Extracellular (non-invaded) hyphal sections are stained cyan; invading hyphae are stained blue (Almeida R, with permission). (D) *C. albicans* hyphae in association with human oral tissue from biopsy (Schmidt-Westhausen A, with permission). A colour version of this figure is available in the plate section at the back of the book.

studies reported the generation of a large number of targeted gene deletion mutants which were defective in hypha formation and attenuated in virulence. Together, these observations lead to the view that hypha formation is required for virulence (Sudbery *et al.*, 2004).

However, many of the genes required for hypha morphogenesis encode regulatory proteins (such as transcription factors), which are also required for the expression of numerous other genes, in addition to their requirement for morphogenesis. Therefore, direct molecular evidence that the yeast-to-hypha transition was a *bona fide* virulence factor was lacking (Gow *et al.*, 2002; Sudbery *et al.*, 2004). To complicate matters further, mutants lacking repressors of hypha morphogenesis (*tup1*Δ and *nrg1*Δ), which are locked in filamentous morphologies and unable to produce yeast, also exhibited attenuated virulence (Braun *et al.*, 2000; Murad *et al.*, 2001b). For an excellent and up-to-date overview of the complex network of signal transduction pathways which govern the yeast to hypha transition in *C. albicans*, readers are directed to Sudbery (Sudbery, 2011).

Not only is the yeast to hypha transition associated with extensive polarized growth which presumably aids tissue invasion via mechanical pressure, but also the expression of virulence factors. Indeed, when *C. albicans* cells are grown under laboratory conditions which stimulate hypha formation, they express a number of secreted and cell surface-associated proteins with direct roles in pathogenicity (Nantel *et al.*, 2002). These include the major adhesins (Hwp1 and Als3), invasins (Als3) and secreted proteases (Sap4–6).

The observed co-expression of the hyphal morphology and virulence-associated genes strengthened the concept of hyphae as the 'pathogenic morphology' and gave rise to the term 'hypha-specific genes'. However, as discussed below (see 'Commensal colonization'), the expression of many of these hypha-associated genes has since been observed by non-hyphal cells. Indeed, Martin *et al.* (2013) have performed transcriptional profiling of *C. albicans* under a range of hypha-inducing conditions, revealing the presence of a 'core set' of eight hypha-associated genes. These were: *ALS3*, *ECE1*, *HGT2*, *HWP1*, *IHD1*, *RBT1*, *DCK1* and orf19.2457.

So is hypha formation a virulence factor? At least during superficial candidiasis this seems to be the case (see below – 'The distinct stages of epithelial infection'); however, for disseminated candidiasis, the jury is still out (Gow *et al.*, 2002; Jacobsen *et al.*, 2012). Although the majority of *C. albicans* mutants with defects in hypha formation also exhibit attenuated virulence in mouse models of disseminated candidiasis, Spiering *et al.* (2010) have reported that an *sfl2*Δ mutant was as virulent as wild type *C. albicans* in a mouse model of disseminated candidiasis, but grew almost exclusively as yeast in the kidneys of infected mice. Observations, such as this, combined with the fact that non-*albicans Candida* species, such as *C. glabrata* and *C. parapsilosis* (which do not form hyphae), now account for up to 40% of cases of disseminated candidiasis (Perlroth *et al.*, 2007), leaves the exact role of hypha morphogenesis in systemic disease an open question.

Commensal colonization

Human pathogenic *Candida* species are not generally associated with environmental niches such as soil and compost. Instead, these yeasts normally exist in close association with warm blooded animals, such as humans. For example, the most prevalent *Candida* species associated with humans, *C. albicans*, is a member of the commensal microbiome of the oral, gastrointestinal and urogenitary mucosae (Kumamoto, 2011). Indeed, the gastrointestinal tracts of most humans are thought to be colonized with *Candida*. Although colonization is, in itself, harmless, commensal colonization is possibly the most important predisposing factor for developing candidiasis (Perlroth *et al.*, 2007). This is because the vast majority of *Candida* infections (both superficial and invasive) are of endogenous origin. Therefore, colonizing (commensal) *Candida* cells represent the reservoir of infection.

Because of this intimate association of *Candida* and humans, it has been proposed that this commensal environment may have 'trained' *Candida* species, allowing them to evolve attributes which subsequently contribute to their pathogenicity during infection (Gow and Hube, 2012; Hube, 2009). This concept of a 'virulence school' has also been proposed for environmentally acquired fungal pathogens, such as *Aspergillus fumigatus* and *Cryptococcus neoformans*. For example, it is postulated that regular engulfment of *C. neoformans* by amoebae in the environment has 'trained' this fungus to survive subsequent exposure to human macrophages during cryptococcosis (Casadevall, 2008; Casadevall and Pirofski, 2007). Analogously, the natural environment of *Candida*, as part of the human microbiome, may also have primed these fungi with attributes that contribute to their pathogenicity during infection. For example, in the 'commensal virulence school' of the human gastrointestinal tract, *Candida* must adhere, to avoid being washed away, compete with bacteria for nutrients, and survive encounters with host immunity – all attributes which contribute to virulence during candidiasis.

Therefore, commensal colonization represents both the major source of infection and the environment in which *Candida* have evolved to become, often devastating, human pathogens. Despite this central importance of commensal colonization for subsequent disease, relatively few studies have investigated the commensal lifestyle of *Candida*. The group of Carol Kumamoto have published a series of interesting studies examining the commensal phase of *C. albicans* using animal models.

Whole genome microarray analysis of *C. albicans* colonizing the murine gastrointestinal tract has revealed that the fungus experiences conditions of hypoxia, alkaline pH and low glucose. Colonizing cells exhibited gene expression patterns characteristic of both exponential and post-exponential phase laboratory conditions. For example, glycolysis and stress response genes were both induced (Rosenbach *et al.*, 2010).

Interestingly, despite growing predominantly in the yeast morphology, colonizing *C. albicans* cells exhibited transcriptional up-regulation of numerous 'hypha-associated' genes, including *HWP1* and *ECE1*, which have been reported in several studies to represent 'markers' of the hyphal morphology (Rosenbach *et al.*, 2010; White *et al.*, 2007). Not only were hypha-associated genes expressed by yeast cells, but their transcriptional induction occurs independent of Efg1, a key regulator of hypha formation (Lo *et al.*, 1997; Stoldt *et al.*, 1997). Indeed, *ECE1*, *RBT1* and *RBT4* were shown to be expressed at high levels by a *cph1Δ/efg1Δ* mutant, lacking two major regulators of morphogenesis, during colonization (White *et al.*, 2007). It should be noted that under *in vitro* laboratory growth conditions, the expression of these genes (as well as the yeast-to-hypha transition) are Efg1 dependent (Harcus *et al.*, 2004). Therefore fundamental differences exist between gene expression and morphology under *in vitro* laboratory growth conditions and during commensal colonization of the gastrointestinal tract.

Unlike it paralogue Efg1, the transcription factor, Efh1 has been shown to be dispensable for *in vitro* growth and morphogenesis (Doedt *et al.*, 2004). In the gastrointestinal tract, however, Efh1 serves an intriguing function: an *efh1Δ* mutant was able to colonize at far higher levels than wild type *C. albicans*; moreover, overexpression of *EFH1* inhibited *C. albicans* colonization. Therefore, Efh1 acts as a negative regulator of colonization (White *et al.*, 2007). This finding hints towards a more complex relationship between *C. albicans* and its host than commonly thought. In addition to the status of the host immune system and microbiome, it would appear that *C. albicans* actively limits its own proliferation during commensalism (Hube, 2004).

Despite decoupling of the expression of hypha-associated genes from Efg1 in the gut, this transcription factor does play an important role in regulating commensal colonization. Pierce *et al.* (2013) used single cell profiling to analyse expression levels of *EFG1*. Using this approach, the authors demonstrated significant differential *EFG1* expression within subpopulations of cells. Such heterogeneity in the expression of this important transcription factor may promote phenotypic variation within a population. Such variation may promote the colonization capacity of certain cells within a population. Indeed, it was shown that cells with low *EFG1* expression colonized the murine gastrointestinal tract more effectively at early time points, whilst high *EFG1* expression permitted continued colonization in the presence of an intact host immune system (Pierce and Kumamoto, 2012). This may be in part due to an Efg1-dependent response to host induced stresses, as it was found that during gastrointestinal colonization, expression of a superoxide dismutase, encoded by *SOD5*, was Efg1-dependent (Pierce *et al.*, 2013).

In summary, colonization of mucosal sites, such as the gastrointestinal tract is a complex process, which is influenced not only by the status of the host immune system and microbiome, but also by specific fungal activities. Future studies are surely required to further investigate this important stage in the lifecycle of *C. albicans* and other *Candida* species.

Surface life: adhesion, epithelial interaction, contact sensing, thigmotropism and biofilm formation

As discussed in the previous section, *C. albicans* is predominantly a commensal of mucosal surfaces and, with the exception of dissemination within the blood, pathogenic phases are also associated with growth on or within tissue. *Candida* species also form biofilms on medical devices, such as catheters, which can serve as reservoirs for disseminated infection. Unsurprisingly then, these fungi are well adapted to sense contact with surfaces, adhere tightly to them, navigate their topological features and form complex multicellular surface-associated communities, such as biofilms.

First contact

Adherence to host cells is an essential first step in candidiasis. Depending on the anatomical niche in question, adhesion can occur directly to host cells, via extracellular host factors, or via other members of the microbiota (i.e. bacteria). *Candida* cells can also adhere to and colonize abiotic surfaces, typically forming biofilms (see below).

A large body of work has examined the adhesion properties of *C. albicans* and the molecular mechanisms underlying this important stage of infection (Hiller *et al.*, 2011; Mayer *et al.*, 2013a,b; ten Cate *et al.*, 2009; Zhu and Filler, 2010). As the fungal cell wall is the primary interface between a fungus and its environment, the role of cell wall proteins in adhesion have been extensively investigated. Cell surface proteins which mediate adhesion are known collectively as adhesins. *C. albicans* encode over a hundred putative GPI-anchored proteins, many of which have the potential to function as adhesins (De Groot *et al.*, 2003; Heilmann *et al.*, 2012; Klis *et al.*, 2011). Of these, members of the Als (agglutinin-like sequence) family and Hwp1 (hyphal wall protein 1) play dominant roles in *C. albicans* adhesion (Hoyer *et al.*, 1998; Liu and Filler, 2011; Staab *et al.*, 1999; Sundstrom, 2002). As discussed in the following sections, adhesin molecules such as Als3 and Hwp1 play critical roles during interactions both with host cells, other *C. albicans* cells, bacteria and with abiotic surfaces.

The distinct stages of epithelial infection: adhesion, invasion and damage

In addition to its commensal lifestyle (see 'Commensal colonization', above) *C. albicans* is a frequent cause of infections at these mucosal sites, such as oral or vaginal candidiasis ('thrush') (Vazquez and Sobel, 2002). Whilst adhesion to mucosal surfaces occurs both during colonization and infection, host cell invasion by the fungus and epithelial damage are considered to be associated with the pathogenic phase. However, it should be noted that certain levels of invasion, and even moderate cell damage may occur asymptomatically during commensal colonization.

A number of studies have used infection models, of varying complexity, to investigate the pathogenicity mechanisms of oral candidiasis, including oral epithelial monolayers, reconstituted human oral epithelium (commonly referred to as RHE) and clinical samples from patients suffering from oral candidiasis. Experimental oral candidiasis models have been employed to dissect the progression of infection into relatively distinct stages (adhesion, invasion and damage) and have identified fungal factors required for all or for discrete stages.

During the first three hours of experimental oral epithelial infection, *C. albicans* yeast cells undergo rapid germ tube formation and this event is closely associated with the expression of cell surface adhesin molecules (Dalle *et al.*, 2010; Park *et al.*, 2005, 2009; Sohn *et al.*, 2006; Wächtler *et al.*, 2012; Zakikhany *et al.*, 2007). Indeed, the major adhesin molecules of *C. albicans* (Als3 and

Hwp1) are often referred to as 'hypha-specific' or 'hypha-associated' proteins. Although, as discussed above, the term 'hypha-specific' may be a misnomer, as genes encoding hypha-associated proteins are expressed by yeast cells (White *et al.*, 2007) and certain hypha-associated proteins have been directly detected on non-hyphal cells (Sosinska *et al.*, 2008).

Indeed, Als3 and Hwp1 both play important roles during these early epithelial interactions as deletion of either *ALS3* or *HWP1* significantly reduces *C. albicans* adhesion to epithelial cells (Staab *et al.*, 1999; Wächtler *et al.*, 2011a; Zhao *et al.*, 2004).

In defined *in vitro* models, Als3- and Hwp1-mediated adhesion is likely due to direct interactions with host receptors on the epithelial cell surface. For example, Als3 directly binds host cell cadherins (such as E-cadherin on oral epithelial cells) (Phan *et al.*, 2007). Hwp1, on the other hand, has been reported to act as a substrate for host transglutaminases, which are thought to covalently crosslink *C. albicans* hyphae to epithelial surfaces (Staab *et al.*, 1999).

In the clinical setting, adhesion of *C. albicans* cells to host tissue is likely to be more complex, with fungal cells associating with soluble host factors (i.e. 'bridging molecules'), and via other microbial cells. Interestingly, Als3 mediates adhesion not only to host cells directly (via cadherins), but also to other *C. albicans* cells, via a complementary adhesion mechanism with Hwp1 (Nobile *et al.*, 2008) and to bacterial cells, by interacting with prokaryotic cell surface components such as the SspA and SspB proteins of *Streptococcus gordonii* (Silverman *et al.*, 2010). Als3 is also able to bind the intracellular iron storage molecule, ferritin (Almeida *et al.*, 2008) (see 'Nutrient acquisition', below) and mediate epithelial invasion by induced endocytosis (Phan *et al.*, 2007). Therefore, Als3 is a multifunctional molecule with a plethora of binding partners and activities (Liu and Filler, 2011).

Following hypha formation and adhesin expression, *C. albicans* can invade epithelial tissue via two different mechanisms: induced endocytosis and active penetration (Zakikhany *et al.*, 2007; Zhu and Filler, 2010). Induced endocytosis is a predominantly host driven process, whilst active penetration relies on properties of the growing hypha.

Induced endocytosis is triggered upon binding of *C. albicans* invasins to host epithelial cell receptors. The first described *C. albicans* invasin was Als3, although, more recently, the cell surface-associated heat shock protein, Ssa1 has also been shown to mediate this process (Phan *et al.*, 2007; Sun *et al.*, 2010; Wächtler *et al.*, 2012). Both invasins bind epithelial cell E-cadherin to facilitate induced endocytosis (note that in the case of endothelial cells, a similar process occurs via interaction with N-cadherin). Als3 and Ssa1 also interact with the epidermal growth factor receptors EGFR and HER2 (Zhu *et al.*, 2012).

These interactions induce epithelial membrane ruffling and pseudopod formation (Fig. 7.1A), ultimately resulting in engulfment of the hypha (Dalle *et al.*, 2010; Zakikhany *et al.*, 2007). Endocytosis relies on reorganization of the actin cytoskeleton, as the microfilament inhibitor, cytochalasin D, blocks this process (Park *et al.*, 2005). Dalle *et al.* (2010) tested the effects of a number of cellular inhibitors on invasion into the human oral epithelial cell line, TR146, and found that treatment with amiloride strongly reduced *C. albicans* invasion into oral epithelial cells, suggesting that induced endocytosis is mediated by macropinocytosis. These authors did not observe alterations in invasion upon inhibition of clathrin-dependent or lipid raft-mediated endocytosis. In a parallel study, Moreno-Ruiz and colleagues investigated the endocytic route of *C. albicans*. Using a combination of live cell imaging and fluorescence microscopy, these authors observed the involvement of E-cadherin, clathrin, dynamin and cortactin in *C. albicans* internalization, suggesting that induced endocytosis is a clathrin-dependent process (Moreno-Ruiz *et al.*, 2009). Interestingly, even killed *C. albicans* hyphae can be internalized by epithelial cells via induced endocytosis (Dalle *et al.*, 2010). Engulfed, killed hyphae become tightly surrounded by host cell membrane and do not elicit damage of the epithelial cells (Wächtler *et al.*, 2012).

The second route of invasion is known as active penetration. In contrast to induced endocytosis, which is a predominantly host-driven process, active penetration can occur in the absence of epithelial cell cytoskeleton rearrangements and can permit fungal invasion of dead epithelial tissue. The properties of active penetration have been investigated by performing *C. albicans*-epithelial infections in the presence of cytochalasin D, which inhibits host cell actin polymerization, and thus blocks induced endocytosis. A combination of scanning (Dalle *et al.*, 2010) and transmission (Wächtler *et al.*, 2012) electron microscopy has demonstrated that active penetration results in a distinct depression in the epithelial cell surface at the site of invasion (Fig. 7.1B) and that, within the epithelial cell, hyphae are located in a broad intracellular space, with only few intimate contacts with epithelial structures. Active penetration is also required for inter-epithelial invasion (Fig. 7.1C).

In contrast to induced endocytosis, where defined receptor–ligand interactions and endocytic pathways have been characterized (above), the molecular mechanisms driving active penetration are somewhat more elusive (Martin *et al.*, 2011b; Zhu and Filler, 2010). It is likely that a combination of hyphal extension, turgor pressure generated at the hyphal tip and the secretion of hydrolytic enzymes are important factors for active penetration. This is supported by molecular studies which have shown that mutants lacking Gpd2 and Gpp1, which are involved in intracellular glycerol synthesis (turgor generation) and Rsr1 and Bud2, which are involved in the maintenance of directed polarized growth, are defective for invasion into enterocytes, a cell type which does not facilitate cadherin dependent induced endocytosis of *C. albicans* (Wächtler *et al.*, 2011a).

In vitro models of candidiasis indicate that active penetration is the dominant invasion mechanism, accounting for approximately 70% of epithelial invasion within the first 3 hours of infection (Wächtler *et al.*, 2012). However, it is likely that both invasion mechanisms contribute to the pathogenesis of oral candidiasis. For example, Zhu *et al.* (2012) demonstrated that interfering with EGFR and HER2 signalling inhibited endocytosis of *C. albicans* and reduced the severity of oral candidiasis in mice. On the other hand, active penetration is likely to be essential for invasion of cornified epithelial cell layers.

Following adhesion to, and then invasion into epithelia, *C. albicans* is able to cause significant damage to the host cells (Dalle *et al.*, 2010; Park *et al.*, 2005; Zakikhany *et al.*, 2007). Although prior adhesion and invasion are prerequisites for subsequent damage, perhaps surprisingly, invasion itself does not cause damage. *C. albicans* mutants lacking Eed1, a regulatory protein, dispensable for initial hypha formation, but essential for hyphal maintenance and elongation, the GTPase Rsr1 and its cognate GTPase-activating protein Bud2, both of which are essential for the maintenance of directed polarized growth, are capable of invading epithelial cells at the same rate as wild-type *C. albicans* during the first three hours of infection; yet these mutants cause very little epithelial damage (Martin *et al.*, 2011a; Wächtler *et al.*, 2011a; Zakikhany *et al.*, 2007). Similarly, treatment of *C. albicans* with low levels of azole drugs at levels which do not block hypha formation permit invasion into epithelial cells, but preclude damage (Wächtler *et al.*, 2011b).

Similar to invasion via active penetration, the exact mechanisms of epithelial damage caused by *C. albicans* require further investigation. It is clear that hypha formation is critical for epithelial cell destruction: this is based on numerous studies which have shown that both clinical isolates and defined mutants with defects in hypha formation on epithelial cells cause less damage to these cell types and hyphae are observed in biopsies from patients suffering from oral candidiasis (Fig. 7.1D). In addition, non-*albicans Candida* species cause less damage in experimental models of epithelial infection (Brunke *et al.*, 2010; Gacser *et al.*, 2007; Li and Dongari-Bagtzoglou, 2007; Moyes *et al.*, 2012). In addition to the initiation of hypha formation, hyphal maintenance and elongation are critical for sustained, inter-epithelial invasion and host cell damage (Wächtler *et al.*, 2011a).

Hyphal morphogenesis is not only critical for direct epithelial damage by *C. albicans*, but also for the recognition of the pathogen by the host. The group of Naglik has recently demonstrated that human oral epithelial cells can discriminate between the two morphologies. Interestingly, epithelial cells are able to tolerate *C. albicans* yeast cells or low numbers of hyphae, but when hyphal burdens reach a certain 'pathogenic threshold', the epithelial cells respond via prolonged MAP kinase signalling and activation of the transcription factor, cFos (Moyes *et al.*, 2010). This 'danger response' triggers the production of proinflammatory cytokines, which would serve to recruit neutrophils to the site of infection. Neutrophils can resolve oral epithelial infection both by directly killing the pathogen and by stimulating epithelial cells to produce TLR4 (Fradin *et al.*, 2005; Schaller *et al.*, 2004; Thompson and Wilton, 1992).

In addition to morphological changes, *C. albicans* is able to secrete a wide range of hydrolytic enzymes with the potential to degrade host substrates and structures (Schaller *et al.*, 2005). These include phospholipases, lipases and proteases. Molecular characterization of the role of secreted hydrolases in oral pathogenicity is complicated by the size of the gene families that encode these enzymes. For example, both the secreted aspartyl proteases and lipases are each encoded by gene families of 10 members (Schaller *et al.*, 2005). However, it is likely that secreted aspartyl proteases are important pathogenicity factors, as treatment with the protease inhibitor, pepstatin A, effectively inhibits damage of oral epithelial cells *in vitro* (Naglik *et al.*, 2008). Indeed, it has even been proposed that the reduction in prevalence of oral candidiasis in HIV patients, which coincided with the introduction of highly active anti retroviral therapy (which includes aspartyl protease inhibitors) is in part due to the inhibition of fungal secreted aspartyl proteases (Munro and Hube, 2002).

Contact sensing and thigmotropism

When *C. albicans* yeast cells come into contact with surfaces, such as epithelia, they rapidly form hyphae. This filamentation itself reinforces adhesion because the dominant adhesins (e.g. Als3 and Hwp1) are highly expressed by hyphal cells.

Hypha formation by *C. albicans* on a surface is not a random event, as germ tubes always appear laterally from the mother cell (Brand and Gow, 2009). Interestingly, the site of germination can be manipulated by applying an electrical field across *C. albicans* cells: germ tubes emerge preferentially towards the cathode (Crombie *et al.*, 1990). Brand *et al.* (2007) took advantage of this galvanotropic response to dissect the molecular mechanisms of germ tube emergence. These authors found that the site of germ tube emergence is dependent on local calcium influx via the voltage-gated calcium channel, Cch1 and that this process is regulated by the calcineurin pathway and the transcription factor, Crz1.

Following germ tube emergence on a surface, the extending hypha can sense and respond to topological features in a process called thigmotropism. Thigmotropism is also a calcium-dependent process. When the hyphal tip encounters topological features such as ridges, membrane perturbation are thought to induce the opening of mechanosensitive ion channels (Kumamoto, 2008). In line with this, the stretch-activated calcium channel, Mid1, is required for hyphal reorientation in response surface topology. The role of calcium influx in this process is supported by the fact that two other calcium channels, Cch1 and Fig1, are also required for correct thigmotropic responses (Brand *et al.*, 2009). Interestingly, thigmotropism relies on the transcription factor, Crz1, but not on upstream components of the calcineurin pathway in *C. dubliniensis* (Chen *et al.*, 2011). Hyphal tip reorientation has also been shown to rely on the Ras-like GTPase, Rsr1, which forms an internal polarity landmark to orchestrate actin cytoskeleton rearrangements (Brand *et al.*, 2008).

Whilst Rsr1, together with its cognate GTPase-activating protein, Bud2, have been shown to play important roles in epithelial pathogenesis (Brand *et al.*, 2008; Wächtler *et al.*, 2011a), the importance of Fig1, Mid1 and Cch1 for interactions with epithelial cells has yet to be determined.

Interestingly, despite this sensitive and orchestrated thigmotropic behaviour of *C. albicans* hyphae, invasion of hyphae into epithelial monolayers was shown to rarely occur at intercellular junctions. Rather, epithelial invasion generally occurs at the apical face (Dalle *et al.*, 2010). This is in contrast to the invasive behaviour of certain plant pathogenic fungi, which seek out 'weak points' such as stomata. It is certainly possible that this preferential invasion into epithelial cells by *C. albicans* (rather than at cell junctions), is a nutrient-acquisition strategy.

Biofilms

Following contact with a surface and adhesion to the substratum, *Candida* cells can form biofilms:

structured, surface associated communities, surrounded by an extracellular matrix. Biofilms can form on both biotic and abiotic surfaces and are often associated with implanted medical devices. These biofilms pose a significant health risk as they can act as a reservoir of infection, seeding the bloodstream, or other normally sterile sites, with fungal cells (see 'Biofilm dispersal', below). Moreover, biofilms are notoriously difficult to treat due to their inherent resistance to antifungal drugs (as well as components of the immune system) (Douglas, 2002). Therefore, biofilm formation is a key pathogenicity attribute of *Candida* species. Here we discuss recent developments that have furthered our understanding of the molecular basis of *C. albicans* biofilm formation and maturation.

Experimental approaches have demonstrated that biofilm formation can be dissected into different stages: adhesion of yeast cells to the substratum; fungal proliferation, cell–cell adhesion and the initiation of extracellular matrix production; biofilm maturation, associated with significant increases in cell density and matrix production; and, finally, dispersal of fungal cells from the mature biofilm (Blankenship and Mitchell, 2006; Douglas, 2003).

The initial stage of biofilm formation – attachment to the substratum – is mediated by a plethora of cell surface adhesin molecules under the control of a complex network of transcriptional regulators (see below). The transcription factor Bcr1, and its upstream regulator, Tec1, play essential roles in the expression of adhesin-encoding genes and the initiation of biofilm formation (Nobile and Mitchell, 2005). The Bcr1 orthologue of *C. parapsilosis* also governs biofilm formation in this species, suggesting a conserved role for this transcription factor in biofilm formation (Ding and Butler, 2007).

Recent studies have made significant advances in our understanding of the molecular mechanisms governing biofilm formation. Nobile and coworkers (Nobile *et al.*, 2012) created a library of 165 transcription factor mutants and screened this for defects in biofilm formation *in vitro*. They identified six genes – *BCR1, TEC1, EFG1* (which were already known to contribute to biofilm formation) and *NDT80, ROB1, BRG1* (which represented newly identified transcription factors) – required for biofilm formation. They went on to analyse the ability of the six transcription factor mutants to form biofilms using a rat venous catheter model and a rat denture model. These two models mimic two very different types of *Candida* biofilms: those formed on indwelling medical devices, such as catheters and those formed on denture material. All six mutants displayed defects in at least one of the animal models, however Bcr1 was more important for catheter-associated than for denture-associated biofilms. Interestingly, the mutant lacking *BRG1* formed robust biofilms on catheters, but was highly defective for biofilm formation on denture material, suggesting niche specific dependency for certain transcription factors.

In order to understand the transcriptional circuitry governing biofilm formation, ChIP-chip analysis was carried out for all six of the transcription factors. This approach identified just over a thousand genes potentially controlled by the six biofilm-regulators. The intergenic regions of many genes (including those of the transcription factors themselves) were bound by multiple transcription factors, suggesting that the biofilm regulatory network is significantly interwoven. Combined with global gene expression analysis of all six transcription factor mutants, hierarchical clustering identified a set of eight core target genes of the biofilm regulatory circuit: orf19.3337, *ALS1, TPO4*, orf19.4000, *EHT1, HYR1, HWP1*, and *CAN2*. Strikingly, overexpression of each of these genes in at least one of the transcription factor mutant backgrounds significantly restored biofilm formation.

Interestingly, a significant number of genes within the biofilm network are young genes, that is they are relatively specific to *C. albicans* and not found in other yeast species. This indicates that the biofilm network in *C. albicans* is relatively recently evolved. Together with other systems biology-based approaches (Finkel *et al.*, 2012; Nobile *et al.*, 2009), significant advances in elucidating the complexity of transcriptional control of biofilm initiation and development have been made.

In addition to contact with the substratum and the concomitant expression of adhesins, a number of other environmental cues have been shown to affect various stages of biofilm formation and maturation, including the production of quorum sensing molecules, alcohols and zinc-sensing. The zinc-responsive transcription factor Zap1 (also known as Csr1) regulates the expression of numerous zinc acquisition genes and is important for growth under zinc limitation (Kim *et al.*, 2008). Interestingly, Nobile *et al.* (2009) also demonstrated a central role for Zap1 in the regulation of biofilm development). They found that a *zap1*Δ/Δ mutant generated higher levels of β-1,3 glucan, which is a major component of biofilm extracellular matrix, indicating that Zap1 is a negative regulator of extracellular matrix production. Micro-array transcriptional profiling and full genome chromatin immunoprecipitation identified a large set of Zap1 target genes including, but not limited to described and predicted zinc acquisition genes. Indeed, overexpression of genes encoding zinc transporters in the *zap1*Δ/Δ mutant was able to restore growth under zinc limited conditions (Kim *et al.*, 2008) and restore the expression of Zap1-regulated genes (Ganguly *et al.*, 2011) but did not affect biofilm matrix production (Nobile *et al.*, 2009) or farnesol production (Ganguly *et al.*, 2011). Instead, Zap1 target genes which are not directly involved in zinc acquisition were found to be involved in β-glucan matrix production. These include *ADH5*, *CSH1* and *IFD6*, which are involved in alcohol metabolism and *GCA1* and *GCA2*, which encode predicted extracellular glucoamylases. Alcohol likely inhibits biofilm formation, as inactivation of the alcohol dehydrogenase, Adh1, resulted in enhanced biofilm formation both *in vitro* and *in vivo* (Mukherjee *et al.*, 2006).

Quorum sensing molecules, such as farnesol also influence the development of biofilms, in particular later phases, such as dispersal (Albuquerque and Casadevall, 2012; Ramage *et al.*, 2002). Dispersal is an important aspect of medically relevant biofilms, as it permits the seeding of infectious cells from biofilms formed on implanted medical devices, such as catheters, into the bloodstream. These events have serious clinical consequences, as they likely contribute to recurrent candidaemia. Despite this, insights into the molecular mechanisms of dispersal of cells from *Candida* biofilms is only a recent development. Uppuluri *et al.* (2010) demonstrated that the morphogenic regulators Ume6, Nrg1 and Pes1 regulate biofilm dispersal and that dispersed yeast cells exhibit increased virulence compared to their planktonic counterpart.

Adaptation and survival

Like other microorganisms, *C. albicans* must tolerate and respond to rapid fluctuations in environmental conditions, including changing pH, oxygen and osmolality; however, as a member of the human microbiome and pathogen, *C. albicans* is also evolutionarily adapted to cope with encounters with components of the immune system. In this section we will discuss the mechanisms by which *C. albicans* deals with environmental stress and their implications in pathogenicity. Three major MAP kinase pathways govern the response to typical environmental insults such as osmotic- or oxidative- stress. These are: Hog1 (high osmolality glycerol), Mkc1 (MAP kinase from *C. albicans*) and Cek1 (*C. albicans* ERK-like kinase). These three pathways integrate multiple environmental signals to initiate an appropriate stress response and also play roles in regulating morphogenesis (de Dios *et al.*, 2010; Monge *et al.*, 2006). The importance of such a robust stress response for pathogenicity is highlighted by a series of studies demonstrating that mutants lacking *HOG1*, *MKC1* or *CEK1* display attenuated virulence (Brown *et al.*, 2007).

Upon osmotic stress, the fungal cell accumulates glycerol to counteract dehydration; oxidative stress triggers the expression of enzymes involved in the detoxification of reactive oxygen species. For example, superoxide dismutases convert superoxide anions to hydrogen peroxide, which is in turn converted to water and molecular oxygen by the action of catalase (Fradin *et al.*, 2005; Martchenko *et al.*, 2004; Wysong *et al.*, 1998). The role of these enzymes in counteracting the oxidative burst of phagocytes is discussed below. Similarly, reactive nitrogen species are detoxified by the nitric oxide dioxygenase, Yhb1 (Ullmann *et al.*, 2004). Thermal stress triggers the expression of heat shock proteins as well as the synthesis of trehalose, which protects the cell from protein unfolding and aggregation (Arguelles, 1997; Leach and Cowen, 2013).

Recent studies, by the groups of Brown and Cowen and others have demonstrated an important role for heat shock proteins in the virulence of *Candida* (Brown *et al.*, 2010; Leach and Cowen, 2013; Leach *et al.*, 2012b). In addition to protecting the cell from thermal stress, the major heat shock protein, Hsp90, plays key roles in governing morphogenesis, biofilm formation cell wall remodelling and antifungal drug resistance (Cowen and Lindquist, 2005; LaFayette *et al.*, 2010; Leach *et al.*, 2012a; Robbins *et al.*, 2011; Shapiro *et al.*, 2009; Singh *et al.*, 2009). Interestingly, the Hsp90 orthologue in *C. glabrata* also regulates resistance to echinocandins (Singh-Babak *et al.*, 2012), suggesting a conserved role for this Hsp in fungi. Indeed, therapeutic antibodies against fungal Hsp90 have been developed through to clinical trials (Herbrecht *et al.*, 2006).

C. albicans encodes a further five major heat shock proteins: Hsp104, Hsp78, two isoforms of Hsp70 (Ssa1 and Ssa2) and Hsp60. *HSP104* has also been shown to contribute to biofilm formation and virulence in a *Caenorhabditis elegans* infection model (Fiori *et al.*, 2012). The role of Hsp78 in virulence has not yet been investigated, however *HSP78* is transcriptionally up-regulated following phagocytosis by macrophages (Lorenz *et al.*, 2004). The two isoforms of Hsp70 have been shown to have multiple functions. Both isoforms mediate binding of the salivary antimicrobial peptide, histatin 5 (Li *et al.*, 2003) (discussed below), whilst Ssa1 functions as a fungal invasin (see above – the distinct stages of epithelial infection) and is required for virulence in both disseminated and mucosal mouse models of infection (Sun *et al.*, 2010). Finally, the role of Hsp60 in pathogenicity has not yet been determined. However, *HSP60* is negatively regulated by Nrg1 and the protein has been detected in hyphae, suggesting a role during hypha morphogenesis (Hernandez *et al.*, 2004; Murad *et al.*, 2001a).

C. albicans also encodes several small heat shock proteins, which have, historically, received less attention than their larger molecular weight counterparts. *HSP12* was shown to be responsive to *in vivo*-relevant stimuli, including physiological CO_2 and pH, but was dispensable for stress tolerance, morphogenesis and virulence in a *Drosophila* infection model (Fu *et al.*, 2012; Sheth *et al.*, 2008). *HSP21*, on the other hand was shown to be essential for growth at elevated temperatures, echinocandin tolerance and for virulence in a mouse model of disseminated candidiasis. It would appear that Hsp21 potentiates stress tolerance via Cek1 MAP kinase signalling and trehalose homeostasis (Mayer *et al.*, 2012b, 2013b).

The anatomical sites faced by *C. albicans* within the human body vary dramatically in pH, from highly acidic in the stomach to mildly acidic in the vagina and on skin to neutral in many other sites. The versatility of *C. albicans* is reflected in its ability to effectively grow at pH 2 through to 10 and this flexibility is under tight regulatory control (Davis, 2009; Vylkova *et al.*, 2011). The response to changes in environmental pH is governed by the transcription factor Rim101. Rim101 is a zinc finger transcriptional activator/repressor that lies downstream of a pH-responsive cascade (Davis, 2009). Changes in environmental pH are sensed at the cell surface by plasma membrane proteins, including Rim21 and Dfg16 (Barwell *et al.*, 2005; Gomez-Raja and Davis, 2012; Thewes *et al.*, 2007). The signal is then transduced by Rim8, Rim13 and Rim20, finally resulting in proteolytic cleavage, via Rim13, and activation of Rim101 (Davis *et al.*, 2000b; Gomez-Raja and Davis, 2012; Li *et al.*, 2004). The Rim101 pathway is essential for hypha morphogenesis in response to changes in pH, but is dispensable for hypha induction in response to other stimuli, such as serum (Davis 2000). Importantly, mutants lacking components of the pH-sensing pathway are attenuated for virulence in mouse models of invasive candidiasis (Davis *et al.*, 2000a; Thewes *et al.*, 2007). Similar observations in other pathogenic fungi (Bignell, 2012), indicate that correct sensing and adaptation to pH is critical for the pathogenicity of fungal pathogens.

As well as efficiently responding to changes in pH within different anatomical niches, *C. albicans* is capable of actively modifying the pH of its surroundings. This is mediated by the metabolism of amino acids, and probably other amine-containing substrates to ammonia (Mayer *et al.*, 2012a; Vylkova *et al.*, 2011). Ammonia is then exported via Ato pumps, resulting in alkalinization of the local environment. As alkalinization is a potent stimulus for the formation of hyphae, this process of active alkalinization has been proposed to serve as a hypha 'auto-induction' mechanism, driving the formation of hyphae in particular anatomical niches, such as following phagocytosis by macrophages (Vylkova *et al.*, 2011).

Indeed, probably the most stressful environment that *C. albicans* finds itself in nature is within host phagocytes, such as macrophages and neutrophils. In these environments, the fungus likely has to deal with multiple stressors at the same time. It has recently been shown that combined environmental stresses are capable of killing *Candida* cells more efficiently than single stresses alone (Kaloriti *et al.*, 2012). A number of studies have investigated the candidicidal mechanisms of macrophages and neutrophils, as well as the fungal processes involved in surviving phagocytosis (Miramon *et al.*, 2013). Neutrophils are considered a vital first line of defence against invasive candidiasis. This is based on the fact that neutrophils exhibit extremely potent candidicidal activity *in vitro* (Fradin *et al.*, 2005) and that neutropenia is a major risk factor for invasive *Candida* infections (Perlroth *et al.*, 2007). Following phagocytosis of a *C. albicans* yeast cell, the neutrophil imposes a combination of extreme nutrient starvation (see 'Proliferation', below) and a barrage of antimicrobial activities in order to block hypha formation and kill the phagocytosed fungus (Fradin *et al.*, 2005; Miramon *et al.*, 2013; Rubin-Bejerano *et al.*, 2003). These activities include the production of antimicrobial peptides and enzymes (such as lactoferrin, elastase and defensins), the 'oxidative burst' and the generation of reactive nitrogen species (Miramon *et al.*, 2013).

In order to counteract these killing mechanisms, *C. albicans* expresses genes to cope with carbohydrate starvation, nitrosative stress and oxidative stress. Indeed, deletion of either *ICL1* (encoding isocitrate lyase, the key enzyme of the glyoxylate cycle), *SOD5* (encoding the major extracellular superoxide dismutase) or *YHB1* (encoding a nitric oxide dioxygenase), increases the sensitivity of *C. albicans* to killing by neutrophils (Fradin *et al.*, 2005; Miramon *et al.*, 2012).

In addition to killing phagocytosed fungal cells by the mechanisms described above, neutrophils can extrude structures called NETs (neutrophil extracellular traps), via a unique form of programmed cell death known as NETosis. NETs are web-like structures of DNA and chromatin, decorated with a range of antimicrobial granule proteins (Urban *et al.*, 2006). However, the major protein component of NETs was found to be calprotectin (Urban *et al.*, 2009). Calprotectin is an antimicrobial peptide which functions via zinc and manganese sequestration (Corbin *et al.*, 2008).

Macrophages also elicit carbohydrate starvation and oxidative stress on phagocytosed *C. albicans* cells and the glyoxylate cycle has been demonstrated to be crucial for the survival of *C. albicans* within macrophages (Lorenz and Fink, 2001; Seider *et al.*, 2010). However, compared to neutrophils, macrophages have been considered relatively poor killers of *C. albicans,* as *in vitro,* yeast cells rapidly germinate into hyphae and can escape the phagocyte (Lorenz *et al.*, 2004). However, recent infection studies using zebra fish suggest that, in this *in vivo* model, macrophages can efficiently inhibit *C. albicans* hypha formation (Brothers *et al.*, 2011; Brothers and Wheeler, 2012). It is likely that, *in vivo,* a combination of context-dependent signals directs the behaviour of macrophages during their interactions *C. albicans*.

The human host also expresses a battery of antimicrobial peptides at different anatomical niches to control *Candida* infections, including cathelicidins, defensins and histatins. Histatin 5 is a member of the histatin family, short histidine-rich peptides, with high anti-candidicidal activity and is a major antimicrobial component of saliva (Oppenheim *et al.*, 1988). Unlike most antimicrobial peptides, which function at the microbial cell surface, histatin 5 must enter the microbial cell (Edgerton *et al.*, 2000).

A series of recent studies has shed light on the elaborate pathway by which histatin 5 targets and kills *C. albicans.* Primarily, histatin 5 binds to the surface of *C. albicans* cells via the cell surface-localized heat shock proteins, Ssa1 and Ssa2 (Li *et al.*, 2003). Following binding to these cell envelope receptors, histatin 5 is internalized via the polyamine transporters, Dur3 and Dur31 (Kumar *et al.*, 2011; Mayer *et al.*, 2012a). Therefore, *C. albicans* actively binds and internalizes this potent antimicrobial peptide, representing a striking example of host–pathogen coevolution. However, *C. albicans* has a number of mechanisms to protect itself in turn from this suicidal tendency. The fungus can shed the plasma membrane glycoprotein, Msb2, from its surface, which inactivates histatin 5 (Puri *et al.*, 2012; Szafranski-Schneider *et al.*, 2012). Alternatively, *C. albicans* can inactivate histatin 5 by proteolytic cleavage with the cell surface-associated secreted aspartic protease, Sap9 (Meiller *et al.*, 2009). Finally, *C. albicans* may actively export histatin 5 via the efflux pump, Flu1 (Li *et al.*, 2013). Interestingly, Flu1 also mediates polyamine export, strengthening the concept that histatin 5 'hijacks' the fungal polyamine import machinery to kill *C. albicans*.

Proliferation

In order to proliferate, either during commensal or pathogenic phases, *C. albicans* must not only survive environmental fluctuations and immune attack, but also acquire sufficient nutrients for growth. Depending on the site of colonization or infection, the availability of nutrients can vary dramatically.

Like its distant relative, *Saccharomyces cerevisiae, C. albicans* can rapidly assimilate sugars such as glucose. However, in contrast to fruits, many anatomical niches do not contain high levels of sugars. Gene expression studies of *C. albicans* infecting different niches have begun to elucidate the carbon sources used by this pathogen (Wilson *et al.*, 2009). During infection of oral epithelium (oral candidiasis), it would appear that *C. albicans* inhabits a glucose-poor environment, indicated by the transcriptional up-regulation of genes encoding components of the glyoxylate cycle (*ICL1* and *MLS1*) and gluconeogenesis (*PCK1*). Importantly, *ICL1* was also identified as up-regulated in clinical samples obtained from HIV patients suffering from oral candidiasis (Zakikhany *et al.*, 2007) and an *icl1*Δ mutant caused significantly less epithelial damage in an *in vitro* model of oral infection (Wächtler *et al.*, 2011a). Genes involved in fatty acid β-oxidation (*FOX2*, *FOX3* and *PXA1*) were also induced, suggesting that lipids may serve as a carbon source during oral infection. In line with this hypothesis, members of the secreted lipase family were also transcriptionally up-regulated (Zakikhany *et al.*, 2007).

The glyoxylate cycle is also of critical importance during interactions with immune cells: *ICL1* is strongly induced following phagocytosis by both macrophages and neutrophils (Barelle *et al.*, 2006; Fradin *et al.*, 2005; Lorenz *et al.*, 2004; Lorenz and Fink, 2001; Miramon *et al.*, 2012) and an *icl1*Δ mutant exhibits increased sensitivity to killing by both of these phagocytes (Lorenz and Fink, 2001; Miramon *et al.*, 2012).

Following dissemination to and growth within the kidney, *C. albicans* also faces a glucose-poor environment, again indicated by the repression of genes involved in glycolysis and the induction of genes involved in fatty acid metabolism and the assimilation of alternative carbon sources (Walker *et al.*, 2009). However, single cell profiling revealed that only sub-populations of cells infecting the kidney-induced expression of glyoxylate cycle (*ICL1*) and gluconeogenesis (*PCK1*) reporter genes (Barelle *et al.*, 2006). In contrast to other sites of infection, the liver would appear to be relatively abundant in sugars as several genes involved in glycolysis and the TCA cycle were shown to be up-regulated in this organ (Thewes *et al.*, 2007).

The carbon sources available to *C. albicans* at these various anatomical niches not only influence growth, but can have profound effects on properties related to pathogenicity. It has long been appreciated that growth of *C. albicans* with different sugars can have strong effects on properties such as adherence (McCourtie and Douglas, 1981). More recently, the group of Alistair Brown has investigated the effect of lactate assimilation on *C. albicans* physiology. Lactate is a physiologically relevant carbon source for *C. albicans*, as it is found at relatively high levels in the gastrointestinal and vaginal mucosae. Compared to growth in glucose, lactate-grown *C. albicans* cells exhibited a significantly altered cell wall, with reduced chitin and mannan (Ene *et al.*, 2012a) and differential expression of cell wall and secreted proteins (Ene *et al.*, 2012b). In addition to these physical alterations, lactate grown cells exhibited increased tolerance to a range of stressors, including osmotic, oxidative and cell wall stress and to antifungal compounds. Lactate, rather than glucose assimilation also impacted a number of processes directly involved in host pathogen interactions: lactate grown cells were more adherent, formed more robust biofilms, stimulated lower production of the protective cytokine, IL-17 and were more virulent in a mouse model of infection (Ene *et al.*, 2012a,b, 2013).

To proliferate, *C. albicans* must also secure sufficient trace metals, such as iron, zinc, manganese and copper, for growth. Microbial high-affinity metal uptake systems are critical for pathogenicity, as the free levels of trace metals in the human host are very low, and are further decreased during infection (e.g. ferriaemia and zincaemia occur during inflammation) (Hood and Skaar, 2012). The 'tug of war' between microbial pathogens and their hosts for iron has been well investigated. Humans sequester iron from invading microbes via a number of sophisticated mechanisms known collectively as 'nutritional immunity'. In response to this, successful pathogens have evolved ways to circumvent this restriction or exploit host iron-binding molecules (Almeida *et al.*, 2009; Hood and Skaar, 2012).

Around two thirds of iron in the human body is bound to haemoglobin. *C. albicans* is able to access the iron bound in haemoglobin in a multi-stage process. Primarily, *C. albicans* hyphae can rosette complement-opsonized erythrocytes (Moors *et al.*, 1992). Hyphae secrete an, as yet uncharacterised, haemolytic factor capable of lysing the red blood cells (Manns *et al.*, 1994). Next, expression of a family of surface-associated haem/haemoglobin-binding proteins encoded by *RBT5*, *RBT51*, *CSA1*, *CSA2* and *PGA7* may facilitate haemoglobin binding at the fungal cell surface. Of these, Rbt5 and Rbt51 have been demonstrated to facilitate iron usage from haemoglobin (Weissman and Kornitzer, 2004). Following binding of haemoglobin to the fungal cell surface via e.g. Rbt5, the complex is internalized to the fungal vacuole via an endocytic mechanism (Weissman *et al.*, 2008). Subsequent iron utilization relies on acidification of secretory components and on the haem oxygenase, Hmx1 (Pendrak *et al.*, 2004). The reductive pathway, which is required for high affinity iron uptake and utilization of iron from certain iron-binding proteins (below) is not required for iron utilization from haemoglobin.

The majority of intracellular iron within the human host is tightly sequestered within the iron storage protein, ferritin. Although ferritin is an extremely stable molecule, *C. albicans* can utilize ferritin as an iron source *in vitro* (Almeida *et al.*, 2008). As *C. albicans* hyphae invade into human epithelial cells they bind ferritin on the hyphal surface. This process is mediated by the multifunctional cell surface protein Als3. Als3 is able to bind ferritin and facilitate its use as an iron source. In addition to Als3-mediated binding, the high-affinity reductive pathway is required for iron assimilation from ferritin (Almeida *et al.*, 2008). The reductive pathway is also required for utilization of iron from transferrin, however the molecular mechanism of *C. albicans*-transferrin binding is not yet known (Knight *et al.*, 2005).

The iron reductive pathway is located in the fungal plasma membrane and sequentially reduces ferric (Fe^{3+}) iron to ferrous (Fe^{2+}) iron via ferric reductase activity and then oxidizes the metal back to ferric iron via multicopper oxidase activity (reviewed in (Almeida *et al.*, 2009). The resultant ferric iron is transported into the cell via iron permeases. The *C. albicans* genome encodes 15 putative ferric reductases, four multicopper oxidases and two plasma membrane iron permeases (Almeida *et al.*, 2009). Genes encoding ferric reductases have been demonstrated to be differentially regulated depending on morphology (Jeeves *et al.*, 2011) or pH, via the pH-responsive transcription factor Rim101 (Baek *et al.*, 2008).

The multicopper oxidase, encoded by *FET3* has been shown to be important for high affinity iron uptake *in vitro*, but dispensable for virulence in a mouse model of systemic candidiasis, probably due to functional redundancy *in vivo* (Eck *et al.*, 1999).

In contrast, the high-affinity iron permease, encoded by *FTR1*, is required for both damage of epithelial cells *in vitro* and for virulence in the mouse model of systemic candidiasis (Almeida *et al.*, 2008; Ramanan and Wang, 2000).

Recently, it has been demonstrated that *C. albicans* can assimilate zinc via a novel 'zincophore' scavenging system. This system is encoded by two syntenic, co-regulated genes: *PRA1* and *ZRT1*. Pra1 is secreted by the fungus and can bind zinc; Zrt1 is a surface associated zinc transporter, which also mediates the reassociation of soluble Pra1 with the fungal cell. Together, this system facilitates assimilation of zinc from host cells (Citiulo *et al.*, 2012; Wilson *et al.*, 2012). Although a *zrt1*Δ mutant has not yet been tested for virulence in murine models, deletion of *PRA1* does not attenuate virulence (Soloviev *et al.*, 2007), suggesting that additional zinc assimilation mechanisms may function *in vivo* (Wilson *et al.*, 2012). In contrast, the mechanisms by which *C. albicans* assimilates other essential trace minerals, such as copper and manganese during infection, remain unexplored.

Summary and outlook

Yeasts such as *C. albicans* thrive at mucosal surfaces as members of the natural microbiome, but can cause superficial infections at these sites, as well as invading to deeper tissues and organs, with often fatal consequences. The technical advances that have propelled *Candida* research into the post-genomic era have led to exciting new insights into the biology of these fascinating microorganisms and the infections that they are responsible for. We have seen how *C. albicans* colonizes and infects diverse host niches, defends itself against the onslaught of immune activities and exploits nutritional resources.

The next decade will be dominated by systems biology, as we attempt to integrate and make sense of ever expanding data-sets, of ever increasing complexity. These global analyses will be accompanied by precise investigations into the interactions of the host, microbiome and pathogen factors which drive the pathogenesis of candidiasis.

We apologize to all colleagues in the field of *Candida* research whose work we were unable to cite due to space limitations.

References

Albuquerque, P., and Casadevall, A. (2012). Quorum sensing in fungi – a review. Med. Mycol. *50*, 337–345.

Almeida, R.S., Brunke, S., Albrecht, A., Thewes, S., Laue, M., Edwards, J.E., Filler, S.G., and Hube, B. (2008). The hyphal-associated adhesin and invasin Als3 of *Candida albicans* mediates iron acquisition from host ferritin. PLoS Pathog. *4*, e1000217.

Almeida, R.S., Wilson, D., and Hube, B. (2009). *Candida albicans* iron acquisition within the host. FEMS Yeast Res. *9*, 1000–1012.

Arguelles, J.C. (1997). Thermotolerance and trehalose accumulation induced by heat shock in yeast cells of *Candida albicans*. FEMS Microbiol. Lett. *146*, 65–71.

Baek, Y.U., Li, M., and Davis, D.A. (2008). *Candida albicans* ferric reductases are differentially regulated in response to distinct forms of iron limitation by the Rim101 and CBF transcription factors. Eukaryot. Cell *7*, 1168–1179.

Barelle, C.J., Priest, C.L., Maccallum, D.M., Gow, N.A., Odds, F.C., and Brown, A.J. (2006). Niche-specific regulation of central metabolic pathways in a fungal pathogen. Cell. Microbiol. *8*, 961–971.

Barwell, K.J., Boysen, J.H., Xu, W., and Mitchell, A.P. (2005). Relationship of *DFG16* to the Rim101p pH response pathway in *Saccharomyces cerevisiae* and *Candida albicans*. Eukaryot. Cell *4*, 890–899.

Bignell, E.M. (2012). Conservation in *Aspergillus fumigatus* of pH-signaling seven transmembrane domain and arrestin proteins, and implications for drug discovery. Annal. New York Acad. Sci. *1273*, 35–43.

Blankenship, J.R., and Mitchell, A.P. (2006). How to build a biofilm: a fungal perspective. Curr. Opin. Microbiol. *9*, 588–594.

Brand, A., and Gow, N.A. (2009). Mechanisms of hypha orientation of fungi. Curr. Opin. Microbiol. *12*, 350–357.

Brand, A., Shanks, S., Duncan, V.M., Yang, M., Mackenzie, K., and Gow, N.A. (2007). Hyphal orientation of *Candida albicans* is regulated by a calcium-dependent mechanism. Curr. Biol. *17*, 347–352.

Brand, A., Vacharaksa, A., Bendel, C., Norton, J., Haynes, P., Henry-Stanley, M., Wells, C., Ross, K., Gow, N.A., and Gale, C.A. (2008). An internal polarity landmark is important for externally induced hyphal behaviors in *Candida albicans*. Eukaryot. Cell 7, 712–720.

Brand, A., Lee, K., Veses, V., and Gow, N.A. (2009). Calcium homeostasis is required for contact-dependent helical and sinusoidal tip growth in *Candida albicans* hyphae. Mol. Microbiol. *71*, 1155–1164.

Braun, B.R., Head, W.S., Wang, M.X., and Johnson, A.D. (2000). Identification and characterization of *TUP1*-regulated genes in *Candida albicans*. Genetics *156*, 31–44.

Brothers, K.M., and Wheeler, R.T. (2012). Non-invasive imaging of disseminated candidiasis in zebrafish larvae. J. Vis. Exp. 30;(65). pii: 4051. doi: 10.3791/4051.

Brothers, K.M., Newman, Z.R., and Wheeler, R.T. (2011). Live imaging of disseminated candidiasis in zebrafish reveals role of phagocyte oxidase in limiting filamentous growth. Eukaryot. Cell *10*, 932–944.

Brown, A.J., Odds, F.C., and Gow, N.A. (2007). Infection-related gene expression in *Candida albicans*. Curr. Opin. Microbiol. *10*, 307–313.

Brown, A.J., Leach, M.D., and Nicholls, S. (2010). The relevance of heat shock regulation in fungal pathogens of humans. Virulence *1*, 330–332.

Brunke, S., and Hube, B. (2012). Two unlike cousins: *Candida albicans* and *C. glabrata* infection strategies. Cell. Microbiol. *15*, 701–708.

Brunke, S., Seider, K., Almeida, R.S., Heyken, A., Fleck, C.B., Brock, M., Barz, D., Rupp, S., and Hube, B. (2010). *Candida glabrata* tryptophan-based pigment production via the Ehrlich pathway. Mol. Microbiol. *76*, 25–47.

Butler, G., Rasmussen, M.D., Lin, M.F., Santos, M.A., Sakthikumar, S., Munro, C.A., Rheinbay, E., Grabherr, M., Forche, A., Reedy, J.L., *et al.* (2009). Evolution of pathogenicity and sexual reproduction in eight *Candida genomes*. Nature *459*, 657–662.

Casadevall, A. (2008). Evolution of intracellular pathogens. Annu. Rev. Microbiol. *62*, 19–33.

Casadevall, A., and Pirofski, L.A. (2007). Accidental virulence, cryptic pathogenesis, martians, lost hosts, and the pathogenicity of environmental microbes. Eukaryot. Cell *6*, 2169–2174.

ten Cate, J.M., Klis, F.M., Pereira-Cenci, T., Crielaard, W., and de Groot, P.W. (2009). Molecular and cellular mechanisms that lead to *Candida* biofilm formation. J. Dent. Res. *88*, 105–115.

Chen, Y.L., Brand, A., Morrison, E.L., Silao, F.G., Bigol, U.G., Malbas, F.F., Jr., Nett, J.E., Andes, D.R., Solis, N.V., Filler, S.G., *et al.* (2011). Calcineurin controls drug tolerance, hyphal growth, and virulence in *Candida dubliniensis*. Eukaryot. Cell *10*, 803–819.

Citiulo, F., Jacobsen, I.D., Miramon, P., Schild, L., Brunke, S., Zipfel, P., Brock, M., Hube, B., and Wilson, D. (2012). *Candida albicans* scavenges host zinc via Pra1 during endothelial invasion. PLoS Pathog. *8*, e1002777.

Corbin, B.D., Seeley, E.H., Raab, A., Feldmann, J., Miller, M.R., Torres, V.J., Anderson, K.L., Dattilo, B.M., Dunman, P.M., Gerads, R., *et al.* (2008). Metal chelation and inhibition of bacterial growth in tissue abscesses. Science *319*, 962–965.

Cowen, L.E., and Lindquist, S. (2005). Hsp90 potentiates the rapid evolution of new traits: drug resistance in diverse fungi. Science *309*, 2185–2189.

Crombie, T., Gow, N.A., and Gooday, G.W. (1990). Influence of applied electrical fields on yeast and hyphal growth of *Candida albicans*. J. Gen. Microbiol. *136*, 311–317.

Dalle, F., Wächtler, B., L'Ollivier, C., Holland, G., Bannert, N., Wilson, D., Labruere, C., Bonnin, A., and Hube, B. (2010). Cellular interactions of *Candida albicans* with human oral epithelial cells and enterocytes. Cell. Microbiol. *12*, 248–271.

Davis, D.A. (2009). How human pathogenic fungi sense and adapt to pH: the link to virulence. Curr. Opin. Microbiol. *12*, 365–370.

Davis, D., Edwards, J.E., Jr., Mitchell, A.P., and Ibrahim, A.S. (2000a). *Candida albicans RIM101* pH response pathway is required for host–pathogen interactions. Infect. Immun. *68*, 5953–5959.

Davis, D., Wilson, R.B., and Mitchell, A.P. (2000b). *RIM101*-dependent and-independent pathways govern pH responses in *Candida albicans*. Mol. Cell Biol. *20*, 971–978.

De Groot, P.W., Hellingwerf, K.J., and Klis, F.M. (2003). Genome-wide identification of fungal GPI proteins. Yeast *20*, 781–796.

Ding, C., and Butler, G. (2007). Development of a gene knockout system in *Candida parapsilosis* reveals a conserved role for *BCR1* in biofilm formation. Eukaryot. Cell *6*, 1310–1319.

de Dios, C.H., Roman, E., Monge, R.A., and Pla, J. (2010). The role of MAPK signal transduction pathways in the response to oxidative stress in the fungal pathogen *Candida albicans*: implications in virulence. Curr. Prot. Pept. Sci. *11*, 693–703.

Doedt, T., Krishnamurthy, S., Bockmuhl, D.P., Tebarth, B., Stempel, C., Russell, C.L., Brown, A.J., and Ernst, J.F. (2004). APSES proteins regulate morphogenesis and metabolism in *Candida albicans*. Mol. Biol. Cell *15*, 3167–3180.

Douglas, L.J. (2002). Medical importance of biofilms in *Candida* infections. Revista iberoamericana de micologia *19*, 139–143.

Douglas, L.J. (2003). *Candida* biofilms and their role in infection. Trends Microbiol. *11*, 30–36.

Dujon, B., Sherman, D., Fischer, G., Durrens, P., Casaregola, S., Lafontaine, I., De Montigny, J., Marck, C., Neuveglise, C., Talla, E., *et al.* (2004). Genome evolution in yeasts. Nature *430*, 35–44.

Eck, R., Hundt, S., Hartl, A., Roemer, E., and Kunkel, W. (1999). A multicopper oxidase gene from *Candida albicans*: cloning, characterization and disruption. Microbiology *145 (Pt 9)*, 2415–2422.

Edgerton, M., Koshlukova, S.E., Araujo, M.W., Patel, R.C., Dong, J., and Bruenn, J.A. (2000). Salivary histatin 5 and human neutrophil defensin 1 kill *Candida albicans* via shared pathways. Antimicrob. Agents Chemother. *44*, 3310–3316.

Ene, I.V., Adya, A.K., Wehmeier, S., Brand, A.C., MacCallum, D.M., Gow, N.A., and Brown, A.J. (2012a). Host carbon sources modulate cell wall architecture, drug resistance and virulence in a fungal pathogen. Cell. Microbiol. *14*, 1319–1335.

Ene, I.V., Heilmann, C.J., Sorgo, A.G., Walker, L.A., de Koster, C.G., Munro, C.A., Klis, F.M., and Brown, A.J. (2012b). Carbon source-induced reprogramming of the cell wall proteome and secretome modulates the adherence and drug resistance of the fungal pathogen *Candida albicans*. Proteomics *12*, 3164–3179.

Ene, I.V., Cheng, S.C., Netea, M.G., and Brown, A.J. (2013). Growth of *Candida albicans* cells on the physiologically relevant carbon source lactate affects their recognition and phagocytosis by immune cells. Infect. Immun. *81*, 238–248.

Finkel, J.S., Xu, W., Huang, D., Hill, E.M., Desai, J.V., Woolford, C.A., Nett, J.E., Taff, H., Norice, C.T., Andes, D.R., *et al.* (2012). Portrait of *Candida albicans* adherence regulators. PLoS Pathog. *8*, e1002525.

Fiori, A., Kucharikova, S., Govaert, G., Cammue, B.P., Thevissen, K., and Van Dijck, P. (2012). The heat-induced molecular disaggregase Hsp104 of *Candida albicans* plays a role in biofilm formation and pathogenicity in a worm infection model. Eukaryot. Cell *11*, 1012–1020.

Fradin, C., De Groot, P., MacCallum, D., Schaller, M., Klis, F., Odds, F.C., and Hube, B. (2005). Granulocytes govern the transcriptional response, morphology and proliferation of *Candida albicans* in human blood. Mol. Microbiol. *56*, 397–415.

Fu, M.S., De Sordi, L., and Muhlschlegel, F.A. (2012). Functional characterization of the small heat shock protein Hsp12p from *Candida albicans*. PloS One *7*, e42894.

Gacser, A., Schafer, W., Nosanchuk, J.S., Salomon, S., and Nosanchuk, J.D. (2007). Virulence of *Candida parapsilosis, Candida orthopsilosis*, and *Candida metapsilosis* in reconstituted human tissue models. Fung. Genet. Biol. *44*, 1336–1341.

Ganguly, S., Bishop, A.C., Xu, W., Ghosh, S., Nickerson, K.W., Lanni, F., Patton-Vogt, J., and Mitchell, A.P. (2011). Zap1 control of cell–cell signaling in *Candida albicans* biofilms. Eukaryot. Cell *10*, 1448–1454.

Gomez-Raja, J., and Davis, D.A. (2012). The beta-arrestin-like protein Rim8 is hyperphosphorylated and complexes with Rim21 and Rim101 to promote adaptation to neutral-alkaline pH. Eukaryot. Cell *11*, 683–693.

Gow, N.A., and Hube, B. (2012). Importance of the *Candida albicans* cell wall during commensalism and infection. Curr. Opin. Microbiol. *15*, 406–412.

Gow, N.A., Brown, A.J., and Odds, F.C. (2002). Fungal morphogenesis and host invasion. Curr. Opin. Microbiol. *5*, 366–371.

Harcus, D., Nantel, A., Marcil, A., Rigby, T., and Whiteway, M. (2004). Transcription profiling of cyclic AMP signaling in *Candida albicans*. Mol. Biol. Cell *15*, 4490–4499.

Heilmann, C.J., Sorgo, A.G., and Klis, F.M. (2012). News from the fungal front: wall proteome dynamics and host–pathogen interplay. PLoS Pathog. *8*, e1003050.

Herbrecht, R., Fohrer, C., and Nivoix, Y. (2006). Mycograb for the treatment of invasive candidiasis. Clin. Infect. Dis. *43*, 1083; author reply 1083–1084.

Hernandez, R., Nombela, C., Diez-Orejas, R., and Gil, C. (2004). Two-dimensional reference map of *Candida albicans* hyphal forms. Proteomics *4*, 374–382.

Hiller, E., Zavrel, M., Hauser, N., Sohn, K., Burger-Kentischer, A., Lemuth, K., and Rupp, S. (2011). Adaptation, adhesion and invasion during interaction of *Candida albicans* with the host--focus on the function of cell wall proteins. Int. J. Med. Microbiol. *301*, 384–389.

Hood, M.I., and Skaar, E.P. (2012). Nutritional immunity: transition metals at the pathogen–host interface. Nat. Rev. Microbiol. *10*, 525–537.

Hoyer, L.L., Payne, T.L., Bell, M., Myers, A.M., and Scherer, S. (1998). *Candida albicans ALS3* and insights into the nature of the *ALS* gene family. Current genetics 33, 451–459.

Hube, B. (2004). From commensal to pathogen: stage- and tissue-specific gene expression of *Candida albicans*. Curr. Opin. Microbiol. *7*, 336–341.

Hube, B. (2009). Fungal adaptation to the host environment. Curr. Opin. Microbiol. *12*, 347–349.

Jacobsen, I.D., Wilson, D., Wächtler, B., Brunke, S., Naglik, J.R., and Hube, B. (2012). *Candida albicans* dimorphism as a therapeutic target. Exp. Rev. Anti-infect. Ther. *10*, 85–93.

Jeeves, R.E., Mason, R.P., Woodacre, A., and Cashmore, A.M. (2011). Ferric reductase genes involved in high-affinity iron uptake are differentially regulated in yeast and hyphae of *Candida albicans*. Yeast *28*, 629–644.

Kaloriti, D., Tillmann, A., Cook, E., Jacobsen, M., You, T., Lenardon, M., Ames, L., Barahona, M., Chandrasekaran, K., Coghill, G., *et al.* (2012). Combinatorial stresses kill pathogenic *Candida* species. Med. Mycol. *50*, 699–709.

Kim, M.J., Kil, M., Jung, J.H., and Kim, J. (2008). Roles of Zinc-responsive transcription factor Csr1 in filamentous growth of the pathogenic Yeast *Candida albicans*. J. Microbiol. Biotech. *18*, 242–247.

Klis, F.M., de Koster, C.G., and Brul, S. (2011). A mass spectrometric view of the fungal wall proteome. Future Microbiol. *6*, 941–951.

Knight, S.A., Vilaire, G., Lesuisse, E., and Dancis, A. (2005). Iron acquisition from transferrin by *Candida albicans* depends on the reductive pathway. Infect. Immun. *73*, 5482–5492.

Kumamoto, C.A. (2008). Molecular mechanisms of mechanosensing and their roles in fungal contact sensing. Nat. Rev. Microbiol. *6*, 667–673.

Kumamoto, C.A. (2011). Inflammation and gastrointestinal *Candida* colonization. Curr. Opin. Microbiol. *14*, 386–391.

Kumar, R., Chadha, S., Saraswat, D., Bajwa, J.S., Li, R.A., Conti, H.R., and Edgerton, M. (2011). Histatin 5 uptake by *Candida albicans* utilizes polyamine transporters Dur3 and Dur31 proteins. J. Biol. Chem. *286*, 43748–43758.

LaFayette, S.L., Collins, C., Zaas, A.K., Schell, W.A., Betancourt-Quiroz, M., Gunatilaka, A.A., Perfect, J.R., and Cowen, L.E. (2010). PKC signaling regulates drug resistance of the fungal pathogen *Candida albicans* via circuitry comprised of Mkc1, calcineurin, and Hsp90. PLoS Pathog. *6*, e1001069.

Leach, M.D., and Cowen, L.E. (2013). Surviving the heat of the moment: a fungal pathogens perspective. PLoS Pathog. *9*, e1003163.

Leach, M.D., Budge, S., Walker, L., Munro, C., Cowen, L.E., and Brown, A.J. (2012a). Hsp90 orchestrates transcriptional regulation by Hsf1 and cell wall remodelling by MAPK signalling during thermal adaptation in a pathogenic yeast. PLoS Pathog. *8*, e1003069.

Leach, M.D., Klipp, E., Cowen, L.E., and Brown, A.J. (2012b). Fungal Hsp90: a biological transistor that tunes cellular outputs to thermal inputs. Nat. Rev. Microbiol. *10*, 693–704.

Leberer, E., Ziegelbauer, K., Schmidt, A., Harcus, D., Dignard, D., Ash, J., Johnson, L., and Thomas, D.Y. (1997). Virulence and hyphal formation of *Candida albicans* require the Ste20p-like protein kinase CaCla4p. Curr. Biol. *7*, 539–546.

Li, L., and Dongari-Bagtzoglou, A. (2007). Oral epithelium-*Candida glabrata* interactions *in vitro*. Oral microbiology and immunology *22*, 182–187.

Li, M., Martin, S.J., Bruno, V.M., Mitchell, A.P., and Davis, D.A. (2004). *Candida albicans* Rim13p, a protease required for Rim101p processing at acidic and alkaline pHs. Eukaryot. Cell *3*, 741–751.

Li, R., Kumar, R., Tati, S., Puri, S., and Edgerton, M. (2013). *Candida albicans* Flu1-Mediated Efflux of Salivary Histatin 5 Reduces Its Cytosolic Concentration and Fungicidal Activity. Antimicrobial agents and chemotherapy *57*, 1832–1839.

Li, X.S., Reddy, M.S., Baev, D., and Edgerton, M. (2003). *Candida albicans* Ssa1/2p is the cell envelope binding protein for human salivary histatin 5. J. Biol. Chem. *278*, 28553–28561.

Liu, Y., and Filler, S.G. (2011). *Candida albicans* Als3, a multifunctional adhesin and invasin. Eukaryot. Cell *10*, 168–173.

Lo, H.J., Kohler, J.R., DiDomenico, B., Loebenberg, D., Cacciapuoti, A., and Fink, G.R. (1997). Nonfilamentous *C. albicans* mutants are avirulent. Cell *90*, 939–949.

Lorenz, M.C., and Fink, G.R. (2001). The glyoxylate cycle is required for fungal virulence. Nature *412*, 83–86.

Lorenz, M.C., Bender, J.A., and Fink, G.R. (2004). Transcriptional response of *Candida albicans* upon internalization by macrophages. Eukaryot. Cell 3, 1076–1087.

Manns, J.M., Mosser, D.M., and Buckley, H.R. (1994). Production of a hemolytic factor by *Candida albicans*. Infect. Immun. *62*, 5154–5156.

Martchenko, M., Alarco, A.M., Harcus, D., and Whiteway, M. (2004). Superoxide dismutases in *Candida albicans*: transcriptional regulation and functional characterization of the hyphal-induced *SOD5* gene. Mol. Biol. Cell *15*, 456–467.

Martin, R., Moran, G.P., Jacobsen, I.D., Heyken, A., Domey, J., Sullivan, D.J., Kurzai, O., and Hube, B. (2011a). The *Candida albicans*-specific gene *EED1* encodes a key regulator of hyphal extension. PloS One *6*, e18394.

Martin, R., Wächtler, B., Schaller, M., Wilson, D., and Hube, B. (2011b). Host–pathogen interactions and virulence-associated genes during *Candida albicans* oral infections. Int. J. Med. Microbiol. *301*, 417–422.

Martin, R., Albrecht-Eckardt, D., Brunke, S., Hube, B., Hunniger, K., and Kurzai, O. (2013). A Core Filamentation Response Network in *Candida albicans* Is Restricted to Eight Genes. PloS One *8*, e58613.

Mayer, F.L., Wilson, D., Jacobsen, I.D., Miramon, P., Grosse, K., and Hube, B. (2012a). The novel *Candida albicans* transporter Dur31 Is a multi-stage pathogenicity factor. PLoS Pathog. *8*, e1002592.

Mayer, F.L., Wilson, D., Jacobsen, I.D., Miramon, P., Slesiona, S., Bohovych, I.M., Brown, A.J., and Hube, B. (2012b). Small but crucial: the novel small heat shock protein Hsp21 mediates stress adaptation and virulence in *Candida albicans*. PloS One 7, e38584.

Mayer, F.L., Wilson, D., and Hube, B. (2013a). *Candida albicans* pathogenicity mechanisms. Virulence *4*, 119–128.

Mayer, F.L., Wilson, D., and Hube, B. (2013b). Hsp21 Potentiates Antifungal Drug Tolerance in *Candida albicans*. PloS One *8*, e60417.

McCourtie, J., and Douglas, L.J. (1981). Relationship between cell surface composition of *Candida albicans* and adherence to acrylic after growth on different carbon sources. Infect. Immun. *32*, 1234–1241.

Meiller, T.F., Hube, B., Schild, L., Shirtliff, M.E., Scheper, M.A., Winkler, R., Ton, A., and Jabra-Rizk, M.A. (2009). A novel immune evasion strategy of *Candida albicans*: proteolytic cleavage of a salivary antimicrobial peptide. PloS One *4*, e5039.

Miramon, P., Dunker, C., Windecker, H., Bohovych, I.M., Brown, A.J., Kurzai, O., and Hube, B. (2012). Cellular Responses of *Candida albicans* to Phagocytosis and the Extracellular Activities of Neutrophils Are Critical to Counteract Carbohydrate Starvation, Oxidative and Nitrosative Stress. PloS One *7*, e52850.

Miramon, P., Kasper, L., and Hube, B. (2013). Thriving within the host: *Candida* spp. interactions with phagocytic cells. Med. Microbiol. Immunol.

Monge, R.A., Roman, E., Nombela, C., and Pla, J. (2006). The MAP kinase signal transduction network in *Candida albicans*. Microbiology *152*, 905–912.

Moors, M.A., Stull, T.L., Blank, K.J., Buckley, H.R., and Mosser, D.M. (1992). A role for complement receptor-like molecules in iron acquisition by *Candida albicans*. The Journal of experimental medicine *175*, 1643–1651.

Moreno-Ruiz, E., Galan-Diez, M., Zhu, W., Fernandez-Ruiz, E., d'Enfert, C., Filler, S.G., Cossart, P., and Veiga, E. (2009). *Candida albicans* internalization by host cells is mediated by a clathrin-dependent mechanism. Cell. Microbiol. *11*, 1179–1189.

Moyes, D.L., Runglall, M., Murciano, C., Shen, C., Nayar, D., Thavaraj, S., Kohli, A., Islam, A., Mora-Montes, H., Challacombe, S.J., *et al.* (2010). A biphasic innate immune MAPK response discriminates between the yeast and hyphal forms of *Candida albicans* in epithelial cells. Cell Host Microb. *8*, 225–235.

Moyes, D.L., Murciano, C., Runglall, M., Kohli, A., Islam, A., and Naglik, J.R. (2012). Activation of MAPK/c-Fos induced responses in oral epithelial cells is specific to *Candida albicans* and *Candida dubliniensis* hyphae. Med. Microbiol. Immunol. *201*, 93–101.

Mukherjee, P.K., Mohamed, S., Chandra, J., Kuhn, D., Liu, S., Antar, O.S., Munyon, R., Mitchell, A.P., Andes, D., Chance, M.R., *et al.* (2006). Alcohol dehydrogenase restricts the ability of the pathogen *Candida albicans* to form a biofilm on catheter surfaces through an ethanol-based mechanism. Infect. Immun. *74*, 3804–3816.

Munro, C.A., and Hube, B. (2002). Anti-fungal therapy at the HAART of viral therapy. Trends Microbiol. *10*, 173–177.

Murad, A.M., d'Enfert, C., Gaillardin, C., Tournu, H., Tekaia, F., Talibi, D., Marechal, D., Marchais, V., Cottin, J., and Brown, A.J. (2001a). Transcript profiling in *Candida albicans* reveals new cellular functions for the transcriptional repressors CaTup1, CaMig1 and CaNrg1. Mol. Microbiol. *42*, 981–993.

Murad, A.M., Leng, P., Straffon, M., Wishart, J., Macaskill, S., MacCallum, D., Schnell, N., Talibi, D., Marechal, D., Tekaia, F., *et al.* (2001b). *NRG1* represses yeast-hypha morphogenesis and hypha-specific gene expression in *Candida albicans*. EMBO J. *20*, 4742–4752.

Naglik, J.R., Moyes, D., Makwana, J., Kanzaria, P., Tsichlaki, E., Weindl, G., Tappuni, A.R., Rodgers, C.A., Woodman, A.J., Challacombe, S.J., *et al.* (2008). Quantitative expression of the *Candida albicans* secreted aspartyl proteinase gene family in human oral and vaginal candidiasis. Microbiology *154*, 3266–3280.

Nantel, A., Dignard, D., Bachewich, C., Harcus, D., Marcil, A., Bouin, A.P., Sensen, C.W., Hogues, H., van het

Hoog, M., Gordon, P., *et al.* (2002). Transcription profiling of *Candida albicans* cells undergoing the yeast-to-hyphal transition. Mol. Biol. Cell *13*, 3452–3465.

Nobile, C.J., and Mitchell, A.P. (2005). Regulation of cell-surface genes and biofilm formation by the *C. albicans* transcription factor Bcr1p. Curr. Biol. *15*, 1150–1155.

Nobile, C.J., Schneider, H.A., Nett, J.E., Sheppard, D.C., Filler, S.G., Andes, D.R., and Mitchell, A.P. (2008). Complementary adhesin function in *C. albicans* biofilm formation. Curr. Biol. *18*, 1017–1024.

Nobile, C.J., Nett, J.E., Hernday, A.D., Homann, O.R., Deneault, J.S., Nantel, A., Andes, D.R., Johnson, A.D., and Mitchell, A.P. (2009). Biofilm matrix regulation by *Candida albicans* Zap1. PloS Biol. 7, e1000133.

Nobile, C.J., Fox, E.P., Nett, J.E., Sorrells, T.R., Mitrovich, Q.M., Hernday, A.D., Tuch, B.B., Andes, D.R., and Johnson, A.D. (2012). A recently evolved transcriptional network controls biofilm development in *Candida albicans*. Cell *148*, 126–138.

Odds, F.C. (1985). Morphogenesis in *Candida albicans*. Crit. Rev. Microbiol. *12*, 45–93.

Oppenheim, F.G., Xu, T., McMillian, F.M., Levitz, S.M., Diamond, R.D., Offner, G.D., and Troxler, R.F. (1988). Histatins, a novel family of histidine-rich proteins in human parotid secretion. Isolation, characterization, primary structure, and fungistatic effects on *Candida albicans*. J. Biol. Chem. *263*, 7472–7477.

Park, H., Myers, C.L., Sheppard, D.C., Phan, Q.T., Sanchez, A.A., E. Edwards., J., and Filler, S.G. (2005). Role of the fungal Ras-protein kinase A pathway in governing epithelial cell interactions during oropharyngeal candidiasis. Cell. Microbiol. 7, 499–510.

Park, H., Liu, Y., Solis, N., Spotkov, J., Hamaker, J., Blankenship, J.R., Yeaman, M.R., Mitchell, A.P., Liu, H., and Filler, S.G. (2009). Transcriptional responses of *Candida albicans* to epithelial and endothelial cells. Eukaryot. Cell *8*, 1498–1510.

Pendrak, M.L., Chao, M.P., Yan, S.S., and Roberts, D.D. (2004). Heme oxygenase in *Candida albicans* is regulated by hemoglobin and is necessary for metabolism of exogenous heme and hemoglobin to alpha-biliverdin. J. Biol. Chem. *279*, 3426–3433.

Perlroth, J., Choi, B., and Spellberg, B. (2007). Nosocomial fungal infections: epidemiology, diagnosis, and treatment. Med. Mycol. *45*, 321–346.

Pfaller, M.A., Diekema, D.J., Gibbs, D.L., Newell, V.A., Ellis, D., Tullio, V., Rodloff, A., Fu, W., Ling, T.A., and and the Global Antifungal Surveillance, G. (2010). Results from the ARTEMIS DISK Global Antifungal Surveillance Study, 1997 to 2007: a 10.5-year analysis of susceptibilities of *Candida* Species to fluconazole and voriconazole as determined by CLSI standardized disk diffusion. J. Clin. Microbiol. *48*, 1366–1377.

Phan, Q.T., Myers, C.L., Fu, Y., Sheppard, D.C., Yeaman, M.R., Welch, W.H., Ibrahim, A.S., Edwards, J.E., Jr., and Filler, S.G. (2007). Als3 is a *Candida albicans* invasin that binds to cadherins and induces endocytosis by host cells. PloS Biol. *5*, e64.

Pierce, J.V., and Kumamoto, C.A. (2012). Variation in *Candida albicans EFG1* expression enables host-dependent changes in colonizing fungal populations. Mbio 3, e00117–00112.

Pierce, J.V., Dignard, D., Whiteway, M., and Kumamoto, C.A. (2013). Normal adaptation of *Candida albicans* to the murine gastrointestinal tract requires Efg1p-dependent regulation of metabolic and host defense genes. Eukaryot. Cell *12*, 37–49.

Puri, S., Kumar, R., Chadha, S., Tati, S., Conti, H.R., Hube, B., Cullen, P.J., and Edgerton, M. (2012). Secreted aspartic protease cleavage of *Candida albicans* Msb2 activates Cek1 MAPK signaling affecting biofilm formation and oropharyngeal candidiasis. PloS One 7, e46020.

Ramage, G., Saville, S.P., Wickes, B.L., and Lopez-Ribot, J.L. (2002). Inhibition of *Candida albicans* biofilm formation by farnesol, a quorum-sensing molecule. Appl. Env. Microbiol. *68*, 5459–5463.

Ramanan, N., and Wang, Y. (2000). A high-affinity iron permease essential for *Candida albicans* virulence. Science *288*, 1062–1064.

Robbins, N., Uppuluri, P., Nett, J., Rajendran, R., Ramage, G., Lopez-Ribot, J.L., Andes, D., and Cowen, L.E. (2011). Hsp90 governs dispersion and drug resistance of fungal biofilms. PLoS Pathog. 7, e1002257.

Rosenbach, A., Dignard, D., Pierce, J.V., Whiteway, M., and Kumamoto, C.A. (2010). Adaptations of *Candida albicans* for growth in the mammalian intestinal tract. Eukaryot. Cell *9*, 1075–1086.

Rubin-Bejerano, I., Fraser, I., Grisafi, P., and Fink, G.R. (2003). Phagocytosis by neutrophils induces an amino acid deprivation response in *Saccharomyces cerevisiae* and *Candida albicans*. Proc. Nat. Acad. Sci. U.S.A. *100*, 11007–11012.

Ruhnke, M. (2002). Skin and mucous membrane infections. In *Candida* and Candidiasis, R Calderone, ed. (ASM Press, Washington DC), pp. 307–325.

Santos, M.A., Gomes, A.C., Santos, M.C., Carreto, L.C., and Moura, G.R. (2011). The genetic code of the fungal CTG clade. Compt. Rend. Biol. *334*, 607–611.

Schaller, M., Boeld, U., Oberbauer, S., Hamm, G., Hube, B., and Korting, H.C. (2004). Polymorphonuclear leukocytes (PMNs) induce protective Th1-type cytokine epithelial responses in an *in vitro* model of oral candidosis. Microbiology *150*, 2807–2813.

Schaller, M., Borelli, C., Korting, H.C., and Hube, B. (2005). Hydrolytic enzymes as virulence factors of *Candida albicans*. Mycoses *48*, 365–377.

Seider, K., Heyken, A., Luttich, A., Miramon, P., and Hube, B. (2010). Interaction of pathogenic yeasts with phagocytes: survival, persistence and escape. Curr. Opin. Microbiol. *13*, 392–400.

Shapiro, R.S., Uppuluri, P., Zaas, A.K., Collins, C., Senn, H., Perfect, J.R., Heitman, J., and Cowen, L.E. (2009). Hsp90 orchestrates temperature-dependent *Candida albicans* morphogenesis via Ras1-PKA signaling. Curr. Biol. *19*, 621–629.

Sheth, C.C., Mogensen, E.G., Fu, M.S., Blomfield, I.C., and Muhlschlegel, F.A. (2008). *Candida albicans HSP12* is co-regulated by physiological CO_2 and pH. Fung. Genet. Biol. *45*, 1075–1080.

Silverman, R.J., Nobbs, A.H., Vickerman, M.M., Barbour, M.E., and Jenkinson, H.F. (2010). Interaction of *Candida albicans* cell wall Als3 protein with *Streptococcus gordonii* SspB adhesin promotes development of mixed-species communities. Infect. Immun. *78*, 4644–4652.

Singh-Babak, S.D., Babak, T., Diezmann, S., Hill, J.A., Xie, J.L., Chen, Y.L., Poutanen, S.M., Rennie, R.P., Heitman, J., and Cowen, L.E. (2012). Global analysis of the evolution and mechanism of echinocandin resistance in *Candida glabrata*. PLoS Pathog. *8*, e1002718.

Singh, S.D., Robbins, N., Zaas, A.K., Schell, W.A., Perfect, J.R., and Cowen, L.E. (2009). Hsp90 governs echinocandin resistance in the pathogenic yeast *Candida albicans* via calcineurin. PLoS Pathog. *5*, e1000532.

Sobel, J.D. (2007). Vulvovaginal candidosis. Lancet *369*, 1961–1971.

Sohn, K., Senyurek, I., Fertey, J., Konigsdorfer, A., Joffroy, C., Hauser, N., Zelt, G., Brunner, H., and Rupp, S. (2006). An *in vitro* assay to study the transcriptional response during adherence of *Candida albicans* to different human epithelia. FEMS Yeast Res. *6*, 1085–1093.

Soloviev, D.A., Fonzi, W.A., Sentandreu, R., Pluskota, E., Forsyth, C.B., Yadav, S., and Plow, E.F. (2007). Identification of pH-regulated antigen 1 released from *Candida albicans* as the major ligand for leukocyte integrin alphaMbeta2. J. Immunol. *178*, 2038–2046.

Sosinska, G.J., de Groot, P.W., Teixeira de Mattos, M.J., Dekker, H.L., de Koster, C.G., Hellingwerf, K.J., and Klis, F.M. (2008). Hypoxic conditions and iron restriction affect the cell-wall proteome of *Candida albicans* grown under vagina-simulative conditions. Microbiology *154*, 510–520.

Spiering, M.J., Moran, G.P., Chauvel, M., Maccallum, D.M., Higgins, J., Hokamp, K., Yeomans, T., d'Enfert, C., Coleman, D.C., and Sullivan, D.J. (2010). Comparative transcript profiling of *Candida albicans* and *Candida dubliniensis* identifies *SFL2*, a *C. albicans* gene required for virulence in a reconstituted epithelial infection model. Eukaryot. Cell *9*, 251–265.

Staab, J.F., Bradway, S.D., Fidel, P.L., and Sundstrom, P. (1999). Adhesive and mammalian transglutaminase substrate properties of *Candida albicans* Hwp1. Science *283*, 1535–1538.

Stoldt, V.R., Sonneborn, A., Leuker, C.E., and Ernst, J.F. (1997). Efg1p, an essential regulator of morphogenesis of the human pathogen *Candida albicans*, is a member of a conserved class of bHLH proteins regulating morphogenetic processes in fungi. EMBO J. *16*, 1982–1991.

Sudbery, P.E. (2011). Growth of *Candida albicans* hyphae. Nat. Rev. Microbiol. *9*, 737–748.

Sudbery, P., Gow, N., and Berman, J. (2004). The distinct morphogenic states of *Candida albicans*. Trends Microbiol. *12*, 317–324.

Sullivan, D.J., Westerneng, T.J., Haynes, K.A., Bennett, D.E., and Coleman, D.C. (1995). *Candida dubliniensis* sp. nov.: phenotypic and molecular characterization of a novel species associated with oral candidosis in HIV-infected individuals. Microbiology *141 (Pt 7)*, 1507–1521.

Sun, J.N., Solis, N.V., Phan, Q.T., Bajwa, J.S., Kashleva, H., Thompson, A., Liu, Y., Dongari-Bagtzoglou, A., Edgerton, M., and Filler, S.G. (2010). Host cell invasion and virulence mediated by *Candida albicans* Ssa1. PLoS Pathog. *6*, e1001181.

Sundstrom, P. (2002). Adhesion in *Candida* spp. Cell. Microbiol. *4*, 461–469.

Szabo, E.K., and MacCallum, D.M. (2011). The contribution of mouse models to our understanding of systemic candidiasis. FEMS Microbiol. Lett. *320*, 1–8.

Szafranski-Schneider, E., Swidergall, M., Cottier, F., Tielker, D., Roman, E., Pla, J., and Ernst, J.F. (2012). Msb2 shedding protects *Candida albicans* against antimicrobial peptides. PLoS Pathog. *8*, e1002501.

Thewes, S., Kretschmar, M., Park, H., Schaller, M., Filler, S.G., and Hube, B. (2007). *In vivo* and *ex vivo* comparative transcriptional profiling of invasive and non-invasive *Candida albicans* isolates identifies genes associated with tissue invasion. Mol. Microbiol. *63*, 1606–1628.

Thompson, H.L., and Wilton, J.M. (1992). Interaction and intracellular killing of *Candida albicans* blastospores by human polymorphonuclear leucocytes, monocytes and monocyte-derived macrophages in aerobic and anaerobic conditions. Clin. Exp. Immunol. *87*, 316–321.

Ullmann, B.D., Myers, H., Chiranand, W., Lazzell, A.L., Zhao, Q., Vega, L.A., Lopez-Ribot, J.L., Gardner, P.R., and Gustin, M.C. (2004). Inducible defense mechanism against nitric oxide in *Candida albicans*. Eukaryot. Cell 3, 715–723.

Uppuluri, P., Chaturvedi, A.K., Srinivasan, A., Banerjee, M., Ramasubramaniam, A.K., Kohler, J.R., Kadosh, D., and Lopez-Ribot, J.L. (2010). Dispersion as an important step in the *Candida albicans* biofilm developmental cycle. PLoS Pathog. *6*, e1000828.

Urban, C.F., Reichard, U., Brinkmann, V., and Zychlinsky, A. (2006). Neutrophil extracellular traps capture and kill *Candida albicans* yeast and hyphal forms. Cell. Microbiol. *8*, 668–676.

Urban, C.F., Ermert, D., Schmid, M., Abu-Abed, U., Goosmann, C., Nacken, W., Brinkmann, V., Jungblut, P.R., and Zychlinsky, A. (2009). Neutrophil extracellular traps contain calprotectin, a cytosolic protein complex involved in host defense against *Candida albicans*. PLoS Pathog. *5*, e1000639.

Vazquez, J.A., and Sobel, J.D. (2002). Mucosal candidiasis. Infect. Dis. Clin. North Am. *16*, 793–820, v.

Vylkova, S., Carman, A.J., Danhof, H.A., Collette, J.R., Zhou, H., and Lorenz, M.C. (2011). The fungal pathogen *Candida albicans* autoinduces hyphal morphogenesis by raising extracellular pH. Mbio *2*, e00055–00011.

Wächtler, B., Wilson, D., Haedicke, K., Dalle, F., and Hube, B. (2011a). From attachment to damage: defined genes of *Candida albicans* mediate adhesion, invasion and damage during interaction with oral epithelial cells. PloS One *6*, e17046.

Wächtler, B., Wilson, D., and Hube, B. (2011b). *Candida albicans* adhesion to and invasion and damage of vaginal epithelial cells: stage-specific inhibition by clotrimazole and bifonazole. Antimicrob. Agents Chemother. *55*, 4436–4439.

Wächtler, B., Citiulo, F., Jablonowski, N., Forster, S., Dalle, F., Schaller, M., Wilson, D., and Hube, B. (2012). *Candida albicans*-epithelial interactions: dissecting the roles of active penetration, induced endocytosis and host factors on the infection process. PloS One *7*, e36952.

Walker, L.A., Maccallum, D.M., Bertram, G., Gow, N.A., Odds, F.C., and Brown, A.J. (2009). Genome-wide analysis of *Candida albicans* gene expression patterns during infection of the mammalian kidney. Fung. Genet. Biol. *46*, 210–219.

Weissman, Z., and Kornitzer, D. (2004). A family of *Candida* cell surface haem-binding proteins involved in haemin and haemoglobin-iron utilization. Mol. Microbiol. *53*, 1209–1220.

Weissman, Z., Shemer, R., Conibear, E., and Kornitzer, D. (2008). An endocytic mechanism for haemoglobin-iron acquisition in *Candida albicans*. Mol. Microbiol. *69*, 201–217.

White, S.J., Rosenbach, A., Lephart, P., Nguyen, D., Benjamin, A., Tzipori, S., Whiteway, M., Mecsas, J., and Kumamoto, C.A. (2007). Self-regulation of *Candida albicans* population size during GI colonization. PLoS Pathog. *3*, e184.

Wilson, D., Thewes, S., Zakikhany, K., Fradin, C., Albrecht, A., Almeida, R., Brunke, S., Grosse, K., Martin, R., Mayer, F., *et al.* (2009). Identifying infection-associated genes of *Candida albicans* in the postgenomic era. FEMS Yeast Res. *9*, 688–700.

Wilson, D., Citiulo, F., and Hube, B. (2012). Zinc exploitation by pathogenic fungi. PLoS Pathog. *8*, e1003034.

Wysong, D.R., Christin, L., Sugar, A.M., Robbins, P.W., and Diamond, R.D. (1998). Cloning and sequencing of a *Candida albicans* catalase gene and effects of disruption of this gene. Infect. Immun. *66*, 1953–1961.

Zakikhany, K., Naglik, J.R., Schmidt-Westhausen, A., Holland, G., Schaller, M., and Hube, B. (2007). In vivo transcript profiling of *Candida albicans* identifies a gene essential for interepithelial dissemination. Cell. Microbiol. *9*, 2938–2954.

Zhao, X., Oh, S.H., Cheng, G., Green, C.B., Nuessen, J.A., Yeater, K., Leng, R.P., Brown, A.J., and Hoyer, L.L. (2004). *ALS3* and *ALS8* represent a single locus that encodes a *Candida albicans* adhesin; functional comparisons between Als3p and Als1p. Microbiology *150*, 2415–2428.

Zhu, W., and Filler, S.G. (2010). Interactions of *Candida albicans* with epithelial cells. Cell. Microbiol. *12*, 273–282.

Zhu, W., Phan, Q.T., Boontheung, P., Solis, N.V., Loo, J.A., and Filler, S.G. (2012). EGFR and HER2 receptor kinase signaling mediate epithelial cell invasion by *Candida albicans* during oropharyngeal infection. Proc. Nat. Acad. Sci. U.S.A. *109*, 14194–14199.

Virulence Characteristics of *Aspergillus fumigatus* 8

Rebecca A. Owens, Grainne O'Keeffe, Karen A. O'Hanlon,
Lorna Gallagher and Sean Doyle

Abstract

Aspergillus fumigatus is an opportunistic fungal pathogen which causes serious disease, broadly termed aspergillosis, in immunocompetent and immunocompromised individuals. Advances in the diagnosis of aspergillosis exploit detection of fungal nucleic acid via PCR and secreted galactomannan by ELISA. New diagnostic strategies are emerging based on detection of secreted low molecular mass metabolites. Drug treatments, which interfere with membrane biosynthesis and integrity, as well as cell wall function, are available to treat infection with *A. fumigatus*. No single characteristic appears to confer complete virulence on *A. fumigatus*, rather, it appears to depend on a combination of survival factors, in part derived from its function as a saprophyte, to facilitate host colonization and fungal growth. These attributes include a robust cell wall, small hydrophobic conidia and biofilm architecture; resistance to oxidative stress, hypoxia and iron starvation; secondary metabolite biosynthesis and secretion, as well as thermotolerance and an ability to secrete degradative enzymes for nutrient uptake. Attenuation of the efficiency of these survival strategies offers possibilities for the development of new classes of antifungal drugs.

General characteristics of *Aspergillus fumigatus*

The saprophytic fungus *Aspergillus fumigatus* is a member of the class Ascomycota, with over 200 species of the genus *Aspergillus* identified to date (Anzai *et al*., 2008). This soil-associated fungus plays an important role in recycling carbon and nitrogen, and is prevalent in the environment (Latgé, 1999). In a laboratory setting, *A. fumigatus* can grow on minimal agar containing a simple carbon source (e.g. glucose), a nitrogen source (e.g. ammonium tartrate) and trace elements (Brakhage and Langfelder, 2002). The ability of *A. fumigatus* to thrive at 37°C enables the pathogenicity of the fungus. The relative thermotolerance of *A. fumigatus* allows it to grow at temperatures up to 55°C, reflective of the presence of an ecological niche in compost heaps, and the conidia can withstand temperatures up to 70°C (Bhabhra and Askew, 2005; Latgé, 1999). *A. fumigatus* produces hydrophobic conidia (spores) that are aerially dispersed and only 2.5–3 μm in diameter (Brakhage and Langfelder, 2002) (Fig. 8.1). These conidia are ubiquitous in the environment and hundreds are inhaled every day (Latgé, 1999). The small size and buoyant nature of these conidia enables them to reach the alveoli of the lungs, where they are generally cleared in healthy individuals (Dagenais and Keller, 2009). A relatively small proportion of *Aspergillus* species are associated with human disease, with *A. fumigatus* demonstrating the highest pathogenicity (Kradin and Mark, 2008). *A. fumigatus* is an opportunistic pathogen, which causes disease in immunocompromised individuals (Ben-Ami *et al*., 2010). A markedly high mortality rate is observed with invasive *Aspergillus*-related disease, ranging from 40% to 95% (Abad *et al*., 2010).

A. fumigatus reproduction is predominantly asexual and is mediated by the dispersion of haploid conidia. The asexual life cycle of *A. fumigatus* is initiated by germination of conidia into septate hyphae which form a network of mycelia (Ward

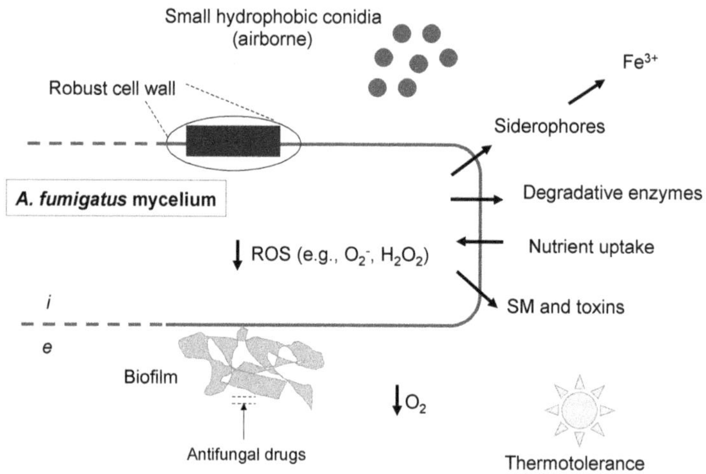

Figure 8.1 Overview of the structural features and virulence factors which influence the opportunistic pathogenicity of *Aspergillus fumigatus*. The rigid cell wall, small hydrophobic conidia and robust biofilm are key structural features which enhance organismal pathogenicity, and contribute to antifungal drug resistance. Toxins, secondary metabolites (SM), siderophores and secreted degradative enzymes also enhance the fungal virulence arsenal. An intricate capacity to grow under hypoxic conditions, dismutate reactive oxygen species (ROS; both intracellularly and host-derived) is also important for the success of *A. fumigatus* as an opportunistic pathogen. Key: i, intracellular; e, extracellular. A colour version of this figure is available in the plate section at the back of the book.

et al., 2005). Conidiophores form on hyphal extensions from the mycelial mass, and produce chains of grey-green conidia for dispersion. Genes regulating sexual development in *A. fumigatus* were identified following the sequencing of the genome, in addition to pheromone-associated genes (Nierman *et al.*, 2005; Paoletti *et al.*, 2005). Subsequently, the presence of a sexual cycle in *A. fumigatus* was experimentally verified (O'Gorman *et al.*, 2009).

Sequencing of the genome of the *A. fumigatus* clinical isolate Af293 revealed the presence of 9926 putative protein-coding genes along eight chromosomes, in a 29.4 megabase genome (Nierman *et al.*, 2005). Subsequently, a second strain of *A. fumigatus*, A1163, was sequenced, and demonstrated the presence of a set of core, highly conserved genes common to both sequenced strains (Fedorova *et al.*, 2008). Comparison of the *A. fumigatus*, *Neosartorya fischeri* and *Aspergillus clavatus* genome sequences revealed a number of genes that are unique to *A. fumigatus*. These include genes involved in secondary metabolism and detoxification, which may contribute to pathogenicity (Fedorova *et al.*, 2008; Nierman *et al.*, 2005).

Data from genome sequencing and annotation is available from a number of online warehouses including the Central *Aspergillus* Data Repository (CADRE) (http://www.cadre-genomes.org.uk/) and the *Aspergillus* Genome Database (AspGD) (http://www.aspgd.org/). These resources combine *in silico* computational gene annotation with manually curated information obtained from experimentation to provide extensive profiling of the genome. CADRE provides a unique identifier for each gene with the nomenclature indicating the respective strain (e.g. AFUA_ indicates *A. fumigatus* Af293, AFUB_ indicates *A. fumigatus* A1163). In the case of *A. fumigatus* Af293, the chromosome number and relative gene locus are also included in the identifier (e.g. AFUA_6G09740 indicates the gene is at locus number 9740 along chromosome 6 in *A. fumigatus* Af293).

Pathogenesis of *Aspergillus fumigatus*

A. fumigatus-related disease

A. fumigatus is responsible for a range of diseases, with host susceptibility closely linked to the immune status of the individual (Latgé, 1999). *A. fumigatus*-associated disease can be classified into three general groups: (a) allergic reactions, (b) colonization with limited invasiveness and (c) invasive infections (Brakhage and Langfelder, 2002). The latter category tends to be observed in immunocompromised individuals, demonstrative of the classification of *A. fumigatus* as an opportunistic pathogen (Abad *et al.*, 2010).

Allergic airway diseases, associated with *A. fumigatus,* include allergic bronchopulmonary aspergillosis (ABPA) and related conditions. ABPA is a hypersensitivity disorder that predominantly affects individuals with cystic fibrosis (CF) and asthma (Knutsen and Slavin, 2011). ABPA can lead to chronic lung damage and deterioration of lung function in this cohort of patients (Chaudhary and Marr, 2011; Kraemer *et al.*, 2006). The major predisposing factor for development of ABPA is the ineffective clearance of inhaled conidia from the lung (Patterson and Strek, 2010; Pihet *et al.*, 2009a). Structural abnormalities associated with chronic lung disease can contribute to the conidial evasion of the host mucociliary clearance mechanism (Chaudhary and Marr, 2011; Thomas *et al.*, 2010). CF is characterized by mutations in the CF transmembrane conductance regulator (CFTR), leading to disruption of chloride channels in many epithelial cells. This results in the development of thick, viscous mucous in the lung, which can impede the clearance of inhaled microorganisms (Pihet *et al.*, 2009a). In the absence of ABPA, lung function is not affected by *A. fumigatus* colonization, which is observed frequently in CF patients (de Vrankrijker *et al.*, 2011). Conidia evading extrusion from the lung must germinate before allergy can be established, as dormant conidia are immunologically inert due to presence of an external hydrophobic rodlet layer (Aimanianda *et al.*, 2009). Damage to pulmonary epithelium following germination permits exposure of *A. fumigatus* antigens to host dendritic cells (DCs). Pulmonary DCs subsequently elicit a Th2-type response with secretion of cytokines and B-cell isotype switching to IgE production (Chaudhary and Marr, 2011). Elevation of total IgE levels and secretion of proinflammatory cytokines is indicative of the allergic response associated with ABPA (Knutsen, 2006; Patterson and Strek, 2010). Diagnostic guidelines for ABPA include the presence of predisposing conditions (e.g. CF), elevation in total IgE, elevation in anti-*A. fumigatus* antibodies and a number of other criteria (Agarwal, 2011).

Aspergilloma (fungus-ball) is an example of the second category of *A. fumigatus*-related disease, characterized by colonization in the absence of extensive invasiveness. *A. fumigatus* colonization of pre-existing cavities in the lung can result in the formation of an aspergilloma. Tuberculosis (TB) is the most common cause of cavitation leading to aspergilloma formation, with 11% of individuals possessing cavities showing radiographical signs of aspergilloma (Kawamura *et al.*, 2000; Zmeili and Soubani, 2007). The aspergilloma consists of a mass of fungal hyphae, inflammatory cells, fibrin mucous and cell debris and the condition is usually non-invasive (Latgé, 1999; Zmeili and Soubani, 2007). Detection and diagnosis of aspergilloma is usually through routine radiography, as the condition is often asymptomatic (Kradin and Mark, 2008; Zmeili and Soubani, 2007). Sporulation of conidia is postulated to occur in addition to mycelial growth, as multiple isogenic, azole-resistant strains were isolated from an aspergilloma during a course of anti-fungal therapy (Camps *et al.*, 2012). Surgical intervention is required for removal of the aspergilloma, and the use of antifungal drugs as an adjunct has been found to be ineffective (Brik *et al.*, 2008; Sagan and Goździuk, 2010).

The third category of disease associated with *A. fumigatus* is classified by invasive disease, which may disseminate and result in systemic infection. Invasive Aspergillosis (IA) is a term used to describe clinical presentations that fall into this group. IA is the most detrimental *Aspergillus*-related disease, with associated mortality rates ranging from 40% to 95% (Abad *et al.*, 2010; Nivoix *et al.*, 2008). Various factors including the immune status of the patient, site of infection and treatment affect these mortality rates (Maertens *et al.*, 2002). Individuals at risk for the development of IA include patients with haematological malignancies (e.g. leukaemia), chronic obstructive pulmonary disease (COPD), solid organ and haematopoietic stem cell transplant (HSCT) recipients, extended corticosteroid use and HIV-positive individuals (Dagenais and Keller, 2009; Gangneux *et al.*, 2010). COPD has also been identified as a risk factor for development of *Aspergillus*-related hypersensitivity and ABPA (Agarwal *et al.*, 2010). Genetic factors also influence the susceptibility of individuals to IA, and individuals with single nucleotide polymorphisms (SNPs) in genes such as plasminogen, IL-10 and mannose-binding lectin (MBL) are pre-disposed to develop IA (Brouard *et al.*, 2005; Crosdale *et al.*, 2001; Sainz *et al.*, 2007; Zaas *et al.*, 2008). Two distinct risk factors exist for the development of IA, neutropenia and corticosteroid-induced immunosuppression. Prolonged neutropenia represents the dominant risk for development of pulmonary IA and the associated disease is characterized by angioinvasion and disseminated fungal growth (Ben-Ami *et al.*, 2010; Kradin and Mark, 2008). Additionally, disruption in neutrophil function [e.g. chronic granulomatous disease (CGD) can result in a similar clinical presentation of IA, although angioinvasion is generally not observed in this subset of patients (Segal and Romani, 2009)]. A different pathology is observed in IA associated with corticosteroid-induced immunosuppression. Non-neutropenic or corticosteroid-related IA is generally non-angioinvasive, with limited fungal development or dissemination. Instead, the condition is characterized by excessive inflammation resulting in damage to tissues due to an aggressive host response (Balloy *et al.*, 2005). Inhaled conidia that evade mucociliary clearance, are generally phagocytosed by macrophages in the lung (Bhatia *et al.*, 2011). In addition to their function as phagocytes, alveolar macrophages modulate the immune response, and elicit the migration of other immune effector cells through the release of chemo- and cytokines. Engulfed conidia are killed in macrophages through the action of reactive oxygen species (ROS) and acidification in the phagolysosome (Ibrahim-Granet *et al.*, 2003; Philippe *et al.*, 2003). While corticosteroids do not affect phagocytosis of conidia by alveolar macrophages, ROS-mediated killing is inhibited. This can lead to germination of phagocytosed conidia in individuals undergoing corticosteroid treatment (Philippe *et al.*, 2003). Circulating neutrophils are recruited to the lung and are important in defence against fungal hyphae. Neutrophils mediate killing of hyphae by oxidative mechanisms following attachment to hyphal surfaces and de-granulation (Feldmesser, 2006; Levitz and Farrell, 1990). Tissue damage resulting from corticosteroid-associated IA is through an excessive influx of neutrophils and associated inflammation (Balloy *et al.*, 2005). Conversely, in neutropenia-associated IA, hyphal growth and extensive fungal development occurs due to the absence of neutrophils (Balloy *et al.*, 2005; Feldmesser, 2006). Diagnosis of IA is often delayed due to the non-specificity of the associated symptoms and limited sensitivity of diagnostic tests, which can contribute to delayed treatment (Segal and Walsh, 2006; Trof *et al.*, 2007). Treatment of IA involves the use of antifungal drug therapy, which includes the azole, echinocandin and polyene classes of drugs (Kontoyiannis, 2012).

Detection and diagnostic strategies for invasive aspergillosis

Historically, diagnosis of IA has been hampered by the presence of a multi-factorial disease, coupled with a wide-range of pre-disposing host factors. Together with the limited array of validated laboratory diagnostic methods, late diagnosis of IA contributes to delayed treatment and correspondingly high mortality rates (Maertens *et al.*, 2007). This phenomenon is exemplified by the high disparity between post-mortem detection and ante-mortem diagnosis of invasive fungal infections (IFIs), and specifically IA (Antinori *et al.*, 2009; Chamilos *et al.*, 2006). The current criteria for diagnosis of IFIs have been outlined by the consensus group of the European Organisation for Research and Treatment of Cancer/Invasive Fungal Infections Cooperative Group and the National Institute of Allergy and Infectious Diseases Mycoses Study Group (EORTC/MSG) (de Pauw *et al.*, 2008). Following the revised EORTC/MSG guidelines diagnosis of IFIs can be delineated into proven, probable or possible categories of disease. Classification is dependent on the presence of host factors (e.g. recent neutropenia or prolonged corticosteroid use), observation of clinical indications (e.g. radiographic findings) and mycological criteria (e.g. direct microscopy or detection of antigens). Mycological evidence of

IFI involves the direct or indirect detection of the causative fungal agent. Direct methods of detection include culture of the fungus from patient specimens, including sputum or bronchoalveolar lavage (BAL) fluid, or observation of fungal elements by microscopy. Indirect mycological detection methods currently approved for use in the EORTC/MSG guidelines are typically only applicable to aspergillosis and candidiasis due to the selectivity of these tests (de Pauw *et al.*, 2008).

Detection of circulating antigens using enzyme-immunoassay (EIA) has contributed to the detection and monitoring of *Aspergillus* infections. Galactomannan (GM) is a heat-stable polysaccharide component of the cell walls of *Aspergillus* and *Penicillium* spp., which is secreted during fungal growth (Latgé *et al.*, 1994). The FDA-approved Platelia© sandwich EIA, for the detection of *Aspergillus* GM, is routinely used as an adjunct diagnostic tool for IA (Maertens *et al.*, 2007). Levels of circulating GM have been considered to be proportional to the relative fungal load, with absence of GM subsidence an indicator of prognosis (Boutboul *et al.*, 2002). An additional cell wall constituent, β-1,3-glucan, is another diagnostic target for IFIs. Since β-1,3-glucan is present on cell walls from most pathogenic fungi, excluding *Cryptococcus* spp. and zygomycetes, it does not specifically indicate IA but instead is a 'pan-fungal' detection strategy (Hope *et al.*, 2005; Maertens *et al.*, 2007; Thornton, 2010). Investigation of the specificity and sensitivity of these indirect detection strategies, currently in routine use for IFI diagnosis, has resulted in widely varying results. Host-factors appear to influence the performance of these antigen assays, with accuracy of GM detection differing between patients with haematological disorders and immunosuppressed individuals (Ku *et al.*, 2012; Pfeiffer *et al.*, 2006). Furthermore, anti-fungal therapy can also reduce the sensitivity of these antigen immunoassays, highlighting the need for further validated tests for IA diagnosis (Marr *et al.*, 2005). Commercial polymerase chain reaction (PCR)-based detection strategies, for use in IA diagnosis, are in development (http://www.lab21.com/), with further validation required before they can be included in the EORTC/MSG guidelines for IFI diagnosis (de Pauw *et al.*, 2008). Efforts have been undertaken to standardize procedures used in PCR-based detection of IA, in addition to reduction in the incidence of false-positives (White *et al.*, 2010, 2011). Emerging real-time quantitative PCR (qPCR) strategies for detection of *Aspergillus* are compliant with stringent guidelines for reporting and may signal the shift towards inclusion of these molecular tests in the IA diagnostic tool-belt (Johnson *et al.*, 2012).

Several novel strategies are under investigation for the diagnosis of IA, including immuno-based methods and detection of *Aspergillus*-specific low molecular mass metabolites (Lewis *et al.*, 2005; Doyle *et al.*, 2012). The use of monoclonal antibodies (MAbs) for detection of alternative targets previously described (i.e. GM, β-1,3-glucan) has been examined by a number of groups with the potential for enhanced selectivity and sensitivity of detection. The development of an *Aspergillus* antigen capture ELISA has been described, using two distinct MAbs for the enhanced capture and detection of an *Aspergillus* antigen (Hao *et al.*, 2008). Combination of an antigen-assay with detection of anti-*Aspergillus* antibodies has also been explored, for the overall improvement in diagnostic capabilities. The gliotoxin oxidoreductase, GliT, has demonstrated immunoreactivity and has been identified both intracellularly and extracellularly in *A. fumigatus* (Kumar *et al.*, 2011; Schrettl *et al.*, 2010). Measurement of anti-GliT antibody levels, has been examined as a putative tool for IA diagnosis in non-neutropenic patients, although coupling of this assay with GM detection was suggested for optimal results (Shi *et al.*, 2012). A lateral flow device has been developed for the detection of an *Aspergillus*-specific antigen, utilizing a MAb directed against an undisclosed external component of *Aspergillus*, which is secreted during active growth (Thornton, 2008). Due to the ease of use, this device represents a significant development in IA diagnostics (Thornton *et al.*, 2012).

Additional studies have been carried out to investigate the potential of low molecular mass fungal metabolites in diagnosis. A novel metabolite-based detection strategy, under preliminary investigation, comprises a breath test for IA, and is based on the detection of an *Aspergillus*-specific volatile organic compound (VOC) (Chambers *et al.*, 2011). The premise of this test is the use of 2-pentylfuran (2PF) as a biomarker of IA, however the technical apparatus required to characterize these samples make this technology incompatible with a clinical setting in its current form (Chambers *et al.*, 2011). Mass spectrometry-based detection of cyclic non-ribosomal peptides has been proposed as a way to detect fungal infection and distinguish between fungal strains, based on distinct cyclic peptide profiles (Jegorov *et al.*, 2006). Additionally, detection of the bis-methylated derivative of gliotoxin (bmGT), was achieved using thin layer chromatography (TLC) and high performance liquid chromatography (HPLC), and was put forward as a more reliable diagnostic candidate than native gliotoxin (Domingo *et al.*, 2012). Further to their potential for use as biomarkers of infection, fungal metabolites can also be employed for detection of fungal growth *in vivo*. Petrik *et al.* (2010) exploited the iron scavenging activity of *A. fumigatus*, by using modified siderophores as a reporting mechanism. Radiolabelled siderophores were selectively taken up by *A. fumigatus in vivo* and accumulation allowed the detection of *A. fumigatus* infection using positron emission tomography (PET). Again this strategy hinged on the principle of using fungal-specific molecules or mechanisms for detection, to eliminate false positives from host interference. Further development of indirect detection methods for IA is of paramount importance, as sensitive and selective techniques would preclude the need for invasive diagnostic procedures.

Anti-fungal therapy for invasive aspergillosis

Therapeutic goals in the treatment of IFI include the restoration of immune function, if applicable, and the reduction of the fungal burden (Traunmüller *et al.*, 2011). Antifungal agents are utilized to achieve the latter, and include polyenes, triazoles and echinocandins for the treatment of IA (Thompson and Patterson, 2008). Due to the relatively close phylogenetic relationship between fungi and humans, unique targets for antifungal therapeutics are limited and the identification of differential mechanisms may expand this base (Denning and Hope, 2010). One such target for antifungal agents is the cell wall, which represents a fungal-specific entity, distinct from the host background (Vandeputte *et al.*, 2012). Additionally ergosterol, a component of the fungal cell membrane, is not found in human cell membranes, with particular classes of drugs targeting these molecules (Beauvais and Latgé, 2001). The polyene class of antifungals, including amphotericin B (AmB) deoxycholate, exploit this unique fungal target to elicit their function. AmB binds to ergosterol in the fungal cell membrane and forms pores, which lead to leakage of potassium ions. The resultant proton gradient ultimately results in fungal cell death (Lemke *et al.*, 2005). AmB was classically the primary antifungal used for the treatment of IA, however substantial side effects, including nephrotoxicity, have led to the development and use of lipid formulations with reduced toxicity (Kleinberg, 2006; Ullmann *et al.*, 2006; Wingard *et al.*, 1999).

Voriconazole, itraconazole and posaconazole are members of the triazole class of antifungals used to treat invasive fungal infection. Triazoles inhibit the cytochrome P450 enzyme resulting in disruption of ergosterol biosynthesis and consequently cell membrane dysfunction and cell death (Thompson and Patterson, 2008). Voriconazole inhibits the action of 14α-lanosterol demethylase, a key step in the production of ergosterol which is required for normal cell membrane function (Denning and Hope, 2010). Some hepatic-related side effects are associated with voriconazole use, principally due to metabolism via host cytochrome P450 enzymes (Johnson and Kauffman, 2003). Despite this, voriconazole has emerged as the primary therapy for IA due to the reduced toxicity profile and enhanced efficacy, relative to AmB (Azie *et al.*, 2012; Herbrecht *et al.*, 2002). Emergence of resistance to azoles has been noted in some cases and this represents a potential concern for long-term use of these therapeutics (Camps *et al.*, 2012; Howard *et al.*, 2006; Trof *et al.*, 2007).

Caspofungin is a member of the echinocandin class of antifungal agents, which function by disrupting cell wall synthesis. Caspofungin is currently recommended for use as a second line or salvage therapy for IA (Kartsonis *et al.*, 2005; Maertens *et al.*, 2004). Recent studies have also indicated the effectiveness of this agent in the primary treatment of IA, with efficacy and favourable toxicity profiles shown in patients with haematological disorders and HSCT recipients (Herbrecht *et al.*, 2010; Jarque *et al.*, 2013). The mechanism of action of caspofungin is the targeted disruption of β-1,3-glucan biosynthesis, through non-competitive inhibition of β-1,3-glucan synthase. Inhibition of β-1,3-glucan biosynthesis, a principle cell wall component, results in destabilization of fungal cell walls and limits fungal growth (Letscher-Bru, 2003). Caspofungin is utilized as a monotherapy for IFIs, in addition to inclusion in combination therapy (Maertens *et al.*, 2010). Combination therapy for the treatment of IA offers a number of theoretical advantages, including reduced risk of resistance, wider target area, and more rapid effect; however, some studies have not conclusively demonstrated the effectiveness of this approach over monotherapy (Garbati *et al.*, 2012; Trof *et al.*, 2007; Tunger *et al.*, 2008). *In vitro* studies suggest positive outcomes upon combinations of azoles with echinocandins, possibly due to differential targets associated with these agents (Jeans *et al.*, 2012).

Immunotherapy represents another strategy for the treatment of IA, focusing on modulation of immune function for enhanced antifungal activity (Carvalho *et al.*, 2012). Administration of recombinant colony stimulating factors (CSFs), including granulocyte-CSF (G-CSF) and granulocyte–macrophage CSF (GM-CSF), has been investigated for potential in the treatment of IFIs. G-CSF stimulates the proliferation and survival of neutrophils and their precursors, and is used in the prevention of chemotherapy-associated febrile neutropenia (Silvestris *et al.*, 2012). GM-CSF promotes neutrophil survival and stimulates neutrophil effector function, in addition to stimulation of macrophage proliferation and activity (Hercus *et al.*, 2012). Therapeutic use of these CSFs in treatment of IFI was expected to restore immune function and consequently enhance fungal clearance; however, results from various *in vitro* studies and clinical reports were contradictory (Lehrnbecher *et al.*, 2011). Post-transplant administration of G-CSF has also been demonstrated to impair immune recovery due to the induction of an inflammatory Th2 response (Volpi *et al.*, 2001). GM-CSF represents a more appropriate agent for use against *Aspergillus*-associated infection due to stimulation of both neutrophils and macrophage activity. Additionally GM-CSF, unlike G-CSF, does not dramatically increase total leucocytes counts, resulting in reduced tissue injury from the inflammatory neutrophil response (Graybill *et al.*, 1998). Use of G-CSF has been shown to shorten recovery times and length of hospitalization when used in the treatment of IA, however no effect was noted regarding mortality rates (Pagano *et al.*, 2010). The most beneficial use of CSFs appears to be in prophylactic treatment for the prevention of infection in high-risk patient cohorts (Falagas *et al.*, 2008). Therefore this immunotherapy represents a promising preventative or adjunct therapy for IA.

Structures and factors contributing to *Aspergillus fumigatus* pathogenicity

Cell wall

Aspergillus fumigatus has been postulated to be an accidental pathogen, lacking sophisticated virulence factors (McCormick *et al.*, 2010a). Indeed, many of the traits of *A. fumigatus*, which contribute to its pathogenicity under opportune conditions, have developed to enable survival in the primary ecological niche of this fungus, the soil. The cell wall provides a physical barrier between the fungal cell and the external environment, affording structural integrity to the hyphae and conidia in addition to physical protection from exogenous stresses (Latgé, 2007) (Fig. 8.1). In *A. fumigatus* the cell wall is composed primarily of polysaccharides, and is a dynamic structure embedded with proteins, that can change in response to environmental stimuli (Abad *et al.*, 2010). The constituents of the cell wall include α-1,3-glucans, β-1,3-glucans, chitins and galactomannans, and disruption of the cell wall integrity increases the sensitivity of *A. fumigatus* to external stresses, and additionally can enhance susceptibility to antifungal agents. Fungi will die if the cell wall is weakened or removed, unless they are osmotically stabilized (Latgé, 2007). The cell wall also bears an aggressive function, as it is a reservoir of toxic and hydrolytic molecules, which are used by the fungus to invade ecological niches (Latgé, 2007). Despite the essential role of the fungal cell wall, the biosynthesis and accurate structure of the cell wall of most fungal species is not fully understood, and this is particularly so amongst filamentous fungi (Latgé and Calderone, 2002; Lesage and Bussey, 2006). The cell wall can be organized into two distinct layers; an inner and an outer layer, whose composition varies between fungal species and growth stage of the fungus. Fibrillar polysaccharides are close to the plasma membrane, while amorphous polysaccharides are located throughout the cell wall but mostly facing outwards (Latgé, 2010). This distinction between the two layers is purely theoretical since it is impossible to analyse the cell wall without prior enzymatic or chemical treatment, thereby altering the arrangement of the layers (Latgé, 2010).

More than 90% of the cell wall is composed of polysaccharide, and the outer layer of the cell wall is an insoluble structure requiring solubilization, usually by means of hot alkali treatment before it can be properly analysed (Latgé, 2007). For most fungal species, the central core of the cell wall is composed of a branched β-1,3,1,6-glucan linked to chitin via a β-1,4 linkage. In *A. fumigatus* and *Saccharomyces cerevisiae*, interchain β-1,6-glycosidic linkages make up 3% and 4% of the total glucan linkages, respectively (Fontaine *et al.*, 2000; Kollar *et al.*, 1995; Nguyen *et al.*, 1998; Pérez and Ribas, 2004). This structural core varies between fungal species, and is thought to be fibrillar and embedded in an unstructured cement, which is alkali-soluble (Latgé, 2007). The glucan–chitin complex is bound covalently to other polysaccharides, and this is variable between pathogens. In *A. fumigatus*, the polysaccharides are composed of galactomannan and β-1,3-1,4-glucan. This is absent in the human-pathogenic yeast *Candida albicans*, which contains mainly β-1,6-glucan, that is absent in *A. fumigatus* (Aimanianda *et al.*, 2009; Fontaine *et al.*, 2000). In *A. fumigatus*, the alkali-soluble fraction is comprised of α-1,3-glucan with 1% interconnecting α-1,4-linkages (Latgé, 2010). β-1,3-glucans are synthesized by a plasma membrane bound glucan synthase complex, and using UDP-glucose as a substrate, linear chains are extruded through the membrane (Beauvais *et al.*, 1993). Fungal β-glucan exhibits important immunostimulatory properties, mediated by the innate immune receptor Dectin-1 (Brown, 2006; Taylor *et al.*, 2007).

Also found in the cell wall of *A. fumigatus* are N-mannans and O-mannans which often terminate in a galactofuranose residue at the non-reducing end (Leitao *et al.*, 2003; Morelle *et al.*, 2005), and are covalently bound to the glucan fibrillar core (Latgé, 2010). The polysaccharide skeleton of the cell wall is coated with cell wall proteins (CWPs). The glycophosphatidylinositol (GPI)-modified proteins represent the major class of CWPs (Eisenhaber *et al.*, 2004; De Groot *et al.*, 2003). GPI-modified proteins possess an N-terminal hydrophobic signal peptide sequence targeting them to the endoplasmic reticulum. A C-terminal hydrophobic domain is present that is cleaved in the endoplasmic reticulum and replaced with a GPI-anchor. The GPI moiety may then be processed and attached to β-1,6-glucan in the cell wall (Romano *et al.*, 2006). The *S. cerevisiae* ECM33 protein has the typical features of a GPI-anchored protein. In *C. albicans*, ECM33 protein is required for normal cell architecture and expression of cell-surface proteins, and ECM33 mutants are reduced in the ability to invade epithelial cells, and are less virulent in a murine model of candidosis (Martinez-Lopez *et al.*, 2004, 2006). The *A. fumigatus* ECM33 homologue is involved in conidial adherence and morphogenesis (Romano *et al.*, 2006). Interestingly, an *AfuECM33* mutant exhibits rapid germination, increased cell–cell adhesion, increased resistance towards the antifungal caspofungin and increased virulence in a mouse model for IA (Romano *et al.*, 2006). Thus, cell wall alteration in the *AfuECM33* mutant may lead to greater exposure of cell-surface proteins involved in adherence, leading to increased cell–cell adhesion, and that a softer, more pliable cell wall allows for faster germination. Reduced synthesis of β-1,3-glucan in the *AfuECM33* mutant may explain the increased resistance to caspofungin, since caspofungin inhibits the β-1,3-glucan synthase (Romano *et al.*, 2006).

An important feature of the fungal cell wall is that it is a highly dynamic structure that is constantly changing during the cell cycle and in response to the environment. The majority of structural changes that occur in the cell wall are associated with the outer layer (Latgé, 2010). *A. fumigatus* conidia are covered by hydrophobins and melanin, while germinating conidia expose α-1,3-glucans, galactomannan, galactosaminogalactan and N-glycosylated proteins including galactomannoproteins on their surface (Latgé, 2010). As noted earlier, the exposure of such galactomannan and associated galactomannoproteins on germinating conidia is the basis of the commercial Platelia© EIA, which has been used widespread for the diagnosis of IA, and detection of β-1,3-glucans is also used as a diagnostic strategy for IA (Thornton, 2010).

Importantly, the fungal cell wall is an essential and major target for antifungal drug discovery and development for several reasons: it accounts for one quarter of the fungal cell; it is a physically rigid structure protecting the fungus from the environment; it is essential for fungal life; and it is composed of unique molecules which do not have equivalents in humans (e.g. β-glucans and chitin) (Beauvais and Latgé, 2001). Given the huge potential for cell wall antifungal drug targets and the importance of the cell wall in innate immune recognition of *A. fumigatus*, a greater understanding of the cell wall, its composition, biosynthesis and regulation is necessary.

Deletion of *afmnt1* and *afpmt1*, involved in cell wall component biosynthesis, increases the sensitivity of the mutants to azoles and hygromycin, respectively (Wagener *et al.*, 2008; Zhou *et al.*, 2007). Additionally, *wsc1* mutants lacking a protein involved in cell wall integrity signalling, are significantly more sensitive to echinocandin treatment (Dichtl *et al.*, 2012). The cell wall constituent, β-glucan is recognized by the host immune system and is a ligand for the Dectin-1 receptor, which activates the inflammatory immune response (Steele *et al.*, 2005). β-glucan is not exposed on the surface of resting conidia, but becomes available for Dectin-1 binding following the initiation of conidia swelling and germination (Hohl *et al.*, 2005). Components of the cell wall also exhibit immunomodulatory mechanisms, which can result in attenuation of the innate immune response. Specifically, α-glucan reduces toll-like receptor (TLR)2 and TLR4-mediated production of interleukin (IL)6, while β-1,3-glucan decreases IL6 production via TLR4 (Chai *et al.*, 2011a). These effects are dose-dependent and more pronounced IL6 attenuation was observed using extracts from germinating conidia compared to resting conidia (Chai *et al.*, 2011a). TLRs recognize specific non-self molecules and trigger immune responses through signalling mechanisms. IL6 triggers the innate immune response and modulation of the production of this molecule demonstrates an immunoevasion mechanism elicited by the cell wall of *A. fumigatus*.

An additional mechanism employed by neutrophils in the defence against *A. fumigatus* is the production of Nuclear Extracellular Traps (NETs), consisting of secreted DNA complexes with antimicrobial granular proteins, which surround the fungal cells (Bruns *et al.*, 2010a; Wartha *et al.*, 2007). Indeed, neutrophils have attracted much attention following the identification of NETs. Once the nuclear membrane is ruptured, the NETs are formed by a mixture of nuclear DNA with granular contents and are then explosively released, in a process associated with cell death, known as NETosis (Brinkmann and Zychlinsky, 2007). Dying polymorphonuclear leucocytes (PMNLs) have been shown to eject nuclear DNA in response to *A. fumigatus*, forming NETs, a dynamic web-like structure embedded with fungicidal proteins which inhibit hyphal growth (Bruns *et al.*, 2010a; McCormick *et al.*, 2010b). NET formation has been observed in human neutrophils upon exposure to *A. fumigatus* and other Aspergilli both *in vivo* and *in vitro* (Bianchi *et al.*, 2009; Bruns *et al.*, 2010a; McCormick *et al.*, 2010b), and has been found to reduce hyphal activity in the respiratory tract after nine hours of co-incubation (Bruns *et al.*, 2010a). Following NETosis, the final NET release from a neutrophil occurs within three hours of DNA release (Bruns *et al.*, 2010a). NET formation was observed upon exposure to live and UV-killed *A. fumigatus* conidia and hyphae, and was associated with the presence of elastase, known to be characteristic of NETs (McCormick *et al.*, 2010b). NETs also trap *A. fumigatus* conidia (Bruns *et al.*, 2010a; Jaillon *et al.*, 2007; McCormick *et al.*, 2010b) and hyphae (Bruns *et al.*, 2010a), and appear to be fungistatic rather than fungicidal (Bruns *et al.*, 2010a; McCormick *et al.*, 2010b). It is suggested that they may help recruit more neutrophils or other immune cells to the infected site (Brakhage *et al.*, 2010), as well as reducing fungal dissemination (Bruns *et al.*, 2010a; McCormick *et al.*, 2010b). NETs are formed following the induction of a ROS signalling cascade in neutrophils which results in breakdown of the nuclear envelope and granular membranes (Fuchs *et al.*, 2007). These findings highlight the importance of neutrophils in the defence against *A. fumigatus*. RodA is responsible for reduced NET formation, with reduced levels or absence correlating with enhanced secretion of NETs by neutrophils (Bruns *et al.*, 2010a). As with melanin, the hydrophobin layer provides a mask for immunogenic cell surface ligands in dormant conidia, allowing spores to evade host defences prior to germination.

Conidia

Dormant *A. fumigatus* conidia are covered by a hydrophobic rodlet layer comprised of a thin layer of specifically arranged RodA hydrophobins (Thau *et al.*, 1994) (Fig. 8.1). The rodlet layer favours the buoyancy and air dispersal of conidia (Beever and Dempsey, 1978). The hydrophobic proteins are characterized by a conserved spacing of eight cysteine residues (Wessels, 1997; Wösten and de Vocht, 2000). The conidia of *A. fumigatus* contain two hydrophobins, RodA and RodB. The RodB protein, although homologous to RodA, is not essential for rodlet formation (Paris *et al.*, 2003). Using rodlet mutants, it was shown that RodA protects *A. fumigatus* conidia against killing by alveolar macrophages (Paris *et al.*, 2003). RodA is covalently bound to the cell wall polysaccharides, indicated by the presence of a GPI-anchoring sequence in the *rodA* gene (AFUA_5G09580) (Latgé, 2007). The rodlet layer renders *A. fumigatus* conidia immunologically silent, and is the likely reason why inhaled spores do not usually initiate inflammatory responses in immunocompetent individuals (Aimanianda *et al.*, 2009). Removal of the RodA protein chemically, by gene deletion, or through germination, resulted in conidia which induced greater immune responses compared to the wild-type (Aimanianda *et al.*, 2009). RodA has recently been shown to prevent *A. fumigatus* conidia from triggering NET formation in human and murine neutrophils, thereby presenting a novel mechanism by which *A. fumigatus* conidia escape neutrophil attack (Bruns *et al.*, 2010a).

The innate immune response directed against *A. fumigatus* involves alveolar macrophages and dendritic cells, resident in the lung, as a first-line of defence. Additionally, recruitment and activation of other leucocytes is required to fend off infiltrating fungal infection. As *A. fumigatus* is an opportunistic pathogen, virulence is associated with impairment in the host defence system. Moreover, *A. fumigatus* has a number of characteristics that enhance immune evasion and enable resistance to components of the host immune response. In addition to the cell wall components, *A. fumigatus* possesses exterior features that impart protection against external stresses. Melanin is located on the external surface of conidia and is responsible for the dark pigment associated with *A. fumigatus* conidia (Schmaler-Ripcke *et al.*, 2009; Tsai *et al.*, 1999). Melanin is synthesized by the dihydroxynaphthalene (DHN)-melanin pathway, directed by a cluster of 6 genes in *A. fumigatus*, which are expressed during conidiation (Tsai *et al.*, 1999) and is a documented virulence factor for *A. fumigatus* (Latgé, 1999). The polyketide synthetase PksP, encoded by the *pksp* gene in *A. fumigatus* (also called *alb1*) mediates one of the first steps of the DHN-melanin pathway. *pskp* mutants did not produce melanin and were found to be less virulent than wild-type strains in murine models of IA (Jahn *et al.*, 1997; Tsai *et al.*, 1998). Recently, it has been shown that melanin also plays a structural role in the conidial cell wall, is essential for correct assembly of the fungal cell wall layers, and the expression of the hydrophobic rodlet layer at the outer cell wall (Pihet *et al.*, 2009b). Melanin has been demonstrated to play a protective role in *A. fumigatus* conidia through a number of mechanisms: (i) protection against UV light, (ii) down-regulation of the complement cascade, (iii) ROS scavenging and (iv) masking cell surface ligands of the innate immune system (e.g. β-1,3-glucan) (Brakhage and Liebmann, 2005; Chai *et al.*, 2011b; Nosanchuk and Casadevall, 2006; Tsai *et al.*, 1998, 1999). Furthermore, following phagocytosis of conidia by macrophages, melanin inhibits phagolysosome acidification and apoptosis, thus prolonging the survival of infected macrophages (Thywißen *et al.*, 2011; Volling *et al.*, 2011).

An additional hydrophobic layer, encases the surface of conidia, and is composed of RodA protein covalently bound to the cell wall. This rodlet or hydrophobin layer aids in the dispersion of conidia, through the associated hydrophobicity, and also decreases the immunogenicity of resting conidia. Conidia from mutants deficient in the RodA protein activated dendritic cells and alveolar macrophages, with associated release of cytokines, while wild-type resting conidia were immunologically silent (Aimanianda *et al.*, 2009; Dagenais *et al.*, 2010).

Biofilm formation

Mowat *et al.* (2007) were the first to conclusively demonstrate that *A. fumigatus* forms biofilms *in vitro* which resulted in a 1000-fold decrease in susceptibility to a range of antifungal drugs, compared to planktonic cells, as determined by metabolic activity assay (Fig. 8.1). Bruns *et al.* (2010b) revealed the presence of gliotoxin, a redox-active molecule, in *A. fumigatus* biofilms, *in vitro*, and speculated that this may be an important factor in protecting the fungus against the host immune response, *in vivo*. It was subsequently shown in *C. albicans* that the molecular chaperone Hsp90 is an important regulator of biofilm formation and that disruption of Hsp90 augmented sensitivity to the azole class of antifungal drugs (Robbins *et al.*, 2011). Moreover, geldanamycin-mediated inhibition of Hsp90 function caused morphological changes in biofilms which increased the antifungal efficacy of both azole and echinocandin drugs against *A. fumigatus* (Robbins *et al.*, 2011). *A. fumigatus* biofilms are mainly composed of galactomannan and α–1,3-glucans, and can be disrupted using enzymes such as alginate lyase (Papi *et al.*, 2012). The involvement of extracellular DNA in biofilm architecture has also been demonstrated by Rajendran *et al.* (2013) who also found that DNAse presence increased the susceptibility of *A. fumigatus* to the antifungals, amphotericin B and caspofungin.

Resistance to oxidative stress

An integral component of *A. fumigatus*, in resistance to effector functions of the host defences, is the plethora of enzymes for detoxification of ROS (Fig. 8.1). Macrophage and neutrophil-mediated killing of conidia and hyphae is achieved in part through the generation of ROS. NADPH oxidase (NOX) is responsible for release of ROS and is essential for defence against fungal infection (Brown *et al.*, 2009). Correspondingly, defects in the NOX enzyme, associated with CGD, results in impaired capacity of neutrophils to kill fungi and leads to increased susceptibility to infections (Segal and Romani, 2009). The mitogen-activated protein kinase (MAPK) family of enzymes play a role in cell signalling and regulate intracellular responses to stress. In *A. fumigatus*, MpkA and SakA are two MAPKs associated with response to ROS and mediate their function through signal transduction cascades (Du *et al.*, 2006; Valiante *et al.*, 2008). MpkA is involved in the regulation of a number of genes encoding antioxidant proteins, including catalases and superoxide dismutase (SOD) (Jain *et al.*, 2011). In *A. nidulans*, the interaction of SakA with the transcription factor AtfA facilitates stress signalling and is involved in the regulation of catalase and peroxiredoxin expression (Lara-Rojas *et al.*, 2011). The transcription factor, Yap1, plays an integral role in the regulation of antioxidant genes in *A. fumigatus* (Aguirre *et al.*, 2006). Upon exposure to oxidative stress, Yap1 accumulates in the nucleus and induces the transcription of genes involved in protecting the cell from stress-induced damage (Kuge *et al.*, 1997). Yap1 targets, identified by comparative proteomics using a *yap1* deletion strain, included catalases, chaperones, peroxidases and a mitochondrial peroxiredoxin (Lessing *et al.*, 2007). While Δ*yap1* displayed no change in pathogenicity relative to the wild-type in a mouse model of invasive pulmonary aspergillosis (IPA), Yap1 was essential for virulence in an ocular keratitis infection model (Lessing *et al.*, 2007; Sixto *et al.*, 2012). This may result from the use of a neutropenic mouse model for IPA infection, as Yap1 is crucial in the defence against neutrophil-mediated killing (Lessing *et al.*, 2007; Sixto *et al.*, 2012). A zinc finger transcription factor, SebA, has recently been identified in *A. fumigatus*, with a role in protection against various forms of stress, including oxidative and heat shock stress (Dinamarco *et al.*, 2012). The deletion strain of *sebA*, demonstrated increased sensitivity to H_2O_2 and paraquat treatments, and *sebA*-dependent regulation of antioxidant enzymes was observed in response to these stimuli (Dinamarco *et al.*, 2012). The fungal response to oxidative stress therefore plays an integral role in the ability to function as a human pathogen (Brown *et al.*, 2009) and further characterization of this response is necessary for the understanding of pathogenicity.

Hypoxic responses

Hypoxic conditions represent a further physiological stress imposed on *A. fumigatus* during pathogenesis (Vödisch *et al.*, 2011) (Fig. 8.1). The sterol-regulatory element binding protein, SrbA, regulates ergosterol biosynthesis, maintains cell polarity and is essential for growth in hypoxic conditions (Willger *et al.*, 2008). Furthermore, deletion of *srbA* results in attenuated virulence in both neutropenic and corticosteroid models of IA, underpinning the role of hypoxia adaptation in the pathogenicity of *A. fumigatus* (Willger *et al.*, 2008, 2009). Large-scale proteomic and transcriptomic investigations have been carried out to identify mechanisms differentially regulated in *A. fumigatus* in response to both short-term and long-term hypoxia (Barker *et al.*, 2012; Vödisch *et al.*, 2011). Long-term exposure to low oxygen conditions correlated with the increase in expression of proteins involved in glycolysis, respiration and secondary metabolite (SM) production. The transcription factor SrbA was activated in response to a requirement for ergosterol biosynthesis, and additionally nitric oxide (NO)-detoxifying flavohaemoprotein was up-regulated, indicating the production of reactive nitrogen species (RNS) may be associated with hypoxia (Vödisch *et al.*, 2011). Similarly, following short-term exposure to hypoxia, ergosterol biosynthesis and flavohemoprotein were also up-regulated, indicating the requirement for these components in the adaption to, and the maintenance of growth in, hypoxic conditions (Barker *et al.*, 2012).

Iron uptake and siderophores

Another prerequisite for the survival of *A. fumigatus* in the host is the ability to obtain and store iron, an essential nutrient and cofactor for a number of enzymes (Haas, 2012) (Fig. 8.1). Conditions encountered by *A. fumigatus* in the lung are iron-limited and the capacity to acquire iron is crucial for growth (Haas, 2012; Schrettl *et al.*, 2007). Two distinct routes exist by which *A. fumigatus* can obtain iron, reductive iron assimilation (RIA) and siderophore-assisted iron uptake, both of which are induced in iron-limiting conditions (Schrettl *et al.*, 2007). The RIA system reduces ferric iron (Fe^{3+}) to ferrous iron (Fe^{2+}) and subsequently sequesters the reduced iron, whereas siderophores are low molecular mass, ferric iron-specific chelators. Siderophores which specifically chelate Fe^{3+} are used to sequester iron from the host environment or store iron internally (Schrettl *et al.*, 2004). Abrogation of siderophore biosynthesis, by disruption of the L-ornithine N^5-monooxygenase gene, *sidA* (Fig. 8.2) results in complete attenuation of virulence in a neutropenic mouse model of IA, demonstrating the essential role of the siderophore system in *A. fumigatus* pathogenicity (Schrettl *et al.*, 2004, 2007).

A. fumigatus produces three hydroxamate-type siderophores: extracellular fusarinine C (FSC) and triacetylfusarinine C (TAFC), and an intracellular ferricrocin (FC) (Schrettl *et al.*, 2007) (Fig. 8.2). The siderophore biosynthetic pathway in *A. fumigatus* has been extensively studied (Schrettl *et al.*, 2007). *sidA* is involved in the first committed step of the biosynthesis of both intracellular and extracellular siderophores in *A. fumigatus*. Here, L-ornithine is converted to N^5-hydroxy-L-ornithine by L-ornithine-N^5-oxygenase encoded by *sidA* (Eisendle *et al.*, 2003). After N^5-hydroxy-L-ornithine is synthesized, the pathway splits to enable either extracellular or intracellular siderophore biosynthesis (Fig. 8.2). For the extracellular siderophore, TAFC, the N^5-hydroxy-L-ornithine is converted to N^5-cis-anhydromevalonyl-N^5-hydroxy-L-ornithine by the N^5-transacylase, encoded by *sidF* using cis-anhydro-mevalonyl-CoA as the substrate. The N^5-cis-anhydromevalonyl-N^5-hydroxy-L-ornithine is converted to FSC by the non-ribosomal peptide synthetase (NRPS), encoded by *sidD*. FSC is then converted into the extracellular siderophore, TAFC, by a N^2-transacetylase, encoded by *sidG*, using acetyl-CoA as the substrate. For the intracellular siderophore, N^5-hydroxy-L-ornithine is converted

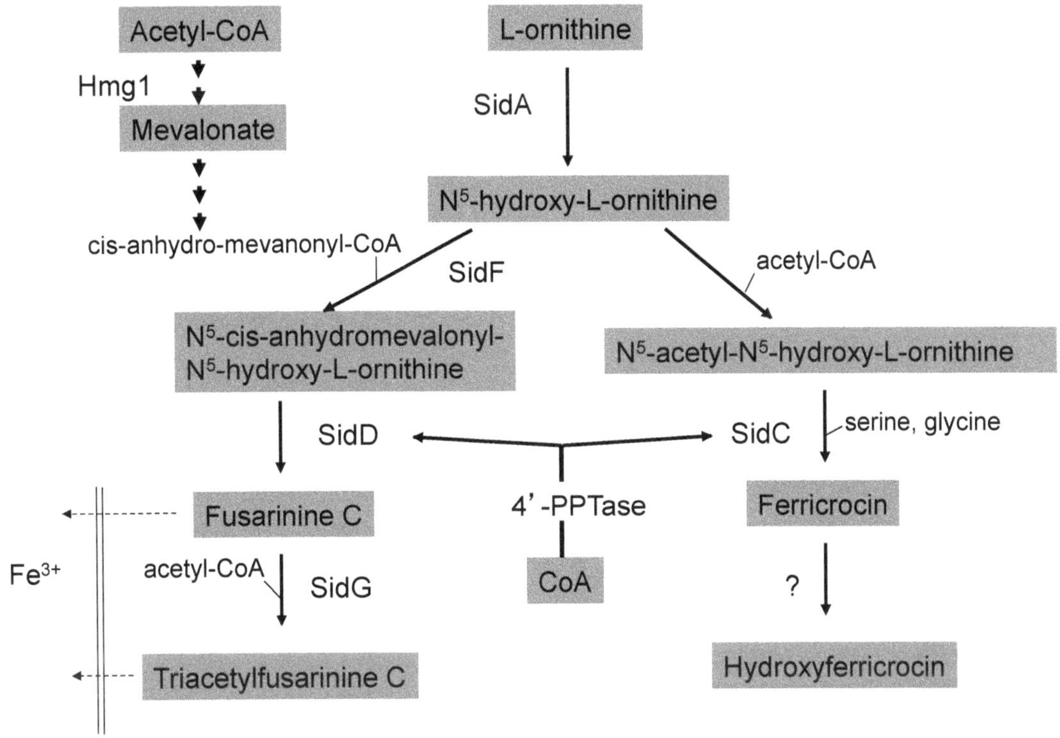

Figure 8.2 Biosynthetic pathway leading to siderophore biosynthesis in *A. fumigatus*. Siderophore biosynthesis is essential for virulence of the organism. SidA and 4′-phosphopantetheinyl transferase (4′-PPTase) are key enzymes in siderophore biosynthesis (Haas, 2012). Using coenzyme A as a substrate, 4′-PPTase post-translationally activates SidC and SidD via 4′-phosphopantetheinylation of a specific serine residue in NRPS thiolation domains. The extracellular siderophores, fusarinine C (FSC) and triacetylfusarinine C (TAFC) are secreted to acquire Fe^{3+} which is essential for fungal growth. From Schrettl *et al.* (2007) with modifications.

to N^5-acetyl-N^5-hydroxy-L-ornithine, by an unidentified N^5-transacylase using acetyl-CoA as the substrate. The N^5-acetyl- N^5-hydroxy-L-ornithine is converted into FC by the NRP synthetase, encoded by *sidC*, using serine and glycine as substrates. FC is converted to hydroxyferricrocin (HFC) by an unknown ferricrocin hydroxylase (Schrettl *et al.*, 2007) (Fig. 8.2). The enzyme SidD and the FC synthetase, SidC were identified by Reiber *et al.* (2005), and SidF and SidG were identified by Schrettl *et al.* (2007). The cis-anhydro-mevalonyl-CoA destined for FSC and TAFC biosynthesis is derived from mevalonate, a product of 3-hydroxy-3-methylglutaryl (HMG) CoA reductase, Hmg1 (Fig. 8.2). Interestingly, Hmg1 overexpression increases TAFC production, while lovastatin, an Hmg1 inhibitor suppresses TAFC production (Yasmin *et al.*, 2012). Consequently, it has been speculated that since siderophore biosynthesis is an essential virulence attribute, that the cholesterol-lowering statins may play a role in suppressing fungal infection in immunocompromised patients (Haas, 2012).

Secondary metabolites as virulence factors

A. fumigatus is capable of producing a range of toxic molecules which may contribute to its success as an opportunistic pathogen. Many of these molecules are SM in *A. fumigatus* and those implicated in pathogenicity, or at least interference with host biochemistry, include verruculogen, ergot alkaloids, gliotoxin and fumitremorgin. Over 200 SM are secreted by *A. fumigatus* and include melanins and ergosterols discussed earlier (Frisvad *et al.*, 2009). Gliotoxin represents the most characterized SM of *A. fumigatus*, with relevance to virulence. The mechanism of action of gliotoxin will be discussed in detail later this chapter, in addition to the effect of gliotoxin on the host defences. Briefly, gliotoxin is a redox-active metabolite, which exerts numerous immunomodulatory mechanisms, contributing to the success of *A. fumigatus* as an opportunistic pathogen (Abad *et al.*, 2010). Helvolic acid is a triterpene which demonstrates ciliostatic activity, albeit with lower activity than gliotoxin and fumagillin (Amitani *et al.*, 1995). A putative cluster involved in helvolic acid biosynthesis has been identified on Chromosome 4 in *A. fumigatus*, however the biosynthetic pathway has not been elucidated to date (Lodeiro *et al.*, 2009; Mitsuguchi *et al.*, 2009). Ergot alkaloids are conidia-associated toxins that interact with monoamine receptors and can affect the nervous and reproductive systems through this antagonism (Coyle *et al.*, 2007). The NRPSs, Pes1 and PesL, have been implicated in the production of the ergot alkaloid, fumigaclavine C (O'Hanlon *et al.*, 2012). Fumigaclavine C has been shown to inhibit the proliferation and activation of T lymphocytes in addition to reducing the production of TNFα *in vivo* and *in vitro* (Zhao *et al.*, 2004). Other mycotoxins associated with conidia have been identified following sporulation and include fumiquinazoline C, tryptoquivaline F, trypacidin, monomethylsulochrin and questin. Of these, trypacidin was observed to have toxic characteristics, and significantly and substantially reduced the cell viability of a human alveolar carcinoma cell line (A549) (Gauthier *et al.*, 2012). Trypacidin triggered cell death through necrosis in both primary and immortal lung cells, with oxidative stress induction implicated in the irreversible death process (Gauthier *et al.*, 2012). Fumagillin produced by *A. fumigatus* has been suggested to have genotoxic effects on mammalian cells *in vivo*, in addition to anti-angiogenic activity (Stanimirovic *et al.*, 2007). Moreover, fumagillin has been identified as an inhibitor of neutrophil function through interruption of the NAPDH oxidase complex assembly (Fallon *et al.*, 2010). Degranulation of neutrophils was also reduced in fumagillin treated cells, which could contribute to the persistence of *A. fumigatus* in the host (Fallon *et al.*, 2010). Pseurotin A and related analogues have demonstrated suppression of IgE production *in vitro* (Ishikawa *et al.*, 2009). Pseurotin A biosynthesis is encoded by a portion of a SM 'supercluster' on chromosome 8, with the sole hybrid polyketide synthase/non-ribosomal peptide synthetase (PKS/NRPS) in *A. fumigatus* essential for pseurotin A production (Maiya *et al.*, 2007; Perrin *et al.*, 2007). The pseurotin precursor molecules, propionyl-coenzyme A, phenylalanine and malonyl-coenzyme A, are utilized to produce a PKS-NRPS bound intermediate, with the remaining genes in the cluster putatively responsible for further modification of this intermediate (Maiya *et al.*, 2007).

Identification of secondary metabolite clusters

SM are predominantly low molecular mass molecules, produced by a range of organisms, which are dispensable for primary growth (Keller *et al.*, 2005). The production of SMs can often be associated with specific morphological stages and SMs may confer selective advantages in the presence of competing organisms (Calvo *et al.*, 2002; Losada *et al.*, 2009). SM can be broadly categorized based on the enzyme classes responsible for their biosynthesis, and include (i) polyketides, (ii) non-ribosomal peptides (iii) terpenes and (iv) indole alkaloids. Biosynthesis of these SMs is carried out by polyketide synthetases (PKSs), NRPSs, terpene cyclases and dimethylallyl tryptophan synthetases (DMATS), respectively. These enzymes catalyse the first step in the biosynthesis of SMs and are referred to as 'backbone enzymes' (Khaldi *et al.*, 2010). Enzymes involved in the biosynthesis of SMs are usually grouped in contiguous clusters in the genome with a proclivity towards telomeric localization (Keller *et al.*, 2005; Nierman *et al.*, 2005). This phenomenon is not observed in primary metabolite biosynthetic pathways, nor in secondary metabolism genes from other kingdoms. The sequencing of the genome of *A. fumigatus* Af293, revealed the presence of 26 SM clusters, including a number of clusters not identified in *A. oryzae* or *A. nidulans* (Nierman *et al.*, 2005). The number of SM gene clusters was subsequently revised to 22 by Perrin *et al.* (2007).

Developments in automated web-based tools, including Secondary Metabolite Unknown Region

Finder (SMURF; http://www.jcvi.org/smurf/index.php), have aided in locating SM clusters in fungi (Khaldi *et al.*, 2010). Additionally, the inclusion of a second sequenced strain of *A. fumigatus*, A1163, has resulted in the total number of SMs in *A. fumigatus* being adjusted to 36 (Fedorova *et al.*, 2008; Sanchez *et al.*, 2012). Characterization of mechanisms required for the production of specific classes of SM has enabled the identification of similar clusters in disparate species. The identification of the gliotoxin biosynthetic cluster in *A. fumigatus* was facilitated by comparison with another previously characterized epipolythiodioxopiperazine (ETP), sirodesmin, from *Leptosphaeria maculans* (Gardiner and Howlett, 2005). Subsequently, both of these biosynthetic clusters were used to identify other ETP clusters from across a range of fungi, and resulted in the detection of a second smaller ETP cluster in *A. fumigatus* (Patron *et al.*, 2007).

Secondary metabolite cluster regulation

Various factors can influence the production of SM and regulation of SM gene cluster expression is controlled by transcription factors. 'Broad'-domain transcription factors can co-regulate the expression of multiple gene clusters, resulting in an integrated response to external stimuli (Keller *et al.*, 2005). Conversely, 'narrow' range transcription factors, are often located within gene clusters and specifically regulate expression of the respective SM pathway. Secondary metabolism is influenced by external environmental factors, including pH, temperature, light and nutrient source, in addition to morphological development (Calvo *et al.*, 2002; Reverberi *et al.*, 2010). In *A. terreus* the production of the polyketide lovastatin, is influenced by the carbon:nitrogen ratio, while the biosynthesis of aflatoxin in *A. flavus* is dependent on temperature (Casas López *et al.*, 2003; O'Brian *et al.*, 2007).

A methyltransferase, LaeA, is a global regulator of secondary metabolism and development in *Aspergillus* species (Bok and Keller, 2004; Sarikaya Bayram *et al.*, 2010). The function of LaeA is putatively executed via methyltransferase activity and regulation of chromatin remodelling (Bok and Keller, 2004; Bok *et al.*, 2005; Keller *et al.*, 2006; Perrin *et al.*, 2007). LaeA was originally identified as a SM regulator following complementation of the loss of sterigmatocystin biosynthesis from *A. nidulans*, and was subsequently recognized as essential for the production of penicillin and conidial pigments (Bok and Keller, 2004). In *A. nidulans*, two of the velvet family of proteins, VeA and VelB, were shown to form a trimeric complex with LaeA and subsequently up-regulate asexual development and secondary metabolism (e.g. sterigmatocystin biosynthesis) (Bayram *et al.*, 2008). Translocation of VeA, and consequently VelB, to the nucleus is inhibited by light and thus interaction with LaeA occurs in the absence of light (Bayram *et al.*, 2008; Sarikaya Bayram *et al.*, 2010). While a similar trimeric interaction (VelB-VeA-LaeA) is noted in *A. fumigatus*, absence of VeA and VelB did not affect biosynthesis of gliotoxin in *A. fumigatus* (Park *et al.*, 2012). A large-scale transcriptional investigation identified multiple SM gene clusters under full or partial LaeA-regulation in *A. fumigatus*, including those involved in ergot alkaloid biosynthesis and the gliotoxin biosynthetic cluster (Perrin *et al.*, 2007). The contribution of SM to the pathogenicity of *A. fumigatus* was evidenced by the reduction in virulence, in a neutropenic model of pulmonary aspergillosis, upon deletion of the *laeA* gene (Bok *et al.*, 2005). An increase in macrophage-mediated conidial phagocytosis of Δ*laeA*, was caused by a reduction in the hydrophobic rodlet layer due to delayed expression of *rodA* (Bok *et al.*, 2005; Dagenais *et al.*, 2010). Moreover, reduced killing of polymorphonuclear neutrophils (PMNs) by Δ*laeA* hyphae was due to reduction in secreted SMs, causing reduction in the PMN-respiratory burst (Bok *et al.*, 2005; Sugui *et al.*, 2007a). During the onset of IA, a significant number of the SM genes regulated by LaeA were increased in expression, including members of the pseurotin and gliotoxin gene clusters (McDonagh *et al.*, 2008).

Gliotoxin

Gliotoxin ($C_{13}H_{14}N_2O_4S_2$) a 326-Da metabolite, is classified as an epipolythiodioxopiperazine (ETP) toxin, and is produced by a range of fungal species including *A. fumigatus*, *A. terreus*, *A. flavus*, *A. oryzae*, *Trichoderma virens* and some

Penicillium species (Patron *et al.*, 2007). ETPs are characterized by the presence of an internal disulfide bridge across a dioxopiperazine ring, formed from a modified cyclic dipeptide (Fox and Howlett, 2008; Gardiner *et al.*, 2005a). The toxicity and reactivity of gliotoxin is mediated by the disulfide bridge, which can cross-link proteins through reaction with thiol-containing cysteine residues (Patron *et al.*, 2007). Additionally, gliotoxin is a redox-active metabolite, generating ROS as it cycles between its oxidized (disulfide) and reduced (dithiol) forms (Gardiner and Howlett, 2005). Gliotoxin has been extensively studied for its toxicity properties towards mammalian cells, with antimicrobial activity also attributed to this molecule (Carberry *et al.*, 2012; Pardo *et al.*, 2006).

Gliotoxin biosynthesis

As mentioned previously, the genes for gliotoxin biosynthesis are clustered together, and the *gli* cluster is comprised of thirteen genes (Gardiner and Howlett, 2005; Schrettl *et al.*, 2010). Functional genomics demonstrated that the NRPS, GliP, is essential for the production of gliotoxin (Cramer *et al.*, 2006; Kupfahl *et al.*, 2006; Spikes *et al.*, 2008) (Fig. 8.3). The initial step in the biosynthesis of gliotoxin was subsequently elucidated, through heterologous expression and purification of GliP from *Escherichia coli* (Balibar and Walsh, 2006). GliP was observed to catalyse a condensation reaction between L-phenylalanine and L-serine, resulting in the formation of a dipeptide (Balibar and Walsh, 2006) (Fig. 8.3). GliP is the 'backbone' enzyme of the *gli* cluster and is multi-modular, a feature of NRPSs. Release of the dipeptide from the GliP enzyme occurs non-enzymatically, due to the absence of a thioesterase domain on the NRPS, and results in a cyclic diketopiperazine molecule (Balibar and Walsh, 2006; Davis *et al.*, 2011).

An integral stage in the formation of gliotoxin is the introduction of the sulfur atoms that confer the signature redox activity to gliotoxin. Since neither phenylalanine nor serine are sulfur-containing amino acids, other molecular species were proposed as potential sulfur donors, including cysteine, methionine and sodium sulfate (Gardiner *et al.*, 2005a; Suhadolnik and

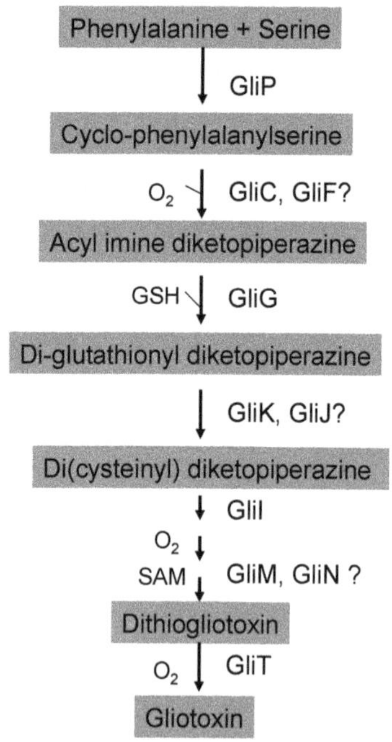

Figure 8.3 Schematic overview of the gliotoxin biosynthetic pathway in *A. fumigatus*. The NRPS GliP forms a diketopiperazine which is activated via GliC and possibly GliF monooxygenases to a reactive acyl imine intermediate. The glutathione s-transferase, GliG, conjugates two GSH molecules to this intermediate, followed by proteolytic modification (possibly via GliJ and GliK), C-S lyase activity (GliI) and N-methylation (GliM or GliN(?)) to yield the dithiol form of gliotoxin. GliT effects disulfide bridge closure to produce gliotoxin.

Chenowath, 1958). A putative glutathione-S-transferase (GST) gene, *gliG*, is situated within the *gli* cluster, and presented a potential route for sulfur incorporation into the gliotoxin precursor (Gardiner *et al.*, 2005a). This hypothesis was confirmed through functional genomics whereby the *gliG* was deleted, resulting in abrogation of gliotoxin biosynthesis and accumulation of an off-pathway shunt metabolite (Davis *et al.*, 2011; Scharf *et al.*, 2011). Davis *et al.* (2011) validated the GST activity of the enzyme using recombinant GliG and demonstrated that GliG does not confer auto-protection against exogenous gliotoxin. Scharf *et al.* (2011) demonstrated the activity of the P450 monooxygenase, GliC, which is required for an hydroxylation reaction prior to the GliG-mediated conjugation of glutathione (GSH) to the gliotoxin precursor

(Fig. 8.3). The function of a number of GSTs relates to detoxification of non-polar compounds (i.e. xenobiotics) through conjugation of GSH, allowing subsequent metabolism of the glutathionylated molecule. While the proposed gliotoxin biosynthetic pathway partially resembles this mechanism, GliG is phylogenetically distinct from these detoxification enzymes (Davis *et al.*, 2011; Scharf *et al.*, 2011). Exposure of the thiol groups of gliotoxin was predicted to be catalysed by GliI utilizing pyridoxyl-5′-phosphate (PLP) as a co-factor (Davis *et al.*, 2011; Fox and Howlett, 2008; Scharf *et al.*, 2011) (Fig. 8.3). Recently, the activity of GliI as a C-S lyase was experimentally verified, and was demonstrated to perform concurrent cleavage for the formation of both thiols in the ETP (Scharf *et al.*, 2012). Identification of a cytosolic, water-soluble intermediate, exclusive to the *gliI* mutant (Δ*gliI*), enabled elucidation of the mechanism of action of this enzyme in gliotoxin biosynthesis (Scharf *et al.*, 2012). The final step in gliotoxin biosynthesis has also been elucidated and involves the oxidation of the dithiol form of gliotoxin to form the disulfide bridge (Scharf *et al.*, 2010; Schrettl *et al.*, 2010) (Fig. 8.3). This action is mediated by the enzyme, GliT, a gliotoxin oxidoreductase. While dithiol to disulfide oxidation is required to complete the biosynthesis of gliotoxin, GliT also exhibits NADPH-dependant gliotoxin reductase activity (Schrettl *et al.*, 2010). This mechanism is key in the auto-protection of *A. fumigatus* from the toxic effects exerted by gliotoxin through the disulfide bond. Indeed deletion of *gliT* resulted in a phenotype that was hypersensitive to exogenous gliotoxin, confirming the role of GliT in self-protection against gliotoxin (Scharf *et al.*, 2010; Schrettl *et al.*, 2010). Additionally, transformation of GliT into *A. nidulans* and *S. cerevisiae* conferred gliotoxin resistance to these typically sensitive species. Hlm1, a functionally homologous enzyme to GliT has been identified in *Streptomyces clavuligerus* and is associated with the production of the antibiotic, holomycin. As with GliT, Hlm1 is responsible for the oxidation of the dithiol in the final step of holomycin biosynthesis, and also confers protection against the toxic effects of this molecule (Li and Walsh, 2011).

The *gli* cluster also contains a selection of genes that are not directly involved in the gliotoxin biosynthetic process, namely the transporter, *gliA*, and the transcriptional regulator, *gliZ*. The *gliA* gene encodes a transporter of the major facilitator superfamily (MFS), while the homologous gene in the sirodesmin cluster of *L. maculans*, *sirA*, encodes an ATP binding cassette (ABC) transporter (Gardiner and Howlett, 2005). MFS transporters are more commonly found in fungal toxin biosynthetic clusters (e.g. aflatoxin) than ABC transporters (Gardiner *et al.*, 2005b). Disruption of the *gliA* homologue, *sirA*, in *L. maculans* resulted in increased secretion of sirodesmin, indicating that SirA is not solely responsible for efflux of sirodesmin from the cell. Despite this observation, Δ*sirA* demonstrated enhanced sensitivity to exogenous sirodesmin and gliotoxin, relative to the parent strain (Gardiner *et al.*, 2005b). Furthermore, complementation of Δ*sirA* with *gliA* resulted in acquired tolerance to gliotoxin but not sirodesmin. This indicates that the MFS transporter, GliA, specifically imparts auto-protection against gliotoxin through its action as an efflux pump.

Regulation of the *gli* cluster expression is controlled by the global transcription factor, LaeA (Bok and Keller, 2004). Additionally, a narrow range transcriptional regulator, GliZ, is present within the cluster and regulates the expression of the cluster genes in response to various stimuli (Bok *et al.*, 2006; Schrettl *et al.*, 2010). GliZ is a Zn_2Cys_6 transcription factor and deletion of the *gliZ* gene results in abolition of gliotoxin biosynthesis due to loss of expression of a number of *gli* cluster genes, including *gliA*, *gliG* and *gliI* (Bok *et al.*, 2006; Schrettl *et al.*, 2010). Exogenous gliotoxin induces *gli* cluster expression, a process mediated by GliZ (Schrettl *et al.*, 2010). In the absence of *gliZ*, gliotoxin is unable to induce expression of cluster components such as *gliG* or *gliA*; however, *gliT* expression is regulated independently of *gliZ* (Schrettl *et al.*, 2010). While deletion of *gliZ* resulted in reduced virulence of *A. fumigatus* in an insect model of infection, Δ*gliZ* demonstrated no significant change in pathogenicity in a murine IA infection model relative to the parent strain (Bok *et al.*, 2006; Schrettl *et al.*, 2010). This is in contrast to deletion of the global SM regulator, *laeA*, which resulted in reduced virulence owing to the disruption of SM biosynthesis (Bok *et al.*, 2005, 2006). Furthermore, supernatants from both Δ*laeA* and Δ*gliZ* triggered significantly less apoptosis of PMN *in vitro*, relative to the respective parent strains (Bok *et al.*, 2006). Recently, differential metabolomic analysis has resulted in the identification of a number of *gliZ*-dependent metabolites (Forseth *et al.*, 2011). While many of these molecules represent shunt metabolites of the gliotoxin biosynthesis pathway, they may elaborate on the mechanisms involved in gliotoxin biosynthesis (Forseth *et al.*, 2011). Several of the sulfurized *gliZ*-dependent metabolites identified were S-methylated, including the previously characterized bisdethiobis(methylthio)gliotoxin, with a methyl group bound to each of the thiol groups on the piperazine ring (Forseth *et al.*, 2011). While the significance of this thiol methylation is not yet understood, this may represent a mechanism by which to reduce the toxicity of gliotoxin and associated intermediates in *A. fumigatus*, through capping of the reactive moieties (Li *et al.*, 2012).

To date, the putative functions of the other members of the *gli* cluster have not been experimentally validated. While both *gliC* and *gliF* genes are predicted to encode cytochrome P450 monooxygenases, the hydroxylation of the diketopiperazine, prior to C–S bond formation, does not appear to involve GliF (Scharf *et al.*, 2011). GliM and GliN have putative assigned functions as an O-methyltransferase and a methyltransferase, respectively, both genes have homologues in the sirodesmin biosynthetic cluster (Gardiner and Howlett, 2005). Transfer of methyl groups to gliotoxin precursors is likely mediated through the methyl donor S-adenosylmethionine (SAM) (Davis *et al.*, 2011) (Fig. 8.3). Additionally, a GliM homologue is also found in a second smaller putative ETP biosynthetic cluster in *A. fumigatus* (Kremer *et al.*, 2007; Patron *et al.*, 2007). A putative dipeptidase, GliJ, may act by cleavage of L-glutamate groups from the bis-glutathionylated intermediate or alternatively may release the cyclic peptide, tethered to the GliP enzyme (Davis *et al.*, 2011; Fox and Howlett, 2008; Scharf *et al.*, 2011). Deletion of *gliK*, a putative γ-glutamylcyclotransferase (Fig. 8.3), abrogates gliotoxin biosynthesis, increases mutant sensitivity to exogenous gliotoxin, and exhibits significantly increased levels of ergothioneine – a redox-active species (Gallagher *et al.*, 2012). An unknown function protein, GliH, is also present in the *gli* cluster and has been demonstrated to be involved in gliotoxin biosynthesis but is not involved in self-protection against gliotoxin (Schrettl *et al.*, 2010).

Gliotoxin: effects on host cells

Gliotoxin has been shown to elicit a range of modulatory mechanisms on immune and non-immune cells, both *in vitro* and *in vivo* (Kwon-Chung and Sugui, 2009; Pardo *et al.*, 2006). Induction of mammalian cell apoptosis has been associated with gliotoxin, and this effect has been demonstrated in a range of cells from the immune system. PMNLs undergo apoptosis following incubation with gliotoxin *in vitro*, in both hypoxic and normoxic conditions (Dyugovskaya *et al.*, 2011). Apoptosis of monocytes, the precursors of macrophages, was also induced by gliotoxin (Orciuolo *et al.*, 2007; Stanzani *et al.*, 2005). Culture supernatants from *A. fumigatus* mutants, deficient in gliotoxin production (e.g. Δ*gliP*, Δ*gliZ*), induced significantly less apoptosis in PMNLs and macrophage-like cells (Bok *et al.*, 2006; Kupfahl *et al.*, 2006; Sugui *et al.*, 2007b). Gliotoxin-induced apoptosis is mediated by a member of the proapoptotic Bcl-2 family, Bak (Pardo *et al.*, 2006). The Bak protein is localized in the mitochondria of mammalian cells and elicits initiation of apoptosis following a conformational change in response to cell-damage stimuli (Griffiths *et al.*, 1999). Subsequent oligomerization of the Bak protein occurs and creates a pore in the mitochondrial membrane resulting in membrane permeability (Dewson *et al.*, 2009). Gliotoxin interacts with the Bak protein to cause a conformational change and consequently leads to ROS production and disruption of the mitochondrial membrane (Pardo *et al.*, 2006). Interestingly oligomerization of the Bak protein involves cysteine linkages (Dewson *et al.*, 2009) and may represent the mode of action for gliotoxin activation of this process. Following activation of Bak, release of other proapoptotic factors, including cytochrome c and apoptosis-inducing factor (AIF), from the mitochondria ultimately results in cell death. This mechanism contributes to the virulence of the fungus, with mice deficient in Bak exhibiting reduced mortality in a corticosteroid model of IA (Pardo *et al.*, 2006).

Another mechanism employed by gliotoxin, to suppress the immune response to *A. fumigatus*, is the inhibition of the NADPH-dependent oxidative burst from PMN cells such as neutrophils. Culture supernatants from gliotoxin-producing strains of *A. fumigatus* inhibited the oxidative burst of human neutrophils, while gliotoxin-deficient mutants had no inhibitory effects (Sugui *et al.*, 2007b). ROS generation is achieved through NADPH oxidase, which generates O_2^- from molecular oxygen. Release of ROS from immune effector cells contributes to the killing of invading microorganisms, and defects in NADPH oxidase (e.g. CGD) result in increased incidence of IFIs (Henriet *et al.*, 2012). The components of NADPH oxidase are dispersed between the cytosol and the membrane in resting cells. Upon stimulation the cytosolic elements ($p47^{phox}$, $p67^{phox}$, $p40^{phox}$ and Rac2) translocate to the membrane and assemble with flavocytochrome b_{558}. Gliotoxin inhibits the assembly and activation of the NADPH oxidase enzyme through disruption of the translocation of $p47^{phox}$, $p67^{phox}$ and $p40^{phox}$ to the membrane (Tsunawaki *et al.*, 2004). Furthermore, gliotoxin directly interacts with flavocytochrome b_{558}, inhibiting electron transport capabilities prior to assembly of the oxidase enzyme (Nishida *et al.*, 2005). These authors postulated that the mechanism of inhibition of flavocytochrome activity was through the reaction of gliotoxin with available cysteines on the protein. In this way gliotoxin prevents the activation of NADPH oxidase but is not effective in inhibiting the activity of the assembled enzyme (Tsunawaki *et al.*, 2004). The action of gliotoxin is mediated through the disulfide bridge on the ETP molecule, and blocking the thiol groups of gliotoxin negated the effect of the metabolite on NADPH activation (Tsunawaki *et al.*, 2004).

Angiogenesis, the formation of new blood vessels from existing ones, is a feature of IA and occurs possibly in response to tissue hypoxia associated with invasion of the pulmonary vasculature (Ben-Ami *et al.*, 2010). Following interaction with *A. fumigatus* hyphae, endothelial cells release proinflammatory cytokines, including TNFα and IL-8, and this event is not affected by the presence or absence of gliotoxin (Chiang *et al.*, 2008). The activation of pro-angiogenic signalling pathways is elicited by these cytokines, resulting in the induction of vascular endothelial growth factor (VEGF) and basic fibroblast growth factor (bFGF) (Yoshida *et al.*, 1997). Cytokine-mediated recruitment of PMN cells to the site of infection results in the release of ROS, and consequently, induction of NF-κβ. Further up-regulation of pro-angiogenic molecules is induced by NF-κβ. ROS, including H_2O_2, derived from NADPH oxidase, are important mediators of angiogenesis (Ushio-Fukai and Alexander, 2004). Gliotoxin performs an anti-angiogenic function through reduction of ROS generation by PMN cells and inhibition of NF-κβ. Gliotoxin also has the potential to act as an antioxidant, reducing intracellular H_2O_2 to H_2O via the mammalian thioredoxin redox system, hence eliminating a potent inducer of NF-κβ and angiogenesis (Choi *et al.*, 2007). By replacing the function of 2-cys peroxiredoxin, gliotoxin reduces H_2O_2 in a dose-dependent manner, thus preventing H_2O_2-induced angiogenesis. Redox-cycling between the oxidized and reduced forms of gliotoxin is integral to this action and involves the transfer of electrons from NADPH to gliotoxin (Choi *et al.*, 2007). This suggests a dual-function for gliotoxin, in production and neutralization of ROS, and further investigation of this dichotomy is required to understand the role of gliotoxin in *A. fumigatus* and pathogenesis. Despite the pro-angiogenic signals elicited by *A. fumigatus* infection, culture filtrates were demonstrated to have potent anti-angiogenic properties *in vitro* (Ben-Ami *et al.*, 2009). This anti-angiogenic activity is attributable to SM in *A. fumigatus*, as evidenced by the loss of this activity from the Δ*laeA* mutant. Using a gliotoxin-deficient mutant, Δ*gliP*, these authors verified that gliotoxin was responsible for approximately 40% of the LaeA-dependent anti-angiogenic activity of *A. fumigatus* (Ben-Ami *et al.*, 2009, 2010).

A number of other immunomodulatory effects are attributable to gliotoxin and may contribute to the virulence of *A. fumigatus in vivo*. These functions include inhibition of T-cell responses, impairment of phagocytosis and reduction in ciliary movement of epithelial cells (Abad *et al.*, 2010). Gliotoxin interferes with human T-cell activation, suppressing the response of these cells to antigens. This action is accomplished by inhibition of antigen presentation as opposed to directly impairing T-cell function (Stanzani *et al.*, 2005). Gliotoxin also impedes the activities of cytotoxic T lymphocytes (CTLs) through blocking the binding of CTLs to target cells (Yamada *et al.*, 2000). Again, the disulfide bridge of gliotoxin is indispensable for this process, as reduced gliotoxin (i.e. dithiol form) is unable to inhibit CTL-mediated cytotoxicity (Yamada *et al.*, 2000). The influence of gliotoxin on phagocytosis has also been highlighted, with low concentrations of gliotoxin significantly inhibiting the phagocytic capacity of human PMN cells (Coméra *et al.*, 2007; Orciuolo *et al.*, 2007). It was also noted that gliotoxin promoted cytoskeleton reorganization, with F-actin collapse noted around nuclei of affected cells, however it was proposed that these events were independent (Coméra *et al.*, 2007). Through inhibition of phagocytosis, gliotoxin could contribute to the persistence of *A. fumigatus* in infected tissue. Additionally, reduction in the rate of ciliary movement, as caused by gliotoxin, could result in impaired clearance of inhaled conidia from the lung and reduced capacity to keep conidia from reaching the epithelium (Amitani *et al.*, 1995).

The variety of host mechanisms affected by gliotoxin is due to the disulfide bridge across the piperazine ring. This confers the ability to cross-link proteins or other molecules via thiol residues and can result in antagonism of active sites or induction of conformational changes, disrupting protein function (Hurne *et al.*, 2000; Srinivasan *et al.*, 2006; Waring *et al.*, 1995). Furthermore, redox cycling between the reduced (dithiol) and oxidized (disulfide) forms of gliotoxin can produce ROS, with deleterious effects (Bernardo *et al.*, 2003; Gardiner *et al.*, 2005a; Kwon-Chung and Sugui, 2009). As discussed, ROS reducing activity is also possible through incorporation into the thioredoxin redox system (Choi *et al.*, 2007), contributing to the multifaceted role of gliotoxin. An increase in the potency of gliotoxin is observed upon reduction of mammalian cell density, as a result of active concentration of the toxin within cells (Bernardo *et al.*, 2003). The oxidized (disulfide) form of gliotoxin can gain entry into cells while the reduced (dithiol) form cannot permeate cells (Bernardo *et al.*, 2003). This mechanism forms the basis of gliotoxin accumulation in cells and the consequent execution of toxic effects. Oxidized gliotoxin, is subsequently reduced by intracellular GSH leading to concentration of reduced gliotoxin within the cell. The ensuing depletion of glutathione leads to reversion of gliotoxin to the oxidized form and efflux is restored. Glutathione-dependent accumulation of gliotoxin correlated with an increase in mammalian cell apoptosis and hence defines a route by which this metabolite exerts its effects (Bernardo *et al.*, 2003). This process allows recycling of gliotoxin from apoptotic cells and increases the efficiency of cytotoxicity. Bernardo *et al.* (2003) also postulated that the accrual of gliotoxin intracellularly could be caused by other cellular reductants, in addition to glutathione, indicating intracellular redox potential as a moderator of gliotoxin susceptibility.

Despite the range of effects of gliotoxin on mammalian cells, the contribution of gliotoxin to the pathogenicity of *A. fumigatus* has been debated. Mouse models infected with gliotoxin-deficient strains of *A. fumigatus* displayed divergent results, making the definition of gliotoxin as a virulence factor tentative. The immunosuppression regime utilized for the IA infection model appears to be of paramount importance in assessing the effect of gliotoxin on the pathogenicity of *A. fumigatus*. Mice treated with cyclophosphamide (inducing neutropenia) coupled with corticosteroids demonstrated comparable mortality rates when infected with gliotoxin-deficient strains of *A. fumigatus* (Δ*gliP*, Δ*gliZ*), relative to the respective gliotoxin-producing parent strains (Bok *et al.*, 2006; Cramer *et al.*, 2006; Kupfahl *et al.*, 2006; Spikes *et al.*, 2008). These results appeared to indicate that gliotoxin did not contribute significantly to the development of IA in the absence of neutrophils. Conversely, when corticosteroids were used alone for immunosuppression, gliotoxin-deficient strains of *A. fumigatus* displayed significantly reduced pathogenicity compared to the wild-type strains (Spikes *et al.*, 2008; Sugui *et al.*, 2007b). This observation correlates with the effects of gliotoxin on neutrophils and suggests that gliotoxin contributes to virulence in a non-neutropenic setting but may be dispensable in a more severely immunocompromised host.

Proteases, thermotolerance and unfolded protein response

As *A. fumigatus* is a saprophytic fungus, the ability to extract nutrients from the host plays an integral part in the ability of the fungus to survive and persist in the lung. Secreted enzymes, including proteases, hydrolases and lipases, allow acquisition of nutrients, in addition to destruction of host barriers to enable invasive growth. A transcriptional regulator of proteases, PrtT, was detected in *A. fumigatus* following the identification of homologues in *A. niger* and *A. oryzae* (Bergmann *et al.*, 2009; Sharon *et al.*, 2009). While PrtT-regulated proteases were essential for the utilization of protein as a nutrient source, these enzymes did not contribute to virulence in neutropenic or corticosteroid mouse models of IA (Bergmann *et al.*, 2009; Sharon *et al.*, 2009).

Through functional genomics investigations a number of genes connected to thermotolerance in *A. fumigatus* were identified, including *afpmt1* (AFUA_3G06450), *afmnt1* (AFUA_5G10760), *cgrA* (AFUA_8G02750), *thtA* (AFUA_1G03992) and *midA* (AFUA_3G10960) (Bhabhra *et al.*, 2004; Chang *et al.*, 2004; Dichtl *et al.*, 2012; Wagener *et al.*, 2008; Zhou *et al.*, 2007). The genes *afmnt1*, *cgrA*, *thtA* and *midA* are essential for growth at 48°C, while conidiation at 50°C is inhibited upon deletion of *afpmt1*. A range of distinctive mechanisms contribute to attenuation of thermotolerance in these deletion strains. CprA plays a role in ribosomal synthesis, and is required for conidial germination at higher temperatures. *Afpmt1* and *afmnt1* code for an α-1,2-mannosyltransferase and an O-mannosyltransferase respectively, with deletion of these genes resulting in loss of cell wall integrity (Wagener *et al.*, 2008; Zhou *et al.*, 2007). This implicates the cell wall as an integral component in resistance to thermal stress. Comparative proteomics has also identified mechanisms that are differentially regulated upon exposure to heat shock. Protein chaperones, proteins involved in carbon and nitrogen metabolism, translation, and proteins involved in defence against oxidative and nitrosative stress, were all significantly upregulated upon temperature shift from 30°C to 48°C (Albrecht *et al.*, 2010). Proteins involved in cell wall and cytoskeleton assembly were also observed to increase in expression following thermal stress, further demonstrating the importance of cell wall integrity in resistance to stress.

The endoplasmic reticulum (ER) provides the location for the majority of the protein folding events in the cell. Proteins are translocated into the lumen of the ER in an unfolded state before, with the aid of ER-resident chaperones, folding enzymes and post-translational modifications, acquisition of 3D conformation (Malhotra and Kaufman, 2007; Richie *et al.*, 2009). The levels of unfolded proteins in the ER needs to be carefully monitored as unfolded proteins can form toxic aggregates that interfere with the function of normal proteins (Richie *et al.*, 2009). Disruption to the redox balance (e.g. under oxidative stress conditions) can also lead to a build up of unfolded proteins which ultimately results in ER stress (Malhotra and Kaufman, 2007). ER stress results from an imbalance between the amount of unfolded proteins in the ER and the capacity of the ER to deal with them (Ron and Walter, 2007). In the event of ER stress, the unfolded protein response (UPR) is initiated (Ron and Walter, 2007). The UPR is a conserved eukaryotic signalling pathway in the ER which transmits information regarding the protein folding capacity of the ER to the nucleus (Richie *et al.*, 2009).

The level of unfolded proteins is detected by a bifunctional protein call Ire1p (Malhotra and Kaufman, 2007). Under non-stressed ER conditions, Ire1p is bound to the protein chaperone BiP and is maintained in an inactive state (Malhotra and Kaufman, 2007). Under ER stress conditions, the accumulated misfolded proteins bind to BiP and Ire1p is released in an active state (Malhotra and Kaufman, 2007). Ire1p possesses both a protein kinase and an endoribonuclease (Ma and Hendershot, 2001; Ron and Walter, 2007). The endoribonuclease excises an intron from the only substrate for Ire1p, *hac1*, and in doing so enables translation of this transcription factor which activates UPR target genes (Ma and Hendershot, 2001; Malhotra and Kaufman, 2007; Ron and Walter, 2007). The endoribonuclease activity of

Ire1p is regulated by the intrinsic kinase module (Ron and Walter, 2007). Higher eukaryotes also contain two additional ER stress transducers; AFT6 and PERK (Malhotra and Kaufman, 2007; Ron and Walter, 2007), the functions of which will not be discussed here.

Activation of the UPR reduces protein synthesis and translocation to the ER, consequently reducing the amount of unfolded proteins entering the ER (Ron and Walter, 2007). In addition to this, up-regulation of UPR target genes involved in the ER protein-folding machinery increases the capacity of the ER to accommodate unfolded proteins (Ron and Walter, 2007). Targets of the UPR also includes genes involved in ER-associated protein degradation (ERAD) (Malhotra and Kaufman, 2007; Ron and Walter, 2007). ERAD is involved in the retro-translocation of unfolded proteins from the ER lumen to the cytosol where they are degraded by the ubiquitin-proteasome system (Goeckeler and Brodsky, 2010). In *A. fumigatus* the UPR is required for virulence and antifungal tolerance (Richie *et al.*, 2009). An *A. fumigatus* deletion strain for *hacA*, the orthologue of *hac1*, displayed increased sensitivity to antifungals, and also had attenuated virulence in different mouse models of IA (Richie *et al.*, 2009).

Concluding remarks

Major advances in our understanding of the biology of *Aspergillus fumigatus* have been forthcoming, especially since the genome sequence of the opportunistic pathogen became available. This, allied to the development of a plethora of new molecular and cellular tools have allowed dissection of the sophisticated and surprising strategies deployed by the organism to cause disease, and avoid the host immune response. New diagnostic strategies and biomarkers are now *in situ*, which, allied to translation of biological research findings into improved therapies, should significantly improve the treatments available to combat the insidious nature of infection caused by *A. fumigatus*, and related *Aspergilli*.

Acknowledgements

Aspergillus fumigatus research in the Corresponding Authors laboratory is funded by a Science Foundation Ireland Principal Investigator Award (SFI/PI/1188).

References

Abad, A., Fernández-Molina, J.V., Bikandi, J., Ramírez, A., Margareto, J., Sendino, J., Luis Hernando, F., Pontón, J., Garaizar, J., and Rementeria, A. (2010). What makes *Aspergillus fumigatus* a successful pathogen? Genes and molecules involved in invasive aspergillosis. Rev. Iberoam. Micol. *27*, 155–182.

Agarwal, R. (2011). Allergic bronchopulmonary aspergillosis: Lessons for the busy radiologist. World J. Radiol. *3*, 178–181.

Agarwal, R., Hazarika, B., Gupta, D., Aggarwal, A.N., Chakrabarti, A., and Jindal, S.K. (2010). *Aspergillus* hypersensitivity in patients with chronic obstructive pulmonary disease: COPD as a risk factor for ABPA? Med. Mycol. *48*, 988–994.

Aguirre, J., Hansberg, W., and Navarro, R. (2006). Fungal responses to reactive oxygen species. Med. Mycol. *44*, 101–107.

Aimanianda, V., Bayry, J., Bozza, S., Kniemeyer, O., Perruccio, K., Elluru, S.R., Clavaud, C., Paris, S., Brakhage, A.A., Kaveri, S.V., *et al.* (2009). Surface hydrophobin prevents immune recognition of airborne fungal spores. Nature *460*, 1117–1121.

Albrecht, D., Guthke, R., Brakhage, A.A., and Kniemeyer, O. (2010). Integrative analysis of the heat shock response in *Aspergillus fumigatus*. BMC Genomics *11*, 32.

Amitani, R., Taylor, G., Elezis, E., Llewellyn-jones, C., Mitchell, J., Kuze, F., Cole, P.J., and Wilson, R. (1995). Purification and characterization of factors produced by *Aspergillus fumigatus* which affect human ciliated respiratory epithelium. Infect. Immun. *63*, 3266–3271.

Antinori, S., Nebuloni, M., Magni, C., Fasan, M., Adorni, F., Viola, A., Corbellino, M., Galli, M., Vago, G., Parravicini, C., *et al.* (2009). Trends in the postmortem diagnosis of opportunistic invasive fungal infections in patients with AIDS: a retrospective study of 1,630 autopsies performed between 1984 and 2002. Am. J. Clin. Pathol. *132*, 221–227.

Anzai, K., Mayuzumi, S., Nakashima, T., Sato, H., Inaba, S., Park, J.-Y., Kuwahara, N., Suzuki, R., Utsumi, N., Yokoyama, F., *et al.* (2008). Comparison of groupings among members of the genus *Aspergillus* based on phylogeny and production of bioactive compounds. Biosci. Biotechnol. Biochem. 72, 2199–2202.

Azie, N., Neofytos, D., Pfaller, M., Meier-Kriesche, H.-U., Quan, S.-P., and Horn, D. (2012). The PATH (Prospective Antifungal Therapy) Alliance® registry and invasive fungal infections: update 2012. Diagn. Microbiol. Infect. Dis. 73, 293–300.

Balibar, C.J., and Walsh, C.T. (2006). GliP, a multimodular nonribosomal peptide synthetase in *Aspergillus fumigatus*, makes the diketopiperazine scaffold of gliotoxin. Biochemistry 45, 15029–15038.

Balloy, V., Huerre, M., Latgé, J., and Chignard, M. (2005). Differences in Patterns of Infection and Inflammation for Corticosteroid Treatment and Chemotherapy in Experimental Invasive Pulmonary Aspergillosis. Infect. Immun. *73*, 494–503.

Barker, B.M., Kroll, K., Vödisch, M., Mazurie, A., Kniemeyer, O., and Cramer, R.A. (2012). Transcriptomic and proteomic analyses of the *Aspergillus fumigatus* hypoxia response using an oxygen-controlled fermenter. BMC Genomics *13*, 62.

Bayram, Ö., Krappmann, S., Ni, M., Bok, J.W., Helmstaedt, K., Yu, J., and Braus, G.H. (2008). VelB/VeA/LaeA Complex Coordinates Light Signal with Fungal Development and Secondary Metabolism. Science *320*, 1504–1506.

Beauvais, A., and Latgé, J.P. (2001). Membrane and cell wall targets in *Aspergillus fumigatus*. Drug Resist. Updates *4*, 38–49.

Beauvais, A., Drake, R., Ng, K., Diaquin, M., and Latgé, J.P. (1993). Characterization of the 1,3-beta-glucan synthase of *Aspergillus fumigatus*. J. Gen. Microbiol. *139*, 3071–3078.

Beever, R.E., and Dempsey, G.P. (1978). Function of rodlets on the surface of fungal spores. Nature *272*, 608–610.

Ben-Ami, R., Lewis, R.E., Leventakos, K., and Kontoyiannis, D.P. (2009). *Aspergillus fumigatus* inhibits angiogenesis through the production of gliotoxin and other secondary metabolites. Blood *114*, 5393–5399.

Ben-Ami, R., Lewis, R.E., and Kontoyiannis, D.P. (2010). Enemy of the (immunosuppressed) state: an update on the pathogenesis of *Aspergillus* fumigatus infection. Br. J. Haematol. *150*, 406–417.

Bergmann, A., Hartmann, T., Cairns, T., Bignell, E.M., and Krappmann, S. (2009). A regulator of *Aspergillus fumigatus* extracellular proteolytic activity is dispensable for virulence. Infect. Immun. *77*, 4041–4050.

Bernardo, P.H., Brasch, N., Chai, C.L.L., and Waring, P. (2003). A novel redox mechanism for the Glutathione-dependent reversible uptake of a fungal toxin in cells. J. Biol. Chem. *278*, 46549–46555.

Bhabhra, R., and Askew, D.S. (2005). Thermotolerance and virulence of *Aspergillus fumigatus*: role of the fungal nucleolus. Med. Mycol. *43*, 87–93.

Bhabhra, R., Miley, M.D., Mylonakis, E., Boettner, D., Fortwendel, J., Panepinto, J.C., Postow, M., Rhodes, J.C., and Askew, D.S. (2004). Disruption of the *Aspergillus fumigatus* Gene Encoding Nucleolar Protein CgrA Impairs Thermotolerant Growth and Reduces Virulence. Infect. Immun. *72*, 4731–4740.

Bhatia, S., Fei, M., Yarlagadda, M., Qi, Z., Akira, S., Saijo, S., Iwakura, Y., Van Rooijen, N., Gibson, G.A., St Croix, C.M., *et al.* (2011). Rapid host defense against *Aspergillus fumigatus* involves alveolar macrophages with a predominance of alternatively activated phenotype. PloS One *6*, e15943.

Bianchi, M., Hakkim, A., Brinkmann, V., Siler, U., Seger, R.A., Zychlinsky, A., and Reichenbach, J. (2009). Restoration of NET formation by gene therapy in CGD controls aspergillosis. Blood *114*, 2619–2622.

Bok, J.W., and Keller, N.P. (2004). LaeA, a Regulator of Secondary Metabolism in *Aspergillus* spp. Eukaryot. Cell *3*, 527–535.

Bok, J.W., Balajee, S.A., Marr, K.A., Andes, D., Nielsen, K.F., Frisvad, J.C., and Keller, N.P. (2005). LaeA, a Regulator of Morphogenetic Fungal Virulence Factors. Eukaryot. Cell *4*, 1574–1582.

Bok, J.W., Chung, D., Balajee, S.A., Marr, K.A., Andes, D., Nielsen, K.F., Frisvad, J.C., Kirby, K.A., and Keller, N.P. (2006). GliZ, a transcriptional regulator of gliotoxin biosynthesis, contributes to *Aspergillus fumigatus* virulence. Infect. Immun. *74*, 6761–6768.

Boutboul, F., Alberti, C., Leblanc, T., Sulahian, A., Gluckman, E., Derouin, F., and Ribaud, P. (2002). Invasive Aspergillosis in Allogeneic Stem Cell Transplant Recipients: Increasing Antigenemia Is Associated with Progressive Disease. Clin. Infect. Dis. *34*, 939–943.

Brakhage, A.A., and Langfelder, K. (2002). Menacing mold: the molecular biology of *Aspergillus fumigatus*. Annu. Rev. Microbiol. *56*, 433–455.

Brakhage, A.A., and Liebmann, B. (2005). *Aspergillus fumigatus* conidial pigment and cAMP signal transduction: significance for virulence. Med. Mycol. *43*, 75–82.

Brakhage, A.A., Bruns, S., Thywissen, A., Zipfel, P.F., and Behnsen, J. (2010). Interaction of phagocytes with filamentous fungi. Curr. Opin. Microbiol. *13*, 409–415.

Brik, A., Salem, A.M., Kamal, A.R., Abdel-Sadek, M., Essa, M., El Sharawy, M., Deebes, A., and Bary, K.A. (2008). Surgical outcome of pulmonary aspergilloma. Eur. J. Cardiothorac. Surg. *34*, 882–885.

Brinkmann, V., and Zychlinsky, A. (2007). Beneficial suicide: why neutrophils. Nat. Rev. Microbiol. *5*, 577–582.

Brouard, J., Knauer, N., Boelle, P.-Y., Corvol, H., Henrion-Caude, A., Flamant, C., Bremont, F., Delaisi, B., Duhamel, J.-F., Marguet, C., *et al.* (2005). Influence of interleukin-10 on *Aspergillus fumigatus* infection in patients with cystic fibrosis. J. Infect. Dis. *191*, 1988–1991.

Brown, G.D. (2006). Dectin-1: a signalling non-TLR pattern-recognition receptor. Nat. Rev. Immunol. *6*, 33–43.

Brown, A.J.P., Haynes, K., and Quinn, J. (2009). Nitrosative and oxidative stress responses in fungal pathogenicity. Curr. Opin. Microbiol. *12*, 384–391.

Bruns, S., Kniemeyer, O., Hasenberg, M., Aimanianda, V., Nietzsche, S., Thywissen, A., Jeron, A., Latgé, J.-P., Brakhage, A.A., and Gunzer, M. (2010a). Production of extracellular traps against *Aspergillus fumigatus in vitro* and in infected lung tissue is dependent on invading neutrophils and influenced by hydrophobin RodA. PLoS Pathogens *6*, e1000873.

Bruns, S., Seidler, M., Albrecht, D., Salvenmoser, S., Remme, N., Hertweck, C., Brakhage, A.A., Kniemeyer, O., and Müller, F.M. (2010b). Functional genomic profiling of *Aspergillus fumigatus* biofilm reveals enhanced production of the mycotoxin gliotoxin. Proteomics. *10*, 3097–3107.

Calvo, A.M., Wilson, R.A., Bok, J.W., and Keller, N.P. (2002). Relationship between Secondary Metabolism and Fungal Development Fungal Development. Microbiol. Mol. Biol. Rev. *66*, 447–459.

Camps, S.M.T., Van der Linden, J.W.M., Li, Y., Kuijper, E.J., Van Dissel, J.T., Verweij, P.E., and Melchers, W.J.G. (2012). Rapid induction of multiple resistance mechanisms in *Aspergillus fumigatus* during azole therapy: a case study and review of the literature. Antimicrob. Agents Chemother. *56*, 10–16.

Carberry, S., Molloy, E., Hammel, S., O'Keeffe, G., Jones, G.W., Kavanagh, K., and Doyle, S. (2012). Gliotoxin effects on fungal growth: Mechanisms and exploitation. Fungal Genet. Biol. *49*, 302–312.

Carvalho, A., Cunha, C., Bistoni, F., and Romani, L. (2012). Immunotherapy of aspergillosis. Clin. Microbiol. Infect. *18*, 120–125.

Casas López, J.L., Sánchez Pérez, J.A., Fernández Sevilla, J.M., Acién Fernández, F., Molina Grima, E., and Chisti, Y. (2003). Production of lovastatin by *Aspergillus terreus*: effects of the C:N ratio and the principal nutrients on growth and metabolite production. Enzyme. Microb. Technol. *33*, 270–277.

Chai, L.Y.A., Vonk, A.G., Kullberg, B.J., Verweij, P.E., Verschueren, I., Van der Meer, J.W.M., Joosten, L.A.B., Latgé, J.-P., and Netea, M.G. (2011a). *Aspergillus fumigatus* cell wall components differentially modulate host TLR2 and TLR4 responses. Microbes Infect. *13*, 151–159.

Chai, L.Y.A., Netea, M.G., Sugui, J., Vonk, A.G., Van de Sande, W.J., Warris, A., Kwon-chung, K.J., and Jan, B. (2011b). *Aspergillus fumigatus* Conidial Melanin Modulates Host Cytokine Response. Immunobiology *215*, 915–920.

Chambers, S.T., Bhandari, S., Scott-Thomas, A., and Syhre, M. (2011). Novel diagnostics: progress toward a breath test for invasive *Aspergillus fumigatus*. Med. Mycol. *49 (Suppl 1)*, S54–61.

Chamilos, G., Luna, M., Lewis, R.E., Bodey, G.P., Chemaly, R., Tarrand, J.J., Safdar, A., Raad, I.I., and Kontoyiannis, D.P. (2006). Invasive fungal infections in patients with hematologic malignancies in a tertiary care cancer center: an autopsy study over a 15-year period (1989–2003). Haematologica *91*, 986–989.

Chang, Y.C., Tsai, H.-F., Karos, M., and Kwon-Chung, K.J. (2004). THTA, a thermotolerance gene of *Aspergillus fumigatus*. Fungal Genet. Biol. *41*, 888–896.

Chaudhary, N., and Marr, K.A. (2011). Impact of *Aspergillus fumigatus* in allergic airway diseases. Clin. Transl. Allergy *1*, 4.

Chiang, L.Y., Sheppard, D.C., Gravelat, F.N., Patterson, T.F., and Filler, S.G. (2008). *Aspergillus fumigatus* stimulates leukocyte adhesion molecules and cytokine production by endothelial cells *in vitro* and during invasive pulmonary disease. Infect. Immun. *76*, 3429–3438.

Choi, H.S., Shim, J.S., Kim, J.-A., Kang, S.W., and Kwon, H.J. (2007). Discovery of gliotoxin as a new small molecule targeting thioredoxin redox system. Biochem. Biophys. Res. Commun. *359*, 523–528.

Coméra, C., André, K., Laffitte, J., Collet, X., Galtier, P., and Maridonneau-Parini, I. (2007). Gliotoxin from *Aspergillus fumigatus* affects phagocytosis and the organization of the actin cytoskeleton by distinct signalling pathways in human neutrophils. Microbes Infect. *9*, 47–54.

Coyle, C.M., Kenaley, S.C., Rittenour, W.R., and Panaccione, D.G. (2007). Association of ergot alkaloids with conidiation in *Aspergillus fumigatus*. Mycologia *99*, 804–811.

Cramer, R.A., Gamcsik, M.P., Brooking, R.M., Najvar, L.K., Kirkpatrick, W.R., Patterson, T.F., Balibar, C.J., Graybill, J.R., Perfect, J.R., Abraham, S.N., *et al.* (2006). Disruption of a nonribosomal peptide synthetase in *Aspergillus fumigatus* eliminates gliotoxin production. Eukaryot. Cell 5, 972–980.

Crosdale, D.J., Poulton, K.V., Ollier, W.E., Thomson, W., and Denning, D.W. (2001). Mannose-binding lectin gene polymorphisms as a susceptibility factor for chronic necrotizing pulmonary aspergillosis. J. Infect. Dis. *184*, 653–656.

Dagenais, T.R.T., and Keller, N.P. (2009). Pathogenesis of *Aspergillus fumigatus* in invasive Aspergillosis. Clin. Microbiol. Rev. *22*, 447–465.

Dagenais, T.R.T., Giles, S.S., Aimanianda, V., Latgé, J.-P., Hull, C.M., and Keller, N.P. (2010). *Aspergillus fumigatus* LaeA-mediated phagocytosis is associated with a decreased hydrophobin layer. Infect. Immun. *78*, 823–829.

Davis, C., Carberry, S., Schrettl, M., Singh, I., Stephens, J.C., Barry, S.M., Kavanagh, K., Challis, G.L., Brougham, D., and Doyle, S. (2011). The role of glutathione S-transferase GliG in gliotoxin biosynthesis in *Aspergillus fumigatus*. Chem. Biol. *18*, 542–552.

Denning, D.W., and Hope, W.W. (2010). Therapy for fungal diseases: opportunities and priorities. Trends Microbiol. *18*, 195–204.

De Vrankrijker, A.M.M., Van der Ent, C.K., Van Berkhout, F.T., Stellato, R.K., Willems, R.J.L., Bonten, M.J.M., and Wolfs, T.F.W. (2011). *Aspergillus fumigatus* colonization in cystic fibrosis: implications for lung function? Clin. Microbiol. Infect. *17*, 1381–1386.

Dewson, G., Kratina, T., Czabotar, P., Day, C.L., Adams, J.M., and Kluck, R.M. (2009). Bak activation for apoptosis involves oligomerization of dimers via their α6 helices. Mol. Cell *36*, 696–703.

Dichtl, K., Helmschrott, C., Dirr, F., and Wagener, J. (2012). Deciphering cell wall integrity signalling in *Aspergillus fumigatus*: identification and functional characterization of cell wall stress sensors and relevant Rho GTPases. Mol. Microbiol. *83*, 506–519.

Dinamarco, T.M., Almeida, R.S., De Castro, P.A., Brown, N.A., Dos Reis, T.F., Ramalho, L.N.Z., Savoldi, M., Goldman, M.H.S., and Goldman, G.H. (2012). Molecular characterization of the putative transcription factor SebA involved in virulence in *Aspergillus fumigatus*. Eukaryot. Cell *11*, 518–531.

Domingo, M.P., Colmenarejo, C., Martínez-Lostao, L., Müllbacher, A., Jarne, C., Revillo, M.J., Delgado, P., Roc, L., Meis, J.F., Rezusta, A., *et al.* (2012). Bis (methyl) gliotoxin proves to be a more stable and reliable marker for invasive aspergillosis than gliotoxin and suitable for use in diagnosis. Diagn. Microbiol. Infect. Dis. *73*, 57–64.

Doyle, J.M., Walshe, K. Gordon, N., Kavanagh, K., and Gallagher, L. Method for detecting infections. European Patent Application. Filed 16 May 2012. EP2012/059133.

Du, C., Sarfati, J., Latgé, J.-P., and Calderone, R. (2006). The role of the *sakA* (Hog1) and *tcsB* (sln1) genes in the oxidant adaptation of *Aspergillus fumigatus*. Med. Mycol. *44*, 211–218.

Dyugovskaya, L., Polyakov, A., Ginsberg, D., Lavie, P., and Lavie, L. (2011). Molecular pathways of spontaneous and TNFalpha-mediated neutrophil apoptosis under intermittent hypoxia. Am. J. Respir. Cell Mol. Biol. *45*, 154–162.

Eisendle, M., Oberegger, H., Zadra, I., and Haas, H. (2003). The siderophore system is essential for viability of *Aspergillus nidulans*: functional analysis of two genes encoding L-ornithine N5-monooxygenase (*sidA*) and a non-ribosomal peptide synthetase (*sidC*). Mol. Microbiol. *49*, 359–375.

Eisenhaber, B., Schneider, G., Wildpaner, M., and Eisenhaber, F. (2004). A sensitive predictor for potential GPI lipid modification sites in fungal protein sequences and its application to genome-wide studies for *Aspergillus nidulans, Candida albicans, Neurospora crassa, Saccharomyces cerevisiae* and *Schizosaccharomyc.* J. Mol. Biol. 337, 243–253.

Falagas, M.E., Vardakas, K.Z., and Samonis, G. (2008). Decreasing the incidence and impact of infections in neutropenic patients: evidence from meta-analyses of randomized controlled trials. Curr. Med. Res. Opin. *24*, 215–235.

Fallon, J.P., Reeves, E.P., and Kavanagh, K. (2010). Inhibition of neutrophil function following exposure to the *Aspergillus fumigatus* toxin fumagillin. J. Med. Microbiol. *59*, 625–633.

Fedorova, N.D., Khaldi, N., Joardar, V.S., Maiti, R., Amedeo, P., Anderson, M.J., Crabtree, J., Silva, J.C., Badger, J.H., Albarraq, A., *et al.* (2008). Genomic islands in the pathogenic filamentous fungus *Aspergillus fumigatus*. PLoS Genetics *4*, e1000046.

Feldmesser, M. (2006). Role of neutrophils in invasive aspergillosis. Infect. Immun. *74*, 6514–6516.

Fontaine, T., Simenel, C., Dubreucq, G., Adam, O., Delepierre, M., Lemoine, J., Vorgias, C.E., Diaquin, M., and Latgé, J.P. (2000). Molecular organization of the alkali-insoluble fraction of *Aspergillus fumigatus* cell wall. J. Biol. Chem. *275*, 27594–27607.

Forseth, R.R., Fox, E.M., Chung, D., Howlett, B.J., Keller, N.P., Frank, C., and Schroeder, F.C. (2011). Identification of cryptic products of the gliotoxin gene cluster using NMR-based comparative metabolomics and a model for gliotoxin biosynthesis. J. Am. Chem. Soc. *133*, 9678–9681.

Fox, E.M., and Howlett, B.J. (2008). Biosynthetic gene clusters for epipolythiodioxopiperazines in filamentous fungi. Mycol. Res. *112*, 162–169.

Frisvad, J.C., Rank, C., Nielsen, K.F., and Larsen, T.O. (2009). Metabolomics of *Aspergillus fumigatus*. Med. Mycol. *47 (Suppl 1)*, S53–71.

Fuchs, T.A., Abed, U., Goosmann, C., Hurwitz, R., Schulze, I., Wahn, V., Weinrauch, Y., Brinkmann, V., and Zychlinsky, A. (2007). Novel cell death program leads to neutrophil extracellular traps. J. Cell Biol. *176*, 231–241.

Gallagher, L., Owens, R.A., O' Keeffe, G., Dolan, S.K., Schrettl, M., Kavanagh, K., Jones, G., and Doyle, S. (2012). The *Aspergillus fumigatus* protein GliK protects against oxidative stress and is essential for gliotoxin biosynthesis. Eukaryot. Cell *11*, 1226–1238.

Gangneux, J.-P., Camus, C., and Philippe, B. (2010). Epidemiology of invasive aspergillosis and risk factors in non neutropaenic patients. Rev. Mal. Respir. *27*, e34–46.

Garbati, M.A., Alasmari, F.A., Al-Tannir, M.A., and Tleyjeh, I.M. (2012). The role of combination antifungal therapy in the treatment of invasive aspergillosis: a systematic review. Int. J. Infect. Dis. *16*, e76–81.

Gardiner, D.M., and Howlett, B.J. (2005). Bioinformatic and expression analysis of the putative gliotoxin biosynthetic gene cluster of *Aspergillus fumigatus*. FEMS Microbiol. Lett. *248*, 241–248.

Gardiner, D.M., Waring, P., and Howlett, B.J. (2005a). The epipolythiodioxopiperazine (ETP) class of fungal toxins: distribution, mode of action, functions and biosynthesis. Microbiology *151*, 1021–1032.

Gardiner, D.M., Jarvis, R.S., and Howlett, B.J. (2005b). The ABC transporter gene in the sirodesmin biosynthetic gene cluster of *Leptosphaeria maculans* is not essential for sirodesmin production but facilitates self-protection. Fungal Genet. Biol. *42*, 257–263.

Gauthier, T., Wang, X., Sifuentes Dos Santos, J., Fysikopoulos, A., Tadrist, S., Canlet, C., Artigot, M.P., Loiseau, N., Oswald, I.P., and Puel, O. (2012). Trypacidin, a spore-borne toxin from *Aspergillus fumigatus*, is cytotoxic to lung cells. PloS One 7, e29906.

Goeckeler, J.L., and Brodsky, J.L. (2010). Molecular chaperones and substrate ubiquitination control the efficiency of endoplasmic reticulum-associated degradation. Diabetes Obes. Metab. *12*, 32–38.

Graybill, J.R., Bocanegra, R., Najvar, L.K., Loebenberg, D., and Luther, M.F. (1998). Granulocyte colony-stimulating factor and azole antifungal therapy in murine aspergillosis: role of immune suppression. Antimicrob. Agents Chemother. *42*, 2467–2473.

Griffiths, G.J., Dubrez, L., Morgan, C.P., Jones, N.A., Whitehouse, J., Corfe, B.M., Dive, C., and Hickman, J.A. (1999). Cell damage-induced conformational changes of the pro-apoptotic protein Bak *in vivo* precede the onset of apoptosis. J. Cell Biol. *144*, 903–914.

De Groot, P.W.J., Hellingwerf, K.J., and Klis, F.M. (2003). Genome-wide identification of fungal GPI proteins. Yeast *20*, 781–796.

Haas, H. (2012). Iron – a key nexus in the virulence of *Aspergillus* fumigatus. Front. Microbiol. *3*, 28.

Hao, W., Pan, Y.-X., Ding, Y.-Q., Xiao, S., Yin, K., Wang, Y.-D., Qiu, L.-W., Zhang, Q.-L., Woo, P.C.Y., Lau, S.K.P., *et al.* (2008). Well-characterized monoclonal antibodies against cell wall antigen of *Aspergillus* species improve immunoassay specificity and sensitivity. CVI *15*, 194–202.

Henriet, S.S., Verweij, P.E., and Warris, A. (2012). *Aspergillus nidulans* and chronic granulomatous disease: a unique host-pathogen interaction. J. Infect. Dis. *206*, 1128–1137.

Herbrecht, R., Denning, D.W., Patterson, T.F., Bennett, J.E., Greene, R.E., Oestmann, J.-W., Kern, W.V., Marr, K.A., Ribaud, P., Lortholary, O., *et al.* (2002). Voriconazole versus amphotericin B for primary therapy of invasive aspergillosis. N. Engl. J. Med. *347*, 408–415.

Herbrecht, R., Maertens, J., Baila, L., Aoun, M., Heinz, W., Martino, R., Schwartz, S., Ullmann, A.J., Meert, L., Paesmans, M., *et al.* (2010). Caspofungin first-line therapy for invasive aspergillosis in allogeneic hematopoietic stem cell transplant patients: an European Organisation for Research and Treatment of Cancer study. Bone Marrow. Transplant. *45*, 1227–1233.

Hercus, T.R., Broughton, S.E., Ekert, P.G., Ramshaw, H.S., Perugini, M., Grimbaldeston, M., Woodcock, J.M., Thomas, D., Pitson, S., Hughes, T., *et al.* (2012). The GM-CSF receptor family: mechanism of activation and implications for disease. Growth Factors *30*, 63–75.

Hohl, T.M., Van Epps, H.L., Rivera, A., Morgan, L.A., Chen, P.L., Feldmesser, M., and Pamer, E.G. (2005). *Aspergillus fumigatus* triggers inflammatory responses by stage-specific beta-glucan display. PLoS Pathogens *1*, e30.

Hope, W.W., Walsh, T.J., and Denning, D.W. (2005). Laboratory diagnosis of invasive aspergillosis. Lancet Infect. Dis. *5*, 609–622.

Howard, S.J., Webster, I., Moore, C.B., Gardiner, R.E., Park, S., Perlin, D.S., and Denning, D.W. (2006). Multi-azole resistance in *Aspergillus fumigatus*. Int. J. Antimicrob. Agents *28*, 450–453.

Hurne, A.M., Chai, C.L.L., and Waring, P. (2000). Inactivation of rabbit muscle creatine kinase by reversible formation of an internal disulfide bond induced by the fungal toxin gliotoxin. J. Biol. Chem. *275*, 25202–25206.

Ibrahim-Granet, O., Philippe, B., Boleti, H., Boisvieux-Ulrich, E., Grenet, D., Stern, M., and Latgé, J.P. (2003). Phagocytosis and intracellular fate of *Aspergillus fumigatus* conidia in alveolar macrophages. Infect. Immun. *71*, 891–903.

Ishikawa, M., Ninomiya, T., Akabane, H., Kushida, N., Tsujiuchi, G., Ohyama, M., Gomi, S., Shito, K., and Murata, T. (2009). Pseurotin A and its analogues as inhibitors of immunoglobulin E production. Bioorg. Med. Chem. Lett. *19*, 1457–1460.

Jahn, B., Koch, A., Schmidt, A., Wanner, G., Gehringer, H., Bhakdi, S., and Brakhage, A.A. (1997). Isolation and characterization of a pigmentless-conidium mutant of *Aspergillus fumigatus* with altered conidial surface and reduced virulence. Infect. Immun. *65*, 5110–5117.

Jaillon, S., Peri, G., Delneste, Y., Frémaux, I., Doni, A., Moalli, F., Garlanda, C., Romani, L., Gascan, H., Bellocchio, S., *et al.* (2007). The humoral pattern recognition receptor PTX3 is stored in neutrophil granules and localizes in extracellular traps. J. Exp. Med. *204*, 793–804.

Jain, R., Valiante, V., Remme, N., Docimo, T., Heinekamp, T., Hertweck, C., Gershenzon, J., Haas, H., and Brakhage, A.A. (2011). The MAP kinase MpkA controls cell wall integrity, oxidative stress response, gliotoxin production and iron adaptation in *Aspergillus fumigatus*. Mol. Microbiol. *82*, 39–53.

Jarque, I., Tormo, M., Bello, J.L., Rovira, M., Batlle, M., Julia, A., Tabares, S., Rivas, C., Capote, F.J., and Sanz, M.A. (2013). Caspofungin for the treatment of invasive fungal disease in hematological patients (ProCAS Study). Med. Mycol. *51*, 150–154.

Jeans, A.R., Howard, S.J., Al-Nakeeb, Z., Goodwin, J., Gregson, L., Warn, P.A., and Hope, W.W. (2012). Combination of voriconazole and anidulafungin for the treatment of triazole resistant *Aspergillus fumigatus* in an *in vitro* model of invasive pulmonary aspergillosis. Antimicrob. Agents Chemother. *56*, 5180–5185.

Jegorov, A., Hajduch, M., Sulc, M., and Havlicek, V. (2006). Nonribosomal cyclic peptides: specific markers of fungal infections. J. Mass Spectrom. *41*, 563–576.

Johnson, L.B., and Kauffman, C.A. (2003). Voriconazole: a new triazole antifungal agent. Clin. Infect. Dis. *36*, 630–637.

Johnson, G.L., Bibby, D.F., Wong, S., Agrawal, S.G., and Bustin, S.A. (2012). A MIQE-Compliant Real-Time PCR Assay for *Aspergillus* Detection. PLoS ONE *7*, e40022.

Kartsonis, N.A., Saah, A.J., Joy Lipka, C., Taylor, A.F., and Sable, C.A. (2005). Salvage therapy with caspofungin for invasive aspergillosis: results from the caspofungin compassionate use study. J. Infect. *50*, 196–205.

Kawamura, S., Maesaki, S., Tomono, K., Tashiro, T., and Kohno, S. (2000). Clinical evaluation of 61 patients with pulmonary aspergilloma. Intern. Med. *39*, 209–212.

Keller, N.P., Turner, G., and Bennett, J.W. (2005). Fungal secondary metabolism – from biochemistry to genomics. Nat. Rev. Microbiol. *3*, 937–947.

Keller, N., Bok, J., Chung, D., Perrin, R.M., and Keats Shwab, E. (2006). LaeA, a global regulator of *Aspergillus* toxins. Med. Mycol. *44*, 83–85.

Khaldi, N., Seifuddin, F.T., Turner, G., Haft, D., Nierman, W.C., Wolfe, K.H., and Fedorova, N.D. (2010). SMURF: genomic mapping of fungal secondary metabolite clusters. Fungal Genet. Biol. *47*, 736–741.

Kleinberg, M. (2006). What is the current and future status of conventional amphotericin B? Int. J. Antimicrob. Agents *27 (Suppl 1)*, 12–16.

Knutsen, A.P. (2006). Genetic and respiratory tract risk factors for aspergillosis: ABPA and asthma with fungal sensitization. Med. Mycol. *44*, 61–70.

Knutsen, A.P., and Slavin, R.G. (2011). Allergic bronchopulmonary aspergillosis in asthma and cystic fibrosis. Clin. Dev. Immunol. *2011*, 843763.

Kollar, R., Petrakova, E., Ashwell, G., Robbins, P.W., and Cabib, E. (1995). Architecture of the yeast cell wall – the linkage between chitin and beta(1- 3)-glucan. J. Biol. Chem. *270*, 1170–1178.

Kontoyiannis, D.P. (2012). Invasive mycoses: strategies for effective management. Am. J. Med. *125*, S25–38.

Kradin, R.L., and Mark, E.J. (2008). The pathology of pulmonary disorders due to *Aspergillus* spp. Arch. Pathol. Lab. Med. *132*, 606–614.

Kraemer, R., Deloséa, N., Ballinari, P., Gallati, S., and Crameri, R. (2006). Effect of allergic bronchopulmonary aspergillosis on lung function in children with cystic fibrosis. Am. J. Respir. Crit. Care. Med. *174*, 1211–1220.

Kremer, A., Westrich, L., and Li, S.-M. (2007). A 7-dimethylallyltryptophan synthase from *Aspergillus fumigatus*: overproduction, purification and biochemical characterization. Microbiology *153*, 3409–3416.

Ku, N.S., Han, S.H., Choi, J.Y., Kim, S.B., Kim, H.-W., Jeong, S.J., Kim, C.O., Song, Y.G., and Kim, J.M. (2012). Diagnostic value of the serum galactomannan assay for invasive aspergillosis: It is less useful in non-haematological patients. Scand. J. Infect. Dis. *44*, 600–604.

Kuge, S., Jones, N., and Nomoto, A. (1997). Regulation of yAP-1 nuclear localization in response to oxidative stress. EMBO J. *16*, 1710–1720.

Kumar, A., Ahmed, R., Singh, P.K., and Shukla, P.K. (2011). Identification of virulence factors and diagnostic markers using immunosecretome of *Aspergillus* fumigatus. J. Protemics *74*, 1104–1112.

Kupfahl, C., Heinekamp, T., Geginat, G., Ruppert, T., Härtl, A., Hof, H., and Brakhage, A.A. (2006). Deletion of the gliP gene of *Aspergillus fumigatus* results in loss of gliotoxin production but has no effect on virulence of the fungus in a low-dose mouse infection model. Mol. Microbiol. *62*, 292–302.

Kwon-Chung, K.J., and Sugui, J.A. (2009). What do we know about the role of gliotoxin in the pathobiology of *Aspergillus fumigatus*? Med. Mycol. *47*, 1–12.

Lara-Rojas, F., Sánchez, O., Kawasaki, L., and Aguirre, J. (2011). *Aspergillus nidulans* transcription factor AtfA interacts with the MAPK SakA to regulate general stress responses, development and spore functions. Mol. Microbiol. *80*, 436–454.

Latgé, J. (1999). *Aspergillus fumigatus* and Aspergillosis. Clin. Microbiol. Rev. *12*, 310–350.

Latgé, J.-P. (2007). The cell wall: a carbohydrate armour for the fungal cell. Mol. Microbiol. *66*, 279–290.

Latgé, J.-P. (2010). Tasting the fungal cell wall. Cell. Microbiol. *12*, 863–872.

Latgé, J., and Calderone, R. (2002). Host–microbe interactions: fungi. Invasive human fungal opportunistic infections. Editorial overview. Curr. Opin. Microbiol. *5*, 355–358.

Latgé, J.P., Kobayashi, H., Debeaupuis, J.P., Diaquin, M., Sarfati, J., Wieruszeski, J.M., Parra, E., Bouchara, J.P., and Fournet, B. (1994). Chemical and immunological characterization of the extracellular galactomannan of *Aspergillus fumigatus*. Infect. Immun. *62*, 5424–5433.

Lehrnbecher, T., Tramsen, L., Koehl, U., Schmidt, S., Bochennek, K., and Klingebiel, T. (2011). Immunotherapy against invasive fungal diseases in stem cell transplant recipients. Immunol. Investig. *40*, 839–852.

Leitao, E.A., Bittencourt, V.C.B., Haido, R.M.T., Valente, A.P., Peter-Katalinic, J., Letzel, M., De Souza, L.M., and Barreto-Bergter, E. (2003). Beta-galactofuranose-containing O-linked oligosaccharides present in the cell wall peptidogalactomannan of *Aspergillus fumigatus* contain immunodominant epitopes. Glycobiology *13*, 681–692.

Lemke, A., Kiderlen, A.F., and Kayser, O. (2005). Amphotericin B. Appl. Microbiol. Biotechnol. *68*, 151–162.

Lesage, G., and Bussey, H. (2006). Cell wall assembly in *Saccharomyces cerevisiae*. Microbiol. Mol. Biol. Rev. *70*, 317–343.

Lessing, F., Kniemeyer, O., Wozniok, I., Loeffler, J., Kurzai, O., Haertl, A., and Brakhage, A.A. (2007). The *Aspergillus fumigatus* transcriptional regulator AfYap1 represents the major regulator for defense against reactive oxygen intermediates but is dispensable for pathogenicity in an intranasal mouse infection model. Eukaryot. Cell *6*, 2290–2302.

Letscher-Bru, V. (2003). Caspofungin: the first representative of a new antifungal class. J. Antimicrob. Chemother. *51*, 513–521.

Levitz, S.M., and Farrell, T.P. (1990). Human neutrophil degranulation stimulated by *Aspergillus fumigatus*. J. Leukoc. Biol. *47*, 170–175.

Lewis, R.E., Wiederhold, N.P., Chi, J., Han, X.Y., Komanduri, K.V., Kontoyiannis, D.P., and Prince, R.A. (2005). Detection of gliotoxin in experimental and human aspergillosis. Infect Immun. *73*, 635–637.

Li, B., and Walsh, C.T. (2011). *Streptomyces clavuligerus* HlmI is an intramolecular disulfide- forming dithiol oxidase in holomycin biosynthesis. Biochemistry *50*, 4615–4622.

Li, B., Forseth, R.R., Bowers, A.A., Schroeder, F.C., and Walsh, C.T. (2012). A Backup Plan for Self-Protection: S-Methylation of Holomycin Biosynthetic Intermediates in *Streptomyces clavuligerus*. Chembiochem *13*, 2521–2526.

Lodeiro, S., Xiong, Q., Wilson, W.K., Ivanova, Y., Smith, M.L., May, G.S., and Matsuda, S.P.T. (2009). Protostadienol biosynthesis and metabolism in the pathogenic fungus *Aspergillus fumigatus*. Org. Lett. *11*, 1241–1244.

Losada, L., Ajayi, O., Frisvad, J.C., Yu, J., and Nierman, W.C. (2009). Effect of competition on the production and activity of secondary metabolites in *Aspergillus species*. Med. Mycol. 47 (Suppl. 1), S88–96.

Ma, Y., and Hendershot, L.M. (2001). The Unfolding Tale of the Unfolded Protein Response. Cell *107*, 827–830.

Maertens, J., Theunissen, K., and Boogaerts, M. (2002). Invasive Aspergillosis: Focus on New Approaches and New Therapeutic Agents. Curr. Med. Chem. *1*, 65–81.

Maertens, J., Raad, I., Petrikkos, G., Boogaerts, M., Selleslag, D., Petersen, F.B., Sable, C.A., Kartsonis, N.A., Ngai, A., Taylor, A., *et al.* (2004). Efficacy and safety of caspofungin for treatment of invasive aspergillosis in patients refractory to or intolerant of conventional antifungal therapy. Clin. Infect. Dis. *39*, 1563–1571.

Maertens, J., Theunissen, K., Lodewyck, T., Lagrou, K., and Van Eldere, J. (2007). Advances in the serological diagnosis of invasive *Aspergillus infections* in patients with haematological disorders. Mycoses *50* (Suppl. 1), 2–17.

Maertens, J., Egerer, G., Shin, W.S., Reichert, D., Stek, M., Chandwani, S., Shivaprakash, M., and Viscoli, C. (2010). Caspofungin use in daily clinical practice for treatment of invasive aspergillosis: results of a prospective observational registry. BMC Infect. Dis. *10*, 182.

Maiya, S., Grundmann, A., Li, X., Li, S.-M., and Turner, G. (2007). Identification of a hybrid PKS/NRPS required for pseurotin A biosynthesis in the human pathogen *Aspergillus fumigatus*. Chembiochem *8*, 1736–1743.

Malhotra, J.D., and Kaufman, R.J. (2007). The endoplasmic reticulum and the unfolded protein response. Semin. Cell Dev. Biol. *18*, 716–731.

Marr, K.A., Laverdiere, M., Gugel, A., and Leisenring, W. (2005). Antifungal therapy decreases sensitivity of the *Aspergillus* galactomannan enzyme immunoassay. Clin. Infect. Dis. *40*, 1762–1769.

Martinez-Lopez, R., Monteoliva, L., Diez-Orejas, R., Nombela, C., and Gil, C. (2004). The GPI-anchored protein CaEcm33p is required for cell wall integrity, morphogenesis and virulence in *Candida albicans*. Microbiology *150*, 3341–3354.

Martinez-lopez, R., Park, H., Myers, C.L., Gil, C., and Filler, S.G. (2006). *Candida albicans* Ecm33p is important for normal cell wall architecture and interactions with host cells. Eukaryot. Cell *5*, 140–147.

McCormick, A., Loeffler, J., and Ebel, F. (2010a). *Aspergillus fumigatus*: contours of an opportunistic human pathogen. Cell. Microbiol. *12*, 1535–1543.

McCormick, A., Heesemann, L., Wagener, J., Marcos, V., Hartl, D., Loeffler, J., Heesemann, J., and Ebel, F. (2010b). NETs formed by human neutrophils inhibit growth of the pathogenic mold *Aspergillus fumigatus*. Microbes Infect. *12*, 928–936.

McDonagh, A., Fedorova, N.D., Crabtree, J., Yu, Y., Kim, S., Chen, D., Loss, O., Cairns, T., Goldman, G., Armstrong-James, D., *et al.* (2008). Sub-telomere directed gene expression during initiation of invasive aspergillosis. PLoS Pathogens *4*, e1000154.

Mitsuguchi, H., Seshime, Y., Fujii, I., Shibuya, M., Ebizuka, Y., and Kushiro, T. (2009). Biosynthesis of steroidal antibiotic fusidanes: functional analysis of oxidosqualene cyclase and subsequent tailoring enzymes from *Aspergillus fumigatus*. J. Am. Chem. Soc. *131*, 6402–6411.

Morelle, W., Bernard, M., Debeaupuis, J., Buitrago, M., Latgé, J., Tabouret, M., and Latgé, J. (2005). Galactomannoproteins of *Aspergillus fumigatus*. Eukaryot. Cell *4*, 1308–1316.

Mowat, E., Butcher, J., Lang, S., Williams, C., and Ramage, G. (2007). Development of a simple model for studying the effects of antifungal agents on multicellular communities of *Aspergillus fumigatus*. J. Med. Microbiol. *56*, 1205–1212.

Nguyen, T.H., Fleet, G.H., and Rogers, P.L. (1998). Composition of the cell walls of several yeast species. Appl. Microbiol. Biotechnol. *50*, 206–212.

Nierman, W.C., Pain, A., Anderson, M.J., Wortman, J.R., Kim, H.S., Arroyo, J., Berriman, M., Abe, K., Archer, D.B., Bermejo, C., *et al.* (2005). Genomic sequence of the pathogenic and allergenic filamentous fungus *Aspergillus fumigatus*. Nature *438*, 1151–1156.

Nishida, S., Yoshida, L.S., Shimoyama, T., Nunoi, H., Kobayashi, T., and Tsunawaki, S. (2005). Fungal Metabolite Gliotoxin Targets Flavocytochrome b 558 in the Activation of the Human Neutrophil NADPH Oxidase. Infect. Immun. *73*, 235–244.

Nivoix, Y., Velten, M., Letscher-Bru, V., Moghaddam, A., Natarajan-Amé, S., Fohrer, C., Lioure, B., Bilger, K., Lutun, P., Marcellin, L., *et al.* (2008). Factors associated with overall and attributable mortality in invasive aspergillosis. Clin. Infect. Dis. *47*, 1176–1184.

Nosanchuk, J.D., and Casadevall, A. (2006). Impact of melanin on microbial virulence and clinical resistance to antimicrobial compounds. Antimicrob. Agents Chemother. *50*, 3519–3528.

Orciuolo, E., Stanzani, M., Canestraro, M., Galimberti, S., Carulli, G., Lewis, R., Petrini, M., and Komanduri, K.V. (2007). Effects of *Aspergillus fumigatus* gliotoxin and methylprednisolone on human neutrophils: implications for the pathogenesis of invasive aspergillosis. J. Leukoc. Biol. *82*, 839–848.

O'Brian, G.R., Georgianna, D.R., Wilkinson, J.R., Yu, J., Abbas, H.K., Bhatnagar, D., Cleveland, T.E., Nierman, W., and Payne, G.A. (2007). The effect of elevated temperature on gene transcription and aflatoxin biosynthesis. Mycologia *99*, 232–239.

O'Gorman, C.M., Fuller, H.T., and Dyer, P.S. (2009). Discovery of a sexual cycle in the opportunistic fungal pathogen *Aspergillus fumigatus*. Nature *457*, 471–474.

O'Hanlon, K.A., Gallagher, L., Schrettl, M., Jöchl, C., Kavanagh, K., Larsen, T.O., Doyle, S., and O' Hanlon, K.A. (2012). Nonribosomal Peptide Synthetase Genes *pesL* and *pes1* Are Essential for Fumigaclavine C Production in *Aspergillus fumigatus*. Appl. Environ. Microbiol. *78*, 3166–3176.

Pagano, L., Caira, M., Candoni, A., Offidani, M., Martino, B., Specchia, G., Pastore, D., Stanzani, M., Cattaneo, C., Fanci, R., *et al.* (2010). Invasive aspergillosis in patients with acute myeloid leukemia: a SEIFEM-2008 registry study. Haematologica *95*, 644–650.

Paoletti, M., Rydholm, C., Schwier, E.U., Anderson, M.J., Szakacs, G., Lutzoni, F., Debeaupuis, J.-P., Latgé, J.-P., Denning, D.W., and Dyer, P.S. (2005). Evidence for sexuality in the opportunistic fungal pathogen *Aspergillus fumigatus*. Curr. Biol. *15*, 1242–1248.

Papi, M., Maiorana, A., Bugli, F., Torelli, R., Posteraro, B., Maulucci, G., De Spirito, M., and Sanguinetti, M. (2012). Detection of biofilm-grown *Aspergillus fumigatus* by means of atomic force spectroscopy: ultrastructural effects of alginate lyase. Microsc Microanal. *18*, 1088–1094.

Pardo, J., Urban, C., Galvez, E.M., Ekert, P.G., Müller, U., Kwon-Chung, J., Lobigs, M., Müllbacher, A., Wallich, R., Borner, C., *et al.* (2006). The mitochondrial protein Bak is pivotal for gliotoxin-induced apoptosis and a critical host factor of *Aspergillus fumigatus* virulence in mice. J. Cell Biol. *174*, 509–519.

Paris, S., Debeaupuis, J., Crameri, R., Charlès, F., Prévost, M.C., Philippe, B., Latgé, J.P., Carey, M., Charle, F., Pre, M.C., *et al.* (2003). Conidial Hydrophobins of *Aspergillus fumigatus*. Appl. Environ. Microbiol. *69*, 1581–1588.

Park, H.-S., Bayram, Ö., Braus, G.H., Kim, S.C., and Yu, J.-H. (2012). Characterization of the velvet regulators in *Aspergillus fumigatus*. Mol. Microbiol. *86*, 937–953.

Patron, N.J., Waller, R.F., Cozijnsen, A.J., Straney, D.C., Gardiner, D.M., Nierman, W.C., and Howlett, B.J. (2007). Origin and distribution of epipolythiodioxopiperazine (ETP) gene clusters in filamentous ascomycetes. BMC Evol. Biol. 7, 174.

Patterson, K., and Strek, M.E. (2010). Allergic bronchopulmonary aspergillosis. Proc. Am. Thorac. Soc. 7, 237–244.

De Pauw, B., Walsh, T.J., Donnelly, J.P., Stevens, D.A., Edwards, J.E., Calandra, T., Pappas, P.G., Maertens, J., Lortholary, O., Kauffman, C.A., *et al.* (2008). Revised Definitions of Invasive Fungal Disease from the European Organization for Research and Treatment of Cancer/Invasive Fungal Infections Cooperative Group and the National Institute of Allergy and Infectious Diseases Mycoses Study Group (EORTC/MSG). Clin. Infect. Dis. *46*, 1813–1821.

Perrin, R.M., Fedorova, N.D., Bok, J.W., Cramer, R.A., Wortman, J.R., Kim, H.S., Nierman, W.C., and Keller, N.P. (2007). Transcriptional regulation of chemical diversity in *Aspergillus fumigatus* by LaeA. PLoS Pathogens 3, e50.

Petrik, M., Haas, H., Dobrozemsky, G., Lass-Flörl, C., Helbok, A., Blatzer, M., Dietrich, H., and Decristoforo, C. (2010). 68Ga-siderophores for PET imaging of invasive pulmonary aspergillosis: proof of principle. J. Nucl. Med. *51*, 639–645.

Pfeiffer, C.D., Fine, J.P., and Safdar, N. (2006). Diagnosis of invasive aspergillosis using a galactomannan assay: a meta-analysis. Clin. Infect. Dis. *42*, 1417–1427.

Philippe, B., Ibrahim-Granet, O., Prevost, M.C., Gougerot-Pocidalo, M.A., Sanchez Perez, M., Van Der Meeren, A., and Latgé, J.P. (2003). Killing of *Aspergillus fumigatus* by alveolar macrophages is mediated by reactive oxidant intermediates. Infect. Immun. *71*, 3034–3042.

Pihet, M., Carrere, J., Cimon, B., Chabasse, D., Delhaes, L., Symoens, F., and Bouchara, J.-P. (2009a). Occurrence and relevance of filamentous fungi in respiratory secretions of patients with cystic fibrosis–a review. Med. Mycol. *47*, 387–397.

Pihet, M., Vandeputte, P., Tronchin, G., Renier, G., Saulnier, P., Georgeault, S., Mallet, R., Chabasse, D., Symoens, F., and Bouchara, J.-P. (2009b). Melanin is an essential component for the integrity of the cell wall of *Aspergillus fumigatus* conidia. BMC Microbiol. *9*, 177.

Pérez, P., and Ribas, J.C. (2004). Cell wall analysis. Methods *33*, 245–251.

Rajendran, R., Williams, C., Lappin, D.F., Millington, O., Martins, M., and Ramage, G. (2013). Extracellular DNA release acts as an antifungal resistance mechanism in mature *Aspergillus fumigatus* biofilms. Eukaryot. Cell. [Epub ahead of print]

Reiber, K., Reeves, E.P., Neville, C.M., Winkler, R., Gebhardt, P., Kavanagh, K., and Doyle, S. (2005). The expression of selected non-ribosomal peptide synthetases in *Aspergillus fumigatus* is controlled by the availability of free iron. FEMS Microbiol. Lett. *248*, 83–91.

Reverberi, M., Ricelli, A., Zjalic, S., Fabbri, A.A., and Fanelli, C. (2010). Natural functions of mycotoxins and control of their biosynthesis in fungi. Appl. Microbiol. Biotechnol. *87*, 899–911.

Richie, D.L., Hartl, L., Aimanianda, V., Winters, M.S., Fuller, K.K., Miley, M.D., White, S., McCarthy, J.W., Latgé, J.-P., Feldmesser, M., *et al.* (2009). A role for the unfolded protein response (UPR) in virulence and antifungal susceptibility in *Aspergillus fumigatus*. PLoS Pathogens 5, e1000258.

Robbins, N., Uppuluri, P., Nett, J., Rajendran, R., Ramage, G., Lopez-Ribot, J.L., Andes, D., and Cowen, L.E. (2011). Hsp90 governs dispersion and drug resistance of fungal biofilms. PLoS Pathog. 7, e1002257.

Romano, J., Nimrod, G., Ben-Tal, N., Shadkchan, Y., Baruch, K., Sharon, H., and Osherov, N. (2006). Disruption of the *Aspergillus* fumigatus ECM33 homologue results in rapid conidial germination, antifungal resistance and hypervirulence. Microbiology *152*, 1919–1928.

Ron, D., and Walter, P. (2007). Signal integration in the endoplasmic reticulum unfolded protein response. Nat. Rev. Mol. Cell Biol. *8*, 519–529.

Sagan, D., and Goździuk, K. (2010). Surgery for pulmonary aspergilloma in immunocompetent patients: no benefit from adjuvant antifungal pharmacotherapy. Ann. Thorac. Surg. *89*, 1603–1610.

Sainz, J., Hassan, L., Perez, E., Romero, A., Moratalla, A., López-Fernández, E., Oyonarte, S., and Jurado, M. (2007). Interleukin-10 promoter polymorphism as risk factor to develop invasive pulmonary aspergillosis. Immunol. Lett. *109*, 76–82.

Sanchez, J.F., Somoza, A.D., Keller, N.P., and Wang, C.C. (2012). Advances in *Aspergillus* secondary metabolite research in the post-genomic era. Nat. Prod. Rep. *29*, 351–371.

Sarikaya Bayram, O., Bayram, O., Valerius, O., Park, H.S., Irniger, S., Gerke, J., Ni, M., Han, K.-H., Yu, J.-H., and Braus, G.H. (2010). LaeA control of velvet family regulatory proteins for light-dependent development and fungal cell-type specificity. PLoS Genetics *6*, e1001226.

Scharf, D.H., Remme, N., Heinekamp, T., Hortschansky, P., Brakhage, A.A., and Hertweck, C. (2010). Transannular Disulfide Formation in Gliotoxin Biosynthesis and Its Role in Self-Resistance of the Human Pathogen *Aspergillus fumigatus*. J. Am. Chem. Soc. *132*, 10136–10141.

Scharf, D.H., Remme, N., Habel, A., Chankhamjon, P., Scherlach, K., Heinekamp, T., Hortschansky, P., Brakhage, A.A., and Hertweck, C. (2011). A dedicated glutathione S-transferase mediates carbon–sulfur bond formation in gliotoxin biosynthesis. J. Am. Chem. Soc. *133*, 12322–12325.

Scharf, D.H., Chankhamjon, P., Schlerlach, K., Heinekamp, T., Roth, M., Brakhage, A.A., and Hertweck, C. (2012). Epidithiol formation by an unprecedented twin carbon-sulfur lyase in the gliotoxin pathway. Agnew. Chem. Int. Edit. *51*, 10064–10068.

Schmaler-Ripcke, J., Sugareva, V., Gebhardt, P., Winkler, R., Kniemeyer, O., Heinekamp, T., and Brakhage, A.A. (2009). Production of pyomelanin, a second type of melanin, via the tyrosine degradation pathway in *Aspergillus fumigatus*. Appl. Environ. Microbiol. *75*, 493–503.

Schrettl, M., Bignell, E., Kragl, C., Joechl, C., Rogers, T., Arst, H.N., Haynes, K., and Haas, H. (2004). Siderophore biosynthesis but not reductive iron assimilation is essential for *Aspergillus fumigatus* virulence. J. Exp. Med. *200*, 1213–1219.

Schrettl, M., Bignell, E., Kragl, C., Sabiha, Y., Loss, O., Eisendle, M., Wallner, A., Arst, H.N., Haynes, K., and Haas, H. (2007). Distinct roles for Intra- and Extracellular Siderophores during *Aspergillus fumigatus* Infection. PLoS Pathogens 3, 1195–1207.

Schrettl, M., Carberry, S., Kavanagh, K., Haas, H., Jones, G.W., O'Brien, J., Nolan, A., Stephens, J., Fenelon, O., and Doyle, S. (2010). Self-protection against gliotoxin-a component of the gliotoxin biosynthetic cluster, GliT, completely protects *Aspergillus fumigatus* against exogenous gliotoxin. PLoS Pathogens *6*, e1000952.

Segal, B.H., and Romani, L.R. (2009). Invasive aspergillosis in chronic granulomatous disease. Med. Mycol. *47 (Suppl 1)*, S282–290.

Segal, B.H., and Walsh, T.J. (2006). Current approaches to diagnosis and treatment of invasive aspergillosis. Am. J. Respir. Crit. Care. Med. *173*, 707–717.

Sharon, H., Hagag, S., and Osherov, N. (2009). Transcription factor PrtT controls expression of multiple secreted proteases in the human pathogenic mold *Aspergillus fumigatus*. Infect. Immun. *77*, 4051–4060.

Shi, L.-N., Li, F.-Q., Lu, J.-F., Kong, X.-X., Wang, S.-Q., Huang, M., Shao, H.-F., and Shao, S.-H. (2012). Antibody specific to thioredoxin reductase as a new biomarker for serodiagnosis of invasive aspergillosis in non-neutropenic patients. Clin. Chim. Acta *413*, 938–943.

Silvestris, N., Del Re, M., Azzariti, A., Maiello, E., Lombardi, L., Cinieri, S., Guarini, A., Brunetti, A.E., Delcuratolo, S., De Vita, F., *et al.* (2012). Optimized granulocyte colony-stimulating factor prophylaxis in adult cancer patients: from biological principles to clinical guidelines. Expert Opin. Ther. Targets *16 (Suppl 2)*, S111–117.

Sixto, M.L.J., Vareechon, C., Cowden, S., Cobb, B.A., Latgé, J., Momany, M., Pearlman, E., and Leal, M.S.J. (2012). Fungal antioxidant pathways promote survival against neutrophils during infection. J. Clin. Invest. *122*, 2482–2498.

Spikes, S., Xu, R., Nguyen, C.K., Chamilos, G., Kontoyiannis, D.P., Jacobson, R.H., Ejzykowicz, D.E., Chiang, L.Y., Filler, S.G., and May, G.S. (2008). Gliotoxin production in *Aspergillus fumigatus* contributes to host-specific differences in virulence. J. Infect. Dis. *197*, 479–486.

Srinivasan, U., Bala, A., Jao, S., Starke, D.W., Jordan, T.W., and Mieyal, J.J. (2006). Selective Inactivation of Glutaredoxin by Sporidesmin and Other Epidithiopiperazinediones. Biochemistry *45*, 8978–8987.

Stanimirovic, Z., Stevanovic, J., Bajic, V., and Radovic, I. (2007). Evaluation of genotoxic effects of fumagillin by cytogenetic tests *in vivo*. Mutat. Res. *628*, 1–10.

Stanzani, M., Orciuolo, E., Lewis, R., Kontoyiannis, D.P., Martins, S.L.R., St John, L.S., and Komanduri, K.V. (2005). *Aspergillus fumigatus* suppresses the human cellular immune response via gliotoxin-mediated apoptosis of monocytes. Blood *105*, 2258–2265.

Steele, C., Rapaka, R.R., Metz, A., Pop, S.M., Williams, D.L., Gordon, S., Kolls, J.K., and Brown, G.D. (2005). The beta-glucan receptor dectin-1 recognizes specific morphologies of *Aspergillus fumigatus*. PLoS Pathogens *1*, e42.

Sugui, J.A., Pardo, J., Chang, Y.C., Müllbacher, A., Zarember, K.A., Galvez, E.M., Brinster, L., Zerfas, P., Gallin, J.I., Simon, M.M., *et al.* (2007a). Role of *laeA* in the Regulation of *alb1*, *gliP*, Conidial Morphology, and Virulence in *Aspergillus fumigatus*. Eukaryot. Cell *6*, 1552–1561.

Sugui, J.A., Pardo, J., Chang, Y.C., Zarember, K.A., Nardone, G., Galvez, E.M., Müllbacher, A., Gallin, J.I., Simon, M.M., and Kwon-Chung, K.J. (2007b). Gliotoxin is a virulence factor of *Aspergillus fumigatus*: *gliP* deletion attenuates virulence in mice immunosuppressed with hydrocortisone. Eukaryot. Cell *6*, 1562–1569.

Suhadolnik, R.J., and Chenowath, R.G. (1958). Biosynthesis of Gliotoxin. 1. Incorporation of Phenylalanine-1- and -2-C^{14}. J. Am. Chem. Soc. *80*, 4391–4392.

Taylor, P.R., Tsoni, S.V., Willment, J.A., Dennehy, K.M., Findon, H., Haynes, K., Steele, C., Botto, M., Gordon, S., and Brown, G.D. (2007). Dectin-1 is required for β-glucan recognition and control of fungal infection. Nat. Immunol. *8*, 31–38.

Thau, N., Monod, M., Crestani, B., Rolland, C., Tronchin, G., Latgé, J.P., and Paris, S. (1994). *rodletless* mutants of *Aspergillus fumigatus*. Infect. Immun. *62*, 4380–4388.

Thomas, B., Rutman, A., Hirst, R.A., Haldar, P., Wardlaw, A.J., Bankart, J., Brightling, C.E., and O'Callaghan, C. (2010). Ciliary dysfunction and ultrastructural abnormalities are features of severe asthma. J. Allergy Clin. Immunol. *126*, 722–729.e2.

Thompson, G.R., and Patterson, T.F. (2008). Pulmonary aspergillosis. Sem. Resp. Crit. Care. M. *29*, 103–110.

Thornton, C.R. (2008). Development of an immunochromatographic lateral-flow device for rapid serodiagnosis of invasive aspergillosis. CVI *15*, 1095–1105.

Thornton, C.R. (2010). Detection of invasive aspergillosis. Adv. Appl. Microbiol. *70*, 187–216.

Thornton, C., Johnson, G., and Agrawal, S. (2012). Detection of invasive pulmonary aspergillosis in haematological malignancy patients by using lateral-flow technology. JoVE e3721.

Thywißen, A., Heinekamp, T., Dahse, H.-M., Schmaler-Ripcke, J., Nietzsche, S., Zipfel, P.F., and Brakhage, A.A. (2011). Conidial dihydroxynaphthalene melanin of the human pathogenic fungus *Aspergillus fumigatus* interferes with the host endocytosis pathway. Front. Microbiol. *2*, 96.

Traunmüller, F., Popovic, M., Konz, K.-H., Smolle-Jüttner, F.-M., and Joukhadar, C. (2011). Efficacy and safety of current drug therapies for invasive aspergillosis. Pharmacology *88*, 213–224.

Trof, R.J., Beishuizen, A., Debets-Ossenkopp, Y.J., Girbes, A.R.J., and Groeneveld, A.B.J. (2007). Management of invasive pulmonary aspergillosis in non-neutropenic critically ill patients. Intensive Car Med. *33*, 1694–1703.

Tsai, H.F., Chang, Y.C., Washburn, R.G., Wheeler, M.H., and Kwon-Chung, K.J. (1998). The developmentally regulated *alb1* gene of *Aspergillus fumigatus*: its role in modulation of conidial morphology and virulence. J. Bacteriol. *180*, 3031–3038.

Tsai, H.F., Wheeler, M.H., Chang, Y.C., and Kwon-Chung, K.J. (1999). A developmentally regulated gene cluster involved in conidial pigment biosynthesis in *Aspergillus fumigatus*. J. Bacteriol. *181*, 6469–6477.

Tsunawaki, S., Yoshida, L.S., Nishida, S., Kobayashi, T., and Shimoyama, T. (2004). Fungal metabolite gliotoxin inhibits assembly of the human respiratory burst NADPH oxidase. Infect. Immun. *72*, 3373–3382.

Tunger, O., Bayram, H., Degerli, K., Dinc, G., and Cetin, B.C. (2008). Comparison of the efficacy of combination and monotherapy with caspofungin and liposomal amphotericin B against invasive candidiasis. Saudi Med. J. *29*, 728–733.

Ullmann, A.J., Sanz, M.A., Tramarin, A., Barnes, R.A., Wu, W., Gerlach, B.A., Krobot, K.J., and Gerth, W.C. (2006). Prospective study of amphotericin B formulations in immunocompromised patients in 4 European countries. Clin. Infect. Dis. *43*, e29–38.

Ushio-Fukai, M., and Alexander, R.W. (2004). Reactive oxygen species as mediators of angiogenesis Signaling Role of NAD(P)H oxidase. Mol. Cell. Biochem. *264*, 85–97.

Valiante, V., Heinekamp, T., Jain, R., Härtl, A., and Brakhage, A.A. (2008). The mitogen-activated protein kinase MpkA of *Aspergillus fumigatus* regulates cell wall signaling and oxidative stress response. Fungal Genet. Biol. *45*, 618–627.

Vandeputte, P., Ferrari, S., and Coste, A.T. (2012). Antifungal resistance and new strategies to control fungal infections. Int. J. Microbiol. *2012*, 713687.

Volling, K., Thywissen, A., Brakhage, A.A., and Saluz, H.P. (2011). Phagocytosis of melanized *Aspergillus* conidia by macrophages exerts cytoprotective effects by sustained PI3K/Akt signalling. Cell. Microbiol. *13*, 1130–1148.

Volpi, I., Perruccio, K., Tosti, A., Capanni, M., Ruggeri, L., Posati, S., Aversa, F., Tabilio, A., Romani, L., Martelli, M.F., *et al.* (2001). Postgrafting administration of granulocyte colony-stimulating factor impairs functional immune recovery in recipients of human leukocyte antigen haplotype-mismatched hematopoietic transplants. Blood *97*, 2514–2521.

Vödisch, M., Scherlach, K., Winkler, R., Hertweck, C., Braun, H.-P., Roth, M., Haas, H., Werner, E.R., Brakhage, A.A., and Kniemeyer, O. (2011). Analysis of the *Aspergillus fumigatus* proteome reveals metabolic changes and the activation of the pseurotin A biosynthesis gene cluster in response to hypoxia. J. Proteome Res. *10*, 2508–2524.

Wagener, J., Echtenacher, B., Rohde, M., Kotz, A., Krappmann, S., Heesemann, J., and Ebel, F. (2008). The putative alpha-1,2-mannosyltransferase AfMnt1 of the opportunistic fungal pathogen *Aspergillus fumigatus* is required for cell wall stability and full virulence. Eukaryot. Cell 7, 1661–1673.

Ward, O.P., Qin, W.M., Dhanjoon, J., Ye, J., and Singh, A. (2005). Physiology and biotechnology of *Aspergillus*. Adv. Appl. Microbiol. *58C*, 1–75.

Waring, P., Sjaarda, A., and Lin, Q.H. (1995). Gliotoxin Inactivates Alcohol Dehydrogenase by either Covalent Modification or Free Radical Damage Mediated by Redox Cycling. Biochem. Pharmacol. *49*, 1195–1201.

Wartha, F., Beiter, K., Normark, S., and Henriques-Normark, B. (2007). Neutrophil extracellular traps: casting the NET over pathogenesis. Curr. Opin. Microbiol. *10*, 52–56.

Wessels, J. (1997). Hydrophobins: proteins that change the nature of the fungal surface. Adv. Microb. Physiol. *38*, 1–45.

White, P.L., Bretagne, S., Klingspor, L., Melchers, W.J.G., McCulloch, E., Schulz, B., Finnstrom, N., Mengoli, C., Barnes, R.A., Donnelly, J.P., *et al.* (2010). *Aspergillus* PCR: one step closer to standardization. Journal of Clinical Microbiology *48*, 1231–1240.

White, P.L., Mengoli, C., Bretagne, S., Cuenca-Estrella, M., Finnstrom, N., Klingspor, L., Melchers, W.J.G., McCulloch, E., Barnes, R.A., Donnelly, J.P., *et al.* (2011). Evaluation of *Aspergillus* PCR protocols for testing serum specimens. J. Clin. Microbiol. *49*, 3842–3848.

Willger, S.D., Puttikamonkul, S., Kim, K.-H., Burritt, J.B., Grahl, N., Metzler, L.J., Barbuch, R., Bard, M., Lawrence, C.B., and Cramer, R.A. (2008). A sterol-regulatory element binding protein is required for cell polarity, hypoxia adaptation, azole drug resistance, and virulence in *Aspergillus* fumigatus. PLoS Pathogens *4*, e1000200.

Willger, S.D., Grahl, N., and Cramer Jr, R.A. (2009). *Aspergillus fumigatus* metabolism: Clues to mechanisms of *in vivo* fungal growth and virulence. Med. Mycol. *47*, S72–79.

Wingard, J.R., Kubilis, P., Lee, L., Yee, G., White, M., Walshe, L., Bowden, R., Anaissie, E., Hiemenz, J., and Lister, J. (1999). Clinical significance of nephrotoxicity in patients treated with amphotericin B for suspected or proven aspergillosis. Clin. Infect. Dis. *29*, 1402–1407.

Wösten, H.A., and De Vocht, M.L. (2000). Hydrophobins, the fungal coat unravelled. Biochim. Biophys. Acta *1469*, 79–86.

Yamada, A., Kataoka, T., and Nagai, K. (2000). The fungal metabolite gliotoxin: immunosuppressive activity on CTL-mediated cytotoxicity. Immunol. Lett. *71*, 27–32.

Yasmin, S., Alcazar-fuoli, L., Gründlinger, M., Puempel, T., Cairns, T., and Blatzer, M. (2012). Mevalonate governs interdependency of ergosterol and siderophore biosyntheses in the fungal pathogen *Aspergillus fumigatus*. Proc. Natl. Acad. Sci. U.S.A. *109*, 2707–2708.

Yoshida, S., Ono, M., Shono, T., Izumi, H., Ishibashi, T., Suzuki, H., and Kuwano, M. (1997). Involvement of Interleukin-8, Vascular Endothelial Growth Factor, and Basic Fibroblast Growth Factor in Tumor Necrosis Factor Alpha-Dependent Angiogenesis. Mol. Cell. Biol. *17*, 4015–4023.

Zaas, A.K., Liao, G., Chien, J.W., Weinberg, C., Shore, D., Giles, S.S., Marr, K.A., Usuka, J., Burch, L.H., Perera, L., *et al.* (2008). Plasminogen alleles influence susceptibility to invasive aspergillosis. PLoS Genetics *4*, e1000101.

Zhou, H., Hu, H., Zhang, L., Li, R., Ouyang, H., Ming, J., and Jin, C. (2007). O-Mannosyltransferase 1 in *Aspergillus fumigatus* (AfPmt1p) is crucial for cell wall integrity and conidium morphology, especially at an elevated temperature. Eukaryot. Cell *6*, 2260–2268.

Zhao, Y., Liu, J., Wang, J., Wang, L., Yin, H., Tan, R., and Xu, Q. (2004). Fumigaclavine C improves concanavalin A-induced liver injury in mice mainly via inhibiting TNF-alpha production and lymphocyte adhesion to extracellular matrices. J. Pharm. Pharmacol. *56*, 775–782.

Zmeili, O.S., and Soubani, A.O. (2007). Pulmonary aspergillosis: a clinical update. QJM *100*, 317–334.

Cryptococcus

9

Rocío García-Rodas, Radames J.B. Cordero and Oscar Zaragoza

Abstract

Cryptococcus neoformans is a cosmopolitan pathogen of major concern among HIV-infected patients and has a significant incidence in developing countries. *Cryptococcus neoformans* blastoconidia or desiccated spores are inhaled and can cause pneumonia and meningitis in immunodeficient individuals, being fatal if not treated. Its sibling species *Cryptococcus gattii* is more restricted to tropical areas and infects mainly immunocompetent individuals. *Cryptococcus* interactions with the human host have been widely studied, in particular its capacity to modulate the immune response, replicate within macrophages, disseminate from the lungs to the central nervous system, and/or remain in a latent state for years. Our current understanding of the virulence determinants that facilitate *Cryptococcus* to cause infection, particularly its polysaccharide capsule and melanin accumulation, as well as other traits that confer adaptation to the host will be reviewed in this chapter.

Introduction

Among pathogenic yeasts, *Cryptococcus* spp. offer a unique model to study fungal virulence. These yeasts are basidiomycetes and the two main species that cause disease in humans are *C. neoformans* and *C. gattii*. Both species are found in the environment and it is well known that these pathogens are not host-specific and can cause disease in a wide variety of organisms; from unicellular predators to plants, nematodes, insects and higher hosts, including both aquatic and terrestrial mammals (Feldmesser *et al.*, 2000; Mylonakis *et al.*, 2002; Staib *et al.*, 1972; Venn-Watson *et al.*, 2012; Warpeha *et al.*, 2013; Xue *et al.*, 2007). Although closely related, *C. neoformans* and *C. gattii* differ in their distribution and epidemiology. While *C. neoformans* has a cosmopolitan distribution, *C. gattii* is more restricted to tropical areas (Litvintseva *et al.*, 2005; Morgan *et al.*, 2006; Sorrell, 2001). However, the main difference between them is that *C. neoformans* is primarily an opportunistic pathogen, while *C. gattii* causes disease mainly in immunocompetent individuals (Sorrell, 2001). In both cases, the main phenotypic feature is the presence of a polysaccharide capsule around the cell wall that provides unique physical, biochemical, and immunological properties. For this reason, the capsule has been the subject of a large number of studies that have clearly established its role as a virulence factor (Chang and Kwon-Chung, 1994; McClelland *et al.*, 2005).

As we will review in the following sections, these species offer a versatile system to study fungal virulence. Firstly, they are significant opportunistic pathogens in susceptible patient populations (Park *et al.*, 2009). In addition, they also offer an excellent model to analyse virulence in environmental organisms, in particular how these organisms acquire virulence traits and how they use them to infect complex hosts such as humans.

In this chapter, we will provide an overview of the main aspects regarding cryptococcal pathogenesis and its usefulness as a model to investigate fungal virulence.

Classification of *Cryptococcus neoformans/gattii*

Since it was first described at the end of the nineteenth century, *Cryptococcus neoformans* has been denominated with many different names (*Saccharomyces neoformans, Torula histolytica* (see review in Kwon-Chung, 2011). In 1949, *C. neoformans* was first divided into three serotypes (A, B, and C) (Evans, 1949), and years later, serotypes D and A/D hybrids were also described (Wilson *et al.*, 1968). *Cryptococcus gattii* was described in 1970, and originally corresponded to serotypes B and C. These serotypes were then grouped as *C. neoformans* variety *gattii*. However, because of the morphological and genetic differences between the two varieties, *C. gattii* was finally recognized as a different species in 2006 (Kwon-Chung and Varma, 2006). Currently, the *C. neoformans/gattii* complex is recognized as two different species with different serotypes and varieties. In *C. neoformans*, serotype A is denominated as *C. neoformans* var. *grubii* (Franzot *et al.*, 1999), and serotype D as *C. neoformans* var. *neoformans*. More recently, some hybrid strains *neoformans/gattii* (A/B and B/D) have been described (Aminnejad *et al.*, 2012; Bovers *et al.*, 2006).

The development of molecular biology techniques has allowed a more detailed description and characterization of *Cryptococcus* species. Using different methodologies (i.e. PCR fingerprinting, AFLP, RFLP, MLST or IGS genotyping), *C. neoformans/gattii* can also be divided into different genotypes (Meyer *et al.*, 2009). *Cryptococcus neoformans* is found in four subgroups: VNI and VNII (AFLP1A and AFLP1B), which correspond to serotype A; VNIII (AFLP 3), which corresponds to A/D hybrids; and VNIV (AFLP 2), which groups serotype D strains. *Cryptococcus gattii* is also divided in four different subgroups: VGI (AFLP4), VGII (AFLP6), VGIII (AFLP5), and VGIV (AFLP7).

Incidence and epidemiology

Cryptococcus neoformans has been known as a pathogenic yeast since the nineteenth century. However, its incidence significantly increased at the end of the twentieth century due to the emergence of HIV. In some areas such as the U.S.A., it was estimated that around 10% of HIV-positive patients suffered from cryptococcosis (Casadevall and Perfect, 1998). The main clinical outcome of the disease is acute meningitis, which is fatal unless diagnosed and treated in the early stages of the disease. The recommended treatment for this infection is an initial therapy based on a combination of amphotericin B plus flucytosine, followed by a maintenance therapy with fluconazole (Perfect *et al.*, 2010). In contrast, echinocandins (caspofungin, micafungin and anidulafungin) are not active against this fungus.

The development of the highly active antiretroviral therapy for AIDS (HAART) has controlled the incidence in developed countries, although its associated early mortality was not affected (Lortholary *et al.*, 2006). Moreover, new groups of patients at risk have been described, such as transplant recipients (Singh *et al.*, 2008).

In developing areas, such as sub-Saharan Africa, Asia or South America, cryptococcosis still has a significant incidence. It is estimated that it causes one million infections and 650,000 deaths per year (Park *et al.*, 2009). Although *C. neoformans* is strongly associated with HIV infection, there are reports indicating high prevalence among apparent immunocompetent individuals in some geographical regions (Chau *et al.*, 2010; Chen *et al.*, 2008), suggesting that there are still unknown factors that predispose humans to get a disseminated disease after being exposed to the pathogen.

Once in the lung, *C. neoformans* is normally controlled by the host's immune response. Classically, it was believed that the pathogen was gradually eliminated. However, in the last years, there is strong evidence that *Cryptococcus neoformans* can cause asymptomatic infections and stay latent in the host during long periods of time (Goldman *et al.*, 1994, 2000; Larsen *et al.*, 1989; Salyer *et al.*, 1974).

Cryptococcus gattii is more restricted to tropical areas, and in contrast to *C. neoformans*, it mainly causes infection in immunocompetent individuals (Byrnes *et al.*, 2010). However, recent studies highlight a relatively high prevalence of molecular types VGIII and VGIV among isolates from HIV-infected patients suffering from cryptococcosis in Africa (Litvintseva *et al.*, 2005) and

North America (Byrnes *et al.*, 2011). In the last decade, *C. gattii* gained importance after an outbreak of infections in Vancouver Island in British Columbia, Canada (Fraser *et al.*, 2005; Fyfe *et al.*, 2008; Hoang *et al.*, 2004; Ngamskulrungroj *et al.*, 2011). This outbreak affected not only humans, but also many farm animals and pets (Stephen *et al.*, 2002). Recent data indicate that the outbreak has expanded through the US Northwest, which raises major concerns about the virulence and incidence of these strains (Byrnes *et al.*, 2009; Byrnes and Marr, 2011; Byrnes and Heitman, 2009; Datta *et al.*, 2009). Concerning the origin of these strains, genotypic analysis revealed evidence of sexual reproduction between two alpha mating-type parents instead of the normal sexual cycle (Fraser *et al.*, 2005), which could explain the capability of these strains to inhabit different geographical niches. In the case of *C. gattii*, the primary disease is pneumonia (Ngamskulrungroj *et al.*, 2012), but it can also disseminate to the brain (Stephen *et al.*, 2002).

Virulence determinants and importance as a human pathogen

Cryptococcus has several well-defined virulence determinants and many mechanisms that are necessary for adaptation to the host. In this section, we will review these traits that are required for virulence and their importance during disease.

Capsule

The most characteristic feature of *C. neoformans* is a polysaccharide capsule that surrounds the cell body. It is not visible under bright field microscopy. However, it can be observed after placing yeasts in a suspension of India ink as a white and uniform halo (Fig. 9.1). The capsule is highly hydrated, and it is estimated that in regular conditions the water content accounts for more than 95% of the wet weight of the capsule (Maxson *et al.*, 2007a). At the molecular level, the main component of the capsule is glucuronoxylomannan (GXM), a very large heteropolymer (kilo- to megadaltons) which consist of a α(1,3)-mannan main chain with β(1,2)-glucuronic acid residues attached to every first mannose forming the basic core with 6-*O*-acetyl and β-D-xylose components (Cherniak *et al.*, 1980; McFadden *et al.*, 2006, 2007). The GXM serotype specificity depends on the amount of *O*-acetylation, the linkage between mannose and the xylose sugars (β-1,2 for serotypes A and D or β-1,2 and β-1,4 for serotypes B and C), and the location and type of xylose substitution (Bhattacharjee *et al.*, 1984; Cherniak *et al.*, 1988).

Nowadays there are a large number of mouse monoclonal antibodies against GXM differing in specificity, affinity and isotype (Casadevall *et al.*, 1992, 1994; Dromer *et al.*, 1987; Eckert and Kozel, 1987; Pirofski *et al.*, 1995; Todaro-Luck *et al.*, 1989; van de Moer *et al.*, 1990). These antibodies can alter the course of the infection,

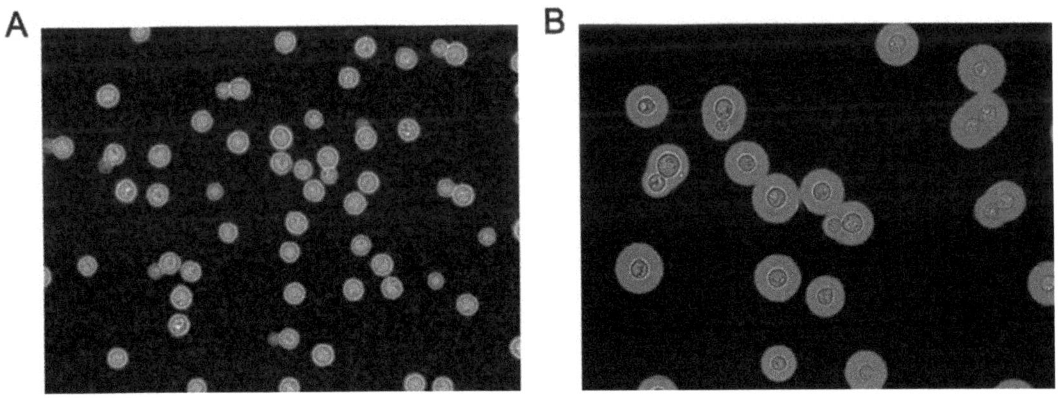

Figure 9.1 Capsule size of *C. neoformans* in different conditions. India ink suspension of *Cryptococcus neoformans* var. *grubii* grown in (A) Sabouraud liquid medium overnight (regular capsule size) and in (B) 10% Sabouraud, pH 7 (enlarged capsule).

and they can behave as protective, neutral or even disease enhancing, depending on the Ab isotype and the pathogen dose (Mukherjee *et al.*, 1992, 1993; Taborda and Casadevall, 2001; Taborda *et al.*, 2003). In addition to immunological studies, these antibodies have also been used to demonstrate the heterogeneity of the capsule structure (McFadden *et al.*, 2007; Todaro-Luck *et al.*, 1989). Some antibodies produce an annular pattern around the capsule, while others produce a punctuate pattern (Fig. 9.2), this difference depends on the strain serotype and Ab isotype and specificity (Cleare and Casadevall, 1998; Dromer *et al.*, 1987; Nussbaum *et al.*, 1997).

Another polysaccharide found in the cryptococcal capsule is galactoxylomannan (GalXM), later revised to glucuronoxylomannogalactan (GXMGal) (Heiss *et al.*, 2009), constituting approximately 8% of the capsular mass (Bose *et al.*, 2003; Vaishnav *et al.*, 1998). This heteropolymer exhibits an average molecular weight of approximately 10^5 Daltons (Cherniak and Sundstrom, 1994; McFadden *et al.*, 2006) and it is composed of a galactopyranose backbone containing four possible 3-O-linked oligosaccharide branches (α (1–3)-Man, α(1–4)-Man, β-galactosidase trisaccharide with variable amounts of β(1–2)-or β(1–3)-xylose side groups) (Bose *et al.*, 2003; Heiss *et al.*, 2009; McFadden *et al.*, 2007; Vaishnav *et al.*, 1998).

The location of GXMGal in the cryptococcal capsule is uncertain. The first studies reported that GXMGal was mostly cell wall associated (Vaishnav *et al.*, 1998; van de Moer *et al.*, 1990). However, recent studies suggest that GXMGal is a transient component of the capsule and that it may be present in vesicular fractions or associated with newly formed capsule (De Jesus *et al.*, 2009). Mutants unable to produce GXMGal have larger capsules (Moyrand *et al.*, 2007), indicating that this component might play an important structural role. More recently, several reports highlighted the importance of GXMGal as virulence factor given its strong anti-inflammatory properties (Chow and Casadevall, 2011; Pericolini *et al.*, 2006; Vecchiarelli *et al.*, 2011).

Mannoproteins (MP) are a minor component of the capsule (Cherniak and Sundstrom, 1994; Doering, 2000). Despite their immunogenic effect during infection (Huang *et al.*, 2002; Levitz *et al.*, 2001; Levitz and Specht, 2006; Mansour *et al.*, 2002), their role in capsule structure has never

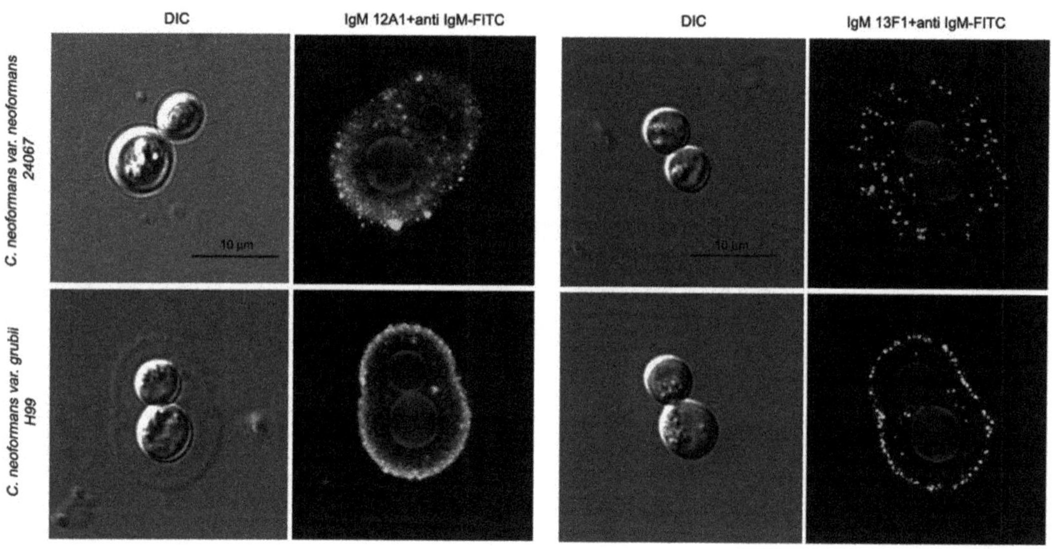

Figure 9.2 Binding patterns of two monoclocal antibodies against cryptococcal capsule. Yeast cells from *C. neoformans* var. *neoformans* (24067) and *C. neoformans* var. *grubii* (H99) were stained with Calcofluor White (blue fluorescence, cell wall) and two IgM mAbs (12A1 and 13F1, green fluorescence, capsule) and observed by confocal microscopy. Scale bar in upper bright field panels apply to the rest of the panels. A colour version of this figure is available in the plate section at the back of the book.

been established. It is thought that they are found in the inner cell wall, and not associated with GXM or GXMGal. (Jesus *et al.*, 2010; Vartivarian *et al.*, 1989). However, very little is known about their capsular localization and role in virulence.

Capsule organization and synthesis

In contrast to bacterial capsules, which are synthesized at the extracellular space by plasma-membrane-associated enzymes (Raetz and Whitfield, 2002; Whitfield, 2006; Whitfield *et al.*, 2003), GXM is synthesized intracellularly (Feldmesser *et al.*, 2001a; Garcia-Rivera *et al.*, 2004), inside the Golgi apparatus (Yoneda and Doering, 2006). GXM traffic to the cell surface involves secretion of vesicles containing polysaccharide that cross the cell wall and release their content outside the cell (Rodrigues *et al.*, 2007). The polysaccharide can bind to the capsule or be released into the medium as exopolysaccharide (Feldmesser *et al.*, 2001a; Garcia-Rivera *et al.*, 2004; Rodrigues *et al.*, 2007). The production of exopolysaccharide is of high physiological relevance since it produces immunomodulation in the host (see review in Zaragoza *et al.*, 2009).

Once exported, the capsular polysaccharide molecules are attached and/or anchored to the cell surface through molecular interactions with cell wall components. Capsule attachment involves the interaction between GXM and cell wall derived α(1,3)-glucans and chitin-derived macromolecules (Reese and Doering, 2003; Rodrigues *et al.*, 2008). Capsular polysaccharides can also bind to each other by self-association through non-covalent bonds involving divalent cation bridges (McFadden *et al.*, 2006; Nimrichter *et al.*, 2007).

Capsular polysaccharide molecules are characterized by having a very large molecular size (diameter) and molecular mass, ranging between 200 nm and 600 nm and 10^7–10^8 g/mol, respectively. Higher-order structure analysis of DMSO-extracted capsular polysaccharide demonstrated macromolecular characteristics consistent with a branched and highly entangled conformation, similar to other branched polysaccharides (i.e.amylopectin, glycogen) (Cordero *et al.*, 2011a). This appears to be an important determinant of capsular polysaccharide biological activity (Cordero *et al.*, 2011a) and implies the existence of biochemical mechanisms responsible for generating such higher order polysaccharide structures.

Different genes involved in capsule biosynthesis have been identified (see seminal reviews by Haynes *et al.*, 2011; O'Meara and Alspaugh, 2012; Zaragoza *et al.*, 2009). The first one, *CAP59*, was cloned by complementation of an acapsular mutant isolated by UV-random mutagenesis (Chang and Kwon-Chung, 1994). The loss of virulence by the acapsular mutant and the following restoration of the phenotype by complementation represented the first molecular evidence for the capsule as a virulence factor (Chang and Kwon-Chung, 1994). Following *CAP59* discovery, three other genes were identified; *CAP64* (Chang *et al.*, 1996); *CAP60* (Chang and Kwon-Chung, 1998) and *CAP10* (Chang and Kwon-Chung, 1999). The expression of the *CAP* genes is affected by conditions that regulate capsule size (Okabayashi *et al.*, 2005). Moreover, Cap64-like proteins Cas3, Cas31, Cas32, Cas33, Cas34, and Cas35 may be involved in assembling the monomers of the GXM or GXMGal, and/or adding modifications to existing polysaccharide molecules (Moyrand *et al.*, 2004).

Genes involved in the addition of β-1,3-linked xylose to the capsule have been described (Castle *et al.*, 2008; Klutts *et al.*, 2007) as well as those responsible of the addition and/or modification of mannose groups to the carbohydrate backbone (Cottrell *et al.*, 2007). Furthermore, a putative novel regulator of capsule expansion was discovered and found to have homology with the Ada2 protein of *Saccharomyces cerevisiae*, and genes downstream of *ADA2* were found to be involved in mating and virulence (Haynes *et al.*, 2011).

Capsule dynamics and plasticity

The cryptococcal capsule is a plastic structure that can change in structure and size depending on the environment or growth conditions. Analysis of the capsule using mAbs has demonstrated that the capsule exhibits inter- and intra-strains differences (McFadden *et al.*, 2007). Even under consistent growth conditions, different cellular reactivities to the same antibody can be discerned, a fact that has been related to age-dependent differences

(Cordero *et al.*, 2011a,b). In this sense, it has been shown that old cells are more resistant to capsular release by γ-radiation than young cells, which indicates that capsule complexity and cross-linking increases with cell age (Cordero *et al.*, 2011b; Maxson *et al.*, 2007b). Changes in capsule structure also occur during infection (Charlier *et al.*, 2005; Garcia-Hermoso *et al.*, 2004; Gates *et al.*, 2004). This process is of extreme importance, since it can be considered as a mechanism of immune evasion. The response elicited by the cells of the immune system (Ab secretion or cytokine production) occurs in response to recognition of specific capsule structures, and thus changes in this structure will enable cryptococcal cells to evade the immune system (Cleare and Casadevall, 1998; Kozel *et al.*, 2003; Mukherjee *et al.*, 1998; Todaro-Luck *et al.*, 1989). Furthermore, it is known that changes in capsule structure are required for dissemination and organ colonization (Charlier *et al.*, 2005; Garcia-Hermoso *et al.*, 2004).

The size of the capsule is variable, not only among strains, but also between individual cells from the same strain (Fig. 9.1, see review in Zaragoza, 2011). This is a dramatic change for the cell, since the cellular dry weight can increase by up to 20% due to accumulation of new polysaccharide on the capsule, and the capsule can account for more than the 95% of the total cell volume (Maxson *et al.*, 2007a). Changes in capsule size are characteristic of the outcome of the interaction between *C. neoformans* and the host and are considered an early morphologic response during infection (Feldmesser *et al.*, 2001a). Moreover, this phenomenon is also observed during *C. neoformans* intracellular parasitism of macrophages (Ma *et al.*, 2006) and during interaction with non-mammalian models, such as *Galleria mellonella* (García-Rodas *et al.*, 2011) and amoebae (Chrisman *et al.*, 2011). There are several factors that induce capsule enlargement such as a CO_2 enriched atmosphere (5%) (Granger *et al.*, 1985). These authors also demonstrated the relationship between capsular enlargement and virulence in *C. neoformans*, which was the first evidence for the role of the capsule in the pathogenesis of this microorganism. Iron limitation also results in capsule enlargement (Vartivarian *et al.*, 1993). Furthermore, mammalian serum, nutrient limitation, basic pH, and mannitol are also known to induce capsule enlargement *in vitro* (Guimaraes *et al.*, 2010; Zaragoza and Casadevall, 2004; Zaragoza *et al.*, 2003a). In addition, it has been shown that capsule enlargement is regulated at the polymer level, through the synthesis of new polysaccharide molecules of larger dimensions (Frases *et al.*, 2009; Yoneda and Doering, 2008),

The capsule can also change in polysaccharide density, becoming denser in the inner regions close to the cell wall (Gates *et al.*, 2004; Maxson *et al.*, 2007b; Pierini and Doering, 2001; Zaragoza *et al.*, 2006). This increase in density can be observed *in vitro* due to ageing (Cordero *et al.*, 2011b; Maxson *et al.*, 2007b) and *in vivo* during infection in mammals and non-vertebrate hosts (García-Rodas *et al.*, 2011; Gates *et al.*, 2004).

Capsule as a virulence factor

The capsule has multiple effects in the host, and for this reason, the capsular polysaccharide is considered the main virulence factor of *C. neoformans*. The capsule inhibits phagocytosis by macrophages (Kozel and Gotschlich, 1982) and other environmental predators such as amoebas (Steenbergen *et al.*, 2001). In addition, the capsule impairs a subset of processes which are dependent on phagocytosis such as T-cell responses (Syme *et al.*, 1999) or antigen processing and presentation by macrophages (Vecchiarelli *et al.*, 1994a,b). GXM modulates the innate and adaptive immune response through different mechanisms (Dong *et al.*, 1999; Dong and Murphy, 1995, 1997; Monari *et al.*, 2005a,b; Retini *et al.*, 1996; Vecchiarelli, 1996, 2000a, b; Vecchiarelli and Monari, 2012) and GXMGal has anti-inflammatory and apoptotic properties (Chow and Casadevall, 2011; Pericolini *et al.*, 2006; Vecchiarelli *et al.*, 2011). For these reasons, the cryptococcal capsular polysaccharides have been considered as a treatment for inflammatory disease, such as rheumatoid arthritis (Monari *et al.*, 2009; Vecchiarelli *et al.*, 2011).

Cryptococcus neoformans is an intracellular facultative pathogen, since it can avoid killing inside phagocytic cells. This yeast is able to exploit the intracellular environment for its replication (Feldmesser *et al.*, 2000; Lee *et al.*, 1995)

and also exit infected macrophages (Alvarez and Casadevall, 2006; Ma *et al.*, 2006) or spread from cell-to-cell (Alvarez and Casadevall, 2007; Ma *et al.*, 2007). Acapsular mutants cannot replicate inside phagocytic cells (Feldmesser *et al.*, 2000), nor can they be extruded from macrophages (Alvarez and Casadevall, 2006; Ma *et al.*, 2006). However, this phenotype can be restored by coating the acapsular mutants with GXM (Alvarez and Casadevall, 2006). These data confirm that the capsule is a key factor to avoid killing and to replicate inside macrophages.

In addition, there is an important accumulation of GXM-containing vesicles after engulfment of *C. neoformans*, which in turn alter macrophage function (Tucker and Casadevall, 2002). Moreover, capsule protects against reactive oxygen species and antifungal compounds (Zaragoza *et al.*, 2008), which provides a mechanism for the protective role of the capsule during intracellular pathogenesis.

Melanin and laccase

Melanin is a negatively charged, dark pigment widely found in nature, with an amorphous, hydrophobic structure that is insoluble in aqueous or organic fluids (Casadevall *et al.*, 2000).

Cryptococcus neoformans accumulates melanin in the presence of diphenolic compounds. The synthesis of melanin by *C. neoformans* requires exogenous dihydroxyphenolic or polyaminobenzene compounds and molecular oxygen (Nurudeen and Ahearn, 1979), which could explain the tropism of *C. neoformans* for the brain (Eisenman *et al.*, 2007). Melanin is synthesized by a phenoloxidase, specifically a laccase, encoded by the *LAC1* gene (Williamson, 1994). The expression of this gene is regulated by multiple factors including iron, copper, and glucose levels (Alspaugh *et al.*, 1997; Jiang *et al.*, 2009). Melanin synthesis from 3,4-dyhydroxyphenylalanine (L-dopa) was validated by electron spin resonance spectroscopy criteria and nowadays it is the most used medium to study melanization (Eisenman *et al.*, 2007). In addition, *lac1* mutants cannot melanise and display a delay in dissemination from the lungs (Noverr *et al.*, 2004; Williamson, 1997). Laccase is localized in the fungal cell wall and melanin may contribute to fungal cell survival by maintaining cell wall integrity (Zhu *et al.*, 2001). In addition, there is a second laccase enzyme, encoded by the *LAC2* gene, which has a cytoplasmatic localization and a 75% nucleotide identity with *LAC1* (Heitman *et al.*, 2011; Missall *et al.*, 2005). Both laccases contribute to *C. neoformans* melanin production, although *LAC2* basal transcript levels are much lower than those of *LAC1* (Pukkila-Worley *et al.*, 2005).

Several studies have associated melanin synthesis and virulence. Melanization occurs both *in vitro* and *in vivo* (Nosanchuk *et al.*, 1999). Mutants unable to produce melanin have reduced virulence compared with wild-type strains (Kwon-Chung *et al.*, 1982; Williamson, 1997). Melanin protects cryptococcal cells from multiple stress factors, such as UV-light and free radicals, which contributes to the survival of the fungal cells in the host (Casadevall *et al.*, 2000; Liu *et al.*, 1999; Noverr *et al.*, 2004; Wang and Casadevall, 1994). Furthermore, melanized cells are resistant to antifungal and antimicrobial compounds suggesting that melanization can result in clinical therapy failure (van Duin *et al.*, 2002). In addition to its protective role, injection of melanin ghosts in mice results in tissue inflammation in organs such as the spleen (Mednick *et al.*, 2005), indicating that melanin itself can alter the host immune response and act as a virulence factor. Moreover, monoclonal antibodies to melanin protect against *C. neoformans* infection in mice (Rosas *et al.*, 2001) which correlates with the observation that *C. neoformans* melanizes in tissues (Nosanchuk *et al.*, 2000).

Development of titan/giant cells, microcells, and pseudohyphae

Although classic morphological changes, such as hypha and pseudohypha formation are rare in *C. neoformans*, it exhibits variations in cellular structure and size, which have important consequences during infection. Some of these changes have been described above, which are related to capsule structure and capsule size. But one striking phenomenon that has been recently described in *C. neoformans* is its ability to form titan/giant cells (Fig. 9.3) (Okagaki *et al.*, 2010; Zaragoza *et al.*, 2010). *In vitro*, *C. neoformans* total cell size can vary from 5–7 μm. However, *in vivo*, cells sizes of

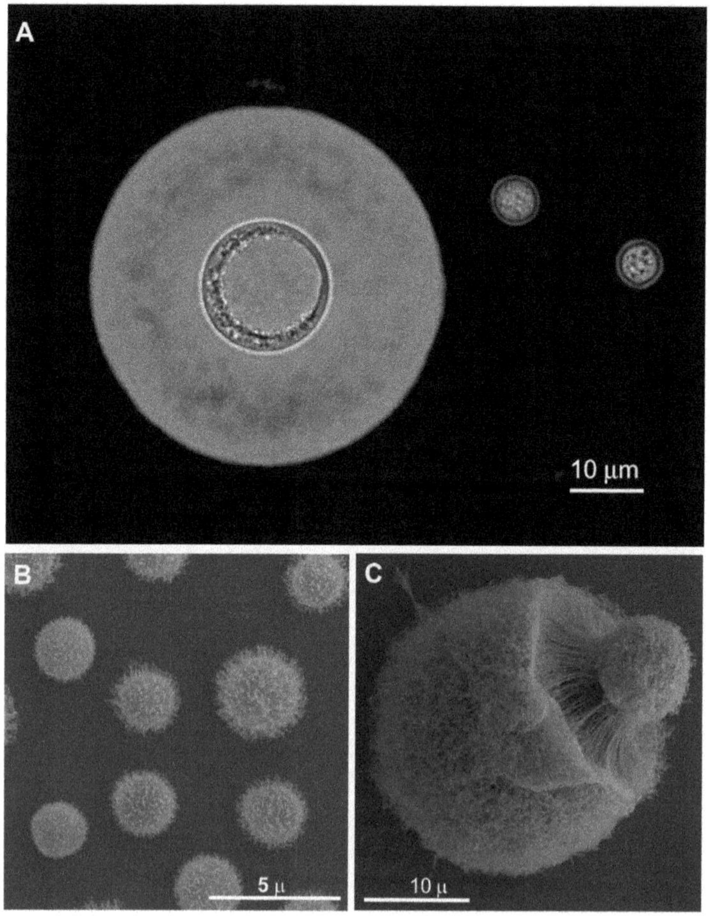

Figure 9.3 Titan/giant cryptococcal cells recovered from mouse lungs. (A) Picture of India ink suspensions of cryptococcal cells isolated from mouse lungs extract showing a titan/giant cell and two regular cryptococcal cells. (B) Scanning electron microscopy of regular and (C) giant/titan cells. Pictures from panels (B) and (C) were obtained from Zaragoza *et al.* (2010).

up to 100 μm have been described (Cruickshank *et al.*, 1973; Feldmesser *et al.*, 2001a; Okagaki *et al.*, 2010; Zaragoza *et al.*, 2010).

These cells show phenotypic differences compared to cells grown *in vitro*, such as a denser capsule (Fig. 9.3C), enlarged cell wall, and polyploidy (Okagaki *et al.*, 2010; Zaragoza *et al.*, 2010). The proportion of titan cells during infection is variable. In regular conditions its frequency is very low (<2–5%), but it increases in conditions of asymptomatic chronic infection (Zaragoza *et al.*, 2010) or when there is a coinfection with MATa and MATα mating types (Okagaki *et al.*, 2010). The signalling pathways involved in the formation of these cells are still uncharacterized, but some elements are already known. For instance, adenylate cyclase is required for cellular enlargement (Zaragoza *et al.*, 2010), indicating that cAMP and PKA signalling pathway are required for the titan cell formation (Choi *et al.*, 2012; Zaragoza *et al.*, 2010). Moreover, titan cell formation is regulated by a G-protein coupled receptor signal transduction pathway associated with pheromone sensing (Okagaki *et al.*, 2010, 2011). Titan cells can be found after the interaction of *C. neoformans* with *Galleria mellonella* (García-Rodas *et al.*, 2011). Moreover, injection of *G. mellonella* with titan cells produces acute killing of the larvae, indicating that these cells are virulent. Their production promotes survival in the host and enhances cryptococcal virulence by inhibiting phagocytosis, not only of themselves, but of regular size cryptococcal cells as well (Crabtree *et al.*, 2012; Okagaki and Nielsen, 2012).

In addition, microforms (cells of a diameter around 1 µm) have also been described during infection of *C. neoformans* (Feldmesser *et al.*, 2001a). However, their importance and mechanism of formation are still unknown.

Cryptococcus neoformans can form hyphae during sexual reproduction and mating, but these forms are rare during vegetative growth or infection. Pseudohyphal forms have been occasionally described in *C. neoformans* after co-incubation with *Acanthamoeba* species, and have been considered as an adaptation mechanism after the interaction with this host (Neilson *et al.*, 1981). Recently, the molecular basis of this process has been elucidated. Pseudohypha development is associated with mutations in the RAM pathway (Magditch *et al.*, 2012). Moreover, it can also occur in response to nitrogen limitation (Lee *et al.*, 2012). The pseudohyphal phenotype is not stable and is avirulent in mammalian models of cryptococcosis (Magditch *et al.*, 2012; Neilson *et al.*, 1981), so its role during infection remains to be elucidated.

Other characteristics that favour virulence

Growth at 37°C

Adaptation to physiological temperature is essential to cause disease in humans. Although *C. neoformans* is an environmental yeast, it can grow at 37°C, which gives this yeast the ability to cause disease in susceptible hosts (reviewed in Perfect, 2005, 2006). Calcineurin is essential for *C. neoformans* survival at 37°C but not at lower temperatures (Odom *et al.*, 1997). Many genes have been described to be necessary for growth at high temperatures, such as *COX1* (cytochrome oxidase c subunit 1) (Toffaletti *et al.*, 2003), *AOX1* (alternative oxidase) (Akhter *et al.*, 2003), *VPH1* (vacuolar ATP-ase) (Erickson *et al.*, 2001), and *ILV2* (an acetolactate synthase) (Kingsbury *et al.*, 2004). In addition, genes from different signalling pathways such as *RAS1*, *MKP1* are also required for growth at 37°C (Alspaugh *et al.*, 2000; Kraus *et al.*, 2003).

Urease

Most cryptococcal strains contain urease, which hydrolyses urea to ammonium and carbamate (Casadevall and Perfect, 1998). This enzyme is a virulence determinant since an *ure1* mutant showed decreased pathogenicity (Cox *et al.*, 2000). In addition, urease promotes lethal meningitis in a pulmonary infection mouse model by polarizing the T-cell helper 2 (Th2) immune response and promoting the accumulation of immature dendritic cells in the lung, impairing the clearance of the infection (Osterholzer *et al.*, 2009). Real-time imaging has recently shown a urease-dependent transmigration of *C. neoformans* in the mouse brain (Olszewski *et al.*, 2004; Shi *et al.*, 2010). This finding supports the idea of urease as a novel virulence factor for *C. neoformans* and, thus, a possible target for therapeutic strategies.

Phospholipases

Two phospholipases have been described in *C. neoformans*: phospholipase B (PLB), which is secreted into the extracellular environment where it hydrolyses host phospholipids (Chen *et al.*, 1997, 2000; Vidotto *et al.*, 1996), and phospholipase C (PLC), which hydrolyses phosphatidylinositol (PI) substrates within the cryptococcal cell and interferes with different cellular functions including Plb1 secretion (Chayakulkeeree *et al.*, 2008). In addition, extracellular phospholipase B promotes the survival and replication of *C. neoformans* in macrophages *in vitro*, confirming its role in cryptococcal pathogenicity (Santangelo *et al.*, 2004). Interestingly, Plb1 is required for dissemination from the lung to regional lymph nodes and for the entry into the bloodstream but it is not essential for the establishment of neurological infections (Santangelo *et al.*, 2004).

Host–pathogen interaction

Cryptococcus neoformans interferes with the immune response at multiple levels, and as a consequence, cryptococcal infection is associated with immunological unresponsiveness. In this section we will provide a brief overview of the most characteristic aspects of the *Cryptococcus*–host interaction.

Immune response elicited by T cells and dendritic cells

The interaction between *C. neoformans* and the immune system is very complex and has been intensely studied (see review in (Wozniak and Levitz, 2011) and (Vecchiarelli, 2000a, b, 2007). In general, the most effective response against the infection is based on Th1 immunity. In this context, $CD4^+$ and $CD8^+$ T cells play a major role in controlling *C. neoformans* infection and for this reason, HIV patients with low $CD4^+$ cells counts are highly susceptible to the infection.

Dendritic cells (DCs) function as sentinels in the innate immune system and are a heterogeneous population present throughout the body (Crowley *et al.*, 1990). Immature dendritic cells phagocytose *C. neoformans* cells in the lungs (Syme *et al.*, 2002). This results in DC maturation, which transforms them into efficient antigen presenting cells (APCs). Dendritic cells migrate to the lymphoid organs where they present antigens to native T cells and produce cytokines that modulate the adaptative immune response (Banchereau *et al.*, 2000). These cytokines can polarize T-cell development or induce tolerance. There is a strong association between development of a Th1 response (Koguchi and Kawakami, 2002; Zhang *et al.*, 2009), promoted by IL-12 and IFN-γ, and protection against cryptococcosis. A Th2 response is considered non-protective for cryptococcosis and is promoted by IL-4, IL-10, IL-13, and the absence of IL-12 (Muller *et al.*, 2007). The role of regulatory T-cells and Th17 cells in cryptococcosis is not very clear, although a Th17 response has been associated with protection (Wozniak *et al.*, 2011; Zhang *et al.*, 2009).

Animal models have shown that an adaptative T-cell mediated immune response is required for protection against cryptococcosis. However, restoring $CD4^+$ T-cell function in patients with impaired T-cell function, such as HIV-infected patients or patients receiving immunosuppressive medication, can result in an 'immune response inflammatory syndrome', which could be life-threatening (Chuck and Sande, 1989; Haddow *et al.*, 2012; Lortholary *et al.*, 2005).

Interaction with phagocytic cells: phagocytosis avoidance and intracellular parasitism

After being inhaled, *Cryptococcus* interacts with alveolar macrophages and neutrophils, and the interaction between this yeast and phagocytic cells has been intensively studied. It is believed that the outcome of this interaction plays a significant role in the disease produced by this pathogen.

Cryptococcus uptake mechanisms

Early studies demonstrated that *C. neoformans* is poorly phagocytosed (Bolaños and Mitchell, 1989b; Levitz and DiBenedetto, 1989). This phenomenon depends on the capsule (Kozel and Gotschlich, 1982), since acapsular mutants are normally phagocytosed. Phagocytosis occurs after binding of macrophage receptors to fungal epitopes that are mainly present at the cell wall. Thus, in the case of *Cryptococcus*, the presence of the capsule impairs this recognition (Small and Mitchell, 1989).

In recent years, at least two antiphagocytic capsule-independent mechanisms have been described. One depends on App1 (antiphagocytic protein 1), which inhibits phagocytosis by binding to complement receptors CR2 and CR3 (Luberto *et al.*, 2003; Stano *et al.*, 2009). The second one depends on Gat201, a pleiotropic virulence determinant, which also plays a role in phagocytosis avoidance (Chun *et al.*, 2011; Liu *et al.*, 2008).

Despite these antiphagocytic mechanisms, yeast cells are found inside macrophages after a few hours of infection (Feldmesser *et al.*, 2000), which implies that there are mechanisms that overcome the antiphagocytic effects described above (reviewed in (Garcia-Rodas and Zaragoza, 2012). These mechanisms rely on the presence of opsonins which bind to the capsule and have specific receptors on the phagocytic cells. The main opsonins described for *C. neoformans* are proteins from the complement system and antibodies.

Cryptococcus neoformans induces complement activation through the alternative pathway, resulting in deposition of iC3b on the capsule (Kozel *et al.*, 1984, 1988, 1989), which facilitates phagocytosis through the complement receptors (CR), in particular CR3 (Cross *et al.*, 1997; Levitz, 2002;

Zaragoza *et al.*, 2003b). Phagocytosis mediated through antibodies consists of binding between the Fc region of the antibody and the Fc receptors of macrophages. This process is influenced by many factors, such as Ab isotype (Saylor *et al.*, 2010; Yuan *et al.*, 1998), epitope localization, and binding properties (Cleare and Casadevall, 1998; Nussbaum *et al.*, 1997).

However, other phagocytic mechanisms which do not depend on opsonisation have also been described. For example, anti-GXM IgM antibodies are opsonic for *C. neoformans*, although this isotype does not bind to any Fc receptor. In this case, the binding of the IgM Ab induces capsule rearrangements that allow binding of some capsular epitopes to CD18, inducing CR-depending phagocytosis (Taborda and Casadevall, 2002). Furthermore, different receptors such as mannose receptor in dendritic cells (Mansour *et al.*, 2006; Syme *et al.*, 2002) or FcRγII in macrophages (Monari *et al.*, 2005b) are known to mediate *C. neoformans* phagocytosis (Kozel and Gotschlich, 1982; Mansour and Levitz, 2002; Small and Mitchell, 1989).

Intracellular fate of cryptococcal cells

Once *C. neoformans* is internalized, multiple outcomes are possible. Macrophages can kill *C. neoformans* after ingestion (Bolaños and Mitchell, 1989a,b; Kitz *et al.*, 1984). However, the yeast can also survive and replicate inside phagocytic cells (Fig. 9.4) (Del Poeta, 2004; Feldmesser *et al.*, 2001b) through several mechanisms (reviewed in (Garcia-Rodas and Zaragoza, 2012; Voelz *et al.*, 2011)). Although *Cryptococcus* does not avoid phagosome–lysosome fusion, the phagolysosomes containing cryptococcal cells become leaky (Tucker and Casadevall, 2002), which suggests that these phagolysosomes have altered functionality. Moreover, *C. neoformans* impairs the production of nitric oxide (Naslund *et al.*, 1995), an important host mechanism for killing phagocytosed pathogens (Alvarez *et al.*, 2009; Diamond and Bennett, 1973; Lee *et al.*, 1995).

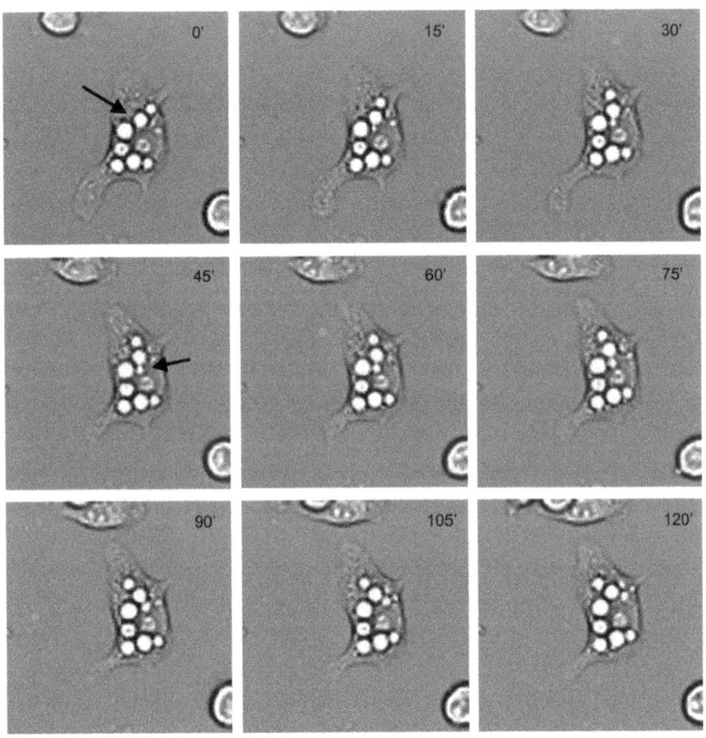

Figure 9.4 Intracellular replication of *C. neoformans* in a murine macrophage. The sequence of images shows the process of budding of *C. neoformans* within a macrophage. The first black arrow indicates the beginning of the budding process. Second black arrow shows the emerging bud.

During intracellular parasitism, there is an accumulation of GXM-containing vesicles that can damage the phagolysosome (Tucker and Casadevall, 2002), leading to polysaccharide release into the cytoplasm where it can interfere with macrophage functionality. In this sense, GXM can bind to phosphofructokinase, which provides a mechanism by which the capsular polysaccharide interferes with the macrophage's metabolism (Grechi *et al.*, 2011). Moreover, capsule enlargement results in phagolysosome dilatation, and thus, dilution of the lysosomal antimicrobial compounds.

Intracellular replication has been extensively observed *in vitro* (Coenjaerts *et al.*, 2006; Diamond and Bennett, 1973; Lee *et al.*, 1995; Mitchell and Friedman, 1972; Tucker and Casadevall, 2002) and occurs at a higher rate than budding of extracellular cells, which indicates that *C. neoformans* has developed adaptation mechanisms to the stress conditions of the phagolysosomes (Levitz *et al.*, 1999). In fact, experimental acidification of the phagosome leads to a decrease intracellular proliferation (Levitz *et al.*, 1997).

The fate of intracellular yeast is very diverse. Massive replication can lead to host cell lysis. Host cell lysis has been linked to phospholipase B1, although its ability to form pores on the phagosome membrane has not been demonstrated (Noverr *et al.*, 2003). Moreover, mechanical rupture of the phagosome due to the rapid intracellular replication of *C. neoformans* and polysaccharide accumulation has been shown as a plausible explanation for this phenomenon (Del Poeta, 2004; Feldmesser *et al.*, 2000).

However, there are additional mechanisms that allow escape from host cells without affecting the viability of the macrophages or the yeast cell. Extrusion or 'vomocytosis' (also called non-lytic exocytosis) happens in both murine and human macrophages (Alvarez and Casadevall, 2006; Ma *et al.*, 2006; Voelz; *et al.*, 2011) and consist of exiting from the phagocytic cells without lysis in a similar process to exocytosis. Real-time imaging experiments confirmed that extrusion depends on live yeast cells and that mAb opsonized cells were more frequently expelled that those opsonized by complement proteins (Alvarez and Casadevall, 2006). This process has also been studied using flow cytometry confirming that it happens both *in vitro* and *in vivo* in a murine model (Nicola *et al.*, 2011). Mechanisms to accomplish this phenomenon are still unknown but the Wiskott–Aldrich syndrome protein (WASP)-Arp2/3 may have a role (Johnston and May, 2010).

Finally, lateral transfer has been observed in *C. neoformans*-infected macrophages. This phenomenon depends on live yeast cells (Alvarez and Casadevall, 2007) and actin (Ma *et al.*, 2007) and could be considered as a strategy of *C. neoformans* for surviving in the host without being exposed to antifungal agents or other immune factors (Ma *et al.*, 2007).

The factors that influence and determine the fate of internalized cells are unknown, but the activation state of the macrophage seems to play a determinant role in these processes, since Th1 cytokines inhibit intracellular fungal proliferation (Voelz *et al.*, 2009).

Cryptococcus gattii can also behave as a facultative intracellular pathogen, and this aspect seems to correlate with the enhanced virulence shown by some of the strains isolated from the outbreak on Vancouver Island (Ma *et al.*, 2009).

Dissemination to the central nervous system (CNS)

Cryptoccocus neoformans neurotropism has been intensively studied, especially since this yeast is able to produce melanin from diphenolic compounds such as L-DOPA and dopamine, which are abundant in the brain. To reach the brain, fungemia, which is observed in 50% of HIV patients with cryptococcosis (Dromer *et al.*, 2007) must be present followed by traversal of the blood–brain barrier (BBB) (Chretien *et al.*, 2002).

The BBB is composed of brain microvascular endothelial cells (BMECs), astrocytes and pericytes (Kim, 2008). Three mechanisms have been proposed to explain traversal of the BBB by *C. neoformans*: 'Trojan horse' (Charlier *et al.*, 2009), transcellular penetration (Chang *et al.*, 2004), and paracellular penetration (Charlier *et al.*, 2005). The Trojan horse mechanism occurs when cryptococcal cells cross the BBB within infected macrophages (Dromer and Levitz, 2011; Kim, 2008). Phagocyte depletion results in reduced disease severity and lower fungal burden, which suggests that in fact macrophages contribute to the dissemination of the microorganisms to brain (Charlier *et al.*, 2009).

Transcellular penetration has been observed after incubation of *C. neoformans* with human BMECs monolayers. In these experiments, it was found that a low percentage of cryptococcal cells adhere and transverse the monolayer, without affecting the integrity of the mammalian cells (Chang *et al.*, 2004). Recent findings indicate that this is a passive mechanism induced when the cryptococcal cells are trapped by the BBB and can occur in the absence of the capsule (Sabiiti and May, 2012). In contrast to this process, *C. neoformans* can also transverse monolayers formed by BMECs through paracellular transport, which causes damage of the endothelial barrier (Charlier *et al.*, 2009; Chen *et al.*, 2003). Supporting this mechanism, histopathological studies have shown disruption of the BBB and the presence of yeast cells close to the damaged basal lamina and endothelial cells (Charlier *et al.*, 2005). In this sense, *C. neoformans* also alters the cytoskeleton of human brain endothelial cells (Chen *et al.*, 2003). Recently, a mouse brain transmigration assessment was used to study the pathogenesis of *C. neoformans* in the central nervous system (CNS) which clearly demonstrated the role of transcellular and paracellular transport (Shi *et al.*, 2010).

As stated above, none of these mechanisms are exclusive and the most probable scenario is that *C. neoformans* has developed different strategies to invade the CNS depending on the host immune state (Casadevall, 2010). Genetic approaches which investigate the transport of different mutants through the BBB support this idea (Tseng *et al.*, 2012). Furthermore, laccase participates in dissemination to the CNS. This enzyme seems to induce changes in the polarization of $CD4^+$ T cells which contributes to CNS dissemination (Qiu *et al.*, 2012).

Latency and reactivation

There is increasing evidence that *C. neoformans* can reside in the host causing latent asymptomatic infections. A screen for anti- *C. neoformans* antibodies in individuals aged from 1 week to 21 years old in New York City provided evidence that infection might occur early in life (Goldman *et al.*, 2001). In addition, there are differences in the reactivity of the sera depending on geographical location, confirming the influence of regional differences in exposure to *C. neoformans* (Davis *et al.*, 2007). Furthermore, an epidemiological study of African patients diagnosed in France suggested that cryptococcal infection can reactivate from latency (Garcia-Hermoso *et al.*, 1999).

In addition to this latent state, there is also evidence of subclinical infection and reactivation following immunosuppression in patients who developed cryptococcosis after transplantation (Saha *et al.*, 2007). The mechanism by which *C. neoformans* remains latent in the host is still unknown. Dormant yeast cells are believed to be localized in subpleural nodules and lymph nodes (Paterson *et al.*, 1998). In men, the prostate could also represent a site for chronic asymptomatic infection (Larsen *et al.*, 1989). Macrophages could also be a site for latency since *C. neoformans* is able to survive and replicate within macrophages (see above). Rats are the best characterized animal model for the study of latency since they are intrinsically resistant to cryptococcosis (Goldman *et al.*, 1994). Rats are unable to eliminate the fungus but elicit a granulomatous response, where cryptococcal cells are usually found inside macrophages (Goldman *et al.*, 2000). Finally, titan/giant cells could also provide a mechanism by which *C. neoformans* could remain in the host for long time periods evading the host immune response (Okagaki *et al.*, 2010; Zaragoza *et al.*, 2010).

Cryptococcus as environmental pathogen

Cryptococcus has been classically associated with pigeon droppings, trees, soil, and water. Although it has not been fully elucidated, it is accepted that it enters in the human host by inhalation and it first resides in the lungs. In recent years, different studies have demonstrated that spores can behave as infective agents, findings that have heightened interest in the sexual cycle in the environment. For instance, *C. neoformans* can complete its sexual cycle in association with plants and pigeon droppings (Nielsen *et al.*, 2007; Xue *et al.*, 2007).

One of the interesting aspects of this fungus is its ability to interact not only with the mammalian host, but also with environmental predators. Nematodes, insects, and amoebae are susceptible to cryptococcal infection and the study of these non-conventional hosts has become of great interest for two main reasons. Firstly, they offer a model to study host–pathogen interactions that reduces the bioethical problems associated with classical animal experimentation (Chamilos *et al.*, 2007; Desalermos *et al.*, 2012). Secondly, they also provide an excellent model to identify conserved infection mechanisms among different hosts. For example, amoebae can phagocytose *C. neoformans* cells and the interaction between these organisms is similar to that observed with mammalian macrophages. *Cryptococcus neoformans* can replicate inside amoebae (Steenbergen *et al.*, 2001), and also be extruded from these cells (Chrisman *et al.*, 2010). This finding suggests that the environmental interaction of *C. neoformans* with amoebas has resulted in the acquisition of mechanisms that allow evasion of killing that may also be used during infection in mammals. Additionally, in this model the capsule also plays an important role in the outcome of the interaction given that acapsular mutants are not able to replicate intracellularly (Steenbergen *et al.*, 2001).

In the case of insects, two different models have been used: the lepidopteran *Galleria mellonella* and the fly *Drosophila melanogaster*. *Galleria mellonella* offers an excellent system to study innate responses (mainly phagocytosis), and more specifically, virulence traits of the pathogen. Genes required for virulence in mammals, such as *CAP59*, *RAS1*, *PKA1* and *GPA1*, are also required for virulence in *G. mellonella* (Mylonakis *et al.*, 2005). *Cryptococus neoformans* induces capsular enlargement and titan/giant cell formation in *G. mellonella* in a similar way as in mammalian models (García-Rodas *et al.*, 2011). In this way, it is also hypothesized that morphogenesis in *C. neoformans* is a virulence trait acquired from environmental interactions that confers persistence in the host. *Drosophila melanogaster* is a more versatile model because it has both adaptive and innate immune responses. In addition, genetic tools are available for this model, so it is possible to investigate the role of specific host elements in the immune response to this pathogen. For instance, a particular role for Toll receptors was elucidated during systemic infection after injection in the hemolymph, but not after ingestion and gastrointestinal localization (Apidianakis *et al.*, 2004). Recently, based on RNAi technology, and using the *D. melanogaster* S2 macrophage-like cell line, autophagy proteins were identified as predominant elements that regulate the intracellular behaviour of *C. neoformans* in these phagocytes (Qin *et al.*, 2011). Autophagy was later confirmed in mammalian cells as a mechanism to control intracellular replication of *C. neoformans* and *Candida albicans* (Nicola *et al.*, 2012) in murine macrophages.

Finally, the nematode *Caenorhabditis elegans* has also been used to study host–cryptococcal interactions (Mylonakis *et al.*, 2002). This model has been useful to elucidate the role of specific pathogen factors in the virulence of both *C. neoformans* and *C. gattii* (Chayakulkeeree *et al.*, 2008; Mylonakis *et al.*, 2004; Ngamskulrungroj *et al.*, 2009; Tang *et al.*, 2005). The availability of genetics and the possibility of large-scale experiments also make this nematode a suitable model to investigate immune responses and antifungal efficacy during *in vivo* infection.

Diagnosis of cryptococcosis

The laboratory diagnosis of cryptococcosis has evolved from microscopic methods and culture to antigen detection. However, sensitivity and specificity varies, not only between methods, but also between type of samples and type of patients.

The cryptococcal capsule is an excellent feature to distinguish this yeast from other medically important fungi. Detection of the capsule by India ink offers a simple and quick technique that detects from 30% to 70% of the cases of meningitis in non-HIV-infected patients and up to 80% in HIV-infected patients, where the fungal burden is usually higher. India ink can be applied to different fluids such as cerebrospinal fluid (CSF), urine, and serum. However, its sensitivity depends on the fungal burden and the experience of the clinician (Diaz and Nguyen, 2011; Saha *et al.*, 2009).

Detection of capsular polysaccharide antigen is the most valuable test for diagnosis of cryptococcosis. Capsular polysaccharide is released in body fluids such as CSF and serum during infection, and thus, it can be detected and quantified with specific rabbit anti-cryptococcal antiserum. There are several formats for capsular antigen testing: LAT assays (latex agglutination test) and EIA (enzyme immunoassay).

The latex agglutination test relies on agglutination of latex particles covered with immunoglobulin against cryptococcal polysaccharide with antigens present in the fluids tested (CSF, serum, etc.). Sensitivity of the technique reaches 100% whereas specificity decreases due to false-positives (Saha *et al.*, 2008, 2009). The most common cause of false positive is the presence of rheumatoid factor (Bennett and Bailey, 1971; Heelan *et al.*, 1991).

The EIA kit detects glucuronoxylomannan. It is more expensive than LAT assays and requires a spectrophotometer (Diaz and Nguyen, 2011). However, it does not require pre-treatment and does not react with rheumatoid factor (Engler and Shea, 1994). Although it offers the same sensitivity as LAT assays, EIA specificity is higher compared to microscopic techniques and LAT assays (Frank *et al.*, 1993; Gade *et al.*, 1991; Saha *et al.*, 2009).

Recently, a lateral flow and enzyme-linked immunoassay (LFA) has been developed for the detection of cryptococcal antigen in serum and cerebrospinal fluid (Jarvis *et al.*, 2011; Lindsley *et al.*, 2011). It consists of a low cost and ready-to-use membrane strip coated with a mixture of two anti-GXM mAbs. The results can be read after 10 minutes and sensitivity and specificity have been shown to be similar to EIA (Hansen *et al.*, 2012). Consequently, the main advantage of the LFA is the feasibility of use in a point-of-care setting in developing countries where resources are limited but cryptococcosis has a high incidence.

The application of real-time PCR for diagnosis allows the detection of pathogen DNA in clinical samples. A multiplex-PCR Luminex xTAG has been employed for detection of clinically important fungi from blood cultures including *C. neoformans* (Balada-Llasat *et al.*, 2012). Recently, a multi-target real time PCR was set up to identify organisms responsible for meningitis using cerebrospinal fluid from patients (Favaro *et al.*, 2012). This method was reported to be quick, reliable, and easy to perform.

Loop-mediated isothermal amplification technique (LAMP) is a recently described technique, which consists of denaturing and replicating the fungal DNA simultaneously without changing temperature levels. This makes it a cost-effective alternative to PCR as there is no need for a thermal cycler but only a thermal block or a water bath. A LAMP technique has been recently developed for *C. neoformans* and *C. gattii* using CSF samples from HIV-infected patients (Rezaei Amirabadi, 2012).

Perspectives and conclusions

The incidence of cryptococcosis remains high in developing areas and its associated mortality is unacceptable. For this reason, further efforts are required to reduce the impact of this disease. The development of cheaper methods to diagnose the disease, such as the LFA, have great potential to facilitate early detection of the disease in developing countries. The treatment of the disease is complicated for different reasons. Firstly, only a limited number of antifungals are effective, such as polyenes and azoles. In the case of *C. neoformans*, echinocandins are not active, so treatment is based on amphotericin B and azoles, especially fluconazole. Conventional amphotericin B is very toxic, and the new liposomal formulations that present reduced toxicity have very limited use in developing areas due to their high cost. Regarding the azoles, fluconazole is the most commonly used drug to treat cryptococcosis given its low cost and its good diffusion across the BBB. Therefore, new antifungal treatments are needed to efficiently control the disease. Moreover, the distribution of the highly active antiretroviral therapy in developing areas or the development of an efficient vaccine against HIV would be key to decrease the incidence of cryptococcosis worldwide.

The design of new antifungal therapies requires the understanding of the virulence mechanisms of the pathogen. In this sense, *C. neoformans* is a unique model, since it possesses a virulence factor (the polysaccharide capsule) that impairs the host immune response. Despite years of investigation, there are still key aspects about the biology of the capsule that remain unknown, specially its synthesis and assembly. The elucidation of the mechanisms that regulate capsule synthesis will allow the design of specific antifungal therapies. Finally, *C. neoformans* offers a singular model to study host–pathogen interaction, because it can range from a latent disease state to a severe disseminated disease. For this reason, more studies are needed to understand the responses that allow persistence in the host, such as the formation of giant/titan cells, intracellular pathogenesis, and dissemination to the brain.

References

Akhter, S., McDade, H.C., Gorlach, J.M., Heinrich, G., Cox, G.M., and Perfect, J.R. (2003). Role of alternative oxidase gene in pathogenesis of *Cryptococcus neoformans*. Infect. Immun. *71*, 5794–5802.

Alspaugh, J.A., Perfect, J.R., and Heitman, J. (1997). *Cryptococcus neoformans* mating and virulence are regulated by the G-protein alpha subunit GPA1 and cAMP. Genes Dev. *11*, 3206–3217.

Alspaugh, J.A., Cavallo, L.M., Perfect, J.R., and Heitman, J. (2000). *RAS1* regulates filamentation, mating and growth at high temperature of *Cryptococcus neoformans*. Mol. Microbiol. *36*, 352–365.

Alvarez, M., and Casadevall, A. (2006). Phagosome extrusion and host-cell survival after *Cryptococcus neoformans* phagocytosis by macrophages. Curr. Biol. *16*, 2161–2165.

Alvarez, M., and Casadevall, A. (2007). Cell-to-cell spread and massive vacuole formation after *Cryptococcus neoformans* infection of murine macrophages. BMC Immunol. *8*, 16.

Alvarez, M., Burn, T., Luo, Y., Pirofski, L.A., and Casadevall, A. (2009). The outcome of *Cryptococcus neoformans* intracellular pathogenesis in human monocytes. BMC Microbiol. *9*, 51.

Aminnejad, M., Diaz, M., Arabatzis, M., Castaneda, E., Lazera, M., Velegraki, A., Marriott, D., Sorrell, T.C., and Meyer, W. (2012). Identification of novel hybrids between *Cryptococcus neoformans* var. *grubii* VNI and *Cryptococcus gattii* VGII. Mycopathologia *173*, 337–346.

Apidianakis, Y., Rahme, L.G., Heitman, J., Ausubel, F.M., Calderwood, S.B., and Mylonakis, E. (2004). Challenge of *Drosophila melanogaster* with *Cryptococcus neoformans* and role of the innate immune response. Eukaryot. Cell. *3*, 413–419.

Balada-Llasat, J.M., LaRue, H., Kamboj, K., Rigali, L., Smith, D., Thomas, K., and Pancholi, P. (2012). Detection of yeasts in blood cultures by the Luminex xTAG fungal assay. J. Clin. Microbiol. *50*, 492–494.

Banchereau, J., Briere, F., Caux, C., Davoust, J., Lebecque, S., Liu, Y.J., Pulendran, B., and Palucka, K. (2000). Immunobiology of dendritic cells. Annu. Rev. Immunol. *18*, 767–811.

Bennett, J.E., and Bailey, J.W. (1971). Control for rheumatoid factor in the latex test for cryptococcosis. Am. J. Clin. Pathol. *56*, 360–365.

Bhattacharjee, A.K., Bennett, J.E., and Glaudemans, C.P. (1984). Capsular polysaccharides of *Cryptococcus neoformans*. Rev. Infect. Dis. *6*, 619–624.

Bolaños, B., and Mitchell, T.G. (1989a). Killing of *Cryptococcus neoformans* by rat alveolar macrophages. J. Med. Vet. Mycol. *27*, 219–228.

Bolaños, B., and Mitchell, T.G. (1989b). Phagocytosis of *Cryptococcus neoformans* by rat alveolar macrophages. J. Med. Vet. Mycol. *27*, 203–217.

Bose, I., Reese, A.J., Ory, J.J., Janbon, G., and Doering, T.L. (2003). A yeast under cover: the capsule of *Cryptococcus neoformans*. Eukaryot. Cell. *2*, 655–663.

Bovers, M., Hagen, F., Kuramae, E.E., Diaz, M.R., Spanjaard, L., Dromer, F., Hoogveld, H.L., and Boekhout, T. (2006). Unique hybrids between the fungal pathogens *Cryptococcus neoformans* and *Cryptococcus gattii*. FEMS Yeast Res. *6*, 599–607.

Byrnes, E.J., and Heitman, J. (2009). *Cryptococcus gattii* outbreak expands into the Northwestern United States with fatal consequences. F1000 Biol. Rep. *1*.

Byrnes, E.J., 3rd, and Marr, K.A. (2011). The Outbreak of *Cryptococcus gattii* in Western North America: Epidemiology and Clinical Issues. Curr. Infect. Dis. Rep. *13*, 256–261.

Byrnes, E.J., 3rd, Bildfell, R.J., Frank, S.A., Mitchell, T.G., Marr, K.A., and Heitman, J. (2009). Molecular evidence that the range of the Vancouver Island outbreak of *Cryptococcus gattii* infection has expanded into the Pacific Northwest in the United States. J. Infect. Dis. *199*, 1081–1086.

Byrnes, E.J., 3rd, Li, W., Lewit, Y., Ma, H., Voelz, K., Ren, P., Carter, D.A., Chaturvedi, V., Bildfell, R.J., May, R.C., *et al.* (2010). Emergence and pathogenicity of highly virulent *Cryptococcus gattii* genotypes in the northwest United States. PLoS Pathog. *6*, e1000850.

Byrnes, E.J., 3rd, Li, W., Ren, P., Lewit, Y., Voelz, K., Fraser, J.A., Dietrich, F.S., May, R.C., Chaturvedi, S., Chaturvedi, V., *et al.* (2011). A diverse population of *Cryptococcus gattii* molecular type VGIII in southern Californian HIV/AIDS patients. PLoS Pathog. *7*, e1002205.

Casadevall, A. (2010). Cryptococci at the brain gate: break and enter or use a Trojan horse? J. Clin. Invest. *120*, 1389–1392.

Casadevall, A., and Perfect, J. (1998). *Cryptococcus neoformans* (ASM Press, Washington DC).

Casadevall, A., Mukherjee, J., Devi, S.J., Schneerson, R., Robbins, J.B., and Scharff, M.D. (1992). Antibodies elicited by a *Cryptococcus neoformans*-tetanus toxoid conjugate vaccine have the same specificity as those elicited in infection. J. Infect. Dis. *165*, 1086–1093.

Casadevall, A., DeShaw, M., Fan, M., Dromer, F., Kozel, T.R., and Pirofski, L.A. (1994). Molecular and idiotypic analysis of antibodies to *Cryptococcus neoformans* glucuronoxylomannan. Infect. Immun. *62*, 3864–3872.

Casadevall, A., Rosas, A.L., and Nosanchuk, J.D. (2000). Melanin and virulence in *Cryptococcus neoformans*. Curr. Opin. Microbiol. *3*, 354–358.

Castle, S.A., Owuor, E.A., Thompson, S.H., Garnsey, M.R., Klutts, J.S., Doering, T.L., and Levery, S.B. (2008). Beta-1,2-xylosyltransferase Cxt1p is solely responsible for xylose incorporation into *Cryptococcus neoformans* glycosphingolipids. Eukaryot. Cell. *7*, 1611–1615.

Chamilos, G., Lionakis, M.S., Lewis, R.E., and Kontoyiannis, D.P. (2007). Role of mini-host models in the study of medically important fungi. Lancet Infect. Dis. *7*, 42–55.

Chang, Y.C., and Kwon-Chung, K.J. (1994). Complementation of a capsule-deficient mutation of *Cryptococcus neoformans* restores its virulence. Mol. Cell. Biol. *14*, 4912–4919.

Chang, Y.C., and Kwon-Chung, K.J. (1998). Isolation of the third capsule-associated gene, CAP60, required for virulence in *Cryptococcus neoformans*. Infect. Immun. *66*, 2230–2236.

Chang, Y.C., and Kwon-Chung, K.J. (1999). Isolation, characterization, and localization of a capsule-associated gene, *CAP10*, of *Cryptococcus neoformans*. J. Bacteriol. *181*, 5636–5643.

Chang, Y.C., Penoyer, L.A., and Kwon-Chung, K.J. (1996). The second capsule gene of *Cryptococcus neoformans*, *CAP64*, is essential for virulence. Infect. Immun. *64*, 1977–1983.

Chang, Y.C., Stins, M.F., McCaffery, M.J., Miller, G.F., Pare, D.R., Dam, T., Paul-Satyaseela, M., Kim, K.S., and Kwon-Chung, K.J. (2004). Cryptococcal yeast cells invade the central nervous system via transcellular penetration of the blood–brain barrier. Infect. Immun. *72*, 4985–4995.

Charlier, C., Chretien, F., Baudrimont, M., Mordelet, E., Lortholary, O., and Dromer, F. (2005). Capsule structure changes associated with *Cryptococcus neoformans* crossing of the blood–brain barrier. Am. J. Pathol. *166*, 421–432.

Charlier, C., Nielsen, K., Daou, S., Brigitte, M., Chretien, F., and Dromer, F. (2009). Evidence of a role for monocytes in dissemination and brain invasion by *Cryptococcus neoformans*. Infect. Immun. *77*, 120–127.

Chau, T.T., Mai, N.H., Phu, N.H., Nghia, H.D., Chuong, L.V., Sinh, D.X., Duong, V.A., Diep, P.T., Campbell, J.I., Baker, S., *et al.* (2010). A prospective descriptive study of cryptococcal meningitis in HIV uninfected patients in Vietnam – high prevalence of *Cryptococcus neoformans* var *grubii* in the absence of underlying disease. BMC. Infect. Dis. *10*, 199.

Chayakulkeeree, M., Sorrell, T.C., Siafakas, A.R., Wilson, C.F., Pantarat, N., Gerik, K.J., Boadle, R., and Djordjevic, J.T. (2008). Role and mechanism of phosphatidylinositol-specific phospholipase C in survival and virulence of *Cryptococcus neoformans*. Mol. Microbiol. *69*, 809–826.

Chen, J., Varma, A., Diaz, M.R., Litvintseva, A.P., Wollenberg, K.K., and Kwon-Chung, K.J. (2008). *Cryptococcus neoformans* strains and infection in apparently immunocompetent patients, China. Emerg. Infect. Dis. *14*, 755–762.

Chen, S.C., Wright, L.C., Santangelo, R.T., Muller, M., Moran, V.R., Kuchel, P.W., and Sorrell, T.C. (1997). Identification of extracellular phospholipase B, lysophospholipase, and acyltransferase produced by *Cryptococcus neoformans*. Infect. Immun. *65*, 405–411.

Chen, S.C., Wright, L.C., Golding, J.C., and Sorrell, T.C. (2000). Purification and characterization of secretory phospholipase B, lysophospholipase and lysophospholipase/transacylase from a virulent strain of the pathogenic fungus *Cryptococcus neoformans*. Biochem. J. *347*, 431–439.

Chen, S.H., Stins, M.F., Huang, S.H., Chen, Y.H., Kwon-Chung, K.J., Chang, Y., Kim, K.S., Suzuki, K., and Jong, A.Y. (2003). *Cryptococcus neoformans* induces alterations in the cytoskeleton of human brain microvascular endothelial cells. J. Med. Microbiol. *52*, 961–970.

Cherniak, R., and Sundstrom, J.B. (1994). Polysaccharide antigens of the capsule of *Cryptococcus neoformans*. Infect. Immun. *62*, 1507–1512.

Cherniak, R., Reiss, E., Slodki, M.E., Plattner, R.D., and Blumer, S.O. (1980). Structure and antigenic activity of the capsular polysaccharide of *Cryptococcus neoformans* serotype A. Mol. Immunol. *17*, 1025–1032.

Cherniak, R., Cheeseman, M.M., Reyes, G.H., Reiss, E., and Todaro, F. (1988). Enhanced binding of capsular polysaccharides of *Cryptococcus neoformans* to polystyrene microtitration plates for enzyme-linked immunosorbent assay. Diagn. Clin. Immunol. *5*, 344–348.

Choi, J., Vogl, A.W., and Kronstad, J.W. (2012). Regulated expression of cyclic AMP-dependent protein kinase A reveals an influence on cell size and the secretion of virulence factors in *Cryptococcus neoformans*. Mol. Microbiol. *85*, 700–715.

Chow, S.K., and Casadevall, A. (2011). Evaluation of *Cryptococcus neoformans* galactoxylomannan–protein conjugate as vaccine candidate against murine cryptococcosis. Vaccine *29*, 1891–1898.

Chretien, F., Lortholary, O., Kansau, I., Neuville, S., Gray, F., and Dromer, F. (2002). Pathogenesis of cerebral *Cryptococcus neoformans* infection after fungemia. J. Infect. Dis. *186*, 522–530.

Chrisman, C.J., Alvarez, M., and Casadevall, A. (2010). Phagocytosis of *Cryptococcus neoformans* by, and nonlytic exocytosis from, *Acanthamoeba castellanii*. Appl. Environ. Microbiol. *76*, 6056–6062.

Chrisman, C.J., Albuquerque, P., Guimaraes, A.J., Nieves, E., and Casadevall, A. (2011). Phospholipids trigger *Cryptococcus neoformans* capsular enlargement during interactions with amoebae and macrophages. PLoS Pathog. *7*, e1002047.

Chuck, S.L., and Sande, M.A. (1989). Infections with *Cryptococcus neoformans* in the acquired immunodeficiency syndrome. N. Engl. J. Med. *321*, 794–799.

Chun, C.D., Brown, J.C., and Madhani, H.D. (2011). A major role for capsule-independent phagocytosis-inhibitory mechanisms in mammalian infection by *Cryptococcus neoformans*. Cell Host Microbe *9*, 243–251.

Cleare, W., and Casadevall, A. (1998). The different binding patterns of two immunoglobulin M monoclonal antibodies to *Cryptococcus neoformans* serotype A and D strains correlate with serotype classification and differences in functional assays. Clin. Diagn. Lab. Immunol. *5*, 125–129.

Coenjaerts, F.E., Hoepelman, A.I., Scharringa, J., Aarts, M., Ellerbroek, P.M., Bevaart, L., Van Strijp, J.A., and Janbon, G. (2006). The Skn7 response regulator of *Cryptococcus neoformans* is involved in oxidative stress signalling and augments intracellular survival in endothelium. FEMS Yeast Res. *6*, 652–661.

Cordero, R.J., Frases, S., Guimaraes, A.J., Rivera, J., and Casadevall, A. (2011a). Evidence for branching in cryptococcal capsular polysaccharides and consequences on its biological activity. Mol. Microbiol. *79*, 1101–1117.

Cordero, R.J., Pontes, B., Guimaraes, A.J., Martinez, L.R., Rivera, J., Fries, B.C., Nimrichter, L., Rodrigues, M.L., Viana, N.B., and Casadevall, A. (2011b). Chronological aging is associated with biophysical and chemical changes in the capsule of *Cryptococcus neoformans*. Infect. Immun. *79*, 4990–5000.

Cottrell, T.R., Griffith, C.L., Liu, H., Nenninger, A.A., and Doering, T.L. (2007). The pathogenic fungus *Cryptococcus neoformans* expresses two functional GDP-mannose transporters with distinct expression patterns and roles in capsule synthesis. Eukaryot. Cell. *6*, 776–785.

Cox, G.M., Mukherjee, J., Cole, G.T., Casadevall, A., and Perfect, J.R. (2000). Urease as a virulence factor in experimental cryptococcosis. Infect. Immun. *68*, 443–448.

Crabtree, J.N., Okagaki, L.H., Wiesner, D.L., Strain, A.K., Nielsen, J.N., and Nielsen, K. (2012). Titan cell production enhances the virulence of *Cryptococcus neoformans*. Infect. Immun. *80*, 3776–3785.

Cross, C.E., Collins, H.L., and Bancroft, G.J. (1997). CR3-dependent phagocytosis by murine macrophages: different cytokines regulate ingestion of a defined CR3 ligand and complement-opsonized *Cryptococcus neoformans*. Immunology *91*, 289–296.

Crowley, M., Inaba, K., and Steinman, R.M. (1990). Dendritic cells are the principal cells in mouse spleen bearing immunogenic fragments of foreign proteins. J. Exp. Med. *172*, 383–386.

Cruickshank, J.G., Cavill, R., and Jelbert, M. (1973). *Cryptococcus neoformans* of unusual morphology. Appl. Microbiol. *25*, 309–312.

Datta, K., Bartlett, K.H., Baer, R., Byrnes, E., Galanis, E., Heitman, J., Hoang, L., Leslie, M.J., MacDougall, L., Magill, S.S., *et al.* (2009). Spread of *Cryptococcus gattii* into Pacific Northwest region of the United States. Emerg. Infect. Dis. *15*, 1185–1191.

Davis, J., Zheng, W.Y., Glatman-Freedman, A., Ng, J.A., Pagcatipunan, M.R., Lessin, H., Casadevall, A., and Goldman, D.L. (2007). Serologic evidence for regional differences in pediatric cryptococcal infection. Pediatr. Infect. Dis. J. *26*, 549–551.

De Jesus, M., Nicola, A.M., Rodrigues, M.L., Janbon, G., and Casadevall, A. (2009). Capsular localization of the *Cryptococcus neoformans* polysaccharide component galactoxylomannan. Eukaryot. Cell. *8*, 96–103.

Del Poeta, M. (2004). Role of phagocytosis in the virulence of *Cryptococcus neoformans*. Eukaryot. Cell. *3*, 1067–1075.

Desalermos, A., Fuchs, B.B., and Mylonakis, E. (2012). Selecting an invertebrate model host for the study of fungal pathogenesis. PLoS Pathog. *8*, e1002451.

Diamond, R.D., and Bennett, J.E. (1973). Growth of *Cryptococcus neoformans* within human macrophages *in vitro*. Infect. Immun. *7*, 231–236.

Diaz, M.R., and Nguyen, M.H. (2011). Diagnostic approach based on capsular antigen, capsule detection, beta-glucan, and DNA analysis. In *Cryptococcus*: From Human Pathogen to Model Yeast, J. Heitman, T.R. Kozel, K.J. Kwon-Chung, J.R. Perfect, and A. Casadevall, eds. (ASM Press, Washington DC), pp. 547–564.

Doering, T.L. (2000). How does *Cryptococcus* get its coat? Trends Microbiol. *8*, 547–553.

Dong, Z.M., and Murphy, J.W. (1995). Effects of the two varieties of *Cryptococcus neoformans* cells and culture filtrate antigens on neutrophil locomotion. Infect. Immun. *63*, 2632–2644.

Dong, Z.M., and Murphy, J.W. (1997). Cryptococcal polysaccharides bind to CD18 on human neutrophils. Infect. Immun. *65*, 557–563.

Dong, Z.M., Jackson, L., and Murphy, J.W. (1999). Mechanisms for induction of L-selectin loss from T lymphocytes by a cryptococcal polysaccharide, glucuronoxylomannan. Infect. Immun. *67*, 220–229.

Dromer, F., and Levitz, S.M. (2011). Invasion of *Cryptococcus* into the central nervous system. In *Cryptococcus*: From Human Pathogen to Model Yeast, J. Heitman, T.R. Kozel, K.J. Kwon-Chung, J.R. Perfect, and A. Casadevall, eds. (ASM Press, Washington DC), pp. 464–471.

Dromer, F., Salamero, J., Contrepois, A., Carbon, C., and Yeni, P. (1987). Production, characterization, and antibody specificity of a mouse monoclonal antibody reactive with *Cryptococcus neoformans* capsular polysaccharide. Infect. Immun. *55*, 742–748.

Dromer, F., Mathoulin-Pelissier, S., Launay, O., and Lortholary, O. (2007). Determinants of disease presentation and outcome during cryptococcosis: the CryptoA/D study. PLoS Med. *4*, e21.

van Duin, D., Casadevall, A., and Nosanchuk, J.D. (2002). Melanization of *Cryptococcus neoformans* and *Histoplasma capsulatum* reduces their susceptibilities to amphotericin B and caspofungin. Antimicrob. Agents Chemother. *46*, 3394–3400.

Eckert, T.F., and Kozel, T.R. (1987). Production and characterization of monoclonal antibodies specific for *Cryptococcus neoformans* capsular polysaccharide. Infect. Immun. *55*, 1895–1899.

Eisenman, H.C., Mues, M., Weber, S.E., Frases, S., Chaskes, S., Gerfen, G., and Casadevall, A. (2007). *Cryptococcus neoformans* laccase catalyses melanin synthesis from both D- and L-DOPA. Microbiology *153*, 3954–3962.

Engler, H.D., and Shea, Y.R. (1994). Effect of potential interference factors on performance of enzyme immunoassay and latex agglutination assay for cryptococcal antigen. J. Clin. Microbiol. *32*, 2307–2308.

Erickson, T., Liu, L., Gueyikian, A., Zhu, X., Gibbons, J., and Williamson, P.R. (2001). Multiple virulence factors of *Cryptococcus neoformans* are dependent on *VPH1*. Mol. Microbiol. *42*, 1121–1131.

Evans, E.E. (1949). An immunologic comparison of 12 strains of *Cryptococcus neoformans* (Torula histolytica). Proc. Soc. Exp. Biol. Med. *71*, 644–646.

Favaro, M., Savini, V., Favalli, C., and Fontana, C. (2013). A multi-target real-time PCR assay for rapid identification of meningitis-associated microorganisms. Mol. Biotechnol. *53*, 74–79.

Feldmesser, M., Kress, Y., Novikoff, P., and Casadevall, A. (2000). *Cryptococcus neoformans* is a facultative intracellular pathogen in murine pulmonary infection. Infect. Immun. *68*, 4225–4237.

Feldmesser, M., Kress, Y., and Casadevall, A. (2001a). Dynamic changes in the morphology of *Cryptococcus neoformans* during murine pulmonary infection. Microbiology *147*, 2355–2365.

Feldmesser, M., Tucker, S., and Casadevall, A. (2001b). Intracellular parasitism of macrophages by *Cryptococcus neoformans*. Trends Microbiol. *9*, 273–278.

Frank, U.K., Nishimura, S.L., Li, N.C., Sugai, K., Yajko, D.M., Hadley, W.K., and Ng, V.L. (1993). Evaluation of an enzyme immunoassay for detection of cryptococcal capsular polysaccharide antigen in serum and cerebrospinal fluid. J. Clin. Microbiol. *31*, 97–101.

Franzot, S.P., Salkin, I.F., and Casadevall, A. (1999). *Cryptococcus neoformans* var. *grubii*: separate varietal status for *Cryptococcus neoformans* serotype A isolates. J. Clin. Microbiol. *37*, 838–840.

Fraser, J.A., Giles, S.S., Wenink, E.C., Geunes-Boyer, S.G., Wright, J.R., Diezmann, S., Allen, A., Stajich, J.E., Dietrich, F.S., Perfect, J.R., *et al.* (2005). Same-sex mating and the origin of the Vancouver Island *Cryptococcus gattii* outbreak. Nature *437*, 1360–1364.

Frases, S., Pontes, B., Nimrichter, L., Viana, N.B., Rodrigues, M.L., and Casadevall, A. (2009). Capsule of *Cryptococcus neoformans* grows by enlargement of polysaccharide molecules. Proc. Natl. Acad. Sci. U.S.A. *106*, 1228–1233.

Fyfe, M., MacDougall, L., Romney, M., Starr, M., Pearce, M., Mak, S., Mithani, S., and Kibsey, P. (2008). *Cryptococcus gattii* infections on Vancouver Island, British Columbia, Canada: emergence of a tropical fungus in a temperate environment. Can. Commun. Dis. Rep. *34*, 1–12.

Gade, W., Hinnefeld, S.W., Babcock, L.S., Gilligan, P., Kelly, W., Wait, K., Greer, D., Pinilla, M., and Kaplan, R.L. (1991). Comparison of the PREMIER cryptococcal antigen enzyme immunoassay and the latex agglutination assay for detection of cryptococcal antigens. J. Clin. Microbiol. *29*, 1616–1619.

Garcia-Hermoso, D., Janbon, G., and Dromer, F. (1999). Epidemiological evidence for dormant *Cryptococcus neoformans* infection. J. Clin. Microbiol. *37*, 3204–3209.

Garcia-Hermoso, D., Dromer, F., and Janbon, G. (2004). *Cryptococcus neoformans* capsule structure evolution *in vitro* and during murine infection. Infect. Immun. *72*, 3359–3365.

Garcia-Rivera, J., Chang, Y.C., Kwon-Chung, K.J., and Casadevall, A. (2004). *Cryptococcus neoformans CAP59* (or Cap59p) is involved in the extracellular trafficking of capsular glucuronoxylomannan. Eukaryot. Cell. *3*, 385–392.

Garcia-Rodas, R., and Zaragoza, O. (2012). Catch me if you can: phagocytosis and killing avoidance by *Cryptococcus neoformans*. FEMS Immunol. Med. Microbiol. *64*, 147–161.

García-Rodas, R., Casadevall, A., Rodríguez-Tudela, J.L., Cuenca-Estrella, M., and Zaragoza, O. (2011). *Cryptococcus neoformans* capsular enlargement and cellular gigantism during *Galleria mellonella* infection. PLoS One *6*, e24485.

Gates, M.A., Thorkildson, P., and Kozel, T.R. (2004). Molecular architecture of the *Cryptococcus neoformans* capsule. Mol. Microbiol. *52*, 13–24.

Goldman, D., Lee, S.C., and Casadevall, A. (1994). Pathogenesis of pulmonary *Cryptococcus neoformans* infection in the rat. Infect. Immun. *62*, 4755–4761.

Goldman, D.L., Lee, S.C., Mednick, A.J., Montella, L., and Casadevall, A. (2000). Persistent *Cryptococcus neoformans* pulmonary infection in the rat is associated with intracellular parasitism, decreased inducible nitric oxide synthase expression, and altered antibody responsiveness to cryptococcal polysaccharide. Infect. Immun. *68*, 832–838.

Goldman, D.L., Khine, H., Abadi, J., Lindenberg, D.J., Pirofski, L., Niang, R., and Casadevall, A. (2001). Serologic evidence for *Cryptococcus neoformans* infection in early childhood. Pediatrics *107*, E66.

Granger, D.L., Perfect, J.R., and Durack, D.T. (1985). Virulence of *Cryptococcus neoformans*. Regulation of capsule synthesis by carbon dioxide. J. Clin. Invest. *76*, 508–516.

Grechi, J., Marinho-Carvalho, M., Zancan, P., Cinelli, L.P., Gomes, A.M., Rodrigues, M.L., Nimrichter, L., and Sola-Penna, M. (2011). Glucuronoxylomannan from *Cryptococcus neoformans* down-regulates the enzyme 6-phosphofructo-1-kinase of macrophages. J. Biol. Chem. *286*, 14820–14829.

Guimaraes, A.J., Frases, S., Cordero, R.J., Nimrichter, L., Casadevall, A., and Nosanchuk, J.D. (2010). *Cryptococcus neoformans* responds to mannitol by increasing capsule size *in vitro* and *in vivo*. Cell Microbiol. *12*, 740–753.

Haddow, L.J., Moosa, M.Y., Mosam, A., Moodley, P., Parboosing, R., and Easterbrook, P.J. (2012). Incidence, clinical spectrum, risk factors and impact of HIV-associated immune reconstitution inflammatory syndrome in south africa. PLoS One *7*, e40623.

Hansen, J., Slechta, E.S., Gates-Hollingsworth, M.A., Neary, B., Barker, A., Bauman, S., Kozel, T.R., and Hanson, K.E. (2013). Large scale evaluation of the immuno-mycologics lateral flow and enzyme-linked immunoassays for the detection of cryptococcal antigen in serum and cerebrospinal fluid. Clin. Vaccine Immunol. *20*, 52–55.

Haynes, B.C., Skowyra, M.L., Spencer, S.J., Gish, S.R., Williams, M., Held, E.P., Brent, M.R., and Doering, T.L. (2011). Toward an integrated model of capsule regulation in *Cryptococcus neoformans*. PLoS Pathog. *7*, e1002411.

Heelan, J.S., Corpus, L., and Kessimian, N. (1991). False-positive reactions in the latex agglutination test for *Cryptococcus neoformans* antigen. J. Clin. Microbiol. *29*, 1260–1261.

Heiss, C., Klutts, J.S., Wang, Z., Doering, T.L., and Azadi, P. (2009). The structure of *Cryptococcus neoformans* galactoxylomannan contains β-D-glucuronic acid. Carbohydr. Res. *344*, 915–920.

Heitman, J., Kozel, T., Kwon-Chung, K., Perferct, J., and Casadevall, A. (2011). *Cryptococcus*. From Human pathogen to model yeast (ASM Press, Washington DC).

Hoang, L.M., Maguire, J.A., Doyle, P., Fyfe, M., and Roscoe, D.L. (2004). *Cryptococcus neoformans* infections at Vancouver Hospital and Health Sciences Centre (1997–2002): epidemiology, microbiology and histopathology. J. Med. Microbiol. *53*, 935–940.

Huang, C., Nong, S.H., Mansour, M.K., Specht, C.A., and Levitz, S.M. (2002). Purification and characterization of a second immunoreactive mannoprotein from *Cryptococcus neoformans* that stimulates T-Cell responses. Infect. Immun. *70*, 5485–5493.

Jarvis, J.N., Percival, A., Bauman, S., Pelfrey, J., Meintjes, G., Williams, G.N., Longley, N., Harrison, T.S., and Kozel, T.R. (2011). Evaluation of a novel point-of-care cryptococcal antigen test on serum, plasma, and urine from patients with HIV-associated cryptococcal meningitis. Clin. Infect. Dis. *53*, 1019–1023.

Jesus, M.D., Nicola, A.M., Chow, S.K., Lee, I.R., Nong, S., Specht, C.A., Levitz, S.M., and Casadevall, A. (2010). Glucuronoxylomannan, galactoxylomannan, and mannoprotein occupy spatially separate and discrete regions in the capsule of *Cryptococcus neoformans*. Virulence *1*, 500–508.

Jiang, N., Sun, N., Xiao, D., Pan, J., Wang, Y., and Zhu, X. (2009). A copper-responsive factor gene CUF1 is required for copper induction of laccase in *Cryptococcus neoformans*. FEMS Microbiol. Lett. *296*, 84–90.

Johnston, S.A., and May, R.C. (2010). The human fungal pathogen *Cryptococcus neoformans* escapes macrophages by a phagosome emptying mechanism that is inhibited by Arp2/3 complex-mediated actin polymerisation. PLoS Pathog. *6*, e1001041.

Kim, K.S. (2008). Mechanisms of microbial traversal of the blood–brain barrier. Nat. Rev. Microbiol. *6*, 625–634.

Kingsbury, J.M., Yang, Z., Ganous, T.M., Cox, G.M., and McCusker, J.H. (2004). *Cryptococcus neoformans* Ilv2p confers resistance to sulfometuron methyl and is required for survival at 37°C and *in vivo*. Microbiology *150*, 1547–1558.

Kitz, D.J., Johnson, C.R., Kobayashi, G.S., Medoff, G., and Little, J.R. (1984). Growth inhibition of *Cryptococcus neoformans* by cloned cultured murine macrophages. Cell Immunol. *88*, 489–500.

Klutts, J.S., Levery, S.B., and Doering, T.L. (2007). A β-1,2-xylosyltransferase from *Cryptococcus neoformans* defines a new family of glycosyltransferases. J. Biol. Chem. *282*, 17890–17899.

Koguchi, Y., and Kawakami, K. (2002). Cryptococcal infection and Th1-Th2 cytokine balance. Int. Rev. Immunol. *21*, 423–438.

Kozel, T.R., and Gotschlich, E.C. (1982). The capsule of *Cryptococcus neoformans* passively inhibits phagocytosis of the yeast by macrophages. J. Immunol. *129*, 1675–1680.

Kozel, T.R., Highison, B., and Stratton, C.J. (1984). Localization on encapsulated *Cryptococcus neoformans* of serum components opsonic for phagocytosis by macrophages and neutrophils. Infect. Immun. *43*, 574–579.

Kozel, T.R., Pfrommer, G.S., Guerlain, A.S., Highison, B.A., and Highison, G.J. (1988). Role of the capsule in phagocytosis of *Cryptococcus neoformans*. Rev. Infect. Dis. *10 (Suppl 2)*, S436–439.

Kozel, T.R., Wilson, M.A., Pfrommer, G.S., and Schlageter, A.M. (1989). Activation and binding of opsonic fragments of C3 on encapsulated *Cryptococcus neoformans* by using an alternative complement pathway reconstituted from six isolated proteins. Infect. Immun. *57*, 1922–1927.

Kozel, T.R., Levitz, S.M., Dromer, F., Gates, M.A., Thorkildson, P., and Janbon, G. (2003). Antigenic and biological characteristics of mutant strains of *Cryptococcus neoformans* lacking capsular O acetylation or xylosyl side chains. Infect. Immun. *71*, 2868–2875.

Kraus, P.R., Fox, D.S., Cox, G.M., and Heitman, J. (2003). The *Cryptococcus neoformans* MAP kinase Mpk1 regulates cell integrity in response to antifungal drugs and loss of calcineurin function. Mol. Microbiol. *48*, 1377–1387.

Kwon-Chung, K.B., Boekhout, T., Wickes, B.L., and Fell, J.W. (2011). Section 1. General principles: systematics of the genus *Cryptococcus* and its type species *C. neoformans*. In *Cryptococcus:* From Human Pathogen to Model Yeast, J. Heitman, T.R. Kozel, K.J. Kwon-Chung, J.R. Perfect, and A. Casadevall, eds. (ASM Press, Washington DC), pp. 3–15.

Kwon-Chung, K.J., and Varma, A. (2006). Do major species concepts support one, two or more species within *Cryptococcus neoformans*? FEMS Yeast Res. *6*, 574–587.

Kwon-Chung, K.J., Polacheck, I., and Popkin, T.J. (1982). Melanin-lacking mutants of *Cryptococcus neoformans* and their virulence for mice. J. Bacteriol. *150*, 1414–1421.

Larsen, R.A., Bozzette, S., McCutchan, J.A., Chiu, J., Leal, M.A., and Richman, D.D. (1989). Persistent *Cryptococcus neoformans* infection of the prostate after successful treatment of meningitis. California Collaborative Treatment Group. Ann. Intern. Med. *111*, 125–128.

Lee, S.C., Kress, Y., Zhao, M.L., Dickson, D.W., and Casadevall, A. (1995). *Cryptococcus neoformans* survive and replicate in human microglia. Lab. Invest. *73*, 871–879.

Lee, S.C., Phadke, S., Sun, S., and Heitman, J. (2012). Pseudohyphal growth of *Cryptococcus neoformans* is a reversible dimorphic transition in response to ammonium that requires Amt1 and Amt2 ammonium permeases. Eukaryot. Cell. *11*, 1391–1398.

Levitz, S.M. (2002). Receptor-mediated recognition of *Cryptococcus neoformans*. Nihon Ishinkin Gakkai Zasshi 43, 133–136.

Levitz, S.M., and DiBenedetto, D.J. (1989). Paradoxical role of capsule in murine bronchoalveolar macrophage-mediated killing of *Cryptococcus neoformans*. J. Immunol. *142*, 659–665.

Levitz, S.M., and Specht, C.A. (2006). The molecular basis for the immunogenicity of *Cryptococcus neoformans* mannoproteins. FEMS Yeast Res. *6*, 513–524.

Levitz, S.M., Harrison, T.S., Tabuni, A., and Liu, X. (1997). Chloroquine induces human mononuclear phagocytes to inhibit and kill *Cryptococcus neoformans* by a mechanism independent of iron deprivation. J. Clin. Invest. *100*, 1640–1646.

Levitz, S.M., Nong, S.H., Seetoo, K.F., Harrison, T.S., Speizer, R.A., and Simons, E.R. (1999). *Cryptococcus neoformans* resides in an acidic phagolysosome of human macrophages. Infect. Immun. *67*, 885–890.

Levitz, S.M., Nong, S., Mansour, M.K., Huang, C., and Specht, C.A. (2001). Molecular characterization of a mannoprotein with homology to chitin deacetylases that stimulates T cell responses to *Cryptococcus neoformans*. Proc. Natl. Acad. Sci. U.S.A. *98*, 10422–10427.

Lindsley, M.D., Mekha, N., Baggett, H.C., Surinthong, Y., Autthateinchai, R., Sawatwong, P., Harris, J.R., Park, B.J., Chiller, T., Balajee, S.A., *et al.* (2011). Evaluation of a newly developed lateral flow immunoassay for the diagnosis of cryptococcosis. Clin. Infect. Dis. *53*, 321–325.

Litvintseva, A.P., Thakur, R., Reller, L.B., and Mitchell, T.G. (2005). Prevalence of clinical isolates of *Cryptococcus gattii* serotype C among patients with AIDS in Sub-Saharan Africa. J. Infect. Dis. *192*, 888–892.

Liu, L., Tewari, R.P., and Williamson, P.R. (1999). Laccase protects *Cryptococcus neoformans* from antifungal activity of alveolar macrophages. Infect. Immun. *67*, 6034–6039.

Liu, O.W., Chun, C.D., Chow, E.D., Chen, C., Madhani, H.D., and Noble, S.M. (2008). Systematic genetic analysis of virulence in the human fungal pathogen *Cryptococcus neoformans*. Cell *135*, 174–188.

Lortholary, O., Fontanet, A., Memain, N., Martin, A., Sitbon, K., and Dromer, F. (2005). Incidence and risk factors of immune reconstitution inflammatory syndrome complicating HIV-associated cryptococcosis in France. AIDS *19*, 1043–1049.

Lortholary, O., Poizat, G., Zeller, V., Neuville, S., Boibieux, A., Alvarez, M., Dellamonica, P., Botterel, F., Dromer, F., and Chene, G. (2006). Long-term outcome of AIDS-associated cryptococcosis in the era of combination antiretroviral therapy. Aids *20*, 2183–2191.

Luberto, C., Martinez-Marino, B., Taraskiewicz, D., Bolanos, B., Chitano, P., Toffaletti, D.L., Cox, G.M., Perfect, J.R., Hannun, Y.A., Balish, E., *et al.* (2003). Identification of App1 as a regulator of phagocytosis and virulence of *Cryptococcus neoformans*. J. Clin. Invest. *112*, 1080–1094.

Ma, H., Croudace, J.E., Lammas, D.A., and May, R.C. (2006). Expulsion of live pathogenic yeast by macrophages. Curr. Biol. *16*, 2156–2160.

Ma, H., Croudace, J.E., Lammas, D.A., and May, R.C. (2007). Direct cell-to-cell spread of a pathogenic yeast. BMC Immunol. *8*, 15.

Ma, H., Hagen, F., Stekel, D.J., Johnston, S.A., Sionov, E., Falk, R., Polacheck, I., Boekhout, T., and May, R.C. (2009). The fatal fungal outbreak on Vancouver Island is characterized by enhanced intracellular parasitism driven by mitochondrial regulation. Proc. Natl. Acad. Sci. U.S.A. *106*, 12980–12985.

McClelland, E.E., Bernhardt, P., and Casadevall, A. (2005). Coping with multiple virulence factors: which is most important? PLoS Pathog. *1*, e40.

McFadden, D.C., De Jesus, M., and Casadevall, A. (2006). The physical properties of the capsular polysaccharides from *Cryptococcus neoformans* suggest features for capsule construction. J. Biol. Chem. *281*, 1868–1875.

McFadden, D.C., Fries, B.C., Wang, F., and Casadevall, A. (2007). Capsule structural heterogeneity and antigenic variation in *Cryptococcus neoformans*. Eukaryot. Cell. *6*, 1464–1473.

Magditch, D.A., Liu, T.B., Xue, C., and Idnurm, A. (2012). DNA mutations mediate microevolution between host-adapted forms of the pathogenic fungus *Cryptococcus neoformans*. PLoS Pathog. *8*, e1002936.

Mansour, M.K., and Levitz, S.M. (2002). Interactions of fungi with phagocytes. Curr. Opin. Microbiol. *5*, 359–365.

Mansour, M.K., Schlesinger, L.S., and Levitz, S.M. (2002). Optimal T cell responses to *Cryptococcus neoformans* mannoprotein are dependent on recognition of conjugated carbohydrates by mannose receptors. J. Immunol. *168*, 2872–2879.

Mansour, M.K., Latz, E., and Levitz, S.M. (2006). *Cryptococcus neoformans* glycoantigens are captured by multiple lectin receptors and presented by dendritic cells. J. Immunol. *176*, 3053–3061.

Maxson, M.E., Cook, E., Casadevall, A., and Zaragoza, O. (2007a). The volume and hydration of the *Cryptococcus neoformans* polysaccharide capsule. Fungal Genet. Biol. *44*, 180–186.

Maxson, M.E., Dadachova, E., Casadevall, A., and Zaragoza, O. (2007b). Radial mass density, charge, and epitope distribution in the *Cryptococcus neoformans* capsule. Eukaryot. Cell. *6*, 95–109.

Mednick, A.J., Nosanchuk, J.D., and Casadevall, A. (2005). Melanization of *Cryptococcus neoformans* affects lung inflammatory responses during cryptococcal infection. Infect. Immun. 73, 2012–2019.

Meyer, W., Aanensen, D.M., Boekhout, T., Cogliati, M., Diaz, M.R., Esposto, M.C., Fisher, M., Gilgado, F., Hagen, F., Kaocharoen, S., *et al.* (2009). Consensus multi-locus sequence typing scheme for *Cryptococcus neoformans* and *Cryptococcus gattii*. Med. Mycol. *47*, 561–570.

Missall, T.A., Moran, J.M., Corbett, J.A., and Lodge, J.K. (2005). Distinct stress responses of two functional laccases in *Cryptococcus neoformans* are revealed in the absence of the thiol-specific antioxidant Tsa1. Eukaryot. Cell. *4*, 202–208.

Mitchell, T.G., and Friedman, L. (1972). *In vitro* phagocytosis and intracellular fate of variously encapsulated strains of *Cryptococcus neoformans*. Infect. Immun. *5*, 491–498.

van de Moer, A., Salhi, S.L., Cherniak, R., Pau, B., Garrigues, M.L., and Bastide, J.M. (1990). An anti-*Cryptococcus neoformans* monoclonal antibody directed against galactoxylomannan. Res. Immunol. *141*, 33–42.

Monari, C., Bistoni, F., Casadevall, A., Pericolini, E., Pietrella, D., Kozel, T.R., and Vecchiarelli, A. (2005a). Glucuronoxylomannan, a microbial compound, regulates expression of costimulatory molecules and production of cytokines in macrophages. J. Infect. Dis. *191*, 127–137.

Monari, C., Pericolini, E., Bistoni, G., Casadevall, A., Kozel, T.R., and Vecchiarelli, A. (2005b). *Cryptococcus neoformans* capsular glucuronoxylomannan induces expression of fas ligand in macrophages. J. Immunol. *174*, 3461–3468.

Monari, C., Bevilacqua, S., Piccioni, M., Pericolini, E., Perito, S., Calvitti, M., Bistoni, F., Kozel, T.R., and Vecchiarelli, A. (2009). A microbial polysaccharide reduces the severity of rheumatoid arthritis by influencing Th17 differentiation and proinflammatory cytokines production. J. Immunol. *183*, 191–200.

Morgan, J., McCarthy, K.M., Gould, S., Fan, K., Arthington-Skaggs, B., Iqbal, N., Stamey, K., Hajjeh, R.A., and Brandt, M.E. (2006). *Cryptococcus gattii* infection: characteristics and epidemiology of cases identified in a South African province with high HIV seroprevalence, 2002–2004. Clin. Infect. Dis. *43*, 1077–1080.

Moyrand, F., Chang, Y.C., Himmelreich, U., Kwon-Chung, K.J., and Janbon, G. (2004). Cas3p belongs to a seven-member family of capsule structure designer proteins. Eukaryot. Cell. *3*, 1513–1524.

Moyrand, F., Fontaine, T., and Janbon, G. (2007). Systematic capsule gene disruption reveals the central role of galactose metabolism on *Cryptococcus neoformans* virulence. Mol. Microbiol. *64*, 771–781.

Mukherjee, J., Scharff, M.D., and Casadevall, A. (1992). Protective murine monoclonal antibodies to *Cryptococcus neoformans*. Infect. Immun. *60*, 4534–4541.

Mukherjee, J., Pirofski, L.A., Scharff, M.D., and Casadevall, A. (1993). Antibody-mediated protection in mice with lethal intracerebral *Cryptococcus neoformans* infection. Proc. Natl. Acad. Sci. U.S.A. *90*, 3636–3640.

Mukherjee, J., Kozel, T.R., and Casadevall, A. (1998). Monoclonal antibodies reveal additional epitopes of serotype D *Cryptococcus neoformans capsular* glucuronoxylomannan that elicit protective antibodies. J. Immunol. *161*, 3557–3568.

Muller, U., Stenzel, W., Kohler, G., Werner, C., Polte, T., Hansen, G., Schutze, N., Straubinger, R.K., Blessing, M., McKenzie, A.N., *et al.* (2007). IL-13 induces disease-promoting type 2 cytokines, alternatively activated macrophages and allergic inflammation during pulmonary infection of mice with *Cryptococcus neoformans*. J. Immunol. *179*, 5367–5377.

Mylonakis, E., Ausubel, F.M., Perfect, J.R., Heitman, J., and Calderwood, S.B. (2002). Killing of *Caenorhabditis elegans* by *Cryptococcus neoformans* as a model of yeast pathogenesis. Proc. Natl. Acad. Sci. U.S.A. *99*, 15675–15680.

Mylonakis, E., Idnurm, A., Moreno, R., El Khoury, J., Rottman, J.B., Ausubel, F.M., Heitman, J., and Calderwood, S.B. (2004). *Cryptococcus neoformans* Kin1 protein kinase homologue, identified through a *Caenorhabditis elegans* screen, promotes virulence in mammals. Mol. Microbiol. *54*, 407–419.

Mylonakis, E., Moreno, R., El Khoury, J.B., Idnurm, A., Heitman, J., Calderwood, S.B., Ausubel, F.M., and Diener, A. (2005). *Galleria mellonella* as a model system to study *Cryptococcus neoformans* pathogenesis. Infect. Immun. *73*, 3842–3850.

Naslund, P.K., Miller, W.C., and Granger, D.L. (1995). *Cryptococcus neoformans* fails to induce nitric oxide synthase in primed murine macrophage-like cells. Infect. Immun. *63*, 1298–1304.

Neilson, J.B., Fromtling, R.A., and Bulmer, G.S. (1981). Pseudohyphal forms of *Cryptococcus neoformans*: decreased survival *in vivo*. Mycopathologia 73, 57–59.

Ngamskulrungroj, P., Himmelreich, U., Breger, J.A., Wilson, C., Chayakulkeeree, M., Krockenberger, M.B., Malik, R., Daniel, H.M., Toffaletti, D., Djordjevic, J.T., *et al.* (2009). The trehalose synthesis pathway is an integral part of the virulence composite for *Cryptococcus gattii*. Infect. Immun. *77*, 4584–4596.

Ngamskulrungroj, P., Price, J., Sorrell, T., Perfect, J.R., and Meyer, W. (2011). *Cryptococcus gattii* virulence composite: candidate genes revealed by microarray analysis of high and less virulent Vancouver island outbreak strains. PLoS One *6*, e16076.

Ngamskulrungroj, P., Chang, Y., Sionov, E., and Kwon-Chung, K.J. (2012). The primary target organ of *Cryptococcus gattii* is different from that of *Cryptococcus neoformans* in a murine model. Mbio. *3*, e00103–00112.

Nicola, A.M., Robertson, E.J., Albuquerque, P., Derengowski Lda, S., and Casadevall, A. (2011). Nonlytic exocytosis of *Cryptococcus neoformans* from macrophages occurs *in vivo* and is influenced by phagosomal pH. Mbio 2, e00167–00111.

Nicola, A.M., Albuquerque, P., Martinez, L.R., Dal-Rosso, R.A., Saylor, C., De Jesus, M., Nosanchuk, J.D., and Casadevall, A. (2012). Macrophage autophagy in immunity to *Cryptococcus neoformans* and *Candida albicans*. Infect. Immun. *80*, 3065–3076.

Nielsen, K., De Obaldia, A.L., and Heitman, J. (2007). *Cryptococcus neoformans* mates on pigeon guano: implications for the realized ecological niche and globalization. Eukaryot. Cell. *6*, 949–959.

Nimrichter, L., Frases, S., Cinelli, L.P., Viana, N.B., Nakouzi, A., Travassos, L.R., Casadevall, A., and Rodrigues, M.L. (2007). Self-aggregation of *Cryptococcus neoformans* capsular glucuronoxylomannan is dependent on divalent cations. Eukaryot. Cell. *6*, 1400–1410.

Nosanchuk, J.D., Valadon, P., Feldmesser, M., and Casadevall, A. (1999). Melanization of *Cryptococcus neoformans* in murine infection. Mol. Cell. Biol. *19*, 745–750.

Nosanchuk, J.D., Rosas, A.L., Lee, S.C., and Casadevall, A. (2000). Melanisation of *Cryptococcus neoformans* in human brain tissue. Lancet 355, 2049–2050.

Noverr, M.C., Cox, G.M., Perfect, J.R., and Huffnagle, G.B. (2003). Role of *PLB1* in pulmonary inflammation and cryptococcal eicosanoid production. Infect. Immun. *71*, 1538–1547.

Noverr, M.C., Williamson, P.R., Fajardo, R.S., and Huffnagle, G.B. (2004). Cn*LAC1* is required for extrapulmonary dissemination of *Cryptococcus neoformans* but not pulmonary persistence. Infect. Immun. 72, 1693–1699.

Nurudeen, T.A., and Ahearn, D.G. (1979). Regulation of melanin production by *Cryptococcus neoformans*. J. Clin. Microbiol. *10*, 724–729.

Nussbaum, G., Cleare, W., Casadevall, A., Scharff, M.D., and Valadon, P. (1997). Epitope location in the *Cryptococcus neoformans* capsule is a determinant of antibody efficacy. J. Exp. Med. *185*, 685–694.

Odom, A., Muir, S., Lim, E., Toffaletti, D.L., Perfect, J., and Heitman, J. (1997). Calcineurin is required for virulence of *Cryptococcus neoformans*. EMBO J. *16*, 2576–2589.

Okabayashi, K., Kano, R., Watanabe, S., and Hasegawa, A. (2005). Expression of capsule-associated genes of *Cryptococcus neoformans*. Mycopathologia *160*, 1–7.

Okagaki, L.H., and Nielsen, K. (2012). Titan cells confer protection from phagocytosis in *Cryptococcus neoformans* infections. Eukaryot. Cell. *11*, 820–826.

Okagaki, L.H., Strain, A.K., Nielsen, J.N., Charlier, C., Baltes, N.J., Chretien, F., Heitman, J., Dromer, F., and Nielsen, K. (2010). Cryptococcal cell morphology affects host cell interactions and pathogenicity. PLoS Pathog. *6*, e1000953.

Okagaki, L.H., Wang, Y., Ballou, E.R., O'Meara, T.R., Bahn, Y.S., Alspaugh, J.A., Xue, C., and Nielsen, K. (2011). Cryptococcal titan cell formation is regulated by G-protein signaling in response to multiple stimuli. Eukaryot. Cell. *10*, 1306–1316.

Olszewski, M.A., Noverr, M.C., Chen, G.H., Toews, G.B., Cox, G.M., Perfect, J.R., and Huffnagle, G.B. (2004). Urease expression by *Cryptococcus neoformans* promotes microvascular sequestration, thereby enhancing central nervous system invasion. Am. J. Pathol. *164*, 1761–1771.

O'Meara, T.R., and Alspaugh, J.A. (2012). The *Cryptococcus neoformans* capsule: a sword and a shield. Clin. Microbiol. Rev. *25*, 387–408.

Osterholzer, J.J., Surana, R., Milam, J.E., Montano, G.T., Chen, G.H., Sonstein, J., Curtis, J.L., Huffnagle, G.B., Toews, G.B., and Olszewski, M.A. (2009). Cryptococcal urease promotes the accumulation of immature dendritic cells and a non-protective T2 immune response within the lung. Am. J. Pathol. *174*, 932–943.

Park, B.J., Wannemuehler, K.A., Marston, B.J., Govender, N., Pappas, P.G., and Chiller, T.M. (2009). Estimation of the current global burden of cryptococcal meningitis among persons living with HIV/AIDS. AIDS *23*, 525–530.

Paterson, D.L., Singh, N., Gayowski, T., and Marino, I.R. (1998). Pulmonary nodules in liver transplant recipients. Medicine (Baltimore) *77*, 50–58.

Perfect, J.R. (2005). *Cryptococcus neoformans*: a sugar-coated killer with designer genes. FEMS Immunol. Med. Microbiol. *45*, 395–404.

Perfect, J.R. (2006). *Cryptococcus neoformans*: the yeast that likes it hot. FEMS Yeast Res. *6*, 463–468.

Perfect, J.R., Dismukes, W.E., Dromer, F., Goldman, D.L., Graybill, J.R., Hamill, R.J., Harrison, T.S., Larsen, R.A., Lortholary, O., Nguyen, M.H., *et al.* (2010). Clinical practice guidelines for the management of cryptococcal disease: 2010 update by the infectious diseases society of america. Clin. Infect. Dis. *50*, 291–322.

Pericolini, E., Cenci, E., Monari, C., De Jesus, M., Bistoni, F., Casadevall, A., and Vecchiarelli, A. (2006). *Cryptococcus neoformans* capsular polysaccharide component galactoxylomannan induces apoptosis of human T-cells through activation of caspase-8. Cell. Microbiol. *8*, 267–275.

Pierini, L.M., and Doering, T.L. (2001). Spatial and temporal sequence of capsule construction in *Cryptococcus neoformans*. Mol. Microbiol. *41*, 105–115.

Pirofski, L., Lui, R., DeShaw, M., Kressel, A.B., and Zhong, Z. (1995). Analysis of human monoclonal antibodies elicited by vaccination with a *Cryptococcus neoformans* glucuronoxylomannan capsular polysaccharide vaccine. Infect. Immun. *63*, 3005–3014.

Pukkila-Worley, R., Gerrald, Q.D., Kraus, P.R., Boily, M.J., Davis, M.J., Giles, S.S., Cox, G.M., Heitman, J., and Alspaugh, J.A. (2005). Transcriptional network of multiple capsule and melanin genes governed by the *Cryptococcus neoformans* cyclic AMP cascade. Eukaryot. Cell. *4*, 190–201.

Qin, Q.M., Luo, J., Lin, X., Pei, J., Li, L., Ficht, T.A., and de Figueiredo, P. (2011). Functional analysis of host factors that mediate the intracellular lifestyle of *Cryptococcus neoformans*. PLoS Pathog. *7*, e1002078.

Qiu, Y., Davis, M.J., Dayrit, J.K., Hadd, Z., Meister, D.L., Osterholzer, J.J., Williamson, P.R., and Olszewski, M.A. (2012). Immune modulation mediated by cryptococcal laccase promotes pulmonary growth and brain dissemination of virulent *Cryptococcus neoformans* in mice. PLoS One 7, e47853.

Raetz, C.R., and Whitfield, C. (2002). Lipopolysaccharide endotoxins. Annu. Rev. Biochem. *71*, 635–700.

Reese, A.J., and Doering, T.L. (2003). Cell wall alpha-1,3-glucan is required to anchor the *Cryptococcus neoformans* capsule. Mol. Microbiol. *50*, 1401–1409.

Retini, C., Vecchiarelli, A., Monari, C., Tascini, C., Bistoni, F., and Kozel, T.R. (1996). Capsular polysaccharide of *Cryptococcus neoformans* induces proinflammatory cytokine release by human neutrophils. Infect. Immun. *64*, 2897–2903.

Rezaei Amirabadi, A., Hassan Shahhosseiny, M., Vand Yousefi, J., and Ghahri, M. (2012). LOOP mediated isothermal AMPlification (LAMP) in diagnosis of neuorocryptococcosis. African J. Biotech. *11*, 3986–3992.

Rodrigues, M.L., Nimrichter, L., Oliveira, D.L., Frases, S., Miranda, K., Zaragoza, O., Alvarez, M., Nakouzi, A., Feldmesser, M., and Casadevall, A. (2007). Vesicular polysaccharide export in *Cryptococcus neoformans* is a eukaryotic solution to the problem of fungal trans-cell wall transport. Eukaryot. Cell. *6*, 48–59.

Rodrigues, M.L., Nakayasu, E.S., Oliveira, D.L., Nimrichter, L., Nosanchuk, J.D., Almeida, I.C., and Casadevall, A. (2008). Extracellular vesicles produced by *Cryptococcus neoformans* contain protein components associated with virulence. Eukaryot. Cell. *7*, 58–67.

Rosas, A.L., Nosanchuk, J.D., and Casadevall, A. (2001). Passive immunization with melanin-binding monoclonal antibodies prolongs survival of mice with lethal *Cryptococcus neoformans* infection. Infect. Immun. *69*, 3410–3412.

Sabiiti, W., and May, R.C. (2012). Capsule independent uptake of the fungal pathogen *Cryptococcus neoformans* into brain microvascular endothelial cells. PLoS One 7, e35455.

Saha, D.C., Goldman, D.L., Shao, X., Casadevall, A., Husain, S., Limaye, A.P., Lyon, M., Somani, J., Pursell, K., Pruett, T.L., *et al.* (2007). Serologic evidence for reactivation of cryptococcosis in solid-organ transplant recipients. Clin. Vaccine Immunol. *14*, 1550–1554.

Saha, D.C., Xess, I., and Jain, N. (2008). Evaluation of conventional and serological methods for rapid diagnosis of cryptococcosis. Indian J. Med. Res. *127*, 483–488.

Saha, D.C., Xess, I., Biswas, A., Bhowmik, D.M., and Padma, M.V. (2009). Detection of *Cryptococcus* by conventional, serological and molecular methods. J. Med. Microbiol. *58*, 1098–1105.

Salyer, W.R., Salyer, D.C., and Baker, R.D. (1974). Primary complex of *Cryptococcus* and pulmonary lymph nodes. J. Infect. Dis. *130*, 74–77.

Santangelo, R., Zoellner, H., Sorrell, T., Wilson, C., Donald, C., Djordjevic, J., Shounan, Y., and Wright, L. (2004). Role of extracellular phospholipases and mononuclear phagocytes in dissemination of cryptococcosis in a murine model. Infect. Immun. *72*, 2229–2239.

Saylor, C.A., Dadachova, E., and Casadevall, A. (2010). Murine IgG1 and IgG3 isotype switch variants promote phagocytosis of *Cryptococcus neoformans* through different receptors. J. Immunol. *184*, 336–343.

Shi, M., Li, S.S., Zheng, C., Jones, G.J., Kim, K.S., Zhou, H., Kubes, P., and Mody, C.H. (2010). Real-time imaging of trapping and urease-dependent transmigration of *Cryptococcus neoformans* in mouse brain. J. Clin. Invest. *120*, 1683–1693.

Singh, N., Dromer, F., Perfect, J.R., and Lortholary, O. (2008). Cryptococcosis in solid organ transplant recipients: current state of the science. Clin. Infect. Dis. *47*, 1321–1327.

Small, J.M., and Mitchell, T.G. (1989). Strain variation in antiphagocytic activity of capsular polysaccharides from *Cryptococcus neoformans* serotype A. Infect. Immun. *57*, 3751–3756.

Sorrell, T.C. (2001). *Cryptococcus neoformans* variety gattii. Med. Mycol. *39*, 155–168.

Staib, F., Thielke, C., Randhawa, H.S., Senska, M., and Kulins, G. (1972). Colonisation of dead plants by *Cryptococcus neoformans*. Zentralbl. Bakteriol. Orig. A. *222*, 115–125.

Stano, P., Williams, V., Villani, M., Cymbalyuk, E.S., Qureshi, A., Huang, Y., Morace, G., Luberto, C., Tomlinson, S., and Del Poeta, M. (2009). App1: an antiphagocytic protein that binds to complement receptors 3 and 2. J. Immunol. *182*, 84–91.

Steenbergen, J.N., Shuman, H.A., and Casadevall, A. (2001). *Cryptococcus neoformans* interactions with amoebae suggest an explanation for its virulence and intracellular pathogenic strategy in macrophages. Proc. Natl. Acad. Sci. U.S.A. *98*, 15245–15250.

Stephen, C., Lester, S., Black, W., Fyfe, M., and Raverty, S. (2002). Multispecies outbreak of cryptococcosis on southern Vancouver Island, British Columbia. Can. Vet. J. *43*, 792–794.

Syme, R.M., Bruno, T.F., Kozel, T.R., and Mody, C.H. (1999). The capsule of *Cryptococcus neoformans* reduces T-lymphocyte proliferation by reducing phagocytosis, which can be restored with anticapsular antibody. Infect. Immun. *67*, 4620–4627.

Syme, R.M., Spurrell, J.C., Amankwah, E.K., Green, F.H., and Mody, C.H. (2002). Primary dendritic cells phagocytose *Cryptococcus neoformans* via mannose receptors and Fcgamma receptor II for presentation to T lymphocytes. Infect. Immun. *70*, 5972–5981.

Taborda, C.P., and Casadevall, A. (2001). Immunoglobulin M efficacy against *Cryptococcus neoformans*: mechanism, dose dependence, and prozone-like effects in passive protection experiments. J. Immunol. *166*, 2100–2107.

Taborda, C.P., and Casadevall, A. (2002). CR3 (CD11b/CD18) and CR4 (CD11c/CD18) are involved in complement-independent antibody-mediated phagocytosis of *Cryptococcus neoformans*. Immunity *16*, 791–802.

Taborda, C.P., Rivera, J., Zaragoza, O., and Casadevall, A. (2003). More is not necessarily better: prozone-like effects in passive immunization with IgG. J. Immunol. *170*, 3621–3630.

Tang, R.J., Breger, J., Idnurm, A., Gerik, K.J., Lodge, J.K., Heitman, J., Calderwood, S.B., and Mylonakis, E. (2005). *Cryptococcus neoformans* gene involved in mammalian pathogenesis identified by a *Caenorhabditis elegans* progeny-based approach. Infect. Immun. *73*, 8219–8225.

Todaro-Luck, F., Reiss, E., Cherniak, R., and Kaufman, L. (1989). Characterization of *Cryptococcus neoformans* capsular glucuronoxylomannan polysaccharide with monoclonal antibodies. Infect. Immun. *57*, 3882–3887.

Toffaletti, D.L., Del Poeta, M., Rude, T.H., Dietrich, F., and Perfect, J.R. (2003). Regulation of cytochrome c oxidase subunit 1 (*COX1*) expression in *Cryptococcus neoformans* by temperature and host environment. Microbiology *149*, 1041–1049.

Tseng, H.K., Liu, C.P., Price, M.S., Jong, A.Y., Chang, J.C., Toffaletti, D.L., Betancourt-Quiroz, M., Frazzitta, A.E., Cho, W.L., and Perfect, J.R. (2012). Identification of genes from the fungal pathogen *Cryptococcus neoformans* related to transmigration into the central nervous system. PLoS One 7, e45083.

Tucker, S.C., and Casadevall, A. (2002). Replication of *Cryptococcus neoformans* in macrophages is accompanied by phagosomal permeabilization and accumulation of vesicles containing polysaccharide in the cytoplasm. Proc. Natl. Acad. Sci. U.S.A. *99*, 3165–3170.

Vaishnav, V.V., Bacon, B.E., O'Neill, M., and Cherniak, R. (1998). Structural characterization of the galactoxylomannan of *Cryptococcus neoformans* Cap67. Carbohydr. Res. *306*, 315–330.

Vartivarian, S.E., Reyes, G.H., Jacobson, E.S., James, P.G., Cherniak, R., Mumaw, V.R., and Tingler, M.J. (1989). Localization of mannoprotein in *Cryptococcus neoformans*. J. Bacteriol. *171*, 6850–6852.

Vartivarian, S.E., Anaissie, E.J., Cowart, R.E., Sprigg, H.A., Tingler, M.J., and Jacobson, E.S. (1993). Regulation of cryptococcal capsular polysaccharide by iron. J. Infect. Dis. *167*, 186–190.

Vecchiarelli, A. (2000a). Cytokines and costimulatory molecules: positive and negative regulation of the immune response to *Cryptococcus neoformans*. Arch. Immunol. Ther. Exp. (Warsz) *48*, 465–472.

Vecchiarelli, A. (2000b). Immunoregulation by capsular components of *Cryptococcus neoformans*. Med. Mycol. *38*, 407–417.

Vecchiarelli, A. (2007). Fungal capsular polysaccharide and T-cell suppression: the hidden nature of poor immunogenicity. Crit. Rev. Immunol. *27*, 547–557.

Vecchiarelli, A., and Monari, C. (2012). Capsular material of *Cryptococcus neoformans*: virulence and much more. Mycopathologia *173*, 375–386.

Vecchiarelli, A., Dottorini, M., Pietrella, D., Monari, C., Retini, C., Todisco, T., and Bistoni, F. (1994a). Role of human alveolar macrophages as antigen-presenting cells in *Cryptococcus neoformans* infection. Am. J. Respir. Cell. Mol. Bio.l *11*, 130–137.

Vecchiarelli, A., Pietrella, D., Dottorini, M., Monari, C., Retini, C., Todisco, T., and Bistoni, F. (1994b). Encapsulation of *Cryptococcus neoformans* regulates fungicidal activity and the antigen presentation process in human alveolar macrophages. Clin. Exp. Immunol. *98*, 217–223.

Vecchiarelli, A., Retini, C., Monari, C., Tascini, C., Bistoni, F., and Kozel, T.R. (1996). Purified capsular polysaccharide of *Cryptococcus neoformans* induces interleukin-10 secretion by human monocytes. Infect. Immun. *64*, 2846–2849.

Vecchiarelli, A., Pericolini, E., Gabrielli, E., Chow, S.K., Bistoni, F., Cenci, E., and Casadevall, A. (2011). *Cryptococcus neoformans* galactoxylomannan is a potent negative immunomodulator, inspiring new approaches in anti-inflammatory immunotherapy. Immunotherapy 3, 997–1005.

Venn-Watson, S., Daniels, R., and Smith, C. (2012). Thirty year retrospective evaluation of pneumonia in a bottlenose dolphin *Tursiops truncatus* population. Dis. Aquat. Organ. *99*, 237–242.

Vidotto, V., Sinicco, A., Di Fraia, D., Cardaropoli, S., Aoki, S., and Ito-Kuwa, S. (1996). Phospholipase activity in *Cryptococcus neoformans*. Mycopathologia *136*, 119–123.

Voelz, K., Lammas, D.A., and May, R.C. (2009). Cytokine signaling regulates the outcome of intracellular macrophage parasitism by *Cryptococcus neoformans*. Infect. Immun. 77, 3450–3457.

Voelz, K., Johnston, S.A., and May, R.C. (2011). Intracellular replication and exit strategies. In *Cryptococcus*: From Human Pathogen to Model Yeast, J. Heitman, T.R. Kozel, K.J. Kwon-Chung, J.R. Perfect, and A. Casadevall, eds. (ASM Press, Washington DC), pp. 441–451.

Wang, Y., and Casadevall, A. (1994). Decreased susceptibility of melanized *Cryptococcus neoformans* to UV light. Appl. Environ. Microbiol. *60*, 3864–3866.

Warpeha, K.M., Park, Y.D., and Williamson, P.R. (2013). Susceptibility of Intact Germinating *Arabidopsis thaliana* to the Human Fungal Pathogen *Cryptococcus*. Appl. Environ. Microbiol. *79*, 2979–2988.

Whitfield, C. (2006). Biosynthesis and assembly of capsular polysaccharides in *Escherichia coli*. Annu. Rev. Biochem. *75*, 39–68.

Whitfield, C., Kaniuk, N., and Frirdich, E. (2003). Molecular insights into the assembly and diversity of the outer core oligosaccharide in lipopolysaccharides from *Escherichia coli* and *Salmonella*. J. Endotoxin Res. *9*, 244–249.

Williamson, P.R. (1994). Biochemical and molecular characterization of the diphenol oxidase of *Cryptococcus neoformans*: identification as a laccase. J. Bacteriol. *176*, 656–664.

Williamson, P.R. (1997). Laccase and melanin in the pathogenesis of *Cryptococcus neoformans*. Front. Biosci. *2*, e99–107.

Wilson, D.E., Bennett, J.E., and Bailey, J.W. (1968). Serologic grouping of *Cryptococcus neoformans*. Proc. Soc. Exp. Biol. Med. *127*, 820–823.

Wozniak, K.L., and Levitz, S.M. (2011). T Cell and Dendritic Cell Immune Responses to *Cryptococcus*. In *Cryptococcus*: From Human Pathogen to Model Yeast,. J. Heitman, T.R. Kozel, K.J. Kwon-Chung, J.R. Perfect, and A. Casadevall, eds. (ASM Press, Washington DC).

Wozniak, K.L., Hardison, S.E., Kolls, J.K., and Wormley, F.L. (2011). Role of IL-17A on resolution of pulmonary *C. neoformans* infection. PLoS One *6*, e17204.

Xue, C., Tada, Y., Dong, X., and Heitman, J. (2007). The human fungal pathogen *Cryptococcus* can complete its sexual cycle during a pathogenic association with plants. Cell Host Microbe *1*, 263–273.

Yoneda, A., and Doering, T.L. (2006). A eukaryotic capsular polysaccharide is synthesized intracellularly and secreted via exocytosis. Mol. Biol. Cell *17*, 5131–5140.

Yoneda, A., and Doering, T.L. (2008). Regulation of *Cryptococcus neoformans* capsule size is mediated at the polymer level. Eukaryot. Cell. 7, 546–549.

Yuan, R.R., Spira, G., Oh, J., Paizi, M., Casadevall, A., and Scharff, M.D. (1998). Isotype switching increases efficacy of antibody protection against *Cryptococcus neoformans* infection in mice. Infect. Immun. *66*, 1057–1062.

Zaragoza, O. (2011). Multiple disguises for the same party: the concepts of morphogenesis and phenotypic variations in *Cryptococcus neoformans*. Front. Microbiol. 2, 181.

Zaragoza, O., and Casadevall, A. (2004). Experimental modulation of capsule size in *Cryptococcus neoformans*. Biol. Proced. Online *6*, 10–15.

Zaragoza, O., Fries, B.C., and Casadevall, A. (2003a). Induction of capsule growth in *Cryptococcus neoformans* by mammalian serum and CO_2. Infect. Immun. *71*, 6155–6164.

Zaragoza, O., Taborda, C.P., and Casadevall, A. (2003b). The efficacy of complement-mediated phagocytosis of *Cryptococcus neoformans* is dependent on the location of C3 in the polysaccharide capsule and involves both direct and indirect C3-mediated interactions. Eur J. Immunol. 33, 1957–1967.

Zaragoza, O., Telzak, A., Bryan, R.A., Dadachova, E., and Casadevall, A. (2006). The polysaccharide capsule of the pathogenic fungus *Cryptococcus neoformans* enlarges by distal growth and is rearranged during budding. Mol. Microbiol. *59*, 67–83.

Zaragoza, O., Chrisman, C.J., Castelli, M.V., Frases, S., Cuenca-Estrella, M., Rodriguez-Tudela, J.L., and Casadevall, A. (2008). Capsule enlargement in *Cryptococcus neoformans* confers resistance to oxidative stress suggesting a mechanism for intracellular survival. Cell. Microbiol. *10*, 2043–2057.

Zaragoza, O., Rodrigues, M.L., De Jesus, M., Frases, S., Dadachova, E., and Casadevall, A. (2009). The capsule of the fungal pathogen *Cryptococcus neoformans*. Adv. Appl. Microbiol. *68*, 133–216.

Zaragoza, O., Garcia-Rodas, R., Nosanchuk, J.D., Cuenca-Estrella, M., Rodriguez-Tudela, J.L., and Casadevall, A. (2010). Fungal cell gigantism during mammalian infection. PLoS Pathog. *6*, e1000945.

Zhang, Y., Wang, F., Tompkins, K.C., McNamara, A., Jain, A.V., Moore, B.B., Toews, G.B., Huffnagle, G.B., and Olszewski, M.A. (2009). Robust Th1 and Th17 immunity supports pulmonary clearance but cannot prevent systemic dissemination of highly virulent *Cryptococcus neoformans* H99. Am. J. Pathol. *175*, 2489–2500.

Zhu, X., Gibbons, J., Garcia-Rivera, J., Casadevall, A., and Williamson, P.R. (2001). Laccase of *Cryptococcus neoformans* is a cell wall-associated virulence factor. Infect. Immun. *69*, 5589–5596.

Dermatophytes as Saprophytes and Pathogens

10

Michel Monod, Bernard Mignon and Peter Staib

Abstract

Dermatophytes infect the *stratum corneum*, nails and hair and are the most common agents of superficial mycoses in humans and animals. At present the genome of seven species has been sequenced. Between 22.5 and 24 Mb, the dermatophyte genomes are smaller in size than those of *Coccidioides* spp., *Histoplasma* spp. and *Aspergillus* spp. They are enriched for particular families of genes encoding secreted proteases, fungal specific kinases and proteins containing the LysM domain that is known to bind chitin. Different tools were recently developed to improve genetic analyses of dermatophytes, including efficient systems for targeted gene inactivation, gene silencing and broad transcriptional profiling techniques. Unexpectedly, gene expression profiles in the skin and hair of infected guinea pigs were found to be very different from those during *in vitro* growth using hard keratin as a substrate. Instead of the major *in vitro* expressed protease genes, others were found to only be activated in the skin of infected animals. In other words, the expression of putative virulence genes in dermatophytoses is more complex than previously assumed and likely depends on the site and type of infection. Further broad transcriptional profiling approaches during infections will give new insights into the pathogenic mechanisms of dermatophytes.

Introduction

Dermatophytes are highly specialized filamentous fungi that are able to digest and grow on keratinized substrates (Ajello, 1974). Pathogenic species are the most common agents of superficial mycoses, infecting the *stratum corneum*, nails and hair almost exclusively. Dermatophytes evolved from soil saprophyte fungi, which acquired the ability to efficiently degrade hard keratin into amino acids and short peptides in the process of recycling soil nitrogen. Species such as *Trichophyton terrestre* and *Trichophyton* (*Keratinomyces*) *ajelloi* very rarely infect animals (Vanbreuseghem *et al.*, 1978) while other species gradually evolved to parasitize keratinous tissues of animals living in close contact with soil. The origin of dermatophytes corresponds to the early development of mammals (Harmsen *et al.*, 1995).

Three broad ecological groups of dermatophyte species are recognized, namely, anthropophilic, zoophilic and geophilic depending on their major reservoir in nature (Georg, 1960; Ajello, 1962; Weitzman and Summerbell, 1995) (Table 10.1). Anthropophilic species naturally colonize humans, while zoophilic species are found predominantly in animals. Geophilic species of dermatophytes are saprophytes existing in the soil without or only sporadically causing disease. They can be easily isolated from hair deposited on loam in a Petri dish (hair baiting technique) (Vanbreuseghem, 1952, Pore and Plunkett, 1967).

Zoophilic dermatophytes may result in zoonoses when humans are exposed to these organisms, and dermatophytosis is considered to be one of the most common zoonotic diseases. Zoophilic dermatophytes apparently lose pathogenicity during serial passage in humans, and thus infections are generally acquired directly from the animal or, less frequently, from the environment rather than from human contact (Marples, 1956). The majority of zoonotic dermatophytoses are caused by four species: *Microsporum canis* (usually derived from pet animals, particularly cats and dogs), *Trichophyton*

Table 10.1 Ecological classification of dermatophytes into geophilic, zoophilic, and anthropophilic species

Geophilic	Zoophilic	Anthropophilic
*M. gypseum** (*A. gypseum*, *A. fulvum*, *A. incurvatum*) (Stockdale 1962, 1964; Weitzman *et al.*, 1986)	*M. canis* (*A. otae*) (Hasegawa and Usui, 1974, 1975; Takashio 1981; Weitzman *et al.*, 1986)	*M. audouinii*
*T. terrestre** (*A. quadrifidum*) (Dawson and Gentles, 1961) (non-pathogenic)	*M. equinum*	*M. ferrugineum*
T. ajelloi (*A. uncinatum*) (Dawson and Gentles, 1959, 1961) (non-pathogenic)	*M. persicolor* (*A. persicolor*) (Stockdale 1967; Weitzman *et al.*, 1986)	*T. interdigitale*
A. curreyi†	*M. nanum* (*A. obtusa*) (Dawson and Gentles, 1961)	*T. tonsurans*
	A. benhamiae (formerly *T. mentagrophytes**) (Ajello and Cheng, 1967)	*T. rubrum*
	T. erinacei	*T. violaceum*
	A. vanbreuseghemii (formerly *T. mentagrophytes**) (Takashio, 1973)	*T. soudanense*
	T. quinckeanum‡ (also named *T. mentagrophytes* var. *quinckeanum*)	*T. schoenleinii*
	T. equinum	*T. concentricum*
	T. simii (*A. simii*) (Stockdale *et al.*, 1965)	*E. floccosum*
	T. verrucosum	
	T. bullosum	
	T. gallinae	

*Species complex.
†In contrast to other *Arthroderma* species, *A. curreyi* is homothallic (Pore and Plunkett, 1967).
‡Species to which the neotype of *T. mentagrophytes* belongs (Graser *et al.*, 1999b; Beguin *et al.*, 2012).
A, *Arthoderma*; *E*, *Epidermophyton*; *M*, *Microsporum*; *T*, *Trichophyton*.

verrucosum (usually derived from cattle), and two species of the *Trichophyton mentagrophytes* complex, *Arthroderma vanbreuseghemii* (usually derived from cats and dogs) and *Arthroderma benhamiae* (usually derived from guinea pigs) (Table 10.2). Infection with anthropophilic species may be also acquired indirectly through contact with a contaminated environment.

Taxonomy of dermatophytes

Species identification

Dermatophytes are ascomycete fungi (Onygenales order), but only anamorphs (or asexual forms) are isolated from infected patients, animals or soil. Mating studies (see section 'Dermatophyte teleomorphs', below) led to the discovery of the perfect states of dermatophyte species and demonstrated that anamorphic states may represent complexes of species. After isolation, dermatophyte species identification is usually performed on the basis of macroscopic and microscopic characteristics of the organism grown in culture, in particular with the observation of micro and macroconidia (Fig. 10.1).

Dermatophyte anamorphs are classified in three genera, *Trichophyton*, *Microsporum*, and *Epidermophyton*. In cultures, *Microsporum* species make spindle-shaped thick-walled rough macroconidia with several transverse septa (Fig. 10.1E). In *Trichophyton* species, macroconidia when present are smooth and blunt ended. *Microsporum* and *Trichophyton* species produce microconidia along the side of the hyphae. Microconidia vary in their number from none or a few in *Trichophyton rubrum* to numerous in species of the *T. mentagrophytes* complex (Fig. 10.1D). *Epidermophyton* has only two recognized species, *E. floccosum* and *E. stockdaleae*, which produce clavate macroconidia and no microconidia, respectively, and which are phylogenetically separated (Kwon-Chung and Bennett; 1992; Gräser *et al.*, 1999a). *Epidermophyton floccosum* is an anthropophilic pathogen closely related to other anthropophilic species of the genus *Trichophyton* (Gräser *et al.*, 1999a). *Epidermophyton stockdaleae* is a

Table 10.2 Zoonotic dermatophytes and their reservoirs

Dermatophyte	Main animal reservoir(s)	Frequency in humans	Geographic distribution
M. canis	Cats, dogs	Common	Worldwide
M. equinum	Horses	Rare	Worldwide
M. persicolor	Rodents, voles	Rare	Europe, USA
M. nanum	Pigs	Rare	Worldwide
A. benhamiae	Guinea pigs	Common	Worldwide
A. vanbreuseghemii	Cats, dogs, rodents (mice, chinchillas)	Common	Worldwide
A. simii	Monkeys, poultry, dogs	Rare outside India	Rare outside India
T. verrucosum	Cattle	Common	Worldwide
T. erinacei	Hedgehogs	Occasional	East Asia, Europe, New Zealand
T. quinckeanum	Mice (favus)	Rare	Worldwide
T. equinum	Horses	Occasional	Worldwide
*T. bullosum**	Horses, donkeys	One case	France, Sudan, Syria, Tunisia
T. gallinae	Chicken	Rare	Worldwide

*Species recently described from one case in a young man who had direct contact with donkeys (Sitterle *et al.*, 2012).
A, *Arthroderma*; *M*, *Microsporum; T*, *Trichophyton*.

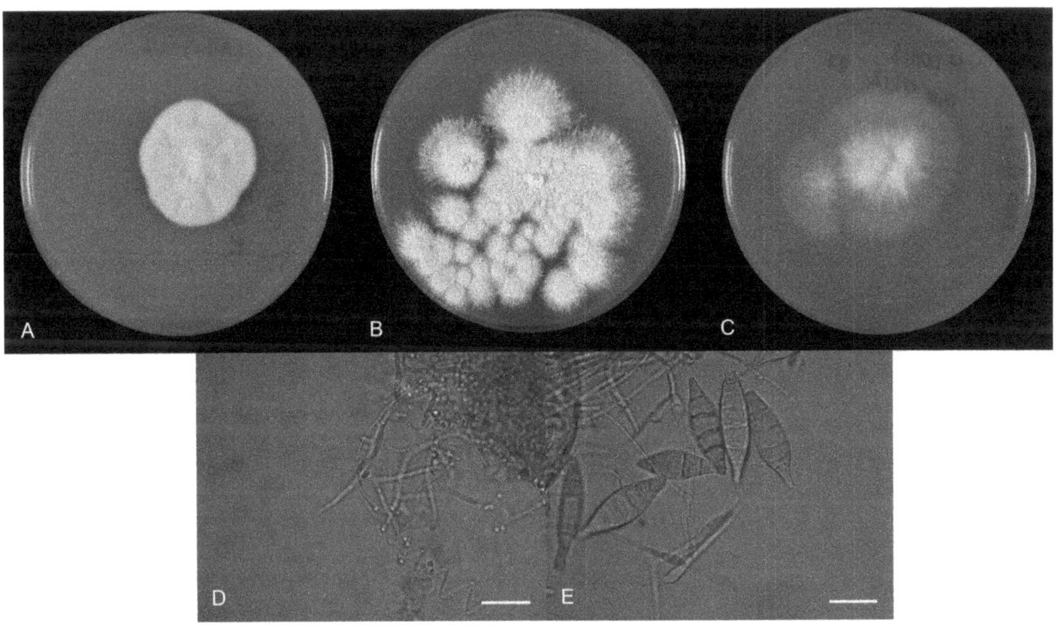

Figure 10.1 Macroscopic and microscopic characteristics of various dermatophytes. (A–C) Cultures of *T. rubrum*, *A. vanbreuseghemii* and *M. canis*, respectively, on Sabouraud agar. (D) *Arthroderma vanbreuseghemii* microspores. (E) *Microsporum canis* spindle-shaped, thick-walled macroconidia (bar in D and E=10 μm). A colour version of this figure is available in the plate section at the back of the book.

non-pathogenic species isolated only once from soil and close to other geophilic dermatophytes (Prochacki and Engelhardt-Zasada, 1974).

Morphological dermatophyte identification often remains difficult or uncertain because there can be variations from one isolate to another.

Therefore, DNA sequencing is particularly useful for accurate species identification (Gräser *et al.*, 1999b; Fréalle *et al.*, 2007; Ninet *et al.*, 2003; Symoens *et al.*, 2011, 2013). The polymorphism of the internal transcribed spacers ITS1 and ITS2 flanking the DNA sequence encoding the 5.8S rDNA is very discriminating and reliable for distinguishing different species. Identification can be performed by comparison of the ITS sequence of the species to be identified with ITS sequences available in the CBS public database (http://www.cbs.knaw.nl/dermatophytes/BioloMICSID.aspx). The 28S rDNA sequences were also found to be suitable for dermatophyte species identification (Ninet *et al.*, 2003) but are less discriminating.

Matrix-assisted laser desorption ionization time of-flight mass spectrometry (MALDI-TOF MS) can also be used now for rapid dermatophyte species identification (Nenoff *et al.*, 2013; de Respinis *et al.*, 2013). Closely related species in the *Microsporum gypseum* and *T. mentagrophytes* species complexes (see section 'Species delineation by mating experiments and DNA sequencing', below) can be discriminated using this technology.

Dermatophyte teleomorphs

Dermatophytes are generally heterothallic fungi, meaning that conjugation for sexual reproduction is only possible through the interaction of individual isolates of different mating types. This is the case for all pathogenic species. One geophilic species, *Arthroderma curreyi* was found to be homothallic (Pore and Plunkett, 1967). For heterothallic species, teleomorphs (or sexual forms) can be obtained by inoculating two isolates of different mating types (designated as either '+' or '−') a short distance apart (3 cm) in the centre of a Petri plate (Fig. 10.2) containing a nutrient-poor agar medium (for instance Sabouraud's medium 1/10) (Takashio, 1972). A good medium for mating experiments also consists of 3.0% (w/v) ground seeds of *Guizotia abyssinica* (from a seed store) and 2.0% (w/v) agar in distilled water (De Vroey, 1964; Symoens *et al.*, 2011). Dermatophyte teleomorphs have been classified in the *Arthroderma* genus in the Ascomycotina subphylum and in the Onygenales order, which contains *Talaromyces* and *Neosartorya*, the teleomorphs of *Penicillium* spp. and *Aspergillus* spp., respectively. Sexual fructifications are cleistothecia (globulose and closed ascocarps; Fig. 10.2B) similar to those of *Talaromyces* and *Neosartorya*. The wall of the cleistothecia of dermatophytes, the peridium, is constituted of loosely interwoven thin-walled hyphae characteristic of this group. Cleistothecia are full of asci with a thin and undifferentiated cell wall (prototunicate asci), containing eight ascospores (Fig. 10.2C). Teleomorphs of *Microsporum* species were previously classified in the *Nannizzia* genus but this has now been shown to be congeneric with *Arthroderma* (Weitzman *et al.*, 1986).

In anthropophilic species, as in a number of zoophilic species, sexual reproduction has not been observed. For instance, all isolates of *T. rubrum*, *T. tonsurans* and *T. interdigitale* were found to be of the − mating type, as determined by

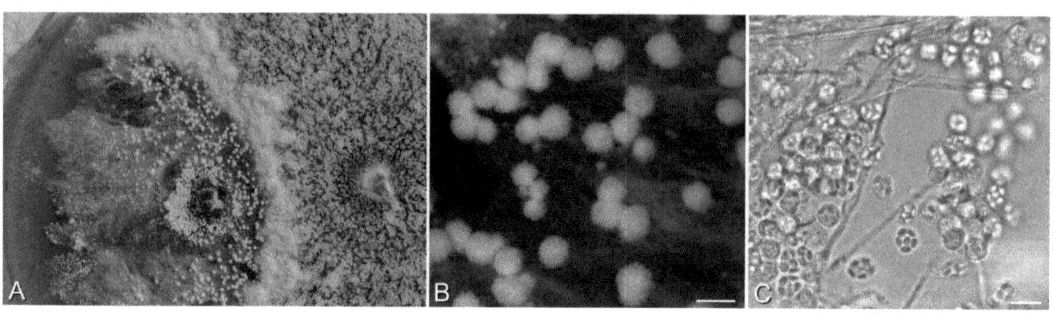

Figure 10.2 *Arthroderma benhamiae* teleomorphs. (A) Mating of a + strain with a − strain of *A. benhamiae*. (B) Binocular microscope observation of cleistothecia (bar=100 μm). (C) Asci and ascospores from squashed cleistothecia (bar=10 μm). A colour version of this figure is available in the plate section at the back of the book.

sexual stimulation and pseudocleistothecia formation using an *A. simii* tester strain of the opposite mating type (Stockdale, 1981). A bias towards one particular mating type has been observed for other species such as *M. canis* (*A. otae*), strains of which are generally of mating type –. In *A. benhamiae*, both mating types were isolated within a group of strains with a particular white phenotype, but only one mating type (–) was detected in a second group of strains characterized by a yellow phenotype (Symoens *et al.*, 2013). A bias towards one particular mating type has also been observed in other pathogenic fungi such as *Cryptococcus neoformans* (Idnurm *et al.*, 2005).

Dermatophytes with a + mating type were found to contain an *HMG* (high mobility group) box coding for the mating type protein MAT1-2, while dermatophytes with a – mating type were found to contain an alpha box coding for the mating type protein MAT1-1. However, in other ascomycete species, the + mating type and the – mating type designate strains with an alpha box and an *HMG* box, respectively. This principle (rule) has been followed in recent literature relative to dermatophyte mating types (Lee *et al.*, 2010; Li *et al.*, 2010).

Species delineation by mating experiments and DNA sequencing

Mating experiments and observation of fertile cleistothecia were particularly helpful to delineate species in the *M. gypseum* and *T. mentagrophytes* species complexes (Stockdale, 1964; Ajello and Cheng 1967; Takashio 1973; Symoens *et al.*, 2011, 2013). In confrontations of dermatophyte strains of opposite mating type belonging to closely related species, colonies join together and show an increase in aerial mycelium along the line where they meet. Cleistothecia initials or pseudocleistothecia without asci and ascospores can be observed. However, interspecific successful matings generating fertile cleistothecia were also reported (Kawasaki *et al.*, 2009; Anzawa *et al.*, 2010).

The *M. gypseum* species complex contains geophilic dermatophytes, which produce numerous spindle shaped macroconidia. In this complex, three separate teleomorph species with no interspecific sexual reaction could be distinguished, *Arthroderma fulvum*, *Arthroderma gypseum* and *Arthroderma incurvatum* (Stockdale, 1964).

The *T. mentagrophytes* species complex contains various species that produce numerous pyriform or round microconidia and for which identifications are often difficult to ascertain. Mating experiments and ribosomal DNA sequencing showed that this complex contains different teleomorph species and other species for which only the anamorph is known (Table 10.1). *Arthroderma simii* was the first species for which cleistothecia were obtained. *Arthroderma simii* grows fast and makes abundant spindle-shaped macroconidia, in contrast to the other species in the *T. mentagrophytes* complex. This species, which is prevalently isolated in India from soil and from various animal samples, including poultry, generally infects skin in humans (Vanbreuseghem *et al.*, 1978) but was recently isolated from a human nail (Beguin *et al.*, 2013). Two other teleomorph species are *A. benhamiae* and *A. vanbreuseghemii*. Both species are zoophilic. *Arthroderma benhamiae* was described by Ajello and Cheng (1967) after mating strains isolated from rodents, and *A. vanbreuseghemii* was described by Takashio (1973) after mating strains which were isolated from humans, mice and chinchillas. The latter species was frequently isolated from hunting European short hair cats and from dogs showing lesions (Drouot *et al.*, 2008). It can be suspected that infections occurred during hunting and that the natural source of *A. vanbreuseghemii* is either soil or rodents. Although fertile cleistothecia were obtained after mating an *A. simii* isolate (KMU 4810=JCM 27961) and a tester strain of *A. vanbreuseghemii* RV27961 (Kawasaki *et al.*, 2009), *A. simii* and *A. vanbreuseghemii* are still considered as separate species.

Mating experiments allowed clear delineation of the anthropophilic species *Trichophyton interdigitale*, which causes non-inflammatory tinea pedis (foot mycosis) and tinea unguium (onychomycosis). No fertile cleistothecia were obtained from confrontations between *T. interdigitale* isolates and strains of zoophilic species of the same *T. mentagrophytes* complex, indicating that *T. interdigitale* should be considered as a humanized species derived from the sexual relative *A. vanbreuseghemii* (Symoens *et al.*, 2011).

Of important note, the recently designated *T. mentagrophytes* neotype (CBS 318.56) (Gräser *et al.*, 1999b) belongs to a different species of *A. benhamiae, A. vanbreuseghemii* and *T. interdigitale*. The ITS and 28S sequences of this neotype strain were found to be identical to those of many strains of *T. mentagrophytes* var. *quinckeanum* or *Trichophyton quinckeanum* (Beguin *et al.*, 2012). This particular species is the agent of mouse and human favus, a special form of tinea capitis characterized by crusts (called scutulum) consisting of mycelium, neutrophils and epidermal cells (Schneider 1954, Blank 1957; Kwon-Chung and Bennett, 1992). *Trichophyton quinckeanum* was used to establish an experimental mouse model of dermatophytosis (Hay *et al.*, 1983). Following the report by Gruby (1842) which first and in French named a fungus 'mentagrophytes' in a case of beard hair sheath infection, *T. mentagrophytes senso stricto* should have been reserved for a dermatophyte other than the one species causing a favus. *Trichophyton mentagrophytes* should have been reserved for the asexual form of either *A. benhamiae* or *A. vanbreuseghemii* (Beguin *et al.*, 2012). The name *T. mentagrophytes* is still routinely used in clinical laboratories and in many publications for these two species. However, according to the rules proposed by the Amsterdam declaration on Fungal Nomenclature (One Fungus=One Name) (Hawksworth *et al.*, 2011), the sexual names *A. benhamiae* and *A. vanbreuseghemii* should always be used.

Signs and diagnosis of dermatophytosis, and dermatophyte isolation from clinical samples

Dermatophytoses vary depending on the causative agent and the body site affected. The disease is often described with the word 'tinea' followed by a term for the particular infected body site (Degreef, 2008). Anthropophilic species (e.g. *T. rubrum, T. interdigitale,* and *Trichophyton tonsurans*), tend to be associated with more chronic infections which are less inflammatory. In contrast, zoophilic and geophilic species of dermatophytes (e.g. *A. benhamiae, A. vanbreuseghemii, Trichophyton erinacei, T. verrucosum, M. canis* and *M. gypseum*) often cause highly inflamed lesions in humans (Fig. 10.3). Animal dermatophytoses caused by zoophilic species are shown on Fig. 10.4.

Dermatophytosis cannot usually be diagnosed on the basis of clinical signs alone but needs both direct microscopic examination of dermatological samples and infecting species identification. Direct microscopic examination using fluorescence techniques is by far the most sensitive technique

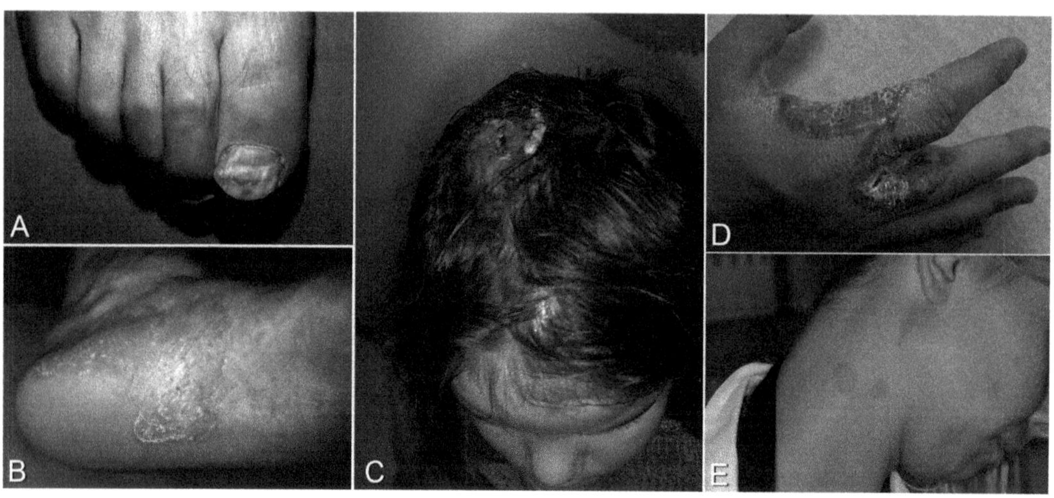

Figure 10.3 Human dermatophytoses. (A) Tinea unguium with *T. rubrum*. (B) Tinea pedis with *T. rubrum*. (C) Tinea capitis (kerion) with *A. benhamiae*, (D) Tinea manum with *T. erinacei*. (E) Tinea corporis with *M. canis*. A colour version of this figure is available in the plate section at the back of the book.

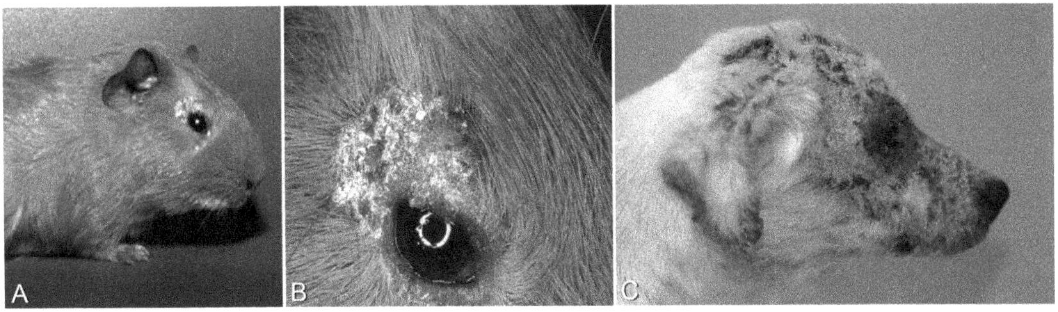

Figure 10.4 Animal dermatophytoses. (A and B) Alopecia and scaling caused by *A. benhamiae* in a guinea pig. (C) Extensive alopecia with *A. vanbreuseghemii* in a dog. A colour version of this figure is available in the plate section at the back of the book.

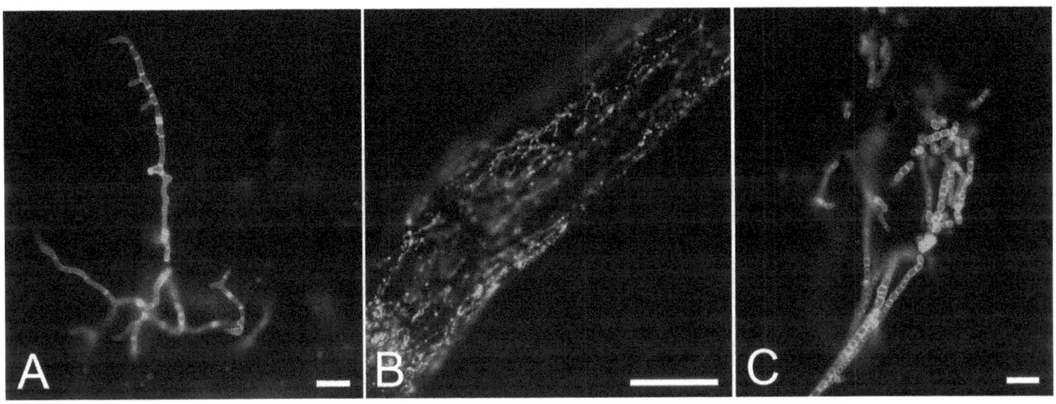

Figure 10.5 Direct mycological examinations of clinical and animal samples. (A) Hyphae in infected nails. (B) Tinea capitis, infected hair. (C) Hyphae and arthrospores from an infected guinea pig (bar in A and C=10 μm, bar in B=40 μm). One portion of each sample was examined in a dissolving solution containing a fluorochrome that specifically binds to vegetal and fungal cell wall polysaccharides (Monod *et al.*, 1989). The sample preparations were observed using a fluorescence microscope using excitation between 400 nm and 440 nm. The fluorescence techniques allow the best contrast of fungal elements in comparison to other techniques such as that using 10% KOH without fungal staining. A colour version of this figure is available in the plate section at the back of the book.

for detecting rare hyphae and spores in dermatological samples (Fig. 10.5) (Monheit *et al.*, 1984; Holländer *et al.*, 1984; Gip and Abelin, 1987; Monod *et al.*, 1989). However, a positive direct examination does not automatically mean that the infecting fungus is a dermatophyte. As an example, although *T. rubrum* and *T. interdigitale* are the main aetiological agents of onychomycosis, PCR-sequencing and restriction fragment length polymorphism (RFLP) assays also revealed *Fusarium* spp. and other less frequently isolated non-dermatophyte fungi as infectious agents in nails (Monod *et al.*, 2006; Verrier *et al.*, 2012).

The infecting fungus is generally identified by fungal cultures. A given dermatological sample is often seeded onto two media. The first consists of Sabouraud agar medium containing an antibiotic, usually chloramphenicol (50 μg/ml), and the second tube contains Sabouraud agar medium with chloramphenicol (50 μg/ml) plus cycloheximide (400 μg/ml). Dermatophytes can be selected from other moulds such as *Alternaria* spp., *Penicillium* spp. or *Aspergillus* spp. on the second medium because they are insensitive to cycloheximide. Since dermatophytes are usually slowly growing fungi, cultures are typically maintained for 2–3 weeks before species identification. Another particularity of dermatophytes is their ability to increase the pH of a growth medium through the secretion of alkaline metabolites (Taplin *et al.*, 1969; Li *et al.*, 2009). Therefore, dermatophytes can also be distinguished from other moulds using a dermatophyte test medium (DTM) that incorporates phenol red in the culture medium as a pH indicator. Dermatophytes turn the yellow colour of the medium red.

Results of fungal cultures may be negative, especially from infected nails, emphasizing the importance of other additional diagnostic techniques for direct *in situ* fungus identification in a collected sample. In the last few years, PCR assays have been developed for direct detection and identification of dermatophytes in nail, skin and hair samples (Brillowska-Dabrowska *et al.*, 2007; Bergmans *et al.*, 2010; Verrier *et al.*, 2012), as well as dermatophytes and non-dermatophyte fungi in nails (Monod *et al.*, 2006; Verrier *et al.*, 2012).

Tools to investigate dermatophyte biology

Dermatophyte genomes

Prior to the availability of the complete genome sequences, collections of dermatophyte open reading frame and protein sequences were revealed from transcriptome analysis (cDNA analysis and EST sequencing). A comprehensive *T. rubrum* Expression Database (TrED) was launched by Wang *et al.* (2004, 2006), where cDNAs and ESTs were obtained from the fungus grown under different biologically interesting and also pathogenicity related conditions (Zaugg *et al.*, 2009).

Currently, seven dermatophyte genomes have been sequenced and annotated. The genomes of two closely related zoophilic species, *A. benhamiae* and *T. verrucosum*, were sequenced at the Hans Knöll Institute in Jena (Germany) (Burmester *et al.*, 2011). The genomes of the anthropophilic species *T. rubrum* and *T. tonsurans*, the zoophilic species *Trichophyton equinum* and *M. canis*, and the geophilic species *M. gypseum* were sequenced by the BROAD Institute (Martinez *et al.*, 2012). The annotated genome sequences are available in the Dermatophyte Comparative Database released by the BROAD Institute (http://www.broadinstitute.org/annotation/genome/dermatophyte_comparative/GenomesIndex.html).

The genomes of the dermatophytes were found to be similar in size (ranging from 22.5 Mb for *T. rubrum* to 24.1 Mb for *T. equinum*) and highly collinear. They are smaller than those of *Coccidioides* spp. (27–29 Mb), *Histoplasma capsulatum* (30–39 Mb) and *Aspergilli* (e.g. 28 Mb and 37 Mb for *A. fumigatus* and *A. oryzae*, respectively). The number of predicted protein-coding genes varies from 7980 in *A. benhamiae* to 8915 in *M. canis* (Martinez *et al.*, 2012). Major distinguishing features of the dermatophytes genomes can be summarized as following:

1. Large gene families encoding secreted endoproteases of the subtilisin, deuterolisin and fungalisin families are expanded in comparison to other fungi such as *Aspergillus* spp. The enrichment in secreted protease families can be *a priori* associated with the ability of dermatophytes to use hard keratin as a growth substrate.
2. Genes encoding enzymes involved in sugar metabolism and plant cell wall breakdown are lacking. For instance, there are no genes encoding glycoside hydrolases and alpha-amylases. These deficiencies attest to the high specialization of dermatophytes and their adaptation to particular substrates other than vegetal debris.
3. The genomes are enriched in several genes encoding proteins containing one or several LysM domains known to specifically bind chitin. Dermatophyte proteins with a LysM domain could mask chitin on the surface of the hyphae as in fungal plant pathogens (de Jonge and Thomma, 2009). This neutralization could help the fungus evade the immune system by preventing its stimulation by accessible chitin (Martinez *et al.*, 2012).
4. The genomes encode novel sets of fungal-specific kinases with unknown specificities.
5. Dermatophytes contain a large number of genes coding for enzymes that synthesize secondary metabolites.
6. Despite species differences in mating ability, genes involved in mating and meiosis are

conserved across species, suggesting the possibility of cryptic mating in species where it has not been previously detected.

Comparative genome analyses of anthropophilic, zoophilic and geophilic species should reveal differences and address factors which are potentially involved in host preferences, pathogenicity and related adaptation mechanisms of dermatophytes during infection. A large number of orthologues were found to be shared by the seven dermatophytes that have been sequenced and annotated (6168 genes or groups of genes with paralogue duplications) whilst the number of unique genes to each species is 801, 937 and 943 in *T. rubrum*, *M. gypseum* and *M. canis*, respectively (Martinez *et al.*, 2012). However, altogether these genes contain little functional information, as recognizable Interpro domains (Hunter *et al.*, 2009) were found in only 4% of the encoded proteins in *T. rubrum*, and 17% in *M. canis*. Forty-eight, 271 and 89 orthologue groups were found to be specific for anthropophilic species (*T. rubrum* and *T. tonsurans*), zoophilic species (*A. benhamiae*, *T. verrucosum*, *T. equinum* and *M. canis*) and *M. gypseum* as a unique geophilic species, respectively. Validation of genome indications and predictions is now one of the major challenges in dermatophyte molecular biology. Table 10.3 presents an inventory of genetics tools which can be applied to investigate dermatophyte biology for each species.

Useful species to investigate basic dermatophyte biology, pathogenicity and related adaptation mechanisms during infection

There are multiple dermatophyte species from different ecological groups. Their peculiar phenotypes may differ greatly during cultivation in terms of growth rate, colony appearance and cellular morphology. Some species, such as *T. verrucosum*, are even difficult to propagate *in vitro*, which makes experimental studies difficult to perform. Depending on the dermatophyte species itself, but also upon individual host characteristics and the part of the body infected, dermatophytoses may differ strongly in terms of symptoms, e.g. when comparing *T. rubrum* induced onychomycosis with highly inflammatory skin lesions caused by *A. benhamiae*. In other words, it appears difficult to address dermatophytes and dermatophytoses as a whole, while at the same time there is an increasing need for appropriate models to study the biological and physiopathological aspects of dermatophytes and dermatophytoses, respectively.

In recent years the zoophilic dermatophyte *A. benhamiae*, which is frequently isolated (Fumeaux *et al.*, 2004, Symoens *et al.*, 2013), was selected as an appropriate organism to investigate dermatophyte biology and virulence factors at a molecular level for several reasons:

1. Compared to other dermatophytes, *A. benhamiae* grows relatively fast under laboratory conditions. This characteristic represents a major advantage, especially when studying pathogenicity related traits, e.g. keratin degradation, at the level of global transcriptomes or proteomes (Zaugg *et al.*, 2009; Staib *et al.*, 2010; Sriranganadane *et al.*, 2011a).
2. *A. benhamiae* produces abundant microconidia, single-nucleated round-oval or pear-shaped cells. Microconidia can be used as a starting material for genetic manipulation, such as efficient gene knock-out, and for inoculation of dermatophytoses models, such as *in vitro* reconstituted epidermis (Tabart *et al.*, 2007; Grumbt *et al.*, 2011a; for review see Grumbt *et al.*, 2011b).
3. Since *A. benhamiae* is zoophilic and causes inflammatory cutaneous infections naturally in guinea pigs (Fig. 10.4), this enabled establishment of an animal model, making possible the analysis of transformant strains *in vivo* too (Staib *et al.*, 2010).
4. *A. benhamiae* is teleomorphic and offers the opportunity to provide insights into basic dermatophyte biology, including cleistothecia formation and sexual development. Fertile cleistothecia can be easily produced by confrontations of wild isolates of opposite mating types (Symoens *et al.*, 2011, 2013). It was recently demonstrated that recombinant progeny could be generated in *M. gypseum* from two parental strains of opposite mating types (Li *et al.*, 2010).

Table 10.3 Species characteristics and genetic tools to investigate dermatophyte biology

	T. rubrum	*A. benhamiae*	*A. vanbreuseghemii* (*T. mentagrophytes*)*	*A. otae* (*M. canis*)
Species characteristics				
Natural host	Human	Guinea pig	Cat, dog, rabbit, mouse (?)	Cat, dog
Growth rate	Quite slow	Relatively fast	Relatively fast	Relatively fast
Microconidia	On poor media	Abundant	Abundant	Low amount
Macroconidia	Some isolates form club shaped macroconidia	Rarely observed	Rarely observed	Most isolates form spindle shaped macroconidia
Strain mating results				
Teleomorph	Unknown	*A. benhamiae*	*A. vanbreuseghemii*	*A. otae*
Mating results		Fertile cleistothecia easily produced in laboratory	Fertile cleistothecia easily produced in laboratory	Fertile cleistothecia Hasegawa and Usui (1975); Takashio (1981)
Genetics tools				
Genome	Martinez *et al*. (2012)	Burmester *et al*. (2011)	Not sequenced	Martinez *et al*. (2012)
Selection markers for transformation				
hph	Fachin *et al*. (2006), Ferreira-Nozawa *et al*. (2006)	Grumbt *et al*. (2011a)	Gonzalez *et al*. (1989), Kaufman *et al*. (2004)	Yamada *et al*. (2005, 2006)
neo		Grumbt *et al*. (2011a)	Yamada *et al*. (2008)	
nat1			Alshahni *et al*. (2010)	
Targeted gene inactivation	Fachin *et al*. (2006), Ferreira-Nozawa *et al*. (2006)	Grumbt *et al*. (2011a)	Yamada *et al*. (2009a,b)	Yamada *et al*. (2006)
Availability of *ku70* mutants to increase the rate of targeted integration of linear DNA cassettes into the genome		Grumbt *et al*. (2011a)	Yamada *et al*. (2009a), Alshahni *et al*. (2011)	
RNA silencing				Vermout *et al*. (2007)
Conditional gene knock-down			Iwata *et al*. (2012)	

*The anamorph name *T. mentagrophytes* was often used instead of *A. vanbreuseghemii* in dermatophyte genetics investigations.
A, *Arthroderma*; *M*, *Microsporum*; *T*, *Trichophyton*.

5 All experimental approaches in *A. benhamiae* are now facilitated by the fact that the genome of *A. benhamiae* has recently been sequenced and annotated (Burmester *et al.*, 2011).

Arthroderma vanbreuseghemii is another interesting species for investigation into dermatophyte biology. Like *A. benhamiae* it grows relatively fast, produces abundant microconidia and is teleomorphic. Many powerful genetics tools were developed in this species (Table 10.3). However, the genome of *A. vanbreuseghemii* has not yet been sequenced. This species naturally causes inflammatory cutaneous infections in cats and dogs (Fig.

10.4), but also in rabbits, which could make the establishment of natural model of infections more difficult or inadequate for ethical reasons.

Transformation and targeted gene inactivation in dermatophytes

The construction and subsequent phenotypic analysis of specific knock-out mutants represents a major advance in the research of pathogenic microorganisms. However, dermatophytes are less amenable to genetic manipulation than many other fungi. This drawback putatively explains why only little is known about dermatophytes at the molecular level. The first successful transformation of a dermatophyte was performed by Gonzalez *et al.* (1989) in *T. mentagrophytes*, following a protoplast/polyethylene glycol (PEG)-mediated protocol. Transformants with random DNA insertions were selected by use of the bacterial hygromycin B phosphotransferase gene *hph* as a marker. The next reported transformation was carried out 15 years later by Kaufman *et al.* (2004), who also used PEG-mediated transformation of *T. mentagrophytes* and transformant selection via *hph*. The latter marker was also successfully applied in *M. canis* (Yamada *et al.*, 2005, 2006) and *T. rubrum* (Fachin *et al.*, 2006; Ferreira-Nozawa *et al.*, 2006) before different alternative drugs/dominant markers were established in *T. mentagrophytes*, i.e. geniticin (G-418)/*Escherichia coli* neomycin phosphotransferase gene *neo* (Yamada *et al.*, 2008) and nourseothricin/*Streptomyces noursei* nourseothricin acetyltransferase gene *nat1* (Alshahni *et al.*, 2010). The *neo* and *hph* markers were also recently proven to be successful in *A. benhamiae* (Grumbt *et al.*, 2011a). Although in most transformation experiments PEG-mediated protoplast transformation was performed to allow DNA transfer, additional techniques have also been employed in dermatophytes, including an *Agrobacterium tumefaciens*-mediated transformation (ATMT) system in *T. mentagrophytes* (Yamada *et al.*, 2009b) and electroporation of germinated conidia in *T. rubrum* (Dobrowolska and Staczek, 2009).

Dermatophytes, similar to other filamentous fungi, are known to support targeted integration of linear DNA cassettes into the genome by homologous recombination only poorly. This drawback was circumvented in diverse fungal species, e.g. in *Neurospora crassa* (Ninomiya *et al.*, 2004) and *Aspergillus* spp. (da Silva Ferreira *et al.*, 2006; Krappmann *et al.*, 2006), by constructing recipient strains which are deficient in the non-homologous end joining (NHEJ) recombination pathway. In this way, mutants were constructed in dermatophytes lacking key factors of NHEJ, e.g. *ku80* and *lig4* mutants in *T. mentagrophytes* (Yamada *et al.*, 2009a; Alshahni *et al.*, 2011), and *ku70* mutants in *A. benhamiae* (Grumbt *et al.*, 2011a). Use of such mutants as parental strains in transformation experiments strongly increases the frequency of targeted insertions by homologous recombination, but at the same time they also display an altered risk of undergoing unforeseen events of genetic variation. However, in a comparison between *A. benhamiae* wild type and the *ku70* mutant, wild-type cells were also found to allow efficient targeted gene deletion. This finding made the species interesting for genetic research in dermatophytes (Grumbt *et al.*, 2011a).

Despite recent advances, up until now only a small number of genes have so far been inactivated in dermatophyte species by site-directed mutagenesis. In addition to the above mentioned genes important for NHEJ, these include *pacC* and *mdr2* in *T. rubrum* (Fachin *et al.*, 2006; Ferreira-Nozawa *et al.*, 2006), *dnr1* in *M. canis* (Yamada *et al.*, 2006), *areA* and *tri m 4* in *T. mentagrophytes* (Yamada *et al.*, 2009a), and *acuE* and *hypA* in *A. benhamiae* (Grumbt *et al.*, 2011a; Heddergott *et al.*, 2012). In *A. benhamiae*, the establishment of two different dominant selection markers, *hph* and *neo*, made possible the specific reconstitution of gene knock-outs for the first time (Grumbt *et al.*, 2011a).

RNA silencing and conditional gene knock-down

RNA interference, first described in the nematode *Caenorhabditis elegans* (Fire *et al.*, 1998), has been widely applied in diverse areas of research as a suitable tool to study gene function. The method appears to be of particular interest in cases where site directed mutagenesis is either difficult or, due to lethality, impossible. In addition, the strategy of gene silencing allows the down regulation of several homologous genes at the same time, a capability that might be advantageous in studying the functions of multiple members of a gene family. In dermatophytes, such factors of interest are, for example, represented by genes encoding the large families of subtilisins (Subs) or metalloproteases (Meps) that are putatively important for dermatophyte pathogenicity. RNA silencing in dermatophytes was established by Vermout *et al.* (2007). *M. canis* transformants have been constructed in which the expression of *dppIV* (encoding dipeptidyl peptidase IV) or *sub3* was suppressed. Testing the latter knock-down strains in a sophisticated model of *in vitro* reconstructed feline epidermis and in *ex vivo* models of human or animal epidermis showed that Sub3 obviously contributed to the adherence of *M. canis* to host tissue (Baldo *et al.*, 2008, 2010; Bagut *et al.*, 2012). In contrast, the epidermal invasiveness and virulence of these transformants was not impaired in a model of cutaneous guinea pig infection (Baldo *et al.*, 2010). A drawback of the RNA interference tool for gene silencing is that the stability of the constructs was found to vary among independently generated transformants. All experiments based on *M. canis sub* gene silencing were performed with a single stable transformant.

Another strategy to characterize essential genes can offer conditional gene knockdown by regulatable promoters. Such a system has recently been established in the dermatophyte *A. vanbreuseghemii* by use of tightly copper-responsive promoters (Iwata *et al.*, 2012). In particular, the *T. rubrum* promoter of the copper transporter gene *ctr4* was proven successful in repressing a fused gene of interest in the presence of copper.

Infection models

A distinctive biological characteristic of pathogenic dermatophytes is their ability to infect immunocompetent hosts, contrasting with most other fungi that are opportunistic pathogens (Vermout *et al.*, 2008b). Anthropophilic and zoophilic dermatophytes can even be considered as obligate parasites. They have no known reproductive cycle in the environment, such as exosaprophytes, e.g. *Aspergillus* spp. They also differ from endo- and episaprophytes, e.g. *Candida* and *Malassezia* species, respectively, which can not only survive but also reproduce for long periods in their host without any pathological significance. With the possible exception of asymptomatic chronic colonization observed for some dermatophyte species in their natural host under certain conditions, dermatophytes are not part of the normal skin flora (Wagner and Sohnle, 1995). Dermatophytoses are contagious infections, eliciting both clinical signs and a complex but misunderstood immune response in their hosts (Mignon *et al.*, 2008). These fungi produce a highly polymorphic spectrum of diseases depending on the fungal species involved, the infected body area, and the hosts' inflammatory and immune responses (Degreef, 2008; Brasch, 2010). This makes the modelling of dermatophytosis very hard to perform. As a consequence, it is necessary to have appropriate infection models to study the various aspects of the host–fungus relationship, especially for the identification of dermatophyte virulence factors (Achterman and White, 2012) and to characterize the hosts' immune responses against these parasitic fungi (Mignon *et al.*, 2008).

So far, most studies of the pathophysiological mechanisms involved in dermatophytoses have been performed *in vivo*. The guinea pig was used as the experimental animal model in more than 80% of these studies. This model of infection currently has several limitations, e.g. the difficulty in inducing clinical signs similar to those caused by most anthropophilic species, including chronic infections, such as those that can be induced by *T. rubrum* in man. For zoophilic species, the guinea pig model has been useful to evaluate the role of several putative virulence factors of both *M. canis* (Baldo *et al.*, 2010) and *A. benhamiae* (Staib *et al.*, 2010; Grumbt *et al.*, 2011a,b) and to assess the immunogenicity of some dermatophyte antigens, including characterized proteases (Mignon *et al.*, 1999; Brouta *et al.*, 2003; Descamps *et al.*, 2003). Regarding studies of the immune response to dermatophytes, both for testing particular hypotheses and to develop efficient and safe vaccines, it is clearly preferable to use the target species when using *in vivo* models. In this context, the concomitant use of the guinea pig and *A. benhamiae* remains attractive for the study of dermatophyte virulence. However, the operating costs and the lack of immunological tools for this animal species render the use of a mouse model more promising for the near future. Despite the existence of an ancient and particular model of mouse favus (Hay *et al.*, 1983) and a recent model of invasive dermatophytosis (Venturini *et al.*, 2012), modern murine models for the study of superficial dermatophyte infections are currently lacking (Shimamura *et al.*, 2012). Animal models not only generate ethical issues, but also require highly qualified researchers.

In vitro and *ex vivo* models are more convenient to dissect the in-depth mechanisms of dermatophyte pathogenicity, such as host-fungus cellular and molecular interactions. Such models have been developed, but to a much smaller extent than those for the study of opportunistic fungal infections. With regard to *ex vivo* models, these include mainly skin explants (Duek *et al.*, 2004; Kaufman *et al.*, 2004, 2007), but also nail fragments (Ferreira-Nozawa *et al.*, 2006; Maranhao *et al.*, 2009; Grumbt *et al.*, 2013) and isolated hair follicles (Rashid *et al.*, 1996), *stratum corneum* sheets (Aljabre *et al.*, 1992a,b) and dead keratinocyte suspensions (Zurita and Hay, 1987; Aljabre *et al.* 1993). Recently, skin explants have been used to study dermatophyte invasion (Mathy *et al.*, 2010) and adherence, allowing identification for the first time of a role for a putative dermatophyte pathogenic factor, namely a subtilisin from *M. canis*, in adherence to corneocytes from many animal species and men (Baldo *et al.*, 2010; Bagut *et al.*, 2012). *In vitro* models include the use of a reconstructed human epidermis (Grumbt *et al.*, 2011a), a reconstructed feline epidermis (Tabart *et al.* 2007; Baldo *et al.*, 2008) and also cultivated cells such as epidermal keratinocytes (Nakamura *et al.*, 2002; Shiraki *et al.*, 2006; Tani *et al.*, 2007; Burmester *et al.*, 2011), macrophages (Campos *et al.*, 2006), dendritic cells (Chung *et al.*, 2009; Heddergott *et al.*, 2012), and transient epidermal cells such as neutrophils (Cambier *et al.*, 2013; Heddergott *et al.*, 2012). All these *in vitro* models are necessary to understand the detailed interactions between putative dermatophyte virulence factors and the different host components that are encountered by a fungus during infection. The recent completion of genome projects for dermatophytes, the development of efficient molecular tools applicable to dermatophytes, e.g. mutant technology, and the growing knowledge about innate and specific immunity will help to design relevant *in vitro* tests and then *in vivo* assays, preferably in genetically unique strains of animals, for use in studies on microbial virulence and host defence mechanisms.

Hard keratin digestion by dermatophytes

Proteases secreted by dermatophytes *in vitro* in a protein medium

Since dermatophytes are almost exclusively localized in keratinized tissues, research on mechanisms of fungal invasion has so far primarily been focused on secreted proteases. All investigated dermatophytes secrete proteolytic activities *in vitro* when grown in a medium containing protein, and in particular keratin, as the sole nitrogen source (Mignon *et al.*, 1998; Brouta *et al.*, 2001, 2002; Descamps *et al.*, 2002; Jousson *et al.*, 2004a,b; Monod *et al.*, 2005). Various dermatophyte-secreted proteases were originally purified from culture supernatants after different chromatographic steps. However, the growth of dermatophytes is rather slow, so it was difficult to obtain enough material for the purification of native proteins in sufficient quantities for further characterization. Two native endoproteases, a subtilisin and a metalloprotease, both from *M. canis*, were purified and demonstrated to be keratinolytic (Mignon *et al.*, 1998; Brouta *et al.*, 2001). Afterwards, numerous dermatophyte proteases were individually generated as recombinant proteins for further characterization (Brouta *et al.*, 2003; Descamps *et al.*, 2003; Jousson *et al.*,2004b; Monod *et al.*, 2005). This experimental approach also avoided the problem of purification, since there would be no contamination from other proteases present in the culture supernatant.

At neutral or alkaline pH, dermatophytes secrete multiple endopeptidases of the S8 and M36 families (Brouta *et al.*, 2001, 2002; Descamps *et al.*, 2002; Jousson *et al.*, 2004a,b), called Sub for subtilisins and Mep for metalloproteases, respectively. In addition, dermatophytes secrete various aminopeptidases, including leucine aminopeptidases (Lap1 and Lap2) and dipeptidyl-peptidases (DppIV and DppV), for which an orthologue exists in *Aspergillus* spp., and a major metallocarboxypeptidase homologous to the human pancreatic carboxypeptidase A for which no orthologue was found in *Aspergillus* spp. (Monod *et al.*, 2005; Vermout *et al.*, 2008a; Zaugg *et al.*, 2008). Up-regulation of the genes encoding Subs, Meps, Laps and Dpps was demonstrated by microarray analysis of gene expression profiles during growth of *T. rubrum* and *A. benhamiae* on soy and keratin protein (Zaugg *et al.*, 2009; Staib *et al.*, 2010). The protein sequences of orthologous proteases are highly conserved across dermatophyte species, but their levels of secretion vary from one species to another (Giddey *et al.*, 2007).

In a protein medium at acidic pH, dermatophytes were found to secrete an aspartic protease of the pepsin family (Pep1) as an endoprotease, as well as tripeptidyl peptidases of the sedolisin family, X-prolyl peptidases and carboxypeptidases of the S10 family, all as exoproteases (Sriranganadane *et al.*, 2011a). In summary, dermatophytes secrete at acidic pH, as well as alkaline pH, proteases similar to those of *Aspergillus* spp. with the exception of a glutamic protease. Genes encoding glutamic proteases (Yabuki *et al.*; 2004; Kataoka *et al.*, 2005) are lacking in dermatophyte genomes (Burmester *et al.*, 2011; Martinez *et al.*, 2012).

Tight regulation of *in vitro* secreted proteolytic activity

Genes encoding major proteases secreted by dermatophytes during *in vitro* growth in a protein medium are controlled by mechanisms similar to those described in *Aspergillus* spp. The expression of *Aspergillus* spp. genes encoding alkaline and acidic proteases secreted in a protein medium depends on the transcription factor AreA (the orthologue of Nit2 in *Neurospora crassa*; Stewart and Vollmer, 1986), which is required for the expression of genes involved in nitrogen metabolism (Kudla *et al.*, 1990; Wilson and Arst, 1998; Christensen *et al.*, 1998; Hensel *et al.*, 1998). In the absence of ammonium and glutamine, AreA activates a large number of genes, including genes coding for secreted proteases. In addition, *Aspergillus* spp. secreted proteolytic activity is controlled by the transcription factor PacC, which acts positively on various genes expressed at neutral and alkaline pH and negatively on another set of genes expressed at acidic pH (Peñalva *et al.*, 2008). In particular, genes encoding neutral and alkaline secreted proteases are activated at alkaline pH while genes encoding acidic secreted proteases are repressed. At acidic pH, PacC is inactive and genes encoding secreted acidic proteases are not repressed, while there is a lack of activation of genes encoding secreted alkaline proteases.

Dermatophyte genomes contain nitrogen regulatory genes that are orthologues of those functionally characterized in *Aspergillus* spp. In *M. canis*, inactivation of *dnr1*, the orthologue of *areA* and *nit2* in *Aspergillus nidulans* and *N. crassa*, respectively (Kudla *et al.*, 1990; Stewart and Vollmer, 1986), resulted in a defect in alkaline proteolytic activity (Yamada *et al.*, 2006). Mutants disrupted in *dnr1* were unable to grow on skimmed milk and bovine serum albumin as sole nitrogen source and showed reduced growth on dog hair *in vitro*. *Microsporum canis dnr1* can even complement an *areA* mutation in *A. nidulans*. In *T. rubrum*, deletion of *pacC* was shown to reduce the level of secreted proteolytic activity at neutral pH and also to limit fungal growth in human nail samples in comparison to the wild-type control strain (Ferreira-Nozawa *et al.*, 2006). Dermatophyte genomes do not contain a gene encoding an orthologue of PrtT that is also specifically required for extracellular proteolytic activity in several species of *Aspergillus* (Punt *et al.*, 2008; Bergman *et al.*, 2009; Sharon *et al.*, 2009).

Deduced pathways of protein degradation into amino acids and short peptides

Endo- and exoproteases secreted by microorganisms, including fungi, cooperate very efficiently to digest protein into oligopeptides and free amino acids that are assimilable via transporters (Fig. 10.6, Payne and Smith, 1994). During this process the main function of endoproteases is to produce a large number of free ends on which exoproteases may act. However, fungal secreted proteases are incapable of digesting structures made of hard keratin by themselves. Efficient proteolysis has to be accompanied by the reduction of cysteine disulfide bridges, which are mainly responsible

Keratinized tissues (insoluble network of different cross-linked proteins)

(i) Secreted sulfite ↓

Reduced polypeptidic chains

(ii) Endoproteases ↓

Large peptides

(iii) Exoproteases (non-specific aminopeptidases and X-prolyl peptidases) ↓

Amino acids and short peptides (substrates for membrane transporters)

Figure 10.6 Schema of hard keratin degradation by dermatophytes. Although the activities of several dermatophyte secreted endoproteases are redundant, there are bottlenecks in the degradation process: (i) Hard keratin cannot be degraded by enzymatic activity alone. Sulfite excretion to reduce disulfide bridges is mandatory. (ii) Large peptides with free ends on which exoproteases may act have to be produced by endoproteases. (iii) Proline cannot be bypassed by non-specific aminopeptideases. In cases of proline rich proteins, degradation of a peptidic chain into amino acids and short peptides (2–5-mers) has to be achieved with the complementary action of a prolylpeptidase.

for the resistant nature of keratin (Kunert, 1992, 2000).

Reduction of cystine disulfide bridges preceding proteolytic activity in hard keratin digestion

During *in vitro* keratin degradation, dermatophytes and filamentous fungi were shown to excrete sulfite as a reducing agent (Kunert 1972a, 1976, 2000; Léchenne *et al.*, 2007). Sulfite is a product of the metabolism of cysteine. In the presence of sulfite, disulfide bonds of the keratin substrate are directly cleaved into cysteine and S-sulfocysteine. The presence of sulfocysteine was confirmed in human hair attacked by *M. gypseum* (Kunert, 1972b). This compound was found both in the free form and in oligopeptides with a molecular mass of 700 to 2500 Da that are the primary products of keratinolysis (Kunert, 1976; Ruffin *et al.*, 1976). As highly specialized fungi, dermatophytes are able to utilize free cysteine added to a nutrient broth and excrete sulfite as a product of the metabolism of cysteine-containing compounds. Once disulfide bridges are directly cleaved *in vivo* by sulfitolysis, reduced proteins become accessible to further digestion by various endo- and exoproteases secreted by the fungi.

Genes encoding *T. rubrum* and *A. benhamiae* sulfite efflux pumps (Ssu1) were cloned and their heterologous expression in *Saccharomyces cerevisiae* resulted in a sulfite resistant phenotype in the yeast (Léchenne *et al.*, 2007). Like the *S. cerevisiae* sulfite transporter SceSsu1, these transporters belong to the Tellurite-resistance/Dicarboxylate Transporter (TDT) family, which also include the *E. coli* tellurite efflux pump and the *Schizosaccharomyces pombe* malate transporter encoded by the genes *tehA* and *maeI*, respectively (Walter *et al.*, 1991; Grobler *et al.*, 1995). The relatively high expression of *T. rubrum ssu1* and *A. benhamiae ssu1* compared to that of *Aspergillus fumigatus ssu1* reflects a characteristic of dermatophytes, which renders these fungi able to degrade keratinized epidermal structures, i.e. *stratum corneum*, hair and nails (Léchenne *et al.*, 2007). Recent investigations on sulfite formation and secretion in *A. benhamiae* have provided molecular evidence for the role of this mechanism in the degradation of hard compact keratin by dermatophytes. *Arthroderma benhamiae* mutants in Ssu1 and cysteine dioxygenase Cdo1, a key cytoplasmic enzyme involved in conversion of cysteine into sulfite, were specifically defective for growth on hair and nails (Grumbt *et al.*, 2013).

Protein digestion

The different steps of protein degradation into amino acids, di- and tripeptides by fungi were first investigated in *Aspergillus fumigatus* and *Aspergillus oryzae* and deduced by homology in dermatophytes (Byun *et al.*, 2001; Sriranganadane *et al.*, 2010, 2011a).

Two different extracellular protein degradation pathways at neutral and acidic pH, respectively, were described in *Aspergillus* spp. At neutral and alkaline pH, proteins are first digested by a subtilisin (Alp for alkaline protease) and a metalloprotease (Mep) (endoproteases) to generate large peptides (Jaton-Ogay *et al.*, 1994). The latter are subsequently digested into amino acids and X-pro dipeptides by the synergic action of Laps and DppIV (Byun *et al.*, 2001) in a manner similar to that described in *Lactobacillus* spp. (O'Cuinn *et al.*, 1999). Laps degrade peptides from their N-terminus until an X-Pro sequence is encountered, which acts as a stop, but in a complementary manner the X-Pro sequences can be removed by DppIV, thus allowing Laps access to the next residue. At acidic pH, proteins are first digested by *Aspergillus* spp. with an aspartic protease and a glutamic protease (endoproteases) to generate large peptides (Sriranganadane *et al.*, 2011b). These peptides are subsequently digested into short peptides by the synergic action of tripeptidylpeptidases of the sedolisin family and prolylpeptidases of the S28 family (Sriranganadane *et al.*, 2010). Seds degrade peptides from their N-termini into tripeptides, but a Pro residue in the P1 or the P′1 position (position 3 or 4 relative to the peptide's N-terminus, respectively) acts as a stop sequence. In a complementary manner, X-X-Pro and X-X-X-Pro sequences can be removed by a prolyl peptidase of the S28 family, thus allowing Seds further sequential proteolysis of the trimmed peptide. In conclusion, both alkaline and acidic sets of proteases contain exoprotease activities capable of cleaving after proline residues that are roadblocks for sequential digestion by non-specific aminopeptidases.

Analysis of the repertoire of secreted exoproteases in dermatophytes suggests basic mechanisms of extracellular proteolysis similar to those described in *Aspergillus* spp. at acidic and neutral pH. In particular, two dermatophyte subtilisins secreted in a keratin medium, Sub3 and Sub4, were found to be more active on keratin azure than the *Aspergillus* spp. secreted subtilisin (Alp) and proteinase K, but less active on other protein sources, especially elastin. These results attested to a specificity of both Sub3 and Sub4 towards hard keratin substrates (Jousson *et al.*, 2004b). The four *T. rubrum* aminopeptidases Lap1, Lap2, DppIV and DppV, showed activities comparable to those of *A. fumigatus* orthologues (Monod *et al.*, 2005). The synergistic action of Lap and DppIV at neutral pH and the synergistic action of Seds and prolylpeptidase at acidic pH seems to be a common strategy in many ascomycete fungi in order to degrade large peptides. However, in comparison to *A. fumigatus*, more proteases were found to be secreted by dermatophytes at neutral as well as acidic pH in a protein medium (Sriranganadane *et al.*, 2011a), suggesting the intervention of specific dermatophyte enzymes within the protein degradation pathways homologous to those previously demonstrated in *Aspergillus* spp. (Byun *et al.*, 2001; Sriranganadane *et al.*, 2010).

Disease mechanisms

The successive events leading to dermatophytosis

The first stage in the development of dermatophytosis in a susceptible individual exposed to a sufficiently high level of inoculation is the adherence of arthrospores to corneocytes. This step probably depends on host as well as several fungal factors, most of which remain unidentified, a fungal keratinolytic subtilisin excepted (see section 'Dermatophyte adherence', below). The second stage, arthrospore germination, is promoted by humidity, warmth, maceration and skin excoriation. Arthrospore germination must occur rapidly to avoid fungal elimination by a constantly desquamating epidermis. The following stage, invasion, consists of the production of hyphae which penetrate the *stratum corneum*. For many years, and even decades, keratinolytic proteases from dermatophytes have been considered major virulence factors involved in this invasion step, but recent data show that the role of other factors should be investigated (see sections 'Dissociation of *in vitro* keratinolytic activity and virulence trends' and 'Virulence trends other than proteolytic activity revealed by transcriptome analysis', below) and that previous concepts should be revisited using the most modern and powerful available molecular tools.

In haired areas, when a mycelial filament in the *stratum corneum* encounters a hair opening, it penetrates the keratinized external root sheath of the hair follicle up to the *infundibulum*, the internal root sheath (the external root sheath is no longer keratinized) and the hair shaft. The fungus grows towards the hair bulb but is maintained in the newly formed keratin. The opposing growth of fungus and hair is balanced such that the aerial portion of the hair is progressively invaded by hyphae and newly formed arthrospores. This makes the hair fragile, often causing it to break several millimetres above the skin surface with release of highly resistant and infective arthrospores into the environment. The dermatophyte can invade other adjacent hair follicles, leading to centrifugal spread of lesions and the distinctive clinical appearance, consisting of well-demarcated alopecic circular patches, as seen in tinea capitis. Some dermatophytes, such as *Microsporum persicolor*, *Epidermophyton floccosum* and *Trichophyton concentricum* are incapable of invading hair, even *in vitro*, and stay confined to interfollicular keratin. In glabrous skin, the infection process is basically similar, leading to centrifugal superficial lesions. In addition to the dermatophyte species and the body site affected, the clinical appearance also depends upon the host's defence reactions (see section 'Host defence mechanisms involved in pathogenesis', below). Lesions are sometimes more inflammatory at the periphery and heal from the centre, giving an annular rather than a circular aspect in glabrous skin and some central hair regrowth in tinea capitis.

Dermatophyte adherence

The kinetics of adherence of dermatophytes to the skin or nail surface has been investigated in several dermatophyte species, using several microscopy techniques and different experimental *in vitro* and *ex vivo* models (reviewed by Vermout *et al.*, 2008b; Baldo *et al.*, 2012a). The studies showed a time-dependent increase in the number of adhering spores, followed by germination and invasion of the *stratum corneum* by hyphae growing in multiple directions. Adherence of arthroconidia to keratinocytes generally begins by 2 hours and, depending on the studies, is maximum or still increasing at 4–6 hours.

Little is known about the factors that mediate adherence of dermatophytes. Like other fungi and parasites, adhesins interacting with host-receptors could be involved in the first steps of infection, including adherence. This has been suggested for *T. mentagrophytes* (Esquenazi *et al.*, 2003) and *T. rubrum* (Esquenazi *et al.*, 2004), given the expression of carbohydrate-specific adhesins at the surface of microconidia of both species that are able to recognize mannose and galactose on Chinese hamster ovary epithelial cells. These interesting studies have unfortunately not been pursued, or repeated using infective arthrospores and keratinocytes. More recently, the role of secreted Subs in the adherence process of arthrospores to epidermis was demonstrated for *M. canis* using specific protease inhibitors, or monoclonal antibodies against the keratinolytic Sub3, and an *in vitro* reconstructed feline epidermis (Baldo *et al.*, 2008). Using a stable *M. canis sub3* RNA-silenced strain (Vermout *et al.*, 2007) and *ex vivo* skin explants, Sub3 was shown to be required for adherence of arthrospores to corneocytes not only in the natural host species, namely the cat (Baldo *et al.*, 2010), but also in many other susceptible animal species and man (Bagut *et al.*, 2012). Whether the proteolytic activity of Sub3 is necessary for the adherence process currently remains unknown (Baldo *et al.*, 2012b).

Dissociation of *in vitro* keratinolytic activity and virulence trends

Dermatophytes specifically adapt to host niches during infections and other environmental conditions. *In vivo* and *in vitro* dermatophyte transcriptomes are different. Using a guinea pig infection model with the zoophilic dermatophyte *A. benhamiae*, which causes highly inflammatory cutaneous infections in humans and rodents, microarray analysis revealed an *in vivo* protease gene expression profile totally different from the pattern elicited during *in vitro* growth on keratin (Staib *et al.*, 2010). None of the genes encoding specific endo- and exoproteases involved in keratin digestion *in vitro* was found to be up-regulated during infection. In contrast, the gene encoding a particular subtilisin (Sub6) was revealed as the most up-regulated *A. benhamiae* sequence, while this gene was not detectably activated during *in vitro* growth on keratin. Of particular note, this gene is the orthologue of that encoding the major allergen Tri r 2 in *T. rubrum* (Woodfolk *et al.*, 1998). *Trichophyton rubrum* Tri r 2 was found to induce dual immune responses and elicit either immediate (IH) or delayed-type (DTH) hypersensitivity skin test reactions in different individuals (Slunt *et al.*, 1996; Woodfolk *et al.*, 1998, 2005). Besides *sub6*, other paralogues (the closely related *sub7* as well as *sub1* and *sub2*) and other genes) were also specifically up-regulated *in vivo*. As a general conclusion, most proteases secreted *in vitro* in a keratin medium do not appear to be involved during establishment of an infection, at least in the tested model. In contrast, other secreted proteases not involved in keratin degradation likely fulfil important functions in virulence.

The importance of secreted acidic proteases has not yet been shown for dermatophytes during infection and during keratin degradation in the environment. Genes encoding acidic proteases as well as many other proteases, which so far remain putative, may be relics in these fungi that originate from soil fungi. However, genome sequences reveal that they were conserved in all species during evolution, in contrast to genes encoding glycosidases (see subsection 'Dermatophyte genomes', above). It is also possible that acidic proteases are secreted at specific steps during an infection. Skin pH varies according to many endogenous and exogenous factors, including host species, and is rather acidic in humans.

Virulence trends other than proteolytic activity revealed by transcriptome analysis

Microarray analysis allowed the identification of genes encoding other candidate pathogenicity-related factors in *A. benhamiae*, such as heat shock proteins, transporters, an opsin-related protein and metabolic enzymes (Staib *et al.*, 2010). Of particular interest was the strong up-regulation of key enzymes in the glyoxylate cycle (i.e. isocitrate lyase and malate synthase). The glyoxylate cycle, which is absent in mammals, was previously found to support the pathogenicity of other microbial pathogens, e.g. *Mycobacterium tuberculosis* and the yeast *Candida albicans* (Lorenz and Fink, 2001; McKinney *et al.*, 2000). In dermatophytes, however, similar to *A. fumigatus* (Schöbel *et al.*, 2007), the analyses of knock-out mutants defective in key enzymes of this metabolic pathway have excluded its contribution to fungal pathogenicity *in vivo*, at least in the analysed animal infection models (Grumbt *et al.*, 2011a). The first broad transcriptional profiling results during dermatophyte infection have given new molecular insights into putative pathogenicity-associated mechanisms involved in dermatophytoses, which render necessary the reinvestigation of dermatophyte virulence factors and host–pathogen interactions.

Nothing is currently known about other secreted hydrolases (e.g. ceramidases and lipases), which are also certainly involved in the degradation of the skin barrier. There is no doubt that they will be investigated in the near future in light of new data on dermatophyte genomics.

Host defence mechanisms involved in pathogenesis

The intact skin represents a physical barrier against infection by virtue of its exposure to ultraviolet radiation, its low humidity, competition with the commensal microflora and the cornification process. However, in cases of infection and despite their superficial localization in skin, dermatophytes induce in their host an adaptive immune response, which may result in recovery and protection against reinfection. As with other infections, specific immunity should be driven by innate immunity, both responses having been much less investigated in dermatophytosis than in opportunistic fungal infections (Romani, 2011).

Adaptive immunity

Numerous studies (reviewed by Calderon, 1989; Woodfolk and Platts-Mills, 1998; Woodfolk, 2005; Mignon *et al.*, 2008; Almeida, 2008; Brasch, 2010) have shown that the efficient and protective response against dermatophytes is a cell-mediated response of the delayed type hypersensitivity (DTH) type characterized by elevated activities of the key cytokines interleukin 12 (IL-12) and interferon-γ (IFN-γ) from the Th1 axis, together with recruitment of macrophages as effector cells. In contrast, chronic persistent infections in humans are usually correlated with poor specific DTH, elevated specific IgE and IgG4, IgE-mediated immediate hypersensitivity (IH), and with the production of Th2 cytokines by mononuclear leucocytes. Clinically, a Th1 immune response against dermatophytes is associated with cutaneous inflammation. The latter is most often helpful to the host defence, as the more inflammatory the lesions, the shorter the time they persist, with a spontaneous resolution possible. The degree of inflammation varies according to both the dermatophyte species and host. Typically, a given dermatophyte induces a more intense inflammatory reaction on a host to which it is not adapted than it does on its natural host. An illustration of this is the highly inflammatory acute lesions, such as kerions, that can be caused in humans by zoophilic and geophilic species, contrasting with the persistent asymptomatic infections that may be induced by the anthropophilic *T. rubrum*. However, a given dermatophyte strain in a given host species may induce various clinical expressions, both in humans and animals. The factors responsible for this differential clinical spectrum remain unknown, but the nature of the specific immune response that the host develops should be critical. As an example, it was shown that two dermatophyte antigens, Tri r 2 from *T. rubrum*, homologous to the protease Sub6, and Tri t 4 from *T. tonsurans*, corresponding to DppV, could induce dual immune responses. These antigens notably elicit either IH in patients suffering from a chronic recurrent infection with low-grade inflammatory lesions, or DTH in individuals having had an acute infection (Slunt *et al.*, 1996; Woodfolk *et al.*, 1998, 2000; Woodfolk and Platts-Mills, 2001).

Innate immunity

Some recent insights into the innate immune response against dermatophytes have been gained. The studies on this topic are instrumental because the receptors, pathways and cells that are implicated have an instructive role on cells of the adaptive immune system.

The first epidermal cells encountered by dermatophytes during the infection process are keratinocytes. Upon exposure to the fungi, they can produce a broad spectrum of cytokines, including the chemo-attractant for polymorphonuclear neutrophils (PMN) CXCL-8 (Shiraki *et al.*, 2006) and the pro-inflammatory TNF-α (Nakamura *et al.*, 2002). Interestingly, zoophilic species stimulate the production of pro-inflammatory cytokines by human keratinocytes more than anthropophilic ones (Shiraki *et al.*, 2006; Tani *et al.*, 2007), and this correlates with the clinical aspects observed in both types of infection. Besides cytokines, human keratinocytes also secrete antimicrobial peptides (AMP), cathelicidins and defensins with potential antifungal activity. Their role in skin defence against dermatophytes was suggested by several authors who showed that human β-defensin (Jensen *et al.*, 2007; Fritz *et al.*, 2012) and cathelicidin LL-37 (Lopez-Garcia *et al.*, 2006) are either fungistatic or fungicidal *in vitro* against *T. rubrum* and that their expression is up-regulated in tinea corporis. In addition, defensin-like peptide genes have been identified in dermatophytes themselves (Zhu *et al.*, 2012).

In close contact with both keratinocytes and infecting dermatophytes are the Langerhans cells (LC), a subtype of dendritic cells (DC), which are professional antigen-presenting cells involved in both initiation and modulation of adaptative anti-dermatophyte immunity. For example, dectin-2, a C-type lectin-like receptor constitutively expressed by mature DC, was shown to be a pattern recognition receptor (PRR) for fungi, including dermatophytes. Dectin-2 binds the hyphal forms rather than spores from both *Microsporum audouinii* and *T. rubrum*, leading *in fine* to the activation of NFκB and the up-regulated secretion of TNF-α and IL-1ra (Sato *et al.*, 2006). By contrast, upon phagocytosis by macrophages, *T. rubrum* spores induce secretion of the anti-inflammatory cytokine IL-10, while other factors related to enhancement of a relevant protective immune response are down-regulated, i.e. class II MHC, CD54 and CD80 co-stimulation molecules, nitric oxide and IL-12 (Campos *et al.*, 2006). Another PRR present on mouse DC and human LC, DC-HIL, may also exert positive and negative regulation in anti-dermatophyte immunity. Upon binding with *M. audouinii* and *T. rubrum* cell walls, DC-HIL is able to both increase the antigen presenting cell properties of DC and produce a potent inhibitory signal to T cells in an antigen-independent manner (Chung *et al.*, 2009). All these results show that both the nature of the fungal components and how they are detected by epidermal sentinel cells, keratinocytes and DC, likely play key roles in both initiation and modulation of the specific immune response to dermatophytes.

In addition to keratinocytes and DC, PMN are important cells involved in innate anti-dermatophyte immunity. PMN are the first leucocytes recruited to the site of infection in mice experimentally infected with *T. quinckeanum* (Hay *et al.*, 1983). They accumulate in skin as early as 4–6 hours after infection, i.e. during the fungal adherence and germination phases. PMN, along with macrophages, are also known to be responsible for the *in fine* elimination of dermatophytes via phagocytosis and the production of superoxide anions (Calderon and Hay, 1987). Another effector mechanism, NET 'Neutrophil Extracellular Trap' formation, was recently demonstrated in PMN in contact with the zoophilic *A. benhamiae* (Heddergott *et al.*, 2012). In addition, PMN may have a role in initiation of the immune response in dermatophytoses, as feline PMN produce pro-inflammatory cytokines following exposure to *M. canis* arthrospores (Cambier *et al.*, 2013).

Future directions

Regarding opportunistic fungal infections, it is currently accepted that diseases can result either from the loss of immune competence or over activation of the inflammatory response (Romani, 2011). While this concept does not fit for dermatophyte infections, current views about the immunology of dermatophytoses, including the Th1-Th2 paradigm (see section 'Adaptive immunity', above), are probably too simplistic. They do not encompass a full understanding of the complex host defence mechanisms or the clinical polymorphism of dermatophytic infections. For a better understanding of host–pathogen interactions and for the establishment of rational strategies to develop new treatments and efficient vaccines against dermatophytes, future molecular studies should integrate not only fungal attributes but also host defence mechanisms, using the most appropriate infection models (see section 'Infection models', above) and the input of qualified immunologists. For example, the role of the Th17 pathway and its associated cytokines in the establishment of a protective immune response against dermatophytes should be investigated as for other fungal infections (Romani, 2011). Another important challenge for the future is to identify both the PRR in epidermal sentinel cells and the dermatophyte components able to interact with these PRR, in order to better understand how dermatophytes modulate the host immune response.

Host susceptibility

Fungal infections depend not only on the fungus but also on host susceptibility. For instance, in *A. fumigatus*, the main cause of invasive aspergillosis in neutropenic patients, polymorphisms in pattern recognition receptors have been associated with increased or decreased susceptibility to infections (Bochud *et al.*, 2008). The first evidence for genetic susceptibility of the host in dermatophyte infection is a study by Zaias *et al.* (1996), which suggests a genetic predisposition for distal subungual onychomycosis caused by *T. rubrum*. The pedigrees of 13 families where all affected children had at least one parent with distal subungual onychomycosis were consistent with an autosomal mode of transmission for the predisposition to infection.

Recently, whole genome genotyping was used to identify susceptibility trends in children infected by *T. tonsurans* in the United States (Abdel-Rahman and Preuett, 2012). SNP analysis allowed the identification of 23 genes associated with a child's susceptibility to superficial dermatophyte infections. This group of genes included 6 genes involved in the recruitment, activation, and migration of leucocytes, eight genes involved in the extracellular matrix and elastin fibre network, and seven genes involved in epidermal development and maintenance and wound repair. The finding of multiple genes encoding proteins that are all important for skin homeostasis is a consistent result of this first pioneer analysis to research the host factors which protect against or predispose to chronic *T. tonsurans* infection.

Antifungal drug resistance

Antifungal resistance in many fungi is often mediated either by overexpression of genes encoding multidrug transporters or by missense mutations in the genes encoding the drug target (Sanglard *et al.*, 1995, 1998; Marichal *et al*; 1999; reviewed in Morschhäuser, 2010). Although azoles and terbinafine have been extensively used to fight dermatophytoses, reports of clinical dermatophyte isolates resistant to antifungal compounds are rare in contrast to reports for *C. albicans* and *A. fumigatus*.

Two dermatophyte ATP-binding-cassette (ABC) transporter genes, called *mdr1* and *mdr2* (for multidrug resistance 1 and 2), have been cloned (Cervelatti *et al.*, 2006; Fachin *et al.*, 2006). Their deduced amino acid sequences show high homology with ABC transporters known to be drug efflux pumps in other fungi. Increased expression levels of both genes in response to stress caused by subinhibitory levels of various compounds including antifungal drugs was observed. Disruption of *mdr2* resulted in an increased sensitivity to terbinafine among various tested antifungal compounds that was two-fold higher than that of the non-mutated strains. However, no resistance mediated by efflux pumps of the ABC transporter family, or by efflux pumps in the major facilitator family, has been documented to date from dermatophyte clinical isolates.

Presently, two cases of terbinafine resistance have been investigated at the biochemical and DNA levels in *T. rubrum* clinical isolates from onychomycoses (Mukherjee *et al.*, 2003; Favre *et al.*, 2004; Osborne *et al.*, 2005, 2006). Resistance was shown to be due to single point mutations in the gene encoding the squalene epoxidase that is targeted by terbinafine. Both mutations introduced missense substitutions in the same region of the enzyme (L393F in case 1 and F397L in case 2) resulting in minimum inhibitory concentrations (MIC) of over 4 μg/ml and 64 μg/ml terbinafine, respectively. For comparison, the MIC of wild *T. rubrum* isolates was found to be between 0.2 ng/ml and 4 ng/ml (Osborne *et al.*, 2003, 2006). *Trichophyton rubrum* strains resistant to terbinafine could be selected *in vitro* at a low frequency by plating colony forming units on a growth medium containing 60 ng terbinafine per ml (Osborne *et al.*, 2003). Cross-resistance of all mutants to various squalene epoxidase inhibitors revealed that the resistance was due to alteration in squalene epoxidase activity.

Although it is possible to isolate dermatophyte strains resistant to antifungals, insensitivity of onychomycosis to standard treatments using terbinafine and azoles is generally attributed to non-dermatophyte fungi (NDF), which are especially difficult to cure. *In situ* identification of the fungus by PCR revealed *Fusarium* sp., *Acremonium* sp. and *Aspergillus* sp. as the sole infectious agents in cases of onychomycoses insensitive to systemic terbinafine and itraconazole treatment (Baudraz-Rosselet *et al.*, 2010). Neither *T. rubrum* nor *T. interdigitale* was identified as an infectious fungus, and insensitivity to treatment was not attributed to resistant dermatophyte strains. A treatment using amphotericin B in a topical application was found to be efficacious for curing NDF onychomycosis (Lurati *et al.*, 2011). In tinea capitis, the therapeutic response varies according to the causative dermatophyte species. The cure rate with terbinafine is excellent in cases of infection by the anthropophilic species *T. violaceum* and *T. soudanense*, but not in cases of infection by zoophilic species such as *M. canis*, *A. benhamiae* and *A. vanbreuseghemii* or the anthropophilic *M. audouinii*

(Baudraz-Rosselet *et al.*, 1996). The clinical efficacy of terbinafine was found to be unrelated to the results of *in vitro* susceptibility tests per species (Mock *et al.*, 1998). Griseofulvin remains the treatment of choice in many cases of tinea capitis. In most cases of tinea corporis, dermatophyte identification is not imperative, as these mycoses respond well to topical standard treatments regardless of the incriminated dermatophyte. As a general conclusion, one must be aware that (1) true dermatophyte resistance to antifungal drugs in patients remains exceptional, and (2) correct species identification may be important in cases of superficial mycoses, especially hair and nail infections, in order to prescribe adequate treatments, and prior to investigation into the possibility of acquired antifungal resistance.

Concluding remarks and perspectives

In comparison to other fungi, dermatophytes are remarkable in more than one feature. As in other parasitic and highly specialized microorganisms, the size of their genomes is reduced and particular gene families are expanded. The emergence of multigenic families is most frequently due to ancient gene duplication processes allowing organisms to better adapt to different environmental conditions. The ability of dermatophytes to degrade hard keratin as saprophytes and to develop in animals as parasites apparently requires the use of combinations of different proteases suitable to each particular situation. For comparison, opportunistic fungi such as *Aspergillus* spp. have no specific large gene families encoding secreted proteases. The secretion of a substantial amount of sulfite generated by the catabolism of cysteine appears to be specific for dermatophytes and revealed to be important for the breakdown of hard keratin structures (Grumbt *et al.*, 2013). Dermatophyte genomes are also characterized by an expansion of genes coding for proteins with a LysM domain. This expansion could be linked to the capacity of the fungus to evade the immune system during infection. The depletion of genes associated with sugar metabolism is another trademark of the high specialization of dermatophytes. For comparison, the genomes of saprophytic fungi in soil, vegetal composts and plant parasitic species encode large batteries of secreted glycosidases.

Now that the genome sequences of dermatophytes are available, new insight into virulence factors of the dermatophytes will be obtained through the analysis of their representative transcriptomes in various conditions, as well *in vivo* during infections as *in vitro*. The real importance of individual factors potentially involved in niche and host preference can now be determined using the different tools recently developed for genetic manipulations of dermatophytes and animal infection models. One must be also aware that the non-invasive character of dermatophytes and the extraordinary variety of associated clinical signs depend not only on factors of the fungi themselves but also on the susceptibility of the host. In this context, whole genome genotyping analysis could be helpful to identify possible genetic host factors that may protect or predispose to an infection, and targets of virulence attributes of dermatophytes.

Acknowledgements

We thank Dr Philippe Hauser (CHUV, Lausanne) and Dr Françoise Symoens (Scientific Institute of Public Health, Brussels) for critical review of the manuscript, and Willy Blanchard for preparing the figures.

References

Abdel-Rahman, S.M., and Preuett, B.L. (2012). Genetic predictors of susceptibility to cutaneous fungal infections: a pilot genome wide association study to refine a candidate gene search. J. Dermatol. Sci. *67*, 147–152.

Achterman, R.R., and White, T.C. (2012). Dermatophyte virulence factors: identifying and analyzing genes that may contribute to chronic or acute skin infections. Int. J. Microbiol. *2012*, Article ID 358305.

Ajello, L. (1962). Present day concepts in the dermatophytes. Mycopathol. Mycol. Appl. *17*, 315–324.

Ajello, L. (1974). Natural history of the dermatophytes and related fungi. Mycopathol. Mycol. Appl. *53*, 93–110.

Ajello, L., and Cheng, S.L. (1967). The perfect state of *Trichophyton mentagrophytes*. Sabouraudia *5*, 230–234.

Aljabre, S.H., Richardson, M.D., Scott, E.M., and Shankland, G.S. (1992a). Dormancy of *Trichophyton mentagrophytes* arthroconidia. J. Med. Vet. Mycol. *30*, 409–412.

Aljabre, S.H., Richardson, M.D., Scott, E.M., and Shankland, G.S. (1992b). Germination of *Trichophyton mentagrophytes* on human stratum corneum *in vitro*. J. Med. Vet. Mycol. *30*, 145–152.

Aljabre, S.H., Richardson, M.D., Scott, E.M., Rashid, A., and Shankland, G.S. (1993). Adherence of arthroconidia and germlings of anthropophilic and zoophilic varieties of *Trichophyton mentagrophytes* to human corneocytes as an early event in the pathogenesis of dermatophytosis. Clin. Exp. Dermatol. *18*, 231–235.

Almeida, S.R. (2008). Immunology of dermatophytosis. Mycopathologia *166*, 277–283.

Alshahni, M.M., Makimura, K., Yamada, T., Takatori, K., and Sawada, T. (2010). Nourseothricin acetyltransferase: a new dominant selectable marker for the dermatophyte *Trichophyton mentagrophytes*. Med. Mycol. *48*, 665–668.

Alshahni, M.M., Yamada, T., Takatori, K., Sawada, T., and Makimura, K. (2011). Insights into a nonhomologous integration pathway in the dermatophyte *Trichophyton mentagrophytes*: efficient targeted gene disruption by use of mutants lacking ligase IV. Microbiol. Immunol. *55*, 34–43.

Anzawa, K., Kawasaki, M., Mochizuki, T., and Ishizaki, H. (2010). Successful mating of *Trichophyton rubrum* with *Arthroderma simii*. Med. Mycol. *48*, 629–634.

Băguţ, E.T., Baldo, A., Mathy, A., Cambier, L., Antoine, N., Cozma, V., and Mignon, B. (2012). Subtilisin Sub3 is involved in adherence of *Microsporum canis* to human and animal epidermis. Vet. Microbiol. *160*, 413–419.

Baldo, A., Tabart, J., Vermout, S., Mathy, A., Collard, A., Losson, B., and Mignon, B. (2008). Secreted subtilisins of *Microsporum canis* are involved in adherence of arthroconidia to feline corneocytes. J. Med. Microbiol. *57*, 1152–1156.

Baldo, A., Mathy, A., Tabart, J., Camponova, P., Vermout, S., Massart, L., Maréchal, F., Galleni, M., and Mignon, B. (2010). Secreted subtilisin Sub3 from *Microsporum canis* is required for adherence to but not for invasion of the epidermis. Br. J. Dermatol. *162*, 990–997.

Baldo, A., Monod, M., Mathy, A., Cambier, L., Bagut, E.T., Defaweux, V., Symoens, F., Antoine, N., and Mignon, B. (2012a). Mechanisms of skin adherence and invasion by dermatophytes. Mycoses *55*, 218–223.

Baldo, A., Chevigné, A., Dumez, M.E., Mathy, A., Power, P., Tabart, J., Cambier, L., Galleni, M., and Mignon, B. (2012b). Inhibition of the keratinolytic subtilisin protease Sub3 from *Microsporum canis* by its propeptide (proSub3) and evaluation of the capacity of proSub3 to inhibit fungal adherence to feline epidermis. Vet. Microbiol. *159*, 479–484.

Baudraz-Rosselet, F., Monod, M., Jaccoud, S., and Frenk, E. (1996). Efficacy of terbinafine treatment of tinea capitis in children varies according to the dermatophyte species. Br. J. Dermatol. *135*, 1011–1012.

Baudraz-Rosselet, F., Ruffieux, C., Lurati, M., Bontems, O., and Monod, M. (2010). Onychomycosis insensitive to systemic terbinafine and azole treatments reveals non-dermatophyte moulds as infectious agents. Dermatology *220*, 164–168.

Beguin, H., Pyck, N., Hendrickx, M., Planard, C., Stubbe, D., and Detandt, M. (2012). The taxonomic status of *Trichophyton quinckeanum* and *T. interdigitale* revisited: a multigene phylogenetic approach. Med. Mycol. *50*, 871–882.

Beguin, H., Goens, K., Hendrickx, M., Planard, C., Stubbe, D., and Detandt, M. (2013). Is *Trichophyton simii* endemic of the Indian subcontinent? Med. Mycol. *51*, 444–448.

Bergmann, A., Hartmann, T., Cairns, T., Bignell, E.M., and Krappmann, S. (2009). A regulator of *Aspergillus fumigatus* extracellular proteolytic activity is dispensable for virulence. Infect. Immun. *77*, 4041–4050.

Bergmans, A.M., van der Ent, M., Klaassen, A., Böhm, N., Andriesse, G.I., and Wintermans, R.G. (2010). Evaluation of a single-tube real-time PCR for detection and identification of 11 dermatophyte species in clinical material. Clin. Microbiol. Infect. *16*, 704–710.

Blank, F. (1957). Favus of mice. Can. J. Microbiol. *3*, 885–896.

Bochud, P.Y., Chien, J.W., Marr, K.A., Leisenring, W.M., Upton, A., Janer, M., Rodrigues, S.D., Li, S., Hansen, J.A., Zhao, L.P., *et al.* (2008). Toll-like receptor 4 polymorphisms and aspergillosis in stem-cell transplantation. N. Engl. J. Med. *359*, 1766–1777.

Brasch, J. (2010). Pathogenesis of tinea. J. Dtsch. Dermatol. Ges. *8*, 780–786.

Brillowska-Dabrowska, A., Saunte, D.M., and Arendrup, M.C. (2007). Five-hour diagnosis of dermatophyte nail infections with specific detection of *Trichophyton rubrum*. J. Clin. Microbiol. *45*, 1200–1204.

Brouta, F., Descamps, F., Fett, T., Losson, B., Gerday, C., and Mignon, B. (2001). Purification and characterization of a 43.5 kDa keratinolytic metalloprotease from *Microsporum canis*. Med. Mycol. *39*, 269–275.

Brouta, F., Descamps, F., Monod, M., Vermout, S., Losson, B., and Mignon, B. (2002). Secreted metalloprotease gene family of *Microsporum canis*. Infect. Immun. *70*, 5676–5683.

Brouta, F., Descamps, F., Vermout, S., Monod, M., Losson, B., and Mignon, B. (2003). Humoral and cellular immune response to a *Microsporum canis* recombinant keratinolytic metalloprotease (r-MEP3) in experimentally infected guinea pigs. Med. Mycol. *41*, 495–501.

Burmester, A., Shelest, E., Glöckner, G., Heddergott, C., Schindler, S., Staib, P., Heidel, A., Felder, M., Petzold, A., Szafranski, K., *et al.* (2011). Comparative and functional genomics provide insights into the pathogenicity of dermatophytic fungi. Genome Biol. *12*, R7.

Byun, T., Kofod, L., and Blinkovsky, A. (2001). Synergistic action of an X-prolyl dipeptidyl aminopeptidase and a non-specific aminopeptidase in protein hydrolysis. J. Agric. Food Chem. *49*, 2061–2063.

Calderon, R.A. (1989). Immunoregulation of dermatophytosis. Crit. Rev. Microbiol. *16*, 339–368.

Calderon, R.A., and Hay, R.J. (1987). Fungicidal activity of human neutrophils and monocytes on dermatophyte fungi, *Trichophyton quinckeanum* and *Trichophyton rubrum*. Immunology *61*, 289–295.

Cambier, L., Mathy, A., Baldo, A., Bagut, E.T., Tabart, J., Antoine, N., and Mignon, B. (2013). Feline polymorphonuclear neutrophils produce pro-inflammatory cytokines following exposure to *Microsporum canis*. Vet. Microbiol. *162*, 800–805.

Campos, M.R., Russo, M., Gomes, E., and Almeida, S.R. (2006). Stimulation, inhibition and death of macrophages infected with *Trichophyton rubrum*. Microbes Infect. *8*, 372–379.

Cervelatti, E.P., Fachin, A.L., Ferreira-Nozawa, M.S., and Martinez-Rossi, N.M. (2006). Molecular cloning and characterization of a novel ABC transporter gene in the human pathogen *Trichophyton rubrum*. Med. Mycol. *44*, 141–147.

Christensen, T., Hynes, M.J., and Davis, M.A. (1998). Role of the regulatory gene *areA* of *Aspergillus oryzae* in nitrogen metabolism. Appl. Environ. Microbiol. *64*, 3232–3237.

Chung, J.S., Yudate, T., Tomihari, M., Akiyoshi, H., Cruz, P.D. Jr, and Ariizumi, K. (2009). Binding of DC-HIL to dermatophytic fungi induces tyrosine phosphorylation and potentiates antigen presenting cell function. J. Immunol. *183*, 5190–5198.

Dawson, C.O., and Gentles, J.C. (1959). Perfect stage of *Keratinomyces ajelloi*. Nature *183*, 1345–1356.

Dawson, C.O., and Gentles, J.C. (1961). The perfect states of *Keratinomyces ajelloi* Vanbreuseghem, *Trichophyton terrestre* Durie and Frey and *Microsporum nanum* Fuentes. Sabouraudia *1*, 49–57.

Degreef, H. (2008). Clinical forms of dermatophytosis (ringworm infection). Mycopathologia *166*, 257–265.

Descamps, F., Brouta, F., Monod, M., Zaugg, C., Baar, D., Losson, B., and Mignon, B. (2002). Isolation of a *Microsporum canis* gene family encoding three subtilisin-like proteases expressed *in vivo*. J. Invest. Dermatol. *119*, 830–835.

Descamps, F., Brouta, F., Vermout, S., Monod, M., Losson, B., and Mignon, B. (2003). Recombinant expression and antigenic properties of a 31.5-kDa keratinolytic subtilisin-like serine protease from *Microsporum canis*. FEMS Immunol. Med. Microbiol. *38*, 29–34.

De Vroey, C. (1964). Formes sexuées des dermatophytes. Production de cleistothèces de *Microsporum gypseum* (Bodin) Guiart et Grigorakis sur divers milieux stériles. Ann. Soc. Belge Méd. Trop. *44*, 831–840.

Dobrowolska, A., and Staczek, P. (2009). Development of transformation system for *Trichophyton rubrum* by electroporation of germinated conidia. Curr. Genet. *55*, 537–542.

Drouot, S., Mignon, B., Fratti, M., Roosje, P., and Monod, M. (2008). Pets as the main source of two zoonotic species of the *Trichophyton mentagrophytes* complex in Switzerland, *Arthroderma vanbreuseghemii* and *Arthroderma benhamiae*. Vet. Dermatol. *20*, 13–18.

Duek, L., Kaufman, G., Ulman, Y., and Berdicevsky, I. (2004). The pathogenesis of dermatophyte infections in human skin sections. J. Infect. *48*, 175–180.

Esquenazi, D., de Souza, W., Alviano, C.S., and Rozental, S. (2003). The role of surface carbohydrates on the interaction of microconidia of *Trichophyton mentagrophytes* with epithelial cells. FEMS Immunol. Med. Microbiol. *35*, 113–123.

Esquenazi, D., Alviano, C.S., de Souza, W., and Rozental, S. (2004). The influence of surface carbohydrates during *in vitro* infection of mammalian cells by the dermatophyte *Trichophyton rubrum*. Res. Microbiol. *155*, 144–153.

Fachin, A.L., Ferreira-Nozawa, M.S., Maccheroni, W. Jr, and Martinez-Rossi, N.M. (2006). Role of the ABC transporter TruMDR2 in terbinafine, 4-nitroquinoline N-oxide and ethidium bromide susceptibility in *Trichophyton rubrum*. J. Med. Microbiol. *55*, 1093–1099.

Favre, B., Ghannoum, M.A., and Ryder, N.S. (2004). Biochemical characterization of terbinafine-resistant *Trichophyton rubrum* isolates. Med. Mycol. *42*, 525–529.

Ferreira-Nozawa, M.S., Silveira, H.C., Ono, C.J., Fachin, A.L., Rossi, A., and Martinez-Rossi, N.M. (2006). The pH signaling transcription factor PacC mediates the growth of *Trichophyton rubrum* on human nail *in vitro*. Med. Mycol. *44*, 641–645.

Fire, A., Xu, S., Montgomery, M.K., Kostas, S.A., Driver, S.E., and Mello, C.C. (1998). Potent and specific genetic interference by double stranded RNA in *Caenorhabditis elegans*. Nature *391*, 806–811.

Fréalle, E., Rodrigue, M., Gantois, N., Aliouat, C.M., Delaporte, E., Camus, D., Dei-Cas, E., Kauffmann-Lacroix, C., Guillot, J., and Delhaes, L. (2007). Phylogenetic analysis of *Trichophyton mentagrophytes* human and animal isolates based on MnSOD and ITS sequence comparison. Microbiology *153*, 3466–3477.

Fritz, P., Beck-Jendroschek, V., and Brasch, J. (2012). Inhibition of dermatophytes by the antimicrobial peptides human β-defensin-2, ribonuclease 7 and psoriasin. J. Med. Mycol. *50*, 579–584.

Fumeaux, J., Mock, M., Ninet, B., Jan, I., Bontems, O., Léchenne, B., Lew, D., Panizzon, R., Jousson, O., and Monod, M. (2004). First report of *Arthroderma benhamiae* in Switzerland. Dermatology *208*, 244–250.

Georg, L.K. (1960). Epidemiology of the dermatophytoses: sources of infection, modes of transmission and epidemicity. Ann. N. Y. Acad. Sci. *89*, 69–77.

Giddey, K., Favre, B., Quadroni, M., and Monod, M. (2007). Closely related dermatophyte species produce different patterns of secreted proteins. FEMS Microbiol. Lett. *267*, 95–101.

Gip, L., and Abelin, J. (1987). Differential staining of fungi in clinical specimens using fluorescent whitening agent (Blankophor). Mykosen *30*, 21–24.

Gonzalez, R., Ferrer, S., Buesa, J., and Ramon, D. (1989). Transformation of the dermatophyte *Trichophyton mentagrophytes* to hygromycin B resistance. Infect. Immun. *57*, 2923–2925.

Gräser, Y., El Fari, M., Vilgalys, R., Kuijpers, A.F., De Hoog, G.S., Presber, W., and Tietz, H. (1999a). Phylogeny and taxonomy of the family *Arthrodermataceae* (dermatophytes) using sequence analysis of the ribosomal ITS region. Med. Mycol. *37*, 105–114.

Gräser, Y., Kuijpers, A.F., Presber, W., and De Hoog, G.S. (1999b). Molecular taxonomy of *Trichophyton mentagrophytes* and *T. tonsurans*. Med. Mycol. *37*, 315–330.

Grobler, J., Bauer, F., Subden, R.E., and Van Vuuren, H.J. (1995). The *mae1* gene of *Schizosaccharomyces pombe* encodes a permease for malate and other C4 dicarboxylic acids. Yeast *11*, 1485–1491.

Gruby, D. (1842). Sur une espèce de mentagre contagieuse résultant du développement d'un nouveau cryptogame dans la racine des poils de la barbe de l'homme. C. R. Acad. des Sc. *15*, 512–514.

Grumbt, M., Defaweux, V., Mignon, B., Monod, M., Burmester, A., Wöstemeyer, J., and Staib, P. (2011a). Targeted gene deletion and *in vivo* analysis of putative virulence gene function in the pathogenic dermatophyte *Arthroderma benhamiae*. Eukaryot. Cell *10*, 842–853.

Grumbt, M., Monod, M., and Staib, P. (2011b). Genetic advances in dermatophytes. FEMS Microbiol. Lett. *320*, 79–86.

Grumbt, M., Monod, M., Yamada, T., Hertweck, C., Kunert, J., and Staib, P. (2013). Keratin degradation by dermatophytes relies on cysteine dioxygenase and a sulfite efflux pump. J. Invest. Dermatol. *133*, 1550–1555.

Harmsen, D., Schwinn, A., Weig, M., Bröcker, E.B., and Heesemann, J. (1995). Phylogeny and dating of some pathogenic keratinophilic fungi using small subunit ribosomal RNA. J. Med. Vet. Mycol. *33*, 299–303.

Hasegawa, A., and Usui, K. (1974). The perfect state of *Microsporum canis*. Nihon Juigaku Zasshi *36*, 447–449.

Hasegawa, A., and Usui, K. (1975). *Nannizzia otae* sp. nov., the perfect state of *Microsporum canis* Bodin. Jpn J. Med. Mycol. *16*, 148–153.

Hawksworth, D.L., Crous, P.W., Redhead, S.A., Reynolds, D.R., Samson, R.A., Seifert, K.A., Taylor, J.W., and Wingfield, M.J. (2011). The Amsterdam declaration on fungal nomenclature. IMA Fungus *2*, 105–112.

Hay, R.J., Calderon, R.A., and Collins, M.J. (1983). Experimental dermatophytosis: the clinical and histopathologic features of a mouse model using *Trichophyton quinckeanum* (mouse favus). J. Invest. Dermatol. *81*, 270–274.

Heddergott, C., Bruns, S., Nietzsche, S., Leonhardt, I., Kurzai, O., Kniemeyer, O., and Brakhage, A.A. (2012). The *Arthroderma benhamiae* hydrophobin HypA mediates hydrophobicity and influences recognition by human immune effector cells. Eukaryot. Cell *11*, 673–682.

Hensel, M., Arst, H.N., Aufauvre-Brown, A., and Holden, D.W. (1998). The role of the *Aspergillus fumigatus areA* gene in invasive pulmonary aspergillosis. Mol. Gen. Genet. *258*, 553–557.

Holländer, H., Keilig, W., Bauer, J., and Rothemund, E. (1984). A reliable fluorescent stain for fungi in tissue sections and clinical specimens. Mycopathologia. *88*, 131–134.

Hunter, S., Apweiler, R., Attwood, T.K., Bairoch, A., Bateman, A., Binns, D., Bork, P., Das, U., Daugherty, L., Duquenne, L., *et al.* (2009). InterPro: the integrative protein signature database. Nucleic Acids Res. *37 (Database issue)*, D211–215.

Idnurm, A., Bahn, Y.S., Nielsen, K., Lin, X.R., Fraser, J.A., and Heitman, J. (2005). Deciphering the model pathogenic fungus *Cryptococcus neoformans*. Nat. Rev. Microbiol. *3*, 753–764.

Iwata, A., Alshahni, M.M., Nishiyama, Y., Makimura, K., Abe, S., and Yamada, T. (2012). Development of a tightly regulatable copper-mediated gene switch system in dermatophytes. Appl. Environ. Microbiol. *78*, 5204–5211.

Jaton-Ogay, K., Paris, S., Huerre, M., Quadroni, M., Falchetto, R., Togni, G., Latgé, J.P., and Monod, M. (1994). Cloning and disruption of the gene encoding an extracellular metalloprotease of *Aspergillus fumigatus*. Mol. Microbiol. *14*, 917–928.

Jensen, J.M., Pfeiffer, S., Akaki, T., Schröder, J.M., Kleine, M., Neumann, C., Proksch, E., and Brasch, J. (2007). Barrier function, epidermal differentiation, and human beta-defensin 2 expression in tinea corporis. J. Invest. Dermatol. *127*, 1720–1727.

de Jonge, R., and Thomma, B.P. (2009). Fungal LysM effectors: extinguishers of host immunity? Trends Microbiol. *17*, 151–157.

Jousson, O., Léchenne, B., Bontems, O., Capoccia, S., Mignon, B., Barblan, J., Quadroni, M., and Monod, M. (2004a). Multiplication of an ancestral gene encoding secreted fungalysin preceded species differentiation in the dermatophytes *Trichophyton* and *Microsporum*. Microbiology *150*, 301–310.

Jousson, O., Léchenne, B., Bontems, O., Mignon, B., Reichard, U., Barblan, J., Quadroni, M., and Monod, M. (2004b). Secreted subtilisin gene family in *Trichophyton rubrum*. Gene *339*, 79–88.

Kataoka, Y., Takada, K., Oyama, H., Tsunemi, M., James, M.N., and Oda, K. (2005). Catalytic residues and substrate specificity of scytalidoglutamic peptidase, the first member of the eqolisin in family (G1) of peptidases. FEBS Lett. *579*, 2991–2994.

Kaufman, G., Horwitz, B.A., Hadar, R., Ullmann, Y., and Berdicevsky, I. (2004). Green fluorescent protein (GFP) as a vital marker for pathogenic development of the dermatophyte *Trichophyton mentagrophytes*. Microbiology *150*, 2785–2790.

Kaufman, G., Horwitz, B.A., Duek, L., Ullman, Y., and Berdicevsky, I. (2007). Infection stages of the dermatophyte pathogen *Trichophyton*: microscopic characterization and proteolytic enzymes. Med. Mycol. *45*, 149–155.

Kawasaki, M., Anzawa, K., Mochizuki, T., Ishizaki, H., and Hemashettar, B.M. (2009). Successful mating of a human isolate of *Arthroderma simii* with a tester strain of *A. vanbreuseghemii*. Nihon Ishinkin Gakkai Zasshi *50*, 15–18.

Krappmann, S., Sasse, C., and Braus, G.H. (2006). Gene targeting in *Aspergillus fumigatus* by homologous recombination is facilitated in a nonhomologous end-joining-deficient genetic background. Eukaryot. Cell *5*, 212–215.

Kudla, B., Caddick, M.X., Langdon, T., Martinez-Rossi, N.M., Bennett, C.F., Sibley, S., Davies, R.W., and Arst, H.N. Jr. (1990). The regulatory gene *areA* mediating nitrogen metabolite repression in *Aspergillus nidulans*. Mutations affecting specificity of gene activation alter a loop residue of a putative zinc finger. EMBO J. *9*, 1355–1364.

Kunert, J. (1972a). Keratin decomposition by dermatophytes: evidence of the sulphitolysis of the protein. Experientia *28*, 1025–1026.

Kunert, J. (1972b). Thiosulphate esters in keratin attacked by dermatophytes *in vitro*. Sabouraudia *10*, 6–13.

Kunert, J. (1976). Keratin decomposition by dermatophytes. II. Presence of S-sulphocysteine and cysteic acid in soluble decomposition products. Z. Allg. Mikrobiol. *16*, 97–105.

Kunert, J. (1992). Effect of reducing agents on proteolytic and keratinolytic activity of enzymes of *Microsporum gypseum*. Mycoses 35, 343–348.

Kunert, J. (2000). Physiology of keratinophilic fungi. In Biology of Dermatophytes and Other Keratinophilic Fungi, R.K.S. Kushwaha and J. Guarro, eds. (Revista Iberoamericana de Micología 17, Bilbao), pp. 77–85.

Kwon-Chung, K.J., and Bennett, J.E. (1992). Medical Mycology (Lea and Febiger, Philadelphia and London).

Léchenne, B., Reichard, U., Zaugg, C., Fratti, M., Kunert, J., Boulat, O., and Monod, M. (2007). Sulphite efflux pumps in *Aspergillus fumigatus* and dermatophytes. Microbiology *153*, 905–913.

Lee, S.C., Ni, M., Li, W., Shertz, C., and Heitman, J. (2010). The evolution of sex: a perspective from the fungal kingdom. Microbiol. Mol. Biol. Rev. *74*, 298–340.

Li, W., Metin, B., White, T.C., and Heitman, J. (2010). Organization and evolutionary trajectory of the mating type (MAT) locus in dermatophyte and dimorphic fungal pathogens. Eukaryot. Cell 9, 46–58.

Li, X.F., Shen, Y.N., Chen, W., Chen, H., Lv, G.X., and Liu, W.D. (2009). A new medium for diagnosis of dermatophyte infection. Eur. J. Dermatol. *19*, 34–37.

Lopez-Garcia, B., Lee, P.H., and Gallo, R.L. (2006). Expression and potential function of cathelicidin antimicrobial peptides in dermatophytosis and tinea versicolor. J. Antimicrob. Chemother. 57, 877–882.

Lorenz, M.C., and Fink, G.R. (2001). The glyoxylate cycle is required for fungal virulence. Nature *412*, 83–86.

Lurati, M., Baudraz-Rosselet, F., Vernez, M., Spring, P., Bontems, O., Fratti, M., and Monod, M. (2011). Efficacious treatment of non-dermatophyte mould onychomycosis with topical amphotericin B. Dermatology 223, 289–292.

McKinney, J.D., Honer zu Bentrup, K., Munoz-Elias, E.J., Miczak, A., Chen, B., Chan, W.T., Swenson, D., Sacchettini, J.C., Jacobs, W.R. Jr, and Russell, D.G. (2000). Persistence of *Mycobacterium tuberculosis* in macrophages and mice requires the glyoxylate shunt enzyme isocitrate lyase. Nature *406*, 735–738.

Maranhão, F.C., Paião, F.G., Fachin, A.L., and Martinez-Rossi, N.M. (2009). Membrane transporter proteins are involved in *Trichophyton rubrum* pathogenesis. J. Med. Microbiol. 58, 163–168.

Marichal, P., Koymans, L., Willemsens, S., Bellens, D., Verhasselt, P., Luyten, W., Borgers, M., Ramaekers, F.C., Odds, F.C., and Bossche, H.V. (1999). Contribution of mutations in the cytochrome P450 14alpha-demethylase (Erg11p, Cyp51p) to azole resistance in *Candida albicans*. Microbiology *145*, 2701–2713.

Marples, M.J. (1956). The ecology of *Microsporum canis* Bodin in New Zealand. J. Hyg. 54, 378–387.

Martinez, D.A., Oliver, B.G., Gräser, Y., Goldberg, J.M., Li, W., Martinez-Rossi, N.M., Monod, M., Shelest, E., Barton, R.C., Birch, E., *et al.* (2012). Comparative genome analysis of *Trichophyton rubrum* and related dermatophytes reveals candidate genes involved in infection. Mbio. 3, e00259–00212.

Mathy, A., Baldo, A., Schoofs, L., Cambier, L., Defaweux, V., Tabart, J., Maréchal, F., Symoens, F., and Mignon, B. (2010). Fungalysin and dipeptidyl-peptidase gene transcription in *Microsporum canis* strains isolated from symptomatic and asymptomatic cats. Vet. Microbiol. *146*, 179–182.

Mignon, B., Swinnen, M., Bouchara, J.P., Hofinger, M., Nikkels, A., Pierard, G., Gerday, C., and Losson, B. (1998). Purification and characterization of a 31.5 kDa keratinolytic subtilisin-like serine protease from *Microsporum canis* and evidence of its secretion in naturally infected cats. Med. Mycol. 36, 395–404.

Mignon, B.R., Leclipteux, T., Focant, C., Nikkels, A.J., Piérard, G.E., and Losson, B.J. (1999). Humoral and cellular immune response to a crude exo-antigen and purified keratinase of *Microsporum canis* in experimentally infected guinea pigs. Med. Mycol. 37, 123–129.

Mignon, B., Tabart, J., Baldo, A., Mathy, A., Losson, B., and Vermout, S. (2008). Immunization and dermatophytes. Curr. Opin. Infect. Dis. *21*, 134–140.

Mock, M., Monod, M., Baudraz-Rosselet, F., and Panizzon, R.G. (1998). Tinea capitis dermatophytes: susceptibility to antifungal drugs tested *in vitro* and *in vivo*. Dermatology *197*, 361–367.

Monheit, J.E., Cowan, D.F., and Moore, D.G. (1984). Rapid detection of fungi in tissues using calcofluor white and fluorescence microscopy. Arch. Pathol. Lab. Med. *108*, 616–618.

Monod, M., Baudraz-Rosselet, F., Ramelet, A.A., and Frenk, E. (1989). Direct mycological examination in dermatology: a comparison of different methods. Dermatologica *179*, 183–186.

Monod, M., Léchenne, B., Jousson, O., Grand, D., Zaugg, C., Stocklin, R., and Grouzmann, E. (2005). Aminopeptidases and dipeptidyl-peptidases secreted by the dermatophyte *Trichophyton rubrum*. Microbiology *151*, 145–155.

Monod, M., Bontems, O., Zaugg, C., Lechenne, B., Fratti, M., and Panizzon, R. (2006). Fast and reliable PCR/sequencing/RFLP assay for identification of fungi in onychomycoses. J. Med. Microbiol. 55, 1211–1216.

Morschhäuser, J. (2010). Regulation of multidrug resistance in pathogenic fungi. Fungal Genet. Biol. *47*, 94–106.

Mukherjee, P.K., Leidich, S.D., Isham, N., Leitner, I., Ryder, N.S., and Ghannoum, M.A. (2003). Clinical *Trichophyton rubrum* strain exhibiting primary resistance to terbinafine. Antimicrob. Agents Chemother. *47*, 82–86.

Nakamura, Y., Kano, R., Hasegawa, A., and Watanabe, S. (2002). Interleukin-8 and tumor necrosis factor andalpha; production in human epidermal keratinocytes induced by *Trichophyton mentagrophytes*. Clin. Diagn. Lab. Immunol. 9, 935–937.

Nenoff, P., Erhard, M., Simon, J.C., Muylowa, G.K., Herrmann, J., Rataj, W., and Gräser, Y. (2013). MALDI-TOF mass spectrometry – a rapid method for the identification of dermatophyte species. Med. Mycol. *51*, 17–24.

Ninet, B., Jan, I., Bontems, O., Lechenne, B., Jousson, O., Panizzon, R., Lew, D., and Monod, M. (2003). Identification of dermatophyte species by 28S ribosomal DNA sequencing with a commercial kit. J. Clin. Microbiol. *41*, 826–830.

Ninomiya, Y., Suzuki, K., Ishii, C., and Inoue, H. (2004). Highly efficient gene replacements in *Neurospora* strains deficient for nonhomologous end-joining. Proc. Natl. Acad. Sci. U.S.A. *101*, 12248–12253.

O'Cuinn, G., Fitzgerald, R., Bouchier, P., and McDonnell, M. (1999). Generation of non-bitter casein hydrolysates by using combinations of a proteinase and aminopeptidases. Biochem. Soc. Trans. *27*, 730–734.

Osborne, C.S., Hofbauer, B., Favre, B., and Ryder, N.S. (2003). *In vitro* analysis of the ability of *Trichophyton rubrum* to become resistant to terbinafine. Antimicrob. Agents Chemother. *47*, 3634–3636.

Osborne, C.S., Leitner, I., Favre, B., and Ryder. N.S. (2005). Amino acid substitution in *Trichophyton rubrum* squalene epoxidase associated with resistance to terbinafine. Antimicrob. Agents Chemother. *49*, 2840–2844.

Osborne, C.S., Leitner, I., Hofbauer, B., Fielding, C.A., Favre, B., and Ryder, N.S. (2006). Biological, biochemical, and molecular characterization of a new clinical *Trichophyton rubrum* isolate resistant to terbinafine. Antimicrob. Agents Chemother. *50*, 2234–2236.

Payne, J.W., and Smith, M.W. (1994). Peptide transport by micro-organisms. Adv. Microb. Physiol. *36*, 1–80.

Peñalva, M.A., Tilburn, J., Bignell, E., and Arst, H.N. (2008). Ambient pH gene regulation in fungi: making connections. Trends Microbiol. *16*, 291–300.

Pore, R.S., and Plunkett, O.A. (1967). Biological species and variation in *Arthroderma*. Mycopathol. Mycol. Appl. *31*, 225–241.

Prochacki, H., and Engelhardt-Zasada, C. (1974). *Epidermophyton stockdalae* Sp. Nov. Mycopathol. Mycol. Appl. *54*, 341–345.

Punt, P.J., Schuren, F.H., Lehmbeck, J., Christensen, T., Hjort, C., and van den Hondel, C.A. (2008). Characterization of the *Aspergillus niger* prtT, a unique regulator of extracellular protease encoding genes. Fungal Genet. Biol. *45*, 1591–1599.

Rashid, A., Hodgins, M.B., and Richardson, M.D. (1996). An *in vitro* model of dermatophyte invasion of the human hair follicle. J. Med. Vet. Mycol. *34*, 37–42.

de Respinis, S., Tonolla, M., Pranghofer, S., Petrini, L., Petrini, O., and Bosshard, P.P. (2013). Identification of dermatophytes by matrix-assisted laser desorption/ionization time-of-flight mass spectrometry. Med. Mycol. *51*, 514–521.

Rippon, J.W., and Garber, E.D. (1969). Dermatophyte pathogenicity as a function of mating type and associated enzymes. J. Invest. Dermatol. *53*, 445–448.

Romani, L. (2011). Immunity to fungal infections. Nat. Rev. Immunol. *11*, 275–288.

Ruffin, P., Andrieu, S., Biserte, G., and Biguet, J. (1976). Sulphitolysis in keratinolysis. Biochemical proof. Sabouraudia *14*, 181–184.

Sanglard, D., Kuchler, K., Ischer, F., Pagani, J.L., Monod, M., and Bille, J. (1995). Mechanisms of resistance to azole antifungal agents in *Candida albicans* isolates from AIDS patients involve specific multidrug transporters. Antimicrob. Agents Chemother. *39*, 2378–2386.

Sanglard, D., Ischer, F., Koymans, L., and Bille, J. (1998). Amino acid substitutions in the cytochrome P-450 lanosterol 14alpha-demethylase (CYP51A1) from azole-resistant *Candida albicans* clinical isolates contribute to resistance to azole antifungal agents. Antimicrob. Agents Chemother. *42*, 241–253.

Sato, K., Yang, X.L., Yudate, T., Chung, J.S., Wu, J., Luby-Phelps, K., Kimberly, R.P., Underhill, D., Cruz, P.D. Jr, and Ariizumi, K. (2006). Dectin-2 is a pattern recognition receptor for fungi that couples with the Fc receptor gamma chain to induce innate immune responses. J. Biol. Chem. *281*, 38854–38866.

Schneider, W. (1954). Favusepidemie durch Feldmäuse. Hautarzt *5*, 348–351.

Schöbel, F., Ibrahim-Granet, O., Ave, P., Latge, J.P., Brakhage, A.A., and Brock, M. (2007). *Aspergillus fumigatus* does not require fatty acid metabolism via isocitrate lyase for development of invasive aspergillosis. Infect. Immun. 75, 1237–1244.

Sharon, H., Hagag, S., and Osherov, N. (2009). Transcription factor PrtT controls expression of multiple secreted proteases in the human pathogenic mold *Aspergillus fumigatus*. Infect. Immun. 77, 4051–4060.

Shimamura, T., Kubota, N., and Shibuya, K. (2012). Animal model of dermatophytosis. J. Biomed. Biotechnol. Article ID 125384 (11 pages).

Shiraki, Y., Ishibashi, Y., Hiruma, M., Nishikawa, A., and Ikeda, S. (2006). Cytokine secretion profiles of human keratinocytes during *Trichophyton tonsurans* and *Arthroderma benhamiae* infections. J. Med. Microbiol. *55*, 1175–1185.

da Silva Ferreira, M.E., Kress, M.R., Savoldi, M., Goldman, M.H., Härtl, A., Heinekamp, T., Brakhage, A.A., and Goldman, G.H. (2006). The akuB(KU80) mutant deficient for nonhomologous end joining is a powerful tool for analyzing pathogenicity in *Aspergillus fumigatus*. Eukaryot. Cell *5*, 207–211.

Sitterle, E., Fréalle, E., Foulet, F., Cabaret, O., Cremer, G., Guillot, J., Delhaes, L., and Botterel, F. (2012). *Trichophyton bullosum*: a new zoonotic dermatophyte species. Med Mycol. *50*, 305–309.

Slunt, J.B., Taketomi, E.A., Woodfolk, J.A., Hayden, M.L., and Platts-Mills, T.A. (1996). The immune response to *Trichophyton tonsurans*: distinct T cell cytokine profiles to a single protein among subjects with immediate and delayed hypersensitivity. J. Immunol. *157*, 5192–5197.

Sriranganadane, D., Waridel, P., Salamin, K., Reichard, U., Grouzmann, E., Neuhaus, J.M., Quadroni, M., and Monod, M. (2010). *Aspergillus* protein degradation pathways with different secreted protease sets at neutral and acidic pH. J. Proteome Res. *9*, 3511–3519.

Sriranganadane, D., Waridel, P., Salamin, K., Feuermann, M., Mignon, B., Staib, P., Neuhaus, J.M., Quadroni, M., and Monod, M. (2011a). Identification of novel secreted proteases during extracellular proteolysis by dermatophytes at acidic pH. Proteomics *11*, 4422–4433.

Sriranganadane, D., Reichard, U., Salamin, K., Fratti, M., Jousson, O., Waridel, P., Quadroni, M., Neuhaus, J.M., and Monod, M. (2011b). Secreted glutamic protease rescues aspartic protease Pep deficiency in *Aspergillus fumigatus* during growth in acidic protein medium. Microbiology *157*, 1541–1550.

Staib, P., Zaugg, C., Mignon, B., Weber, J., Grumbt, M., Pradervand, S., Harshman, K., and Monod, M.

(2010). Differential gene expression in the pathogenic dermatophyte *Arthroderma benhamiae in vitro* versus infection. Microbiology *156*, 884–895.

Stewart, V., and Vollmer, S.J. (1986). Molecular cloning of *nit-2*, a regulatory gene required for nitrogen metabolite repression in *Neurospora crassa*. Gene *46*, 291–295.

Stockdale, P.M. (1962). *Nannizzia incurvata* gen. nov., sp. nov., a perfect state of *Microsporum gypseum* (Bodin) Guiart et Grigorakis. Sabouraudia *1*, 41–48.

Stockdale, P.M. (1964). The *Microsporum gypseum* complex (*Nannizzia incurvata* Stockd., *N. gypseum* (Nann.) comb.nov., *N. fulva* sp.nov.). Sabouraudia 3, 114–126.

Stockdale, P.M. (1967). *Nannizzia persicolor* sp. nov., the perfect state of *Trichophyton persicolor* Sabouraud. Sabouraudia *5*, 355–359.

Stockdale, P.M. (1981). Sexual stimulation between *Arthroderma simii* and other dermatophytes. In Sexuality and Pathogenicity of Fungi, R. Vanbreuseghem, and C. DeVroey, eds. (Masson, Paris), pp. 57–68.

Stockdale, P.M., Mackenzie, D.W., and Austwick, P.K. (1965). *Arthroderma simii* sp. nov., the perfect state of *Trichophyton simii* (Pinoy) comb. nov. Sabouraudia *4*, 112–123.

Symoens, F., Jousson, O., Planard, C., Fratti, M., Staib, P., Mignon, B., and Monod, M. (2011). Molecular analysis and mating behaviour of the *Trichophyton mentagrophytes* species complex. Int. J. Med. Microbiol. *301*, 260–266.

Symoens, F., Jousson, O., Packeu, A., Fratti, M., Staib, P., Mignon, B., and Monod, M. (2013). The dermatophyte species *Arthroderma benhamiae*: intraspecies variability and mating behavior. J. Med. Microbiol. *62*, 377–385.

Tabart, J., Baldo, A., Vermout, S., Nusgens, B., Lapiere, C., Losson, B., and Mignon, B. (2007). Reconstructed interfollicular feline epidermis as a model for *Microsporum canis* dermatophytosis. J. Med. Microbiol. *56*, 971–975.

Takashio, M. (1972). Sexual reproduction of some *Arthroderma* and *Nannizzia* on diluted Sabouraud agar with or without salts. Mykosen *15*, 11–17.

Takashio, M. (1973). A new sexual state of the *Trichophyton mentagrophytes* complex, *Arthroderma vanbreuseghemii* sp. nov. Ann. Parasitol. Hum. Comp. *48*, 713–732.

Takashio, M. (1981). Production, maintenance and utilization of tester strains by sexual crossing of dermatophytes. A propos of *Nannizia otae*. In Sexuality and Pathogenicity of Fungi, R. Vanbreuseghem and C. DeVroey, eds. (Masson, Paris), pp. 57–68.

Tani, K., Adachi, M., Nakamura, Y., Kano, R., Makimura, K., Hasegawa, A., Kanda, N., and Watanabe, S. (2007). The effect of dermatophytes on cytokine production by human keratinocytes. Arch. Dermatol. Res. *299*, 381–387.

Taplin, D., Zaias, N., Rebell, G., and Blank, H. (1969). Isolation and recognition of dermatophytes on a new medium (DTM). Arch. Dermatol. *99*, 203–209.

Vanbreuseghem, R. (1952). Technique biologique pour l'isolement des dermatophytes du sol. Ann. Soc. Belge Méd. Trop. *32*, 173–178.

Vanbreuseghem, R., De Vroey, C., and Takashio, M. (1978). Guide Pratique de Mycologie Médicale et Vétérinaire (Masson, Paris).

Venturini, J., Alvares, A.M., Camargo, M.R., Marchetti, C.M., Fraga-Silva, T.F., Luchini, A.C., and Arruda, M.S. (2012). Dermatophyte–host relationship of a murine model of experimental invasive dermatophytosis. Microbes Infect. *14*, 1144–1151.

Vermout, S., Tabart, J., Baldo, A., Monod, M., Losson, B., and Mignon, B. (2007). RNA silencing in the dermatophyte *Microsporum canis*. FEMS Microbiol. Lett. *275*, 38–45.

Vermout, S., Baldo, A., Tabart, J., Losson, B., and Mignon, B. (2008a). Secreted dipeptidyl peptidases as potential virulence factors for *Microsporum canis*. FEMS Immunol. Med. Microbiol. *54*, 299–308.

Vermout, S., Tabart, J., Baldo, A., Mathy, A., Losson, B., and Mignon, B. (2008b). Pathogenesis of dermatophytosis. Mycopathologia *166*, 267–275.

Verrier, J., Pronina, M., Peter, C., Bontems, O., Fratti, M., Salamin, K., Schürch, S., Gindro, K., Wolfender, J.L., Harshman, K., and Monod, M. (2012). Identification of infectious agents in onychomycoses by PCR-terminal restriction fragment length polymorphism. J. Clin. Microbiol. *50*, 553–561.

Wagner, D.K., and Sohnle, P.G. (1995). Cutaneous defenses against dermatophytes and yeasts. Clin. Microbiol. Rev. *8*, 317–335.

Walter, E.G., Weiner, J.H., and Taylor, D.E. (1991). Nucleotide sequence and overexpression of the tellurite-resistance determinant from the IncHII plasmid pHH1508a. Gene *101*, 1–7.

Wang, L., Ma, L., Leng, W., Yang, J., Zhu, J., Dong, J., Xue, Y., Zhe, W., Li, R., and Jin, Q. (2004). Analysis of part of the *Trichophyton rubrum* ESTs. Sci. China Life Sci. *47*, 389–395.

Wang, L., Ma, L., Leng, W., Liu, T., Yu, L., Yang, J., Yang, L., Zhang, W., Zhang, Q., Dong, J., *et al.* (2006). Analysis of the dermatophyte *Trichophyton rubrum* expressed sequence tags. BMC Genomics 7, 255.

Weitzman, I., McGinnis, M.R., Padhye, A.A., and Ajello, L. (1986). The genus *Arthroderma* and its later synonym *Nannizzia*. Mycotaxon 25, 505–518.

Weitzman, I., and Summerbell, R.C. (1995). The dermatophytes. Clin. Microbiol. Rev. *8*, 240–259.

Wilson, R.A., and Arst, H.N. (1998). Mutational analysis of AREA, a transcriptional activator mediating nitrogen metabolite repression in *Aspergillus nidulans* and a member of the "streetwise" GATA family of transcription factors. Microbiol. Mol. Biol. Rev. *62*, 586–596.

Woodfolk, J.A. (2005). Allergy and dermatophytes. Clin. Microbiol. Rev. *18*, 30–43.

Woodfolk, J.A., and Platts-Mills, T.A. (1998). The immune response to dermatophytes. Res. Immunol. *149*, 436–445.

Woodfolk, J.A., and Platts-Mills, T.A. (2001). Diversity of the human allergen-specific T cell repertoire associated with distinct skin test reactions: delayed-type hypersensitivity-associated major epitopes induce Th1- and Th2-dominated responses. J. Immunol. *167*, 5412–5419.

Woodfolk, J.A., Wheatley, L.M., Piyasena, R.V., Benjamin, D.C., and Platts-Mills, T.A. (1998). Trichophyton antigens associated with IgE antibodies and delayed type hypersensitivity – Sequence homology to two families of serine proteinases. J. Biol. Chem. *273*, 29489–29496.

Woodfolk, J.A., Sung, S.S., Benjamin, D.C., Lee, J.K., and Platts-Mills, T.A. (2000). Distinct human T cell repertoires mediate immediate and delayed-type hypersensitivity to the *Trichophyton* antigen, Tri r 2. J. Immunol. *165*, 4379–4387.

Yabuki, Y., Kubota, K., Kojima, M., Inoue, H., and Takahashi, K. (2004). Identification of a glutamine residue essential for catalytic activity of aspergilloglutamic peptidase by site-directed mutagenesis. FEBS Lett. *569*, 161–164.

Yamada, T., Makimura, K., Uchida, K., and Yamaguchi, H. (2005). Reproducible genetic transformation system for two dermatophytes, *Microsporum canis* and *Trichophyton mentagrophytes*. Med. Mycol. *43*, 533–544.

Yamada, T., Makimura, K., and Abe, S. (2006). Isolation, characterization, and disruption of *dnr1*, the *areA*/*nit-2*-like nitrogen regulatory gene of the zoophilic dermatophyte, *Microsporum canis*. Med. Mycol. *44*, 243–252.

Yamada, T., Makimura, K., Hisajima, T., Ito, M., Umeda, Y., and Abe, S. (2008). Genetic transformation of the dermatophyte *Trichophyton mentagrophytes* based on the use of G418 resistance as a dominant selectable marker. J. Dermatol. Sci. *49*, 53–61.

Yamada, T., Makimura, K., Hisajima, T., Ishihara, Y., Umeda, Y., and Abe, S. (2009a). Enhanced gene replacements in Ku80 disruption mutants of the dermatophyte, *Trichophyton mentagrophytes*. FEMS Microbiol. Lett. *298*, 208–217.

Yamada, T., Makimura, K., Satoh, K., Umeda, Y., Ishihara, Y., and Abe, S. (2009b). *Agrobacterium tumefaciens*-mediated transformation of the dermatophyte, *Trichophyton mentagrophytes*: an efficient tool for gene transfer. Med. Mycol. *47*, 485–494.

Zaias, N., Tosti, A., Rebell, G., Morelli, R., Bardazzi, F., Bieley, H., Zaiac, M., Glick, B., Paley, B., Allevato, M., and Baran, R. (1996). Autosomal dominant pattern of distal subungual onychomycosis caused by *Trichophyton rubrum*. J. Am. Acad. Dermatol. *34*, 302–304.

Zaugg, C., Jousson, O., Léchenne, B., Staib, P., and Monod, M. (2008). *Trichophyton rubrum* secreted and membrane-associated carboxypeptidases. Int. J. Med. Microbiol. *298*, 669–682.

Zaugg, C., Monod, M., Weber, J., Harshman, K., Pradervand, S., Thomas, J., Bueno, M., Giddey, K., and Staib, P. (2009). Gene expression profiling in the human pathogenic dermatophyte *Trichophyton rubrum* during growth on proteins. Eukaryot. Cell *8*, 241–250.

Zhu, S., Gao, B., Harvey, P.J., and Craik, D.J. (2012). Dermatophytic defensin with antiinfective potential. Proc. Natl. Acad. Sci. U.S.A. *109*, 8495–8500.

Zurita, J., and Hay, R.J. (1987). Adherence of dermatophyte microconidia and arthroconidia to human keratinocytes *in vitro*. J. Invest. Dermatol. *89*, 529–534.

Pathogenesis Mechanisms of *Histoplasma capsulatum*

11

Chad A. Rappleye

Abstract

The dimorphic fungal pathogen *Histoplasma capsulatum* causes respiratory and systemic disease in both immunocompromised and immunocompetent individuals. In mammalian hosts, *Histoplasma* grows as pathogenic yeasts which survive and replicate within phagocytes. Genetic studies have identified a signalling kinase and 3 transcription factors that regulate the thermally induced yeast-phase differentiation process. The mechanisms that facilitate *Histoplasma* pathogenesis (summarized in Fig. 11.1) centre on three primary tasks facing intracellular pathogens: (1) concealment of molecular signatures to avoid detection by phagocytes and activation of immune defences, (2) detoxification of antimicrobial defence molecules produced by phagocytes, and (3) acquisition of essential nutrients for growth and replication within the nutrient-limited phagosome. Demonstrated virulence factors accomplishing these required aspects of pathogenesis include synthesis of α-glucan which masks immunostimulatory yeast cell wall β-glucans, production and secretion of antioxidant molecules (extracellular superoxide dismutase and catalase) to defend specifically against phagocyte-derived reactive oxygen, and secretion of multiple factors for iron acquisition (siderophores and iron reductases). In addition, *Histoplasma* yeasts secrete several novel factors with undefined functions but which are likely linked to pathogenesis by virtue of their specific expression only by pathogenic-phase cells and their extracellular localization. Together, these mechanisms enable *Histoplasma* yeasts to successfully establish infections in spite of fully functional immune defences of the host.

Histoplasmosis

Histoplasma is the aetiological agent of the respiratory and systemic disease histoplasmosis. Most *Histoplasma* infections occur in endemic regions of North, Central, and South America as well as Africa. In the Ohio and Mississippi river valleys of North America, up to 80% of residents show serological evidence of infection (Edwards *et al.*, 1969) indicating widespread exposure to the fungus. It has been estimated that *Histoplasma* causes up to 200,000 infections annually in the United States (Ajello, 1971). *Histoplasma* is present in bird- and bat guano-rich soils. Infection by *Histoplasma* results from inhalation of conidia, which are produced by the mycelial form (mould) and become airborne upon environmental disturbance. Consistent with this, many outbreaks of histoplasmosis can be traced to disruption of a focal source contaminated with *Histoplasma* mould (Chamany *et al.*, 2004; Cottle *et al.*, 2013; Morgan *et al.*, 2003; O'Keefe *et al.*, 2012; Safranek and Beecham, 2004; Warren and Westman, 2008).

Clinically, lung infection by *Histoplasma* ranges from asymptomatic to severe, dependent upon the host's immune status and the number of conidia inhaled (Rippon, 1988). Unlike opportunistic fungal pathogens for which infections only occur in individuals with significantly impaired immune function (notably lack of innate immunity), *Histoplasma* can successfully infect and cause disease in immunocompetent as well as immunocompromised individuals. Thus, *Histoplasma* is more appropriately classified as a primary, not an opportunistic fungal pathogen. As evidence of this, outbreaks of histoplasmosis primarily involve otherwise healthy individuals (Chamany *et al.*, 2004; Morgan *et al.*, 2003; O'Keefe *et al.*, 2012; Safranek

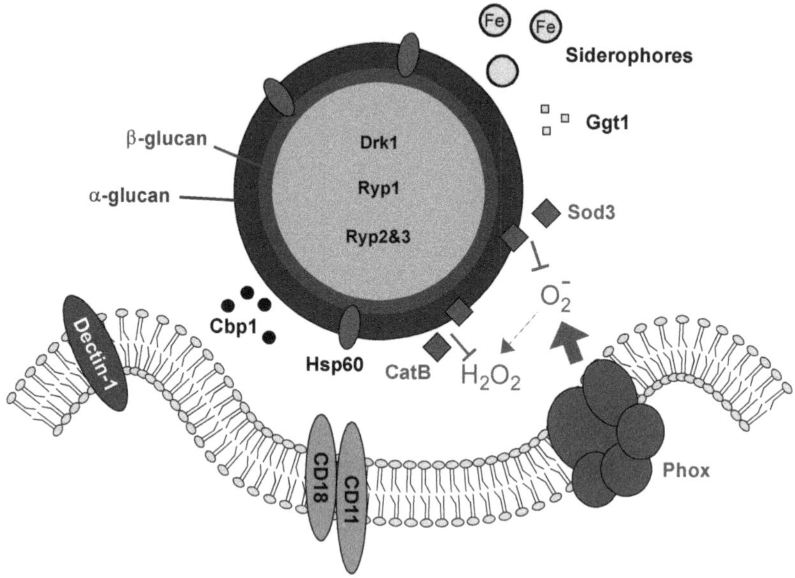

Figure 11.1 Pathogenesis mechanisms at the macrophage–pathogen interface that facilitate Histoplasma virulence. Defined virulence factors are schematically shown and coloured according to their function (adhesion and phagocytosis=grey; concealment=blue/green; reactive oxygen (ROS) detoxification=red; iron acquisition=yellow). Virulence strategies may operate at the macrophage cell surface during initial binding of yeasts to macrophages or they may function in the Histoplasma-containing phagosome (for simplicity, pathogenesis strategies are schematically shown in the figure at the macrophage cell surface). Adherence and phagocytosis of Histoplasma yeasts into macrophages is mediated by yeast cell wall localized Hsp60 (grey) binding to CD18-family receptors (e.g. CD18/CD11) in the macrophage plasma membrane. Simultaneously, Histoplasma yeasts avoid detection by the macrophage β-glucan receptor Dectin-1 (green, host receptor) by concealing fungal β-glucans (green) beneath an α-glucan polysaccharide layer (blue) in the yeast cell wall. Histoplasma yeasts eliminate phagocyte-produced ROS by production of extracellular oxidative stress defence factors (red). Assembly of the multimeric phagocyte NADPH oxidase complex (Phox) in the plasma membrane or the phagosomal membrane leads to production of antimicrobial superoxide (O_2^-) and peroxide (H_2O_2). These toxic ROS molecules are eliminated by the extracellular *Histoplasma* superoxide dismutase (Sod3) and the extracellular catalase (CatB), respectively. *Histoplasma* yeasts have multiple strategies for essential iron acquisition (yellow). Host iron is brought to the phagosome via transferrin. Histoplasma yeasts secrete iron chelating siderophores and a gamma-glutamyl transferase (Ggt1) which reduces ferric to ferrous iron and allows for iron import into yeast cells. The Cbp1 protein (black) is secreted by Histoplasma yeasts and is critical for pathogenesis, although the role it plays is unknown. The majority of these mechanisms are specific for yeast phase growth. Within the yeast cell, a signalling kinase (Drk1) and transcription factors (Ryp1, Ryp2, and Ryp3) act to promote differentiation of Histoplasma cells into the pathogenic yeast phase in response to mammalian temperature. A colour version of this figure is available in the plate section at the back of the book.

and Beecham, 2004). Epidemiological surveys in the United States indicate that most hospitalizations due to histoplasmosis are not associated with other co-morbidities, further underscoring the pathogenic potential of *Histoplasma* and the risk to immunocompetent people (Chu *et al.*, 2006). Nevertheless, immunocompromised people are significantly more at risk for more severe histoplasmosis. In immunocompetent individuals, control and clearance of *Histoplasma* requires activation of $CD4^+$ T-cells. In individuals lacking functional T-cells (e.g. individuals with AIDS), yeasts will become disseminated to extrapulmonary sites causing the most deadly form of histoplasmosis. Although activation of cell-mediated immunity in immunocompetent hosts most often leads to elimination of disease symptoms, infections may not be completely cleared as latent *Histoplasma* cells can persist, constituting a reservoir of organisms that can seed reactivation histoplasmosis (Allen and Deepe, 2006; Durkin *et al.*, 2001). In addition, histoplasmosis also complicates immunomodulatory therapies for individuals, particularly for those receiving TNFα blocking drugs (Hage *et al.*, 2010; Jain *et al.*, 2006; Lee *et al.*, 2002; Wood *et al.*, 2003).

Histoplasma phylogenetics

The species designation *Histoplasma capsulatum* refers to a complex of cryptic species. Based on concordance of multiple gene sequence genealogies, *Histoplasma capsulatum* strains are separated into six major phylogenetic clades which are correlated with their geographic distribution: two in North America (North American class 1 (NAm1) and North American class 2 (NAm2)), three in Latin America [Latin America group A (LAmA), Latin America group B (LAmB), and a Panama clade], and an African clade which includes the *Histoplasma capsulatum* variety *duboisii* (Kasuga *et al.*, 1999, 2003). Infections with isolates from these phylogenetic groups result in differences in clinical disease manifestation including variations in the tendency to cause cutaneous lesions versus restriction to pulmonary tissue (Durkin *et al.*, 2004; Karimi *et al.*, 2002). The majority of the dissection of virulence relies on two clinical isolates representing the NAm2 and the Panama groups since techniques for molecular genetic manipulation have progressed the most for these two strains. These two strains differ in their biochemical composition as well as their overall virulence with the NAm2 strain being more virulent than the Panama strain in mouse models of histoplasmosis (Edwards *et al.*, 2011a; Mayfield and Rine, 2007; Tewari and Berkhout, 1972). Furthermore, the production of and requirement for different secreted factors differs between these two groups (Edwards and Rappleye, 2011). Thus, the different phylogenetic lineages of *Histoplasma capsulatum* should more appropriately be viewed as distinctly separate species of *Histoplasma*.

Molecular genetics of *Histoplasma*

Genomes

Genome sequencing efforts have produced reference genome sequences for two North American strains (WU24 representing NAm1 and G217B representing NAm2), a Panama strain (G186A), and two African strains (H143 and H88) (The Broad Institute, http://www.broadinstitute.org/annotation/genome/histoplasma_capsulatum/MultiHome.html; The Genome Institute, http://genome.wustl.edu/genomes/detail/histoplasma-capsulatum/). *Histoplasma* is haploid. Conidia are asexually produced by the mycelia but a mating type locus (MAT locus) is present in the genome. Strains with different MAT alleles are correlated with opposite mating types in clinical strains; G217B has the MAT1-1 allele whereas G186A has the MAT1-2 allele (Bubnick and Smulian, 2007). The genomes of most strains have approximately 28–30 megabases, with the exception being the NAm2 strain (e.g. G217B) that has a genome size of 41 megabases. The genome size differences are primarily due to differing amounts of repetitive DNA including retrotransposons and cryptons (Goodwin *et al.*, 2003). For example, of the 41 megabases in the G217B genome, roughly 30% is repetitive DNA making the non-repetitive portion (roughly 30 megabases) more consistent with the genomes of the other *Histoplasma* strains. Thus, approximately 28–30 megabases can be considered the 'core' *Histoplasma* genome, which is predicted by bioinformatics analyses to encode roughly 9000 genes. The organization of the non-repetitive portion of the genomes of G186A and G217B has a high degree of synteny with the 'extra' DNA located intergenically as clusters of repetitive sequence. In coding regions, the nucleotide sequence identity is 97% on average (99% amino acid identity) suggesting differential gene regulation, rather than amino acid polymorphisms, contributes more than gene content to phenotypic differences between strains.

Genetic tools and '-omic' approaches

Below, the molecular mechanisms that promote *Histoplasma* pathogenesis are discussed. Much of the current knowledge has resulted from advances in molecular genetic methodologies. The ability to delete or deplete specific gene functions is essential for performing functional tests and establishing the roles of individual factors. *Histoplasma* is particularly recalcitrant to traditional gene knock-out methods that rely on homologous recombination. Nevertheless, by combining positive and negative selections, some success in allelic replacement has been realized (Cooper and Woods, 2009; Hilty *et al.*, 2011; Holbrook *et al.*, 2013; Hwang *et al.*, 2008; Marion *et al.*, 2006; Rappleye *et al.*, 2004; Sebghati *et al.*, 2000; Tian and Shearer, 2002; Woods *et al.*, 1998; Youseff *et al.*, 2012). However the process is laborious and often unsuccessful. RNA interference (RNAi) has proven to be a more efficient and more successful means of testing gene function (Bohse and Woods, 2007a; Cooper and Woods, 2009; Edwards *et al.*, 2011a; Hilty *et al.*, 2011; Holbrook *et al.*, 2011; Hwang *et al.*, 2012; Laskowski-Peak *et al.*, 2012; Marion *et al.*, 2006; Nemecek *et al.*, 2006; Nguyen and Sil, 2008; Rappleye *et al.*, 2004; Youseff *et al.*, 2012; Zarnowski *et al.*, 2008a) and has been further developed with a fluorescent sentinel to visually monitor gene depletion (Edwards *et al.*, 2011a; Youseff and Rappleye, 2012). Insertional mutagenesis exploiting transformation of fungi by *Agrobacterium tumefaciens* is being increasingly applied to both forward genetics for discovery of new genes as well as reverse genetics to interrupt candidate genes (Edwards *et al.*, 2011a,b; Hilty *et al.*, 2008; Kemski *et al.*, 2013; Laskowski and Smulian, 2010; Marion *et al.*, 2006; Nguyen and Sil, 2008; Sullivan *et al.*, 2002; Webster and Sil, 2008; Youseff *et al.*, 2009; Zemska and Rappleye, 2012). Recent advances in platform technologies such as gene expression profiling (Hwang *et al.*, 2003, 2012; Nittler *et al.*, 2005) and proteomics (Albuquerque *et al.*, 2008; Holbrook *et al.*, 2011; Winters *et al.*, 2008) and the candidate gene identification strategies they facilitate complement the parallel development of gene inactivation techniques. These approaches are accelerating the discovery and characterization of factors important for the pathogenic biology of *Histoplasma*.

Histoplasma dimorphism and differentiation

Histoplasma is a true dimorphic fungal pathogen, existing exclusively either as filamentous mycelia or as budding yeasts. This dimorphism is regulated primarily by temperature with *Histoplasma* adopting the mycelial phase at ambient temperatures and the yeast form at 37°C. This has critical relevance to infection since conidia landing in the environment will germinate into mycelia, but conidia inhaled into a mammalian lung will germinate into the pathogenic yeast form. Consequently, disease is associated with yeast-form *Histoplasma* cells. However, the dimorphism of *Histoplasma* goes beyond simple morphology. Adoption of the yeast-form is accompanied by dramatic changes in gene expression compared to mycelia, including determinants of virulence (Abidi *et al.*, 1998; Colonna-Romano *et al.*, 1998; Hwang *et al.*, 2003; Johnson *et al.*, 2002; Kügler *et al.*, 2000; Nguyen and Sil, 2008; Patel *et al.*, 1998; Webster and Sil, 2008). This conversion to yeast and expression of the yeast-phase gene regulon is absolutely necessary for infection and disease establishment. *Histoplasma* cells locked in the mycelial form, either pharmacologically or through genetic mutation, are unable to cause disease (Medoff *et al.*, 1986; Nemecek *et al.*, 2006; Nguyen and Sil, 2008). Thus, the yeast-form is exclusively the pathogenic phase, the mycelia themselves are non-pathogenic, and conidia are the infectious propagules.

Recent studies have begun to dissect the genetic circuitry controlling differentiation of *Histoplasma* cells as pathogenic yeasts. Differentiation signals (i.e. temperature) are in part transmitted by a dimorphism regulating kinase, Drk1. Depletion of Drk1 in *Histoplasma* prevents them from growing as yeast cells at 37°C as well as blocks expression of yeast-phase-specific virulence factors (Nemecek *et al.*, 2006). Drk1 is also required for yeast differentiation of the closely related fungal species, *Blastomyces dermatitidis*, indicating this kinase pathway is conserved among the thermally dimorphic fungi. Differentiation signals culminate in the alteration of gene transcription and studies have identified at least 4 transcriptional regulators that control yeast phase differentiation and function. Forward genetic screens for mutants unable to grow as yeasts at 37°C identified the *RYP1*, *RYP2*, and *RYP3* genes (Nguyen and Sil, 2008; Webster and Sil, 2008). The Ryp1 protein is homologous to the Wor1 differentiation factor that controls white-opaque phenotype switching in *Candida albicans*. Ryp1 binds to promoter regions upstream of yeast-phase genes and activates the expression of approximately 740 yeast-phase genes (Nguyen and Sil, 2008). Ryp2 and Ryp3 are homologues of the VelvetA family proteins VosA and VelB, respectively, which control development of *Aspergillus nidulans* cells as spores. The Ryp1, Ryp2, and Ryp3 proteins show inter-dependence for their own transcription suggesting they may act as a switch complex to control the yeast-phase regulon (Webster and Sil, 2008). *Histoplasma* also encodes a third VelvetA family transcription factor named Vea1 which is homologous to *Aspergillus* VeA. In addition to defects in the formation of filamentous structures of the mycelial phase, depletion of Vea1 impairs differentiation of mycelia into yeast and reduces virulence in mice. While not absolutely essential for yeast-phase differentiation, Vea1 does contribute to regulation of some yeast phase processes necessary for pathogenesis (Laskowski-Peak *et al.*, 2012). VelvetA family member proteins are conserved in filamentous and dimorphic fungi and, consistent with this, *Histoplasma* cells deficient for Ryp2, Ryp3, or Vea1 show mycelial-phase abnormalities. In the dimorphic fungi, the VelvetA-family proteins have been co-opted for additional roles to also control yeast-phase differentiation. The mechanism by which the actual yeast-phase differentiation-inducing thermal signal is transduced into Drk1 activity and ultimately activation of the Ryp-dependent transcriptional program remains to be determined.

Entry into phagocytes

The small size of *Histoplasma* conidia enables them to penetrate into the alveoli of the lung where *Histoplasma* conidia and yeast become phagocytosed by alveolar macrophages. Within the mammalian host, *Histoplasma* is almost exclusively an intracellular pathogen. During infection, *Histoplasma* yeasts are observed within phagocytes and only rarely found extracellularly. Intracellular localization has both benefits and disadvantages for an infecting pathogen. On the one hand, residence within host cells provides protection from soluble immune defence molecules such as antibodies and complement proteins. Within the lung specifically, residence within phagocytes protects *Histoplasma* from the antifungal activity of surfactant proteins (McCormack *et al.*, 2003). Both SP-A and SP-D have direct fungicidal activity on *Histoplasma* through permeabilization of the yeast cells. The effectiveness of SP-A against *Histoplasma* yeast is further indicated by the increased susceptibility of SP-A knockout mice to *Histoplasma* infection *in vivo* (McCormack *et al.*, 2003). However, the intracellular shelter from surfactant provided by intracellular residence is only advantageous if *Histoplasma* yeast survive the initial encounter with these normally fungicidal effector cells of the innate immune system.

Initial adherence of *Histoplasma* yeast to macrophages is mediated by binding to macrophage β-integrins, chiefly the CD18-family complement receptors (CR): LFA-1 (CD11a + CD18), CR3/Mac-1 (CD11b + CD18), or CR4 (CD11c + CD18). Blocking antibodies to the CD18 subunit prevents yeast association with macrophages whereas blocking individual α-subunits (CD11a, CD11b, or CD11c) has a less detrimental effect on adhesion due to CD11 subunit redundancy in macrophage surface receptors (Bullock and Wright, 1987; Lin *et al.*, 2010; Newman *et al.*, 1990). Although *Histoplasma* uses complement receptors for adhesion to macrophages, this association is not dependent on prior opsonization of yeast with complement proteins. Biochemical studies identified heat-shock protein 60 (Hsp60) as the fungal molecule that mediates yeast interaction with CD18-family receptors (Long *et al.*, 2003). Normally Hsp60 is an intracellular stress-response factor and its localization to the yeast cell surface is somewhat surprising. Nevertheless, vaccination with Hsp60 is protective to mice (Deepe *et al.*, 1996; Gomez *et al.*, 1995) and antibodies to *Histoplasma* Hsp60 readily opsonize yeast cells and promote Fc-dependent phagocytosis (Guimarães *et al.*, 2009), providing further evidence of the surface localization of Hsp60. Attachment of *Histoplasma* yeast to other phagocytes is facilitated by different adhesin proteins and phagocyte receptors. For example, yeast-expressed cyclophilin A interacts with VLA-5 receptors on dendritic cells (Gildea *et al.*, 2001; Gomez *et al.*, 2008). No specific adhesin–receptor pair has been determined yet for the interaction of yeasts with PMNs.

Binding of yeast to CD18-family receptors not

only facilitates adherence to macrophages but also benefits *Histoplasma* by triggering internalization of yeasts. Complement receptors are phagocytic receptors in contrast to signalling receptors on the surface of macrophages. However, unlike Fc-mediated phagocytosis of antibody-opsonized particles, complement receptor-mediated phagocytosis is generally non-inflammatory unless accompanied by other stimulating signals (Aderem, 2003; Aderem and Underhill, 1999). Thus, while interacting with the macrophage surface, *Histoplasma* must avoid detection by macrophage-expressed signalling pattern recognition receptors (PRRs), which would activate phagocytic antifungal defence mechanisms.

Avoiding detection

To minimize activation of pro-inflammatory host responses, most *Histoplasma* yeasts block cell wall β-glucans from detection by concealing the β-glucan polysaccharides beneath a layer of α-glucan. Dectin-1 is the major PRR for fungal β-glucans and is responsible for stimulation of antifungal responses of phagocytes in response to β-glucan-containing fungal cells (Drummond and Brown, 2011; Goodridge *et al.*, 2009). Dectin-1 is necessary for pro-inflammatory cytokine production by macrophages in response to *Histoplasma* yeasts (Lin *et al.*, 2010; Rappleye *et al.*, 2007). However, most *Histoplasma* strains synthesize α-linked glucan in addition to β-glucans. The importance of α-glucan was first suggested by the isolation of 'smooth' variants of *Histoplasma* yeasts that spontaneously lost production of α-glucan, and the demonstration that, in contrast to the parent yeast, these variants have significantly attenuated virulence (Eissenberg *et al.*, 1997; Klimpel and Goldman, 1987, 1988). A clue to the role of α-glucan in virulence came from investigation of the distribution of these glucans in the cell wall, which showed the α-glucans overlie the immunostimulatory β-glucans (Rappleye *et al.*, 2007). This spatial organization enables the α-glucans to effectively mask the β-glucans, thereby preventing their detection by Dectin-1. Yeast cells that are unable to synthesize α-glucan are readily detected by Dectin-1 and consequently are severely attenuated in virulence *in vivo* (Rappleye *et al.*, 2004, 2007). α-glucan synthesis requires the function of α-(1,3)-glucan synthase (Ags1) as well as an α-(1,4)-amylase (Amy1) (Camacho *et al.*, 2012; Marion *et al.*, 2006; Rappleye *et al.*, 2004). The current model for α-glucan biosynthesis posits that phosphorylated glucan monomers (synthesized by UTP-glucose-1-phosphate uridyltransferase (*UGP1*); (Marion *et al.*, 2006)) are incorporated either into α-(1,3)-glucan polymers by Ags1 or into short α-(1,4)-glucan oligomers by Amy1 that can act to seed new glucan chains (modelled after Grün *et al.*, 2005).

Interestingly, one phylogenetic group of *Histoplasma* (NAm2) lacks detectable α-glucan yet remains fully virulent. Although the reference genome of this group encodes the *AGS1* gene, the gene is poorly transcribed due to disruption of the promoter by repetitive DNA elements (Edwards *et al.*, 2011a). Complete inactivation of the *AGS1* gene in this lineage does not adversely affect the virulence, confirming that NAm2 strains do not require α-glucan for pathogenesis (Edwards *et al.*, 2011a). This *Histoplasma* species must express an alternate means of minimizing β-glucan detection, although the mechanism is currently undefined. This again highlights the diversity of pathogenesis mechanisms that exist among the different *Histoplasma* phylogenetic groups. Whether by α-glucan or another mechanism, preventing Dectin-1 recognition of yeasts facilitates stealthy entry of *Histoplasma* into macrophages thereby enhancing yeast survival.

Histoplasma yeast cells not only prevent Dectin-1 recognition but also do not trigger macrophage-surface signalling receptors such as Toll-like receptors (TLRs). TLR2 and TLR4 are the most common receptors that recognize fungal cells. *In vitro* studies in which TLRs are blocked indicate that TLR2 and TLR4 do not contribute to macrophage recognition of *Histoplasma* yeast (Lin *et al.*, 2010). Furthermore, TLR2 does not influence the IFNβ response of macrophages to *Histoplasma* conidia (Inglis *et al.*, 2010). One possibility is that *Histoplasma* yeast and conidia simply do not have TLR ligands. However, biochemical fractions of *Histoplasma* yeast cell walls can stimulate lipid body formation by macrophages (which is correlated with pro-inflammatory leukotriene production) in a TLR2-dependent manner indicating the cell wall contains molecules recognized by TLR2 (Sorgi *et al.*, 2009). Thus, a more likely explanation is that the cell wall has TLR-ligands that are recognized in disorganized cell wall material but that the spatial organization of intact yeast cells obscures their recognition in much the same way that α-glucan masks the cell wall β-glucans. It is unknown whether α-glucan might be the cell wall factor that conceals the TLR ligands. The potential role of TLRs in *Histoplasma* pathogenesis *in vivo* remains undetermined.

Neutralization of host defences

In addition to avoidance of detection to facilitate survival in macrophages, *Histoplasma* yeasts must also negate antifungal reactive oxygen species (ROS) produced by phagocytes. Production of ROS by phagocytes, either into the *Histoplasma*-containing phagosome or into the extracellular space, in response to pathogens or as part of the phagocytosis process results from assembly of the phagocyte NADPH-oxidase (Phox) complex. Polymorphonuclear leucocytes (PMNs/neutrophils) are the major source of ROS and PMNs readily release ROS in response to *Histoplasma* yeasts (Kurita *et al.*, 1991a,b; Schaffner *et al.*, 1986; Schnur and Newman, 1990). Macrophages on the other hand only produce significant ROS after cytokine activation (Eissenberg and Goldman, 1987; Wolf and Massof, 1990; Wolf *et al.*, 1987, 1989). The ability to survive phagocyte-produced ROS is a characteristic of primary versus opportunistic fungal pathogens; whereas *Candida* and *Aspergillus* are readily killed by PMNs, *Histoplasma* yeasts maintain their viability (Schaffner *et al.*, 1986). The survival of *Histoplasma* yeasts contrasts with that of *Histoplasma* mycelia indicating that mitigation of ROS effects is solely a pathogenic-phase trait.

Extracellular superoxide dismutase

Histoplasma yeast cells negate the toxic effects of phagocyte-derived superoxide by production of an extracellular superoxide dismutase (Sod3). Sod3 is produced specifically by *Histoplasma* yeast cells (not mycelia) and the protein is localized to the cell surface via a GPI-anchor motif in the C-terminus. Sod3 is also secreted into the soluble extracellular environment (Youseff *et al.*, 2012). As the charge on superoxide prevents it from passing through biological membranes, the extracellular localization of Sod3 perfectly positions it to combat extracellular superoxide molecules such as that produced by phagocytic cells. Deletion of the *SOD3* gene reduces yeast viability when challenged with superoxide *in vitro* and reduces yeast survival when yeasts interact with both PMNs and activated macrophages (Youseff *et al.*, 2012). Consistent with these results, loss of Sod3 attenuates *Histoplasma in vivo*, particularly at early stages of infection during PMN infiltration. Preventing the ability of the host cells to produce superoxide (i.e. knockout of genes encoding Phox components), restores the virulence of yeasts lacking Sod3 indicating that Sod3 specifically acts on the ROS products of the host Phox complex (Youseff *et al.*, 2012). Thus, Sod3 promotes *Histoplasma* pathogenesis by dismutation of host-derived superoxide (Youseff *et al.*, 2012). Intracellular superoxide dismutase, Sod1, does not contribute to detoxification of phagocyte-derived superoxide and does not contribute to pathogenesis *in vivo*, showing the spatial specificity of this virulence mechanism for combating extracellular superoxides produced by the host.

Catalases

In addition to Sod3, *Histoplasma* yeasts express a yeast-phase-specific extracellular catalase (CatB) to defend yeasts against host-derived ROS. CatB is the same protein as M-antigen, one of the early exoantigens adopted for diagnosis of histoplasmosis (Hamilton *et al.*, 1990; Zancopé-Oliveira *et al.*, 1999). Like Sod3, CatB is a yeast-specific factor in the Panama group (Holbrook *et al.*, 2011). In the NAm2 class, CatB is also expressed in the mycelial phase (Johnson *et al.*, 2002). Development of CatB-antibodies by patients (Bauman and Smith, 1976; Bradley *et al.*, 1974) confirms CatB is expressed *in vivo* as does gene expression analysis (Holbrook *et al.*, 2011). As with CatB, *Histoplasma* yeasts have both cell-associated and secreted CatB, although the proportions *in vitro* differ between *Histoplasma* phylogenetic groups (Guimarães *et al.*, 2008; Holbrook *et al.*, 2013). In addition to the gene encoding CatB, the *Histoplasma* genome encodes two additional catalase genes: *CATA* and

CATP. *CATA* is expressed by mycelia and encodes an 80 kDa catalase whereas *CATP* is constitutively expressed by yeast and mycelia and encodes an intracellular 60 kDa catalase enzyme. Because of its specificity for the pathogenic yeast phase and its potential to destroy peroxides, CatB has been assumed to be central to *Histoplasma* virulence.

Contrary to assumptions, deletion of the *CATB* gene from *Histoplasma* yeasts does not impair survival of yeasts when co-cultured with phagocytes nor does CatB loss attenuate virulence *in vivo*. This appears to be due to functional redundancy between CatB and the intracellular CatP catalase (Holbrook *et al.*, 2013). Production of either CatB or CatP is sufficient to protect *Histoplasma* yeasts from host-produced ROS and only deletion of both catalase genes renders *Histoplasma* less virulent in cultured phagocytes and *in vivo* (Holbrook *et al.*, 2013). If intracellular CatP is sufficient, why then do yeasts produce an extracellular catalase? Expression of dual catalases may provide a more effective means of protecting ROS-sensitive cellular targets. Due to its extracellular localization, CatB would be the first to encounter host-derived ROS and would destroy the majority of it before it has an opportunity to affect intracellular targets. CatP, as an intracellular catalase, would provide a second line of defence against any peroxides that escape extracellular destruction by CatB and diffuse into the cell due to their uncharged nature. This two-fold defence strategy provides an effective means of protecting *Histoplasma* yeasts from ROS defences of the host.

The combined activities of Sod3 and the dual catalases efficiently eliminate host-derived reactive oxygen, thereby contributing to the greater virulence of *Histoplasma* yeast compared to opportunistic fungal pathogens. Without either of these mechanisms, *Histoplasma* yeasts are attenuated in virulence. Loss of CatB from yeasts already lacking Sod3 does not further decrease yeast survival against PMNs nor does it further impair lung infection beyond that due to loss of Sod3 (Holbrook *et al.*, 2013). The primary regulation of Sod3 and CatB appears to be differentiation into the yeast phase rather than response to oxidative stresses. As superoxide and peroxide molecules cause lethal damage to cellular constituents (lipids, proteins, iron–sulfur clusters, etc.), pre-formed oxidative stress defences are essential for combating ROS when it is generated by the host. If yeasts relied on stress signals to induce expression of these antifungal oxidants, the yeasts would be eliminated by host cells before synthesis of the defences could be completed. Interestingly, survival of *Histoplasma* against PMNs in the absence of these antioxidant molecules is reduced to levels similar to the survival of wild type *Candida* yeast (Holbrook *et al.*, 2013; Youseff *et al.*, 2012) suggesting that Sod3 and the catalases account for, at least in part, the ability of *Histoplasma* to establish infections even in immunocompetent hosts with functional phagocytes and immune defences.

RNS defence

Host defences against fungal infections also include production of reactive nitrogen species (RNS), particularly by cytokine-activated phagocytic cells. Pro-inflammatory cytokines including IFNγ, TNFα, IL-1β, IL-12, IL-17, and GM-CSF are critical for the ability of hosts to ultimately control *Histoplasma* infections (Allendoerfer and Deepe, 1998, 2000; Clemons *et al.*, 2000; Deepe and Gibbons, 2006, 2009; Deepe and McGuinness, 2006; Deepe *et al.*, 1999; Wu-Hsieh *et al.*, 1992; Zhou *et al.*, 1995). Impaired production of RNS is associated with the loss of IFNγ, TNFα, and GM-CSF and is correlated with increased susceptibility of mice to *Histoplasma* infection. Similarly, pharmacologic inhibition of RNS production by phagocytes decreases their ability to control *Histoplasma* yeasts (Lane *et al.*, 1994; Nakamura *et al.*, 1994). These data indicate that *Histoplasma* yeasts must contend with RNS during infection of mammalian hosts. The effect of RNS on *Histoplasma* is only fungistatic, not fungicidal, suggesting *Histoplasma* can mitigate the lethal but not the inhibitory effects of RNS. Gene expression profiling of yeasts treated with RNS identified 59 genes that are up-regulated in response to nitrosative stress (Nittler *et al.*, 2005). Whether any of these factors are responsible for *Histoplasma* resistance to RNS remains to be determined.

Melanin

Since melanins are scavengers of free radicals, synthesis of melanin by *Histoplasma* yeast cells may provide an additional, non-enzymatic, defence against reactive oxidants produced by the host. Environmental fungi are notorious for synthesizing various pigments, including melanin, which help ameliorate environmental stresses (e.g. UV- and ionizing radiation, heavy metals, etc.). When grown in media with phenolic compounds, *Histoplasma* yeasts can synthesize melanin, which becomes localized to the yeast cell wall (Nosanchuk *et al.*, 2002). Infection with *Histoplasma* stimulates antibody responses to fungal melanin, and yeast cells in infected tissue react with anti-melanin antibodies indicating melanin is synthesized *in vivo* (Nosanchuk *et al.*, 2002). In *Cryptococcus neoformans*, inability to synthesize melanin causes attenuation of virulence (Kwon-Chung *et al.*, 1982; Rhodes *et al.*, 1982) and this correlates with increased sensitivity to reactive oxygen and nitrogen (Wang *et al.*, 1995). The precise role of melanin in *Histoplasma* pathogenesis *in vivo* and whether melanin can provide an additional protection against phagocyte-derived ROS and RNS awaits testing of yeast unable to synthesize melanin.

Stress response

As an intracellular pathogen, *Histoplasma* must also contend with non-ROS- and non-RNS-based stresses during infection. Growth at mammalian temperatures imposes one level of stress typically not encountered by the non-pathogenic mycelia phase. The *Histoplasma* genome encodes members of the Hsp60, Hsp70, and Hsp90 heat shock response proteins and shifting *Histoplasma* cells from 25°C to 37°C induces expression of Hsp60 and the Hsp90-family member Hsp82 (Caruso *et al.*, 1987; Minchiotti *et al.*, 1992; Shearer *et al.*, 1987) suggesting these proteins may contribute to adaptation to the higher temperatures of the mammalian host. The role of Hsp82 in *Histoplasma* pathogenesis was demonstrated by isolation of a hypomorphic mutant in which a T-DNA insertion in the promoter decreases *HSP82* transcription (Edwards *et al.*, 2011b). The residual *HSP82* expression is sufficient for growth of *Histoplasma* yeast in culture at 37°C. However, yeasts with diminished Hsp82 do not efficiently recover from febrile temperatures and are significantly attenuated in their ability to infect mice (Edwards *et al.*, 2011b). Diminished Hsp82 levels also increase sensitivity to hydrogen peroxide treatment *in vitro* indicating that Hsp82 helps alleviate multiple infection-related stresses in addition to thermal stress. Although suggested by expression studies, the role of Hsp82 in the temperature-induced dimorphic transition remains to be determined.

Acquisition of nutrients for intracellular growth

Growth within macrophages, while providing shelter from external immune defences, constrains growth due to nutrient limitation in the phagosome. Unlike many intracellular bacterial pathogens that escape from their intracellular compartment to grow in the nutrient-rich host cell cytosol, *Histoplasma* yeasts remain within the phagosome. The phagosome is comparatively nutrient-poor, and thus *Histoplasma* pathogenesis involves the synthesis and acquisition of nutrients to support growth and replication within host cells.

Iron

Histoplasma yeasts require iron for cellular processes essential for life and thus must acquire this element from the host. IFNγ, produced by activated T-cells as part of cell mediated immunity, limits *Histoplasma* replication and administration of iron chelators mimics this effect (Lane *et al.*, 1991; Newman *et al.*, 1994, 1995). Application of additional iron can reverse some of the IFNγ-triggered restriction of intramacrophage *Histoplasma* growth. This pro-inflammatory cytokine is thus linked to iron restriction by the host, demonstrating iron is crucial for intracellular *Histoplasma* growth (Lane *et al.*, 1991; Newman *et al.*, 1995). Within the phagosome, transferrin-bound-iron is the most likely source of iron for *Histoplasma*. The iron atoms bound to transferrin are released by the drop in pH associated with normal phagosome maturation. However, acidification of the phagosome also poses a danger to *Histoplasma* yeasts due to activation of acidic lysosomal enzymes. The potential lethality of the acidified phagosomal lumen is indicated by the decline in intramacrophage viability of *Histoplasma* yeasts deficient for the HMG-CoA lyase (Hcl1); *hcl1*-mutant yeast overproduce acidic metabolic byproducts causing phagosome acidification, which results in yeast death in macrophages but not in acidified liquid media (Isaac *et al.*, 2013). *Histoplasma* yeasts balance the need for iron release from transferrin and avoidance of lysosomal hydrolases by maintaining the phagosomal pH around 6.0 (Eissenberg *et al.*, 1993; Strasser *et al.*, 1999). This pH is thought to allow release of one of the two iron atoms from transferrin yet is not acidic enough for activation of lysosomal enzymes.

Siderophores

Production and secretion of iron-scavenging siderophores also enables *Histoplasma* to obtain limited iron in the host. *Histoplasma* yeasts secrete hydroxamate-type siderophores under iron limiting conditions (coprogen B, fusarinine, and dimerum acid; (Burt, 1982; Howard *et al.*, 2000)). These siderophores can obtain iron from host molecules such as transferrin and haemin in a pH-independent manner (Foster, 2002; Timmerman and Woods, 2001). Deletion of the *SID1* gene that encodes L-ornithine monooxygenase, the first step in siderophore synthesis, reduces yeast growth under iron-limitation *in vitro* and impairs yeast replication in cultured macrophages by 50% to 75% (Hilty *et al.*, 2011; Hwang *et al.*, 2008). These data are consistent with a role for siderophores in sustaining intramacrophage growth of yeasts. Conflicting evidence exists for the requirement of siderophores *in vivo*, however. Loss of siderophore synthesis in the genetically tractable Panama strain does not impair infection and replication of *Histoplasma* in murine lungs until 15 days post infection (Hwang *et al.*, 2008). On the other hand, RNAi-based depletion of siderophores in the NAm2 strain reduces lung burdens nearly 100-fold at seven days post-infection (Hilty *et al.*, 2011). The fact the Panama strain has genes encoding a high-affinity iron transport system (*FET3* and *FTR1*) which are absent in NAm2 strains suggests the operation of alternative iron acquisition mechanisms between strains may explain some of the phenotypic variations (Hilty *et al.*, 2011).

Although iron is essential, excess iron is detrimental to cells, and consequently iron homeostasis requires regulation and storage mechanisms. Iron-limitation induces changes in both transcription and protein profiles of *Histoplasma* yeasts (Hwang *et al.*, 2008; Winters *et al.*, 2008). The *SID1* gene is part of an enzyme cluster responsible for siderophore biosynthesis which is up-regulated under iron-limiting conditions (Hwang *et al.*, 2008). Transcription of this cluster is repressed under iron-replete conditions by the iron-responsive GATA transcription factor Sre1 (Chao *et al.*, 2008; Hwang *et al.*, 2012). Iron homeostasis also relies upon yeast vacuolar pH as loss of the vacuolar proton ATPase subunit Vma1 or treatment with chemicals that block vacuolar acidification (e.g. bafilomycin) impairs yeast growth under iron-limiting conditions (Hilty *et al.*, 2008). The phenotypes due to loss of Vma1 are reversed by addition of exogenous iron or siderophores.

Iron reductases

Histoplasma yeast preferentially utilize ferrous iron (Zarnowski *et al.*, 2008b) implying iron reduction is a third mechanism by which yeasts can obtain iron from host molecules such as transferrin. Consistent with the need to reduce ferric ions to promote release from host iron-carrying molecules and from secreted siderophores, the ferric-reducing systems are each extracellular. Biochemical studies demonstrated that yeasts produce three extracellular ferric reductases: two soluble and one cell-associated (Timmerman and Woods, 1999, 2001). These *Histoplasma* reducing factors can utilize transferrin and haemin as substrates, reducing the bound iron to the ferrous state, which promotes its release (Timmerman and Woods, 1999). One of the secreted systems depends on reduced glutathione as a co-factor and has been identified as a gamma-glutamyl transferase (Ggt1; Zarnowski *et al.*, 2008a). Ggt1 catalyses a two-step, reduced thiol-dependent but pH-independent reduction of ferric iron (Zarnowski *et al.*, 2008a). Consistent with the role of iron reduction in iron assimilation in macrophages, depletion of Ggt1 function impairs the ability of *Histoplasma* yeasts to replicate in phagocytes. The requirement for Ggt1 *in vivo* remains to be determined.

Secreted pathogenic-phase factors

In order to promote a favourable outcome from the host–pathogen interaction, *Histoplasma* yeasts secrete several novel factors. Many of these factors have, as yet, unidentified functions, but genetic studies indicate they are critical for *Histoplasma* virulence. Expression of each of these factors is linked to pathogenic-phase cells arguing their function is specifically geared to increasing the fitness of this phase. As fitness of yeasts equates to successful survival and infection of the host, these pathogenic-phase secreted factors are closely linked to pathogenesis.

Cbp1

Cbp1 was the first virulence factor demonstrated for *Histoplasma* (Sebghati *et al.*, 2000). Yeast cells lacking the *CBP1* gene are significantly attenuated in their ability to establish respiratory infections and they are unable to parasitize macrophages in culture (Edwards *et al.*, 2011a; Sebghati *et al.*, 2000). Cbp1 is a small (approximately 10 kDa) protein produced in copious amounts by all strains of *Histoplasma* yeasts (Batanghari and Goldman, 1997). Of secreted proteins, Cbp1 is by far the most abundant, and one of the most highly transcribed genes in the genome. This transcription and synthesis is completely restricted to the pathogenic phase (Batanghari *et al.*, 1998; Kügler *et al.*, 2000; Patel *et al.*, 1998). Originally described as a calcium-binding protein based on its staining by calcium-specific stains and labelling with calcium isotopes (Batanghari and Goldman, 1997), Cbp1 has nM affinity for calcium (Beck *et al.*, 2008). Structural studies of Cbp1 demonstrate that Cbp1 is a homodimer covalently held in a tightly compact structure by intramolecular disulfide bonds (Beck *et al.*, 2008). As such, the Cbp1 protein is highly resistant to proteolysis, which suits it well for functioning in the lysosomal hydrolase-rich phagosome. Consistent with this, Cbp1 is secreted into the phagosome during infection (Kügler *et al.*, 2000). Although its role remains undefined, structural homologies provide some clues to its function. The alpha helices of Cbp1 have tertiary structural similarity, but not primary amino acid homology, to saposin family proteins (Beck *et al.*, 2009). As saposins are lipid-binding molecules, this may suggest that Cbp1's purpose is to interact with host membranes or glycolipids in the macrophage phagosome. Although the precise mechanism is still under investigation, Cbp1 clearly has a role in transforming the phagosome into an environment permissive to yeast survival and replication.

Yps3

Histoplasma yeast produce and secrete the Yps3 protein, a protein with sequence homology to the *Blastomyces dermatitidis* adhesin Bad1. The homology between Yps3 and Bad1 includes the C-terminal EGF-domain that mediates attachment to the yeast cell wall via binding to chitin (Bohse and Woods, 2005; Brandhorst *et al.*, 2003). Yps3 diverges from Bad1 in the extent and length of internal tandem repeats. Whereas Bad1 has 24 copies of a 34 amino acid motif, Yps3 is much smaller and has a variable number (2–20 copies) of the 5 amino acid motif YPTDK (Bohse and Woods, 2007b). NAm2 strains have two copies and the Panama strains have 11 copies of this repeat. Functionally, conservation between Yps3 and Bad1 includes secretion and cell association of the protein but Yps3 does not appear to have the adhesin function of Bad1. While strains from all lineages of *Histoplasma* have the *YPS3* gene, production of Yps3 is restricted to the NAm2 phylogenetic group (Bohse and Woods, 2007b; Keath and Abidi, 1994; Keath *et al.*, 1989). The expression differences result from promoter activity differences since replacement of the *YPS3* promoter with a constitutive promoter restores production of Yps3 in *Histoplasma* cells that normally do not synthesize it (Bohse and Woods, 2007b). The absence of Yps3 in Panama strains does not adversely affect the yeast *in vitro* or *in vivo*. Depletion of Yps3 from a NAm2 strain does not hinder pulmonary colonization, but does impair the ability of yeasts to establish disseminated infections (Bohse and Woods, 2007a). The differences in Yps3 among *Histoplasma* strains indicate the phylogenetic groups have both conserved and distinct pathogenic mechanisms.

Peptidases

Histoplasma yeasts also secrete peptidases such as type IV dipeptidyl peptidases. The *Histoplasma* genome encodes two type IV family members (DppIVA and DppIVB; Cooper and Woods, 2009). Gene depletion studies demonstrate that only DppIVB is responsible for the secreted peptidase activity (Cooper and Woods, 2009). In cultured macrophages, *Histoplasma* yeasts unable to produce the DppIVB peptidase are unable to trigger phagocyte death suggesting this secreted factor will be important *in vivo*; however, the animal studies have not been performed to determine this. It is also unknown whether DppIVB is specific for the pathogenic-phase or if it is also produced by non-pathogenic phase cells. This peptidase may play a nutritional role in breaking down proteins so that peptides and amino acids can be taken up by yeasts (and possibly mycelia) for metabolism or the peptidase may have a more pathogenesis-related role in cleaving host molecules (e.g. cytokines, collagen, etc.) like other dipeptidyl peptidases. DppIVB-deficient yeast cells still grow normally in rich media providing some arguments against only a nutritional role for DppIVB (Cooper and Woods, 2009).

Since secreted proteins are likely factors that would interface with host cells and contribute to virulence, more systematic efforts have been made to identify and catalogue the proteins comprising the secreted proteome of pathogenic-phase *Histoplasma* cells (Holbrook *et al.*, 2011). This proteomic approach has resulted in identification of 33 major proteins secreted by *Histoplasma* yeasts. The known virulence factors, Cbp1, CatB, and Sod3 proteins were also identified in this proteome. In addition, newly identified secreted factors include glycanases (an exoglucanase, Exg1; an endoglucanase, Eng1, and a chitinase, Chs1), several dehydrogenases, a thioredoxin reductase-like protein (Trl1) distinct from intracellular thioredoxin reductases, and at least five novel culture filtrate proteins (Cfp) (Holbrook *et al.*, 2011). Consistent with their isolation in the extracellular fraction, these proteins all have recognizable signal sequences that direct them into the eukaryotic secretory pathway. The glycanases likely contribute to yeast cell wall remodelling during growth and division. Cfp4 is a heavily glycosylated protein produced in large amounts by multiple strains of *Histoplasma*. No functional role has been determined for these secreted proteins yet, but the expression of the glycanases, Trl1, and Cfp4 is specific for the pathogenic phase and their expression is maintained during infection (Holbrook *et al.*, 2011). These characteristics (secretion and yeast-phase specificity) are shared with known virulence factors (i.e. Sod3, CatB, and Cbp1), suggesting the new proteins identified are strong candidates for factors that promote *Histoplasma* pathogenesis. *Histoplasma* yeasts also secrete vesicles, which may be a second means of delivering proteins to the extracellular environment (Albuquerque *et al.*, 2008). These secreted vesicles carry over 300 proteins, the vast majority of which are cytosolic factors. It is unknown whether vesicle secretion and their cargo contribute to *Histoplasma* virulence or whether there is any specificity in the cargo transported.

Conclusions

The virulence of *Histoplasma* in mammals contrasts with that of opportunistic pathogens indicating that *Histoplasma* expresses pathogenic mechanisms that counteract immune defences. Opportunistic fungal pathogens such as *Candida albicans* and *Aspergillus fumigatus* are readily eliminated or controlled by host immune cells, especially phagocytes that form the frontline defence against fungi. Consequently, invasive candidiasis and invasive aspergillosis are infections restricted to hosts in which phagocytes are either depleted or non-functional. *Histoplasma*, on the other hand, easily survives the interaction with phagocytes, actually invading these cells and establishing an intracellular environment permissive for *Histoplasma*. The factors promoting pathogenesis are largely restricted to the yeast phase of *Histoplasma* consistent with their purpose to specifically facilitate survival of this form.

Histoplasma pathogenesis involves three major tasks: (1) avoidance of detection by the host, (2) survival of immune defences, and (3) acquisition of necessary nutrients. *Histoplasma* interacts with phagocytes via the CD18-family phagocytic receptors to promote its internalization into macrophages. Simultaneously, *Histoplasma* avoids triggering signalling-type pattern recognition receptors (PRRs) on the macrophage surface by concealing pathogen-associated molecular patterns (PAMPs). The most clearly defined example of this is the shielding of cell wall β-glucans by an outer α-glucan polysaccharide. In contrast to *Histoplasma* yeast, the cell walls of *Candida* and germinating *Aspergillus* conidia are readily detected by macrophage PRRs including TLR2, TLR4, mannose receptor and Dectin-1 (Gantner *et al.*, 2005; Gersuk *et al.*, 2006; Gow *et al.*, 2007; Heinsbroek *et al.*, 2008; Hohl *et al.*, 2005; Luther *et al.*, 2007; Meier *et al.*, 2003; Mezger *et al.*, 2008; Netea *et al.*, 2002, 2006; Steele *et al.*, 2005; Taylor *et al.*, 2007; van de Veerdonk *et al.*, 2009; Wells *et al.*, 2008; Werner *et al.*, 2009). The second mechanism elevating *Histoplasma* pathogenesis above opportunistic fungi is the ability to negate antifungal ROS. *Histoplasma* yeasts produce extracellular superoxide dismutase and extracellular catalase which can efficiently destroy phagocyte-produced reactive oxygen. These defences are constitutively produced by *Histoplasma* yeasts and are thus immediately available to counteract ROS as *Histoplasma* interacts with phagocyte cells during infection. Extracellular antioxidant defences appear to be a characteristic of many primary pathogens that survive invasion of phagocytic cells (e.g. *Mycobacterium tuberculois, Salmonella*, and *Francisella tularensis*). Intracellular residence also protects yeasts from the antifungal effects of surfactant proteins present in the lung. *Histoplasma* yeasts are thus characterized by both the ability to avoid host defence mechanisms as well as detoxify antifungal products released by the host.

Once *Histoplasma* has accomplished survival and has been taken up by macrophages, progression of the infection requires acquisition of essential nutrients for growth and replication. Iron is one of the major limitations imposed on intracellular *Histoplasma* yeasts as evidenced by the number of mechanisms defined to obtain this critical element. Multiple strategies, which vary between phylogenetic groups of *Histoplasma*, can scavenge iron from host molecules such as holotransferrin. These strategies include high-affinity iron binding siderophores and promoting the release of transferrin-bound iron through pH modulation and/or iron reduction. Beyond iron, we know little about the other metabolic requirements of yeasts within the nutrient poor phagosome. Various novel secreted factors such as Cbp1 and the Cfp culture-filtrate proteins will require additional characterization to define the roles they might play in pathogenesis. Continued functional studies on these and other candidate factors will continue to reveal the mechanisms that contribute to *Histoplasma*'s success as an intracellular primary fungal pathogen.

References

Abidi, F.E., Roh, H., and Keath, E.J. (1998). Identification and characterization of a phase-specific, nuclear DNA binding protein from the dimorphic pathogenic fungus *Histoplasma capsulatum*. Infect. Immun. *66*, 3867–3873.

Aderem, A. (2003). Phagocytosis and the Inflammatory Response. J. Infect. Dis. *187*, S340–S345.

Aderem, A., and Underhill, D.M. (1999). Mechanisms of phagocytosis in macrophages. Ann. Rev. Immunol. *17*, 593–623.

Ajello, L. (1971). Distribution of *Histoplasma capsulatum* in the United States. In Histoplasmosis, L. Ajello, E.W. Chick, and M.F. Furcolow, eds. (Charles C. Thomas Publisher, Springfield, IL), pp. 103–122.

Albuquerque, P.C., Nakayasu, E.S., Rodrigues, M.L., Frases, S., Casadevall, A., Zancope-Oliveira, R.M., Almeida, I.C., and Nosanchuk, J.D. (2008). Vesicular transport in *Histoplasma capsulatum*: an effective mechanism for trans-cell wall transfer of proteins and lipids in ascomycetes. Cell. Microbiol. *10*, 1695–1710.

Allen, H.L., and Deepe, G.S. (2006). B Cells and CD4–CD8– T cells are key regulators of the severity of reactivation histoplasmosis. J. Immunol. *177*, 1763–1771.

Allendoerfer, R., and Deepe, G.S., Jr (1998). Blockade of endogenous TNF-α exacerbates primary and secondary pulmonary histoplasmosis by differential mechanisms. J. Immunol. *160*, 6072–6082.

Allendoerfer, R., and Deepe, G.S., Jr (2000). Regulation of infection with *Histoplasma capsulatum* by TNFR1 and -2. J. Immunol. *165*, 2657–2664.

Batanghari, J.W., and Goldman, W.E. (1997). Calcium dependence and binding in cultures of *Histoplasma capsulatum*. Infect. Immun. *65*, 5257–5261.

Batanghari, J.W., Deepe, G.S., Jr, Di Cera, E., and Goldman, W.E. (1998). *Histoplasma* acquisition of calcium and expression of *CBP1* during intracellular parasitism. Mol. Microbiol. *27*, 531–539.

Bauman, D.S., and Smith, C.D. (1976). Comparison of immunodiffusion and complement fixation tests in the diagnosis of histoplasmosis. J. Clin. Microbiol. *2*, 77–80.

Beck, M.R., DeKoster, G.T., Hambly, D.M., Gross, M.L., Cistola, D.P., and Goldman, W.E. (2008). Structural features responsible for the biological stability of *Histoplasma*'s virulence factor CBP. Biochemistry *47*, 4427–4438.

Beck, M.R., DeKoster, G.T., Cistola, D.P., and Goldman, W.E. (2009). NMR structure of a fungal virulence factor reveals structural homology with mammalian saposin B. Mol. Microbiol. 72, 344–353.

Bohse, M.L., and Woods, J.P. (2005). Surface localization of the Yps3p protein of *Histoplasma capsulatum*. Eukaryot. Cell 4, 685–693.

Bohse, M.L., and Woods, J.P. (2007a). RNA interference-mediated silencing of the *YPS3* gene of *Histoplasma capsulatum* reveals virulence defects. Infect. Immun. 75, 2811–2817.

Bohse, M.L., and Woods, J.P. (2007b). Expression and interstrain variability of the *YPS3* gene of *Histoplasma capsulatum*. Eukaryot. Cell 6, 609–615.

Bradley, G., Pine, L., Reeves, M.W., and Moss, C.W. (1974). Purification, composition, and serological characterization of histoplasmin-H and M antigens. Infect. Immun. 9, 870–880.

Brandhorst, T., Wüthrich, M., Finkel-Jimenez, B., and Klein, B. (2003). A C-terminal EGF-like domain governs BAD1 localization to the yeast surface and fungal adherence to phagocytes, but is dispensable in immune modulation and pathogenicity of *Blastomyces dermatitidis*. Mol. Microbiol. 48, 53–65.

Bubnick, M., and Smulian, A.G. (2007). The *MAT1* locus of *Histoplasma capsulatum* is responsive in a mating type-specific manner. Eukaryot. Cell 6, 616–621.

Bullock, W.E., and Wright, S.D. (1987). Role of the adherence-promoting receptors, CR3, LFA-1, and p150,95, in binding of *Histoplasma capsulatum* by human macrophages. J. Exp. Med. 165, 195–210.

Burt, W.R. (1982). Identification of coprogen B and its breakdown products from *Histoplasma capsulatum*. Infect. Immun. 35, 990–996.

Camacho, E., Sepulveda, V.E., Goldman, W.E., San-Blas, G., and Niño-Vega, G.A. (2012). Expression of Paracoccidioides brasiliensis AMY1 in a *Histoplasma capsulatum amy1* mutant, relates an α-(1,4)-amylase to cell wall α-(1,3)-glucan synthesis. PLoS ONE 7, e50201.

Caruso, M., Sacco, M., Medoff, G., and Maresca, B. (1987). Heat shock 70 gene is differentially expressed in *Histoplasma capsulatum* strains with different levels of thermotolerance and pathogenicity. Mol. Microbiol. 1, 151–158.

Chamany, S., Mirza, S.A., Fleming, J.W., Howell, J.F., Lenhart, S.W., Mortimer, V.D., Phelan, M.A., Lindsley, M.D., Iqbal, N.J., Wheat, L.J., *et al.* (2004). A large histoplasmosis outbreak among high school students in Indiana, 2001. Pediatr. Infect. Dis. J. 23, 909–914.

Chao, L.Y., Marletta, M.A., and Rine, J. (2008). Sre1, an iron-modulated GATA DNA-binding protein of iron-uptake genes in the fungal pathogen *Histoplasma capsulatum*. Biochemistry 47, 7274–7283.

Chu, J.H., Feudtner, C., Heydon, K., Walsh, T.J., and Zaoutis, T.E. (2006). Hospitalizations for endemic mycoses: a population-based national study. Clin. Infect. Dis. 42, 822–825.

Clemons, K.V., Darbonne, W.C., Curnutte, J.T., Sobel, R.A., and Stevens, D.A. (2000). Experimental histoplasmosis in mice treated with anti-murine interferon-γ antibody and in interferon-γ gene knockout mice. Microbes Infect. 2, 997–1001.

Colonna-Romano, S., Porta, A., Franco, A., Kobayashi, G.S., and Maresca, B. (1998). Identification and isolation by DDRT-PCR of genes differentially expressed by *Histoplasma capsulatum* during macrophages infection. Microb. Pathog. 25, 55–66.

Cooper, K.G., and Woods, J.P. (2009). Secreted dipeptidyl peptidase IV activity in the dimorphic fungal pathogen *Histoplasma capsulatum*. Infect. Immun. 77, 2447–2454.

Cottle, L.E., Gkrania-Klotsas, E., Williams, H.J., Brindle, H.E., Carmichael, A.J., Fry, G., and Beeching, N.J. (2013). A multinational outbreak of histoplasmosis following a biology field trip in the Ugandan rainforest. J. Travel. Med. 20, 83–87.

Deepe, G.S., Jr, and Gibbons, R.S. (2006). T cells require tumor necrosis factor-α to provide protective immunity in mice infected with *Histoplasma capsulatum*. J. Infect. Dis. 193, 322–330.

Deepe, G.S., Jr, and Gibbons, R.S. (2009). Interleukins 17 and 23 influence the host response to *Histoplasma capsulatum*. J. Infect. Dis. 200, 142–151.

Deepe, G.S., Jr, and McGuinness, M. (2006). Interleukin-1 and host control of pulmonary histoplasmosis. J. Infect. Dis. 194, 855–864.

Deepe, G.S., Gibbons, R., Brunner, G.D., and Gomez, F.J. (1996). A protective domain of heat-shock protein 60 from *Histoplasma capsulatum*. J. Infect. Dis. 174, 828–834.

Deepe, G.S., Jr, Gibbons, R., and Woodward, E. (1999). Neutralization of endogenous granulocyte–macrophage colony-stimulating factor subverts the protective immune response to *Histoplasma capsulatum*. J. Immunol. 163, 4985–4993.

Drummond, R.A., and Brown, G.D. (2011). The role of Dectin-1 in the host defence against fungal infections. Curr. Opin. Microbiol. 14, 392–399.

Durkin, M., Kohler, S., Schnizlein-Bick, C., LeMonte, A., Connolly, P., Goldberg, J., Garringer, T., and Wheat, L.J. (2001). Chronic infection and reactivation in a pulmonary challenge model of histoplasmosis. J. Infect. Dis. 183, 1822–1824.

Durkin, M.M., Connolly, P.A., Karimi, K., Wheat, E., Schnizlein-Bick, C., Allen, S.D., Alves, K., Tewari, R.P., and Keath, E. (2004). Pathogenic differences between North American and Latin American strains of *Histoplasma capsulatum* var. *capsulatum* in experimentally infected mice. J. Clin. Microbiol. 42, 4370–4373.

Edwards, J.A., and Rappleye, C.A. (2011). *Histoplasma* mechanisms of pathogenesis – one portfolio does not fit all. FEMS Microbiol. Letts. 324, 1–9.

Edwards, J.A., Alore, E.A., and Rappleye, C.A. (2011a). The yeast-phase virulence requirement for α-glucan synthase differs among *Histoplasma capsulatum* chemotypes. Eukaryot. Cell 10, 87–97.

Edwards, J.A., Zemska, O., and Rappleye, C.A. (2011b). Discovery of a role for Hsp82 in *Histoplasma* virulence through a quantitative screen for macrophage lethality. Infect. Immun. 79, 3348–3357.

Edwards, L.B., Acquaviva, F.A., Livesay, V.T., Cross, F.W., and Palmer, C.E. (1969). An atlas of sensitivity to tuberculin, PPD-B, and histoplasmin in the United States. Am. Rev. Respir. Dis. *99 (Suppl)*,1–132.

Eissenberg, L.G., and Goldman, W.E. (1987). *Histoplasma capsulatum* fails to trigger release of superoxide from macrophages. Infect. Immun. *55*, 29–34.

Eissenberg, L.G., Goldman, W.E., and Schlesinger, P.H. (1993). *Histoplasma capsulatum* modulates the acidification of phagolysosomes. J. Exp. Med. *177*, 1605–1611.

Eissenberg, L.G., Moser, S.A., and Goldman, W.E. (1997). Alterations to the cell wall of *Histoplasma capsulatum* yeasts during infection of macrophages or epithelial cells. J. Infect. Dis. *175*, 1538–1544.

Foster, L.-A.A. (2002). Utilization and cell-surface binding of hemin by *Histoplasma capsulatum*. Can. J. Microbiol. *48*, 437–442.

Gantner, B.N., Simmons, R.M., and Underhill, D.M. (2005). Dectin-1 mediates macrophage recognition of *Candida albicans* yeast but not filaments. EMBO J. *24*, 1277–1286.

Gersuk, G.M., Underhill, D.M., Zhu, L., and Marr, K.A. (2006). Dectin-1 and TLRs permit macrophages to distinguish between different *Aspergillus fumigatus* cellular states. J. Immunol. *176*, 3717–3724.

Gildea, L.A., Morris, R.E., and Newman, S.L. (2001). *Histoplasma capsulatum* yeasts are phagocytosed via very late antigen-5, killed, and processed for antigen presentation by human dendritic cells. J. Immunol. *166*, 1049–1056.

Gomez, F.J., Allendoerfer, R., and Deepe, G.S. (1995). Vaccination with recombinant heat shock protein 60 from *Histoplasma capsulatum* protects mice against pulmonary histoplasmosis. Infect. Immun. *63*, 2587–2595.

Gomez, F.J., Pilcher-Roberts, R., Alborzi, A., and Newman, S.L. (2008). *Histoplasma capsulatum* cyclophilin A mediates attachment to dendritic cell VLA-5. J. Immunol. *181*, 7106–7114.

Goodridge, H.S., Wolf, A.J., and Underhill, D.M. (2009). β-glucan recognition by the innate immune system. Immunol. Rev. *230*, 38–50.

Goodwin, T.J.D., Butler, M.I., and Poulter, R.T.M. (2003). Cryptons: a group of tyrosine-recombinase-encoding DNA transposons from pathogenic fungi. Microbiology *149*, 3099–3109.

Gow, N.A.R., Netea, M.G., Munro, C.A., Ferwerda, G., Bates, S., Mora-Montes, H.M., Walker, L., Jansen, T., Jacobs, L., Tsoni, V., *et al.* (2007). Immune recognition of *Candida albicans* β-glucan by dectin-1. J. Infect. Dis. *196*, 1565–1571.

Grün, C.H., Hochstenbach, F., Humbel, B.M., Verkleij, A.J., Sietsma, J.H., Klis, F.M., Kamerling, J.P., and Vliegenthart, J.F.G. (2005). The structure of cell wall α-glucan from fission yeast. Glycobiology *15*, 245–257.

Guimarães, A.J., Hamilton, A.J., De M. Guedes, H.L., Nosanchuk, J.D., and Zancopé-Oliveira, R.M. (2008). Biological function and molecular mapping of M antigen in yeast phase of *Histoplasma capsulatum*. PLoS ONE 3, e3449.

Guimarães, A.J., Frases, S., Gomez, F.J., Zancopé-Oliveira, R.M., and Nosanchuk, J.D. (2009). Monoclonal antibodies to heat shock protein 60 alter the pathogenesis of *Histoplasma capsulatum*. Infect. Immun. *77*, 1357–1367.

Hage, C.A., Bowyer, S., Tarvin, S.E., Helper, D., Kleiman, M.B., and Joseph Wheat, L. (2010). Recognition, diagnosis, and treatment of histoplasmosis complicating tumor necrosis factor blocker therapy. Clin. Infect. Dis. *50*, 85–92.

Hamilton, A.J., Bartholomew, M.A., Figueroa, J., Fenelon, L.E., and Hay, R.J. (1990). Evidence that the M antigen of *Histoplasma capsulatum* var. *capsulatum* is a catalase which exhibits cross-reactivity with other dimorphic fungi. J. Med. Vet. Mycol. *28*, 479–485.

Heinsbroek, S.E.M., Taylor, P.R., Martinez, F.O., Martinez-Pomares, L., Brown, G.D., and Gordon, S. (2008). Stage-specific sampling by pattern recognition receptors during *Candida albicans* phagocytosis. PLoS Pathog. *4*, e1000218.

Hilty, J., Smulian, A.G., and Newman, S.L. (2008). The *Histoplasma capsulatum* vacuolar ATPase is required for iron homeostasis, intracellular replication in macrophages and virulence in a murine model of histoplasmosis. Mol. Microbiol. *70*, 127–139.

Hilty, J., George Smulian, A., and Newman, S.L. (2011). *Histoplasma capsulatum* utilizes siderophores for intracellular iron acquisition in macrophages. Med. Mycol. *49*, 633–642.

Hohl, T.M., Van Epps, H.L., Rivera, A., Morgan, L.A., Chen, P.L., Feldmesser, M., and Pamer, E.G. (2005). *Aspergillus* fumigatus triggers inflammatory responses by stage-specific β-glucan display. PLoS Pathog. *1*, e30.

Holbrook, E.D., Edwards, J.A., Youseff, B.H., and Rappleye, C.A. (2011). Definition of the extracellular proteome of pathogenic-phase *Histoplasma capsulatum*. J. Proteome Res. *10*, 1929–1943.

Holbrook, E., Smolnycki, K., Youseff, B., and Rappleye, C. (2013). Redundant catalases detoxify phagocyte reactive oxygen and facilitate *Histoplasma* pathogenesis. Infect. Immun. *81*, 2334–2346.

Howard, D.H., Rafie, R., Tiwari, A., and Faull, K.F. (2000). Hydroxamate siderophores of *Histoplasma capsulatum*. Infect. Immun. *68*, 2338–2343.

Hwang, L., Hocking-Murray, D., Bahrami, A.K., Andersson, M., Rine, J., and Sil, A. (2003). Identifying phase-specific genes in the fungal pathogen *Histoplasma capsulatum* using a genomic shotgun microarray. Mol. Biol. Cell *14*, 2314–2326.

Hwang, L.H., Mayfield, J.A., Rine, J., and Sil, A. (2008). *Histoplasma* requires *SID1*, a member of an iron-regulated siderophore gene cluster, for host colonization. PLoS Pathog *4*, e1000044.

Hwang, L.H., Seth, E., Gilmore, S.A., and Sil, A. (2012). *SRE1* regulates iron-dependent and -independent pathways in the fungal pathogen *Histoplasma capsulatum*. Eukaryot. Cell *11*, 16–25.

Inglis, D.O., Berkes, C.A., Murray, D.R.H., and Sil, A. (2010). Conidia but not yeast cells of the fungal pathogen *Histoplasma capsulatum* trigger a type I Interferon innate immune response in murine macrophages. Infect. Immun. *78*, 3871–3882.

Isaac, D.T., Coady, A., Van Prooyen, N., and Sil, A. (2013). The 3-hydroxy-methylglutaryl coenzyme A lyase *HCL1* is required for macrophage colonization by human fungal pathogen *Histoplasma capsulatum*. Infect. Immun. *81*, 411–420.

Jain, V.V., Evans, T., and Peterson, M.W. (2006). Reactivation histoplasmosis after treatment with anti-tumor necrosis factor alpha in a patient from a nonendemic area. Respir. Med. *100*, 1291–1293.

Johnson, C.H., Klotz, M.G., York, J.L., Kruft, V., and McEwen, J.E. (2002). Redundancy, phylogeny and differential expression of *Histoplasma capsulatum* catalases. Microbiology *148*, 1129–1142.

Karimi, K., Wheat, L.J., Connolly, P., Cloud, G., Hajjeh, R., Wheat, E., Alves, K., Lacaz Cd, C. da S., and Keath, E. (2002). Differences in histoplasmosis in patients with acquired immunodeficiency syndrome in the United States and Brazil. J. Infect. Dis. *186*, 1655–1660.

Kasuga, T., Taylor, J.W., and White, T.J. (1999). Phylogenetic relationships of varieties and geographical groups of the human pathogenic fungus *Histoplasma capsulatum* Darling. J. Clin. Microbiol. *37*, 653–663.

Kasuga, T., White, T.J., Koenig, G., McEwen, J., Restrepo, A., Castañeda, E., Da Silva Lacaz, C., Heins-Vaccari, E.M., De Freitas, R.S., Zancopé-Oliveira, R.M., *et al.* (2003). Phylogeography of the fungal pathogen *Histoplasma capsulatum*. Mol. Ecol. *12*, 3383–3401.

Keath, E.J., and Abidi, F.E. (1994). Molecular cloning and sequence analysis of *yps-3*, a yeast-phase-specific gene in the dimorphic fungal pathogen *Histoplasma capsulatum*. Microbiology *140*, 759–767.

Keath, E.J., Painter, A.A., Kobayashi, G.S., and Medoff, G. (1989). Variable expression of a yeast-phase-specific gene in *Histoplasma capsulatum* strains differing in thermotolerance and virulence. Infect. Immun. *57*, 1384–1390.

Kemski, M.M., Stevens, B., and Rappleye, C.A. (2013). Spectrum of T-DNA integrations for insertional mutagenesis of *Histoplasma capsulatum*. Fungal Biol. *117*, 41–51.

Klimpel, K.R., and Goldman, W.E. (1987). Isolation and characterization of spontaneous avirulent variants of *Histoplasma capsulatum*. Infect. Immun. *55*, 528–533.

Klimpel, K.R., and Goldman, W.E. (1988). Cell walls from avirulent variants of *Histoplasma capsulatum* lack α-(1,3)-glucan. Infect. Immun. *56*, 2997–3000.

Kügler, S., Young, B., Miller, V.L., and Goldman, W.E. (2000). Monitoring phase-specific gene expression in *Histoplasma capsulatum* with telomeric GFP fusion plasmids. Cell. Microbiol. *2*, 537–547.

Kurita, N., Brummer, E., Yoshida, S., Nishimura, K., and Miyaji, M. (1991a). Antifungal activity of murine polymorphonuclear neutrophils against *Histoplasma capsulatum*. J. Med. Vet. Mycol. *29*, 133–143.

Kurita, N., Terao, K., Brummer, E., Ito, E., Nishimura, K., and Miyaji, M. (1991b). Resistance of *Histoplasma capsulatum* to killing by human neutrophils. Evasion of oxidative burst and lysosomal-fusion products. Mycopathologia *115*, 207–213.

Kwon-Chung, K.J., Polacheck, I., and Popkin, T.J. (1982). Melanin-lacking mutants of *Cryptococcus neoformans* and their virulence for mice. J. Bacteriol. *150*, 1414–1421.

Lane, T.E., Wu-Hsieh, B.A., and Howard, D.H. (1991). Iron limitation and the gamma interferon-mediated antihistoplasma state of murine macrophages. Infect. Immun. *59*, 2274–2278.

Lane, T.E., Wu-Hsieh, B.A., and Howard, D.H. (1994). Antihistoplasma effect of activated mouse splenic macrophages involves production of reactive nitrogen intermediates. Infect. Immun. *62*, 1940–1945.

Laskowski, M.C., and Smulian, A.G. (2010). Insertional mutagenesis enables cleistothecial formation in a non-mating strain of *Histoplasma capsulatum*. BMC Microbiol. *10*, 49.

Laskowski-Peak, M.C., Calvo, A.M., Rohrssen, J., and George Smulian, A. (2012). *VEA1* is required for cleistothecial formation and virulence in *Histoplasma capsulatum*. Fungal Genet. Biol. *49*, 838–846.

Lee, J.-H., Slifman, N.R., Gershon, S.K., Edwards, E.T., Schwieterman, W.D., Siegel, J.N., Wise, R.P., Brown, S.L., Udall, J.N., Jr, and Braun, M.M. (2002). Life-threatening histoplasmosis complicating immunotherapy with tumor necrosis factor α antagonists infliximab and etanercept. Arthritis Rheum. *46*, 2565–2570.

Lin, J.-S., Huang, J.-H., Hung, L.-Y., Wu, S.-Y., and Wu-Hsieh, B.A. (2010). Distinct roles of complement receptor 3, Dectin-1, and sialic acids in murine macrophage interaction with *Histoplasma* yeast. J. Leukoc. Biol. *88*, 95–106.

Long, K.H., Gomez, F.J., Morris, R.E., and Newman, S.L. (2003). Identification of heat shock protein 60 as the ligand on *Histoplasma capsulatum* that mediates binding to CD18 receptors on human macrophages. J. Immunol. *170*, 487–494.

Luther, K., Torosantucci, A., Brakhage, A.A., Heesemann, J., and Ebel, F. (2007). Phagocytosis of *Aspergillus fumigatus* conidia by murine macrophages involves recognition by the dectin-1 β-glucan receptor and Toll-like receptor 2. Cell. Microbiol. *9*, 368–381.

McCormack, F.X., Gibbons, R., Ward, S.R., Kuzmenko, A., Wu, H., and Deepe, G.S. (2003). Macrophage-independent fungicidal action of the pulmonary collectins. J. Biol. Chem. *278*, 36250–36256.

Marion, C.L., Rappleye, C.A., Engle, J.T., and Goldman, W.E. (2006). An α-(1,4)-amylase is essential for α-(1,3)-glucan production and virulence in *Histoplasma capsulatum*. Mol. Microbiol. *62*, 970–983.

Mayfield, J.A., and Rine, J. (2007). The genetic basis of variation in susceptibility to infection with *Histoplasma capsulatum* in the mouse. Genes Immun. *8*, 468–474.

Medoff, G., Sacco, M., Maresca, B., Schlessinger, D., Painter, A., Kobayashi, G.S., and Carratu, L. (1986). Irreversible block of the mycelial-to-yeast phase transition of *Histoplasma capsulatum*. Science *231*, 476–479.

Meier, A., Kirschning, C.J., Nikolaus, T., Wagner, H., Heesemann, J., and Ebel, F. (2003). Toll-like receptor (TLR) 2 and TLR4 are essential for *Aspergillus*-induced activation of murine macrophages. Cell. Microbiol. *5*, 561–570.

Mezger, M., Kneitz, S., Wozniok, I., Kurzai, O., Einsele, H., and Loeffler, J. (2008). Proinflammatory response of immature human dendritic cells is mediated by dectin-1 after exposure to *Aspergillus fumigatus* germ tubes. J. Infect. Dis. *197*, 924–931.

Minchiotti, G., Gargano, S., and Maresca, B. (1992). Molecular cloning and expression of *hsp82* gene of the dimorphic pathogenic fungus *Histoplasma capsulatum*. Biochim. Biophys. Acta *1131*, 103–107.

Morgan, J., Cano, M.V., Feikin, D.R., Phelan, M., Monroy, O.V., Morales, P.K., Carpenter, J., Weltman, A., Spitzer, P.G., Liu, H.H., *et al.* (2003). A large outbreak of histoplasmosis among American travelers associated with a hotel in Acapulco, Mexico, spring 2001. Am. J. Trop. Med. Hyg. *69*, 663–669.

Nakamura, L.T., Wu-Hsieh, B.A., and Howard, D.H. (1994). Recombinant murine gamma interferon stimulates macrophages of the RAW cell line to inhibit intracellular growth of *Histoplasma capsulatum*. Infect. Immun. *62*, 680–684.

Nemecek, J.C., Wüthrich, M., and Klein, B.S. (2006). Global control of dimorphism and virulence infFungi. Science *312*, 583–588.

Netea, M.G., Van Der Graaf, C.A.A., Vonk, A.G., Verschueren, I., Van Der Meer, J.W.M., and Kullberg, B.J. (2002). The role of toll-like receptor (TLR) 2 and TLR4 in the host defense against disseminated candidiasis. J. Infect. Dis. *185*, 1483–1489.

Netea, M.G., Gow, N.A.R., Munro, C.A., Bates, S., Collins, C., Ferwerda, G., Hobson, R.P., Bertram, G., Hughes, H.B., Jansen, T., *et al.* (2006). Immune sensing of *Candida albicans* requires cooperative recognition of mannans and glucans by lectin and Toll-like receptors. J. Clin. Invest. *116*, 1642–1650.

Newman, S.L., Bucher, C., Rhodes, J., and Bullock, W.E. (1990). Phagocytosis of *Histoplasma capsulatum* yeasts and microconidia by human cultured macrophages and alveolar macrophages. Cellular cytoskeleton requirement for attachment and ingestion. J. Clin. Invest. *85*, 223–230.

Newman, S.L., Gootee, L., Brunner, G., and Deepe, G.S. (1994). Chloroquine induces human macrophage killing of *Histoplasma capsulatum* by limiting the availability of intracellular iron and is therapeutic in a murine model of histoplasmosis. J. Clin. Invest. *93*, 1422–1429.

Newman, S.L., Gootee, L., Stroobant, V., Goot, H. van der, and Boelaert, J.R. (1995). Inhibition of growth of *Histoplasma capsulatum* yeast cells in human macrophages by the iron chelator VUF 8514 and comparison of VUF 8514 with deferoxamine. Antimicrob. Agents Chemother. *39*, 1824–1829.

Nguyen, V.Q., and Sil, A. (2008). Temperature-induced switch to the pathogenic yeast form of *Histoplasma capsulatum* requires Ryp1, a conserved transcriptional regulator. PNAS *105*, 4880–4885.

Nittler, M.P., Hocking-Murray, D., Foo, C.K., and Sil, A. (2005). Identification of *Histoplasma capsulatum* transcripts induced in response to reactive nitrogen species. Mol. Biol. Cell *16*, 4792–4813.

Nosanchuk, J.D., Gómez, B.L., Youngchim, S., Díez, S., Aisen, P., Zancopé-Oliveira, R.M., Restrepo, A., Casadevall, A., and Hamilton, A.J. (2002). *Histoplasma capsulatum* synthesizes melanin-like pigments *in vitro* and during mammalian infection. Infect. Immun. *70*, 5124–5131.

O'Keefe, A., Frederick, J., and Harmon, B. (2012). Histoplasmosis outbreak among day camp attendees-Nebraska, June 2012. Morb. Mortal. Wkly. Rep. *61*, 747–748.

Patel, J.B., Batanghari, J.W., and Goldman, W.E. (1998). Probing the yeast phase-specific expression of the *CBP1* gene in *Histoplasma capsulatum*. J. Bacteriol. *180*, 1786–1792.

Rappleye, C.A., Engle, J.T., and Goldman, W.E. (2004). RNA interference in *Histoplasma capsulatum* demonstrates a role for α-(1,3)-glucan in virulence. Mol. Microbiol. *53*, 153–165.

Rappleye, C.A., Eissenberg, L.G., and Goldman, W.E. (2007). *Histoplasma capsulatum* α-(1,3)-glucan blocks innate immune recognition by the β-glucan receptor. PNAS *104*, 1366–1370.

Rhodes, J.C., Polacheck, I., and Kwon-Chung, K.J. (1982). Phenoloxidase activity and virulence in isogenic strains of *Cryptococcus neoformans*. Infect. Immun. *36*, 1175–1184.

Rippon, J. (1988). Histoplasmosis. In Medical Mycology: The Pathogenic Fungi and the Pathogenic Actinomycetes (W.B. Saunders Co., Philadelphia), pp. 381–423.

Safranek, T., and Beecham, B. (2004). Outbreak of histoplasmosis among industrial plant workers-Nebraska, 2004. Morb. Mortal. Wkly. Rep. 53, 1020–1022.

Schaffner, A., Davis, C.E., Schaffner, T., Markert, M., Douglas, H., and Braude, A.I. (1986). *In vitro* susceptibility of fungi to killing by neutrophil granulocytes discriminates between primary pathogenicity and opportunism. J. Clin.Invest. *78*, 511–524.

Schnur, R.A., and Newman, S.L. (1990). The respiratory burst response to *Histoplasma capsulatum* by human neutrophils. Evidence for intracellular trapping of superoxide anion. J. Immunol. *144*, 4765–4772.

Sebghati, T.S., Engle, J.T., and Goldman, W.E. (2000). Intracellular parasitism by *Histoplasma capsulatum*: fungal virulence and calcium dependence. Science *290*, 1368–1372.

Shearer, G., Jr, Birge, C.H., Yuckenberg, P.D., Kobayashi, G.S., and Medoff, G. (1987). Heat-shock proteins induced during the mycelial-to-yeast transitions of strains of *Histoplasma capsulatum*. J. Gen. Microbiol. *133*, 3375–3382.

Sorgi, C.A., Secatto, A., Fontanari, C., Turato, W.M., Belangér, C., Medeiros, A.I. de, Kashima, S., Marleau, S., Covas, D.T., Bozza, P.T., *et al.* (2009). *Histoplasma capsulatum* cell wall β-glucan induces lipid body formation through CD18, TLR2, and Dectin-1 receptors: correlation with leukotriene B_4 generation and role in HIV-1 infection. J. Immunol. *182*, 4025–4035.

Steele, C., Rapaka, R.R., Metz, A., Pop, S.M., Williams, D.L., Gordon, S., Kolls, J.K., and Brown, G.D. (2005). The β-glucan receptor dectin-1 recognizes specific morphologies of *Aspergillus fumigatus*. PLoS Pathog. *1*, e42.

Strasser, J.E., Newman, S.L., Ciraolo, G.M., Morris, R.E., Howell, M.L., and Dean, G.E. (1999). Regulation of the macrophage vacuolar ATPase and phagosome–lysosome fusion by *Histoplasma capsulatum*. J. Immunol. *162*, 6148–6154.

Sullivan, T.D., Rooney, P.J., and Klein, B.S. (2002). *Agrobacterium tumefaciens* integrates transfer DNA into single chromosomal sites of dimorphic fungi and yields homokaryotic progeny from multinucleate yeast. Eukaryot. Cell *1*, 895–905.

Taylor, P.R., Tsoni, S.V., Willment, J.A., Dennehy, K.M., Rosas, M., Findon, H., Haynes, K., Steele, C., Botto, M., Gordon, S., *et al.* (2007). Dectin-1 is required for β-glucan recognition and control of fungal infection. Nat. Immunol. *8*, 31–38.

Tewari, R.P., and Berkhout, F.J. (1972). Comparative pathogenicity of albino and brown types of *Histoplasma capsulatum* for mice. J. Infect. Dis. *125*, 504–508.

The Broad Institute (2009). *Histoplasma capsulatum* Database. Available online: www.broadinstitute.org/annotation/genome/histoplasma_capsulatum/MultiHome.html

The Genome Institute (2011). *Histoplasma capsulatum* Genome. Available online: http://genome.wustl.edu/genomes/detail/histoplasma-capsulatum/

Tian, X., and Shearer, G., Jr (2002). The mold-specific *MS8* gene is required for normal hypha formation in the dimorphic pathogenic fungus *Histoplasma capsulatum*. Eukaryot. Cell *1*, 249–256.

Timmerman, M.M., and Woods, J.P. (1999). Ferric reduction is a potential iron acquisition mechanism for *Histoplasma capsulatum*. Infect. Immun. *67*, 6403–6408.

Timmerman, M.M., and Woods, J.P. (2001). Potential role for extracellular glutathione-dependent ferric reductase in utilization of environmental and host ferric compounds by *Histoplasma capsulatum*. Infect. Immun. *69*, 7671–7678.

Van de Veerdonk, F.L., Marijnissen, R.J., Kullberg, B.J., Koenen, H.J.P.M., Cheng, S.-C., Joosten, I., Van den Berg, W.B., Williams, D.L., Van der Meer, J.W.M., Joosten, L.A.B., *et al.* (2009). The macrophage mannose receptor induces IL-17 in response to *Candida albicans*. Cell Host Microbe 5, 329–340.

Wang, Y., Aisen, P., and Casadevall, A. (1995). *Cryptococcus neoformans* melanin and virulence: mechanism of action. Infect. Immun. *63*, 3131–3136.

Warren, K., and Westman, A. (2008). Outbreak of histoplasmosis among travelers returning from El Salvador--Pennsylvania and Virginia, 2008. Morb. Mortal. Wkly. Rep. *57*, 1349–1353.

Webster, R.H., and Sil, A. (2008). Conserved factors Ryp2 and Ryp3 control cell morphology and infectious spore formation in the fungal pathogen *Histoplasma capsulatum*. PNAS *105*, 14573–14578.

Wells, C.A., Salvage-Jones, J.A., Li, X., Hitchens, K., Butcher, S., Murray, R.Z., Beckhouse, A.G., Lo, Y.-L.-S., Manzanero, S., Cobbold, C., *et al.* (2008). The macrophage-inducible C-type lectin, mincle, is an essential component of the innate immune response to *Candida albicans*. J. Immunol. *180*, 7404–7413.

Werner, J.L., Metz, A.E., Horn, D., Schoeb, T.R., Hewitt, M.M., Schwiebert, L.M., Faro-Trindade, I., Brown, G.D., and Steele, C. (2009). Requisite role for the dectin-1 β-glucan receptor in pulmonary defense against *Aspergillus fumigatus*. J. Immunol. *182*, 4938–4946.

Winters, M.S., Spellman, D.S., Chan, Q., Gomez, F.J., Hernandez, M., Catron, B., Smulian, A.G., Neubert, T.A., and Deepe, G.S. (2008). *Histoplasma capsulatum* proteome response to decreased iron availability. Proteome Science *6*, 36.

Wolf, J.E., and Massof, S.E. (1990). *In vivo* activation of macrophage oxidative burst activity by cytokines and amphotericin B. Infect. Immun. *58*, 1296–1300.

Wolf, J.E., Kerchberger, V., Kobayashi, G.S., and Little, J.R. (1987). Modulation of the macrophage oxidative burst by *Histoplasma capsulatum*. J. Immunol. *138*, 582–586.

Wolf, J.E., Abegg, A.L., Travis, S.J., Kobayashi, G.S., and Little, J.R. (1989). Effects of *Histoplasma capsulatum* on murine macrophage functions: inhibition of macrophage priming, oxidative burst, and antifungal activities. Infect. Immun. *57*, 513–519.

Woods, J.P., Retallack, D.M., Heinecke, E.L., and Goldman, W.E. (1998). Rare homologous gene targeting in *Histoplasma capsulatum*: disruption of the *URA5Hc* gene by allelic replacement. J. Bacteriol. *180*, 5135–5143.

Wood, K.L., Hage, C.A., Knox, K.S., Kleiman, M.B., Sannuti, A., Day, R.B., Wheat, L.J., and Twigg, H.L., 3rd (2003). Histoplasmosis after treatment with anti-tumor necrosis factor-α therapy. Am. J. Respir. Crit. Care Med. *167*, 1279–1282.

Wu-Hsieh, B.A., Lee, G.S., Franco, M., and Hofman, F.M. (1992). Early activation of splenic macrophages by tumor necrosis factor α is important in determining the outcome of experimental histoplasmosis in mice. Infect. Immun. *60*, 4230–4238.

Youseff, B.H., and Rappleye, C.A. (2012). RNAi-based gene silencing using a GFP sentinel system in *Histoplasma capsulatum*. In Host–Fungus Interactions, A.C. Brand, and D.M. MacCallum, eds. (Humana Press, New York), pp. 151–164.

Youseff, B.H., Dougherty, J.A., and Rappleye, C.A. (2009). Reverse genetics through random mutagenesis in *Histoplasma capsulatum*. BMC Microbiol. *9*, 236.

Youseff, B.H., Holbrook, E.D., Smolnycki, K.A., and Rappleye, C.A. (2012). Extracellular superoxide dismutase protects *Histoplasma* yeast cells from host-derived oxidative stress. PLoS Pathog *8*, e1002713.

Zancopé-Oliveira, R.M., Reiss, E., Lott, T.J., Mayer, L.W., and Deepe, G.S. (1999). Molecular cloning, characterization, and expression of the M antigen of *Histoplasma capsulatum*. Infect. Immun. *67*, 1947–1953.

Zarnowski, R., Cooper, K.G., Brunold, L.S., Calaycay, J., and Woods, J.P. (2008a). *Histoplasma capsulatum* secreted γ-glutamyltransferase reduces iron by generating an efficient ferric reductant. Mol. Microbiol. *70*, 352–368.

Zarnowski, R., Dobrzyn, A., Ntambi, J.M., and Woods, J.P. (2008b). Ferrous, but not ferric, iron maintains homeostasis in *Histoplasma capsulatum* triacylglycerides. Curr. Microbiol. 57, 153–157.

Zemska, O., and Rappleye, C.A. (2012). *Agrobacterium*-mediated insertional mutagenesis in *Histoplasma capsulatum*. In Host-Fungus Interactions, A.C. Brand, and D.M. MacCallum, eds. (Humana Press, New York), pp. 51–66.

Zhou, P., Sieve, M.C., Bennett, J., Kwon-Chung, K.J., Tewari, R.P., Gazzinelli, R.T., Sher, A., and Seder, R.A. (1995). IL-12 prevents mortality in mice infected with *Histoplasma capsulatum* through induction of IFN-γ. J. Immunol. *155*, 785–795.

Blastomyces dermatitidis and Blastomycosis

12

Gregory M. Gauthier

Abstract

The aetiological agent of blastomycosis, *Blastomyces dermatitidis,* belongs to a group of ascomycete fungi that exhibit thermal dimorphism. In the soil (22–25°C), these fungi grow as filamentous mould that produce infectious conidia. Following the disruption of soil by activities such as construction, aerosolized conidia and mould fragments are inhaled into the lungs of a host (37°C) and convert into pathogenic yeast. In the yeast form, *B. dermatitidis* is capable of evading immune defences by altering the composition of the fungal cell wall and impairing the host's cytokine response. Approximately 25–40% of patients infected with *B. dermatitidis* develop disseminated disease, which frequently involves the skin or bone. Diagnosis of blastomycosis requires a high degree of clinical suspicion coupled with the use of culture and non-culture diagnostics. Treatment requires the use of polyene or azole antifungals.

Historical perspective

Thomas C. Gilchrist first described blastomycosis in 1894 and characterized it as a 'protozoan dermatitidis' (Gilchrist, 1894). Two years later, a histological analysis of cutaneous blastomycosis was published and was referred to as 'blastomycetic dermatitis' (Gilchrist, 1896). In 1896 and 1898, a second case of blastomycetic dermatitis was described and the pathologic agent was named *Blastomyces dermatitidis* (Gilchrist and Stokes, 1896, 1898). Thermal dimorphism in *B. dermatitidis* was described by Walter W. Hamburger in 1907. He analysed four clinical isolates and described the growth of *B. dermatitidis* as mould at room temperature and budding yeast at 37°C (Hamburger, 1907). Increased awareness of blastomycosis resulted in the publication of several case reports and case series describing its clinical characteristics and potential for invasive disease (Montgomery and Ormsby, 1908; Stober, 1914). In the 1950s, Drs Schwarz and Baum established that the primary portal of entry for *B. dermatitidis* was the lung (Schwarz and Baum, 1951). In the mid-1960s, McDonough and Lewis described the sexual form of *B. dermatitidis* (McDonough and Lewis, 1967). *B. dermatitidis* was successfully isolated from the soil in association with a large outbreak in northern Wisconsin in the mid-1980s (Klein *et al.*, 1986). Since the 1990s, investigation of the pathogenesis and immunology of *B. dermatitidis* at the molecular level has led to several important discoveries including the identification of the essential virulence factor *Blastomyces* adhesin-1 (*BAD*-1; formerly WI-1), a dimorphism regulating kinase (*DRK1*) that is critical for transition to yeast at 37°C, and a GATA transcription factor that promotes the conversion to mould at 22°C (Brandhorst *et al.*, 1999; Nemecek *et al.*, 2006; Gauthier *et al.*, 2010). In 2009, the genome of *B. dermatitidis* (strain SLH14081) was sequenced and annotated by the Broad Institute (www.broadinstitute.org). Since then, additional *B. dermatitidis* strains have been sequenced and annotated. Deep RNA sequencing to characterize the transcriptome is ongoing.

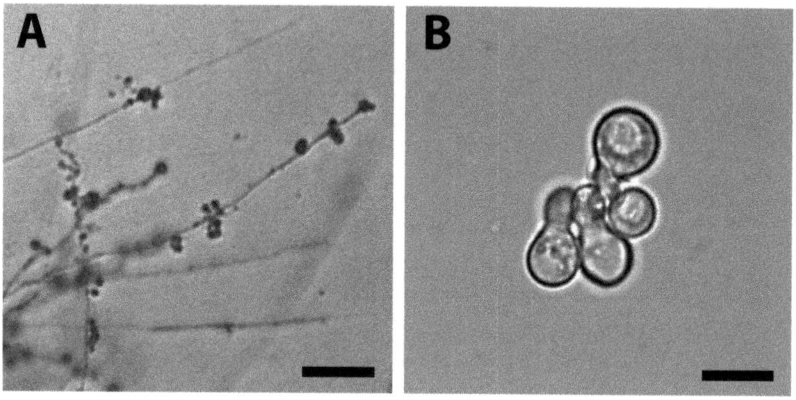

Figure 12.1 *Blastomyces dermatitidis* mould at 22°C (A) and yeast at 37°C (B). Scale bar is 10 μm.

Mycology

Blastomyces dermatitidis is a thermally dimorphic ascomycete, which grows as yeast at 37°C and mould at 22°C to 25°C (Fig. 12.1). The yeast form (8–20 μm) grows in mammalian tissue and is characterized by a broad-based bud (4–5 μm) and doubly retractile cell wall (Wolf *et al.*, 1975). Although the morphologic appearance is unique among the dimorphic fungi, there is potential for giant yeast forms (28–40 μm) to be confused with *Coccidioides* (Wu *et al.*, 2005; Walker *et al.*, 2002). At environmental temperature, *B. dermatitidis* grows as septate hyphae (1–2 μm diameter) that are capable of producing asexual spores known as conidia (Winn *et al.*, 2006; Wolf *et al.*, 1975). Conidia (4–5 μm) are attached to hyphae by short conidiophores (Winn *et al.*,2006; Wolf *et al.*, 1975). Macroscopically, yeast colonies are off-white in pigmentation and individual colonies can have a smooth or rough appearance. Mould colonies initially appear off-white or tan (Wolf *et al.*, 1975). In contrast to yeast, the morphologic features of *B. dermatitidis* mould are not pathognomonic and definitive identification requires molecular confirmation or transition to the yeast form at 37°C. The conversion between mould and yeast is reversible and can be induced in the laboratory by shifting the temperature between 22°C–25°C and 37°C.

The genomes of clinical and environmental *B. dermatitidis* isolates have been sequenced, annotated, and made publicly available (www.broadinstitute.org/annotation/genome/blastomyces_dermatitidis/MultiHome.html). The *B. dermatitidis* genome is 66–75 megabases (Mb) in size and contains 9522–10,104 genes. Approximately 17–19% of the genome comprises coding sequences.

Population analysis using PCR restriction fragment length polymorphism (RFLP) and random amplified polymorphic DNA (RAPD) suggests that genetic heterogeneity exists among different strains of *B. dermatitidis* (McCullough *et al.*, 2000; Yates-Siilata *et al.*, 1995, Meece *et al.*, 2010). Analysis of 27 genes by multilocus microsatellite typing divided 112 *B. dermatitidis* isolates (87 human, 16 veterinarian, 8 environmental, ATCC 26199) into two distinct genetic groups (Meece *et al.*, 2011). Group 1 isolates were relatively homogeneous, whereas group 2 displayed a high degree of allelic diversity (Meece *et al.*, 2011). A separate analysis of 227 human clinical *B. dermatitidis* isolates also clustered into two groups – group 1 isolates were more likely to cause pneumonia, whereas group 2 isolates were associated with disseminated disease. Moreover, group 1 isolates were also linked to outbreaks of blastomycosis (Meece *et al.*, 2013). Analysis of seven genes (*CHS2, DRK1, FADS, PYRF, TUB1, ARF6,* and *ITS2*) in 78 *B. dermatitidis* isolates (67 patient, 2 canine, 9 environmental) by genealogical concordance phylogenetic species recognition (GCPSR) method divided *Blastomyces* into two distinct phylogenetic species (Brown *et al.*, 2013). On the basis of this analysis, Brown and colleagues have proposed a novel species, *Blastomyces gilchristii* sp. nov. (Brown *et al.*, 2013). Both phylogenetic species are found in Wisconsin, Minnesota, and northwestern Ontario; however, unlike *B. dermatitidis*, *B. gilchristii* was not identified in central and Southern Ontario. Analysis of more *B. dermatitidis* strains from a broader geographic distribution is needed to ascertain the degree of geographic overlap or separation for the two phylogenetic species.

Sexual reproduction and mating (*MAT*) loci

B. dermatitidis is considered the asexual or anamorph form, whereas *Ajellomyces dermatitidis* is the sexual or teleomorph form (McDonough and Lewis, 1967; McDonough and Lewis, 1968; Kwon-Chung, 1971; Li *et al.*, 2013). In the laboratory, pairing of opposite mating types (+ and –) results in cleistothecia formation after 3–4 weeks of incubation at 22°C (Fig. 12.2) (Li *et al.*, 2013). Cleistothecia are fruiting bodies derived from mating hyphae and are characterized by the presence of spiral hyphae and asci (McDonough and Lewis, 1967, 1968; Kwon-Chung, 1973; Glick and Kwon-Chung, 1973; Li *et al.*, 2013). Each ascus contains eight ascospores, which can germinate as hyphae at 25°C or convert to virulent yeast when inoculated into mice (37°C) (McDonough and Lewis, 1967, 1968; Li *et al.*, 2013). Only heterothallic mating (pairing of + and –) has been demonstrated for *A. dermatitidis* (Li *et al.*, 2013). Pairing of the same mating types (i.e. homothallic mating) has not resulted in the development of fertile cleistothecia (Kwon-Chung, 1971; Li *et al.*, 2013). Serial passage of cultures in the laboratory can result in the loss of the ability to mate (McDonough *et al.*, 1973).

Similar to *H. capsulatum*, *C. immitis*, and *C. posadasii*, the genome of *B. dermatitidis* contains one of the two mating type (*MAT*) loci, *MAT1-1* or *MAT1-2* (Fraser *et al.*, 2007; Li *et al.*, 2013). None of the sequenced genomes has both mating type loci, which is observed with homothallic fungi such as *A. nidulans* (Fraser *et al.*, 2007; Li *et al.*, 2013, Li *et al.*, 2010). For the 4 sequenced *B. dermatitidis* strains, one contains the *MAT1-1* locus (+, strain ATCC18188), and three possess the *MAT1-2* locus (–, strains ATCC26199, SLH14081, ER-3) (Li *et al.*,2013). *MAT1-1* is 5 kilobases (kb) and is characterized by a gene that encodes an α-box domain transcription factor. *MAT1-2* is 15–38 kb in size and is characterized by a high-mobility group (*HMG*) gene, which also encodes a transcription factor. Size variation in *MAT1-2* loci is due to the presence of retrotransposons (Li *et al.*, 2013). Architectural arrangement of the *MAT* locus is conserved and consists of *SLA2* upstream of the α-box or HMG domain gene, and *APN2* and *COX13* downstream (Li *et al.*, 2010, 2013). *SLA2* is involved with cytoskeletal assembly, *APN2* encodes a DNA lyase, and *COX13* encodes a cytochrome C oxidase subunit VIa. Except for affecting the size of the *MAT* locus, the influence of retrotransposons on mating is unknown. In *H. capsulatum* and *P. brasiliensis*, transcription of *MAT1-1* and *MAT2-2* loci can be induced by in the presence of α-type

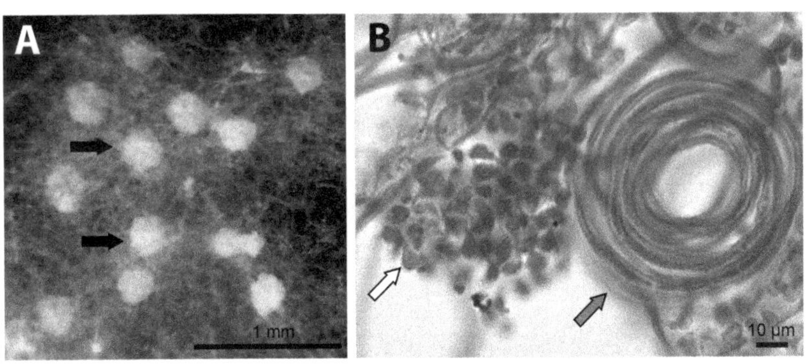

Figure 12.2 Mating structures of *Ajellomyces dermatitidis*. (A) Cleistothecia (black arrows). (B) Spiral hyphae (grey arrow) and asci (white arrow) in a crush preparation of a cleistothecium. Photographs are courtesy of Drs. Bruce S. Klein and Thomas D. Sullivan.

and a-type pheromones, respectively (Bubnick and Smulian, 2007; Teixeira *et al.*, 2013). Although not yet experimentally confirmed, *B. dermatitidis* mating is also likely to be induced by pheromones.

Environmental and human isolates of *B. dermatitidis* have a near equal distribution of + and – mating types (McDonough *et al.*, 1973; Meece *et al.*, 2011). In the early 1970s, McDonough and colleagues characterized the mating type for 129 *B. dermatitidis* clinical isolates from North America (21 American states and three Canadian provinces) and demonstrated a 1:1 distribution of + and – mating types (McDonough *et al.*, 1973). Moreover, patients can be simultaneously infected with both mating types (McDonough *et al.*, 1977). In 2011, Meece and colleagues analysed human, veterinary, and environmental isolates, which were primarily from Wisconsin. For the human isolates, there was a near equal distribution of + and – mating types (Meece *et al.*, 2011). The majority of veterinary and environmental isolates were of the – mating type; however, sample sizes were too small to be definitive (Meece *et al.*, 2011). A 1:1 ratio of mating types has been described for other dimorphic fungi including *Coccidioides* species (human and soil isolates) and *P. brasiliensis* (human and soil isolates) (Mandel *et al.*, 2007; Torres *et al.*, 2010). Despite a 1:1 mating type ratio for *H. capsulatum* soil isolates, the ratio for clinical isolates ranges from 1:2.5 to 1:7 (Kwon-Chung *et al.*, 1974; Gaur *et al.*, 1980; Kwon-Chung *et al.*, 1984; Kwon-Chung *et al.*, 1973). This discrepancy may be related to impaired conversion of mould to yeast for + mating type strains, which was observed by Kwon-Chung, but not by Gaur and colleagues (Kwon-Chung, 1973; Guar *et al.*, 1980).

Geographic distribution, ecology, and epidemiology of *Blastomyces dermatitidis*

Blastomyces dermatitidis is endemic to North America and is restricted to the Midwest, south-central, and southeastern regions of the United States and several Canadian provinces including Saskatchewan, Manitoba, Ontario, and Quebec (Fig. 12.3). *B. dermatitidis* is not uniformly distributed in the endemic region, rather it grows in an ecologic niche characterized by forested, sandy soils with acidic pH that are located nearby water and contain decaying vegetation (Klein *et al.*, 1986). *B. dermatitidis* can also grow in bird and animal excreta. Knowledge about the ecology of *B. dermatitidis* has been primarily gleaned from outbreaks. Human activities that result in soil disruption, such as construction of homes or roads, boating and canoeing, fishing, exploration of beaver dams and underground forts, and use of a community compost pile, have all been associated with increased risk of *B. dermatitidis*

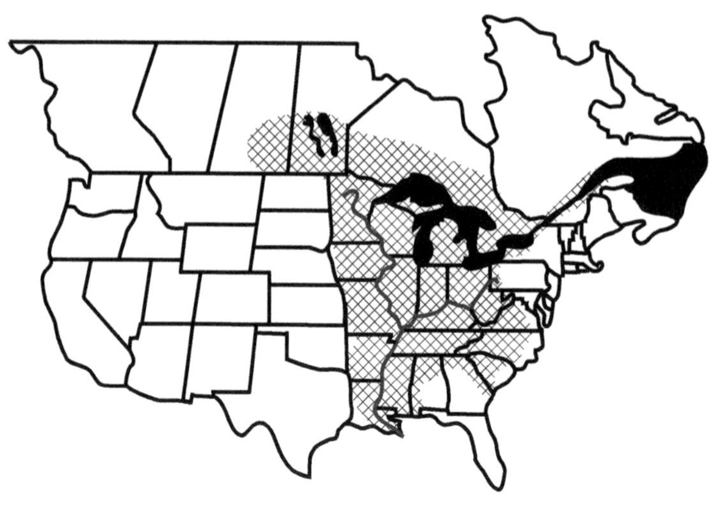

Figure 12.3 Geographic distribution of *B. dermatitidis* in North America. The endemic area is cross-hatched.

infection. Although the majority of cases of blastomycosis are sporadic, 15 well-defined outbreaks have been reported in North America (Smith *et al.*, 1955; Tosh F.E. *et al.*, 1974; Centers for Disease Control and Prevention, 1976; Kitchen *et al.*, 1977; Cockerill *et al.*, 1984; Klein *et al.*, 1986, 1987a; Armstrong *et al.*, 1987; Baumgardner *et al.*, 1991, 2002; Frye and Seifer, 1991; De Groote *et al.*, 2000; Proctor *et al.*, 2002; MacDonald *et al.*, 2006; Pfister *et al.*, 2011).

The epidemiology of blastomycosis is based on retrospective studies and passive surveillance. The annual incidence in the endemic region ranges from 0.2 to 1.94 cases per 100,000 persons (Carlos *et al.*, 2010; Hussein *et al.*, 2009; Morris *et al.*, 2006; Cano *et al.*, 2003; Crampton *et al.*, 2002; Centers for Disease Control and Prevention, 1996; Manetti *et al.*, 1991; Klein and Davis, 1985; Furcolow *et al.*, 1970). Moreover, certain geographic areas are hyperendemic: Kenora, Ontario (117.2 human cases/100,000 population), Vilas County, WI (40.4/100,000), Eagle River, WI (101.3/100,000), Washington Parish, LA (6.8/100,000), and central and south-central Mississippi (> 5/100,000) (Baumgardner *et al.*, 1992; Dwight *et al.*, 2000; Lowry *et al.*, 1989, Chapman *et al.*, 1997). These epidemiological data probably underestimate the true incidence of blastomycosis because reliable serological or skin tests are not available. In addition, > 50% of infected persons develop subclinical or asymptomatic illness (Klein *et al.*, 1986). Therefore, epidemiological data are limited to clinically apparent infections that are diagnosed and reported.

Outside North America, autochthonous cases of culture-proven blastomycosis have been reported from Africa and India (Baily *et al.*, 1991; Shukla *et al.*, 2009). *B. dermatitidis* is not endemic to Central America, South America, Europe, Asia or Australia. Nearly 100 human cases of blastomycosis have been described in African countries (Baily *et al.*,1991; Carman *et al.*, 1989; Frean *et al.*, 1989; McCullough *et al.*, 2000; Shukla *et al.*, 2009). In India, fewer than 12 clinical cases have been described. The ecological niche in Africa or India has not been characterized and *B. dermatitidis* has never been isolated from soil outside North America. Moreover, there have been no human or veterinary outbreaks to define the geographic distribution of *B. dermatitidis* in Africa or India. The clinical presentation often mimics other pulmonary and cutaneous infections, which makes diagnosis difficult. Limitations in public health infrastructure and diagnostic capabilities also impede knowledge about the epidemiology of blastomycosis in Africa and India.

When compared to North American strains, African *B. dermatitidis* isolates display a decreased ability to convert to yeast after an increase in temperature from 25°C to 37°C, unique media requirements for maintenance of yeast growth *in vitro*, and a less antigenically complex cell surface (Klein *et al.*, 1997). Moreover, the essential virulence factor, *Blastomyces* adhesin-1 (BAD1; formerly WI-1) is absent from the genome and cell surface for serotype 2 African strains of *B. dermatitidis* (Klein *et al.*, 1997). In contrast, BAD1 is widely distributed in serotype 1 African strains as well as North American strains (Klein *et al.*,1997).

Pathogenesis

Overview

Blastomyces dermatitidis is considered a primary fungal pathogen because it causes disease in persons with intact or impaired immune defences. Features of this pathogen that promote its growth in human tissue include its ability to survive core human body temperature (37°C), convert to budding yeast, express yeast-phase specific virulence factors, and evade killing by host immune cells.

The phase transition

The conversion between mould and yeast, which is known as the phase transition, is essential for the pathogenesis of *B. dermatitidis*. In the environment (22–25°C), *B. dermatitidis* grows as a filamentous mould that produces infectious conidia. Following disruption of soil, aerosolized conidia and mycelial fragments inhaled into the lungs of a mammalian host (37°C) convert to pathogenic yeast and cause infection (Gauthier and Klein, 2008). The transition between mould and yeast is reversible and can be induced in the laboratory by shifting the temperature between 22–25°C and 37°C, respectively. Similar to other dimorphic fungi such as *H. capsulatum* or *P. brasiliensis*, temperature is the predominant environmental stimulus that promotes the phase transition in *B. dermatitidis* (Gauthier *et al.*, 2010; Nemecek *et al.*, 2006). Other stimuli such as carbon dioxide and steroid hormones (oestradiol, progesterone) have minimal impact on *B. dermatitidis*. Carbon dioxide tension (5% CO_2) is important for the conversion to and maintenance of yeast morphology for *Coccidioides* spp. and *H. capsulatum*, but is dispensable for *B. dermatitidis* (Pine, 1954; Klotz *et al.*, 1984). Progesterone and 17β-oestradiol can bind to *Coccidioides* species and accelerate the growth of arthroconidia, spherules, and endospores (Drutz *et al.*, 1981; Powell *et al.*, 1983). In contrast, binding of progesterone and 17β-oestradiol *B. dermatitidis* is minimal (Powell *et al.*, 1983).

The phase transition is a complex event that involves global changes in metabolism, cell wall composition and plasma membrane lipid content (Gauthier and Klein, 2008). During the conversion from mould to yeast, *B. dermatitidis* goes through three metabolic stages (Medoff *et al.*, 1987). Stage I consists of partial uncoupling of oxidative phosphorylation, a decrease in adenosine triphosphate (ATP) levels, a progressive decline in respiration, and a reduction in mitochondrial cytochrome components (cytochrome b, aa3, c). In stage II, respiration stops. To bypass the blocked cytochrome system, exogenous cysteine is funnelled through a sulfhydryl shunt pathway to provide energy. This requires the uptake of exogenous cysteine. In the absence of sulfhydryl compounds, substrates such as succinate, pyruvate, and α-ketoglutarate cannot be used for respiration. In stage III, respiration resumes, ATP and cytochrome concentrations increase, and the cell converts to yeast morphology.

The cell wall is a dynamic structure that undergoes remodelling during the shift between mould and yeast forms. The mycelial cell wall comprises glucose (4–15%), mannose (3–6%), chitin (25–30%), and glucan (60% α-(1,3)-glucan; 40% β-(1,3)-glucan) (Kanetsuna and Carbonell, 1971; Domer, 1971). The yeast cell wall contains glucose (17–19%), mannose (1%), chitin (32%), and glucan (95% α-(1,3)-glucan; 5% β-(1,3)-glucan) (Kanetsuna and Carbonell, 1971; Domer, 1971). In addition, the mycelial cell wall contains higher concentrations of amino acids and sulfhydryl compounds than the yeast cell wall (Roy and Landau, 1972a,b). The changes in cell wall glucan composition associated with the phase transition may contribute to virulence and immune evasion. *B. dermatitidis* yeast lacking α-(1,3)-glucan have reduced virulence in a murine model of pulmonary infection (Hogan and Klein, 1994). Similarly, reduction or complete loss of α-(1,3)-glucan synthase activity renders *H. capsulatum* avirulent (Rappleye *et al.*, 2004). The decrease in β-(1,3)-glucan from 60% to 5% may limit recognition of this carbohydrate on yeast cells by dectin-1 receptors on innate immune cells and mannose-binding lectin (MBL-A, MbL-C) (Gauthier and Klein, 2008; Koneti *et al.*, 2008). Moreover, the limited amount of β-(1,3)-glucan in the yeast cell wall precludes the use of (1,3)-β-D-glucan test for diagnosis and echinocandins for treatment of blastomycosis (Chapman *et al.*, 2008; Girouard *et al.*, 2007).

Blastomyces dermatitidis biosynthesizes a variety of lipids including triglycerides, digylcerides, sterols, free fatty acids, and phospholipids (Domer and Hamilton, 1971). Yeast cells typically contain 5–10% lipid; however, some strains can have higher lipid content (Peck and Hauser, 1938; Di Salvo and Denton, 1963; Kanetsuna *et al.*, 1969; Al-Doory and Larsh, 1962). The lipid content in mycelia is ≥2-fold higher compared to yeast in some, but not all, strains of *B. dermatitidis* (Di Salvo and Denton, 1963; Kanetsuna *et al.*, 1969). *B. dermatitidis* strains Ga-1, KL-1, and BD64 contain 8.5%, 15.8%, and 5.5% lipid in yeast cells, and 20%, 44.8%, and 8.9–9.6% lipid for mycelia, respectively (Di Salvo and Denton, 1963; Kanetsuna *et al.*, 1969). In contrast, no substantial differences in lipid content between mould and yeast were detected for Found and McCarty strains (Domer and Hamilton, 1971). Although the impact of lipids on the morphologic switch in *B. dermatitidis* is unknown, evidence in *H. capsulatum* indicates that lipid content may influence the phase transition. Exposure of *H. capsulatum* mycelia to exogenous unsaturated fatty acids (oleic acid) prolonged the transition to yeast following an increase in temperature to 37°C. In contrast, the addition of saturated fatty acid (palmitic acid) accelerated the transition from mould to yeast (Maresca and Kobayashi, 1993).

Melanin pigments are found on the cell surface of *B. dermatitidis* yeast and conidia, but not mycelia (Nosanchuk *et al.*, 2004). Melanins are negatively charged, hydrophobic compounds that are biosynthesized by most human pathogenic fungi (Taborda *et al.*, 2008). *B. dermatitidis* yeast produce melanin pigments when grown on medium containing phenolic compounds such as L-DOPA (L-3,4-dihydroxyphenylalanine) or *in vivo* during infection (Nosanchuk *et al.*, 2004). In contrast to yeast, *B. dermatitidis* conidia can synthesize melanin in the absence of L-DOPA. Similar findings have been described for *H. capsulatum*, *P. brasiliensis*, *S. schenkii*, and *P. marneffei* yeast cells and conidia (Gomez *et al.*, 2001; Nosanchuk *et al.*, 2004; Morris-Jones *et al.*, 2003; Youngchim *et al.*, 2005). Collectively, these data suggest *B. dermatitidis* and other dimorphic fungi synthesize L-DOPA melanin (or eumelanin) as yeast and 1,8-dihydroxynapthalene melanin as conidia (Taborda *et al.*, 2008). The production of melanin may contribute towards the pathogenesis of dimorphic fungi by increasing the resistance to phagocytosis, impairing lectin binding to the fungal cell wall, and impairing macrophage killing (da Silva *et al.*, 2006). In addition, melanization reduces the rate at which *B. dermatitidis, H. capsulatum, and P. brasiliensis* are killed by amphotericin B deoxycholate (Nosanchuk *et al.*, 2004; van Duin *et al.*, 2002; da Silva *et al.*, 2006).

Molecular regulation of the phase transition and morphology

The development of molecular tools such as *Agrobacterium tumefaciens*-mediated DNA transfer has enabled the discovery of genes that regulate the phase transition including *DRK1* (dimorphism regulating kinase-1) and *SREB* (siderophore biosynthesis repressor in *Blastomyces*). *DRK1* and *SREB* were identified by a forward genetic approach that involved screening T-DNA insertional mutants for specific defects such as reduction of *BAD1* transcription or aberrant morphology, respectively. *A. tumefaciens*-mediated transformation of *B. dermatitidis* results in random insertion of the T-DNA into chromosomal DNA and homokaryotic progeny (Sullivan *et al.*, 2002). The majority of transformants (>80%) contain a single site of T-DNA insertion and episomal plasmids are not formed (Sullivan *et al.*, 2002). In addition to insertional mutagenesis, *A. tumefaciens*-mediated DNA transfer can be used to knock-out genes by allelic replacement and reduce transcript abundance by RNA interference (Nemecek *et al.*, 2006; Gauthier *et al.*,2010; Marty and Gauthier, 2013). Gene deletion in *B. dermatitidis* is inefficient with a frequency ≤ 1–2% and as low as 0.04% (Nemecek *et al.*, 2006; Brandhorst *et al.*, 1999; Gauthier *et al.*, 2010). This inefficiency plus the lack of sequenced genome hindered a reverse genetic approach to study pathogenesis until the mid-2000s. The development of a green fluorescent protein (GFP) sentinel RNA interference (RNAi) system and a sequenced, annotated genome has facilitated the characterization of *BYS1* (*Blastomyces* yeast-phase-specific 1) and septin-encoding genes (Krajaejun *et al.*, 2007, 2010; Marty and Gauthier, 2013). In this system, a *B. dermatitidis* strain that has been engineered to express GFP is transformed with a silencing plasmid using *A. tumefaciens*-mediated DNA transfer. The silencing plasmid produces a hairpin with nucleotide sequences that correspond to the gene of interest and GFP. Thus, candidate gene and *GFP* are simultaneously silenced. The degree of gene silencing correlates with the decrease in fluorescent intensity (Krajaejun *et al.*, 2007).

B. dermatitidis DRK1 encodes a hybrid histidine kinase that regulates the temperature-dependent conversion from mould to yeast. *DRK1* is a homologue of *Saccharomyces cerevisiae SLN1*, which functions as a response regulator in the high-osmolarity glycerol (HOG) pathway (Winkler *et al.*, 2002). The HOG pathway facilitates adaptation of *S. cerevisiae* to hyperosmotic conditions and heat shock (Winkler *et al.*, 2002). The protein encoded by *B. dermatitidis DRK1* is predicted to contain histidine kinase A and receiver domains important for phosphotransfer, a HATPase_c domain that binds ATP, and 6 HAMP domains (Nemecek *et al.*, 2006; Li *et al.*, 2010). Heterologous expression of *B. dermatitidis* DRK1 in a *S. cerevisiae SLN1* null mutant complemented the growth defect of the mutant strain (Nemecek *et al.*, 2006). Moreover, *B. dermatitidis DRK1* had similar levels of histidine kinase activity as *S. cerevisiae SLN1* (Nemecek *et al.*, 2006). Targeted deletion of *DRK1* in *B. dermatitidis* strain SLH14081 resulted in hyphal growth at 22°C and 37°C, increased sensitivity to calcofluor and Congo red, and reduced transcription of genes encoding α-(1,3)-glucan (*AGS1*) and *Blastomyces adhesin-1* (*BAD1*). In addition, deletion of *DRK1* abolished the production of conidia at 22°C (Nemecek *et al.*, 2006). Retransformation of *DRK1Δ* with an intact *DRK1* gene copy complemented these defects. The failure of *DRK1Δ* to produce conidia and hyphal growth at 37°C precluded virulence studies with the null mutant. To overcome this, *DRK1* transcription was silenced in strains SLH14081 and 60636 using RNAi. *DRK1*-silenced cells grew as pseudohyphae at 37°C and sporulated weakly at 22°C. Mice intratracheally infected with 1 × 10^4 wild-type conidia died within 4 weeks of infection, whereas those infected with *DRK1*-silenced conidia survived (Nemecek *et al.*, 2006).

Homologues of *DRK1* have been identified in other dimorphic fungi including *S. schenckii*, *H. capsulatum*, and *C. immitis* (Nemecek *et al.*, 2006; Li *et al.*, 2010; Hou *et al.*, 2013). In *S. schenckii*, *DRK1* had a 24-fold higher transcript abundance in yeast than mould (Hou *et al.*, 2013). Silencing *DRK1* transcription in *H. capsulatum* strains 186*ura5*AR and HcKD resulted in pseudohyphal growth at 37°C, reduction in *CBP1* (calcium-binding protein 1) and *AGS1* transcript, increased sensitivity to calcofluor, reduced production of conidia, and loss of brown pigmentation of mycelia and conidia (Nemecek *et al.*, 2006). In *H. capsulatum*, *CBP1* encodes an essential virulence factor that promotes survival within macrophages (Sebghati *et al.*, 2000). *AGS1* encodes an enzyme critical for biosynthesis of cell wall α-(1,3)-glucan, which is involved with shielding β-(1,3)-glucan from recognition by dectin-1 receptors (Rappleye *et al.*, 2004). Mice intratracheally infected with 1 × 10^8 wild-type *H. capsulatum* conidia died within 16 days of infection, whereas those infected with *DRK1*-silenced conidia survived (Nemecek *et al.*, 2006). Collectively, these *in vivo* studies using *B. dermatitidis* and *H. capsulatum DRK1*-silenced strains offered the first genetic proof that conversion of mould to yeast was required for virulence.

SREB encodes a GATA transcription factor that promotes the conversion from yeast to mould and regulates genes involved with the biosynthesis and uptake of siderophores (Gauthier *et al.*, 2010). SREB is predicted to contain two zinc finger domains (Cys-X_2-CysX_{17}-Cys-X_2-Cys) separated by a cysteine-rich region and a C-terminus with a coiled-coil domain (Gauthier *et al.*, 2010). The cysteine-rich region contains 4 cysteine residues that bind iron (Chao *et al.*, 2008). Deletion of *SREB* by homologous recombination resulted in cells that failed to complete the conversion from yeast to mould following a drop in temperature from 37°C to 22°C. Moreover, *SREB* null mutants (*SREBΔ*) did not accumulate substantial biomass or expand by radial growth at 22°C when compared to the wild-type isolate. The defects in growth and morphology were present only at 22°C and were independent of exogenous iron concentrations. *SREBΔ* was viable at 22°C and successfully converted to yeast when temperature was increased to 37°C (Gauthier *et al.*, 2010).

In addition to affecting the phase transition, *SREB* impacts iron homeostasis by inhibiting the production and uptake of siderophores under iron-replete conditions (Gauthier *et al.*, 2010). Exogenous iron concentrations influence *SREB* transcript abundance; it is increased under iron-rich conditions and decreased under iron-poor conditions. *H. capsulatum SRE1* and *A. nidulans SREA*, which are homologues of *SREB*, are also regulated at the transcription level (Chao *et al.*, 2008; Haas *et al.*, 1999). In contrast, *SREB* homologues in *Neurospora crassa*, *Ustilago maydis*, *Schizosaccharomyces pombe*, *Candida albicans* and *Cryptococcus neoformans* are constitutively expressed (Zhou *et al.*, 1998; Voisard *et al.*, 1993; Pelletier *et al.*, 2002; Lan *et al.*, 2004; Jung *et al.*, 2006). Under iron-replete conditions (10 μM $FeSO_4$), *SREBΔ* failed to properly repress genes involved with the biosynthesis and uptake of siderophores (dimerum acid and coprogen). This, in turn, resulted in yellow pigmentation of *SREBΔ* colonies and discoloration of the surrounding media (Gauthier *et al.*, 2010).

Homologues of *SREB* in *H. capsulatum* (*SRE1*) and *Cryptococcus neoformans* (*CIR1*) also regulate biologically disparate processes. Silencing *H. capsulatum SRE1* by RNAi resulted in disordered iron homeostasis under iron-replete conditions and impaired conversion to mould at room temperature (Hwang *et al.*, 2012). *C. neoformans CIR1* is unique among GATA transcription factors that regulate iron homeostasis because it contains a single zinc finger (Jung *et al.*, 2006). Gene expression microarray analysis demonstrated that *CIR1* regulates genes involved with melanin biosynthesis, ferric reduction, siderophore uptake, capsule formation, cell wall integrity, ergosterol biosynthesis, thermal tolerance, and mating (Jung *et al.*, 2006; Jung and Kronstad, 2011). In contrast to *SREBΔ*, *CIR1Δ* mutants display a growth defect at 37°C and are avirulent in a murine model of infection (Jung *et al.*, 2006).

Blastomyces yeast phase-specific 1 (*BYS1*) was identified while screening a cDNA library for genes with high transcript abundance in yeast but not mould (Burg and Smith, 1994). Following a drop in temperature from 37°C and 25°C, *BYS1* transcript abundance rapidly declined. Conversely, *BYS1* transcription increased within 12 hours of shifting temperature from 25°C to 37°C (Burg and Smith, 1994). Knocking down *BYS1* transcript by RNAi affected yeast morphology *in vitro* and *in vivo* (Krajaejun *et al.*, 2010). During prolonged incubation at 37°C on 7H10 agar medium, *BYS1*-silenced strains developed a rough colony appearance and grew as pseudohyphae rather than yeast. In experimental pulmonary infection, a subpopulation of cells silenced for *BYS1* grew pseudohyphal when compared to wild-type (3.2% versus < 0.6%). Silencing *BYS1* did not alter virulence, affect α-(1,3)-glucan composition of the cell wall, production of BAD1, or increase sensitivity to calcofluor or Congo red (Krajaejun *et al.*, 2010).

Septin-encoding genes in *B. dermatitidis* affect the morphology of yeast and mould, but are dispensable for the phase transition (Krajaejun *et al.*, 2007; Marty and Gauthier, 2013). Septins are GTP-binding proteins that form complexes at the sites of bud formation and hyphal tips in yeast and mould, respectively (Longtine and Bi, 2003; Juvvadi *et al.*, 2011; Krajaejun *et al.*, 2007). Septin complexes serve as a scaffold for proteins involved with cell polarity and cytokinesis (Longtine and Bi, 2003). They can also create a barrier during cytokinesis to prevent back diffusion of proteins from the daughter to the mother cell (Longtine and Bi, 2003). Five septin genes have been identified in the *B. dermatitidis* genome and include *CDC3*, *CDC10*, *CDC11*, *CDC12*, and *ASPE*. *B. dermatitidis* lacks genes for accessory (*SHS1/SEP7*) or sporulation-regulated (*SPR3*, *SPR8*) septins (Marty and Gauthier, 2013). Targeted knockdown of *CDC3*, *CDC10*, *CDC11*, and *CDC12* using a GFP sentinel RNAi system resulted in aberrant yeast morphology at 37°C when compared to controls (Krajaejun *et al.*, 2007; Marty and Gauthier, 2013). The bean-like, large ovoid, and elongated morphologies exhibited by these knock-down strains suggested a defect in budding. In addition, the growth rates of these silenced strains were sharply reduced when compared to controls, which was indicative of defective cytokinesis. Following a drop in temperature to 22°C, *CDC3*, *CDC10*, *CDC11*, and *CDC12* knockdown strains converted to mould characterized by curved, knobby-appearing hyphae. These strains reverted to aberrant yeast when the temperature was increased from 22°C to 37°C. Silencing of *ASPE* did not affect morphology of yeast or mould (Marty and Gauthier, 2013).

Blastomyces adhesin-1

Blastomyces adhesin-1 (BAD1; formerly WI-1) was discovered in the late 1980s in an effort to identify reliable immune markers to facilitate the serological diagnosis of blastomycosis (Klein and Jones, 1990). It is a 120 kDa molecule that is secreted into the extracellular milieu and binds back to the yeast cell surface (Brandhorst and Klein, 2000). An individual yeast cell is estimated to have 4.7×10^6 BAD1 molecules on the cell surface (Klein and Jones, 1990). During the course of experimental pulmonary infection, the concentration of BAD1 in alveolar fluid increases from 5 ng/ml to 42 ng/ml between days 2 and 14, respectively (Finkel-Jimenez *et al.*,2002). BAD1 is distinct from other known cell wall antigens including A antigen and *Blastomyces* alkali- and water-soluble antigen (BAWS) (Klein and Jones, 1990). Analysis of the BAD1 amino acid composition shows BAD1 to be rich in tryptophan, aspartic acid, lysine, tyrosine, and cysteine (Hogan *et al.*, 1995). Persons infected with *B. dermatitidis* develop BAD-1 specific antibody and T-cell responses to BAD1 (Klein *et al.*, 1992, 1993).

The transcription and translation of *BAD1* is phase regulated (Rooney *et al.*, 2001). Cell surface immunofluorescence staining, Western blot analysis of cell surface protein extracts and culture supernatants, and Northern blot hybridization demonstrate BAD1 protein and transcript is present in cells growing as yeast, but not as mycelia (Rooney *et al.*, 2001). BAD1 is also not detectable on the surface of conidia by immunofluorescence staining (Rooney *et al.*, 2001). Within 6 hours of a shift in temperature from 25°C to 37°C, BAD1 becomes detectable on the cell surface and during the morphologic shift localizes to yeast cells that develop and bud from hyphae (Rooney *et al.*, 2001). The promoter region of BAD1 is critical for its transcriptional regulation in the yeast phase. Several regions in the BAD1 promoter including box A (−175 to −237), box B (−325 to −360), and TCC motif (−20) have homology to the promoter of *YPS3*, which encodes a yeast-phase specific protein in *H. capsulatum* (Rooney *et al.*, 2001; Rooney and Klein, 2004). Functional mapping of the BAD1 promoter using truncated constructs fused to a *LacZ* reporter (P_{BAD1}-*LacZ*) demonstrated that box A is necessary for transcriptional activation of BAD1 in the yeast phase (Rooney and Klein, 2004). In contrast, box B was not required to activate transcription. Putative transcription factor binding sites within box A include CRE and Myb motifs. Replacement of either motif results in reduction of BAD1 transcription. This indicates the potential for transcription factors that bind CRE or Myb motifs to activate BAD1 transcription (Rooney and Klein, 2004).

BAD1 is a complex, hydrophilic protein that is characterized by 41 (10 degenerate; 31 highly conserved) copies of a tandem repeat, a 150 amino acid N-terminus with a Cardin-Weintraub motif (xBBxBx), and a 103-amino acid C-terminus with an epidermal growth factor (EGF)-like domain (Hogan *et al.*, 1995; Brandhorst *et al.*, 2005, 2013). The tandem repeats, which are 24 amino acids in length, are responsible for the biological function of the molecule (Hogan *et al.*, 1995; Brandhorst and Klein, 2000; Brandhorst *et al.*, 2013). Each repeat contains two conserved cysteine residues that form a disulfide bridge resulting in a 17 amino acid loop. Each highly conserved repeat contains a tryptophan-rich motif (WxxWxxW) that mediates binding to heparan sulfate, but not other extracellular matrix components such as laminin or collagen IV. BAD-1 binds heparin with high avidity and moderate affinity (Brandhorst *et al.*, 2013). The C-terminal EGF-like domain facilitates binding of BAD1 to cell wall chitin; loss of the C-terminus (ΔCterm-BAD1) impairs binding to the yeast cell surface and chitin granules, but does not impair virulence in a murine model of pulmonary infection (Brandhorst *et al.*, 2003).

Targeted deletion of *BAD1* by allelic replacement has allowed for in-depth dissection of *B. dermatitidis* virulence. *BAD1* null mutants (*BAD1Δ*) in strains ATCC 26199 and 60915 were avirulent in a murine model of pulmonary infection when compared to wild type (Brandhorst *et al.*, 1999). Mice inoculated with 1×10^2, 1×10^3, or 1×10^4 *BAD1Δ* yeast survived 72 days; whereas mice infected with wild-type or complemented *B. dermatitidis* strains died within 30 days (Brandhorst *et al.*, 1999). Histopathology demonstrated grossly normal lungs that contained few granulomas with yeast for mice infected with *BAD1Δ*. In contrast, there was widespread pulmonary inflammation and abundant yeast cells in mice infected with wild type or complemented *B. dermatitidis* strains. *BAD1Δ* mutants are viable, grow as yeast at 37°C, and do not exhibit defects in the phase transition or growth rate (Brandhorst *et al.*, 1999; Rooney *et al.*, 2001).

Potential mechanisms for loss of virulence in *BAD1Δ* strains include impaired adhesion and failure to alter the host response. BAD1 facilitates binding of yeast to macrophages via CR3 (CD11b/CD18) and CD14 receptors, and tissue via heparan sulfate (Newman *et al.*, 1995; Brandhorst *et al.*, 2013). Yeast devoid of BAD1 bind poorly to macrophages *in vitro* and to lung tissue *ex vivo* (Brandhorst *et al.*, 1999). BAD1 blunts the host response to infection by inhibiting activation of $CD4^+$ T-lymphocytes and production of tumour necrosis factor-α (TNF-α) by innate immune cells. Inhibition of T cell activation, which results in reduced IL-17 and INF-γ, is mediated through BAD1 interacting with CD47 on the $CD4^+$ cell surface (Brandhorst *et al.*2013).

Alteration in TNF-α is mediated by both BAD1 bound to the yeast surface and BAD1 released into the extracellular milieu (soluble BAD1).

BAD1 on the yeast cell surface induces production of transforming growth factor-β (TGF-β) in macrophages and neutrophils, which inhibit the production of TNF-α (Finkel-Jimenez *et al.*, 2001, 2002). TGF-β is postulated to act in an autocrine or paracrine manner to block TNF-α production (Finkel-Jimenez *et al.*, 2002). Experimental evidence supporting this mechanism includes (i) increased TGF-β concentration for wild type versus *BAD1Δ* when cocultured with phagocytes; (ii) increased TNF-α production following antibody neutralization of TGF-β; and (iii) suppression of TNF-α when recombinant TGF-β was added to *BAD1Δ* cocultures (Finkel-Jimenez *et al.*, 2002). Moreover, during murine infection, pulmonary concentration of TGF-β was 2- to 3-fold higher for wild type versus *BAD1Δ* strains, and the rise in TGF-β corresponded with a decline in TNF-α (Finkel-Jimenez *et al.*, 2002).

Soluble BAD1 (sBAD1) is also capable of inhibiting TNF-α production in phagocytes; however, unlike BAD1 bound to the yeast cell wall, the mechanism for TNF-α suppression is independent of TGF-β (Finkel-Jimenez *et al.*, 2002). When exposed to sBAD1, macrophages rapidly internalize it by receptor-mediated endocytosis (Brandhorst *et al.*, 2004). After several hours, sBAD1 is widely distributed throughout the cytoplasm in late endosomes (it is currently unknown if phagolysome fusion is affected). Endocytosis of sBAD1 predominantly occurs by binding to CR3 receptors (Brandhorst *et al.*, 2004). Treatment of macrophages with monodansyl cadaverine (blocks binding) or cytochalasin D (blocks internalization but not binding) impairs sBAD1 inhibition of TNF-α production. CR3-BAD1 binding is important for TNF-α suppression both *in vitro* and *in vivo*. CR3 deficient murine macrophages produce abundant TNF-α in the presence of *B. dermatitidis* yeast (Brandhorst *et al.*, 2004). Similarly, CR3-deficient mice have 6-fold higher concentration of pulmonary TNF-α, reduced burden of infection when analysed by CFU and histopathology, less lung inflammation, and well organized granulomas when compared with wild type or CD14 deficient mice (Brandhorst *et al.*, 2004).

In addition to serving as an adhesin that is essential for virulence, BAD1 also binds calcium (Brandhorst *et al.*, 2005). Equilibrium dialysis using $^{45}Ca^{2+}$ demonstrated that BAD1 binds calcium with high capacity (27 Ca^{2+} ions/BAD1 molecule) and low affinity ($K_d = 41.3 \pm 6.8\,\mu M$) (Brandhorst *et al.*, 2005). The precise biologic impact of calcium binding to BAD1 has not been fully deciphered; however, it appears to affect attachment of BAD1 to the yeast cell surface and promote growth under calcium limited conditions (Brandhorst *et al.*, 2005; Brandhorst and Klein, 2000). Depletion of calcium *in vitro* facilitates release of BAD1 from the yeast cell surface. This phenomenon has been used to enhance the yield of BAD1 during isolation and purification (Brandhorst *et al.*, 2005; Brandhorst and Klein, 2000).

The host response to *Blastomyces dermatitidis*

Innate and adaptive immune responses are required to combat *B. dermatitidis* infection, whereas humoral immunity is dispensable. In the lung, alveolar macrophages phagocytize and kill a large percentage of inhaled conidia and are capable of inhibiting the conversion of conidia to yeast (Drutz and Frey, 1985; Sugar *et al.*, 1995; Sugar and Picard, 1991). Similar to macrophages, neutrophils can ingest and kill conidia (Drutz and Frey, 1985). Conidia that survive the innate immune response by macrophages and neutrophils germinate into yeast, which are more difficult to kill. Yeast suppress TNF-α production and bock $CD4^+$ T-lymphocyte activation (Brandhorst *et al.*, 2004; Finkel-Jimenez *et al.*, 2001, 2002; Brandhorst *et al.*, 2013). In addition, *B. dermatitidis* yeast are poorly killed by activated macrophages (25–37.5% killing) and neutrophils, relatively resistant to damage by reactive oxygen species, and suppress nitric oxide (NO) production in macrophages (Brummer and Stevens, 1987; Brummer *et al.*, 1988, 1992; Rocco *et al.*, 2011). Despite detailed investigation of NO inhibition, the underlying mechanism remains to be fully deciphered (Rocco *et al.*, 2011). *B.*

dermatitidis yeast do not detoxify NO, alter uptake of arginine (the substrate for NO production), or reduce the transcription and translation of inducible nitric oxide synthase (iNOS) (Rocco *et al.*, 2011). NO inhibition requires direct contact of viable yeast with macrophages (Rocco *et al.*, 2011).

The development of adaptive immunity, which is mediated by T-lymphocytes, is important for control of *B. dermatitidis* infection. The cytokine responses (interferon-γ, TNF-α, and interleukin-17) coordinated by Th1 and Th17 T-lymphocytes activate macrophage and neutrophil fungicidal mechanisms and facilitate clearance of infection (Bradsher *et al.*, 1987; Wüthrich *et al.*, 2011a). Moreover, following infection, patients develop cell-mediated immunity against *B. dermatitidis* that can persist for at least 2 years (Klein *et al.*, 1990).

The *BAD1* knockout strain has been used as live-attenuated vaccine to delineate immune mechanisms against *B. dermatitidis*. Subcutaneous administration of the vaccine strain protects 100% of mice against *B. dermatitidis* pulmonary infection and induces sterilizing immunity (Wüthrich *et al.*, 2000). In the subcutaneous tissue, migratory monocyte-derived dendritic cells (DCs) ingest and carry the vaccine strain to the skin draining lymph nodes. At this site, the antigen is transferred to lymph node resident DCs, which then prime naïve $CD4^+$ T lymphocytes and induce their differentiation into protective Th1 and Th17 cells (Ersland *et al.*, 2010; Wüthrich *et al.*, 2011a). Pulmonary infection in vaccinated mice results in migration of Th1 and Th17 cells to the lung to orchestrate host defence against *B. dermatitidis* (Wüthrich *et al.*, 2011a; Wüthrich *et al.*, 2002). The role of Th17 cells in mediating vaccine immunity is conserved for *H. capsulatum* and *C. posadasii* (Wüthrich *et al.*, 2011a). Moreover, mice depleted of $CD4^+$ T lymphocytes can mount a protective immune response against lethal pulmonary infection by induction of TNF-α, INF-γ, and IL-17 producing $CD8^+$ T lymphocytes (Nanjappa *et al.*, 2012; Wüthrich *et al.*, 2003). Subcutaneous vaccine administration is essential for induction of protective immunity. Mucosal administration of the vaccine strain (i.e. intranasal delivery) fails to induce protective T-cell immunity because it results in cleavage of CCL7 by increased lung matrix metalloproteinase 2 (MMP2), which impairs recruitment and maturation of monocyte-derived DCs (Wüthrich *et al.*, 2012). Subcutaneous vaccination in dogs, which have a 10-fold higher incidence of blastomycosis than humans, demonstrated this live-attenuated vaccine to be immunogenic and safe (Wüthrich *et al.*, 2011b).

Clinical manifestations of blastomycosis

The clinical manifestations of blastomycosis range from asymptomatic infection to acute pneumonia to disseminated disease. Several medical conditions can increase the risk for severe clinical disease such as immunosuppression due to solid organ transplantation, AIDS, and treatment with medications that block the activity of tumour necrosis factor alpha (TNF-α) (Gauthier *et al.*, 2007; Pappas *et al.*, 1992; Smith and Kauffman, 2009). Inhalation of aerosolized conidia and mycelia represents the primary route of infection. A less common mechanism for transmission is traumatic inoculation of the skin, which is often related to laboratory accidents (Zampogna *et al.*, 2003; Gray and Baddour, 2002; Larson *et al.*, 1983). Although transmission by transplantation of an infected organ has not been reported for *B. dermatitidis*, this route has been documented for *Histoplasma capsulatum* and *Coccidioides* species (Martín-Dávila *et al.*, 2008). Onset of symptomatic blastomycosis develops 3 weeks to 3.5 months following inhalation of *B. dermatitidis* (Klein *et al.*, 1986, 1987a).

Pneumonia is the most common clinical presentation of blastomycosis and symptomatic disease occurs in 50% of exposed individuals (Klein *et al.*, 1986). Symptoms often resemble community-acquired pneumonia and include fever, chills, cough with sputum production, shortness of breath, chest pain, and malaise (Fanella *et al.*, 2011; Powell and Schuit, 1979; Chapman *et al.*, 2008; Chapman and Sullivan, 2009). Misdiagnosis of bacterial pneumonia is common and delays initiation of antifungal therapy. Importantly, diagnostic delays can lead to disease progression including respiratory failure and extrapulmonary dissemination (Chapman *et al.*, 2008; Gauthier *et al.*, 2007; Meyer *et al.*, 1993; Vasquez *et al.*, 1998).

Extrapulmonary dissemination occurs in 25–40% of persons with pulmonary infection (Chapman *et al.*, 2008). The most frequently affected organs are the skin and bone (Chapman *et al.*, 2008; Saccante and Woods, 2010; Chapman and Sullivan, 2009). The appearance of cutaneous blastomycosis is heterogeneous and includes ulcerative papules, nodules, ulcers, and verrucous lesions (Fig. 12.4A–D) (Saccente and Woods, 2010; Saucier and Gauthier, 2012). Infection of the bone results in destruction and most frequently affects the long bones (i.e. tibia, humerus), skull, ribs, and vertebral bodies (Fanella *et al.*, 2011; Oppenheimer *et al.*, 2007; Saccente *et al.*, 1998). Infection of the vertebral bones can be complicated by abscess formation in the soft tissue and extension into adjacent vertebral discs (Saccente *et al.*, 1998). Progressive destruction can cause vertebral body fracture and spinal cord compression. Dissemination to the central nervous system (CNS), which results in brain abscess or meningitis is uncommon (< 5–10%) in immunocompetent patients and is due to haematogenous spread or direct invasion from scalp or skull infection (Bariola *et al.*, 2010; Fanella *et al.*, 2011; Lemos *et al.*, 2000).

Blastomycosis develops in less than 0.2% of patients immunosuppressed by solid organ transplantation (Gauthier *et al.*, 2007). Similar to immunocompetent patients, transplant associated blastomycosis causes pneumonia; however, it is frequently complicated by respiratory failure or acute respiratory distress syndrome (Gauthier *et al.*, 2007; Grim *et al.*, 2012). The frequency of extrapulmonary dissemination (33–50%) is similar to the general population (25–40%) (Baumgardner *et al.*, 1992; Gauthier *et al.*, 2007; Grim *et al.*, 2012). Mortality ranges from 33% to 38%, but in patients with acute respiratory distress syndrome mortality increases to 67% (Gauthier *et al.*, 2007; Grim *et al.*, 2012; Pappas *et al.*, 1993).

Blastomycosis in patients with other forms of immunosuppression such as AIDS and malignancy (hematologic and solid organ) is relatively uncommon (Pappas *et al.*, 1992, 1993; Torres *et al.*, 2002). In patients with AIDS, the lung is the most commonly infected organ (87%). Extrapulmonary dissemination occurs in 53% and has a high propensity for invading the CNS (40%) (Pappas *et al.*, 1992). Mortality attributable to

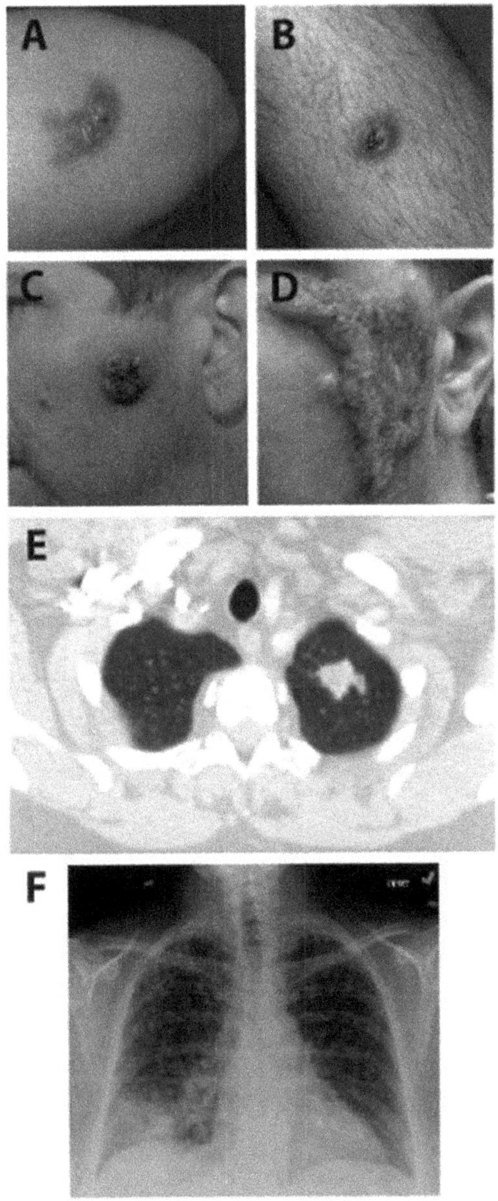

Figure 12.4 Clinical manifestations of blastomycosis. (A) Cutaneous plaque and (B) ulcer in a patient with disseminated blastomycosis. (C and D) Verrucous lesion on the face that expanded in size during a period of several months. (E) Lung mass on a CT scan in a patient with pulmonary blastomycosis. (F) Miliary pattern of pulmonary blastomycosis on a chest roentgenogram (X-ray).

blastomycosis is 53% (Pappas *et al.*, 1992). In patients diagnosed with malignancy, pneumonia is the most common clinical manifestation and disseminated disease occurs in 50% (Pappas *et al.*, 1993). Mortality is high and is attributable to either the underlying malignancy or blastomycosis (Pappas *et al.*, 1992).

Diagnosis of blastomycosis

The diagnosis of blastomycosis requires a high degree of suspicion because its clinical and radiographic manifestations can mimic other diseases including community-acquired pneumonia, mycobacterial infection, and malignancy. A delay from symptom onset to diagnosis often exceeds 2 months (Fanella *et al.*, 2011; Schutze *et al.*, 1996; Steele and Abernathy, 1983). Evaluation for fungal pneumonia including blastomycosis should be performed in patients with pulmonary infections or cutaneous lesions that fail to respond to antibiotic therapy. Moreover, chest radiographic abnormalities in conjunction with skin lesions on physical examination are highly suggestive of blastomycosis. The diagnosis of blastomycosis in a household pet, which is frequently a dog, suggests a common source of exposure for both human and pet (Armstrong *et al.*, 1987; Baumgardner and Paretsky, 1999; Baumgardner *et al.*, 1991; Bromel and Sykes, 2005; Gilor *et al.*, 2006; Menges *et al.*, 1965). Thus, the diagnosis of canine blastomycosis should increase the suspicion for human blastomycosis (Sarosi *et al.*, 1979). Airborne transmission of *B. dermatitidis* between animals and humans does not occur; however, primary cutaneous blastomycosis has been reported from bite injuries of an infected animal (Gnann *et al.*, 1983; Gray and Baddour, 2002).

The presence of broad-based, budding yeast (8–20 μm) with a doubly refractive cell wall when visualized by microscopy can provide a presumptive diagnosis of blastomycosis. Stains such as 10% potassium hydroxide, calcoflour white, Gomori methenamine silver (GMS), periodic acid–Schiff (PAS), or Papanicolaou facilitate identification of *B. dermatitidis* from clinical specimens (e.g. sputum, bronchoalveolar fluid, tissue) (Saccente and Woods, 2010). Histopathological analysis of infected tissue often reveals neutrophil infiltration and non-caseating granulomas (pyogranulomas) (Saccente and Woods, 2010, Saucier and Gauthier, 2012).

Growth of *B. dermatitidis* in culture from sputum, skin, bone or other clinical specimens allows for definitive diagnosis. *B. dermatitidis* can grow on many types of culture media including *Histoplasma* macrophage medium (HMM), 3M minimal medium, 7H10, Sabouraud dextrose agar, potato dextrose agar, and brain-heart infusion agar (Gauthier et al., 2010; Marty and Gauthier, 2013; Saccente and Woods, 2010). Most clinical laboratories use incubator temperatures ranging from 25°C to 30°C. This promotes *B. dermatitidis* growth as mould. Because the morphologic appearance of *B. dermatitidis* mycelia is not pathognomonic, additional confirmatory tests are required. Chemiluminescent DNA probes that target specific rRNA sequences are frequently used (Saccente and Woods, 2010). Conversion from mould to yeast following an increase in temperature to 37°C is rarely performed because the conversion can be slow. The role of PCR for the diagnosis of blastomycosis remains to be determined and no commercial test is available.

A highly sensitive antigen test that detects a galactomannan component in the *Blastomyces* cell wall has largely supplanted traditional serological tests such as complement fixation and immunodiffusion (Bariola *et al.*, 2011; Connolly *et al.*, 2012; Klein *et al.*, 1987b). Urine, serum, bronchoalveolar lavage fluid, and CSF specimens can be used for antigen testing. The sensitivity for urine antigen testing ranges from 85.1% to 92.9% and is influenced by the burden of infection (Bariola *et al.*, 2011; Connolly *et al.*, 2012). In contrast, the sensitivity for complement fixation and immunodiffusion range from 9–28% (Klein *et al.*, 1987b). The galactomannan used for detection is not unique to *B. dermatitidis* and there is substantial cross-reactivity with other dimorphic fungi including *H. capsulatum*, *P. brasiliensis*, and *P. marneffei* (Durkin *et al.*, 2004). Although these cross-reactions can occasionally result in diagnostic confusion between *B. dermatitidis* and *H. capsulatum*, this does not substantially impact clinical management because treatment recommendations for blastomycosis and histoplasmosis are similar (Chapman *et al.*, 2008;

Wheat *et al.*, 2007). Diagnostic confusion related to cross-reactions with *P. brasiliensis* or *P. marneffei* are rare because these fungi are not endemic to North America. The *Blastomyces* antigen test can be used to monitor the response to therapy (Mongkolrattanothai *et al.*, 2006; Tarr *et al.*, 2007).

Chest radiographic findings for patients with blastomycosis are heterogeneous and can mimic other infections or malignancy. Consolidation is the most common radiographic pattern and is indistinguishable from bacterial pneumonia (Alkrinawi *et al.*, 1995; Klein *et al.*, 1986, 1987; Powell *et al.*, 1979; Fang *et al.*, 2007). Other radiographic findings include nodules, interstitial infiltrates, cavities, mass lesions and military patterns (Fig. 12.4E and F) (Fang *et al.*, 2007). Cavitary disease and mass lesions can mimic tuberculosis or lung cancer, respectively.

Treatment of blastomycosis

The severity of infection, presence or absence of dissemination to the CNS, integrity of the immune system, and pregnancy influences antifungal selection and duration of therapy. Antifungal therapy is recommended for all patients diagnosed with blastomycosis (Chapman *et al.*, 2008). Treatment recommendations for immunocompetent patients with moderate to severe blastomycosis (pulmonary or disseminated) include induction therapy with lipid amphotericin B for 1 to 2 weeks followed by oral itraconazole for 12 months (Chapman *et al.*, 2008). Similarly, patients who are immunosuppressed (e.g. organ transplant, HIV, malignancy) require liposomal amphotericin B induction therapy followed by 12 months of oral itraconazole. Patients with CNS blastomycosis require a prolonged period of induction therapy (4–6 weeks) with lipid amphotericin B followed by at least 12 months of an oral azole (itraconazole, fluconazole, or voriconazole) (Chapman *et al.*, 2008). Patients with mild-to-moderate infection (pulmonary or disseminated) do not require induction therapy with lipid amphotericin B and can be treated with itraconazole for 6 to 12 months (Chapman *et al.*, 2008). Azole antifungals should be avoided in pregnant women because they can result in skeletal and craniofacial abnormalities in the developing fetus and increase the risk for spontaneous abortion (Bar-Oz *et al.*, 2000; Desantis *et al.*, 2009; Posaconazole, 2008; Voriconazole, 2011; Pursley *et al.*, 1996). To minimize fetal harm, lipid amphotericin B is recommended (Chapman *et al.*, 2008; Cohen, 1987).

Lipid formulations of amphotericin B are administered intravenously and are fungicidal against *B. dermatitidis*. Adverse effects include nephrotoxicity, urinary wasting of electrolytes (potassium, magnesium, bicarbonate), and infusion reactions (Saliba and Dupont, 2008). Most patients require replacement of potassium and magnesium along with frequent monitoring of electrolytes and renal function (Chapman *et al.*, 2008). In contrast to amphotericin, azoles are fungistatic and route of administration can be oral (itraconazole, voriconazole, fluconazole, posaconazole) or intravenous (voriconazole, fluconazole). Itraconazole is the primary azole used for treatment of blastomycosis; however, voriconazole and posaconazole have excellent *in vitro* activity and clinical experience with these agents is promising (Bakleh *et al.*, 2005; Dismukes *et al.*, 1992; Panicker *et al.*, 2006; Proia and Harnish, 2012; Ta *et al.*, 2009). Once steady-state levels are achieved, monitoring of drug levels for itraconazole, voriconazole, and posaconazole is recommended (Andes *et al.*, 2009; Chapman *et al.*, 2008). Fluconazole is less efficacious than itraconazole and is no longer recommended for treatment of pulmonary and non-CNS disseminated blastomycosis (Chapman *et al.*, 2008; Pappas *et al.*, 1995, 1997).

The newest class of antifungal drugs, the echinocandins, which inhibit the biosynthesis of β-(1,3)-glucan synthase have poor activity against *B. dermatitidis* yeast and are not recommended for treatment (Chapman *et al.*, 2008). Thus, antifungals that target *B. dermatitidis* are limited to two classes, polyenes (i.e. amphotericin B formulations) and azoles. Despite the rise of invasive fungal infections, the discovery of new antifungals has been slow. The discovery of *DRK1*, which encodes a group III hybrid histidine kinase in *B. dermatitidis*, inspired the development of a *Saccharomyces cerevisiae* reporter system to screen 19,762 small molecules in a high-throughput fashion to identify novel antifungal compounds

(Tebbets *et al.*, 2012). Type III hybrid histidine kinases are widely distributed in the fungal kingdom, are important for virulence and morphogenesis, and are absent in mammals including humans (Tebbets *et al.*, 2012). Using this system, two compounds, 2-nitro-5-(4-nitrostyryl)furan and 2-(methylthio)-5-[(5-nitro-1,3-thiazol-2-yl) thio]-1,3,4-thiadiazole, were identified to have activity against *Aspergillus fumigatus, Fusarium solani, Rhizopus oryzae, Candida albicans* including fluconazole resistant strains, *Cryptococcus neoformans,* and *Cryptococcus gattii* (Tebbets *et al.*, 2012). Moreover, these two compounds were synergistic with fluconazole against *C. albicans* biofilms (Tebbets *et al.*, 2012). Gene expression microarray analysis indicates exposure of *C. albicans* to 2-nitro-5-(4-nitroStyryl)furan affects the oxidative stress response, whereas exposure to 2-(methylthio)-5-[(5-nitro-1,3-thiazol-2-yl) thio]-1,3,4-thiadiazole induces genes involved with heavy metal stress (Tebbets *et al.*, 2012). In addition, this screen also identified a group of 3 macrotetrolides (monactin, dinactin, trinactin) derived from *Streptomyces* with activity against *C. albicans, C. krusei, C. neoformans,* and *C. gattii* (Tebbets *et al.*, 2013). These macrotetrolides demonstrated potent activity against *C. albicans* biofilms (Tebbets *et al.*, 2013). Although the screen was designed to identify group III histidine kinase inhibitors, the target(s) of these compounds has yet to be deciphered (Tebbets *et al.*, 2012, 2013).

Acknowledgements

I would like to thank Dr Thomas Sullivan, Dr Marcel Wüthrich and Dr Bruce Klein for their helpful discussions on *B. dermatitidis* mating and immunology. I thank Amber J. Marty for her assistance in proofreading this book chapter.

References

Al-Doory, Y., and Larsh, H.W. (1962). Quantitative studies of total lipids in pathogenic fungi. Appl. Microbiol. *10*, 492–495.

Alkrinawi, S., Reed, M.H., and Pasterkamp, H. (1995). Pulmonary blastomycosis in children: findings on chest radiographs. Am. J. Roentgenol. *165*, 651–654.

Andes, D., Pascual, A., and Marchetti, O. (2009). Antifungal therapeutic drug monitoring: established and emerging indications. Antimicrob. Agents Chemother. *53*, 23–34.

Armstrong, C.W., Jenkins, S.R., Kaufman, L., Kerkering, T.M., Rouse, B.S., Miller, G.B. Jr. (1987). Common-source outbreak of blastomycosis in hunters and their dogs. J. Infect. Dis. *155*, 568–570.

Baily, G.G., Robertson, V.J., Neill, P., Garrido, P., and Levy, L.F. (1991). Blastomycosis in Africa: clinical features, diagnosis, and treatment. Rev. Infect. Dis. *13*, 1005–1008.

Bakleh, M., Aksamit, A.J., Tleyjeh, I.M., and Marshall, W.F. (2005). Successful treatment of cerebral blastomycosis with voriconazole. Clin. Infect. Dis. *40*, e69-e71.

Bariola, J.R., Perry, P., Pappas, P.G., Proia, L., Shealey, W., Wright, P.W., Sizemore, J.M., Robinson, M., and Bradsher, R.W. (2010). Blastomycosis of the central nervous system: a multicenter review of diagnosis and treatment in the modern era. Clin. Infect. Dis. *50*, 797–804.

Bariola, J.R., Hage, C.A., Durkin, M., Bensadoun, E., Gubbins, P.O., Wheat, L.J., and Bradsher, R.W. (2011). Detection of *Blastomyces dermatitidis* antigen in patients with newly diagnosed blastomycosis. Diagn. Microbiol. Infect. Dis. *69*, 187–191.

Bar-Oz, B., Moretti, M.E., Bishai, R., Mareels, G., Van Tittelboom, T., Verspeelt, J., and Koren, G. (2000). Pregnancy outcome after in utero exposure to itraconazole: a prospective cohort study. Am. J. Obstet. Gynecol. *183*, 617–620.

Baumgardner, D.J., and Burdick, J.S. (1991). An outbreak of human and canine blastomycosis. Rev. Infect. Dis. *13*, 898–905.

Baumgardner, D.J., and Paretsky, D.P. (1999). The *in vitro* isolation of *Blastomyces dermatitidis* from a woodpile in north central Wisconsin, USA. Med. Mycol. *37*, 163–168.

Baumgardner, D.J., Buggy, B.P., Mattson, B.J., Burdick, J.S., and Ludwig, D. (1992). Epidemiology of blastomycosis in a region of high endemicity in north central Wisconsin. Clin. Infect. Dis. *15*, 629–635.

Baumgardner, D.J., Egan, G., Giles, S., and Laundre, B. (2002). An outbreak of blastomycosis on a United States Indian reservation. Wilderness Environ. Med. *13*, 250–252.

Bradsher, R.W., Balk, R.A., and Jacobs, R.F. (1987). Growth inhibition of *Blastomyces dermatitidis* in alveolar and peripheral macrophages from patients with blastomycosis. Am. Rev. Respir. Dis. *135*, 412–417.

Brandhorst, T., and Klein, B.S. (2000). Cell wall biogenesis of *Blastomyces dermatitidis*. J. Biol. Chem. *275*, 7925–7934.

Brandhorst, T.T., Wüthrich, M., Warner, T., and Klein, B. (1999). Targeted gene disruption reveals an adhesin indispensable for pathogenicity of *Blastomyces dermatitidis*. J. Exp. Med. *19*, 12078–12016.

Brandhorst, T.T., Wüthrich, M., Finkel-Jimenez, B., and Klein, B.S. (2003). A C-terminal EGF-like domain governs BAD1 localization to the yeast surface and fungal adherence to phagocytes, but is dispensable in immune modulation and pathogenicity of *Blastomyces dermatitidis*. Mol. Microbiol. *48*, 53–65.

Brandhorst, T.T., Wüthrich, M., Finkel-Jimenez, B., Warner, T., and Klein, B.S. (2004). Exploiting type

3 complement receptor for TNA-alpha suppression, immune evasion, and progressive pulmonary infection. J. Immunol. *173*, 7444–7453.

Brandhorst, T.T., Gauthier, G.M., Stein, R.A., and Klein, B.S. (2005). Calcium binding by the essential virulence factor BAD-1 of *Blastomyces dermatitidis*. J. Biol. Chem. *280*, 42156–42163.

Brandhorst, T.T., Roy, R., Wüthrich, M., Nanjappa, S., Filutowicz, H., Galles, K., Tonelli, M., McCaslin, D.R., Satyshur, K., and Klein, B.S. (2013). Structure and function of a fungal adhesion that binds heparin and mimics thrombospondin-1 by blocking T-cell activation and effector function. PLoS. Pathog. *9*, e1003464.

Bromel, C., and Sykes, J.E. (2005). Epidemiology, diagnosis, and treatment of blastomycosis in dogs and cats. Clin. Tech. Small Animal Pract. *20*, 233–239.

Brown, E.M., McTaggart, L.R., Zhang, S.X., Low, D.E., Stevens, D.A., and Richardson, S.E. (2013). Phylogenetic analysis reveals a cryptic species *Blastomyces gilchristii*, sp. nov. within the human pathogenic fungus *Blastomyces dermatitidis*. PLoS One. *8*, e59237.

Brummer, E., and Stevens, D.A. (1987). Fungicidal mechanisms of activated macrophages: evidence for nonoxidative mechanisms for killing of *Blastomyces dermatitidis*. Infect. Immune. *55*, 3221–3224.

Brummer E., Hanson, L.H., Restrepo, A., and Stevens, D.A. (1988). *In vivo* and *In vitro* activation of pulmonary macrophages for enhanced killing of *Paracoccidioides brasiliensis* or *Blastomyces dermatitidis*. J. Immunol. *140*, 2786–2789.

Brummer, E., Kurita, N., Yoshida, S., Nishimura, K., and Miyaji, M. (1992). A basis for resistance of *Blastomyces dermatitidis* killing by human neutrophils: inefficient generation of myeloperoxidase system products. J. Med. Vet. Mycol. *30*, 233–243.

Bubnick, M., and Smulian, A.G. (2007). The *MAT1* locus of *Histoplasma capsulatum* is responsive in a mating-type specific manner. Eukaryot. Cell. *6*, 616–621.

Burg, E.F., and Smith, L.H. (1994). Cloning and characterization of *bys1*, a temperature-dependent cDNA specific to the yeast phase of the pathogenic dimorphic fungus *Blastomyces dermatitidis*. Infect. Immune. *62*, 2521–2528.

Cano, M.V., Ponce-de-Leon, G.F., Tippen, S., Lindsley, M.D., Warwick, M., and Hajjeh, R.A. (2003). Blastomycosis in Missouri: epidemiology and risk factors for endemic disease. Epidemiol. Infect. *131*, 907–914.

Carlos, W.G., Rose, A.S., Wheat, L.J., Norris, S., Sarosi, G.A., Knox, K.S., and Hage, C.A. (2010). Blastomycosis in Indiana: digging up more cases. Chest. *138*, 1377–1382.

Carman, W.F., Frean, J.A., Crewe-Brown, H.H., Culligan, G.A., and Young, C.N. (1989). Blastomycosis in Africa. A review of known cases diagnosed between 1951 and 1987. Mycopathologia. *107*, 25–32.

Centers for Disease Control and Prevention. (1976). Blastomycosis – North Carolina. Morb. Mortal. Wkly. Rep. 25, 205–206.

Centers for Disease Control and Prevention. (1996). Blastomycosis – Wisconsin, 1986–1995. Morb. Mortal. Wkly. Rep. *45*, 601–603.

Chao, L.Y., Marletta, M.A., and Rine, J. (2008). Sre1, an iron-modulated GATA DNA-binding protein of iron-uptake genes in the fungal pathogen *Histoplasma capsulatum*. Biochemistry. *274*, 4613–4619.

Chapman, S.W., and Sullivan, D.C. (2009). *Blastomyces dermatitidis*. In Mandell, Douglas, and Bennett's Principles and Practice of Infectious Diseases, 7th Edition, G.L. Mandell, J.E. Bennett, and R. Dolin, eds. (Churchill Livingstone, Philadelphia, PA), pp. 3319–3332.

Chapman, S.W., Lin,A.C., Hendricks, K.A., Nolan, R.L., Currier, M.M., Morris, K.R., and Turner, H.R. (1997). Endemic blastomycosis in Mississippi: epidemiologic and clinical studies. Semin. Respir. Infect. *12*, 219–218.

Chapman, S.W., Dumukes, W.E., Proia, L.A., Bradsher, R.W., Pappas, P.G., Threlkeid, M.G., and Kauffman, C.A. (2008). Clinical practice guidelines for the management of blastomycosis: 2008 update by the Infectious Diseases Society of America. Clin. Infect. Dis. *46*, 1801–1812.

Cockerill, F.R. 3rd, Roberts, G.D., Rosenblatt, J.E., Utz, J.P., and Utz, D.C. (1984). Epidemic of pulmonary blastomycosis (Namekagon fever) in Wisconsin canoeists. Chest. *86*, 688–692.

Cohen, I. (1987). Absence of congenital infection and teratogenesis in three children born to mothers with blastomycosis and treated with amphotericin B during pregnancy. Pediatr. Infect. Dis. J. *6*, 76–77.

Connolly, P., Hage, C.A., Bariola, J.R., Bensadoun, E., Rodgers, M., Bradsher, R.W., and Wheat, L.J. (2012). *Blastomyces dermatitidis* antigen detection by quantitative enzyme immunoassay. Clin. Vaccine. Immunol. *19*, 53–56.

Crampton, T.L., Light, R.B., Berg, G.M., Meyers, M.P., Schroeder, G.C., Hershfield, E.S., and Embil, J.M. (2002). Epidemiology and clinical spectrum of blastomycosis diagnosed at Manitoba hospitals. Clin. Infect. Dis. *34*, 1310–1316.

De Groote, M.A., Bjerke, R., Smith, H., and Rhodes III, L.V. (2000). Expanding epidemiology of blastomycosis: clinical features and investigation of 2 cases in Colorado. Clin. Infect. Dis. *30*, 582–584.

Desantis, M., Di Gianantonio, E., Cesari, E., Abrosini, G., Straface, G., and Clementi, M. (2009). First-trimester itraconazole exposure and pregnancy outcome: a prospective cohort study of women contacting teratology information services in Italy. Drug. Saf. *32*, 239–244.

Di Salvo, A.F., and Denton, J.F. (1963). Lipid content of four strains of *Blastomyces dermatitidis* of different mouse virulence. J. Bacteriol. *85*, 927–931.

Dismukes, W.E., Bradsher, R.W. Jr., Cloud, G.C., Kauffman, C.A., Chapman, S.W., George, R.B. Stevens, D.A., Girard, W.M., Saag, M.S., and Bowles-Patton, C. (1992). Itraconazole therapy for blastomycosis and histoplasmosis. Am. J. Med. *93*, 489–497.

Domer, J.E. (1971). Monosaccharide and chitin content of cell walls of *Histoplasma capsulatum* and *Blastomyces dermatitidis*. J. Bacteriol. *107*, 870–877.

Domer, J.E., and Hamilton, J.G. (1971). The readily extracted lipids of *Histoplasma capsulatum* and *Blastomyces dermatitidis*. Biochem. Biophys. Acta. *231*, 465–478.

Drutz, D.J., and Frey, C.L. (1985). Intracellular and extracellular defenses of human phagocytes against *Blastomyces dermatitidis* conidia and yeasts. J. Lab. Clin. Med. *105*, 737–750.

Drutz, D.J., Huppert, M., Sun, S.H., and McGuire W.L. (1981). Human sex hormones stimulate the growth and maturation of *Coccidioides immitis*. Infect. Immun. *32*, 897–907.

van Duin, D., Cadadevall, A., and Nosanchuk, J.D. (2002). Melanization of *Cryptococcus neoformans* and *Histoplasma capsulatum* reduces their susceptibilities to amphotericin B and caspofungin. Antimicrob. Agents Chemother. *46*, 3394–3400.

Durkin, M., Witt, J., LeMonte, A., Wheat, B., and Connolly, P. (2004). Antigen assay with the potential to aid in diagnosis of blastomycosis. J. Clin Microbiol. *42*, 4873–4875.

Dwight, P.J., Naus, M., Sarsfield, P., and Limerick, B. (2000). An outbreak of human blastomycosis: the epidemiology of blastomycosis in the Kenora Catchment region of Ontario, Canada. Can. Commun. Dis. Rep. *26*, 82–92.

Ersland, K., Wüthrich, M., and Klein, B.S. (2010). Dynamic interplay among monocyte-derived, dermal and resident lymph node dendritic cells during the generation of vaccine immunity to fungi. Cell. Host. Microbe. *7*, 747–487.

Fang, W., Washington, L., and Kumar, N. (2007). Imaging manifestations of blastomycosis: a pulmonary infection with potential dissemination. Radiographics. *27*, 641–655.

Fanella, S., Skinner, S., Trepman, E., and Embil, J.M. (2011). Blastomycosis in children and adolescents: a 30-year experience from Manitoba. Med. Mycol. *49*, 627–632.

Finkel-Jimenez, B., Wüthrich, M., Brandhorst, T., and Klein, B.S. (2001). The WI-1 adhesin blocks phagocyte TNF-a production imparting pathogenicity on *Blastomyces dermatitidis*. J. Immunol. *166*, 2665–2673.

Finkel-Jimenez, B., Wüthrich, M., and Klein, B.S. (2002). BAD1, an essential virulence factor of *Blastomyces dermatitidis*, suppresses host TNF-α production through TGF-β–dependent and –independent mechanisms. J. Immunol. *168*, 5746–5755.

Fraser, J.A., Stajich, J.E., Tarcha, E.J., Cole, G.T., Inglis, D.O., Sil, A., and Heitman, J. (2007). Evolution of the mating type locus: insights gained from the dimorphic fungal pathogens *Histoplasma capsulatum*, *Coccidioides immitis*, and *Coccidioides posadasii*. Eukaryot. Cell. *6*, 622–629.

Frean, J.A., Carman, W.F., Crewe-Brown, H.H., Culligan, G.A., and Young, C.N. (1989). *Blastomyces dermatitidis* infections in the RSA. S. Afr. Med. J. *76*, 13–16.

Frye, M.D., and Seifer, F.D. (1991). An outbreak of blastomycosis in eastern Tennessee Mycopathologia. *116*, 15–21.

Furcolow, M.L., Chick, E.W., Busey, J.F., and Menges, R.W. (1970). Prevalence and incidence studies of human and canine blastomycosis. I. Cases in the United States, 1885–1968. Am. Rev. Respir. Dis. *102*, 60–67.

Gaur, P.K., and Lightward, R.W. (1980). Mating types and convertibility among soil isolates of *Histoplasma capsulatum*. Mycologia. *72*, 404–406.

Gauthier, G.M., and Klein, B.S. (2008). Insights into fungal morphogenesis and immune evasion. Microbe. *3*, 416–423.

Gauthier, G.M., Safdar, N., Klein, B.S., and Andes, D.R. (2007). Blastomycosis in solid organ transplant recipients. Transpl. Infect. Dis. *9*, 310–317.

Gauthier, G.M., Sullivan, T.D., Gallardo, S.S., Brandhorst, T.T., Vanden Wymelenberg, A.J., Cuomo, C.A., Suen, G., Currie, C.R., and Klein, B.S. (2010). SREB, a GATA transcription factor that directs disparate fates in *Blastomyces dermatitidis* including morphogenesis and siderophore biosynthesis. PLoS Pathog *6*, e1000846.

Gilchrist, TC. (1894). Protozoan dermatitis. Journal of Cutaneous and Genito-Urinary Diseases. *12*, 496.

Gilchrist, T.C. (1896). A case of blastomycetic dermatitis in man. Johns Hopkins Hosp. Rep. *1*, 269–283.

Gilchrist, T.C., and Stokes, W.R. (1896). The presence of an *Oidium* in the tissues of a case of pseudo–lupus vulgaris: Preliminary report. Johns Hopkins Hosp. Bull. *7*, 129–133.

Gilchrist, T.C., and Stokes, W.R. (1898). A case of pseudo-lupus vulgaris caused by a *Blastomyces*. J. Exp. Med. 3, 53–78.

Gilor, C., Graves, T., Barger,, A., and O'Dell-Anderson, K. (2006). Clinical aspects of natural infection with *Blastomyces dermatitidis* in cats: 8 cases (1991–2005). J. Am. Vet. Med. Assoc. *229*, 96–99.

Girouard, G., Lachance, C., and Pelletier, R. (2007). Observations of (1,3)-beta-D-glucan detection as a diagnostic tool in endemic mycosis caused by *Histoplasma* or *Blastomyces*. J. Med. Microbiol. *56*, 1001–1002.

Glick, A.D., and Kwon-Chung, K.J. (1973). Ultrastructural comparison of coils and ascospores of *Emmonsiella capsulate* and *Ajellomyces dermatitidis*. Mycologia. *65*, 216–220.

Gnann, J.W. Jr., Bressler, G.S., Bodet, C.A.3rd., and Avent, C.K. (1983). Human blastomycosis after a dog bite. Ann. Intern. Med. *98*, 48–49.

Gomez, B.L., Nosanchuk, J.D., Diez, S., Youngchim, S., Aisen, P., Cano, L.E., Restrepo, A., Cadadevall, A., and Hamilton, A.J. (2001). Detection of melanin-like pigments in the dimorphic fungal pathogen *Paracoccidioides brasiliensis in vitro* and during infection. Infect. Immun. *69*, 5760–5767.

Gray, N.A., and Baddour, L.M. (2002). Cutaneous inoculation blastomycosis. Clin. Infect. Dis. *34*, e44–49.

Grim, S.A., Proia, L., Miller, R., Alhyraba, M., Costas-Chavarri, A., Oberholzer, J., and Clark, N.M. (2012). A multicenter study of histoplasmosis and blastomycosis after solid organ transplantation. Transpl. Infect. Dis. *14*, 17–23.

Haas, H., Zadra, I., Stoffler, G., and Angermayr, K. (1999). The *Aspergillus nidulans* GATA factor SREA is involved with regulation of siderophore biosynthesis and control of iron uptake. J. Biol. Chem. *274*, 4613–4619.

Hamburger W.W. (1907). A comparative study of four strains of organisms isolated from four cases of generalized blastomycosis. J. Infect. Dis. *4*, 201–209.

Hogan, L.H., and Klein, B.S. (1994). Altered expression of surface alpha-1,3-glucan in genetically related strains of *Blastomyces dermatitidis* that differ in virulence. Infect. Immun. *62*, 3543–3546.

Hogan, L.H., Josvai, S., and Klein, B.S. (1995). Genomic cloning, characterization, and functional analysis of the major surface adhesin WI-1 on *Blastomyces dermatitidis* yeasts. J. Biol. Chem. *270*, 30725–20732.

Hou, B., Zhang, Z., Zheng, F., and Liu, X. (2013). Molecular cloning, characterization, and differential expression of DRK1 in *Sporothrix schenckii*. Int. J. Mol. Med. *31*, 99–104.

Hussein, R., Khan, S., Levy, F., Mehta, J.B., and Sarubbi, F.A. (2009). Blastomycosis in the mountainous region of northeast Tennessee. Chest. *135*, 1019–1023.

Hwang, L.H., Seth, E., Gilmore, S.A., and Sil, A. (2012). *SRE1* regulates iron-dependent and –independent pathways in the fungal pathogen *Histoplasma capsulatum*. Eukaryot. Cell. *11*, 16–25.

Jung, W.H., and Kronstad, J.W. (2011). The iron-responsive, GATA-type transcription factor Cir1 influences mating in *Cryptococcus neoformans*. Mol. Cells. *31*, 73–77.

Jung, W.H., Sham, A., White, R., and Kronstad, J.W. (2006). Iron regulation of the major virulence factors in the AIDS-associated pathogen *Cryptococcus neoformans*. PLoS. Biol. *4*, e410.

Juvvadi, P.R., Fortwendel, J.R., Rogg, L.E., and Steinbach, W.J. (2011). Differential localization of septins during growth of the human fungal pathogen *Aspergillus fumigatus* reveal novel functions. Biochem. Biophys. Res. Commum. *415*, 238–243.

Kanetsuna, F., Carbonell, L.M., Moreno, R.E., and Rodriguez, J. (1969). Cell wall composition of the yeast and mycelial forms of *Paracoccidioides brasiliensis*. J. Bacteriol. 97, 1036–1041.

Kanetsuna, F., and Carbonell, L.M. (1971). Cell wall composition of the yeastlike and mycelial forms of *Blastomyces dermatitidis*. J. Bacteriol. *106*, 946–948.

Kitchen, M.S., Reiber, C.D., and Eastin, G.B. (1977). An urban epidemic of North American blastomycosis. Am. Rev. Respir. Dis. *115*, 1063–1066.

Klein, B.S., and Davis, J.P. (1985). A laboratory-based surveillance of human blastomycosis in Wisconsin between 1973 and 1982. Am. J. Epidemiol. *122*, 897–903.

Klein, B.S., and Jones, J.M. (1990). Isolation, purification, and radiolabeling of a novel 120-kD surface protein on *Blastomyces dermatitidis* yeasts to detect antibody in infected patients. J. Clin. Invest. *85*, 152–161.

Klein, B.S., Vergeront, J.M., Weeks, R.J., Kumar, U.N., Mathai, G., Varkey, B., Kaufman, L., Bradsher, R.W., Stoebig, J.F., and Davis, J.P. (1986). Isolation of *Blastomyces dermatitidis* in soil associated with a large outbreak of blastomycosis in Wisconsin. N. Engl. J. Med. *314*, 529–534.

Klein, B.S., Vergeront, J.M., Di Salvo, A.F., Kaufman, L., and Davis, J.P. (1987a). Two outbreaks of blastomycosis along rivers in Wisconsin: isolation of *Blastomyces dermatitidis* from riverbank soil and evidence of its transmission along waterways. Am. Rev. Respir. Dis. *136*, 1333–1338.

Klein, B.S., Vergeront, J.M., Kaufman, L., Bradsher, R.W., Kumar, U.N., Mathai, G., Varkey, B., and Davis, J.P. (1987b). Serological tests for blastomycosis: assessments during a large point-source outbreak in Wisconsin. J. Infect. Dis. *155*, 262–268.

Klein, B.S., Bradsher, R.W., Vergeront, J.M., and Davis, J.P. (1990). Development of long-term specific cellular immunity after acute *Blastomyces dermatitidis* infection: assessments following a large point-source outbreak in Wisconsin. J. Infect. Dis. *161*, 97–101.

Klein, B.S., Sondel, P.M., and Jones, J.M. (1992). WI-1, a novel 120-kilodalton surface protein on *Blastomyces dermatitidis* yeast cells, is a target antigen of cell-mediated immunity in human blastomycosis. Infect. Immun. *60*, 4291–4300.

Klein, B.S., Hogan, L.H., and Jones, J.M. (1993). Immunologic recognition of a 25-amino acid repeat arrays in tandem on a major antigen of *Blastomyces dermatitidis*. J. Clin. Invest. *92*, 330–337.

Klein, B.S., Aizenstein, B.D., and Hogan, L.H. (1997). African strains of *Blastomyces dermatitidis* that do not express surface adhesin WI-1. Infect. Immunol. *65*, 1505–1509.

Klotz, S.A., Drutz, D.J., Huppert, M., Sun, S.H., and DeMarsh, P.L. (1984). The critical role of CO_2 in the morphogenesis of *Coccidioides immitis* in cell-free subcutaneous chambers. J. Infect. Dis. *150*, 127–134.

Koneti, A., Linke, M.J., Brummer, E., and Stevens, D.A. (2008). Evasion of innate immune responses: evidence for mannose binding lectin inhibition of tumor necrosis factor alpha production by macrophages in response to *Blastomyces dermatitidis*. Infect. Immun. *76*, 994–1002.

Krajaejun, T., Gauthier, G.M., Rappleye, C.M., Sullivan, T.D., and Klein, B.S. (2007). Development and application of a green fluorescent protein sentinel system for identification of RNA interference in *Blastomyces dermatitidis* illuminates the role of septin in morphogenesis and sporulation. Eukaryot. Cell. *6*, 1299–1309.

Krajaejun, T., Wüthrich, M., Gauthier, G.M., Warner, T.F., Sullivan, T.D., and Klein, B.S. (2010). Discordant influence of *Blastomyces dermatitidis* yeast-phase-specific gene *BYS1* on morphogenesis and virulence. Infect. Immun. *78*, 2522–2528.

Kwon-Chung, K.J. (1971). Genetic analysis on the incompatibility system of *Ajellomyces dermatitidis*. Sabouraudia. *9*, 231–238.

Kwon-Chung, K.J. (1973). Studies on *Emmonsiella capsulate* I. Heterothallism and development of the ascocarp. Mycologia. *65*, 109–121.

Kwon-Chung, K.J., Weeks, R.J., and Larsh, H.W. (1974). Studies on *Emmonsiella capsulate* (*Histoplasma capsulatum*). II: distribution of the two mating types in 13 endemic states of the United States. Am. J. Epidemiol. *99*, 44–49.

Kwon-Chung, K.J., Bartlett, M., and Wheat, L.J. (1984). Distribution of the two mating types among *Histoplasma capsulatum* isolates obtained from an urban histoplasmosis outbreak. Sabourdia. *22*, 155–157.

Lan, C.Y., Rodarte, G., Murillo, L.A., Jones, T., Davis, R.W., Dungan, J., Newport, G., and Agabian, N. (2004). Regulatory networks affected by iron availability in *C. albicans*. Mol. Microbiol. *53*, 1451–1469.

Larson, D.M., Eckman, M.R., Alber, R.L., and Goldschmidt, V.G. (1983). Primary cutaneous (inoculation) blastomycosis: An occupational hazard to pathologists. Am. J. Clin. Pathol. *79*, 253–255.

Lemos, L.B., Guo, M., and Baliga, M. (2000). Blastomycosis: organ involvement and etiologic diagnosis. A review of 123 patients from Mississippi. Ann. Diagn. Pathol. *4*, 391–406.

Li, D., Agrellos, O.A., and Calderone, R. (2010). Histidine kinases keep fungi safe and vigorous. Curr. Opin. Microbiol. *13*, 424–430.

Li, W., Metin, B., White, T.C., and Heitman, J. (2010). Organization and evolutionary trajectory of the mating type (*MAT*) locus in dermatophyte and dimorphic fungal pathogens. Eukaryot. Cell. *9*, 46–58.

Li, W., Sullivan, T.D., Walton, E., Averette, A.F., Sakthikumar, S., Cuomo, C.A., Klein, B.S., and Heitman, J. (2013). Identification of the mating-type (*MAT*) locus that controls sexual reproduction of *Blastomyces dermatitidis*. Eukaryot. Cell. *12*, 109–117.

Longtine, M.S., and Bi, E. (2003). Regulation of septin organization and function in yeast. Trends. Cell. Biol. *13*, 403–409.

Lowry, P.W., Kelso, K.Y., and McFarland, L.M. (1989). Blastomycosis in Washington Parish, Louisiana, 1976–1985. Am. J. Epidemiol. *130*, 151–159.

McCullough, M.J., DiSalvo, A.F., Clemons, K.V., Park, P., and Stevens, D.A. (2000). Molecular epidemiology of *Blastomyces dermatitidis*. Clin. Infect. Dis. *30*, 328–335.

MacDonald, P.D., Langley, R.L., Gerkin, S.R., Torok, M.R., and MacCormack, J.N. (2006). Human and canine pulmonary blastomycosis, North Carolina, 2001–2002. Emerg. Infect. Dis. *12*, 1242–1244.

McDonough, E.S., and Lewis, A.L. (1967). *Blastomyces dermatitidis*: production of the sexual stage. Science. *156*, 528–529.

McDonough, E.S., and Lewis, A.L. (1968). The ascigerous stage of *Blastomyces dermatitidis*. Mycologia. *60*, 76–83.

McDonough, E.S., McNamara, W.J., Chan, D.M., and Wallenfang, M.A. (1973). Geographic distribution of "+" and "-" isolates of *Blastomyces dermatitidis* in North America. Am. J. Epidemiol. *98*, 63–67.

McDonough, E.S., Chan, D.M., and McNamara, W.J. (1977). Dual infection by "+" and "-" mating types of *Ajellomyces* (*Blastomyces*) *dermatitidis*. Am. J. Epidemiol. *106*, 67–71.

Mandel, M.A., Barker, B.M., Kroken, S., Rounsley, S.D., and Orbach, M.J. (2007). Genomic and population analyses of mating type loci in *Coccidioides* species reveal evidence for sexual reproduction and gene acquisition. Eukaryot. Cell. *6*, 1189–1199.

Manetti A.C. (1991). Hyperendemic urban blastomycosis. Am. J. Public. Health. *81*, 633–636.

Maresca, B., and Kobayashi, G. (1993). Changes in membrane fluidity modulate heat shock gene expression and produced attenuated strains in the dimorphic fungus *Histoplasma capsulatum*. Arch. Med. Res. *24*, 247–249.

Martín-Dávila, P., Fortún, J., López-Vélez, R., Norman, F., Montes de Oca, M., Zamarrón, P., González, M.I., Moreno, A., Pumarola, T., Garrido, G., *et al.* (2008). Transmission of tropical and geographically restricted infections during solid organ transplantation. Clin. Microbiol. Rev. *21*, 60–96.

Marty, A.J., and Gauthier, G.M. (2013). *Blastomyces dermatitidis* septins CDC3, CDC10, and CDC12 impact the morphology of yeast and hyphae, but are not required for the phase transition. Med. Mycol. *51*, 93–102.

Medoff, G., Painter, A., and Kobayashi, G.S. (1987). Mycelial- to Yeast-phase transitions of the dimorphic fungi *Blastomyces dermatitidis* and *Paracoccidioides brasiliensis*. J. Bacteriol. *169*, 4055–4060.

Meece, J.K., Anderson, J.L., Klein, B.S., Sullivan, T.D., Foley, S.L., Baumgardner, D.J., Brummitt, C.F., and Reed, K.D. (2010). Genetic diversity in *Blastomyces dermatitidis*: implications for PCR detection in clinical and environmental samples. Med. Mycol. *48*, 285–290.

Meece, J.K., Anderson, J.L., Fisher, M.C., Henk, D.A., Sloss, B.L., and Reed, K.D. (2011). Population genetic structure of clinical and environmental isolates of *Blastomyces dermatitidis*, based on 27 polymorphic microsatellite markers. Appl. Environ. Microbiol. *77*, 5123–5131.

Meece, J.K., Anderson, J.L., Gruszka, S., Sloss, B.L., Sullivan, B., and Reed, K.D. (2013). Variation in clinical phenotype of human infection among genetic groups of *Blastomyces dermatitidis*. J. Infect. Dis. *207*, 814–822.

Menges, R.W., Furcolow, M.L., Selby, L.A., Ellis, H.R., and Habermann, R.T. (1965). Clinical and epidemiologic studies on seventy-nine canine blastomycosis cases in Arkansas. Am. J. Epidemiol. *81*, 164–179.

Meyer, K.C., McManus, E.J., and Maki, D.G. (1993). Overwhelming pulmonary blastomycosis associated with acute respiratory distress syndrome. N. Engl. J. Med. *329*, 1231–1236.

Mongkolrattanothai K., Peev, M., Wheat, L.J., and Marcinak, J. (2006). Urine antigen detection of blastomycosis in pediatric patients. Pediatr. Infect. Dis. J. *25*, 1076–1078.

Montgomery, F.H., and Ormsby, O.S. (1908). Systemic blastomycosis: Its etiologic, pathologic and clinical features as established by a critical survey and summary of twenty-two cases, seven previously unpublished; the relation of blastomycosis to coccidioidal granuloma. Arch. Intern. Med. *2*, 1–41.

Morris, S., Brophy, J., Richardson, S.E., Summerbell, R., Parkin, P.C., Jamieson, F., Limerick, B., Wiebe, L., and Ford-Jones, E.L. (2006). Blastomycosis in Ontario, 1994–2003. Emerg. Infect. Dis. *12*, 274–279.

Morris-Jones R., Youngchin, S., Gomez, B.L., Aisen, P., Hay, R.J., Nosanchuk, J.D., Casadevall, A., and Hamilton, A.J. (2003). Synthesis of melanin-like pigments by *Sporothrix schenckii in vitro* and during mammalian infection. Infect. Immun. *71*, 4026–4033.

Nanjappa, S.W., Heninger, E., Wüthrich, M., Gasper, D.J., and Klein, B.S. (2012). Tc17 cells mediate vaccine immunity against lethal fungal pneumonia in immune deficient hosts lacking CD4+ T cells. PLoS Pathog. *8*, e1002771.

Nemecek, J.C., Wüthrich, M., and Klein, B.S. (2006). Global control of dimorphism and virulence in fungi. Science. *312*, 583–588.

Newman, S.L., Chaturvedi, S., and Klein, B.S. (1995). The WI-1 antigen of *Bastomyces dermatitidis* yeasts mediate binding to human macrophage CD11b/CD18 (CR3) and CD14. J. Immunol. *154*, 753–761.

Nosanchuk, J.D., van Duin, D., Mandal, P., Aisen, P., Legendre, A.M., and Casadevall, A. (2004). *Blastomyces dermatitidis* produces melanin *in vitro* and during infection. FEMS Microbiol. Lett. *239*, 187–193.

Oppenheimer, M., Embil, J., Black, B., Wiebe, L., Limerick, B., MacDonald, K., and Trepman, E.(2007). Blastomycosis of bones and joints. South. Med. J. *100*, 570–578.

Panicker, J., Walsh, T., and Kamani, N. (2006). Recurrent central nervous system blastomycosis in an immunocompetent child treated successfully with sequential liposomal amphotericin B and voriconazole. Pediatr. Infect. Dis. J. *25*, 377–379.

Pappas, P.G., Pottage, J.C., Powderly, W.G., Fraser, V.J., Stratton, C.W., McKenzie, S., Tapper, M.L., Chmel, H., Bonebrake, F.C., Blum, R., *et al.* (1992). Blastomycosis in patients with the Acquired Immunodeficiency Syndrome. Ann. Intern. Med. *116*, 847–853.

Pappas, P.G., Threlkeld, M.G., Bedsole, G.D., Cleveland, K.O., Gelfand, M.S., and Dismukes, W.E. (1993). Blastomycosis in immunocompromised patients. Medicine (Baltimore) 72, 311–325.

Pappas, P.G., Bradsher, R.W., Chapman, S.W., Kauffman, C.A., Dine, A., Cloud, G.A., and Dismukes, W.E. (1995). Treatment of blastomycosis with fluconazole: A pilot study. Clin. Infect. Dis. *20*, 267–271.

Pappas, P.G., Bradsher, R.W., Kauffman, C.A., Cloud, G.A., Thomas, C.J., Newman, C., and Dismukes, W.E. (1997). Treatment of blastomycosis with higher doses of fluconazole. Clin. Infect. Dis. 25, 200–205.

Peck, R.L., and Hauser, C.R. (1938). Chemical studies of certain pathogenic fungi. I. The lipids of *Blastomyces dermatitidis*. J. Am. Chem. Soc. *60*, 2599–2603.

Pelletier, B., Beaudoin, J., Mukai, Y., and Labbe, S. (2002). Fep1, an iron sensor regulating iron transporter gene expression in *Schizosaccharomyces pombe*. J. Biol. Chem. *277*, 22950–22958.

Pfister, J.R., Archer, J.R., Hersil, S., Boers, T., Reed, K.D., Meece, J.K., Anderson, J.L., Burgess, J.W., Sullivan, T.D., Klein, B.S., *et al.* (2011). Non-rural point source blastomycosis outbreak near a yard waste collection site. Clin. Med. Res. *9*, 57–65.

Pine, L. (1954). Studies on the growth of *Histoplasma capsulatum*. I. Growth of the yeast phase in liquid media. J. Bacteriol. *68*, 671–679.

Posaconazole [Package Insert]. (2008). Schering Corporation, Kenilworth, NJ.

Powell, B.L., Drutz, D.J., Hupert, M., and Sun, S.H. (1983). Relationship of progesterone- and estradiol-binding proteins in *Coccidioides immitis* to coccidioidal dissemination in pregnancy. Infect. Immun. *40*, 478–485.

Powell, D.A., and Schuit, K.E. (1979). Acute pulmonary blastomycosis in children: Clinical course and follow-up. Pediatrics. *63*, 736–740.

Proctor, M.E., Klein, B.S., Jones, J.M., and Davis, J.P. (2002). Cluster of pulmonary blastomycosis in a rural community: Evidence for multiple high-risk environmental foci following a sustained period of diminished precipitation. Mycopathologia *153*, 113–120.

Proia, L.A., and Harnish, D.O. (2012). Successful use of posaconazole for treatment of blastomycosis. Antimicrob. Agents. Chemother. *56*, 4029.

Pursley, T.J., Blomquist, I.K., Abraham, J., Anderson, H.F., and Bartley, J.A. (1996). Fluconazole-induced congenital abnormalities in three infants. Clin. Infect. Dis. *22*, 336–340.

Rappleye, C.A., Engle, J.T., and Goldman, W.E. (2004). RNA interference in *Histoplasma capsulatum* demonstrates a role for α-(1,3)-glucan in virulence. Mol. Microbiol. *53*, 153–165.

Rocco, N.M., Carmen, J.C., and Klein, B.S. (2011). *Blastomyces dermatitidis* yeast cells inhibit nitric oxide production by alveolar macrophage inducible nitric oxide synthase. Infect. Immun. *79*, 2385–2395.

Rooney, P.J., and Klein, B.S. (2004). Sequence elements necessary for transcriptional activation of *BAD1* in the yeast phase of *Blastomyces dermatitidis*. Eukaryot. Cell. *3*, 785–794.

Rooney, P.J., Sullivan, T.D., and Klein, B.S. (2001). Selective expression of the virulence factor BAD1 upon morphogenesis to the pathogenic yeast form of *Blastomyces dermatitidis*: evidence for transcriptional regulation by a conserved mechanism. Mol. Microbiol. *39*, 875–889.

Roy, I., and Landau, J.W. (1972a). Composition of the alkali resistant cell wall material of dimorphic *Blastomcyes dermatitidis*. Sabourdia. *10*, 107–112.

Roy, I., and Landau, J.W. (1972b). Protein constituents of cell walls of the dimorphic phases of *Blastomyces dermatitidis*. Can. J. Microbiol. *18*, 473–478.

Saccente, M., and Woods, G.L. (2010). Clinical and laboratory update on blastomycosis. Clin. Microbiol. Rev. *23*, 367–381.

Saccente, M., Abernathy, R.S., Pappas, P.G., Shah, H.R., and Bradsher, R.W. (1998). Vertebral blastomycosis with paravertebral abscess: Report of eight cases and review of the literature. Clin. Infect. Dis. *26*, 413–418.

Saliba F., and Dupont, B. (2008). Renal impairment and amphotericin B formulations in patients with invasive fungal infections. Med. Mycol. *46*, 97–112.

Sarosi, G.A., Eckman, M.R., Davies, S.F., and Laskey, W.K. (1979). Canine blastomycosis as a harbinger of human disease. Ann. Intern. Med. *91*, 733–735.

Saucier, J., and Gauthier, G.M. (2012). Verrucous lesions and ectropion in an immunocompetent individual. Clin. Infect. Dis. *55*, 1390–1391, 1426–1428.

Schutze, G.E., Hickerson, S.L., Fortin, E.M., Schellhase, D.E., Darville, T., Gubbins, P.O., and Jacobs, R.F. (1996). Blastomycosis in children. Clin. Infect. Dis. *22*, 496–502.

Schwarz, J., and Baum, G.L. (1951). Blastomycosis. Am. J. Clin. Pathol. *21*, 999–1029.

Sebghati T.S., Engle, J.T., and Goldman, W.E. (2000). Intracellular parasitism by *Histoplasma capsulatum*: fungal virulence and calcium dependence. Science. *290*, 1368–1372.

Shukla, S., Singh, S., Jain, M., Kumar Singh, S., Chander, R., and Kawatra, N. (2009). Paediatric cutaneous blastomycosis: a rare case diagnosed on FNAC. Diagn. Cytopathol. 37, 119–121.

da Silva, M.B., Marques, A.F., Nosanchuk, J.D., Cadadevall, A., Travassos, L.R., and Taborda, C.P. (2006). Melanin in the dimorphic fungal pathogen *Paracoccidioides brasiliensis*: effects on phagocytosis, intracellular resistance, and drug susceptibility. Microbes. Infect. *8*, 197–205.

Smith, J.A., and Kauffman, C.A. (2009). Endemic fungal infections in patients receiving tumour necrosis factor-alpha inhibitor therapy. Drugs. *69*, 1403–1415.

Smith, J.G. Jr, Harris, J.S., Conant, N.F., and Smith, D.T. (1955). An epidemic of North American blastomycosis. J. Am. Med. Assoc. *158*, 641–646.

Steele, R.W., and Abernathy, R.S. (1983). Systemic blastomycosis in children. Pediatr. Infect. Dis. J. *2*, 304–307.

Stober, A.M. (1914). Systemic blastomycosis: a report of its pathological, bacteriological, and clinical features. Arch. Intern. Med. *13*, 509–556.

Sugar, A.M., and Picard, M. (1991). Macrophage- and oxidant-mediated inhibition of the ability of live *Blastomyces dermatitidis* conidia to transform to the pathogenic yeast phase: implication of the pathogenesis of dimorphic fungal infections. J. Infect. Dis. *163*, 371–375.

Sugar, A.M., Picard, M., Wagner, R., and Kornfeld, H. (1995). Interactions between human bronchoalveolar macrophages and *Blastomyces dermatitidis* conidia: Demonstration of fungicidal and fungistatic effects. J. Infect. Dis. *171*, 1559–1562.

Sullivan, T.D., Rooney, P.J., and Klein, B.S. (2002). *Agrobacterium tumefaciens* integrates transfer DNA into single chromosomal sites of dimorphic fungi and yields homokaryotic progeny from multinucleate yeast. Eukaryot. Cell. *1*, 895–905.

Ta, M., Flowers, S.A., and Rogers, P.D. (2009). The role of voriconazole in the treatment of central nervous system blastomycosis. Ann. Pharmacother. *43*, 1696–1700.

Taborda, C.P., da Silva, M.B., Nosanchuk, J.D., and Travassos, L.R. (2008). Melanin as a virulence factor of *Paracoccidioides brasiliensis* and other dimorphic pathogenic fungi: a minireview. Mycopathologia. *165*, 331–339.

Tarr, M., Marcinak, J., Mongkolrattanothai, K., Burns, J.L., Wheat, L.J., Durkin, M., and Ismail, M. (2007). *Blastomyces* antigen detection for monitoring progression of blastomycosis in a pregnant adolescent. Infect. Dis. Obstet. Gynecol. *2007*, 89059.

Tebbets, B., Stewart, D., Lawry, S., Nett, J., Nantel, A., Andes, D., and Klein, B.S. (2012). Identification and characterization of antifungal compounds using a *Saccharomyces cerevisiae* reporter bioassay. PLoS One. *7*, e36021.

Tebbets, B., Yu, Z., Stewart, D., Zhao, L.X., Jiang, Y., Xu, L.H., Andes, D., Shen, B., and Klein, B.S. (2013). Identification of antifungal natural products via *Saccharomyces cerevisiae* bioassay: insights into macrotetrolide drug spectrum, potency, and mode of action. Med. Mycol. *51*, 280–289.

Teixeira, M.D.L., Theodoro, R.C., da Silveira Derengowsky, L., Nicola, A.M., Bagagli, E., and Felipe, M.S. (2013). Molecular and morphological data supports existence of a sexual cycle in species of the genus *Paracoccidioides*. Eukaryot. Cell. *12*, 380–389.

Torres, H., Rivero, G., and Kontoyiannis, D. (2002). Endemic mycoses in a cancer hospital. Medicine (Baltimore). *81*, 201–212.

Torres, I., Garcia, A.M., Hernandez, O., Gonzalez, A., McWeen, J.G., Restrepo, A., and Arango, M. (2010). Presence and expression of the mating type locus in *Paracoccidioides brasiliensis* isolates. Fungal. Genet. Biol. *47*, 373–380.

Tosh, F.E., Hammerman, K.J., Weeks, R.J., and Sarosi, G.A. (1974). A common source epidemic of North American blastomycosis. Am. Rev. Respir. Dis. *109*, 525–529.

Vasquez, J.E., Mehta, J.B., Agrawal, R., and Sarubbi, F.A. (1998). Blastomycosis in northeast Tennessee. Chest. *114*, 436–443.

Voisard, C., Wang, J., McEvoy, J.L., Xu, P., and Leong, S.A. (1993). *Urbs1*, a gene regulating siderophore biosynthesis in *Ustilago maydis*, encodes a protein similar to the erythroid transcription factor GATA-1. Mol. Cell. Biol. *13*, 7091–7100.

Voriconazole [Package Insert]. (2011). Roerig, Division of Pfizer Inc, New York.

Walker, K., Skelton, H., and Smith, K. (2002). Cutaneous lesions showing giant yeast forms of *Blastomyces dermatitidis*. J. Cutan. Pathol. *29*, 616–618.

Wheat, L.J., Freifeld, A.G., Kleiman, M.B., Baddley, J.W., McKinsey, D.S., Loyd, J.E., and Kauffman, C.A. (2007). Clinical practice guidelines for the management of patients with histoplasmosis: 2007 update by the Infectious Diseases Society of America. Clin. Infect. Dis. 45, 807–825.

Winkler, A., Arkind, C., Mattison, C.P., Burkholder, A., Knoche, K., and Ota, I. (2002). Heat stress activates the yeast high-osmolarity glycerol mitogen-activated protein kinase pathway, and protein tyrosine phosphatases are essential under heat stress. Eukaryot. Cell. *1*, 163–173.

Winn, W., Allen, S., Janda, W., Koneman, E., Procop, G., Schreckenberger, P., and Woods, G. (eds). (2006). Mycology. In Koneman's Color atlas and textbook of diagnostic microbiology, ed 6. (Lippincott Williams and Wilkens, Philadelphia), pp. 1194–1198.

Wolf, P.L., Russell, B., and Shimoda, A. (eds). (1975). Section X: Identification of dimorphic fungi causing systemic mycosis. In Practical Clinical Microbiology and Mycology: Techniques and Interpretations. (John Wiley and Sons, New York), pp. 486–488.

Wu, S.J., Valyi-Nagy, T., Engelhard, H.H., Do, M.A., and Janda, W.M. (2005). Secondary intracerebral blastomycosis with giant yeast forms. Mycopathologia. *160*, 253–257.

Wüthrich, M., Filutowicz, H.I., and Klein, B.S. (2000). Mutation of the WI-1 yields an attenuated *Blastomyces dermatitidis* strain that induces host resistance. J. Clin. Invest. *106*, 1381–1389.

Wüthrich, M., Filutowicz, H.I., Warner, T., and Klein, B.S. (2002). Requisite elements in vaccine immunity to *Blastomyces dermatitidis*: plasticity uncovers vaccine potential in immune-deficient hosts. J. Immunol. *169*, 6969–6976.

Wüthrich, M., Filutowicz, H.I., Warner, T., Deepe, G.S., and Klein, B.S. (2003). Vaccine immunity to pathogenic fungi overcomes the requirement for CD4 help in exogenous antigen presentation to $CD8^+$ T cells: implications for vaccine development in immune-deficient hosts. J. Exp. Med. *197*, 1405–1416.

Wüthrich M., Gern, B., Hung, C.Y., Ersland, K., Rocco, N., Pick-Jacobs, J., Galles, K., Filutowicz, H., Warner, T., Evans, M., Cole, G., and Klein, B. (2011a). Vaccine-induced protection against 3 systemic mycoses endemic to North America requires Th17 cells in mice. J. Clin. Invest. *121*, 554–568.

Wüthrich, M., Krajaejun, T., Shearn-Bochsler, V., Bass, C., Filutowicz, H.I., Legendre, A.M., and Klein, B.S. (2011b). Safety, tolerability, and immunogenicity of a recombinant, genetically engineered, live-attenuated vaccine against canine blastomycosis. Clin. Vacc. Immun. *18*, 783–789.

Wüthrich, M., Ersland, K., Sullivan, T., Galles, K., and Klein, B.S. (2012). Fungi subvert vaccine T cell priming at the respiratory mucosa by preventing chemokine-induced influx of inflammatory monocytes. Immunity. *36*, 680–692.

Yates-Siilata, K.E., Sander, D.M., and Keath, E.J. (1995). Genetic diversity in clinical isolates of the dimorphic fungus *Blastomyces dermatitidis* detected by a PCR-based random amplified polymorphic DNA assay. J. Clin. Microbiol. 33, 2171–2175.

Youngchim, S., Hay, R.J., and Hamilton, A.J. (2005). Melanization of *Pencillium marneffei in vitro* and *in vivo*. Microbiology *151*, 291–299.

Zampogna, J., Hoy, M., and Ramos-Caro, F. (2003). Primary cutaneous North American blastomycosis in an immunosuppressed child. Pediatr. Dermatol. *20*, 128–130.

Zhou, L.W., Haas, H., and Marzluf, G.A. (1998). Isolation and characterization of a new gene, *sre*, which encodes a GATA-type regulatory protein that controls iron transport in *Neurospora crassa*. Mol. Gen. Genet. *259*, 532–540.

New Insights into Pathogenesis of *Pneumocystis* Pneumonia

13

Jakrapun Pupaibool and Andrew H. Limper

Abstract

Pneumocystis pneumonia (PCP) remains a major cause of morbidity and mortality in HIV-infected patients. In addition, the prevalence of PCP is increasing in non-HIV patients with suppressed immune status, following the use of potent immunosuppressive therapies. Our understanding of *Pneumocystis* is based on immunosuppressed animal models. Investigations of the *Pneumocystis* life cycle, pathogenesis, and host immune response provide attractive targets for the development of novel anti-*Pneumocystis* agents. Sulfamethoxazole–trimethoprim remains the first-line regimen for treatment and prophylaxis of PCP. Many aspects of *Pneumocystis* biology remain unanswered or controversial, such as respiratory colonization, mode of transmission, clinical significance of mutations to currently available anti-*Pneumocystis* agents, and prophylaxis and therapy in non-HIV immunocompromised patients. There is a critical need to continue investigating this unique medically important pathogenic fungus in order to allow us to develop more effective strategies to prevent and treat *Pneumocystis* infection.

Introduction

Pneumocystis pneumonia (PCP) in humans is caused by *Pneumocystis jirovecii*, a medically important fungus causing pulmonary infection in a wide variety of immunocompromised patients. While the incidence of PCP has decreased since the introduction of highly active anti-retroviral therapy (HAART), it still remains one of the leading causes of mortality in HIV-infected patients (Hanna *et al.*, 2007; Sullivan *et al.*, 2007). In addition, there are increasing numbers of non-HIV infected patients with suppressed immune status, resulting in growing incidence of PCP in these populations (Sepkowitz, 2002; Thomas and Limper, 2004). Advances in molecular technologies and new techniques over the past decade have dramatically improved our understanding of many aspects of these challenging organisms and how they cause disease. However, there remain several areas of scientific uncertainty around this important opportunistic fungal pathogen that still require further study. In particular, the lack of a robust long-term *ex vivo* cultivation method for *Pneumocystis* has remained a major obstacle to understanding this pathogen. This chapter summarizes the current knowledge of the biology, life cycle, pathogenesis, and immune response during *Pneumocystis* infection, and further captures recent developments in PCP diagnosis as well as new targets for potential novel anti-*Pneumocystis* therapies.

Taxonomy and nomenclature

In 1909, *Pneumocystis* was first described in the lungs of guinea pigs by Carlos Chagas, who believed these organisms to be a new form of *Trypanosoma cruzi* (Bartlett and Smith, 1991). In 1910, Antonio Carini discovered similar cysts in the lungs of rats (Gajdusek, 1976). He sent the specimens to the Institut Pasteur in Paris, where the pulmonary cysts were concluded to be a new organism unrelated to *Trypanosoma*. The organism was subsequently named '*Pneumocystis carinii*' in honour of Dr Carini. However, it continued to be classified as a protozoan until 1953 when Giese first proposed that *Pneumocystis* may be a form of yeast (Carmona and Limper, 2011). In 1970, the fungal nature of *Pneumocystis* was again addressed by Vavra and Kucera, based upon their ultrastructural studies (Vavra and Kucera, 1970). The controversy had remained until 1988 when Edman *et al.* (1988) sequenced the organism's ribosomal RNA and proved the fungal nature of *Pneumocystis*.

Pneumocystis carinii was originally thought to be a single species that caused infection in a variety of mammalian hosts. However, in 1972, antigenic differences between rat and human isolates of *P. carinii* were first reported (Kim *et al.*, 1972). Further investigation utilizing polymerase chain reaction (PCR) and DNA sequencing techniques revealed that there was distinct genetic diversity within organisms of the genus *Pneumocystis* (Hong *et al.*, 1990; Wakefield *et al.*, 1988, 1990). Of considerable interest was the observation that specific *Pneumocystis* species each infect a single species of mammalian host (Table 13.1). Furthermore, *Pneumocystis* derived from one host species cannot proliferate or cause pneumonia when transferred to a different mammalian host species. For example, *P. carinii* isolates that infect rats cannot infect humans, and vice versa (Gigliotti *et al.*, 1993). The reason for this stringent host specificity remains unclear. This has led to renaming of *Pneumocystis* that infects humans initially to *P. carinii, humanus* in 1988 (Hughes and Gigliotti, 1988). Subsequently, in 1994, the name was changed to *P. carinii* f. sp. *hominis* (The *Pneumocystis* Workshop, 1994) Interestingly, '*P. jiroveci*' was proposed in 1976 by Frankel to honour to the Czech parasitologist, Otto Jirovec (Frenkel, 1976), but it was not more widely accepted until 2002 when Stringer *et al.* (2002) proposed the same nomenclature.

The name proposed by Frankel in 1976 was based on the International Code of Zoological Nomenclature (ICZN), as at that time *Pneumocystis* was still classified as a protozoan. Stringer *et al.* used the same name for their publication in 2002, by which time it was clear that *Pneumocystis* was a member of the fungi. In 2009, Stringer *et al.* (2009) corrected the error in his previous publication that the species epithet should be modified from '*jiroveci*' to '*jirovecii*' according to International Code of Botanical Nomenclature (ICBN) (Redhead *et al.*, 2006). The correct name and classification under ICBN for this fungal pathogen is *Pneumocystis jirovecii* (Table 13.2).

Table 13.1 *Pneumocystis* species that infect specific mammalian hosts

Mammalian host	*Pneumocystis*
Human	*P. jirovecii*
Rat	*P. carinii*, *P. wakefieldiae*
Mouse	*P. murina*
Rabbit	*P. oryctolagus*

Table 13.2 Currently accepted human-derived *Pneumocystis jirovecii* taxonomy

Kingdom	Fungi
Phylum	Ascomycota
Class	Pneumocystidomycetes
Order	Pneumocystidales
Family	Pneumocystidaceae
Genus	*Pneumocystis*
Species	*Pneumocystis jirovecii*

Pneumocystis life cycle

Life cycle stages of *Pneumocystis* species

Various life cycles for *Pneumocystis* spp. organisms have been proposed since the initial 1984 publication of ultrastructural details of *P. carinii* by Yoshida *et al.* (1984). Their hypotheses were based on morphologic observation from transmission electron microscopy (TEM) and related studies. The inability to cultivate *Pneumocystis* outside the host lung prevents investigators from actually observing fungal growth or stage differentiation *in vitro*. Consistent with most proposed life cycles are two major forms in the *Pneumocystis* life cycle, the smaller trophic form and the cyst form. The sporocyte, a form between the trophic form and the cyst form, and the precyst are other forms of the *Pneumocystis* life cycle that have also been proposed (Aliouat-Denis *et al.*, 2009). The term cyst persists widely in the literature, though recent studies have used the term Ascus, in recognition of the classification of the *Pneumocystis* within the ascomycetous fungi (Cushion, 2010).

The trophic forms represent more than 90% of the total population of *Pneumocystis* in most host lungs. These forms are believed to be haploid, and are thin-walled, mononuclear cells, measuring 1–4 μm in diameter with cytoplasmic projections used to attach to alveolar epithelial cells (Aliouat-Denis *et al.*, 2008; de Souza and Benchimol, 2005; Thomas and Limper, 2007). Diploid trophic forms are also present in small numbers (Wyder *et al.*, 1998) and can result from the conjugation of two trophic forms (Cushion, 2004). After a proposed mating process, these organisms enter meiosis and develop into the cyst forms with eight nuclei (Burgess *et al.*, 2008, 2009). Since the trophic forms predominate over the cyst forms by more than 8:1, it is believed that the trophic forms can also undergo asexual reproduction by haploid mitosis (Cushion, 2004; de Souza and Benchimol, 2005; Thomas and Limper, 2007). The trophic form can be identified by different stains including Gram–Weigert, Wright–Giemsa, modified Wright–Giemsa (Diff-Quik), and Papanicolaou (Table 13.3). *Pneumocystis* trophic and cyst forms were shown to be detectable in respiratory samples using immunofluorescence in 1996 (Calderon *et al.*, 1996).

The cyst form is a thick, smooth-surface structure measuring 8–10 μm in diameter (Carmona and Limper, 2011). β-glucan is a major component of the cyst wall (Carmona *et al.*, 2012; Lebron *et al.*, 2003; Vassallo *et al.*, 2000). A mature cyst contains 8 intracystic nuclei (spores) which gives rise to eight haploid trophic forms upon rupture of the wall, in a process known as excystment (Aliouat-Denis *et al.*, 2009; Carmona and Limper, 2011). β-glucanases and other proteins are required for degradation of the cyst walls during the excystment process (Villegas *et al.*, 2010). This results in the release of the trophic forms which can then complete the life cycle. The cyst form is stainable by calcofluor white, Grocott-Gomori methenamine silver (GMS), cresyl echt violet, and toluidine blue O (Table 13.3).

Control of the *Pneumocystis* life cycle

Since *Pneumocystis* species cannot be continuously cultured in defined media, factors that regulate the life cycle remain poorly defined. That said, accumulating evidence implicates that adherence of *Pneumocystis* to the host epithelial cells in short term co-cultures strongly promotes growth of the organisms (Cushion *et al.*, 1985; Kottom *et al.*, 2011a; Limper *et al.*, 1997). Firm adhesion of *Pneumocystis* to host lung alveolar epithelial cells requires active cytoskeletal re-arrangement of the trophic forms to conform to the epithelial cell surface (Fig. 13.1) (Limper and Martin, 1990).

More recently, genes and molecules controlling *Pneumocystis* cell-cycle transitions have been studied using heterologous expression in *Schizosaccharomyces pombe* and *Saccharomyces cerevisiae*. The cell-cycle control of *Pneumocystis* is regulated by cell division cycle (Cdc) molecules, cell-cycle inhibitors, and Cdc-related cyclins. Cdc2 cyclin-dependent kinase (CDK) regulates transition from G1 to S phase, and from G2 to mitosis. Cdc2 is activated by Cdc25, which is a

Table 13.3 Life cycle stages of *Pneumocystis* species

Characteristics	Trophic forms	Cyst forms
Shape	Irregular, thin wall, cytoplasmic projections	Round, thick wall, smooth surface
Size (μm)	1–4	8–10
Number of nuclei	1, 2 (rarely)	2, 4, 8 (mature cyst)
Stains	Gram-Weigert, Wright–Giemsa, modified Wright–Giemsa (Diff-Quik), Papanicolaou, immunofluorescence with monoclonal antibody	Calcofluor white, Grocott–Gomori methenamine silver (GMS), cresyl echt violet, toluidine blue O, immunofluorescence with monoclonal antibody

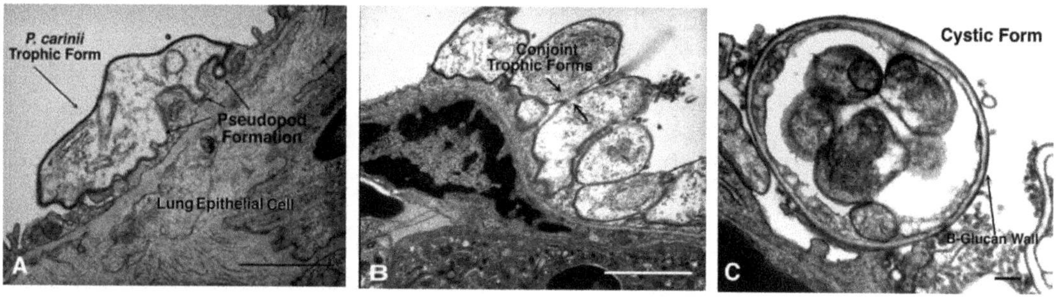

Figure 13.1 Transmission electron micrographic analysis of *Pneumocystis* life forms in the rat. (A) Trophic forms adherent to lung epithelium. (B) Conjoint trophic forms on the lung surface (C) Cystic form, which contains trophic form progeny. The internal bars of each panel represent 1 mm in length for scale. (Adapted from Limper *et al.*, 1997.)

mitotic phosphatase. Cdc25 also restores cell-cycle progression and DNA damage checkpoints (Gustafson *et al.*, 2001). Cdc2 kinase has greater activity in the trophic forms as compared to the cyst from, while Cdc13 B-type cyclin activity was found to be higher in the cyst forms (Thomas *et al.*, 1998). In addition, replication and repair in *Pneumocystis* DNA appears to involve a fungal type PcRtt109 histone transacetylase (Kottom *et al.*, 2011b). There are certainly additional unknown cell cycle molecules needed to regulate the growth of the two forms of *Pneumocystis* species.

Pathogenesis of *Pneumocystis* pneumonia

Transmission of *Pneumocystis* pneumonia

Despite the inability to propagate this fungal pathogen *in vitro*, *Pneumocystis* infections are widely found in mammalian hosts. This allows us to conduct experimental studies using immune suppressed animal models of infection. Immunodeficient mice and macaques become infected with *Pneumocystis* when co-housed with infected animals of the same species (Powles *et al.*, 1992; Vogel *et al.*, 1993). Results from mouse infection models have suggested that the cyst forms are the agents of transmission. Treatment of immune suppressed mice with the echinocandin anidulafungin significantly reduced the numbers of cyst forms and also reduced the ability of mice to transmit the infection to other susceptible hosts (Cushion *et al.*, 2010).

Following the Second World War, there were epidemics of 'plasma cell interstitial pneumonia' (PCP) in malnourished and premature infants who stayed in crowded institutions in many European countries (Dutz *et al.*, 1976; Goetz and Peller, 1974). More recently, there has also been a cluster of PCP cases occurring in transplant and cancer patients co-housed in the same hospital (Singer *et al.*, 1975). Another report proposed transmission of *P. jirovecii* from AIDS patients to renal transplant recipients who attended an HIV clinic (Chave *et al.*, 1991). These observations suggest person-to-person transmission of *Pneumocystis* (Table 13.4). However, it remains unclear whether air-borne transmission is the only mode of PCP acquisition in these patients. As such, there has been insufficient evidence to strongly recommend air-borne isolation in hospitalized patients with PCP. However, most authorities recommend that persons who are immunosuppressed and at risk for PCP should not share the same room with a patient with active PCP.

Another possible mode of transmission of PCP is from environmental acquisition. *Pneumocystis* DNA was identified from environmental samples, including ambient air (Wakefield, 1994, 1996) and pond water (Casanova-Cardiel and Leibowitz, 1997). The DNA in the environment could possibly have come from infected animals residing in that area. Choukri *et al.* (2010) quantified *P. jirovecii* specific DNA in the air

Table 13.4 Acquisition of *Pneumocystis* pneumonia

Possible modes of transmission	Possible mechanisms of acquisition
Person-to-person spread	Air-borne transmission
Environmental acquisition	Ecological niche outside the mammalian hosts
Reactivation of a latent infection	Exposure and respiratory colonization early in life

surrounding patients with PCP. This suggested that *Pneumocystis* organism forms were shed into the surrounding air. However, the simple presence of DNA is not a completely reliable surrogate marker for viable organisms, and it unfortunately remains impossible to culture *Pneumocystis* from environmental specimens.

Pneumocystis colonization

Another proposed mode of active PCP acquisition is through reactivation of latent infections. Serological reactivity to *Pneumocystis* occurs during the first few months of healthy infants' lives, an observation that supports exposure to the organism at extremely early time points in their lives. By the age of 4 years, more than 80% of children exhibit significant titres of circulating anti-*Pneumocystis* antibodies (Pifer *et al.*, 1978).

In more recent years, PCR detection of *Pneumocystis* DNA in respiratory specimens from non-HIV immunocompetent adults was published in several studies (Morris *et al.*, 2002; Nevez *et al.*, 1997; Sing *et al.*, 2001). However, with some of the nested PCR methodologies, *Pneumocystis* DNA was detected in very low levels in respiratory secretions, supporting the concept of colonization with this organism (Wissmann *et al.*, 2010). Whether respiratory colonization of PCP is temporary or persistent is not entirely clear, though transient colonization is evidenced from previous published studies. The overall host risk factors and environmental conditions necessary for PCP colonization have yet to be fully determined.

However, in some studies smoking has been shown to be a risk factor for *Pneumocystis* colonization (Morris *et al.*, 2004a; Vidal *et al.*, 2006). A host colonized with *Pneumocystis* can develop clinical PCP when their immune status is suppressed. Alternatively, the colonized host can possibly serve as reservoirs of *Pneumocystis* and sources of infection for other susceptible immunocompromised individuals. Molecular evidence by Rivera *et al.* (2008) supports the transmission of PCP from colonized immunocompetent carriers to infants. Moreover, carriage of PCP containing genetic mutations of the dihydropteroate synthase (DHPS) gene (Calderon *et al.*, 2004) can contribute to the spread of potentially sulfa-resistant strains to susceptible patients with no previous sulfamethoxazole exposure. The clinical significance of such mutations will be discussed later in this chapter.

Pneumocystis colonization in individuals with chronic pulmonary diseases, including chronic obstructive pulmonary disease (COPD), interstitial lung diseases (ILD), cystic fibrosis (CF), and lung cancers, has been reported. The colonization rate in COPD patients ranged from 5–55% (Calderon *et al.*, 2007; Maskell *et al.*, 2003; Nevez *et al.*, 2006). Additionally, the colonization rate seems to correlate with severity of COPD (Morris *et al.*, 2004b). The inflammatory response to β-glucan, an important *Pneumocystis* cyst wall component, was proposed to affect COPD development and progression by provoking immune reactions through secretion of interleukin-6 (IL-6), IL-8, tumour necrotic factor-α (TNF-α), and macrophage inflammatory protein-2 (MIP-2), and other pathways (Lebron *et al.*, 2003; Varela *et al.*, 2003). The prevalence of *Pneumocystis* colonization in CF patients was 1–22% (Gal *et al.*, 2010; Respaldiza *et al.*, 2005; Sing *et al.*, 2001), while it was reported in up to 23–40% in patients with ILD (Shimizu *et al.*, 2009; Vidal *et al.*, 2006). Patients with small cell lung cancer seemed to have a higher rate of colonization compared to individuals with non-small cell lung cancer (de la Horra *et al.*, 2004; Mori *et al.*, 2010) with a colonization rate of 100% found in one small study (de la Horra *et al.*, 2004). Although *Pneumocystis* colonization seems to correlate with and potentially may have modifying clinical consequences in patients with chronic pulmonary diseases, it is difficult to prove direct causality. Further studies are needed to determine the exact mechanisms and effects of *Pneumocystis* colonization in patients with chronic lung disorders.

Histopathological features of *Pneumocystis* pneumonia

The original descriptions of histopathological features and organism–host interactions were based on appearances of tissue in haematoxylin–eosin stain sections. Such examinations reveal foamy eosinophilic exudates in the alveolar spaces on lung histopathology. These exudates contain abundant organisms, host and organism derived proteins, and inflammatory cells consisting of macrophages, neutrophils, and lymphocytes.

Pneumocystis surface antigens

Studies of freshly derived organism have identified a number of antigens on the surface of *Pneumocystis*. The major surface glycoprotein also termed glycoprotein A, is a heavily glycosylated, mannose-rich major surface glycoprotein (MSG/gpA) complex that represents a predominant surface component of the organism and that has been shown to participate in the attachment process to alveolar epithelial cells (Gigliotti *et al.*, 1988; Linke *et al.*, 1989; Lundgren *et al.*, 1993; O'Riordan *et al.*, 1995; Vuk-Pavlovic *et al.*, 2001). MSG/gpA-related proteins and Prt1 proteins may participate in evading host defence mechanisms during the establishment of lung infection. The Prt1 protein also possesses protease activity that is believed to participate in post-translational processing, resulting in biologically active *Pneumocystis* surface antigens (Lugli *et al.*, 1999). Kex1, is a single copy encoded serine endoprotease, that also is expressed by *Pneumocystis*, and which has been postulated as a protective target in vaccine development (Kutty and Kovacs, 2003; Zheng *et al.*, 2001).

Pneumocystis cell wall components

The *Pneumocystis* cyst wall is comprised of complex carbohydrates including β-glucans, mannoproteins, chitins, and other structural proteins. β(1,3/1,6)-D-glucan is the main structural component of the *Pneumocystis* cell wall (Douglas, 2001). *Pneumocystis* Gsc1 (PcGsc1) polymerizes uridine 5′-diphosphoglucose into the β(1,3)-D-glucan backbone. The activity of the gene encoding this protein was found to be active only in cyst forms of the organism (Kottom and Limper, 2000). β(1,6)-D-glucan side chains are subsequently added onto the core β(1,3)-D-glucan polymers. PcEng2 β-1,3-endoglucanase is an additional protein participating in *Pneumocystis* cell wall remodelling (Kottom and Limper, 2000; Villegas *et al.*, 2010). PcPhr1, provides further activity in glucan cross-linking, further strengthening the *Pneumocystis* cell wall (Kottom *et al.*, 2001). The β(1,3/1,6)-D-glucan cell wall mesh provides rigidity and stability of the cyst forms, but it also provokes inflammatory cascades and induces an immune response in infected host lungs. β(1,3/1,6)-D-glucan-induced lung inflammations results in severe hypoxia and acute respiratory failure (Carmona *et al.*, 2006; Lebron *et al.*, 2003). Accordingly, supplemental corticosteroid therapy in addition to anti-*Pneumocystis* agents have been successful in improving the outcome of patients with moderately severe PCP (Montaner *et al.*, 1990).

Serum β(1,3)-D-glucan has been proposed as a useful biomarker for the diagnosis of invasive fungal infection. Given that β(1,3)-D-glucan is a non-specific antigen for *Pneumocystis* and many other fungi, a positive test result does not indicate which fungi is the causative pathogen. The test is mainly useful for excluding invasive fungal infection in high-risk patients. Initial clinical studies to validate β(1,3)-D-glucan for diagnosis of invasive fungal infections did not include patients with PCP. Serum β(1,3)-D-glucan was subsequently found to be significantly higher in patients with PCP (Desmet *et al.*, 2009; Marty *et al.*, 2007; Tasaka *et al.*, 2007). The results from one retrospective study suggested a higher cut-off value for diagnosis of PCP. At a cut-off value of 80 pg/ml, the sensitivity and specificity were 100% and 89.3%. Using the cut-off value of 100 pg/ml, the sensitivity remained at 100%, while the specificity increased to 96.4% (Desmet *et al.*, 2009).

Adherence of *Pneumocystis* to alveolar epithelial cells

As mentioned above, the adherence of *Pneumocystis* to lung cells represents an important feature of the organism's life cycle, which stimulates growth of the fungus in the host. Accordingly, it is not surprising that several mechanisms have evolved to mediate attachment of *Pneumocystis* cells to host alveolar epithelial cells (AEC). The trophic forms of *Pneumocystis* adhere predominantly to Type I AECs and to a lesser extent to Type II AECs (Yoshida *et al.*, 1984). Binding of the trophic forms to AEC is facilitated by extracellular matrix (ECM), particularly fibronectin and vitronectin, through PcInt1, an integrin-like molecule, which is expressed on the surface of the trophic forms (Kottom *et al.*, 2008; Limper *et al.*, 1993). The production of surfactant phospholipids in the ECM is reduced during active *Pneumocystis* infection, leading to increased lung stiffness and difficult ventilation during PCP.

Pneumocystis carinii genome

The *Pneumocystis carinii* genome sequencing project and advances in molecular biology approaches have improved our understanding of PCP pathogenesis and *Pneumocystis* biology. The *P. carinii* genome is comprised of approximately 8 million DNA base pairs with high adenine and thymine contents of approximately 65% (Stringer and Cushion, 1998; Thomas *et al.*, 1999). Characterization and functional analysis of *Pneumocystis* genes using *Schizosaccharomyces pombe* and *Saccharomyces cerevisiae* models have revealed that *Pneumocystis* genome sequences and functions are homologous to these experimental yeast organisms.

Intracellular signalling pathways

The binding of *Pneumocystis* to AEC and ECM, particularly collagen, fibronectin and vitronectin, stimulates signalling pathways that centre on the activation of the *Pneumocystis* PcSte20 and PcCbk1 kinases. PcSte20 kinase and PcCbk1 kinase are up-regulated and activated following the attachment process (Kottom *et al.*, 2011a). Subsequently, proliferation, mating/meiosis, and cell wall remodelling processes of the organism are believed to be stimulated (Kottom *et al.*, 2003; Kottom and Limper, 2004). PcSte20 exhibits activity in mating and proliferation, while PcCbk1 exhibits activity in regulating cell wall remodelling enzymes. PcSte20 activation requires the presence of a PcCdc42 protein, a Rho family GTPase, which acts as an immediate upstream binding partner necessary for PcSte20 activity (Krajicek *et al.*, 2010). Other additional mechanisms are also involved in regulating PcCdc42–PcSte20 complexes. For instance, preliminary observations suggest that PcCdc42-PcSte20 might be regulated by binding of PcCdc42 with GTPase activating proteins (GAPs) as occurs in other fungal organisms (Limper *et al.*, 1997; Smith *et al.*, 2002).

Host immune response to *Pneumocystis* infection

Macrophages

Alveolar macrophages are a major component of the innate immune response that mediate clearance of *Pneumocystis* from the lungs. MSG/gpA on *Pneumocystis* binds to mannose receptors on alveolar macrophages, leading to internalization of *Pneumocystis* cysts. In addition, the interaction of β(1,3)-D-glucan and dectin-1 β-glucan receptor is another binding mechanism of *Pneumocystis* cysts to macrophages (Steele *et al.*, 2003). Following binding, *Pneumocystis* organisms are phagocytized into a phagosome. Macrophages are further stimulated by inflammatory cytokines such as IFN-γ, and are responsible for direct killing of *Pneumocystis*. Mannose receptor down regulation and impairment of phagocytosis were found in patients with AIDS (Koziel *et al.*, 1998a). Surfactant protein A (SP-A) was shown to variably augment binding of *Pneumocystis* to mannose receptor (Koziel *et al.*, 1998b; Williams *et al.*, 1996), while SP-D was shown to inhibit uptake of *Pneumocystis* (Yong *et al.*, 2003). Antibody-mediated phagocytosis through Fc receptors was also reported (Taylor *et al.*, 1992). The interaction of β(1,3)-D-glucan and dectin-1 receptor on alveolar macrophages mediates uptake and killing of *Pneumocystis* through reactive oxygen species (Steele *et al.*, 2003).

Neutrophils

Immunocompromised patients with PCP were reported to have high numbers of neutrophils in their bronchoalveolar lavage (BAL) fluid (Fleury *et al.*, 1985). β(1,3/1,6)-D-glucan was reported to activate the transcription factor activator protein-1 (AP-1) and nuclear factor-κB (NF-κB) in the AEC. IL-8, a potent chemoattractant factor for neutrophils has been shown to be released following *Pneumocystis* β-glucan mediated inflammatory activation (Carmona *et al.*, 2010). IL-17, an additional cytokine released by T-helper 17 (Th17) cells, can also stimulate alveolar epithelial cells to release IL-8 (Jones and Chan, 2002; Kawaguchi *et al.*, 2001). Phagocytosis of *Pneumocystis* by neutrophils was demonstrated to be enhanced by a soluble mannose receptor binding domain fused with the Fc portion of immunoglobulin G (IgG) (Stehle *et al.*, 2000). In humans, high neutrophil levels correlate with poor oxygenation and worse patient outcome (Limper *et al.*, 1989). Interestingly in murine models depletion of neutrophils did not significantly alter measures of lung injury (Swain *et al.*, 2004).

$CD4^+$ T cells

The essential adaptive immune response to *Pneumocystis* infection is mediated through $CD4^+$ T cells. The risk of infection increases when the $CD4^+$ T cell counts are less than 200 cells/μl (Phair *et al.*, 1990; Shellito *et al.*, 1990). However, the exact mechanisms of clearance of *Pneumocystis* by these cells have not well delineated. Th1 cells produce various cytokines including IFN-γ, while Th2 cells produce IL-4, IL-5, IL-13, and promote antibody production. Shellito *et al.* (2000) demonstrated in mouse model that both types of $CD4^+$ T cells were increased in the lungs after infection, with a higher response of Th2 cells. Theus *et al.* (1998) found that both IFN-γ and IL-4 levels from peripheral blood mononuclear cells (PBMC) were low in AIDS patients. The production of anti-*Pneumocystis* IgG was shown to be mediated by $CD4^+$ T cells through the CD40–CD40L pathway (Wiley and Harmsen, 1995).

$CD8^+$ T cells

Previous studies have revealed that severe combined immunodeficient (SCID) mice with PCP were unable to clear the infection following adoptive transfer of $CD8^+$ T cells (Roths and Sidman, 1992). Furthermore, mice depleted of $CD8^+$ T cells were not susceptible to PCP (Hanano *et al.*, 1996), but mice depleted of both $CD4^+$ T cells and $CD8^+$ T cells developed more severe infection than mice depleted of $CD4^+$ T cells alone (Beck *et al.*, 1996). Mice depleted of $CD4^+$ T cells with PCP were shown to have influx of $CD8^+$ T cells in their lungs (Beck *et al.*, 1991). These might indicate that $CD8^+$ T cells are crucial for *Pneumocystis* clearance when $CD4^+$ T cells are depleted. However, $CD8^+$ T cells have also been demonstrated to be strongly responsible for causing pulmonary inflammation and severe pulmonary damage (Wright *et al.*, 1999).

γδ T cells

Mice depleted of γδ T cells were not found to be susceptible to PCP (Agostini *et al.*, 1995). However, γδ T cells do play a role in host defence of mice depleted of αβ T cells (Hanano and Kaufmann, 1999). HIV-infected patients with PCP have been shown to exhibit elevated numbers of γδ T cells in their blood and BAL fluid (Agostini *et al.*, 1995; Kagi *et al.*, 1993). Regulation of inflammation induced by $CD8^+$ T cells was further demonstrated to be another function of γδ T cells (Steele *et al.*, 2002).

B cells

B-cell-deficient mice were also observed to be susceptible to PCP (Marcotte *et al.*, 1996). HIV-infected patients have been shown to have lower anti-*Pneumocystis* IgG compared to healthy individuals (Laursen and Andersen, 1998). Transfer of *Pneumocystis* monoclonal antibodies or hyperimmune *Pneumocystis* antisera conferred passive protection and promoted clearance of PCP in animal models (Bartlett *et al.*, 1998; Gigliotti *et al.*, 2002; Gigliotti and Hughes, 1988). B cells release cytokines and immunoglobulin M (IgM), which facilitates dendritic cell migration and phagocytosis by phagocytic cells (Rapaka *et al.*, 2010). B cells have a critical role in priming and activating $CD4^+$ T cells in response to PCP (Lund *et al.*, 2006). Further, our group has recently reported cases of PCP that have developed in patients receiving the B cell suppressive agent rituximab alone (Martin-Garrido *et al.*, 2012). Therefore, accumulating evidence does indicate that B-cell function contributes to host defence against *Pneumocystis*.

Infection in non-HIV population

The first published cases of PCP in humans were in 1942 by van der Meer and Brug, who reported *Pneumocystis* as the cause of pneumonia in a

3-month-old infant and in two autopsy cases (a 4-month-old infant and a 21-year-old adult) (Hughes, 2003). In 1952, Vanek and Jirovec reported *Pneumocystis* as the cause of interstitial plasma cell pneumonia in neonates (Vanek and Jirovec, 1952). However, interstitial plasma cell pneumonia in infants was first described in the 1920s in Germany but the causative agent was not known at that time (Gajdusek, 1976). Subsequently, *Pneumocystis* infection in adults was reported in patients with hematologic malignancies receiving corticosteroids and other immunosuppressive agents (Limper *et al.*, 1989; Yale and Limper, 1996).

The prevalence of PCP increased dramatically and awareness of this opportunistic infection was heightened as a major cause of pneumonia in HIV-infected patients during the 1980s (Masur *et al.*, 1981). With the introduction of HAART in the 1990s, the prevalence of PCP in HIV-infected patients has decreased. However, PCP still remains a major AIDS-defining illness in HIV-infected patients. In addition, growing numbers of immunosuppressed patients without HIV infection, such as patients with cancer, patients undergoing solid organ and hematologic stem cell transplantation have rendered PCP as an important cause of pneumonia in these populations over the past decade.

In addition to cancer and transplant patients receiving chemotherapy, patients with rheumatoid arthritis (RA) and other rheumatologic and inflammatory conditions receiving immunomodulatory agents including higher doses of corticosteroids, cytotoxic agents and novel immune suppressants have also exhibited increased risk of developing mycobacterial and fungal infections, including *Pneumocystis* pneumonia. Novel immune suppressive agents implicated with PCP include TNF-α receptor antibody (etanercept), anti-TNF-α antibody (infliximab, adalimumab, golimumab), anti-B7 protein antibody (abatacept), IL-1 receptor antibody (anakinra), and IL-6 receptor antibody (tocilizumab). The incidence of PCP has been shown to be highest with infliximab (Mori and Sugimoto, 2012). Currently, there is no universal consensus for the optimal PCP prophylaxis in non-HIV immunocompromised patients.

Non-HIV immunocompromised patients with PCP tend to mount a more robust inflammatory response compared to HIV-infected patients with PCP. As a result, they have been observed to have less organism burden and higher neutrophil counts in BAL fluid, and they tend to present more acutely with fever, chills, lower arterial oxygenation, impaired gas exchange, and respiratory failure requiring mechanical ventilatory support (Festic *et al.*, 2005; Limper *et al.*, 1989). Of interest, the organism burden in BAL fluid was not shown to be correlated with poor oxygenation (Limper *et al.*, 1989). This suggested instead that robust host inflammatory responses to *Pneumocystis* contribute to lung injury and the clinical manifestations of PCP. These observations set the framework for the use of adjunctive corticosteroid therapy in the treatment of PCP. Corticosteroid therapy has been shown to be beneficial for PCP with a partial arterial oxygen tension less than 70 mmHg or an alveolar–arterial oxygen gradient greater than 35 mmHg (National Institutes of Health—University of California Expert Panel for Corticosteroids as Adjunctive Therapy for *Pneumocystis* Pneumonia, 1990). Unfortunately, prospective studies for the use of adjunctive corticosteroids in PCP have been performed only in HIV-infected patients. The benefits for non-HIV immunocompromised patients are not fully known, though a retrospective analysis does suggest potential benefit in these patients as well (Pareja *et al.*, 1998).

Molecular targets of anti-*Pneumocystis* agents

Ivady and Paldy were the first to report that pentamidine isethionate, an agent used at that time for the treatment of African trypanosomiasis, was effective for treating interstitial plasma cell pneumonia, the entity that was later identified as *Pneumocystis* pneumonia (Ivady and Paldy, 1957). Pentamidine served as the mainstay of treatment for PCP until 1974 when Hughes *et al.* demonstrated that sulfamethoxazole–trimethoprim (SMX-TMP) was effective against *Pneumocystis* in rodent models (Hughes *et al.*, 1974) (Table 13.5). Subsequently, they demonstrated the usefulness of SMX-TMP for treatment and prophylaxis of PCP in children with acute leukaemia (Hughes *et al.*, 1975, 1977). Since 1975, SMX-TMP has been considered the drug of choice for the treatment and prophylaxis of PCP. Both SMX-TMP and pentamidine have significant adverse reactions, such as bone marrow suppression and hypoglycaemia. Later studies revealed that *Pneumocystis* requires *de novo* synthesis of thymidine for DNA synthesis since it does not possess a thymidine salvage pathway (Vestereng and Kovacs, 2004). Other agents involved in this pathway have also been used for the treatment of PCP. Trimetrexate, which targets dihydrofolate reductase (DHFR) in a fashion similar to TMX, has been used in combination with leucovorin in clinical studies, but relapses after treatment have been frequently observed (Fig. 13.2) (Allegra *et al.*, 1988; Sattler *et al.*, 1990). Dapsone has also been employed in combination with TMP in HIV-infected patients since 1986 (Leoung *et al.*, 1986), but methaemoglobinaemia was found be one of the major adverse reactions (Reiter and Cimoch, 1987). The molecular target of dapsone is dihydropteroate synthase (DHPS).

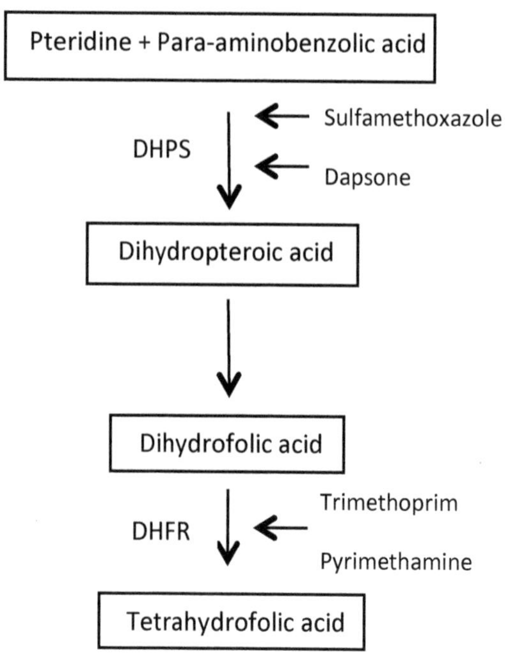

Figure 13.2 Folic acid synthesis pathway.

Because of significant adverse reactions with many of these agents, which ultimately target DNA synthesis, other agents focusing on different targets have been tested in animal models and in humans. The combination of clindamycin and primaquine has been demonstrated to be effective against PCP in humans (Kay and DuBois, 1990; Ruf and Pohle, 1989; Toma *et al.*, 1989). In addition, atovaquone, which inhibits mitochondrial cytochrome b complex in parasites, has also been used for treatment and prophylaxis of PCP. Atovaquone has limitations in that the oral formulation is the only available mode of delivery and this route requires fatty-rich meals to optimize the absorption. Pentamidine, dapsone, clindamycin-primaquine, and atovaquone are all considered to be the second-line regimens because of drug toxicities and/or high rates of relapse and recurrence.

Basic studies on *Pneumocystis* biology, its life cycle, *Pneumocystis*-host cell interactions, and the *Pneumocystis* genome project have allowed investigators to develop a variety of possible targets for novel anti-*Pneumocystis* agents with less adverse reactions and side-effects to the human host. The most recently developed new anti-fungal agents are the echinocandins. They inhibit synthesis of β(1,3)-D-glucan, which is an essential component of fungal cell wall, including the cyst wall of *Pneumocystis* (Kottom and Limper, 2000). Their side effects are less damaging compared to other anti-fungal agents such as azoles and polyenes, since β(1,3)-D-glucan synthase is not present in mammalian host cells. However, echinocandins cannot eradiate *Pneumocystis* infection given the fact that these compounds target the cyst forms while sparing the trophic forms to survive and repopulate once therapy is discontinued (Cushion *et al.*, 2010). The clinical efficacy of these drugs in humans therefore remains quite controversial (Hof and Schnulle, 2008; Kamboj *et al.*, 2006; Waters and Nelson, 2007) and accordingly, echinocandins are not considered the first line of PCP treatment. In addition, they should not be given as a monotherapy (Beltz *et al.*, 2006; Hof and Schnulle, 2008; Utili *et al.*, 2007). Currently, SMX-TMP is still the recommended first-line regimen for patients with PCP.

Table 13.5 Clinically available anti-*Pneumocystis* medications and their molecular targets*

Anti-*Pneumocystis* agent	Molecular target of action	Dosage for treatment	Resistance
Sulfamethoxazole	Dihydropteroate synthase	75–100 mg/kg/day PO/i.v. in combination with TMP	DHPS mutations
Dapsone	Dihydropteroate synthase	100 mg/day PO in combination with TMP	DHPS mutations
Trimethoprim	Dihydrofolate reductase	15–20 mg/kg/day PO/i.v. in combination with SMX or dapsone	DHFR mutations
Pentamidine	No specific target know, possible inhibitation DNA synthesis	3–4 mg/kg/day i.v./IM (aerosolized form is used for prophylaxis only)	Unknown
Clindamycin	Protein synthesis at the ribosomal level	600–900 mg PO/i.v. TID in combination with primaquine	Unknown
Primaquine	Unknown	30 mg/day PO in combination with clindamycin	Unknown
Atovaquone	Mitochondrial cytochrome b complex	750 mg PO BID with fatty food	Cytochrome b mutations

*The clinical use of echinocandin agents for the treatment of PCP remains controversial, and current data indicates that the agents are not first line agents, and further should not be used alone to treat PCP.

Mechanisms of drug resistance

Triazoles

Triazoles inhibit 14α-lanosterol demethylase, which is a cytochrome P450 enzyme encoded by the *ERG11* gene that mediates the first step of the ergosterol synthesis pathway. *Pneumocystis* does not synthesize ergosterol and it is intrinsically resistant to the clinically used triazoles (i.e. fluconazole, itraconazole, voriconazole, posaconazole, and ravuconazole). However, evidence has demonstrated active sterol synthesis in *Pneumocystis* (Cushion *et al.*, 2007; Kaneshiro *et al.*, 1994) and inhibition of particular enzymes in the sterol synthesis pathway has been shown to reduce the viability of *Pneumocystis* (Kaneshiro *et al.*, 2000). Interestingly, sequencing of the *Pneumocystis carinii PCERG11* gene predicts a naturally occurring resistance to the azoles currently used in clinical practice (Morales *et al.*, 2003). Further studies are needed to prove whether modified triazoles that may inhibit this enzyme might represent potential agents with anti-*Pneumocystis* activity.

Polyenes

Amphotericin B and nystatin are polyene antifungal agents currently in clinical use. Their target of action is ergosterol in the fungal cell membrane; however, it can bind to cholesterol of the human cell membrane causing significant cytotoxicity. The reasons that polyenes are not active against *Pneumocystis* are similar to the resistance mechanisms discussed for the triazoles above.

Sulfamethoxazole and dapsone

In *Pneumocystis*, the gene encoding DHPS is a part of the folic acid synthesis (*Pcfas*) gene (Volpe *et al.*, 1992). The gene also encodes for dihydroneopterinaldolase (DHNA) and hydroxymethyl-dihydropteridinepyrophosphokinase (PPPK). This gene is not present in mammalian cells. Mutations of this gene have been detected after the worldwide use of SMX-TMP beginning in the 1990s (Helweg-Larsen *et al.*, 1999; Kazanjian *et al.*, 1998, 2000). Furthermore, these mutations have not been found in non-human *Pneumocystis* species (Demanche *et al.*, 2002; Lane *et al.*, 1997). These observations suggest that DHPS mutations result from selective mutation following the use of sulfonamide. The most common mutations in *Pneumocystis* DHPS are amino acids substitutions at position 55 (alanine or threonine), 57 (serine for proline), or both, which involve the substrate binding sites (Ma *et al.*, 1999; Takahashi *et al.*, 2002). Whether these mutations confer resistance to sulfonamides and/

or dapsone is unknown, since *in vitro* culture of *Pneumocystis* is not available. Furthermore, clinical outcomes of DHPS gene mutations have been controversial and many patients have responded to therapy with SMX-TMP even in the presence of such mutations (Helweg-Larsen *et al.*, 1999; Kazanjian *et al.*, 2000; Ma *et al.*, 1999, 2002; Navin *et al.*, 2001; Takahashi *et al.*, 2000; Visconti *et al.*, 2001; Yoon *et al.*, 2013).

Trimethoprim

TMP inhibits DHFR, which is also present in mammalian cells. It was observed that DHFR mutations were not associated with prior SMX-TMX or dapsone use (Ma *et al.*, 1999; Takahashi *et al.*, 2002). Mutations of DHFR in *Pneumocystis* are generally not located in highly conserved regions that confer resistance to TMP or pyrimethamine (Huang *et al.*, 2004).

Atovaquone

Atovaquone is considered one of the second line agents for treatment and prophylaxis of PCP given its lower efficacy (Helweg-Larsen *et al.*, 2009). Mutations in the cytochrome b gene, resulting in alterations of coenzyme Q binding site have been observed (Kessl *et al.*, 2004; Srivastava *et al.*, 1999). Prophylaxis failure was reported by Helweg-Larsen *et al.* (Hauser *et al.*, 2001); however, the association of the mutations and clinical failure is still unknown. When used for treatment, atovaquone is reserved for patients with mild to moderate disease. It is also useful for PCP prophylaxis (Thomas and Limper, 2004).

Echinocandins

Caspofungin, micafungin, and anidulafungin are the echinocandin agents currently available for clinical use. Anidulafungin appears to be the most active of these agents against *Pneumocystis* in laboratory based animal studies (Cushion and Collins, 2011). However, echinocandins are only active against the cyst forms of *Pneumocystis* since PcGsc1, the *Pneumocystis* β(1,3)-glucan synthetic enzyme is only minimally expressed in the trophic forms (Kottom and Limper, 2000). The trophic forms, which do not possess much β-glucan, are therefore resistant to these agents, resulting in relapse or recurrence after treatment with echinocandins has been withdrawn. Mutations in the regions of the *FKS1* family of genes that mediate β(1,3)-glucan synthesis in other fungi have been demonstrated to confer resistance to all agents in this group (Perlin, 2007). The clinical significance of such mutations in PCP and other fungal infections has yet to be determined. Therefore, at the present time, echinocandins are not considered first line agents, and their clinical use is controversial and, if used at all for PCP treatment, should be employed only in combination with other agents active against *Pneumocystis*.

Conclusions

Pneumocystis pneumonia remains a major cause of morbidity and mortality in HIV-infected patients. The prevalence of PCP is increasing in non-HIV patients with suppressed immune status, following the use of potent immunosuppressive and immune modulating agents. Current knowledge of *Pneumocystis* is based on the use of immunosuppressed animal models and advances in molecular technology. Better understanding of the *Pneumocystis* life cycle, pathogenesis of the disease, and the host immune response provide attractive targets to develop novel anti-*Pneumocystis* agents. sulfamethoxazole–trimethoprim remains the first-line regimen for treatment and prophylaxis of PCP. However, many aspects of *Pneumocystis* remain unanswered or controversial, such as respiratory colonization, mode of transmission, clinical significance of mutations to currently available anti-*Pneumocystis* agents, and prophylaxis therapy in non-HIV immunocompromised patients. Therefore, it is important that studies of this unique medically important pathogenic fungus are continued in order to allow us to develop more effective strategies to prevent and treat *P. jirovecii* infection.

References

Agostini, C., Zambello, R., Trentin, L., and Semenzato, G. (1995). T lymphocytes with gamma/delta T-cell receptors in patients with AIDS and *Pneumocystis carinii* pneumonia. AIDS 9, 203–204.

Aliouat-Denis, C.M., Chabe, M., Demanche, C., Aliouat el, M., Viscogliosi, E., Guillot, J., Delhaes, L., and Dei-Cas, E. (2008). *Pneumocystis* species, co-evolution and pathogenic power. Infect. Genet. Evol. *8*, 708–726.

Aliouat-Denis, C.M., Martinez, A., Aliouat el, M., Pottier, M., Gantois, N., and Dei-Cas, E. (2009). The *Pneumocystis* life cycle. Mem. Inst. Oswaldo Cruz *104*, 419–426.

Allegra, C.J., Chabner, B.A., Tuazon, C.U., Ogata-Arakaki, D., Baird, B., Drake, J.C., and Masur, H. (1988). Treatment of *Pneumocystis* carinii pneumonia with trimetrexate in acquired immunodeficiency syndrome (AIDS). Semin. Oncol. *15*, 46–49.

Bartlett, M.S., and Smith, J.W. (1991). *Pneumocystis carinii*, an opportunist in immunocompromised patients. Clin. Microbiol. Rev. *4*, 137–149.

Bartlett, M.S., Angus, W.C., Shaw, M.M., Durant, P.J., Lee, C.H., Pascale, J.M., and Smith, J.W. (1998). Antibody to *Pneumocystis carinii* protects rats and mice from developing pneumonia. Clin. Diagn. Lab. Immunol. *5*, 74–77.

Beck, J.M., Warnock, M.L., Curtis, J.L., Sniezek, M.J., Arraj-Peffer, S.M., Kaltreider, H.B., and Shellito, J.E. (1991). Inflammatory responses to *Pneumocystis carinii* in mice selectively depleted of helper T lymphocytes. Am. J. Respir. Cell.Mol. Biol. *5*, 186–197.

Beck, J.M., Newbury, R.L., Palmer, B.E., Warnock, M.L., Byrd, P.K., and Kaltreider, H.B. (1996). Role of $CD8^+$ lymphocytes in host defense against *Pneumocystis* carinii in mice. J. Lab. Clin. Med. *128*, 477–487.

Beltz, K., Kramm, C.M., Laws, H.J., Schroten, H., Wessalowski, R., and Gobel, U. (2006). Combined trimethoprim and caspofungin treatment for severe *Pneumocystis jiroveci* pneumonia in a five year old boy with acute lymphoblastic leukemia. Klin. Padiatr. *218*, 177–179.

Burgess, J.W., Kottom, T.J., and Limper, A.H. (2008). *Pneumocystis carinii* exhibits a conserved meiotic control pathway. Infect. Immun. *76*, 417–425.

Burgess, J.W., Kottom, T.J., Villegas, L.R., Lamont, J.D., Baden, E.M., Ramirez-Alvarado, M., and Limper, A.H. (2009). The *Pneumocystis* meiotic PCRan1p kinase exhibits unique temperature-regulated activity. Am. J. Respir. Cell Mol. Biol. *41*, 714–721.

Calderon, E.J., Regordan, C., Medrano, F.J., Ollero, M., and Varela, J.M. (1996). *Pneumocystis carinii* infection in patients with chronic bronchial disease. Lancet *347*, 977.

Calderon, E., de la Horra, C., Medrano, F.J., Lopez-Suarez, A., Montes-Cano, M.A., Respaldiza, N., Elvira-Gonzalez, J., Martin-Juan, J., Bascunana, A., and Varela, J.M. (2004). *Pneumocystis jiroveci* isolates with dihydropteroate synthase mutations in patients with chronic bronchitis. Eur. J. Clin. Microbiol. Infect. Dis. *23*, 545–549.

Calderon, E.J., Rivero, L., Respaldiza, N., Morilla, R., Montes-Cano, M.A., Friaza, V., Munoz-Lobato, F., Varela, J.M., Medrano, F.J., and Horra Cde, L. (2007). Systemic inflammation in patients with chronic obstructive pulmonary disease who are colonized with *Pneumocystis jiroveci*. Clin. Infect. Dis. *45*, e17–19.

Carmona, E.M., and Limper, A.H. (2011). Update on the diagnosis and treatment of *Pneumocystis* pneumonia. Ther. Adv. Respir. Dis. *5*, 41–59.

Carmona, E.M., Vassallo, R., Vuk-Pavlovic, Z., Standing, J.E., Kottom, T.J., and Limper, A.H. (2006). *Pneumocystis* cell wall β-glucans induce dendritic cell costimulatory molecule expression and inflammatory activation through a Fas-Fas ligand mechanism. J. Immunol. *177*, 459–467.

Carmona, E.M., Lamont, J.D., Xue, A., Wylam, M., and Limper, A.H. (2010). *Pneumocystis* cell wall β-glucan stimulates calcium-dependent signaling of IL-8 secretion by human airway epithelial cells. Respir. Res. *11*, 95.

Carmona, E.M., Kottom, T.J., Hebrink, D.M., Moua, T., Singh, R.D., Pagano, R.E., and Limper, A.H. (2012). Glycosphingolipids mediate *Pneumocystis* cell wall β-glucan activation of the IL-23/IL-17 axis in human dendritic cells. Am. J. Respir. Cell Mol. Biol. *47*, 50–59.

Casanova-Cardiel, L., and Leibowitz, M.J. (1997). Presence of *Pneumocystis carinii* DNA in pond water. J. Eukaryot. Microbiol. *44*, 28S.

Chave, J.P., David, S., Wauters, J.P., Van Melle, G., and Francioli, P. (1991). Transmission of *Pneumocystis carinii* from AIDS patients to other immunosuppressed patients: a cluster of *Pneumocystis carinii* pneumonia in renal transplant recipients. AIDS *5*, 927–932.

Choukri, F., Menotti, J., Sarfati, C., Lucet, J.C., Nevez, G., Garin, Y.J., Derouin, F., and Totet, A. (2010). Quantification and spread of *Pneumocystis jirovecii* in the surrounding air of patients with *Pneumocystis* pneumonia. Clin. Infect. Dis. *51*, 259–265.

Cushion, M.T. (2004). *Pneumocystis*: unraveling the cloak of obscurity. Trends Microbiol *12*, 243–249.

Cushion, M.T. (2010). Are members of the fungal genus *Pneumocystis* (a) commensals; (b) opportunists; (c) pathogens; or (d) all of the above? PLoS Pathog. *6*, e1001009.

Cushion, M.T., and Collins, M.S. (2011). Susceptibility of *Pneumocystis* to echinocandins in suspension and biofilm cultures. Antimicrob. Agents Chemother. *55*, 4513–4518.

Cushion, M.T., Ruffolo, J.J., Linke, M.J., and Walzer, P.D. (1985). *Pneumocystis carinii*: growth variables and estimates in the A549 and WI-38 VA13 human cell lines. Exp. Parasitol. *60*, 43–54.

Cushion, M.T., Smulian, A.G., Slaven, B.E., Sesterhenn, T., Arnold, J., Staben, C., Porollo, A., Adamczak, R., and Meller, J. (2007). Transcriptome of *Pneumocystis carinii* during fulminate infection: carbohydrate metabolism and the concept of a compatible parasite. PLoS One 2, e423.

Cushion, M.T., Linke, M.J., Ashbaugh, A., Sesterhenn, T., Collins, M.S., Lynch, K., Brubaker, R., and Walzer, P.D. (2010). Echinocandin treatment of *Pneumocystis* pneumonia in rodent models depletes cysts leaving trophic burdens that cannot transmit the infection. PloS One *5*, e8524.

Demanche, C., Guillot, J., Berthelemy, M., Petitt, T., Roux, P., and Wakefield, A.E. (2002). Absence of mutations associated with sulfa resistance in *Pneumocystis carinii* dihydropteroate synthase gene from non-human primates. Med. Mycol. *40*, 315–318.

Desmet, S., Van Wijngaerden, E., Maertens, J., Verhaegen, J., Verbeken, E., De Munter, P., Meersseman, W., Van Meensel, B., Van Eldere, J., and Lagrou, K. (2009). Serum (1–3)-β-D-glucan as a tool for diagnosis of

Pneumocystis jirovecii pneumonia in patients with human immunodeficiency virus infection or hematological malignancy. J. Clin. Microbiol. 47, 3871–3874.

Douglas, C.M. (2001). Fungal β(1,3)-D-glucan synthesis. Med. Mycol. 39 *(Suppl 1)*, 55–66.

Dutz, W., Post, C., Vessal, K., and Kohout, E. (1976). Endemic infantile *Pneumocystis carinii* infection: the Shiraz study. Natl. Cancer. Inst. Monogr. *43*, 31–40.

Edman, J.C., Kovacs, J.A., Masur, H., Santi, D.V., Elwood, H.J., and Sogin, M.L. (1988). Ribosomal RNA sequence shows *Pneumocystis carinii* to be a member of the fungi. Nature 334, 519–522.

Festic, E., Gajic, O., Limper, A.H., and Aksamit, T.R. (2005). Acute respiratory failure due to *Pneumocystis* pneumonia in patients without human immunodeficiency virus infection: outcome and associated features. Chest *128*, 573–579.

Fleury, J., Escudier, E., Pocholle, M.J., Carre, C., and Bernaudin, J.F. (1985). Cell population obtained by bronchoalveolar lavage in *Pneumocystis carinii* pneumonitis. Acta Cytol. *29*, 721–726.

Frenkel, J.K. (1976). *Pneumocystis jiroveci* n. sp. from man: morphology, physiology, and immunology in relation to pathology. Natl. Cancer Inst. Monogr. *43*, 13–30.

Gajdusek, D.C. (1976). *Pneumocystis carinii* as the cause of human disease: historical perspective and magnitude of the problem: introductory remarks. Natl. Cancer. Inst. Monogr. *43*, 1–11.

Gal, S.L., Hery-Arnaud, G., Ramel, S., Virmaux, M., Damiani, C., Totet, A., and Nevez, G. (2010). *Pneumocystis jirovecii* and cystic fibrosis in France. Scand. J Infect. Dis. *42*, 225–227.

Gigliotti, F., and Hughes, W.T. (1988). Passive immunoprophylaxis with specific monoclonal antibody confers partial protection against *Pneumocystis carinii* pneumonitis in animal models. J. Clin. Invest. *81*, 1666–1668.

Gigliotti, F., Ballou, L.R., Hughes, W.T., and Mosley, B.D. (1988). Purification and initial characterization of a ferret *Pneumocystis carinii* surface antigen. J. Infect. Dis. *158*, 848–854.

Gigliotti, F., Harmsen, A.G., Haidaris, C.G., and Haidaris, P.J. (1993). *Pneumocystis carinii* is not universally transmissible between mammalian species. Infect. Immun. *61*, 2886–2890.

Gigliotti, F., Haidaris, C.G., Wright, T.W., and Harmsen, A.G. (2002). Passive intranasal monoclonal antibody prophylaxis against murine *Pneumocystis carinii* pneumonia. Infect. Immun. *70*, 1069–1074.

Goetz, O., and Peller, P. (1974). *Pneumocystis* carnii pneumonia. Med. Klin. *69*, 1774–1778.

Gustafson, M.P., Thomas, C.F., Jr., Rusnak, F., Limper, A.H., and Leof, E.B. (2001). Differential regulation of growth and checkpoint control mediated by a Cdc25 mitotic phosphatase from *Pneumocystis carinii*. J. Biol. Chem. *276*, 835–843.

Hanano, R., and Kaufmann, S.H. (1999). Effect on parasite eradication of *Pneumocystis carinii*-specific antibodies produced in the presence or absence of CD4(+) alpha-beta T lymphocytes. Eur. J. Immunol. *29*, 2464–2475.

Hanano, R., Reifenberg, K., and Kaufmann, S.H. (1996). Naturally acquired *Pneumocystis carinii* pneumonia in gene disruption mutant mice: roles of distinct T-cell populations in infection. Infect. Immun. *64*, 3201–3209.

Hanna, D.B., Gupta, L.S., Jones, L.E., Thompson, D.M., Kellerman, S.E., and Sackoff, J.E. (2007). AIDS-defining opportunistic illnesses in the HAART era in New York City. AIDS Care *19*, 264–272.

Hauser, P.M., Sudre, P., Nahimana, A., Francioli, P., and Study, G. (2001). Prophylaxis failure is associated with a specific *Pneumocystis carinii* genotype. Clin. Infect. Dis. *33*, 1080–1082.

Helweg-Larsen, J., Benfield, T.L., Eugen-Olsen, J., Lundgren, J.D., and Lundgren, B. (1999). Effects of mutations in *Pneumocystis carinii* dihydropteroate synthase gene on outcome of AIDS-associated *P. carinii* pneumonia. Lancet *354*, 1347–1351.

Helweg-Larsen, J., Benfield, T., Atzori, C., and Miller, R.F. (2009). Clinical efficacy of first- and second-line treatments for HIV-associated *Pneumocystis jiroveci* pneumonia: a tri-centre cohort study. J Antimicrob. Chemother. *64*, 1282–1290.

Hof, H., and Schnulle, P. (2008). *Pneumocystis jiroveci* pneumonia in a patient with Wegener's granulomatosis treated efficiently with caspofungin. Mycoses *51 (Suppl 1)*, 65–67.

Hong, S.T., Steele, P.E., Cushion, M.T., Walzer, P.D., Stringer, S.L., and Stringer, J.R. (1990). *Pneumocystis carinii* karyotypes. J. Clin. Microbiol. *28*, 1785–1795.

de la Horra, C., Varela, J.M., Fernandez-Alonso, J., Medrano, F.J., Respaldiza, N., Montes-Cano, M.A., and Calderon, E.J. (2004). Association between human-*Pneumocystis* infection and small-cell lung carcinoma. Eur. J. Clin. Invest. *34*, 229–235.

Huang, L., Crothers, K., Atzori, C., Benfield, T., Miller, R., Rabodonirina, M., and Helweg-Larsen, J. (2004). Dihydropteroate synthase gene mutations in *Pneumocystis* and sulfa resistance. Emerg. Infect. Dis. *10*, 1721–1728.

Hughes, W.T. (2003). *Pneumocystis carinii* vs. *Pneumocystis jiroveci*: another misnomer (response to Stringer *et al.*). Emerg. Infect. Dis. *9*, 276–277; author reply 277–279.

Hughes, W.T., and Gigliotti, F. (1988). Nomenclature for *Pneumocystis carinii*. J. Infect. Dis. *157*, 432–433.

Hughes, W.T., McNabb, P.C., Makres, T.D., and Feldman, S. (1974). Efficacy of trimethoprim and sulfamethoxazole in the prevention and treatment of *Pneumocystis carinii* pneumonitis. Antimicrob. Agents Chemother. *5*, 289–293.

Hughes, W.T., Feldman, S., and Sanyal, S.K. (1975). Treatment of *Pneumocystis carinii* pneumonitis with trimethoprim-sulfamethoxazole. Can. Med. Assoc. J *112*, 47–50.

Hughes, W.T., Kuhn, S., Chaudhary, S., Feldman, S., Verzosa, M., Aur, R.J., Pratt, C., and George, S.L. (1977). Successful chemoprophylaxis for *Pneumocystis carinii* pneumonitis. N. Engl. J. Med. *297*, 1419–1426.

Ivady, G., and Paldy, L. (1957). New method for the therapy of interstitial plasma cell pneumonia in premature infants by pentavalent stiblum and aromatic diamidines. Orv. Hetil. *98*, 1201–1204.

Jones, C.E., and Chan, K. (2002). Interleukin-17 stimulates the expression of interleukin-8, growth-related oncogene-alpha, and granulocyte-colony-stimulating factor by human airway epithelial cells. Am. J. Respir. Cell Mol. Biol 26, 748–753.

Kagi, M.K., Fierz, W., Grob, P.J., and Russi, E.W. (1993). High proportion of gamma-delta T cell receptor positive T cells in bronchoalveolar lavage and peripheral blood of HIV-infected patients with *Pneumocystis carinii* pneumonias. Respiration *60*, 170–177.

Kamboj, M., Weinstock, D., and Sepkowitz, K.A. (2006). Progression of *Pneumocystis jiroveci* pneumonia in patients receiving echinocandin therapy. Clin. Infect Dis. *43*, e92–94.

Kaneshiro, E.S., Ellis, J.E., Jayasimhulu, K., and Beach, D.H. (1994). Evidence for the presence of "metabolic sterols" in *Pneumocystis*: identification and initial characterization of *Pneumocystis carinii* sterols. J. Eukaryot. Microbiol *41*, 78–85.

Kaneshiro, E.S., Collins, M.S., and Cushion, M.T. (2000). Inhibitors of sterol biosynthesis and amphotericin B reduce the viability of *Pneumocystis carinii* f. sp. *carinii*. Antimicrob. Agents Chemother. *44*, 1630–1638.

Kawaguchi, M., Kokubu, F., Kuga, H., Matsukura, S., Hoshino, H., Ieki, K., Imai, T., Adachi, M., and Huang, S.K. (2001). Modulation of bronchial epithelial cells by IL-17. J Allergy Clin. Immunol. *108*, 804–809.

Kay, R., and DuBois, R.E. (1990). Clindamycin/primaquine therapy and secondary prophylaxis against *Pneumocystis carinii* pneumonia in patients with AIDS. South. Med. J. *83*, 403–404.

Kazanjian, P., Locke, A.B., Hossler, P.A., Lane, B.R., Bartlett, M.S., Smith, J.W., Cannon, M., and Meshnick, S.R. (1998). *Pneumocystis carinii* mutations associated with sulfa and sulfone prophylaxis failures in AIDS patients. AIDS *12*, 873–878.

Kazanjian, P., Armstrong, W., Hossler, P.A., Burman, W., Richardson, J., Lee, C.H., Crane, L., Katz, J., and Meshnick, S.R. (2000). *Pneumocystis carinii* mutations are associated with duration of sulfa or sulfone prophylaxis exposure in AIDS patients. J. Infect. Dis. *182*, 551–557.

Kessl, J.J., Hill, P., Lange, B.B., Meshnick, S.R., Meunier, B., and Trumpower, B.L. (2004). Molecular basis for atovaquone resistance in *Pneumocystis jirovecii* modeled in the cytochrome bc(1) complex of *Saccharomyces* cerevisiae. J. Biol. Chem. *279*, 2817–2824.

Kim, H.K., Hughes, W.T., and Feldman, S. (1972). Studies of morphology and immunofluorescence of *Pneumocystis carinii*. Proc. Soc. Exp. Biol. Med. *141*, 304–309.

Kottom, T.J., and Limper, A.H. (2000). Cell wall assembly by *Pneumocystis carinii*. Evidence for a unique gsc-1 subunit mediating β-1,3-glucan deposition. J. Biol Chem. *275*, 40628–40634.

Kottom, T.J., and Limper, A.H. (2004). *Pneumocystis carinii* cell wall biosynthesis kinase gene *CBK1* is an environmentally responsive gene that complements cell wall defects of cbk-deficient yeast. Infect. Immun. *72*, 4628–4636.

Kottom, T.J., Thomas, C.F., Jr., and Limper, A.H. (2001). Characterization of *Pneumocystis carinii PHR1*, a pH-regulated gene important for cell wall Integrity. J. Bacteriol. *183*, 6740–6745.

Kottom, T.J., Kohler, J.R., Thomas, C.F., Jr., Fink, G.R., and Limper, A.H. (2003). Lung epithelial cells and extracellular matrix components induce expression of *Pneumocystis carinii STE20*, a gene complementing the mating and pseudohyphal growth defects of *STE20* mutant yeast. Infect. Immun. *71*, 6463–6471.

Kottom, T.J., Kennedy, C.C., and Limper, A.H. (2008). *Pneumocystis* PCINT1, a molecule with integrin-like features that mediates organism adhesion to fibronectin. Mol. Microbiol. *67*, 747-761.

Kottom, T.J., Burgess, J.W., and Limper, A.H. (2011a). *Pneumocystis carinii* interactions with lung epithelial cells and matrix proteins induce expression and activity of the PcSte20 kinase with subsequent phosphorylation of the downstream cell wall biosynthesis kinase PcCbk1. Infect. Immun. *79*, 4157–4164.

Kottom, T.J., Han, J., Zhang, Z., and Limper, A.H. (2011b). *Pneumocystis carinii* expresses an active Rtt109 histone acetyltransferase. Am. J. Respir. Cell Mol. Biol. *44*, 768–776.

Koziel, H., Eichbaum, Q., Kruskal, B.A., Pinkston, P., Rogers, R.A., Armstrong, M.Y., Richards, F.F., Rose, R.M., and Ezekowitz, R.A. (1998a). Reduced binding and phagocytosis of *Pneumocystis carinii* by alveolar macrophages from persons infected with HIV-1 correlates with mannose receptor down-regulation. J. Clin. Invest. *102*, 1332–1344.

Koziel, H., Phelps, D.S., Fishman, J.A., Armstrong, M.Y., Richards, F.F., and Rose, R.M. (1998b). Surfactant protein-A reduces binding and phagocytosis of *Pneumocystis carinii* by human alveolar macrophages *in vitro*. Am. J. Resp. Cell Mol. Biol. *18*, 834–843.

Krajicek, B.J., Kottom, T.J., Villegas, L., and Limper, A.H. (2010). Characterization of the PcCdc42 small G protein from *Pneumocystis carinii*, which interacts with the PcSte20 life cycle regulatory kinase. Am. J. Physiol. Lung Cell Mol. Physiol. *298*, L252–260.

Kutty, G., and Kovacs, J.A. (2003). A single-copy gene encodes Kex1, a serine endoprotease of *Pneumocystis jiroveci*. Infect. Immun. *71*, 571–574.

Lane, B.R., Ast, J.C., Hossler, P.A., Mindell, D.P., Bartlett, M.S., Smith, J.W., and Meshnick, S.R. (1997). Dihydropteroate synthase polymorphisms in *Pneumocystis carinii*. J. Infect. Dis. *175*, 482–485.

Laursen, A.L., and Andersen, P.L. (1998). Low levels of IgG antibodies against *Pneumocystis carinii* among HIV-infected patients. Scand. J. Infect. Dis. *30*, 495–499.

Lebron, F., Vassallo, R., Puri, V., and Limper, A.H. (2003). *Pneumocystis carinii* cell wall β-glucans initiate macrophage inflammatory responses through NF-kappaB activation. J. Biol. Chem. *278*, 25001–25008.

Leoung, G.S., Mills, J., Hopewell, P.C., Hughes, W., and Wofsy, C. (1986). Dapsone-trimethoprim for *Pneumocystis carinii* pneumonia in the acquired immunodeficiency syndrome. Ann. Intern. Med. *105*, 45–48.

Limper, A.H., and Martin, W.J., 2nd (1990). *Pneumocystis carinii*: inhibition of lung cell growth mediated by parasite attachment. J. Clin. Invest. *85*, 391–396.

Limper, A.H., Offord, K.P., Smith, T.F., and Martin, W.J., 2nd (1989). *Pneumocystis carinii* pneumonia. Differences in lung parasite number and inflammation in patients with and without AIDS. Am. Rev. Respir. Dis. *140*, 1204–1209.

Limper, A.H., Standing, J.E., Hoffman, O.A., Castro, M., and Neese, L.W. (1993). Vitronectin binds to *Pneumocystis carinii* and mediates organism attachment to cultured lung epithelial cells. Infect. Immun. *61*, 4302–4309.

Limper, A.H., Thomas, C.F., Jr., Anders, R.A., and Leof, E.B. (1997). Interactions of parasite and host epithelial cell cycle regulation during *Pneumocystis carinii* pneumonia. J. Lab. Clin. Med. *130*, 132–138.

Linke, M.J., Cushion, M.T., and Walzer, P.D. (1989). Properties of the major antigens of rat and human *Pneumocystis carinii*. Infect. Immun. *57*, 1547–1555.

Lugli, E.B., Bampton, E.T., Ferguson, D.J., and Wakefield, A.E. (1999). Cell surface protease PRT1 identified in the fungal pathogen *Pneumocystis carinii*. Mol. Microbiol. *31*, 1723–1733.

Lund, F.E., Hollifield, M., Schuer, K., Lines, J.L., Randall, T.D., and Garvy, B.A. (2006). B cells are required for generation of protective effector and memory CD4 cells in response to *Pneumocystis* lung infection. J. Immunol. *176*, 6147–6154.

Lundgren, B., Koch, C., Mathiesen, L., Nielsen, J.O., and Hansen, J.E. (1993). Glycosylation of the major human *Pneumocystis carinii* surface antigen. APMIS *101*, 194–200.

Ma, L., Borio, L., Masur, H., and Kovacs, J.A. (1999). *Pneumocystis carinii* dihydropteroate synthase but not dihydrofolate reductase gene mutations correlate with prior trimethoprim-sulfamethoxazole or dapsone use. J. Infect. Dis. *180*, 1969–1978.

Ma, L., Kovacs, J.A., Cargnel, A., Valerio, A., Fantoni, G., and Atzori, C. (2002). Mutations in the dihydropteroate synthase gene of human-derived *Pneumocystis carinii* isolates from Italy are infrequent but correlate with prior sulfa prophylaxis. J. Infect. Dis. *185*, 1530–1532.

Marcotte, H., Levesque, D., Delanay, K., Bourgeault, A., de la Durantaye, R., Brochu, S., and Lavoie, M.C. (1996). *Pneumocystis carinii* infection in transgenic B cell-deficient mice. J. Infect. Dis. *173*, 1034–1037.

Martin-Garrido, I., Carmona, E.M., Specks, U., and Limper, A.H. (2013). *Pneumocystis* pneumonia in patients treated with rituximab. Chest *144*, 258–265.

Marty, F.M., Koo, S., Bryar, J., and Baden, L.R. (2007). (1→3)β-D-glucan assay positivity in patients with *Pneumocystis* (*carinii*) *jiroveci* pneumonia. Ann. Intern. Med. *147*, 70–72.

Maskell, N.A., Waine, D.J., Lindley, A., Pepperell, J.C., Wakefield, A.E., Miller, R.F., and Davies, R.J. (2003). Asymptomatic carriage of *Pneumocystis jiroveci* in subjects undergoing bronchoscopy: a prospective study. Thorax *58*, 594-597.

Masur, H., Michelis, M.A., Greene, J.B., Onorato, I., Stouwe, R.A., Holzman, R.S., Wormser, G., Brettman, L., Lange, M., Murray, H.W., *et al.* (1981). An outbreak of community-acquired *Pneumocystis carinii* pneumonia: initial manifestation of cellular immune dysfunction. N. Engl. J. Med. *305*, 1431–1438.

Montaner, J.S., Lawson, L.M., Levitt, N., Belzberg, A., Schechter, M.T., and Ruedy, J. (1990). Corticosteroids prevent early deterioration in patients with moderately severe *Pneumocystis carinii* pneumonia and the acquired immunodeficiency syndrome (AIDS). Ann. Intern. Med. *113*, 14–20.

Morales, I.J., Vohra, P.K., Puri, V., Kottom, T.J., Limper, A.H., and Thomas, C.F., Jr. (2003). Characterization of a lanosterol 14 alpha-demethylase from *Pneumocystis carinii*. Am. J. Respir. Cell Mol. Biol. *29*, 232–238.

Mori, H., Ohno, Y., Ito, F., Endo, J., Yanase, K., Funaguchi, N., Bai La, B.L., and Minatoguchi, S. (2010). Polymerase chain reaction positivity of *Pneumocystis jirovecii* during primary lung cancer treatment. Jpn. J. Clin. Oncol. *40*, 658–662.

Mori, S., and Sugimoto, M. (2012). *Pneumocystis jirovecii* infection: an emerging threat to patients with rheumatoid arthritis. Rheumatology *51*, 2120–2130.

Morris, A., Beard, C.B., and Huang, L. (2002). Update on the epidemiology and transmission of *Pneumocystis carinii*. Microbes Infect. *4*, 95–103.

Morris, A., Kingsley, L.A., Groner, G., Lebedeva, I.P., Beard, C.B., and Norris, K.A. (2004a). Prevalence and clinical predictors of *Pneumocystis* colonization among HIV-infected men. AIDS *18*, 793–798.

Morris, A., Sciurba, F.C., Lebedeva, I.P., Githaiga, A., Elliott, W.M., Hogg, J.C., Huang, L., and Norris, K.A. (2004b). Association of chronic obstructive pulmonary disease severity and *Pneumocystis* colonization. Am. J. Respir. Crit. Care Med. *170*, 408–413.

National Institutes of Health – University of California Expert Panel for Corticosteroids as Adjunctive Therapy for *Pneumocystis* Pneumonia. (1990). Consensus statement on the use of corticosteroids as adjunctive therapy for *Pneumocystis* pneumonia in the acquired immunodeficiency syndrome. N. Engl. J. Med. *323*, 1500–1504.

Navin, T.R., Beard, C.B., Huang, L., del Rio, C., Lee, S., Pieniazek, N.J., Carter, J.L., Le, T., Hightower, A., and Rimland, D. (2001). Effect of mutations in *Pneumocystis carinii* dihydropteroate synthase gene on outcome of P carinii pneumonia in patients with HIV-1: a prospective study. Lancet *358*, 545–549.

Nevez, G., Jounieaux, V., Linas, M.D., Guyot, K., Leophonte, P., Massip, P., Schmit, J.L., Seguela, J.P., Camus, D., Dei-Cas, E., *et al.* (1997). High frequency of *Pneumocystis carinii* sp.f. *hominis* colonization in HIV-negative patients. J. Eukaryot. Microbiol. *44*, 36S.

Nevez, G., Magois, E., Duwat, H., Gouilleux, V., Jounieaux, V., and Totet, A. (2006). Apparent absence of *Pneumocystis jirovecii* in healthy subjects. Clin. Infect. Dis. *42*, e99–101.

O'Riordan, D.M., Standing, J.E., and Limper, A.H. (1995). *Pneumocystis carinii* glycoprotein A binds macrophage mannose receptors. Infect. Immun. *63*, 779–784.

Pareja, J.G., Garland, R., and Koziel, H. (1998). Use of adjunctive corticosteroids in severe adult non-HIV *Pneumocystis carinii* pneumonia. Chest *113*, 1215–1224.

Perlin, D.S. (2007). Resistance to echinocandin-class antifungal drugs. Drug Resist. Update *10*, 121–130.

Phair, J., Munoz, A., Detels, R., Kaslow, R., Rinaldo, C., and Saah, A. (1990). The risk of *Pneumocystis carinii* pneumonia among men infected with human immunodeficiency virus type 1. Multicenter AIDS Cohort Study Group. N. Engl. J. Med. *322*, 161–165.

Pifer, L.L., Hughes, W.T., Stagno, S., and Woods, D. (1978). *Pneumocystis carinii* infection: evidence for high prevalence in normal and immunosuppressed children. Pediatrics *61*, 35–41.

Powles, M.A., McFadden, D.C., Pittarelli, L.A., and Schmatz, D.M. (1992). Mouse model for *Pneumocystis carinii* pneumonia that uses natural transmission to initiate infection. Infect. Immun. *60*, 1397–1400.

Rapaka, R.R., Ricks, D.M., Alcorn, J.F., Chen, K., Khader, S.A., Zheng, M., Plevy, S., Bengten, E., and Kolls, J.K. (2010). Conserved natural IgM antibodies mediate innate and adaptive immunity against the opportunistic fungus *Pneumocystis murina*. J. Exp. Med. *207*, 2907–2919.

Redhead, S.A., Cushion, M.T., Frenkel, J.K., and Stringer, J.R. (2006). *Pneumocystis* and *Trypanosoma cruzi*: nomenclature and typifications. J. Eukaryot. Microbiol. *53*, 2–11.

Reiter, W.M., and Cimoch, P.J. (1987). Dapsone-induced methemoglobinemia in a patient with *P. jirovecii* pneumonia and AIDS. N. Engl. J. Med. *317*, 1740–1741.

Respaldiza, N., Montes-Cano, M.A., Dapena, F.J., de la Horra, C., Mateos, I., Medrano, F.J., Calderon, E., and Varela, J.M. (2005). Prevalence of colonisation and genotypic characterisation of *Pneumocystis jirovecii* among cystic fibrosis patients in Spain. Clin. Microbiol. Infect. *11*, 1012–1015.

Rivero, L., de la Horra, C., Montes-Cano, M.A., Rodriguez-Herrera, A., Respaldiza, N., Friaza, V., Morilla, R., Gutierrez, S., Varela, J.M., Medrano, F.J., *et al.* (2008). *Pneumocystis jirovecii* transmission from immunocompetent carriers to infant. Emerg. Infect. Dis. *14*, 1116–1118.

Roths, J.B., and Sidman, C.L. (1992). Both immunity and hyperresponsiveness to *Pneumocystis carinii* result from transfer of $CD4^+$ but not $CD8^+$ T cells into severe combined immunodeficiency mice. J. Clin. Invest. *90*, 673–678.

Ruf, B., and Pohle, H.D. (1989). Clindamycin/primaquine for *Pneumocystis carinii* pneumonia. Lancet *334*, 626–627.

Sattler, F.R., Allegra, C.J., Verdegem, T.D., Akil, B., Tuazon, C.U., Hughlett, C., Ogata-Arakaki, D., Feinberg, J., Shelhamer, J., Lane, H.C., *et al.* (1990). Trimetrexate-leucovorin dosage evaluation study for treatment of *Pneumocystis carinii* pneumonia. J. Infect. Dis. *161*, 91–96.

Sepkowitz, K.A. (2002). Opportunistic infections in patients with and patients without Acquired Immunodeficiency Syndrome. Clin. Infect. Dis. *34*, 1098–1107.

Shellito, J., Suzara, V.V., Blumenfeld, W., Beck, J.M., Steger, H.J., and Ermak, T.H. (1990). A new model of *Pneumocystis carinii* infection in mice selectively depleted of helper T lymphocytes. J. Clin. Invest. *85*, 1686–1693.

Shellito, J.E., Tate, C., Ruan, S., and Kolls, J. (2000). Murine $CD4^+$ T lymphocyte subsets and host defense against *Pneumocystis carinii*. J. Infect. Dis. *181*, 2011–2017.

Shimizu, Y., Sunaga, N., Dobashi, K., Fueki, M., Fueki, N., Makino, S., and Mori, M. (2009). Serum markers in interstitial pneumonia with and without *Pneumocystis jirovecii* colonization: a prospective study. BMC Infect. Dis. *9*, 47.

Sing, A., Geiger, A.M., Hogardt, M., and Heesemann, J. (2001). *Pneumocystis carinii* carriage among cystic fibrosis patients, as detected by nested PCR. J. Clin. Microbiol. *39*, 2717–2718.

Singer, C., Armstrong, D., Rosen, P.P., and Schottenfeld, D. (1975). *Pneumocystis carinii* pneumonia: a cluster of eleven cases. Ann. Intern. Med. *82*, 772–777.

Smith, G.R., Givan, S.A., Cullen, P., and Sprague, G.F., Jr. (2002). GTPase-activating proteins for Cdc42. Eukaryot. Cell *1*, 469–480.

de Souza, W., and Benchimol, M. (2005). Basic biology of *Pneumocystis carinii*: a mini review. Mem. Inst. Oswaldo Cruz *100*, 903–908.

Srivastava, I.K., Morrisey, J.M., Darrouzet, E., Daldal, F., and Vaidya, A.B. (1999). Resistance mutations reveal the atovaquone-binding domain of cytochrome b in malaria parasites. Mol. Microbiol. *33*, 704–711.

Steele, C., Zheng, M., Young, E., Marrero, L., Shellito, J.E., and Kolls, J.K. (2002). Increased host resistance against *Pneumocystis carinii* pneumonia in $\gamma\delta$ T-cell-deficient mice: protective role of gamma interferon and CD8(+) T cells. Infect Immun *70*, 5208–5215.

Steele, C., Marrero, L., Swain, S., Harmsen, A.G., Zheng, M., Brown, G.D., Gordon, S., Shellito, J.E., and Kolls, J.K. (2003). Alveolar macrophage-mediated killing of *Pneumocystis carinii* f. sp. *muris* involves molecular recognition by the Dectin-1 β-glucan receptor. J. Exp. Med. *198*, 1677–1688.

Stehle, S.E., Rogers, R.A., Harmsen, A.G., and Ezekowitz, R.A. (2000). A soluble mannose receptor immunoadhesin enhances phagocytosis of *Pneumocystis carinii* by human polymorphonuclear leukocytes *in vitro*. Scand. J. Immunol. *52*, 131–137.

Stringer, J.R., and Cushion, M.T. (1998). The genome of *Pneumocystis carinii*. FEMS Immunol. Med. Microbiol. *22*, 15–26.

Stringer, J.R., Beard, C.B., Miller, R.F., and Wakefield, A.E. (2002). A new name (*Pneumocystis jiroveci*) for *Pneumocystis* from humans. Emerg. Infect. Dis. *8*, 891–896.

Stringer, J.R., Beard, C.B., and Miller, R.F. (2009). Spelling *Pneumocystis jirovecii*. Emerg. Infect. Dis. *15*, 506.

Sullivan, P.S., Denniston, M., McNaghten, A., Buskin, S.E., Broyles, S.T., and Mokotoff, E.D. (2007). Use of a population-based survey to determine incidence of AIDS-defining opportunistic illnesses among HIV-positive persons receiving medical care in the United States. AIDS Res. Ther. *4*, 17.

Swain, S.D., Wright, T.W., Degel, P.M., Gigliotti, F., and Harmsen, A.G. (2004). Neither neutrophils nor reactive oxygen species contribute to tissue damage during *Pneumocystis* pneumonia in mice. Infect. Immun. *72*, 5722–5732.

Takahashi, T., Hosoya, N., Endo, T., Nakamura, T., Sakashita, H., Kimura, K., Ohnishi, K., Nakamura, Y., and Iwamoto, A. (2000). Relationship between mutations in dihydropteroate synthase of *Pneumocystis* carinii f. sp. *hominis* isolates in Japan and resistance to sulfonamide therapy. J. Clin. Microbiol. *38*, 3161–3164.

Takahashi, T., Endo, T., Nakamura, T., Sakashitat, H., Kimurat, K., Ohnishit, K., Kitamura, Y., and Iwamoto, A. (2002). Dihydrofolate reductase gene polymorphisms in *Pneumocystis carinii* f. sp. *hominis* in Japan. J. Med. Microbiol. *51*, 510–515.

Tasaka, S., Hasegawa, N., Kobayashi, S., Yamada, W., Nishimura, T., Takeuchi, T., and Ishizaka, A. (2007). Serum indicators for the diagnosis of *Pneumocystis* pneumonia. Chest *131*, 1173–1180.

Taylor, M.B., Phillips, M., and Easmon, C.S. (1992). Opsonophagocytosis of *Pneumocystis carinii*. J. Med. Microbiol. *36*, 223–228.

The *Pneumocystis* Workshop. (1994). Revised nomenclature for *Pneumocystis carinii*. J. Eukaryot. Microbiol. *41*, 121S-122S.

Theus, S.A., Sawhney, N., Smulian, A.G., and Walzer, P.D. (1998). Proliferative and cytokine responses of human T lymphocytes isolated from human immunodeficiency virus-infected patients to the major surface glycoprotein of *Pneumocystis carinii*. J. Infect. Dis. *177*, 238–241.

Thomas, C.F., Jr., and Limper, A.H. (2004). *Pneumocystis* pneumonia. N. Engl. J. Med. *350*, 2487–2498.

Thomas, C.F., Jr., and Limper, A.H. (2007). Current insights into the biology and pathogenesis of *Pneumocystis* pneumonia. Nat. Rev. Microbiol. *5*, 298–308.

Thomas, C.F., Anders, R.A., Gustafson, M.P., Leof, E.B., and Limper, A.H. (1998). *Pneumocystis carinii* contains a functional cell-division-cycle Cdc2 homologue. Am. J. Respir. Cell Mol. Biol. *18*, 297–306.

Thomas, C.F. Jr., Leof, E.B., and Limper, A.H. (1999). Analysis of *Pneumocystis carinii* introns. Infect. Immun. *67*, 6157–6160.

Toma, E., Fournier, S., Poisson, M., Morisset, R., Phaneuf, D., and Vega, C. (1989). Clindamycin with primaquine for *Pneumocystis carinii* pneumonia. Lancet *333*, 1046–1048.

Utili, R., Durante-Mangoni, E., Basilico, C., Mattei, A., Ragone, E., and Grossi, P. (2007). Efficacy of caspofungin addition to trimethoprim-sulfamethoxazole treatment for severe *Pneumocystis* pneumonia in solid organ transplant recipients. Transplantation *84*, 685–688.

Vanek, J., and Jirovec, O. (1952). [Parasitic pneumonia. Interstitial plasma cell pneumonia of premature, caused by *Pneumocystis carinii*]. Zentralbl. Bakteriol. Parasitenkd. Infektionskr. Hyg. *158*, 120–127.

Varela, J.M., Respaldiza, N., Sanchez, B., de la Horra, C., Montes-Cano, M., Rincon, M., Dapena, J., Gonzalez-Becerra, C., Medrano, F.J., and Calderon, E. (2003). Lymphocyte response in subjects with chronic pulmonary disease colonized by *Pneumocystis jirovecii*. J. Eukaryot. Microbiol. *50 (Suppl)*, 672–673.

Vassallo, R., Standing, J.E., and Limper, A.H. (2000). Isolated *Pneumocystis carinii* cell wall glucan provokes lower respiratory tract inflammatory responses. J. Immunol. *164*, 3755–3763.

Vavra, J., and Kucera, K. (1970). *Pneumocystis carinii* delanoe, its ultrastructure and ultrastructural affinities. J. Protozool. *17*, 463–483.

Vestereng, V.H., and Kovacs, J.A. (2004). Inability of *Pneumocystis* organisms to incorporate bromodeoxyuridine suggests the absence of a salvage pathway for thymidine. Microbiology *150*, 1179–1182.

Vidal, S., de la Horra, C., Martin, J., Montes-Cano, M.A., Rodriguez, E., Respaldiza, N., Rodriguez, F., Varela, J.M., Medrano, F.J., and Calderon, E.J. (2006). *Pneumocystis jirovecii* colonisation in patients with interstitial lung disease. Clin. Microbiol. Infect. *12*, 231–235.

Villegas, L.R., Kottom, T.J., and Limper, A.H. (2010). Characterization of PCEng2, a β-1,3-endoglucanase homolog in *Pneumocystis carinii* with activity in cell wall regulation. Am. J. Respir. Cell Mol. Biol. *43*, 192–200.

Visconti, E., Ortona, E., Mencarini, P., Margutti, P., Marinaci, S., Zolfo, M., Siracusano, A., and Tamburrini, E. (2001). Mutations in dihydropteroate synthase gene of *Pneumocystis carinii* in HIV patients with *Pneumocystis carinii* pneumonia. Int. J. Antimicrob. Agents *18*, 547–551.

Vogel, P., Miller, C.J., Lowenstine, L.L., and Lackner, A.A. (1993). Evidence of horizontal transmission of *Pneumocystis carinii* pneumonia in simian immunodeficiency virus-infected rhesus macaques. J. Infect. Dis. *168*, 836–843.

Volpe, F., Dyer, M., Scaife, J.G., Darby, G., Stammers, D.K., and Delves, C.J. (1992). The multifunctional folic acid synthesis *fas* gene of *Pneumocystis carinii* appears to encode dihydropteroate synthase and hydroxymethyldihydropterin pyrophosphokinase. Gene *112*, 213–218.

Vuk-Pavlovic, Z., Standing, J.E., Crouch, E.C., and Limper, A.H. (2001). Carbohydrate recognition domain of surfactant protein D mediates interactions with *Pneumocystis carinii* glycoprotein A. Am J Respir Cell Mol Biol *24*, 475–484.

Wakefield, A.E. (1994). Detection of DNA sequences identical to *Pneumocystis carinii* in samples of ambient air. J. Eukaryot. Microbiol. *41*, 116S.

Wakefield, A.E. (1996). DNA sequences identical to *Pneumocystis carinii* f. sp. *carinii* and *Pneumocystis carinii* f. sp. *hominis* in samples of air spora. J. Clin. Microbiol. *34*, 1754–1759.

Wakefield, A.E., Hopkin, J.M., Burns, J., Hipkiss, J.B., Stewart, T.J., and Moxon, E.R. (1988). Cloning of DNA from *Pneumocystis carinii*. J. Infect. Dis. *158*, 859–862.

Wakefield, A.E., Pixley, F.J., Banerji, S., Sinclair, K., Miller, R.F., Moxon, E.R., and Hopkin, J.M. (1990). Detection of *Pneumocystis carinii* with DNA amplification. Lancet *336*, 451–453.

Waters, L., and Nelson, M. (2007). The use of caspofungin in HIV-infected individuals. Expert Opin. Investig. Drugs *16*, 899–908.

Wiley, J.A., and Harmsen, A.G. (1995). CD40 ligand is required for resolution of *Pneumocystis carinii* pneumonia in mice. J. Immunol. *155*, 3525–3529.

Williams, M.D., Wright, J.R., March, K.L., and Martin, W.J., 2nd (1996). Human surfactant protein A enhances attachment of *Pneumocystis carinii* to rat alveolar macrophages. Am. J. Respir. Cell Mol. Biol. *14*, 232–238.

Wissmann, G., Morilla, R., Friaza, V., Calderon, E., and Varela, J.M. (2010). [Human reservoirs of *Pneumocystis*]. Enferm. Infecc. Microbiol. Clin. *28*, 38–43.

Wright, T.W., Gigliotti, F., Finkelstein, J.N., McBride, J.T., An, C.L., and Harmsen, A.G. (1999). Immune-mediated inflammation directly impairs pulmonary function, contributing to the pathogenesis of *Pneumocystis carinii* pneumonia. J. Clin. Invest. *104*, 1307–1317.

Wyder, M.A., Rasch, E.M., and Kaneshiro, E.S. (1998). Quantitation of absolute *Pneumocystis carinii* nuclear DNA content. Trophic and cystic forms isolated from infected rat lungs are haploid organisms. J. Eukaryot. Microbiol. *45*, 233–239.

Yale, S.H., and Limper, A.H. (1996). *Pneumocystis carinii* pneumonia in patients without acquired immunodeficiency syndrome: associated illness and prior corticosteroid therapy. Mayo Clin. Proc. *71*, 5–13.

Yong, S.J., Vuk-Pavlovic, Z., Standing, J.E., Crouch, E.C., and Limper, A.H. (2003). Surfactant protein D-mediated aggregation of *Pneumocystis carinii* impairs phagocytosis by alveolar macrophages. Infect. Immun. *71*, 1662–1671.

Yoon, C., Subramanian, A., Chi, A., Crothers, K., Meshnick, S.R., Taylor, S.M., Beard, C.B., Jarlsberg, L.G., Lawrence, G.G., Avery, M., *et al.* (2013). Dihydropteroate synthase mutations in *Pneumocystis* pneumonia: impact of applying different definitions of prophylaxis, mortality endpoints and mutant in a single cohort. Med. Mycol. *51*, 568–575.

Yoshida, Y., Matsumoto, Y., Yamada, M., Okabayashi, K., Yoshikawa, H., and Nakazawa, M. (1984). *Pneumocystis carinii*: electron microscopic investigation on the interaction of trophozoite and alveolar lining cell. Zentralbl. Bakteriol. Mikrobiol. Hyg. A *256*, 390–399.

Zheng, M.Q., Shellito, J.E., Marrero, L., Zhong, Q., Julian, S., Ye, P., Wallace, V., Schwarzenberger, P., and Kolls, J.K. (2001). CD4(+) T cell-independent vaccination against *Pneumocystis carinii* in mice. J. Clin. Invest. *108*, 1469–1474.

Paracoccidioides Mechanisms of Pathogenesis and Virulence

14

Juliana Alves Parente, Clayton Luiz Borges, Maristela Pereira, Alexandre Melo Bailão, Rosely Maria Zancopé Oliveira and Célia Maria de Almeida Soares

Abstract

Members of the genus *Paracoccidioides* are the causative agents of one of the most prevalent systemic mycoses in Latin America, namely, paracoccidioidomycosis (PCM). Studies have reported different degrees of pathogenicity and virulence among the members of this genus and provided evidence of variation in host susceptibility to infection. The identification and validation of virulence factors has been hindered by difficulties in the genetic manipulation of this pathogen, although studies have demonstrated that *Paracoccidioides* pathogenicity is facilitated through a number of virulence attributes, including fungal dimorphism, the composition of the cell surface and adherence to host tissues, secretion of molecules comprising the virulence arsenal of the fungus, competition with the host for essential nutrients, such as iron, and the pathogen's consequent adaptation to the host environment. This chapter provides information on the current knowledge of the virulence of the genus *Paracoccidioides*.

Introduction

In recent years, a significant increase in the incidence of fungal infections in humans has been described (Lass-Flörl, 2009; Prado *et al.*, 2009). Of the fungi regarded as human pathogens, the members of the genus *Paracoccidioides* are frequently recovered from fungal infections. Phylogenetic and population studies of *Paracoccidioides* species indicated that the genus comprises at least four phylogenetic lineages: S1 (species 1 from Brazil, Venezuela, Peru, Paraguay and Argentina), PS2 (phylogenetic species from Brazil and Venezuela), and PS3 (phylogenetic species from Colombia), which are all included in the *P. brasiliensis* species (Matute *et al.*, 2006; Carrero *et al.*, 2008), and *Pb*01, which is included in *P. lutzii* (Teixeira *et al.*, 2009). Recombination events have been demonstrated within the *Paracoccidioides* species (Matute *et al.*, 2006), indicative of sexual reproduction, although a teleomorphic form of *Paracoccidioides* has not been isolated. Nevertheless, a sexual reproduction cycle has been suggested based on the presence of two idiomorphic mating-type genes in *Paracoccidioides* (Torres *et al.*, 2010) and the functionality of the *Paracoccidioides* mating α-pheromone-receptor system in the heterologous host *Saccharomyces cerevisiae* (Gomez-Rezende *et al.*, 2012).

Many aspects of *Paracoccidioides* biology remain unknown, particularly the virulence factors that contribute to the pathogenesis of this fungus. These factors are of interest as they could be exploited for the development of vaccines and targeted therapeutic strategies. The release of the total genome sequences of *Paracoccidioides* for isolates *Pb*01, *Pb*3 and *Pb*18 at the Dimorphic Fungal Database of the Broad Institute (http://www.broadinstitute.org/annotation/genome/dimoprh_collab/MultiHome.html) has provided a more detailed insight into the biology and pathogenesis of these medically important fungi. The *Paracoccidioides* genomes comprise five chromosomes with 40–75 kb of mitochondrial DNA and nuclear genome sizes of 29–33 Mb with 8130, 7610 and 8042 genes for the isolates *Pb*01, *Pb*3 and *Pb*18, respectively (Desjardins *et al.*, 2011).

Virulence factors are considered as both processes and substances/molecules for the initiation and maintenance of *Paracoccidioides* diseases in the host; however, the roles of the majority of the fungus-specific virulence factors in causing damage to the host have not been fully characterized. This lack of characterization primarily reflects difficulties in employing reverse genetic techniques to study this parasite. In fact, genetic manipulation in *Paracoccidioides* has only been recently initiated, and gene expression approaches based on anti-sense RNA (aRNA) technology have facilitated the targeted down regulation and identification of genes related to fungal pathogenesis and virulence.

Paracoccidioidomycosis

The genus *Paracoccidioides* represents the aetiological agents of paracoccidioidomycosis (PCM), which occurs as an active disease in up to 2% of infected individuals (Restrepo-Moreno and Greer, 1983; San Blas *et al.*, 2002). The disease is one of the most prevalent systemic mycoses in Latin America (Colombo *et al.*, 2011; Laniado-Laborin, 2007), and a number of cases have also been reported outside of the South American continent, both in individuals from endemic areas and in travellers returning from those regions (Buitrago and Cuenca-Estrella, 2012).

Paracoccidioides has been randomly isolated from soil in the environment (Bagagli *et al.*, 2008). As a rule, *Paracoccidioides* infections are acquired through the inhalation of air-borne fungal particles after soil revolving. For this reason, PCM is not a contagious disease. The fungus is typically associated with acid soils in agricultural areas in tropical and humid regions (Bonifaz *et al.*, 2011). PCM is a typical occupational infection of farmers and outdoor workers, primarily affecting men of 30–40 years in age. Infection of the host with *Paracoccidioides* occurs after the inhalation of conidia or mycelial fragments that reach the terminal bronchioles and alveoli, where the organism converts to the pathogenic yeast form, producing an infection that hematogenously spreads to other tissues of the body. Exceptionally, the fungus could also be acquired through traumatic skin inoculation (Castro *et al.*, 1975), where the fungus adheres to the skin and mucous membranes (Mendes-Giannini *et al.*, 2008).

The establishment, spread and severity of this fungal disease depend on factors inherent to the fungus itself, such as virulence, antigenic composition, environmental conditions and, particularly, those factors that elicit an effective immune response in the host (Parise-Fortes *et al.*, 2011). Infected persons typically show no symptoms or exhibit mild auto-limited illnesses that are not recognized as PCM, and in endemic areas, the infection rate can be high as 90% (Colombo *et al.*, 2011). In contrast, the rate of the active disease is lower, suggesting that infected individuals remain infected, unless the host-parasite balance is changed, allowing the growth of the fungus and progression to clinically active disease. Symptomatic PCM, according to its natural course and clinical manifestations, is classified as acute, sub acute and chronic clinical forms (Franco, 1987). In the acute and subacute forms, both genders are affected, with the rapid evolution of the disease primarily affecting the organs of the monocyte–macrophage system, leading to a significant deterioration of the general health of the patient. In most cases, high titres of specific antibodies and severe depression of the cellular immunity are presented. The late reactivation of a quiescent pulmonary complex or metastatic focus, associated with factors that might compromise the immune system, results in chronic PCM, also known as 'adult' (Parise-Fortes *et al.*, 2011), which is more common and less severe than the acute form (Marquez *et al.*, 2005). This form is typically observed in adult males over 30 years in age and takes months or years to become established, eventually presenting as pulmonary and tegumental (cutaneous and/or mucosal) damage.

Paracoccidioides presents a complex antigenic structure, and the immunodominant antigen of this fungus is a cell wall surface glycoprotein of 43 kDa (Gp43). This antigen is associated with virulence factors and/or *Paracoccidioides* evasion of the host defence mechanisms. The innate immune response and adaptive immunity are important for antifungal protective effects (Travassos and Taborda, 2012). Innate response cells, such as natural killer cells (NK), neutrophils, monocytes and macrophages play a central role in the resistance to *Paracoccidioides*. Subsequently, activation of phagocytes through interactions of the parasite surface molecules with Toll-like receptors (TLRs) and lectin C-like receptors (CLR) (Parise-Fortes *et al.*, 2011) and elicitation of antibodies and T cell protective responses, particularly the Th1 response, occurs. Thus, cytokines and chemokines are produced. IFN-γ-activated macrophages exhibit increased fungistatic and fungicidal activities (Parise-Fortes *et al.*, 2011). Immunoregulation in PCM results in the formation of granulomas in healthy individuals with an intact immune system. The humoral immune response against *Paracoccidioides*, characterized by the production of high titres of antibodies, is not effective and has been observed in patients infected with more severe forms of PCM (Mamoni *et al.*, 2002).

Although the clinical manifestations of PCM have been well described, there is significant overlap with symptoms with other diseases, and the diagnosis cannot be achieved based on clinical information alone. The definitive diagnosis requires the isolation of *Paracoccidioides* on specific culture media or the visualization of the yeast form through the direct examination of clinical specimens using specific fungal staining techniques (Bonifaz *et al.*, 2011). Yeasts are readily observed using these techniques, varying from young, recently separated buds measuring 2 to 10 μm in size to mature cells of up to 30 μm or more in size with multipolar budding from the mother cell. Despite the usual high sensitivity of direct microscopic examination, atypical forms of the fungus might be mistaken for those of other fungi (Telles and Martins, 2011). The gold standard for diagnosis is the growth of the organism; however, cultures typically take 2–4 weeks to grow at 37°C and up to 2 months when incubated at 25°C (Bonifaz *et al.*, 2011; Telles and Martins, 2011). In addition, the conversion from the mould to the yeast phase at 37°C is necessary for the accurate diagnosis of *Paracoccidioides*. A combination of non-culture and culture-based methods are used to improve the diagnosis of *Paracoccidioides* infection and guide therapy for PCM. Serological methods typically have a rapid turnaround time, and the detection of either antibodies or antigens provides information indicative of the current disease (Camargo, 2008). The selection of each test for diagnosis primarily depends on the clinical manifestation of the disease and host factors. However, enzyme-based immunoassays are more amenable to automation for high-throughput testing, but exhibit cross-reactivity with other fungal species (Camargo, 2008; Telles and Martins, 2011). Molecular assays have been developed to improve the diagnosis of PCM further, particularly for the detection of disease at an early stage, and enhance the specificity of the diagnosis. The polymerase chain reaction (PCR), in particular, has presented greater usefulness in rapidly detecting invasive fungal infections with high sensitivity. However, these methodologies are in the early stages of application for the routine diagnosis of *Paracoccidioides* (Telles and Martins, 2011).

Trimethoprim-sulfamethoxazole, amphotericin B and itraconazole, are the antifungal agents most commonly used in the treatment of PCM (Shikanai-Yasuda, 2006). Trimethoprim–sulfamethoxazole is an inexpensive and relatively well-tolerated treatment, although the prolonged time of administration causes frequent self-exclusion of the patient from treatment (Mendes *et al.*, 1994). Amphotericin B is used in the severe disseminated PCM and in immunocompromised patients (Bernard and Duarte, 2000). The Brazilian Society of Tropical Medicine suggested that itraconazole is the best alternative for first-line therapy of PCM (Shikanai-Yasuda *et al.*, 2006). Itraconazole is the drug of choice for mild to moderate acute and chronic infections, producing a 90% cure and 0–15% relapse rate (Shikanai-Yasuda *et al.*, 2002). The formulation of itraconazole has improved antifungal efficacy, fewer undesirable effects and extended dosing intervals (Amaral *et al.*, 2009; Khandare and Minko, 2006).

The use of immunomodulatory and immunosuppressive compounds has been suggested as an adjuvant to chemotherapy (Naranjo *et al.*, 2011; Massuda *et al.*, 2011; Bernard *et al.*, 2012). Thus, the combination of chemotherapy and antigen administration has been studied to increase the PCM treatment success. Based on the protein amino acid sequence, the T-cell epitope for Gp43 was mapped to a 15-mer peptide known as P10 (Taborda *et al.*, 1998). Mice previously immunized with P10 (using complete Freund's adjuvant – CFA) and infected with *Paracoccidioides* cells presented a 200-fold reduction of yeast cells recovered in the lung, suggesting that P10 can be used as a potential effective vaccine against PCM (Travassos *et al.*, 2004). In addition, immunization with a plasmid expressing P10 might be a powerful treatment for experimental PCM, with potential as an effective treatment in human patients (Rittner *et al.*, 2012). Another important candidate molecule for PCM treatment is the antigenic protein Pb27. The recombinant protein (rPb27) was tested by immunization of mice, and the fungal infection was deeply reduced in the lungs, liver and spleen, suggesting the potential of Pb27 as a prophylactic in PCM treatment (Reis *et al.*, 2008). The potential use of Pb27 as an adjuvant to chemotherapy was also reported. The combined therapy of rPb27 with fluconazole controlled PCM in the liver and spleen of infected mice (Fernandes *et al.*, 2011). A D-mannose binding lectin from *Artocarpus heterophyllus* seeds, referred to as ArtinM, has been implicated in modulating immunity against several intracellular pathogens. This molecule was tested in PCM experimental infection, eliciting a significant Th1 immune response mediated through IL-12 in therapeutic protocols, thereby reducing the fungal recovery in infected mice (Coltri *et al.*, 2008, 2010). The proteins Pb27 and ArtinM have been tested in experimental immunizations in mice with successful results and are also potential targets for PCM immunization in patients (Fernandes *et al.*, 2011; Coltri *et al.*, 2010).

Patients commonly abandon treatment as a result of the high costs of drugs, treatment length, lack of trained personnel and the limited access of rural workers to the healthcare system (Yasuda, 2005). In addition, the drug toxicity and *Paracoccidioides* resistance to sulfonamides, azoles and amphotericin B have hindered the treatment of PCM (Mundargi *et al.*, 2008; Khandare and Minko, 2006; Yasuda, 2005, Hahn *et al.*, 2003; Ollague *et al.*, 2000). Thus, the identification of new antifungal compounds to treat PCM is of relevance. Studies have identified natural products, such as thichothecenes obtained from *Fusarium* sp., which inhibit *Paracoccidioides* growth (Campos *et al.*, 2011), and ajoenes obtained from *Allium sativum*, which increases the Th1-type cytokine responses in treated mice (Thomaz *et al.*, 2008). In addition, oenothein B obtained from the plant *Eugenia uniflora* has been implicated in the inhibition of the β(1,3)-glucan synthase, causing fungal cell-wall rupture (Santos *et al.*, 2007). Chemotherapeutic agents acting on specific biochemical steps have also been investigated, and chemically synthesized derivatives of the azalanosterol were identified as potent inhibitors of the enzyme (S)-adenosyl-L-methionine: D^{24}-sterol methyltransferase (SMT), which catalyses sterol C-methylation in the sterol biosynthetic pathway. The SMT inhibitors act as anti-proliferative agents against *Paracoccidioides* and are potential targets for rational drug design (Pereira *et al.*, 2010; Visbal *et al.*, 2005).

Dimorphism

Dimorphism is an important characteristic of the *Paracoccidioides* genus, as the fungus is encountered in the mycelial form in the environment and a yeast cell form in the host. A temperature of 37°C triggers the dimorphic switch to the yeast cell form in the host (Carbonell and Rodriguez, 1965). The morphological differentiation that occurs during the mycelia or conidia to yeast transition is considered essential to successful infection. Thus, *Paracoccidioides* virulence is primarily attributed to the ability of this microorganism to transit between filamentous to unicellular yeast forms during infection. Indeed, the transition between these morphotypes is critical for pathogenesis, as strains that are unable to transit are not virulent (Borba and Schaffer, 2002). Because the transition of inhaled mycelia propagules to yeast cells is crucial for the establishment of *Paracoccidioides* infection, understanding the molecular aspects related to this transition is important to elucidate the potential virulence factors in this fungus.

The importance of the morphological transition to the establishment of infection in many dimorphic fungi (Nemecek *et al.*, 2006) has been well established, and the pathways and proteins that regulate this process might represent candidate virulence factors. In eukaryotes, many cell-signalling processes are mediated through guanine-nucleotide binding proteins, known as G proteins. Studies have indicated a direct interaction between the β-subunit of the G protein and the fungal adenylyl cyclase of *Paracoccidioides*, the cellular levels of which are increased during the fungal transition from mycelium to yeast cells, consistent with the regulation of the morphological transition through changing cAMP levels (Chen *et al.*, 2007) as demonstrated in other fungi (Rocha *et al.*, 2001). Other important signalling protein classes include the Ras protein family, which possess a GTP-binding domain and exhibits GTPase activity. The Ras proteins are activated in response to several extracellular stimuli (Wiesmüller and Wittinghofer, 1994). *Paracoccidioides* possesses two Ras proteins (named Ras1 and Ras2) attached to the membrane through farnesylation. The incubation of *Paracoccidioides* mycelia with farnesylation inhibitors blocks the mycelium to yeast cell transition, suggesting the importance of this post-translational modification to Ras activity and dimorphism in *Paracoccidioides* (Fernandes *et al.*, 2008). The blocking of farnesylation also inhibits the conversion from yeast to pseudohyphae in *Candida albicans* (McGeady *et al.*, 2002). The Ca^{2+}/calmodulin-dependent protein calcineurin has been described as essential for adaptation to environmental stresses, growth, morphogenesis, and pathogenesis in many fungal species (Stie and Fox, 2008), such as in the dimorphic fungus *Ustilago maydis* where the knockout of the calcineurin gene reduces the dimorphic transition and drastically reduces fungal virulence in plants (Egan *et al.*, 2009). The calcineurin protein is relevant to *Paracoccidioides* dimorphism, as the calcineurin inhibitor Cyclosporin A (CsA) blocks the mycelium to yeast cell transition and causes abnormal morphology during the cultivation of both mycelium and yeast cells. The calcineurin inhibition through CsA unbalances intracellular calcium levels, thereby reducing cell viability in both mycelium and yeast cells (Campos *et al.*, 2008).

Heat shock proteins (Hsps) are a family of large conserved proteins and some have been characterized in *Paracoccidioides*, such as Hsp60, Hsp87, Hsp70, Hsp100 (also named ClpB) and Hsp90. The expression of Hsp60, Hsp70, Hsp87, Hsp90 and Hsp100 is increased during the fungus dimorphic transition from mycelia to yeast cells (Salem-Izacc *et al.*, 2001; Diez *et al.*, 2002; Silva *et al.*, 1999, Goldman *et al.*, 2003, Jesuíno *et al.*, 2002; Tamayo *et al.*, 2013, Nunes *et al.*, 2005). The inhibition of Hsp90 through geldanamycin (GDA) reduces the mycelium to yeast transition rate, consistent with the involvement of the protein in fungal dimorphism (Matos *et al.*, 2012). The transcript encoding Hsp90 is induced during the dimorphic transition and when yeast cells are treated with hydrogen peroxide (H_2O_2) and menadione, suggesting the importance of Hsp90 to dimorphism and the oxidative stress response (Nicola *et al.*, 2008). The *Paracoccidioides* Hsp90 function was evaluated by antisense RNA techniques, and the silenced strains presented a lower yeast growth rate compared with wild-type cells. The yeast cell viability in the Hsp90 knockdown strain was reduced after incubation with macrophages, suggesting the importance of this protein during infection. In addition, the Hsp90-silenced strain was more sensitive to oxidative treatment with H_2O_2 and acid culturing conditions, indicating the importance of Hsp90 to cell integrity and survival under stress conditions, which was consistent with previously published results (Tamayo *et al.*, 2013).

The metabolic differences between mycelium and yeast cells are deeply related to the niche adaptations of the morphotypes. Thus, to understand the biochemical aspects of each morphotype facilitates the elucidation of the virulence attributes of the yeast phase. Transcriptional analyses through sequencing cDNA libraries obtained from the mycelium and yeast cells of *Paracoccidioides Pb*01 have revealed that transcripts related to aerobic metabolism are up regulated in mycelium, while those related to anaerobic metabolism are induced in yeast cells. Several enzymes that provide substrates for the glyoxylate cycle are up-regulated in the yeast cells, such as isocitrate lyase, which generates oxaloacetate, aminotransferases, which generate acetyl-CoA, and acetamidase, which forms acetate from acetamide, suggesting the importance of the glyoxylate cycle to yeast cell maintenance (Felipe *et al.*, 2003, 2005). The glyoxylate cycle is an important carbon source for *Paracoccidioides* yeast cells also in the host, as transcripts encoding enzymes of the glyoxylate cycle are induced in yeast cells recovered from murine macrophages (Derengowski *et al.*, 2008).

The induction of virulence factors during the dimorphic transition in fungi is important, as an increase in temperature is induced through exposure to host conditions, leading to the adaptation of the fungal cell to the host (Klein and Tebbets, 2007). Thus, many studies have focused on the global analysis of gene expression to elucidate the transition of mycelia/conidia to yeast cells and describe potential virulence factors. The transcriptome analysis of the early mycelium to yeast cell transition in *Paracoccidioides* (*Pb*01) was performed. The transcriptome analysis revealed that a relevant aspect of the fungal transition was the synthesis/remodelling of the cell wall/membrane, as suggested by the induction of transcripts encoding enzymes, such as α-glucosidase, α(1,3)-glucan synthase and chitinases, putatively reflecting cell wall reorganization during the mycelia to yeast transition. Transcripts encoding catalases, proteases, haemolysin and urease were significantly induced during the transition, suggesting that *Paracoccidioides* responds to the new conditions encountered in the host milieu (Bastos *et al.*, 2007). The transcriptional analysis of *Paracoccidioides* during the mycelium to yeast transition was also performed with *Pb*18 using the microarray technique. Transcripts encoding proteins related to cell wall organization, such as chitin synthases, were induced, and the down-regulation of transcripts encoding endochitinases was also observed, reflecting the high chitin content in yeast cells compared with mycelium. Notably, a 4-hydroxyl-phenyl pyruvate dioxygenase (4-HPPD), which is involved in aromatic amino acid catabolism, was highly over expressed during the differentiation of mycelium to yeast. The incubation of mycelium with the 4-HPPD-inhibitor NTBC [2-(2-nitro-4-trifluoromethylbenzoyl)-cyclohexane-1,3-dione] blocks mycelial growth and the mycelium to yeast transition. These data demonstrate that 4-HPPD might be an important molecule in the mycelium to yeast transition, suggesting that 4-HPPD contributes to morphological transitions in the host (Nunes *et al.*, 2005).

One important aspect that influences the dimorphism of *Paracoccidioides* is the hormone 17β-oestradiol encountered in adult women. *Paracoccidioides* presents a high-affinity oestrogen-binding protein expressed in both mycelium and yeast cells (Stover *et al.*, 1986), and the interaction of the fungal cell with this hormone inhibits the mycelia/conidia transition to yeast cells *in vitro* (Restrepo *et al.*, 1984; Salazar *et al.*, 1988) and during experimental infection in mice (Aristizabal *et al.*, 1998). Consistently, epidemiological studies of PCM patients indicate that the disease is approximately 11 to 30 times more common in adult males (Blotta *et al.*, 1999). Microarray assays were employed to study the effect of 17β-oestradiol on fungal dimorphism (Shankar *et al.*, 2011). Genes related to the modulation of the heat shock response, cell wall maintenance/remodelling, energy metabolism, transport and detoxification, cell signalling, and retro-transposable elements were induced. These results suggest that blocking of the fungal transition by this hormone is mediated via global transcriptional alterations. The transcript encoding the Hsp40 protein is significantly induced in fungal cells in response to 17β-oestradiol. *S. cerevisiae* mutants lacking Hsp40 displayed high basal levels of steroid receptor (Johnson and Craig, 2000), suggesting that Hsp40 modulates the expression of the hormone-binding protein. It has been suggested that the high expression of Hsp40 in *Paracoccidioides* during 17β-oestradiol treatment might negatively modulate steroid receptor expression, reducing fungal cell binding to the hormone (Shankar *et al.*, 2011).

The development of proteomic approaches has increased our knowledge of the metabolic aspects of the *Paracoccidioides* transition. Thus, the global analysis of *Paracoccidioides Pb*01 mycelia, the mycelia to yeast transition and the yeast proteome was performed using 2-DE in combination with mass spectrometry (Rezende *et al.*, 2011). The results showed the up regulation of several metabolic processes. One major change detected in these studies was the accumulation of the enzymes of the glycolytic pathway and alcohol dehydrogenase at 22 hours after the mycelia to yeast transition, consistent with transcriptional studies indicating a shift towards anaerobic metabolism in yeast cells (Felipe *et al.*, 2005). Proteomic analysis also demonstrated induced Hsp levels during the transition from mycelia to yeast cells and the yeast phase, reflecting the fungal response to the temperature shift. During the dimorphic transition it was possible to identify induced metabolic pathways, such as glycolysis, and proteins that remain up-regulated during the yeast phase, including those related to glycolysis and pentose phosphate pathways. Notably, the induction of the pentose phosphate pathway during the mycelium to yeast transition and the yeast phase can be associated with the recycling of $NADP^+$ to NADPH, protecting the fungal cells against reactive oxygen intermediates, which is potentially important in the parasitic phase to avoid macrophage-induced ROS damage (Rezende *et al.*, 2011).

Candidate virulence factors at the *Paracoccidioides* yeast cell surface

The cell wall is the first barrier to make contact with the host. During the mycelia to yeast cell transition, remodelling of the cell wall occurs, which involves a change in the type of glucan utilized. The mycelia are predominantly enriched with β-glucan compared with yeast cells that present α(1,3)-glucan (San-Blas, 1993). The relationship between α(1,3)-glucan with fungal virulence has long been speculated, as reduced virulence occurs concomitantly with the loss of externally arranged α(1,3)-glucan (San-Blas *et al.*, 1976; San-Blas and San-Blas, 1977; Kashino *et al.*, 1990). In *Paracoccidioides*, the final step of the α(1,3)-glucan synthesis is catalysed by α(1,3)-glucan synthase (Ags1) which is highly expressed during the yeast phase (Sorais *et al.*, 2010). The activity of Ags1 was evaluated through complementation analysis of a *Histoplasma capsulatum* AMY1 mutant (encoding α-1,4-amylase). The mutant complemented with the Ags1 protein from *Paracoccidioides* exhibited a restored ability to synthesize α(1,3)-glucan. In addition, the *H. capsulatum* Amy1 mutant complemented with Ags1 exhibited a restored ability to infect macrophages, suggesting the importance of the Ags1 as a virulence factor in *Paracoccidioides* (Camacho *et al.*, 2012).

Morphogenesis and cell-wall modelling has been associated with a network of glycoproteins and polysaccharides, including chitin. Reduced glycosylation of yeast cells produces smaller mother cells compared to their fully glycosylated counterparts with a drastic reduction of cell budding, reflecting an impairment of growth and morphogenesis. This fact was evidenced when the chitinase paracoccin was underglycosylated. When exhibiting low levels of glycosylation, paracoccin accumulates at intracellular locations, loses the ability to bind to laminin and presents reduced N-acetyl-β-D-glucosaminidase activity. In addition, yeast cells treated with the N-glycosylation inhibitor tunicamycin are smaller than control cells and present a drastic reduction of cell budding, which reduces the proliferation rate of yeast cells in the host (Almeida *et al.*, 2011). Thus, an optimal dynamic and structure of cell wall carbohydrates is required for yeast cell adhesion and pathogenesis.

The initial contact of pathogenic fungi with host cells is essential for successful tissue colonization and the initiation of the infectious process. Thus, the factors related to fungal attachment have been the primary target of many scientific investigations (Long *et al.*, 2003; Barbosa *et al.*, 2006; Rappleye and Goldman, 2006; Verstrepen and Klis, 2006; Upadhyay *et al.*, 2009; Murciano *et al.*, 2012). Pathogenic fungi display a large variety of molecules at their surface, which facilitate adherence to different cell types and interaction with several ligands present in various host sites, such as the extracellular matrix, body fluids and basement membranes (Tronchin *et al.*, 2008). PCM development depends on the binding of fungal cells to host components. Therefore, the ability of *Paracoccidioides* to adhere to host tissues and mucosal surfaces is a crucial step in the establishment of the infection (Brito *et al.*, 1973; Hanna *et al.*, 2000). It has been shown that this fungus binds to and invades different host cells (Brummer *et al.*, 1989; Mendes-Giannini *et al.*, 2008). In addition, during *Paracoccidioides* dissemination, the fungus adheres to endothelial cells and crosses basement membranes to invade different tissues (Franco, 1987; Mendes-Giannini *et al.*, 2008). Thus, this pathogenic fungus must exhibit adhesion in order to establish infection and to disseminate and colonize a wide range of host tissues.

Evidence suggests that adhesion is an important step for *Paracoccidioides* pathogenesis. Precoating yeast cells with laminin before infection of mice reduces the histopathology of a high virulence strain and reduces the fungal burden of a low virulence strain (Andre *et al.*, 2004). Adherence varies among *Paracoccidioides* strains with differential virulence potentials; highly virulent strains exhibit enhanced adhesion ability *in vitro* (Hanna *et al.*, 2000). Studies have shown that ECM components present a specific fluorescence distribution on the fungal cell surface, indicating the presence of specific binding sites for different ECM proteins (Mendes-Giannini *et al.*, 2000). The yeast cells interact with several ECM proteins, such as laminin, fibronectin, fibrinogen and type I and type IV collagens (Mendes-Giannini *et al.*, 2006). Conidia also interact with fibronectin, fibrinogen and laminin and this association mediates the adhesion of conidia to pulmonary epithelial cells (Caro *et al.*, 2008). Binding assays demonstrated that cell-free extracts obtained from yeast cells, contain several proteins that attach to laminin, fibronectin or type I collagen (Mendes-Giannini *et al.*, 2006). *Paracoccidioides* use different strategies to promote adhesion to host cells/tissues, enlisting proteins, lipids and carbohydrate structures to the arsenal of adhesive capabilities, which facilitate the colonization of a wide range of host niches.

As the binding of *Paracoccidioides* cells to ECM components is important to PCM establishment, many efforts have been done to characterize adhesins of this pathogen. The Gp43, present on the cell wall, was the first laminin-binding molecule identified in this fungus (Vicentini *et al.*, 1994). The glycoprotein Gp43 also mediated the adherence to macrophages and epithelial Vero cells (Almeida *et al.*, 1998; Hanna *et al.*, 2000). Competition assays showed that synthetic peptides derived from this protein inhibited interactions between Gp43 and Vero cells (Mendes-Giannini *et al.*, 2006). Purified Gp43 also reduces the adhesion and invasion of *Paracoccidioides* to epithelial cells (Andreotti *et al.*, 2005). Anti-Gp43 antibodies inhibited the phagocytosis of yeast cells by macrophages (Almeida *et al.*, 1998). Taken together, these data show that Gp43 plays a significant role in the *Paracoccidioides* adherence to extracellular matrix components, acting as an important element in fungal adhesion.

Other proteins mediate the *Paracoccidioides* adhesion to the ECM and host cells. An adhesin of 30 kDa was isolated from a strain with a high adhesion capacity. This protein presented the ability to bind to laminin, and the pre-treatment of A549 pneumocytes with the recombinant molecule dramatically inhibited the adherence of yeast cells to A549 pneumocytes (Andreotti *et al.*, 2005). In addition, two proteins of 19 and 32 kDa in *Paracoccidioides* whole-cell extracts presented the ability to attach to fibronectin, fibrinogen and laminin. These proteins were localized on the cell surface of yeast cells, mycelia and conidia of *Paracoccidioides* (Gonzalez *et al.*, 2005). Studies identified the 32 kDa molecule as a member of the haloacid dehalogenase (Had) super family. Knockdown mutants for the Had32 protein were obtained using anti-sense RNA technology. Silenced strains showed a reduced ability to adhere to epithelial cells. Infection experiments showed that this adhesin is important for *Paracoccidioides* pathogenesis, as mutants were less virulent in a mouse model of infection (Hernandez *et al.*, 2010).

Enzymes of carbohydrate metabolism have been described as adhesins at the *Paracoccidioides* surface. The glycolytic enzyme glyceraldehyde-3-phosphate dehydrogenase (Gapdh) is localized to the cell wall of the yeast form in addition to the expected cytoplasm localization. Gapdh binds to fibronectin, laminin and type I collagen. Moreover, the anti-Gapdh polyclonal antibodies and the recombinant protein promoted the inhibition of adherence and internalization of yeast cells to A549 pneumocytes, indicating that the protein localized to the cell wall contributes to the initial infection of *Paracoccidioides* (Barbosa *et al.*, 2006). Triose phosphate isomerase (Tpi), another glycolytic enzyme of *Paracoccidioides*, interacts with fibronectin and laminin, as revealed through Far Western assays. Tpi was localized to the cell wall of *Paracoccidioides* and bound to A549 pneumocytes and Vero cells. The association of yeast cells with A549 pneumocytes was hindered using recombinant molecules and anti-Tpi antibodies, indicating that Tpi might be included in the adhesin arsenal of *Paracoccidioides* (Pereira *et al.*,

2007). Enolase (Eno) was identified as another adhesin of *Paracoccidioides*. Antibodies raised against Eno inhibited fungal adhesion to epithelial cells (Donofrio *et al.*, 2009). Studies revealed that Eno binds to laminin, types I and IV collagen and fibrinogen. The interaction between Eno and epithelial cells is partially dependent on the RGD motif and the accumulation of the protein on the yeast surface enhances the *Paracoccidioides* infection of A549 cell (Marcos *et al.*, 2012). Eno was further characterized as a plasminogen-binding protein. Notably, this protein binds to and activates plasminogen in plasmin, most likely contributing to the pathogenesis of the fungus (Nogueira *et al.*, 2010). Thus, during infection, yeast cells might use the plasmin degradation on the cell surface to penetrate and spread throughout host tissues. Malate synthase (Mls), an enzyme of the glyoxylate cycle, was identified on the surface of *Paracoccidioides* yeast cells. The use of polyclonal antibodies against this enzyme blocked the infection of this pathogen in A549 pneumocytes cells (Silva Neto *et al.*, 2009). These observations strongly demonstrated that *Paracoccidioides* presents a complex pool of proteins with adhesive characteristics that mediate the interaction of this pathogen with different host cells and tissues.

The influence of ECM components on the *Paracoccidioides* gene expression profile has been investigated in the presence of type I collagen and fibronectin using a representational difference analysis (RDA) strategy. The results showed the induced expression of 127 genes, putatively affected through the presence of ECM components. A copper transporter and enolase were identified among the induced genes as molecules that bind to different ECM components. In addition, it was demonstrated the capacity of enolase to bind in addition to collagen type I, type IV and fibronectin, to J774 macrophages as demonstrated using immunofluorescence assays (Bailão *et al.*, 2012a).

Many proteins are associated with the cell wall through glycosylphosphatidylinositol anchors (GPI). The Dfg5 protein (defective for filamentous growth), first described in *S. cerevisiae*, was characterized in *Paracoccidioides*. Dfg5 is a cell wall, GPI-anchored protein, primarily expressed during the yeast phase. This protein bids to many ECM proteins, such as laminin, fibronectin and type I and IV collagen, and reacts with the sera of PCM patients, suggesting that this protein is relevant to *Paracoccidioides* infection (Castro *et al.*, 2008). The GPI-anchored O-glycoprotein *Pb*Pga1 is preferentially expressed during the fungus yeast phase and was characterized in *Paracoccidioides* as a molecule that reacts with the sera of PCM patients. *Pb*Pga1 induces TNF-α release and nitric oxide (NO) production in macrophages, suggesting a role for this protein in both humoral and innate immune responses. *Pb*ga1 is localized to the bud surface and the bud neck, suggesting the importance of this protein in yeast cell proliferation in the host (Valim *et al.*, 2012).

Proteins, lipids and carbohydrates play roles on *Paracoccidioides*-host cells interactions. The fungus uses different strategies to promote adhesion to host cells/tissues, employing proteins, lipids and carbohydrate structures in an arsenal of adhesive molecules that facilitate the colonization of a wide range of host niches. Membrane rafts are cholesterol and sphingolipid-enriched membrane domains that functions in cell–pathogen interactions. Fluorescence microscopy using a specific marker for membrane raft-specific lipids revealed that these structures are located in the *Paracoccidioides* yeast-epithelial cell contact region, indicating that membrane raft regions are composed of fungus-mammalian cell-interacting molecules. Furthermore, the depletion of cholesterol in pulmonary epithelial cells, as performed through incubations with methyl-β-cyclodextrin, strongly inhibited the *Paracoccidioides* association with epithelial cells (Maza *et al.*, 2008). *Paracoccidioides* also adhere to plates coated with glycosphingolipids (GSL). It was demonstrated that the fungus also binds to GSL-enriched fractions from fibroblast cultures, suggesting that this microorganism uses host glycostructures as fixing points during infection. Consistently, antibodies raised against the glycosides of membrane glycosphingolipids reduced fungal adhesion to fibroblasts (Ywazaki *et al.*, 2011), strongly suggesting that the adhesion *Paracoccidioides* mechanism is also mediated through carbohydrates or glycoconjungates.

Molecules secreted from *Paracoccidioides*

Microorganisms secrete/excrete molecules for nutrient acquisition, colonization, cell signalling, adhesion, virulence factors, detoxification, and the modulation of host immune responses, among other functions (Bonin-Debs *et al.*, 2004; Carberry *et al.*, 2006; Fernandez-Acero *et al.*, 2007; Yajima and Kav 2006; Seidler, 2013). *Paracoccidioides* secrete molecules related to important functions in host–pathogen interactions, including proteins (Vallejo *et al.*, 2012a; Weber *et al.*, 2012), lipids (Vallejo *et al.*, 2012b) and carbohydrates (Albuquerque *et al.*, 2012) that likely play roles in facilitating the establishment of adhesion, dissemination and infection.

Some secreted proteins in *Paracoccidioides* have been characterized. The secreted Gp43 of *Paracoccidioides* (Straus *et al.*, 1996) interacts with human monocytes, affecting the innate immune response and the production of pro- and anti-inflammatory cytokines, which influences in the establishment of infection (Nakaira-Takahagi *et al.*, 2011). A serine protease of *Paracoccidioides,* is an N-glycosylated enzyme induced in yeast cells that infect murine macrophages (Parente *et al.*, 2010; Parente *et al.*, 2005), suggesting that this protein might be involved in the pathogenesis of *Paracoccidioides* (Parente *et al.*, 2005). Other important proteases identified as secreted molecules in *Paracoccidioides* include a serine-thiol protease, which cleaves laminin and fibronectin *in vitro* (Carmona *et al.*, 1995). This protease activity is modulated through neutral polysaccharides (Matsuo *et al.*, 2006). The interaction of this protease with endogenous polysaccharides likely increases the thermotolerance of this protein to the host and facilitates fungal penetration in tissues (Matsuo *et al.*, 2007).

Members of the *Paracoccidioides* genus harbour genes encoding predicted secreted proteins (297, 291 and 322 genes in *Pb*01, *Pb*03 and *Pb*18, respectively) as previously described (Desjardins *et al.*, 2011). For example, the important virulence factor in *H. capsulatum* Cbp1 (calcium binding protein 1) (Rappleye and Goldman, 2006) has homologues in the three sequenced genomes of the *Paracoccidioides* genus, *Pb*01, *Pb*03 and *Pb*18 (http://www.broadinstitute.org/annotation/genome/paracoccidioides_brasiliensis/MultiHome.html) (Rappleye and Goldman 2006; Desjardins *et al.*, 2011). Several genes encoding proteases and peptidases have also been identified in the *Paracoccidioides* genome database (http://www.broadinstitute.org/annotation/genome/paracoccidioides_brasiliensis/MultiHome.html). For example, *Paracoccidioides* secrete ubiquitinyl hydrolase and ATP-dependent metalloendopeptidases, which are also present in other pathogenic and dimorphic fungi, acting in the establishment of infection (Desjardins *et al.*, 2011; Rappleye and Goldman 2006). These proteases might be relevant to fungal invasion and pathogenesis.

In the post-genomic era, proteomic studies have been performed to describe the collection of proteins secreted from two members of the *Paracoccidioides* complex (Weber *et al.*, 2012; Vallejo *et al.*, 2012a). In both studies, the authors identified proteins secreted through conventional and non-conventional secretory mechanisms. The proteins secreted from *Paracoccidioides Pb*18 were analysed in vesicular and vesicular-free fractions, and 85 proteins from vesicles, 140 proteins from vesicle-free fractions and 120 proteins common between the two fractions were identified. Bioinformatics analyses revealed that 72 proteins could also be identified in *Pb*03 and *Pb*01. From the proteins common to both fractions, a diversified collection of molecules was described, including glyceraldehyde 3-phosphate dehydrogenase, triose phosphate isomerase, and enolase, which have previously been implicated in the virulence of *Paracoccidioides* (Pereira *et al.*, 2007; Barbosa *et al.*, 2006; Nogueira *et al.*, 2010). In addition, Gp43 was detected in the *Paracoccidioides Pb*18 secretome. This protein possesses antigenic and immunomodulatory properties and is effective as a vaccine, as this protein confers protection to experimentally infected mice, as previously described (Rittner *et al.*, 2012; Mayorga *et al.*, 2012; Konno *et al.*, 2012).

The mycelia and yeast secretomes of *Pb*01 were characterized using proteomics and two-dimensional gel electrophoresis to identify differentially expressed proteins in the mycelia and yeast cells of each *Paracoccidioides* morphotype (Weber *et al.*, 2012). A total of 160 proteins were identified, including 30 and 24 proteins preferentially expressed in mycelia and yeast cells, respectively. The protein formamidase was identified as preferentially secreted in mycelia. This cell surface protein is antigenic (Borges *et al.*, 2005, 2010), and the alkalinization of the surrounding milieu by ammonia produced by this enzyme presumably damages host tissues, facilitating the initial steps of the infection process, as described in other pathogenic organisms (Bury-Moné *et al.*, 2004). However, the yeast parasitic form of *Paracoccidioides* contains preferentially expressed detoxification and stress-related proteins compared with mycelia. Among the yeast-secreted proteins related to detoxification, glutathione S-transferase and superoxide dismutase were implicated in resistance against oxidative stress in host tissues and may act to protect the fungus under these conditions. The heat shock protein Hsp60 was also preferentially secreted from the yeast form. Hsp60 induces a protective immune response against PCM, and vaccination with Hsp60 reduced the fungal burden in a mice model of infection (Soares *et al.*, 2008). Bioinformatics analyses of adhesin motifs revealed that 21% of the identified proteins in the secretomes of both the mycelia and yeast cells of *Pb*01 are predicted as adhesin molecules, including glyceraldehyde 3-phosphate dehydrogenase, triose phosphate isomerase and enolase, thus confirming previous descriptions (Pereira *et al.*, 2007; Barbosa *et al.*, 2006; Nogueira *et al.*, 2010). The inhibition of protein secretion in yeast cells using brefeldin-A significantly inhibited the macrophage/fungus interaction and reduced *Paracoccidioides* cell recovery from macrophages, suggesting that the secreted molecules somehow act in the host pathogen interactions and quantitatively influence the recovery of *Paracoccidioides* in this *ex vivo* model of infection. Notably, as demonstrated through ultra-performance liquid chromatography (UPLC) and mass spectrometry, proteins detected in the secretome of yeast cells, such as enolase (Nogueira *et al.*, 2010), glyceraldehyde 3-phosphate dehydrogenase (Barbosa *et al.*, 2006), Hsp60 (Soares *et al.*, 2008) and malate dehydrogenase (Silva Neto *et al.*, 2009), were also detected during fungal infection in macrophages, strongly indicating a role for these proteins in host–parasite interactions and the pathogenesis of *Paracoccidioides* (Weber *et al.*, 2012).

The *Paracoccidioides Pb*01 and *Pb*18 present some similarities and differences in the characterized yeast secretomes (Weber *et al.*, 2012; Vallejo *et al.*, 2012a). Among the 160 proteins/isoforms characterized in *Pb*01, 63 proteins were common between the isolates. A number of the common proteins were identified as enzymes involved in the metabolism of carbohydrates, lipids and heat shock proteins. For example, pyruvate kinase, 2-methylcitrate synthase and Hsp70, were common between *Pb*01 and *Pb*18. However, 47 proteins/isoforms were exclusively identified in *Pb*01, including the stress-related proteins glutathione reductase, glutathione s-transferase and Hsp60. The differential profile of the secreted proteins could reflect the genomic differences described in the three sequenced *Paracoccidioides* strains (Desjardins *et al.*, 2011) and the differential patterns of virulence and pathogenicity identified in different isolates of *Paracoccidioides* (Kurokawa *et al.*, 2005).

The lipidomic analysis of extracellular vesicles released from the yeast pathogenic phase of *Paracoccidioides* was performed using mass spectrometry (Vallejo *et al.*, 2012b). Lipids are important molecules in *Paracoccidioides*–host interactions that, along with enriched fractions of fatty acid/triglycerides from the fungus, induce intense granulomatous reactions in mice (Silva 1985; Alves *et al.*, 1987). The characterization of the lipid composition of *Paracoccidioides* secreted vesicles might elucidate the biogenesis of these molecules and highlight certain aspects of the secretion mechanism. The lipidomic analysis of *Paracoccidioides* extracellular vesicles facilitated the identification of monohexosylceramides, phospholipid species (phosphatidylcholine, phosphatidylethanolamine, phosphatidic acid, phosphatidylserine, phosphatidylinositol and phosphatidylglycerol), and sterols (brassicasterol, ergosterol and lanosterol). These lipids, together with other molecules secreted by *Paracoccidioides*, might contribute to the virulence and pathogenesis of this fungus. Diphosphatidylglycerol (DPG) has been identified as one of the lipids associated with the virulence of *Paracoccidioides*, as this lipid was primarily identified in a virulent strain and was barely detected in an avirulent strain, reflecting the increased metabolic status of virulent strains, as DPG is generally associated with mitochondria or cellular organelles of high metabolic activity (Manocha, 1980).

Carbohydrates released from fungi are required for maintaining cell wall architecture, the modulation of the host immune response, interactions with host cells, the evasion of host defences and the virulence of the pathogen (Rodrigues *et al.*, 2011; Casadevall and Pirofski, 2006; Casadevall *et al.*, 1998). Glucuronoxylomannan (GXM) is an important virulence factor, which is a heteropolysaccharide produced in *Cryptococcus*, associated with the cell wall. The production and association of GXM with the cell wall requires α(1,3)-glucans (Zaragoza *et al.*, 2009; Reese and Doering 2003), and this event is an important step in the formation of the *Cryptococcus* capsule, a virulence factor of the *Cryptococcus* spp.(Araujo *et al.*, 2012; Jesus *et al.*, 2010). *Paracoccidioides* is a non-encapsulated fungus that possesses cell wall-associated fibrillar material (Kanetsuna and Carbonell, 1970; Carbonell, 1967). *Paracoccidio*ides synthesizes and secretes α(1,3)-glucans, which are associated with fungal virulence (San-Blas and San-Blas, 1977) and are required for the association of the GXM heteropolysaccharides in *Cryptococcus*, as previously described. Furthermore, *Paracoccidioides Pb*18 presents homologues of all of the genes encoding the proteins required for polysaccharide export and GXM synthesis, similar to *Cryptococcus*. Additional *in silico* analyses of the Broad genome database (http://www.broadinstitute.org/annotation/genome/paracoccidioides_brasiliensis/MultiHome.html) demonstrated that *Pb*01 and *Pb*03 also contain the genes described above (Clayton Borges, personal communication). The polysaccharides secreted from *Pb*18 are recognized by monoclonal antibodies against the GXM from *Cryptococcus*, as demonstrated through the immunofluorescence analysis of *Paracoccidioides* yeast cells incubated with anti-GXM antibodies. *Paracoccidioides* vesicles were recognized by monoclonal antibodies against the GXM from *Cryptococcus*, suggesting that *Paracoccidioides* polysaccharides are also exported to the extracellular space in these vesicles. Interestingly, an acapsular mutant of *Cryptococcus* incorporated molecules derived from *Paracoccidioides* in the cell wall, resulting in the formation of an capsule-like structure (Albuquerque *et al.*, 2012). Thus, it is likely that the *Paracoccidioides* glycans share similar properties with the molecules of *Cryptococcus*, and might be similarly important for the pathogenesis and virulence of *Paracoccidioides*.

Adaptive response to the host and expression of presumptive virulence attributes

An important aspect of pathogenesis is how microorganisms modulate or adapt to the host environment. The essential nutrient iron lies at the competitive interface between the mammalian host and most microbial pathogens. Hosts have evolved mechanisms to efficiently acquire iron and reduce the availability of this molecule to pathogens. Thus, the mammalian host sequesters iron through a variety of mechanisms in response to microbial infection (Nairz *et al.*, 2010). Studies have demonstrated that iron limitation is a host defence against *Paracoccidioides*. The intracellular transition of conidia to yeast cells of *Paracoccidioides* in murine macrophages was inhibited when the cells were treated with the iron chelator deferoxamine, and this inhibitory effect was reversed through holotransferin treatment, indicating that iron restriction is one of the mechanisms through which activated macrophages control the intracellular differentiation of ingested *Paracoccidioides* (Cano *et al.*, 1994). In this sense, the iron supplementation of macrophages infected with *Paracoccidioides* conidia restores fungal growth and the capacity to transit to yeast cells (Gonzalez *et al.*, 2007). Moreover, low iron conditions have been associated with the susceptibility of *Paracoccidioides* to the antimicrobial actions of monocytes (Dias-Melício *et al.*, 2005). The administration of exogenous iron results in increased fungal load in mouse tissues infected with *Paracoccidioides* (Parente *et al.*, 2011).

To survive and proliferate in the competitively iron-limited environment of the host, *Paracoccidioides* must possess effective iron-acquisition mechanisms, and some have been described (Silva *et al.*, 2011; Bailão *et al.*, 2012b). A potential iron acquisition mechanism involving ferric reduction has been proposed (Silva *et al.*, 2011; Bailão *et al.*, 2012b). *In silico* analysis has revealed that the genome of this fungus contains genes that encode redundant ferric reductase homologues, and experiments have demonstrated a significant increase in the expression of these genes upon iron restriction (Parente *et al.*, 2011; Silva *et al.*, 2011; Bailão *et al.*, 2012b). *Paracoccidioides* also exhibits glutathione-dependent ferrireductase activity (Zarnovski and Woods, 2005), and transcriptional analysis had shown that the over expression of a transcript encoding an orthologue of the high-affinity zinc/iron permease (*ZTR1*) in yeast cells derived from infected liver cells supports the assumption that iron plays a role in fungal survival in the host (Bailão *et al.*, 2006).

Another iron acquisition mechanism, mediated through small chelating agents that bind ferric iron, the siderophores, has been described in *Paracoccidioides*. The major evidence of iron capture through siderophores is derived from the stimulation of fungal growth in the presence of coprogen B and dimeric acid (Castaneda *et al.*, 1988). An analysis of the *Paracoccidioides* genomes revealed the presence of genes putatively involved in siderophore biosynthesis and uptake (Silva *et al.*, 2011; Bailão *et al.*, 2012b). There are additional data showing that *Paracoccidioides* utilize iron through siderophores. Under iron-limited conditions, the fungus induces the expression of *Paracoccidioides* orthologues to known iron-dependent transcripts, such as those encoding the siderophore biosynthetic enzyme SidA and the siderophore transporter Sit1 (Parente *et al.*, 2011).

Phagocytic cells synthesize an array of toxic chemicals that promote fungal killing (Bogdan *et al.*, 2000). The ability to respond to several stress conditions might increase *Paracoccidioides* survival during infection. This possibility is reflected, for example, in the ability of detoxifying enzymes for the promotion of fungus survival. Thus, an alternative oxidase (AOX)-silenced strain of *Paracoccidioides*, generated using antisense RNA, presented reduced viability in the presence of H_2O_2 and after incubation with macrophages compared with the wild-type strain. In addition, the AOX-silenced mutants presented a reduction in the fungal burden in a murine infection model (Hérnandez *et al.*, 2011).

Macrophages and polymorphonuclear neutrophils (PMN) phagocytose pathogens and produce reactive oxygen species (ROS), superoxide radical ($O2^-$), H_2O_2 and reactive nitrogen species (RNS) derived from nitric oxide (NO) (Test and Weiss 1986; MacMicking *et al.*, 1997). The production of ROS from macrophage oxidases (particularly NADPH oxidase 2) is a defence response to microbial infections, a process known as oxidative burst. The acidic pH and ROS production within phagosomes accelerates the pathogen elimination (Kotsias *et al.*, 2013). The principal RNS produced by macrophages is NO, which synthesized through nitric oxide synthase from the amino acid L-arginine via two sequential reactions, producing NO. NO plays role as a mediator in several biological processes, including signalling, cytotoxicity, immunoregulation and antimicrobial activities (Fang, 1997). During *Paracoccidioides* infection, the ROS and RNS response occurs in macrophages (Bannwart *et al.*, 2010; Nishikaku *et al.*, 2009; Meloni-Bruneri *et al.*, 1996). The production of $O2^-$ in polymorphonuclear neutrophils (PMNs) is strongly correlated with the ability of these cells to kill *Paracoccidioides* yeast cells. In addition, there is a direct correlation between reduced ROS production in the PMN and reduced fungicidal activity during infection with *Paracoccidioides* yeast cells (Meloni-Bruneri *et al.*, 1996). Studies of the bronchoalveolar fluids of PCM patients have shown that H_2O_2 levels are increased in macrophages from PCM patients compared with those obtained from healthy control patients, demonstrating that the host response includes the production of ROS, generating an oxidative stress (Calvi *et al.*, 2003).

Studies have resulted in advances in the understanding of the oxidative stress response in *Paracoccidioides*. Transcriptional analyses of yeast cells derived from infected macrophages showed that transcripts related to oxidative stress, such as superoxide dismutase, are up-regulated during *Paracoccidioides* murine macrophage infection. The *Paracoccidioides* response to the macrophage internalization includes the negative regulation of transcripts related to glycolysis, suggesting a carbon-depleted environment inside the macrophage (Tavares *et al.*, 2007). The ability of ROS to induce the oxidative stress response in *Paracoccidioides* reflects the ability of the fungus to induce the expression of catalases in the presence of hydrogen peroxide (Chagas *et al.*, 2008). An analysis of the oxidative stress induced proteome showed that *Paracoccidioides* yeast cells respond to ROS by activating the overexpression of detoxification mechanisms that include catalases, peroxidases, superoxide dismutases and components of the thioredoxin system. A dynamic rearrangement of the metabolic flux to the pentose phosphate pathway leading to the generation of the reduced electron carrier NADPH has been observed (Grossklaus *et al.*, 2013).

The *Paracoccidioides* survival in the host might also be associated with the production of melanin produced in fungi through the polymerization of phenolic and/or indolic compounds by oxidases, such as laccases (Henson *et al.*, 1999). The melanin production in *Paracoccidioides* was first reported in yeast cells growing in the presence of L-DOPA. Melanin-like particles were also recovered from infected mouse tissues, suggesting the molecule as a virulence factor in this fungus. Laccase-like activity was detected in the protein extracts of *Paracoccidioides*, which was associated with the enzymatic synthesis of melanin in yeast cells (Gómez *et al.*, 2001). Melanized *Paracoccidioides* yeast cells presented an increased ability to resist to phagocytosis (Silva *et al.*, 2006). In *Paracoccidioides*, melanin has been associated with ROS and RNS resistance, as melanization increases the survival rate of yeast cells after incubation with nitric oxide, H_2O_2 and hypochlorite (Silva *et al.*, 2009).

Concluding remarks

From its saprobic existence in the soil to its infection in the host, *Paracoccidioides* adapts to a wide range of environmental conditions. *Paracoccidioides* exhibits several virulence attributes, such as dimorphism, adherence, secretion of potential virulence determinants and adaptation to the host. A number of genes/proteins have been potentially associated with the virulence attributes of *Paracoccidioides*, as described above. An

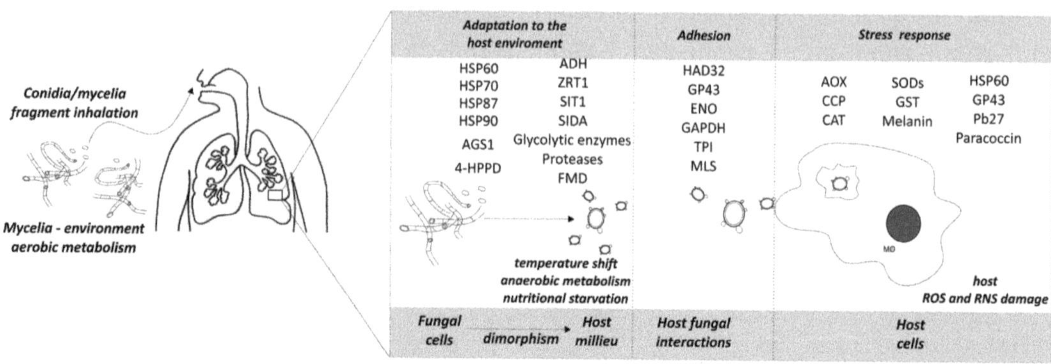

Figure 14.1 Representation of updated potential virulence factors of *Paracoccidioides*. The fungal adaptation to the host is likely mediated through a temperature shift and might include cell wall remodelling mediated through enzymes, such a Ags1 and cell-defence mechanisms, such as those mediated through Hsps. Nutritional adaptations might be performed through glycolytic enzymes, Adh, proteases, and molecules related to iron acquisition, such as SidA and Sit1. The fungal establishment might be mediated through adhesins and secreted molecules (Had32, Gp43, Eno, Gapdh, Tpi, Mls). The fungal response to the stress conditions imposed by the host includes the prevention of RNS and ROS damage (Aox, Ccp, Cat, Sods, Gst and melanin) and the modulation of the host immune response (Hsp60, Gp43, Pb27 and paracoccin). Hsp60 (Heat shock protein 60 kDa), Hsp70 (Heat shock protein 70 kDa), Hsp87 (heat shock protein 87 kDa), Hsp90 (heat shock protein 90 kDa) Ags1 (α-1,3-glucan synthase), Adh (alcohol dehydrogenase), Ztr1 (high-affinity zinc transporter), Sit1 (siderophore transporter), SidA (siderophore biosynthetic enzyme), Fmd (formamidase), Had32 (haloacid dehalogenase), Eno (enolase), Gapdh (glyceraldehyde-3-phosphate dehydrogenase), Tpi (triose phosphate isomerase), Mls (Malate synthase), Gp43 (Glycoprotein of 43 kDa), Proteases (serine proteases), Aox (Alternative oxidase), Ccp (cytochrome C peroxidase), Cat (catalases), Sod (superoxide dismutase), Gst (glutathione-S-transferase) and Pb27 (antigenic protein 27 kDa).

updated description of these genes and proteins is provided in Fig. 14.1. The fungal adaptation to the host initiates after the inhalation of conidia/mycelia propagules by the host. The adaptation to temperature includes cell wall remodelling with the synthesis of chitin and α(1,3)-glucan and the overexpression of Hsps (such as Hsp60, Hsp70, Hsp87, Hsp90). Several mechanism are related to adaptation to the host environment and the expression of virulence factors has been described in *Paracoccidioides*, such as adaptations to a more anaerobic environment in host tissues, including the over expression of glycolytic enzymes, alcohol dehydrogenase and the acquisition of micronutrients, particularly those related to iron acquisition. Proteins that bind to ECM components (enolase, haloacid dehalogenase, glyceraldehyde-3-phosphate dehydrogenase, Gp43, malate synthase and triose phosphate isomerase) are attached to the fungal cell wall or secreted, likely facilitating fungal colonization in the host. After fungal infection, the defence cells of the host are activated, resulting in the exposure of the fungus to stress conditions, including oxidative and nitrosative stresses. The fungus reacts, producing molecules, such as catalases, cytochrome c peroxidase, alternative oxidase, superoxide dismutases, glutathione-S-transferase and melanin, to reduce the effects of ROS and RNS. A number of fungal molecules modulate the host immune response, such as Hsp60, Gp43, Pb27 and paracoccin (Fig. 14.1).

Thus, our knowledge of *Paracoccidioides* is growing, although a major hindrance to the progression of these studies is the absence of efficient molecular techniques that enable targeted gene inactivation in this fungus. While few virulence determinants have been unequivocally identified to date, it is likely that these factors will be identified in the near future.

References

Albuquerque, P.C., Cordero, R.J., Fonseca, F.L., Peres da Silva, R., Ramos, C.L., Miranda, K.R., Casadevall, A., Puccia, R., Nosanchuk, J.D., Nimrichter, L., *et al.* (2012). A *Paracoccidioides brasiliensis* glycan shares serologic and functional properties with cryptococcal glucuronoxylomannan. Fungal Genet. Biol. *49*, 943–954.

Almeida, F.B.R., Carvalho, F.C., Mariano, V.S., Alegre, A.C., Silva Rdo, N., Hanna, E.S., and Roque-Barreira, M.C. (2011). Influence of N-glycosylation on the morphogenesis and growth of *Paracoccidioides brasiliensis* and on the biological activities of yeast proteins. PLoS One. *6*, e29216.

Almeida, S.R., Unterkircher, C.S., and Camargo, Z.P. (1998). Involvement of the major glycoprotein (Gp43) of *Paracoccidioides brasiliensis* in attachment to macrophages. Med. Mycol. *36*, 405–411.

Alves, L.M., Figueiredo, F., Brandao Filho, S.L., Tincani, I., and Silva, C.L. (1987). The role of fractions from *Paracoccidioides brasiliensis* in the genesis of inflammatory response. Mycopathologia. *97*, 3–7.

Amaral, A.C., Bocca, A.L., Ribeiro, A.M., Nunes, J., Peixoto, D.L., Simioni, A.R., Primo, F.L., Lacava, Z.G., Bentes, R., Titze-de-Almeida, R., *et al.* (2009). Amphotericin B in poly(lactic-co-glycolic acid) (PLGA) and dimercaptosuccinic acid (DMSA) nanoparticles against paracoccidioidomycosis. J Antimicrob. Chemother. *63*, 526–533.

Andre, D.C., Lopes, J.D., Franco, M.F., Vaz, C.A., and Calich, V.L. (2004). Binding of laminin to *Paracoccidioides brasiliensis* induces a less severe pulmonary paracoccidioidomycosis caused by virulent and low-virulence isolates. Microbes Infect. *6*, 549–558.

Andreotti, P.F., Monteiro da Silva, J.L., Bailao, A.M., Soares, C.M.A., Benard, G., Soares, C.P., and Mendes-Giannini, M.J. (2005). Isolation and partial characterization of a 30 kDa adhesin from *Paracoccidioides brasiliensis*. Microbes Infect. *7*, 875–881.

Araujo, G.S., Fonseca, F.L., Pontes, B., Torres, A., Cordero, R.J., Zancopé-Oliveira, R.M., Casadevall, A., Viana, N.B., Nimrichter, L., Rodrigues, M.L., *et al.* (2012). Capsules from pathogenic and non-pathogenic *Cryptococcus* spp. manifest significant differences in structure and ability to protect against phagocytic cells. PLoS One. *7*, e29561.

Aristizabal, B.H., Clemons, K.V., Stevens, D.A., and Restrepo, A. (1998). Morphological transition of *Paracoccidioides brasiliensis* conidia to yeast cells: *in vivo* inhibition in females. Infect. Immun. *66*, 5587–5591.

Bagagli E., Theodoro, R.C., Bosco, S.M., McEwen J.G. (2008). *Paracoccidioides brasiliensis*: phylogenetic and ecological aspects. Mycopathologia. *165*, 197–207.

Bailão, A.M., Schrank, A., Borges, C.L., Dutra, V., Molinari-Madlum, E.E.W.I., Felipe, M.S.S., Mendes-Giannini, M.J., Martins, W.S., Pereira, M., and Soares, C.M.A. (2006). Differential gene expression by *Paracoccidioides brasiliensis* in host interaction conditions: representational difference analysis identifies candidate genes associated with fungal pathogenesis. Microbes Infect. *8*, 2686–2697.

Bailão, A.M., Nogueira, S.V., Bonfim, S.M.R.C., Castro, K.P., Silva, J.F., Mendes Giannini, M.J., Pereira, M., and Soares, C.M.A. (2012a). Comparative transcriptome analysis of *Paracoccidioides brasiliensis* during *in vitro* adhesion to type I collagen and fibronectin: identification of potential adhesins. Res. Microbiol. *163*, 182–191.

Bailão, E.F.L.C., Parente, A.F., Parente, J.A., Silva-Bailão, M.G., Castro, K.P., Kmetzsch, L., Staats, C.C., Schrank, A., Vainstein, M.H., Borges, C.L., Bailão, A.M., and

Soares, C.M.A. (2012b). Metal acquisition and homeostasis in fungi. Curr. Fungal Infect. Rep. *6*, 257–266..

Bannwart, C.F., Martins, R.A., Nakaira-Takahashi, E., Dias-Melício, L.A., Soares, A.M., and Peraçoli, M.T. (2010). Interleukin-15 augments oxidative metabolism and fungicidal activity of human monocytes against *Paracoccidioides brasiliensis*. Mem. Inst. Oswaldo Cruz. *105*, 866–872.

Barbosa, M.S., Bao, S.N., Andreotti, P.F., Faria, F.P., Felipe, M.S.S., Feitosa, L.S., Mendes-Giannini, M.J., and Soares, C.M. (2006). Glyceraldehyde-3-phosphate dehydrogenase of *Paracoccidioides brasiliensis* is a cell surface protein involved in fungal adhesion to extracellular matrix proteins and interaction with cells. Infect. Immun. *74*, 382–389.

Bastos, K.P., Bailão, A.M., Borges, C.L., Faria, F.P., Felipe, M.S.S., Silva, M.G., Martins, W.S., Fiúza, R.B., Pereira, M., and Soares, C.M.A. (2007). The transcriptome analysis of early morphogenesis in *Paracoccidioides brasiliensis* mycelium reveals novel and induced genes potentially associated to the dimorphic process. BMC Microbiol. *7*, 29.

Bernard, G., and Duarte, A.J. (2000). Paracoccidioidomycosis: a model for evaluation of the effects of human immunodeficiency virus infection on the natural history of endemic tropical diseases. Clin. Infect. Dis. *31*, 1032–1039.

Bernard, G., Campos, A.F., Netto, L.C., Gonçalves, L.G., Machado, L.R., Mimicos, E.V., França, F.O., and Gryschek, R.C. (2012). Treatment of severe forms of paracoccidioidomycosis: is there a role for corticosteroids? Med. Mycol. *50*, 641–648.

Blotta, M.H., Mamoni, R.L., Oliveira, S.J., Nouér, S.A., Papaiordanou, P.M., Goveia, A., and Camargo, Z.P. (1999). Endemic regions of paracoccidioidomycosis in Brazil: a clinical and epidemiologic study of 584 cases in the southeast region. Am. J. Trop. Med. Hyg. *61*, 390–394.

Bogdan, C., Röllinghoff, M., and Diefenbach, A. (2000). Reactive oxygen and reactive nitrogen intermediates in innate and specific immunity. Curr. Opin. Immunol. *12*, 64–76.

Bonifáz, A., Vazquez-González, D., and Perusquía-Ortiz, A.M. (2011). Endemic systemic mycoses: coccidioidomycosis, histoplasmosis, paracoccidioidomycosis and blastomycosis. J. Dtsch. Dermatol. Ges. *9*, 705–715.

Bonin-Debs, A.L., Boche, I., Gille, H., and Brinkmann, U. (2004). Development of secreted proteins as biotherapeutic agents. Expert. Opin. Biol. Ther. *4*, 551–558.

Borba, C.M., and Schäffer, G.M. (2002). *Paracoccidioides brasiliensis*: virulence and an attempt to induce the dimorphic process with fetal calf serum. Mycoses. *45*, 174–179.

Borges, C.L., Pereira, M., Felipe, M.S.S., Faria, F.P., Gomez, F.J., Deepe Jr., G.S., and Soares, C.M.A. (2005). The antigenic and catalytically active formamidase of *Paracoccidioides brasiliensis*: protein characterization, cDNA and gene cloning, heterologous expression and functional analysis of the recombinant protein. Microbes Infect. *7*, 66–77.

Borges, C.L., Parente, J.A., Barbosa, M.S., Santana, J.M., Bao, S.N., de Sousa, M.V., and Soares, C.M.A. (2010). Detection of a homotetrameric structure and protein–protein interactions of *Paracoccidioides brasiliensis* formamidase lead to new functional insights. FEMS Yeast Res. *10*, 104–113.

Brito, T., Furtado, J.S., Castro, R.M., and Manini, M. (1973). Intraepithelial parasitism as an infection mechanism in human paracoccidioidomycosis (South American blastomycosis). Virchows. Arch. A. Pathol. Pathol. Anat. *361*, 129–138.

Brummer, E., Hanson, L.H., Restrepo, A., and Stevens, D.A. (1989). Intracellular multiplication of *Paracoccidioides brasiliensis* in macrophages: killing and restriction of multiplication by activated macrophages. Infect. Immun. *57*, 2289–2294.

Buitrago, M.J., and Cuenca-Estrella, M. (2012). Epidemiologia actual y diagnóstico de laboratório de lãs micosis endêmicas em España. Enferm. Infecc. Microbiol. Clin. *30*, 407–413.

Bury-Moné, S., Thiberge, J.M., Contreras, M., Maitournam, A., Labigne, A., and De Reuse, H. (2004). Responsiveness to acidity via metal ion regulators mediates virulence in the gastric pathogen *Helicobacter pylori*. Mol. Microbiol. *53*, 623–638.

Calvi, S.A., Soares, A.M., Peraçoli, M.T., Franco, M., Ruiz Jr, R.L., Marcondes-Machado, J., Fecchio, D., Mattos, M.C., and Mendes, R.P. (2003). Study of bronchoalveolar lavage fluid in paracoccidioidomycosis: cytopathology and alveolar macrophage function in response to gamma interferon; comparison with blood monocytes. Microbes Infect. *5*, 1373–1379.

Camacho, E., Sepulveda, V.E., Goldman, W.E., San-Blas, G., and Niño-Vega, G.A. (2012). Expression of *Paracoccidioides brasiliensis* Amy1 in a Histoplasma capsulatum amy1 mutant, relates an α-(1,4)-amylase to cell wall α-(1,3)-glucan synthesis. PLoS One. *7*, e50201.

Camargo, Z.P. (2008). Serology of paracoccidioidomycosis. Mycopathologia. *165*, 289–302.

Campos, C.B., Di Benedette, J.P., Morais, F.V., Ovalle, R., and Nobrega, M.P. (2008). Evidence for the role of calcineurin in morphogenesis and calcium homeostasis during mycelium-to-yeast dimorphism of *Paracoccidioides brasiliensis*. Eukaryot. Cell. *7*, 1856–1864.

Campos, F.F., Johann, S., Cota, B.B., Alves, T.M., Rosa, L.H., Caligiorne, R.B., Cisalpino, P.S., Rosa, C.A., and Zani, C.L. (2011). Antifungal activity of trichothecenes from *Fusarium* sp. against clinical isolates of *Paracoccidioides brasiliensis*. Mycoses. *54*, e122–129.

Cano, L.E., Gomez, B., Brummer, E., Restrepo, A., and Stevens, D.A. (1994). Inhibitory effect of deferoxamine or macrophage activation on transformation of *Paracoccidioides brasiliensis* conidia ingested by macrophages: reversal by holotransferrin. Infect. Immun. *62*, 1494–1496.

Carberry, S., Neville, C.M., Kavanagh, K.A., and Doyle, S. (2006). Analysis of major intracellular proteins of *Aspergillus fumigatus* by MALDI mass spectrometry: identification and characterisation of an elongation factor 1B protein with glutathione transferase activity. Biochem. Biophys. Res. Commun. *341*, 1096–1104.

Carbonell, L.M. (1967). Cell wall changes during the budding process of *Paracoccidioides brasiliensis* and *Blastomyces dermatitidis*. J. Bacteriol. *94*, 213–223.

Carbonell, L.M., and Rodriguez, J. (1965). Transformation of mycelial and yeast forms of *Paracoccidioides brasiliensis* in cultures and in experimental inoculations. J. Bacteriol. *90*, 504–510.

Carmona, A.K., Puccia, R., Oliveira, M.C., Rodrigues, E.G., Juliano, L., and Travassos, L.R. (1995). Characterization of an exocellular serine-thiol proteinase activity in *Paracoccidioides brasiliensis*. Biochem. J. *309*, 209–214.

Caro, E., Gonzalez, A., Munoz, C., Uran, M.E., Restrepo, A., John Hamilton, A., and Cano, L.E. (2008). Recognition of laminin by *Paracoccidioides brasiliensis* conidia: a possible mechanism of adherence to human type II alveolar cells. Med. Mycol. *46*, 795–804.

Carrero, L.L., Niño-Vega, G., Teixeira, M.M., Carvalho, M.J., Soares, C.M.A., Pereira, M., Jesuino, R.S., McEwen, J.G., Mendoza, L., Taylor, J.W., *et al.* (2008). New *Paracoccidioides brasiliensis* isolate reveals unexpected genomic variability in this human pathogen. Fungal Genet. Biol. *45*, 605–612.

Casadevall, A., and Pirofski, L.A. (2006). Polysaccharide–containing conjugate vaccines for fungal diseases. Trends. Mol. Med. *12*, 6–9.

Casadevall, A., Cleare, W., Feldmesser, M., Glatman-Freedman, A., Goldman, D.L., Kozel, T.R., Lendvai, N., Mukherjee, J., Pirofski, L.A., Rivera, J., *et al.* (1998). Characterization of a murine monoclonal antibody to *Cryptococcus neoformans* polysaccharide that is a candidate for human therapeutic studies. Antimicrob. Agents. Chemother. *42*, 1437–1446.

Castaneda, E., Brummer, E., Perlman, A.M., McEwen, J.G., and Stevens, D.A. (1988). A culture medium for *Paracoccidioides brasiliensis* with high plating efficiency, and the effect of siderophores. J. Med. Vet. Mycol. *26*, 351–358.

Castro, R.M., Cuce, L.C., and Netto, C.F. (1975). Paracoccidioidomycosis. Accidental inoculation "in anima nobile." Report of a case. Med. Cutan. Ibero Lat. Am. 3, 289–292.

Castro, N.S., Barbosa, M.S., Maia, Z.A., Báo, S.N., Felipe, M.S.S., Santana, J.M., Mendes-Giannini, M.J., Pereira, M., and Soares, C.M.A. (2008). Characterization of *Paracoccidioides brasiliensis* PbDfg5p, a cell-wall protein implicated in filamentous growth. Yeast. *25*, 141–154.

Chagas, R.F., Bailão, A.M., Pereira, M., Winters, M.S., Smullian, A.G., Deepe Jr, G.S., and Soares, C.M.A. (2008). The catalases of *Paracoccidioides brasiliensis* are differentially regulated: protein activity and transcript analysis. Fungal. Genet. Biol. *45*, 1470–1478.

Chen, D., Janganan, T.K., Chen, G., Marques, E.R., Kress, M.R., Goldman, G.H., Walmsley, A.R., and Borges-Walmsley, M.I. (2007). The cAMP pathway is important for controlling the morphological switch to the pathogenic yeast form of *Paracoccidioides brasiliensis*. Mol. Microbiol. *65*, 761–779.

Colombo, A.L., Tobón, A., Restrepo, A., Queiroz-Telles, F., and Nucci, M. (2011). Epidemiology of endemic sustemic fungal infections in Latin America. Med. Mycol. *49*, 785–798.

Coltri, K.C., Oliveira, L.L., Pinzan, C.F., Vendruscolo, P.E., Martinez, R., Goldman, M.H., Panunto-Castelo, A., and Roque-Barreira, M.C. (2008). Therapeutic administration of KM+ lectin protects mice against *Paracoccidioides brasiliensis* infection via interleukin-12 production in a toll-like receptor 2-dependent mechanism. Am. J. Pathol. *173*, 423–432.

Coltri, K.C., Oliveira, L.L., Ruas, L.P., Vendruscolo, P.E., Goldman, M.H., Panunto-Castelo, A., and Roque-Barreira, M.C. (2010). Protection against *Paracoccidioides brasiliensis* infection conferred by the prophylactic administration of native and recombinant ArtinM. Med. Mycol. *48*, 792–799.

Derengowski, L.S., Tavares, A.H., Silva, S., Procópio, L.S., Felipe, M.S.S., and Silva-Pereira, I. (2008). Up-regulation of glyoxylate cycle genes upon *Paracoccidioides brasiliensis* internalization by murine macrophages and *in vitro* nutritional stress condition. Med. Mycol. *46*, 125–134.

Desjardins, C.A., Champion, M.D., Holder, J.W., Muszewska, A., Goldberg, J., Bailão, A.M., Brigido, M.M., Ferreira, M.E., Garcia, A.M., Grynberg, M., *et al.* (2011). Comparative genomic analysis of human fungal pathogens causing paracoccidioidomycosis. PLoS Genet. *7*, e1002345.

Dias-Melicio, L.A., Calvi, S.A., Peraçoli, M.T., and Soares, A.M. (2005). Inhibitory effect of deferoxamine on *Paracoccidioides brasiliensis* survival in human monocytes: reversal by holotransferrin not by apotransferrin. Rev. Inst. Med. Trop. Sao Paulo. *47*, 263–266.

Díez, S., Gómez, B.L., Restrepo, A., Hay, R.J., and Hamilton, A.J. (2002). *Paracoccidioides brasiliensis* 87-kilodalton antigen, a heat shock protein useful in diagnosis: characterization, purification, and detection in biopsy material via immunohistochemistry. J. Clin. Microbiol. *40*, 359–365.

Donofrio, F.C., Calil, A.C., Miranda, E.T., Almeida, A.M., Benard, G., Soares, C.P., Veloso, S.N., Soares, C.M.A., and Mendes Giannini, M.J. (2009). Enolase from *Paracoccidioides brasiliensis*: isolation and identification as a fibronectin-binding protein. J. Med. Microbiol. *58*, 706–713.

Egan, J.D., García-Pedrajas, M.D., Andrews, D.L., and Gold, S.E. (2009). Calcineurin is an antagonist to PKA protein phosphorylation required for postmating filamentation and virulence, while PP2A is required for viability in *Ustilago maydis*. Mol. Plant Microbe Interact. *22*, 1293–1301.

Fang, F.C. (1997). Perspectives series: host/pathogen interactions. Mechanisms of nitric oxide–related antimicrobial activity. J. Clin. Invest. *99*, 2818–2825.

Felipe, M.S., Andrade, R.V., Petrofeza, S.S., Maranhaom, A.Q., Torres, F.A., Albuquerque, P., Arraes, F.B., Arruda, M., Azevedo, M.O., Baptista, A.J., *et al.* (2003). Transcriptome characterization of the dimorphic and pathogenic fungus *Paracoccidioides brasiliensis* by EST analysis. Yeast. *20*, 263–271.

Felipe, M.S.S., Andrade, R.V., Arraes, F.B., Nicola, A.M., Maranhão, A.Q., Torres, F.A., Silva-Pereira, I., Poças-Fonseca, M.J., Campos, E.G., Moraes, L.M., *et al.* (2005). Transcriptional profiles of the human pathogenic fungus *Paracoccidioides brasiliensis* in mycelium and yeast cells. J. Biol. Chem. *280*, 24706–24714.

Fernandes, L., Paes, H.C., Tavares, A.H., Silva, S.S., Dantas, A., Soares, C.M.A., Torres, F.A., and Felipe, M.S.S. (2008). Transcriptional profile of *ras1* and *ras2* and the potential role of farnesylation in the dimorphism of the human pathogen *Paracoccidioides brasiliensis*. FEMS Yeast Res. *8*, 300–310.

Fernandes, V.C., Martins, E.M., Boeloni, J.N., Coitinho, J.B., Serakides, R., and Goes, A.M. (2011). Additive effect of rPb27 immunization and chemotherapy in experimental paracoccidioidomycosis. PLoS One. *6*, e17885.

Fernandez-Acero, F.J., Jorge, I., Calvo, E., Vallejo, I., Carbu, M., Camafeita, E., Garrido, C., Lopez, J.A., Jorrin, J., and Cantoral, J.M. (2007). Proteomic analysis of phytopathogenic fungus *Botrytis cinerea* as a potential tool for identifying pathogenicity factors, therapeutic targets and for basic research. Arch. Microbiol. *187*, 207–215.

Franco, M. (1987). Host–parasite relationships in paracoccidioidomycosis. J. Med. Vet. Mycol. *25*, 5–18.

Goldman, G.H., dos Reis Marques, E., Duarte Ribeiro, D.C., de Souza Bernardes, L.A., Quiapin, A.C., Vitorelli, P.M., Savoldi, M., Semighini, C.P., de Oliveira, R.C., Nunes, L.R., *et al.* (2003). Expressed sequence tag analysis of the human pathogen *Paracoccidioides brasiliensis* yeast phase: identification of putative homologues of *Candida albicans* virulence and pathogenicity genes. Eukaryot. Cell. *2*, 34–48.

Gomes-Rezende, J.A., Gomes-Alves, A.G., Menino, J.F., Coelho, M.A., Ludovico, P., Gonçalves, P., Sturme, M.H., and Rodrigues, F. (2012). Functionality of the *Paracoccidioides* mating α-pheromone-receptor system. PLoS One. *7*, e47033.

Gómez, B.L., Nosanchuk, J.D., Díez, S., Youngchim, S., Aisen, P., Cano, L.E., Restrepo, A., Casadevall, A., and Hamilton, A.J. (2001). Detection of melanin-like particles in the dimorphic fungal pathogen *Paracoccidioides brasiliensis in vitro* and during infection. Infect. Immun. *69*, 5760–5767.

Gonzalez, A., Gomez, B.L., Restrepo, A., Hamilton, A.J., and Cano, L.E. (2005). Recognition of extracellular matrix proteins by *Paracoccidioides brasiliensis* yeast cells. Med. Mycol. *43*, 637–645.

Gonzalez, A., Restrepo, A., and Cano, L.E. (2007). Role of iron in the nitric oxide-mediated fungicidal mechanism of IFN-gamma-activated murine macrophages against *Paracoccidioides brasiliensis* conidia. Rev. Inst. Med. Trop. Sao Paulo. *49*, 11–16.

Grossklaus, D.A., Bailão, A.M., Rezende, T.C.V., Borges, C.L., Oliveira, M.A.P., Parente, J.A., and Soares, C.M.A. (2013). Response to oxidative stress in *Paracoccidioides* yeast cells as determined by proteomic analysis. Microbes. Infect. *15*, 347–364.

Hahn, R.C., Morato-Conceição, Y.T., Santos, N.L., Ferreira, J.F., and Hamdan, J.S. (2003). Disseminated paracoccidioidomycosis: correlation between clinical and *in vitro* resistance to ketoconazole and trimethoprim sulphamethoxazole. Mycoses *46*, 342–347.

Hanna, S.A., Silva, J.L.M., and Giannini, M.J. (2000). Adherence and intracellular parasitism of *Paracoccidioides brasiliensis* in Vero cells. Microbes Infect. *2*, 877–884.

Henson, J.M., Butler, M.J., and Day, A.W. (1999). The dark side of the mycelium: Melanins of phytopathogenic fungi. Annu. Rev. Phytopathol. *37*, 447–471.

Hernández, O., Almeida, A.J., Gonzalez, A., Garcia, A.M., Tamayo, D., Cano, L.E., Restrepo, A., and McEwen, J.G. (2010). A 32-kilodalton hydrolase plays an important role in *Paracoccidioides brasiliensis* adherence to host cells and influences pathogenicity. Infect. Immun. *78*, 5280–5286.

Hernández, O., Gonzalez, A., Almeida, A.J., Tamayo, D., Garcia, A.M., Restrepo, A., and McEwen, J.G. (2011). Alternative oxidase mediates pathogen resistance in *Paracoccidioides brasiliensis* infection. PLoS Negl. Trop. Dis. *5*, e1353.

Jesuino, R.S., Azevedo, M.O., Felipe, M.S.S., Pereira, M., and Soares, C.M.A. (2002). Characterization of a chaperone ClpB homologue of *Paracoccidioides brasiliensis*. Yeast. *19*, 963–972.

Jesus, M.D., Nicola, A.M., Chow, S.K., Lee, I.R., Nong, S., Specht, C.A., Levitz, S.M., and Casadevall, A. (2010). Glucuronoxylomannan, galactoxylomannan, and mannoprotein occupy spatially separate and discrete regions in the capsule of *Cryptococcus neoformans*. Virulence. *1*, 500–508.

Johnson, J.L., and Craig, E.A. (2000). A role for the Hsp40 Ydj1 in repression of basal steroid receptor activity in yeast. Mol. Cell. Biol. *20*, 3027–3036.

Kanetsuna, F., and Carbonell, L.M. (1970). Cell wall glucans of the yeast and mycelial forms of *Paracoccidioides brasiliensis*. J. Bacteriol. *101*, 675–680.

Kashino, S.S., Singer-Vermes, L.M., Calich, V.L., and Burger, E. (1990). Alterations in the pathogenicity of one *Paracoccidioides brasiliensis* isolate do not correlative with its *in vitro* growth. Mycopathologia. *111*, 173–180.

Khandare, J., and Minko, T. (2006). Polymer–drug conjugates: progress in polymeric prodrugs. Prog. Polym. Sci. *31*, 359–397.

Klein, B.S., and Tebbets, B. (2007). Dimorphism and virulence in fungi. Curr. Opin. Microbiol. *10*, 314–319.

Konno, F.T., Maricato, J., Konno, A.Y., Guereschi, M.G., Vivanco, B.C., Feitosa, S.L., Mariano, M., and Lopes, J.D. (2012). *Paracoccidioides brasiliensis* Gp43-derived peptides are potent modulators of local and systemic inflammatory response. Microbes Infect. *14*, 517–527.

Kotsias, F., Hoffmann, E., Amigorena, S., and Savina, A. (2013). Reactive oxygen species production in the phagosome: impact on antigen presentation in dendritic cells. Antioxid. Redox. Signal. *18*, 714–729.

Kurokawa, C.S., Lopes, C.R., Sugizaki, M.F., Kuramae, E.E., Franco, M.F., and Peraçoli, M.T. (2005). Virulence profile of ten *Paracoccidioides brasiliensis* isolates: association with morphologic and genetic patterns. Rev. Inst. Med. Trop. Sao Paulo. *47*, 257–262.

Laniado-Laborin, R. (2007). Coccidioidomycosis and other endemic mycosis in Mexico. Rev. Iberoam. Micol. *24*, 249–258.

Lass-Flörl, C. (2009). The changing face of epidemiology of invasive fungal disease in Europe. Mycoses *52*, 197–205.

Long, K.H., Gomez, F.J., Morris, R.E., and Newman, S.L. (2003). Identification of heat shock protein 60 as the ligand on *Histoplasma capsulatum* that mediates binding to CD18 receptors on human macrophages. J. Immunol. *170*, 487–494.

McGeady, P., Logan, D.A., and Wansley, D.L. (2002). A protein-farnesyl transferase inhibitor interferes with the serum-induced conversion of *Candida albicans* from a cellular yeast form to a filamentous form. FEMS Microbiol. Lett. *213*, 41–44.

MacMicking, J., Xie, Q.W., and Nathan, C. (1997). Nitric oxide and macrophage function. Ann. Rev. Immunol. *15*, 323–350.

Mamoni, R.L., Nouér, S.A., Oliveira, S.J., Musatti, C.C., Rossi, C.L., Camargo, Z.P., and Blotta, M.H. (2002). Enhanced production of specific IgG4, IgE, IgA and TGF-β in sera from patients with the juvenile form of paracoccidioidomycosis. Med. Mycol. *40*, 153–159.

Manocha, M.S. (1980). Lipid composition of *Paracoccidioides brasiliensis*: comparison between the yeast and mycelial forms. Sabouraudia. *18*, 281–286.

Marcos, C.M., Silva, J.F., Oliveira, H.C., Silva, R.A.M., Mendes-Giannini, M.J., and Fusco-Almeida, A.M. (2012). Surface-expressed enolase contributes to the adhesion of *Paracoccidioides brasiliensis* to host cells. FEMS Yeast Res. *12*, 557–570.

Marquez, A.S., Vicentini, A.P., Ono, M.A., Watanabe, M.A., Camargo, Z.P., and Itano, E.N. (2005). Reactivity of antibodies from patients with acute and chronic paracoccidioidomycosis to a high molecular mass antigen from *Paracoccidioides brasiliensis*. J. Clin. Lab. Anal. *19*, 199–204.

Massuda, T.Y., Nagashima, L.A., Leonello, P.C., Kaminami, M.S., Mantovani, M.S., Sano, A., Uno, J., Venancio, E.J., Camargo, Z.P., and Itano, E.N. (2011). Cyclosporin: A treatment and decreased fungal load/antigenemia in experimental murine paracoccidioidomycosis. Mycopathologia. *171*, 161–169.

Matos, T.G., Morais, F.V., and Campos, C.B. (2012). Hsp90 regulates *Paracoccidioides brasiliensis* proliferation and ROS levels under thermal stress and cooperates with calcineurin to control yeast to mycelium dimorphism. Med. Mycol. *51*, 413–421.

Matsuo, A.L., Tersariol, I.I., Kobata, S.I., Travassos, L.R., Carmona, A.K., and Puccia, R. (2006). Modulation of the exocellular serine-thiol proteinase activity of *Paracoccidioides brasiliensis* by neutral polysaccharides. Microbes Infect. *8*, 84–91.

Matsuo, A.L., Carmona, A.K., Silva, L.S., Cunha, C.E., Nakayasu, E.S., Almeida, I.C., Juliano, M.A., and Puccia, R. (2007). C-Npys (S-3-nitro-2-pyridinesulfenyl) and peptide derivatives can inhibit a serine-thiol proteinase activity from *Paracoccidioides brasiliensis*. Biochem. Biophys. Res. Commun. *355*, 1000–1005.

Matute, D.R., McEwen, J.G., Puccia, R., Montes, B.A., San-Blas, G., Bagagli, E., Rauscher, J.T., Restrepo, A., Morais, F., Niño-Vega, G., and Taylor, J.W. (2006). Cryptic speciation and recombination in the fungus *Paracoccidioides brasiliensis* as revealed by gene genealogies. Mol. Biol. Evol. *23*, 65–73.

Mayorga, O., Munoz, J.E., Lincopan, N., Teixeira, A.F., Ferreira, L.C., Travassos, L.R., and Taborda, C.P. (2012). The role of adjuvants in therapeutic protection against paracoccidioidomycosis after immunization with the P10 peptide. Front. Microbiol. 3, 154.

Maza, P.K., Straus, A.H., Toledo, M.S., Takahashi, H.K., and Suzuki, E. (2008). Interaction of epithelial cell membrane rafts with *Paracoccidioides brasiliensis* leads to fungal adhesion and Src-family kinase activation. Microbes Infect. *10*, 540–547.

Meloni-Bruneri, L.H., Campa, A., Abdalla, D.S., Calich, V.L., Lenzi, H.L., and Burger, E. (1996). Neutrophil oxidative metabolism and killing of *Paracoccidioides brasiliensis* after air pouch infection of susceptible and resistant mice. J. Leukoc. Biol. *59*, 526–533.

Mendes, R.P., Negroni, R., and Arechavala, A. (1994). Treatment and control of cure. In Paracoccidioidomycosis, Franco, M., Lacaz, CS., Restrepo-Moreno, A., Del Negro, G. eds (CRC Press, Boca Raton, FL). pp. 373–392.

Mendes-Giannini, M.J., Taylor, M.L., Bouchara, J.B., Burger, E., Calich, V.L., Escalante, E.D., Hanna, S.A., Lenzi, H.L., Machado, M.P., Miyaji, M.I *et al.* (2000). Pathogenesis II: fungal responses to host responses: interaction of host cells with fungi. Med. Mycol. *38*, 113–123.

Mendes-Giannini, M.J., Andreotti, P.F., Vincenzi, L.R., Silva, J.L., Lenzi, H.L., Benard, G., Zancope-Oliveira, R., Guedes, H.L.M., and Soares, C.P. (2006). Binding of extracellular matrix proteins to *Paracoccidioides brasiliensis*. Microbes Infect. *8*, 1550–1559.

Mendes-Giannini, M.J., Silva, J.L.M., Silva, J.F., Donofrio, F.C., Miranda, E.T., Andreotti, P.F., and Soares, C.P. (2008). Interactions of *Paracoccidioides brasiliensis* with host cells: recent advances. Mycopathologia *165*, 237–248.

Mundargi, R.C., Babu, V.R., Rangaswamy, V., Patel, P., and Aminabhavi, T.M. (2008). Nano/micro technologies for delivering macromolecular therapeutics using poly(D,L-lactide-co-glycolide) and its derivatives. J. Control Release. *125*, 193–209.

Murciano, C., Moyes, D.L., Runglall, M., Tobouti, P., Islam, A., Hoyer, L.L., and Naglik, J.R. (2012). Evaluation of the role of *Candida albicans* agglutinin-like sequence (Als) proteins in human oral epithelial cell interactions. PLoS One. 7, e33362.

Nairz, M., Schroll, A., Sonnweber, T., and Weiss, G. (2010). The struggle for iron – a metal at the host–pathogen interface. Cell Microbiol. *12*, 1691–1702.

Nakaira-Takahagi, E., Golim, M.A., Bannwart, C.F., Puccia, R., and Peracoli, M.T. (2011). Interactions between TLR2, TLR4, and mannose receptors with Gp43 from *Paracoccidioides brasiliensis* induce cytokine production by human monocytes. Med. Mycol. *49*, 694–703.

Naranjo, T.W., Lopera, D.E., Diaz-Granados, L.R., Duque, J.J., Restrepo, A.M., and Cano, L.E. (2011). Combined itraconazole-pentoxifylline treatment promptly reduces lung fibrosis induced by chronic pulmonary paracoccidioidomycosis in mice. Pulm. Pharmacol. Ther. *24*, 81–91.

Nemecek, J.C., Wüthrich, M., and Klein, B.S. (2006). Global control of dimorphism and virulence in fungi. Science. *312*, 583–588.

Nicola, A.M., Andrade, R.V., Dantas, A.S., Andrade, P.A., Arraes, F.B., Fernandes, L., Silva-Pereira, I., and Felipe, M.S. (2008). The stress responsive and morphologically regulated *hsp90* gene from *Paracoccidioides brasiliensis* is essential to cell viability. BMC Microbiol. *8*, 158.

Nishikaku, A.S., Molina, R.F., Ribeiro, L.C., Scavone, R., Albe, B.P., Cunha, C.S., and Burger, E. (2009). Nitric oxide participation in granulomatous response induced by *Paracoccidioides brasiliensis* infection in mice. Med. Microbiol. Immunol. *198*, 123–135.

Nogueira, S.V., Fonseca, F.L., Rodrigues, M.L., Mundodi, V., Abi-Chacra, E.A., Winters, M.S., Alderete, J.F., and Soares, C.M.A. (2010). *Paracoccidioides brasiliensis* enolase is a surface protein that binds plasminogen and mediates interaction of yeast forms with host cells. Infect. Immun. *78*, 4040–4050.

Nunes, L.R., Oliveira, R.C., Leite, D.B., Silva, V.S., Marques, E.R., Ferreira, M.E.S., Ribeiro, D.C., Bernardes, L.A.S., Goldman, M.H., Puccia, R., *et al.* (2005). Transcriptome analysis of *Paracoccidioides brasiliensis* cells undergoing mycelium-to-yeast transition. Eukaryot. Cell. *4*, 2115–2128.

Ollague, J.M., Zurita, A.M., and Calero, G. (2000). Paracoccidioidomycosis (South American blastomycosis) successfully treated with terbinafine: first case report. Br. J. Dermatol. *143*, 188–191.

Parente, A.F., Bailão, A.M., Borges, C.L., Parente, J.A., Magalhães, A.D., Ricart, C.A., and Soares, C.M.A. (2011). Proteomic analysis reveals that iron availability alters the metabolic status of the pathogenic fungus *Paracoccidioides brasiliensis*. PLoS One. *6*, e22810.

Parente, J.A., Costa, M., Pereira, M., and Soares, C.M.A. (2005). Transcriptome overview of *Paracoccidioides brasiliensis* proteases. Genet. Mol. Res. *4*, 358–371.

Parente, J.A., Salem-Izacc, S.M., Santana, J.M., Pereira, M., Borges, C.L., Bailao, A.M., and Soares, C.M.A. (2010). A secreted serine protease of *Paracoccidioides brasiliensis* and its interactions with fungal proteins. BMC Microbiol. *10*, 292.

Parise-Fortes, M.R., Miot, H.A., Kurokawa, C.S., Marques, M.E.A., and Marques, A.S. (2011). Immunology of paracoccidioidomycosis. An. Bras. Dermatol. *86*, 516–525.

Pereira, L.A., Bao, S.N., Barbosa, M.S., Silva, J.L., Felipe, M.S.S., Santana, J.M., Mendes-Giannini, M.J., and Soares, C.M.A. (2007). Analysis of the *Paracoccidioides brasiliensis* triosephosphate isomerase suggests the potential for adhesin function. FEMS Yeast Res. *7*, 1381–1388.

Pereira, M., Song, Z., Santos-Silva, L.K., Richards, M.H., Nguyen, T.T., Liu, J., Soares, C.M.A., Cruz, A.H.S., Ganapathy, K., and Nes, W.D. (2010). Cloning, mechanistic and functional analysis of a fungal sterol C24-methyltransferase implicated in brassicasterol biosynthesis. Biochim. Biophys. Acta. *1801*, 1163–1174.

Prado, M., Silva, M.B., Laurenti, R., Travassos, L.R., and Taborda, C.P. (2009). Mortality due to systemic mycoses as a primary cause of death or in association with AIDS in Brazil: a review from 1996 to 2006. Mem. Inst. Oswaldo Cruz. *104*, 513–521.

Rappleye, C.A., and Goldman, W.E. (2006). Defining virulence genes in the dimorphic fungi. Annu. Rev. Microbiol. *60*, 281–303.

Reese, A.J., and Doering, T.L. (2003). Cell wall alpha-1,3-glucan is required to anchor the *Cryptococcus neoformans* capsule. Mol. Microbiol. *50*, 1401–1409.

Reis, B.S., Fernandes, V.C., Martins, E.M., Serakides, R., and Goes, A.M. (2008). Protective immunity induced by r*Pb*27 of *Paracoccidioides brasiliensis*. Vaccine. *26*, 5461–5469.

Restrepo, A. (1985). The ecology of *Paracoccidioides brasiliensis*: a puzzle still unsolved. Sabouraudia. *23*, 323–334.

Restrepo, A., Salazar, M.E., Cano, L.E., Stover, E.P., Feldman, D., and Stevens, D.A. (1984). Estrogens inhibit mycelium-to-yeast transformation in the fungus *Paracoccidioides brasiliensis*: implications for resistance of females to paracoccidioidomycosis. Infect. Immun. *46*, 346–353.

Restrepo-Moreno, A., and Greer, D.L. (1983). Paracoccidioidomycosis. In Occupational Mycoses, A.F. Di-Salvo, ed. (Lea & Febiger, Philadelphia), pp. 43–64.

Rezende, T.C., Borges, C.L., Magalhães, A.D., Sousa, M.V., Ricart, C.A., Bailão, A.M., and Soares, C.M.A. (2011). A quantitative view of the morphological phases of *Paracoccidioides brasiliensis* using proteomics. J. Proteomics 75, 572–587.

Rittner, G.M., Munoz, J.E., Marques, A.F., Nosanchuk, J.D., Taborda, C.P., and Travassos, L.R. (2012). Therapeutic DNA vaccine encoding peptide P10 against experimental paracoccidioidomycosis. PLoS Negl. Trop. Dis. *6*, e1519.

Rocha, C.R., Schröppel, K., Harcus, D., Marcil, A., Dignard, D., Taylor, B.N., Thomas, D.Y., Whiteway, M., and Leberer, E. (2001). Signaling through adenylyl cyclase is essential for hyphal growth and virulence in the pathogenic fungus *Candida albicans*. Mol. Biol. Cell. *12*, 3631–3643.

Rodrigues, M.L., Nimrichter, L., Cordero, R.J., and Casadevall, A. (2011). Fungal polysaccharides: biological activity beyond the usual structural properties. Front. Microbiol. 2, 171.

Salazar, M.E., Restrepo, A., and Stevens, D.A. (1988). Inhibition by estrogens of conidium-to-yeast conversion in the fungus *Paracoccidioides brasiliensis*. Infect. Immun. *56*, 711–713.

Salem-Izacc, S.M., Gomez, F.J., Jesuino, R.S., Fonseca, C.A., Felipe, M.S.S., Deepe, G.S., and Soares, C.M.A. (2001). Molecular cloning, characterization and expression of the heat shock protein 60 gene from the human pathogenic fungus *Paracoccidioides brasiliensis*. Med. Mycol. *39*, 445–455.

San-Blas, G. (1993). Biochemical and physiological aspects in the dimorphism of *Paracoccidioides brasiliensis*. Arch. Med. Res. *24*, 267–268.

San-Blas, F., San-Blas, G., and Cova, L.J. (1976). A morphological mutant of *Paracoccidioides brasiliensis* strain IVIC Pb9. Isolation and wall characterization. J. Gen. Microbiol. 93, 209–218.

San-Blas, G., and San-Blas, F. (1977). *Paracoccidioides brasiliensis*: cell wall structure and virulence. A review. Mycopathologia. *62*, 77–86.

San-Blas, G., Nino-Vega, G., and Iturriaga, T. (2002). *Paracoccidioides brasiliensis* and paracoccidioidomycosis: molecular approaches to morphogenesis, diagnosis, epidemiology, taxonomy and genetics. Med Mycol. *40*, 225–242.

Santos, G.D., Ferri, P.H., Santos, S.C., Bao, S.N., Soares, C.M.A., and Pereira, M. (2007). Oenothein B inhibits the expression of PbFKS1 transcript and induces morphological changes in *Paracoccidioides brasiliensis*. Med. Mycol. *45*, 609–618.

Seidler, N.W. (2013). GAPDH, as a virulence factor. Adv. Exp. Med. Biol. *985*, 149–178.

Shankar, J., Wu, T.D., Clemons, K.V., Monteiro, J.P., Mirels, L.F., and Stevens, D.A. (2011). Influence of 17β-estradiol on gene expression of *Paracoccidioides* during mycelia-to-yeast transition. PLoS One. *6*, e28402.

Shikanai-Yasuda, M.A., Benard, G., Higaki, Y., Del Negro, G.M., Hoo, S., Vaccari, E.H., Gryschek, R.C., Segurado, A.A., Barone, A.A., and Andrade, D.R. (2002). Randomized trial with itraconazole, ketoconazole and sulfadiazine in paracoccidioidomycosis. Med. Mycol. *40*, 411–417.

Shikanai-Yasuda, M.A., Queiroz-Telles, F., Mendes, R.P., Colombo, A.L., and Moretti, M.L. (2006). Guidelines in Paracoccidioidomycosis. Rev. Soc. Bras. Med. Trop. *39*, 297–310.

Silva, C.L. (1985). Granulomatous reaction induced by lipids isolated from *Paracoccidioides brasiliensis*. Trans. R. Soc. Trop. Med. Hyg. *79*, 70–73.

Silva, M.B., Marques, A.F., Nosanchuk, J.D., Casadevall, A., Travassos, L.R., and Taborda, C.P. (2006). Melanin in the dimorphic fungal pathogen *Paracoccidioides brasiliensis*: effects on phagocytosis, intracellular resistance and drug susceptibility. Microbes Infect. *8*, 197–205.

Silva, M.B., Thomaz, L., Marques, A.F., Svidzinski, A.E., Nosanchuk, J.D., Casadevall, A., Travassos, L.R., and Taborda, C.P. (2009). Resistance of melanized yeast cells of *Paracoccidioides brasiliensis* to antimicrobial oxidants and inhibition of phagocytosis using carbohydrates and monoclonal antibody to CD18. Mem. Inst. Oswaldo Cruz. *104*, 644–648.

Silva, M.G., Schrank, A., Bailão, E.F., Bailão, A.M., Borges, C.L., Staats, C.C., Parente, J.A., Pereira, M., Salem-Izacc, S.M., Mendes-Giannini, M.J., *et al.* (2011). The homeostasis of iron, copper, and zinc in *Paracoccidioides brasiliensis, Cryptococcus neoformans* var. Grubii, and *Cryptococcus gattii*: a comparative analysis. Front. Microbiol. *2*, 49.

Silva, S.P., Borges-Walmsley, M.I., Pereira, I.S., Soares, C.M.A., Walmsley, A.R., and Felipe, M.S.S. (1999). Differential expression of an *hsp70* gene during transition from the mycelial to the infective yeast form of the human pathogenic fungus *Paracoccidioides brasiliensis*. Mol. Microbiol. *31*, 1039–1050.

Silva Neto, B.R., Silva, J.F., Mendes-Giannini, M.J., Lenzi, H.L., Soares, C.M.A., and Pereira, M. (2009). The malate synthase of *Paracoccidioides brasiliensis* is a linked surface protein that behaves as an anchorless adhesin. BMC Microbiol. *9*, 272.

Soares, R.B., Gomez, F.J., Soares, C.M.A., Deepe Jr, G.S. (2008). Vaccination with heat shock protein 60 induces a protective immune response against experimental *Paracoccidioides brasiliensis* pulmonary infection. Infect. Immun. *76*, 4214–4221.

Sorais, F., Barreto, L., Leal, J.A., Bernabé, M., San-Blas, G., and Niño-Veja, G.A. (2010). Cell wall glucan synthases and GTPases in *Paracoccidioides brasiliensis*. Med. Mycol. *48*, 35–47.

Stie, J., and Fox, D. (2008). Calcineurin regulation in fungi and beyond. Eukaryot. Cell. *7*, 177–186.

Stover, E.P., Schär, G., Clemons, K.V., Stevens, D.A., and Feldman, D. (1986). Estradiol-binding proteins from mycelial and yeast-form cultures of *Paracoccidioides brasiliensis*. Infect. Immun. *51*, 199–203.

Straus, A.H., Freymuller, E., Travassos, L.R., and Takahashi, H.K. (1996). Immunochemical and subcellular localization of the 43 kDa glycoprotein antigen of *Paracoccidioides brasiliensis* with monoclonal antibodies. J. Med. Vet. Mycol. *34*, 181–186.

Taborda, C.P., Juliano, M.A., Puccia, R., Franco, M., and Travassos, L.R. (1998). Mapping of the T-cell epitope in the major 43-kilodalton glycoprotein of *Paracoccidioides brasiliensis* which induces a Th-1 response protective against fungal infection in BALB/c mice. Infect. Immun. *66*, 786–793.

Tamayo, D., Muñoz, J.F., Torres, I., Almeida, A.J., Restrepo, A., McEwen, J.G., and Hernández, O. (2013). Involvement of the 90kDa heat shock protein during adaptation of *Paracoccidioides brasiliensis* to different environmental conditions. Fungal Genet. Biol. *51*, 34–41.

Tavares, A.H., Silva, S.S., Dantas, A., Campos, E.G., Andrade, R.V., Maranhão, A.Q., Brígido, M.M., Passos-Silva, D.G., Fachin, A.L., Teixeira, S.M., *et al.* (2007). Early transcriptional response of *Paracoccidioides brasiliensis* upon internalization by murine macrophages. Microbes Infect. *9*, 583–590.

Teixeira, M.M., Theodoro, R.C., Carvalho, M.J., Fernandes, L., Paes, H.C., Hahn, R.C., Mendoza, L., Bagagli, E., San-Blas, G., and Felipe, M.S.S. (2009). Phylogenetic analysis reveals a high level of speciation in the *Paracoccidioides* genus. Mol. Phylogenet. Evol. *52*, 273–283.

Telles, F.R.R., and Martins, M.L. (2011). Laboratorial diagnosis of paracoccidioidomycosis and new insights for the future of fungal diagnosis. Talanta. *85*, 2254–2264.

Test, S.T., and Weiss, S.J. (1986). The generation and utilization of chlorinated oxidants by human neutrophils. Adv. Free Rad. Biol. Med. *2*, 91–116.

Thomaz, L., Apitz-Castro, R., Marques, A.F., Travassos, L.R., and Taborda, C.P. (2008). Experimental paracoccidioidomycosis: alternative therapy with ajoene, compound from Allium sativum, associated with sulfamethoxazole/trimethoprim. Med. Mycol. *46*, 113–118.

Torres, I., García, A.M., Hernández, O., González, A., McEwen, J.G., Restrepo, A., and Arango, M. (2010). Presence and expression of the mating type locus in

Paracoccidioides brasiliensis isolates. Fungal Genet. Biol. *47*, 373–380.

Travassos, L.R. and Taborda, C.P. (2012) Paracoccidioidomycosis vaccine. Hum. Vaccin. Immunother. *8*, 1450–1453.

Travassos, L.R., Taborda, C.P., Iwai, L.K., Cunha Neto, E., and Puccia, R. (2004). The Gp43 from *Paracoccidioides brasiliensis*: A major diagnostic antigen and vaccine candidate. In The Mycota XII, Human Fungal Pathogens, J.E. Domer, and G.S. Kobayashi, eds. (Springer-Verlag Berlin-Heidelberg, Berlin) pp 279–296.

Tronchin, G., Pihet, M., Lopes-Bezerra, L.M., and Bouchara, J.P. (2008). Adherence mechanisms in human pathogenic fungi. Med Mycol. *46*, 749–772.

Upadhyay, S.K., Mahajan, L., Ramjee, S., Singh, Y., Basir, S.F., and Madan, T. (2009). Identification and characterization of a laminin-binding protein of *Aspergillus fumigatus*: extracellular thaumatin domain protein (AfCalAp). J. Med. Microbiol. *58*, 714–722.

Valim, C.X., Basso, L.R. Jr., Almeida, F.B.R., Reis, T.F., Damásio, A.R., Arruda, L.K., Martinez, R., Roque-Barreira, M.C., Oliver, C., Jamur, M.C., *et al.* (2012). Characterization of *Pb*Pga1, an antigenic GPI-protein in the pathogenic fungus *Paracoccidioides brasiliensis*. PLoS One. 7, e44792.

Vallejo, M.C., Nakayasu, E.S., Matsuo, A.L., Sobreira, T.J., Longo, L.V., Ganiko, L., Almeida, I.C., and Puccia, R. (2012a). Vesicle and vesicle-free extracellular proteome of *Paracoccidioides brasiliensis*: comparative analysis with other pathogenic fungi. J. Proteome Res. *11*, 1676–1685.

Vallejo, M.C., Nakayasu, E.S., Longo, L.V., Ganiko, L., Lopes, F.G., Matsuo, A.L., Almeida, I.C., and Puccia, R. (2012b). Lipidomic analysis of extracellular vesicles from the pathogenic phase of *Paracoccidioides brasiliensis*. PLoS One. 7, e39463.

Verstrepen, K.J., and Klis, F.M. (2006). Flocculation, adhesion and biofilm formation in yeasts. Mol. Microbiol. *60*, 5–15.

Vicentini, A.P., Gesztesi, J.L., Franco, M.F., Souza, W., Moraes, J.Z., Travassos, L.R., and Lopes, J.D. (1994). Binding of *Paracoccidioides brasiliensis* to laminin through surface glycoprotein Gp43 leads to enhancement of fungal pathogenesis. Infect. Immun. *62*, 1465–1469.

Visbal, G., San-Blas, G., Murgich, J., and Franco, H. (2005). *Paracoccidioides brasiliensis*, paracoccidioidomycosis, and antifungal antibiotics. Curr. Drug. Targets Infect. Disord. *5*, 211–226.

Weber, S.S., Parente, A.F., Borges, C.L., Parente, J.A., Bailao, A.M., and Soares, C.M.A. (2012). Analysis of the secretomes of *Paracoccidioides* mycelia and yeast cells. PLoS One. 7, e52470.

Wiesmüller L., and Wittinghofer, F. (1994). Signal transduction pathways involving Ras. Mini review. Cell Signal. *6*, 247–267.

Yajima, W., and Kav, N.N. (2006). The proteome of the phytopathogenic fungus *Sclerotinia sclerotiorum*. Proteomics. *6*, 5995–6007.

Yasuda, M.A. (2005). Pharmacological management of paracoccidioidomycosis. Expert Opin. Pharmacother. *6*, 385–397.

Ywazaki, C.Y., Maza, P.K., Suzuki, E., Takahashi, H.K., and Straus, A.H. (2011). Role of host glycosphingolipids on *Paracoccidioides brasiliensis* adhesion. Mycopathologia. *171*, 325–332.

Zaragoza, O., Rodrigues, M.L., Jesus, M., Frases, S., Dadachova, E., and Casadevall, A. (2009). The capsule of the fungal pathogen *Cryptococcus neoformans*. Adv. Appl. Microbiol. *68*, 133–216.

Zarnowski, R., and Woods, J.P. (2005). Glutathione-dependent extracellular ferric reductase activities in dimorphic zoopathogenic fungi. Microbiology. *151*, 2233–2240.

Index

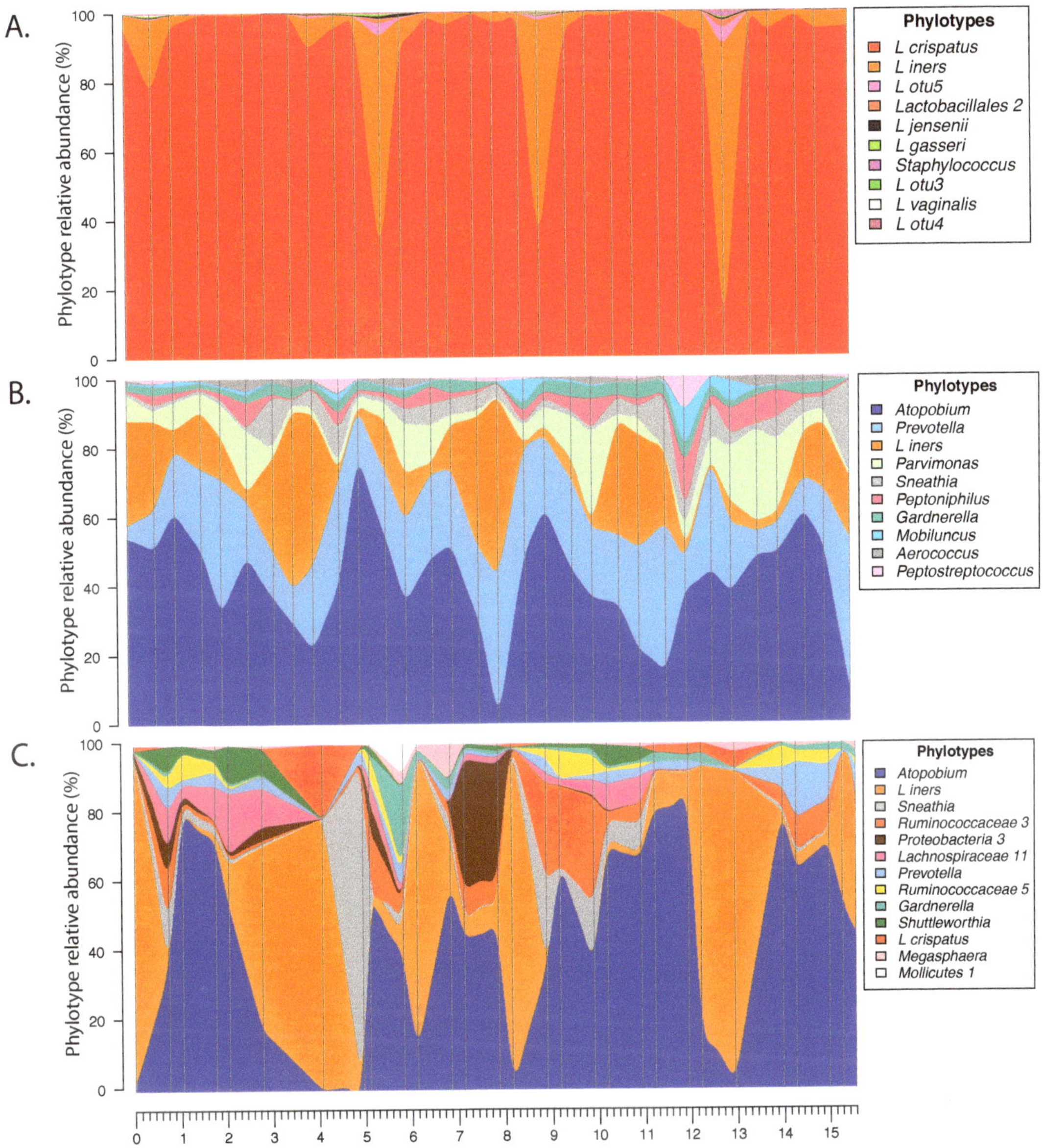

Figure 1.1 Temporal dynamics of vaginal bacterial communities from three different women over 16 weeks. Samples were collected twice-weekly and the community composition was determined by HTS of 16S rRNA gene sequences. The abundances of particular bacteria are depicted as interpolated bar graphs. All phylotypes are coloured as indicated on the right of each bar graph. (A) Example of a highly stable community dominated by *L. crispatus*. (B) Example of a moderately stable community with no or very little *Lactobacillus* sp. (C) Example of an unstable community. Graphs represent sequencing data generated by Gajer *et al.* (2012). Reproduced from Bradford *et al.* (2013) with kind permission from Springer Science Business Media B.V.

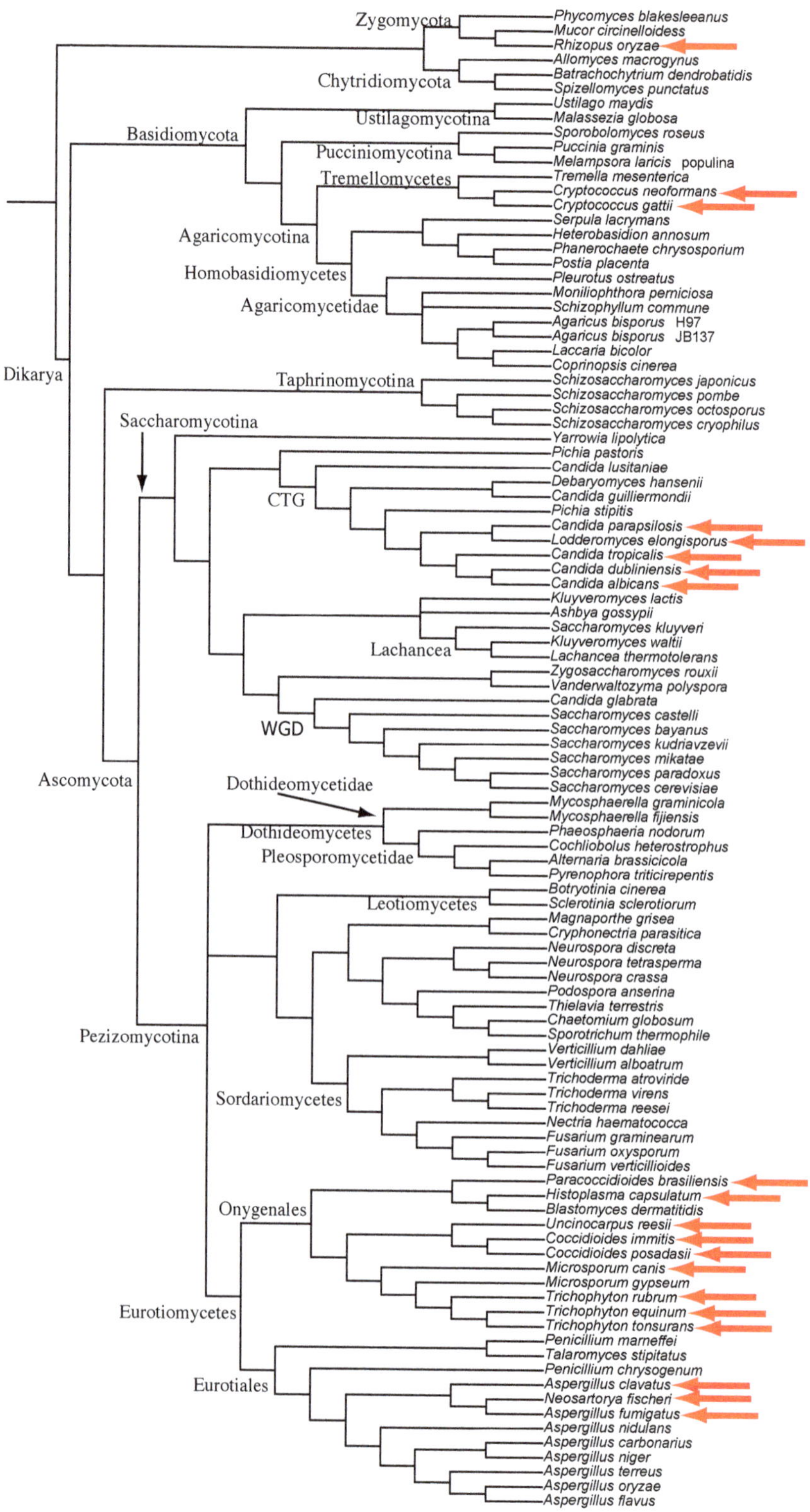

Figure 2.1 Fungal tree of life modified from Medina *et al.* (2011). A number of species discussed in the main text are highlighted with red arrows.

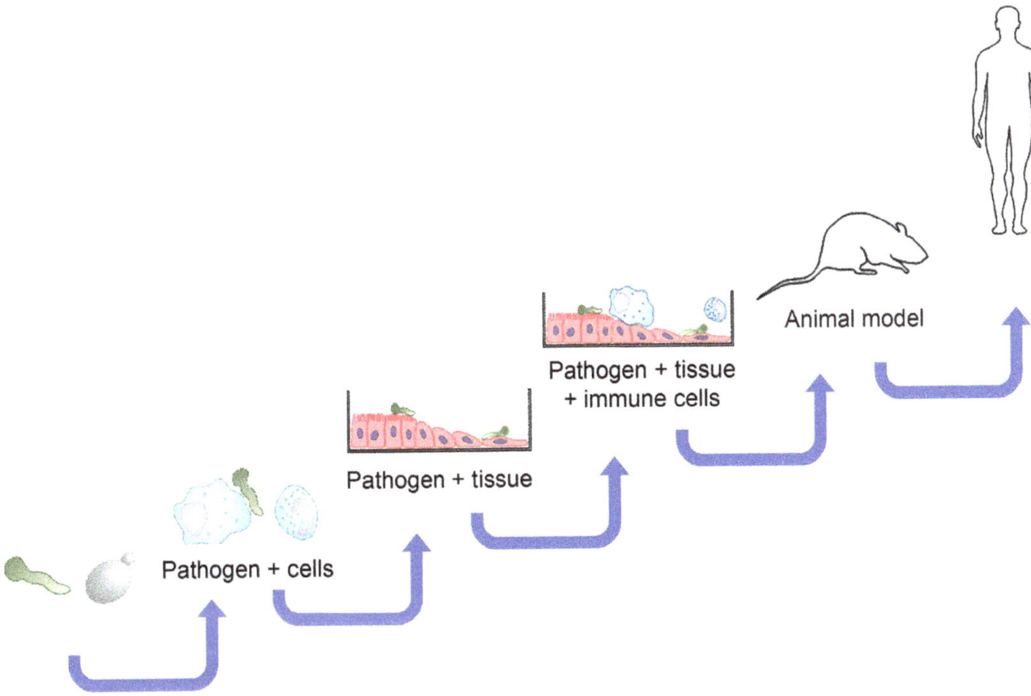

Figure 3.1 Increasing levels of complexity in infection biology studies.

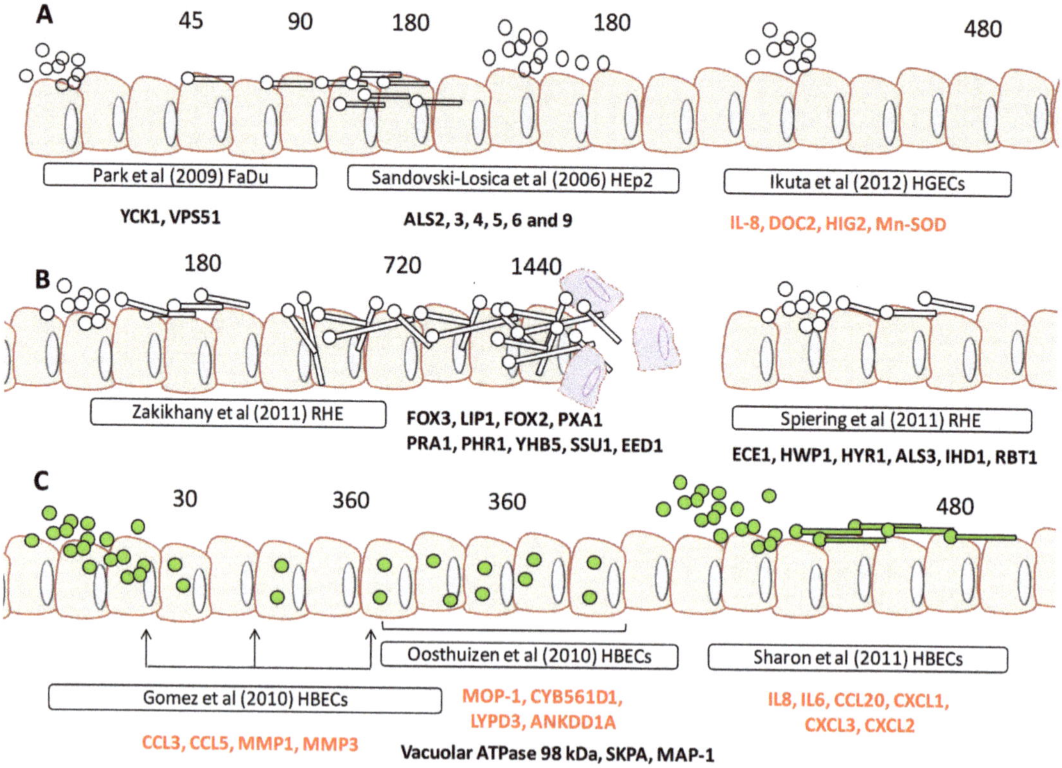

Figure 4.1 Fungus–epithelial interactions subjected to transcriptomic analyses. Overview of studies discussed in this Chapter, including key transcriptomic responses of host (red) and pathogen (black) and morphogenetic responses of fungal cells to host contact. Time-points of analyses (minutes) are indicated above each scheme. Citations and cell types are indicated below each scheme. (A) *Candida*–epithelial interactions (B) *Candida*-reconstituted human epithelium interactions (C) *Aspergillus*–epithelial interactions. Gomez *et al.* (2010) and Oosthuizen *et al.* (2010) studies differ by nature of the host cells studied, respectively targeting only those having internalized *A. fumigatus* conidia or assessing transcriptional responses at the level of the entire monolayer.

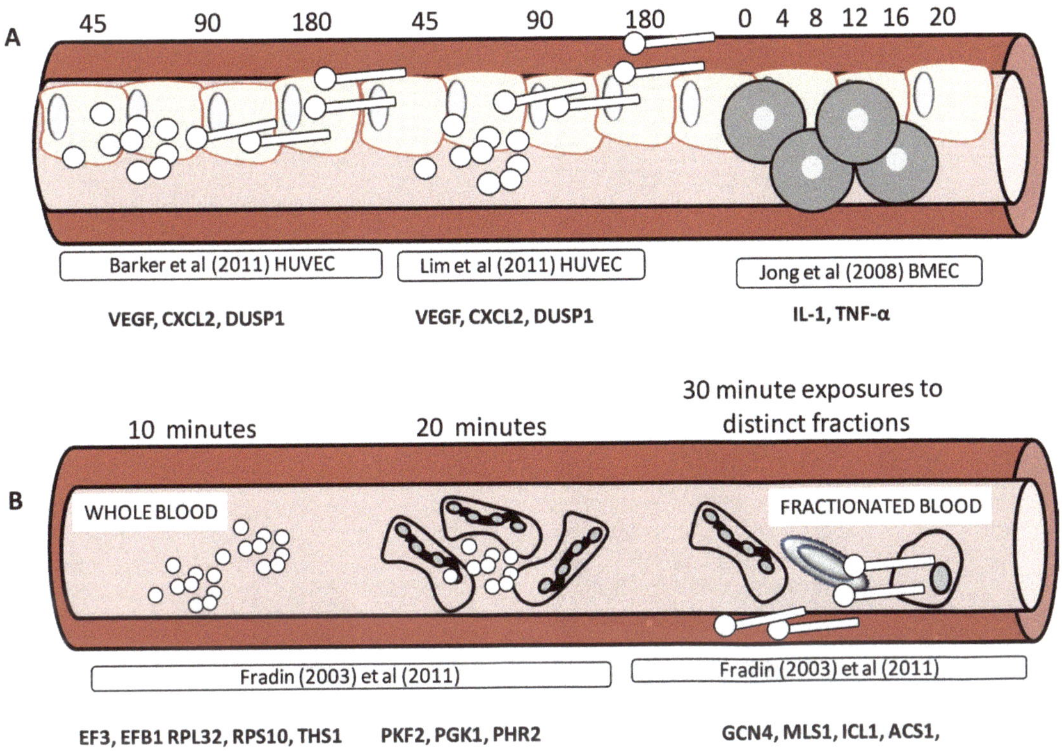

Figure 4.2 Fungus–endothelial and fungus–blood interactions. Overview of studies discussed in this chapter, including key transcriptomic responses of and pathogen (black) and morphogenetic responses of fungal cells to host contact. Time-points of analyses (minutes) are indicated above each scheme. Citations and cell types are indicated below each scheme. (A) *Candida*– and *Cryptococcus*–endothelial interactions (B) *Candida*–blood interactions.

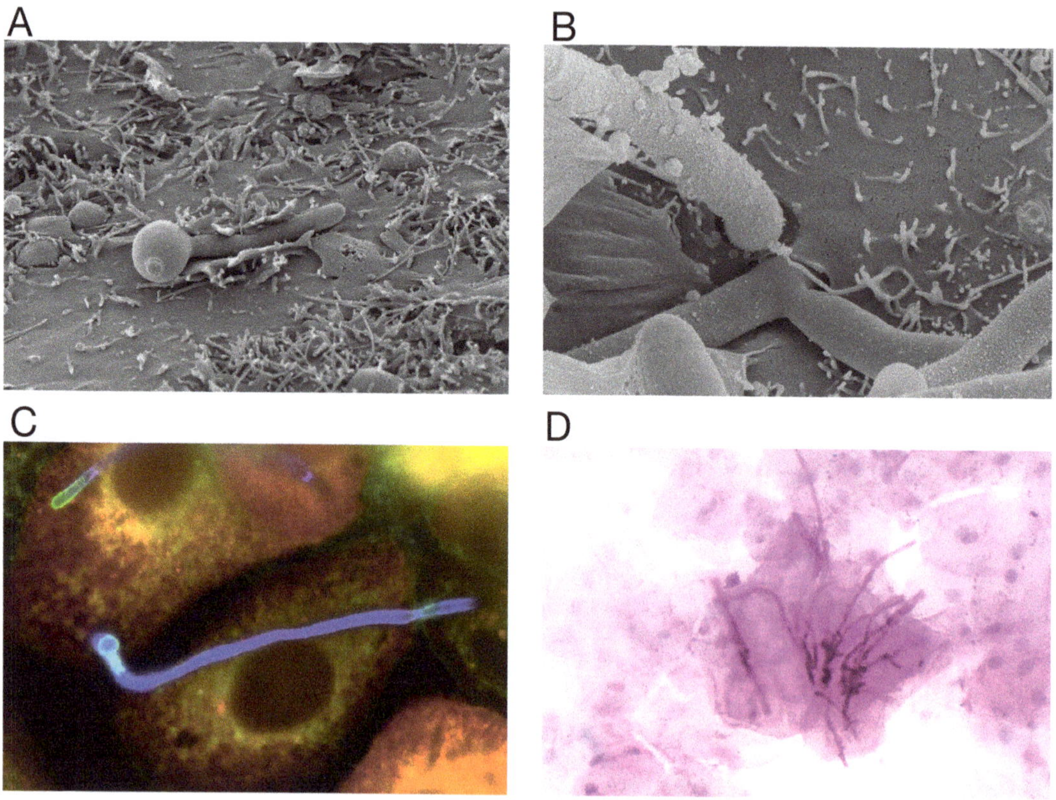

Figure 7.1 *Candida albicans*–epithelial interactions. (A and B) Scanning electron microscopy images of *C. albicans* infecting human oral epithelial cells, 1 hour (A) and 24 hours (B) post infection. Note the presence of epithelial pseudopod-like structures beginning to engulf the *C. albicans* germ tube (A) (Zakikhany K, Holland G, Özel M, with permission). (C) Differential staining reveals inter-epithelial invasion by *C. albicans* hyphae. Extracellular (non-invaded) hyphal sections are stained cyan; invading hyphae are stained blue (Almeida R, with permission). (D) *C. albicans* hyphae in association with human oral tissue from biopsy (Schmidt-Westhausen A, with permission).

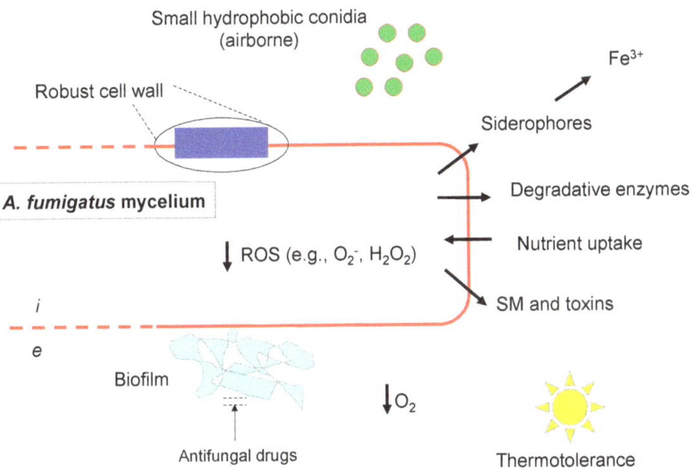

Figure 8.1 Overview of the structural features and virulence factors which influence the opportunistic pathogenicity of *Aspergillus fumigatus*. The rigid cell wall, small hydrophobic conidia and robust biofilm are key structural features which enhance organismal pathogenicity, and contribute to antifungal drug resistance. Toxins, secondary metabolites (SM), siderophores and secreted degradative enzymes also enhance the fungal virulence arsenal. An intricate capacity to grow under hypoxic conditions, dismutate reactive oxygen species (ROS; both intracellularly and host-derived) is also important for the success of *A. fumigatus* as an opportunistic pathogen. Key: i, intracellular; e, extracellular.

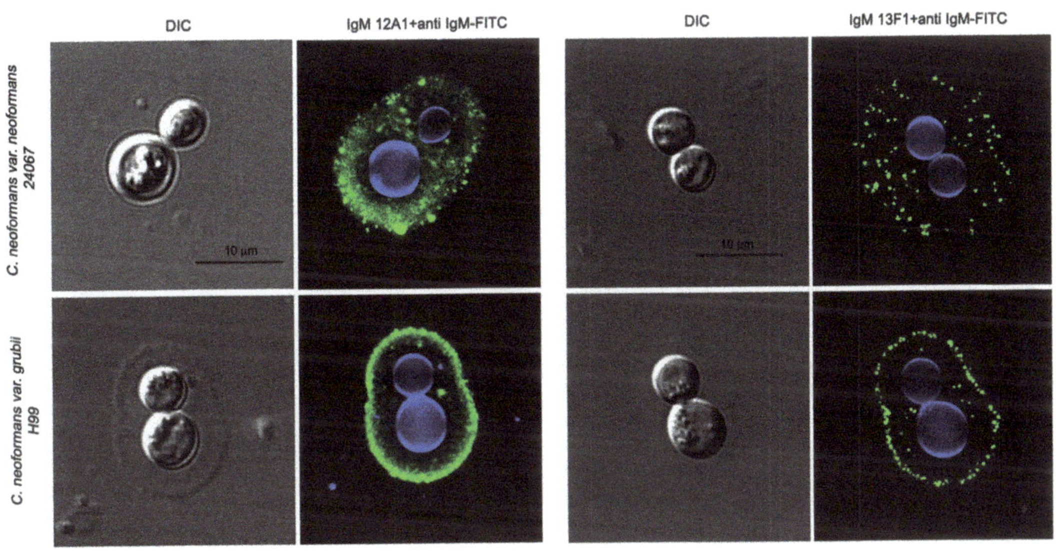

Figure 9.2 Binding patterns of two monoclocal antibodies against cryptococcal capsule. Yeast cells from *C. neoformans* var. *neoformans* (24067) and *C. neoformans* var. *grubii* (H99) were stained with Calcofluor White (blue fluorescence, cell wall) and two IgM mAbs (12A1 and 13F1, green fluorescence, capsule) and observed by confocal microscopy. Scale bar in upper bright field panels apply to the rest of the panels.

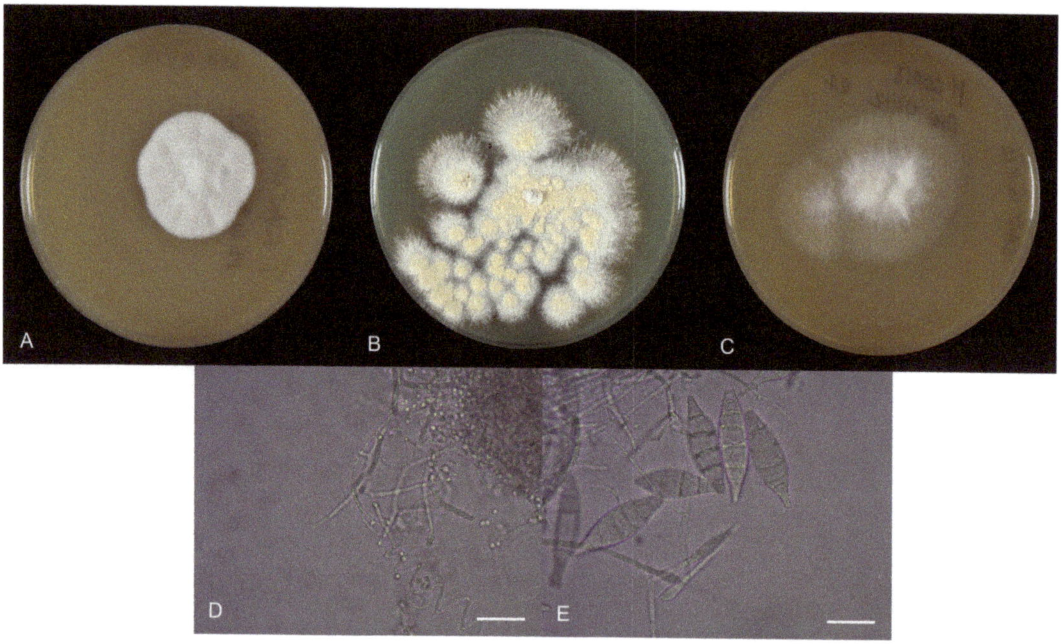

Figure 10.1 Macroscopic and microscopic characteristics of various dermatophytes. (A–C) Cultures of *T. rubrum*, *A. vanbreuseghemii* and *M. canis*, respectively, on Sabouraud agar. (D) *Arthroderma vanbreuseghemii* microspores. (E) *Microsporum canis* spindle-shaped, thick-walled macroconidia (bar in D and E=10 μm).

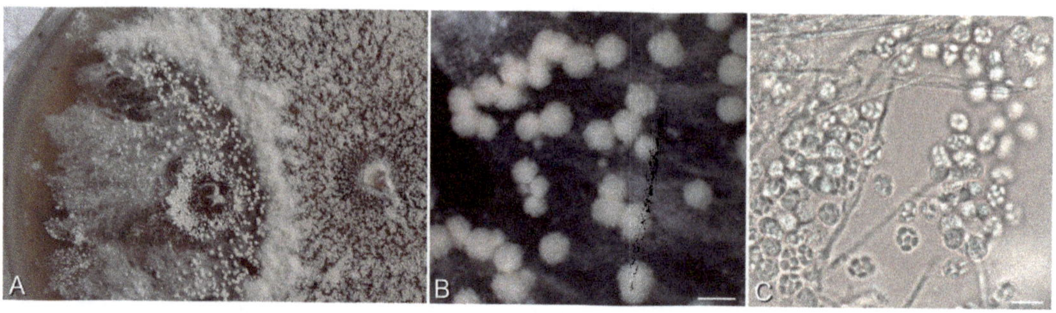

Figure 10.2 *Arthroderma benhamiae* teleomorphs. (A) Mating of a+strain with a – strain of *A. benhamiae*. (B) Binocular microscope observation of cleistothecia (bar=100 μm). (C) Asci and ascospores from squashed cleistothecia (bar=10 μm).

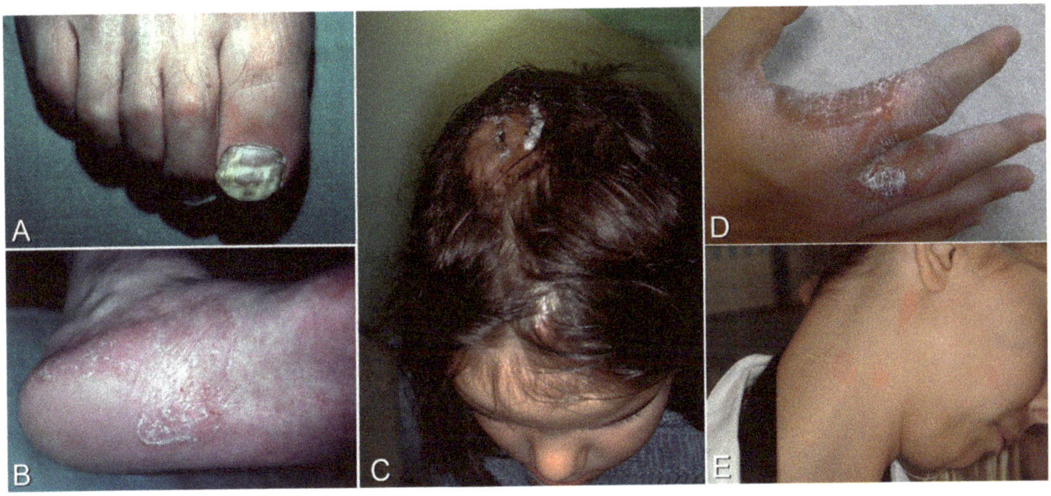

Figure 10.3 Human dermatophytoses. (A) Tinea unguium with *T. rubrum*. (B) Tinea pedis with *T. rubrum*. (C) Tinea capitis (kerion) with *A. benhamiae*, (D) Tinea manum with *T. erinacei*. (E) Tinea corporis with *M. canis*.

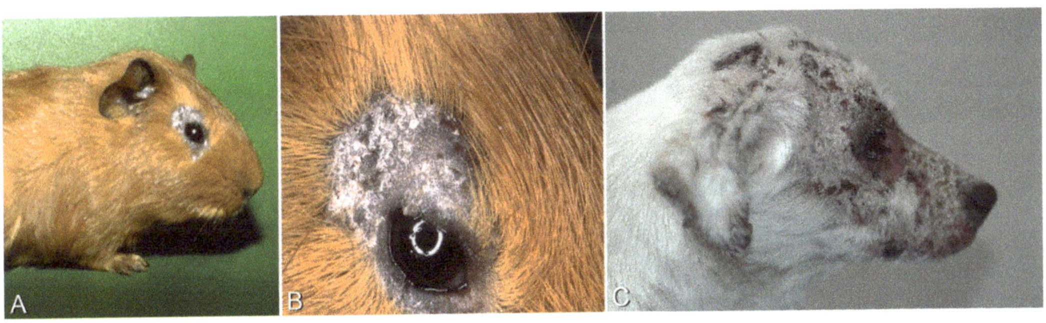

Figure 10.4 Animal dermatophytoses. (A and B) Alopecia and scaling caused by *A. benhamiae* in a guinea pig. (C) Extensive alopecia with *A. vanbreuseghemii* in a dog.

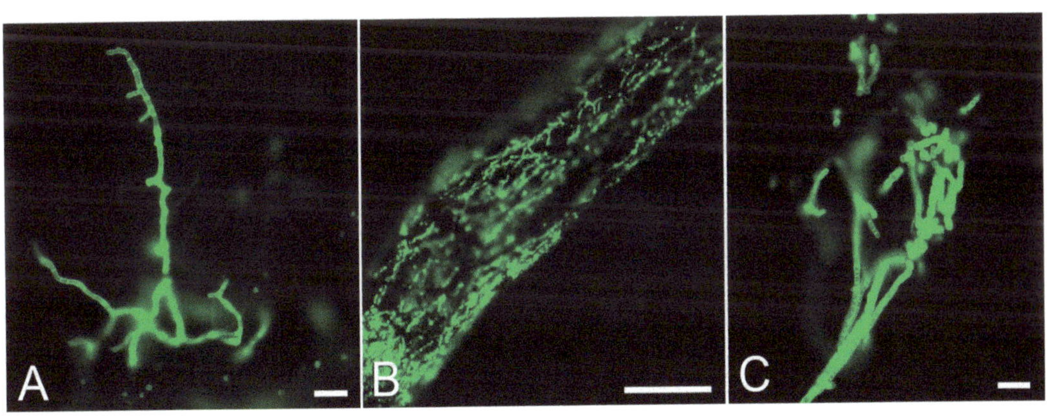

Figure 10.5 Direct mycological examinations of clinical and animal samples. (A) Hyphae in infected nails. (B) Tinea capitis, infected hair. (C) Hyphae and arthrospores from an infected guinea pig (bar in A and C=10 µm, bar in B=40 µm). One portion of each sample was examined in a dissolving solution containing a fluorochrome that specifically binds to vegetal and fungal cell wall polysaccharides (Monod *et al.*, 1989). The sample preparations were observed using a fluorescence microscope using excitation between 400 nm and 440 nm. The fluorescence techniques allow the best contrast of fungal elements in comparison to other techniques such as that using 10% KOH without fungal staining.

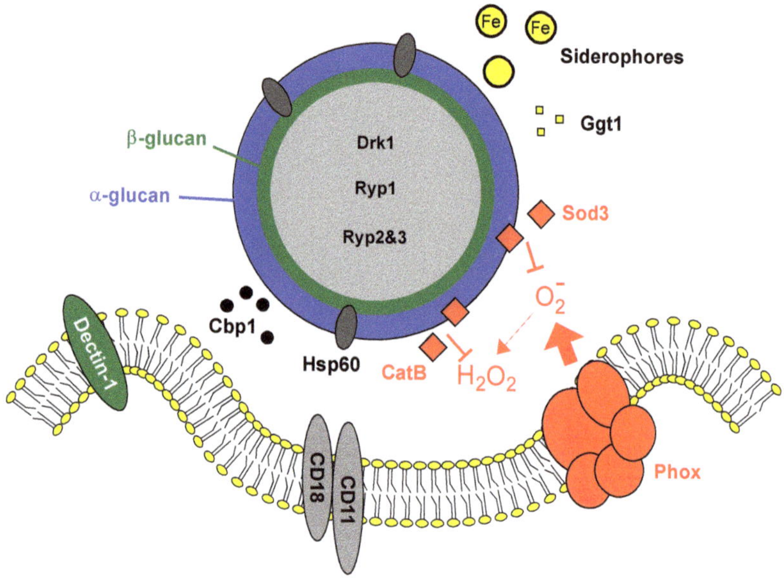

Figure 11.1 Pathogenesis mechanisms at the macrophage–pathogen interface that facilitate Histoplasma virulence. Defined virulence factors are schematically shown and coloured according to their function (adhesion and phagocytosis=grey; concealment=blue/green; reactive oxygen (ROS) detoxification=red; iron acquisition=yellow). Virulence strategies may operate at the macrophage cell surface during initial binding of yeasts to macrophages or they may function in the Histoplasma-containing phagosome (for simplicity, pathogenesis strategies are schematically shown in the figure at the macrophage cell surface). Adherence and phagocytosis of Histoplasma yeasts into macrophages is mediated by yeast cell wall localized Hsp60 (grey) binding to CD18-family receptors (e.g. CD18/CD11) in the macrophage plasma membrane. Simultaneously, Histoplasma yeasts avoid detection by the macrophage β-glucan receptor Dectin-1 (green, host receptor) by concealing fungal β-glucans (green) beneath an α-glucan polysaccharide layer (blue) in the yeast cell wall. Histoplasma yeasts eliminate phagocyte-produced ROS by production of extracellular oxidative stress defence factors (red). Assembly of the multimeric phagocyte NADPH oxidase complex (Phox) in the plasma membrane or the phagosomal membrane leads to production of antimicrobial superoxide (O_2^-) and peroxide (H_2O_2). These toxic ROS molecules are eliminated by the extracellular *Histoplasma* superoxide dismutase (Sod3) and the extracellular catalase (CatB), respectively. *Histoplasma* yeasts have multiple strategies for essential iron acquisition (yellow). Host iron is brought to the phagosome via transferrin. Histoplasma yeasts secrete iron chelating siderophores and a gamma-glutamyl transferase (Ggt1) which reduces ferric to ferrous iron and allows for iron import into yeast cells. The Cbp1 protein (black) is secreted by Histoplasma yeasts and is critical for pathogenesis, although the role it plays is unknown. The majority of these mechanisms are specific for yeast phase growth. Within the yeast cell, a signalling kinase (Drk1) and transcription factors (Ryp1, Ryp2, and Ryp3) act to promote differentiation of Histoplasma cells into the pathogenic yeast phase in response to mammalian temperature.

www.ingramcontent.com/pod-product-compliance
Lightning Source LLC
LaVergne TN
LVHW071336210926
848486LV00003B/157
* 9 7 8 1 9 0 8 2 3 0 4 4 7 *